Engineering Electromagnetic Fields and Waves

Engineering Electromagnetic Fields and Waves

Carl T. A. Johnk

Professor of Electrical Engineering
University of Colorado, Boulder

JOHN WILEY & SONS

New York London Sydney Toronto

Library of Congress Cataloging in Publication Data:

Johnk, Carl Theodore Adolf, 1919–

Engineering electromagnetic fields and waves.

Includes bibliographical references.
1. Electric engineering. 2. Electromagnetic fields.
3. Electromagnetic waves. I. Title.

TK145.J56 621.3 74–13567
ISBN 0–471–44289–5

Printed in the United States of America

10 9 8 7 6 5 4

To
Jeanette
and
the Boys

Preface

It might indeed be difficult to justify another text on undergraduate electromagnetic fields, except that this work grew out of some specific needs in the electrical engineering curriculum at the University of Colorado that might conceivably extend to similar needs existing elsewhere. Until a few years ago, the junior-level two-semester sequence in introductory field and wave-transmission theory covered static electric and magnetic fields during the first semester (employing a more conventional fields text written from the "historical" point of view), and then (using another text, or texts), the second semester covered principally the theory and applications of transmission lines and waveguides, plus an introduction to antenna theory and radiation. The problems arising from this scheme lay in the fact that in using two or more textbooks, the differences in approach and in symbolic notations contributed to a substantial loss of time required in relearning some of the ideas that should have carried over smoothly from the first semester's work. Moreover, in developing static fields from the historical approach (via the experimental Coulomb and Ampère laws), the understanding of the underlying Maxwell's equations was deferred until nearly the end of the first semester, judged by this author to be a major disadvantage.

The present text represents an effort to allay these difficulties. It is the culmination of several preliminary versions, used in the classroom for more than seven years at the University of Colorado and by colleagues at three other universities. Important features of this book might be summarized as follows:

1. Maxwell's equations are postulated for free space at the outset and then developed for material regions, along with the boundary conditions, within

the first three chapters. Applications to both static and time-varying problems illustrate this development. A full treatment of electrostatic and magnetostatic fields as special cases is then offered in Chapters 4 and 5, permitting a smooth transition to quasi-static time-varying fields in those chapters. This much material ordinarily comprises one semester's work.

2. Prerequisites assumed of the student are a freshman- or sophomore-level course covering differential and integral calculus. Vector analysis concepts as needed throughout this text are presented in the first two chapters. While the principal applications of vector ideas in this text involve the rectangular, circular cylindrical, and spherical coordinate systems, the generalized coordinate system is chosen for developing the concepts of dot and cross products as well as the gradient, divergence, and curl operators. Some years of experience with various approaches to vector analysis presentations as well as observations of students' responses have convinced the author that introducing these subjects in generalized form proves a time-saver by embracing the various coordinate systems within a single treatment.

3. In Chapters 4 and 5, on static and quasi-static electric and magnetic fields, the topics of capacitance and inductance are given detailed treatments. Besides the usual approaches via energy and voltage, capacitance is attacked by use of the flux-plotting method and extended to the capacitance-conductance analog. Self- and mutual inductances are given more than average consideration, with their energy definitions developed for both nonlinear and linear devices, together with seven detailed examples worked out in Chapter 5.

4. Chapters 6–11 are suitable for a second semester emphasizing waveguides and transmission lines plus an introduction to antennas. Chapter 6 represents a departure from the conventional manner in which transmission-line methods are usually broached. The student is introduced simultaneously to simple boundary-value problems of electromagnetics and to the analysis of wave-transmission systems via the problem of reflected and transmitted plane waves at normal incidence in a multilayered system. This generic plane-wave approach emphasizes the universality of impedance, reflection coefficient and Smith chart concepts. It has been found to provide additional insight into the wave structures of transmission lines with reflections, a topic considered in detail later in Chapters 9 and 10, while developing basic abilities in handling reflection and transmission problems of radio-wave and optical systems.

5. Chapter 7 gives an in-depth treatment of the real-time and complex forms of the Poynting theorem relative to electromagnetic energy and power, with applications to plane waves in lossless and lossy regions. A thoroughgoing discussion of rectangular hollow waveguide modes is found in Chapter 8, including the concept of group velocity and wall-loss attenuation. The TEM

waves of two-conductor transmission lines are described in Chapter 9, using static-field theory developed in Chapters 4 and 5 to derive line parameters for the lossless and lossy cases. Chapter 10 continues with reflections on transmission lines, drawing from the background of Chapter 6 and including additional applications of the Smith chart, in both impedance and admittance forms, to standing-wave and impedance-matching problems. A consideration of time-domain nonsinusoidal wave reflections on lossless lines rounds out the chapter.

6. In Chapter 11, several aspects of antenna radiation are covered in greater depth than in most texts at this level. These include the Green's theorem to develop the radiation integral, Pocklington's theorem for current distributions on thin wires, the radiation from a center-fed dipole, and the use of the equivalence theorem to find the radiation fields of aperture sources such as horn antennas and lasers.

7. An effort has been made to achieve a balance between the depth of presentation of the theoretical background and the applications via solved problems. Numerous worked-out examples throughout the text provide the student with a useful self-study aid, while giving the instructor greater flexibility in his classroom presentations.

A new book must invariably draw from the works of many authors; here a clear indebtedness to the authors listed in the references should be mentioned. Special gratitude to two of my former teachers, J. L. Glathart and E. C. Jordan, from whom much of my early encouragement was derived, is acknowledged.

While preparing the several earlier versions of this book, many discussions with colleagues and students were of inestimable benefit. Comments by Robert Bond, Ivar Pearson, James Lindsay, Ray King, Paul Klock, David Chang, and Ezekiel Bahar were most helpful, as well as those of anonymous reviewers. The development of this text has been quite rewarding, due in great part to the unflagging spirit of my students, whose remarks have been greatly appreciated. Also acknowledged are the encouragement and support of F. S. Barnes, whose leadership, vision, and indefatigability as Chairman of the Electrical Engineering Department have added materially to this text.

Special thanks go to Mrs. Charlotte Beeson and Mrs. Marie Krenz for their excellent typing efforts. Lastly, the author would like to thank any readers who forward corrections or suggestions for improvements.

Boulder, Colorado *Carl T. A. Johnk*

Contents

CHAPTER 5

CHAPTER 6

CHAPTER 7

CHAPTER 8

Mode Theory of Waveguides

CHAPTER 9

TEM Waves on Two-Conductor Transmission Lines

CHAPTER 10

Analysis of Reflective Transmission Lines

CHAPTER 11

Radiation from Antennas in Free Space

APPENDIX

CHAPTER

1

Vector Analysis and Electromagnetic Fields in Free Space

The important beginnings of vector analysis as a branch of mathematics date back to the middle of the nineteenth century, and since that time it has developed into an essential ingredient of the background of the physical scientist. The object of the treatment of vector analysis contained in the first two chapters is to serve the needs of the remainder of the book. Scalar and vector products, and certain integral processes involving vectors are developed, providing a groundwork for the Lorentz force effects which define the electric and magnetic fields, and for the postulated Maxwell integral relations among these fields in free space. Attention is focused on the generalized orthogonal coordinate system, with examples framed in the more common cartesian, circular cylindrical, and spherical systems.

1-1 Scalar and Vector Fields

A field is taken to mean a mathematical function of space and time. Fields can be classified as *scalar* or *vector* fields. A scalar field is a function having, at each instant in time, an assignable magnitude at every point of a region in space. Thus, the temperature field $T(x, y, z, t)$ inside the block of material of Figure 1-1(*a*) is a scalar field. To each point $P(x, y, z)$ there exists a corresponding temperature $T(x, y, z, t)$ at any instant t in time. The velocity of a fluid moving inside the pipe shown in Figure 1-1(*b*) illustrates a vector field.

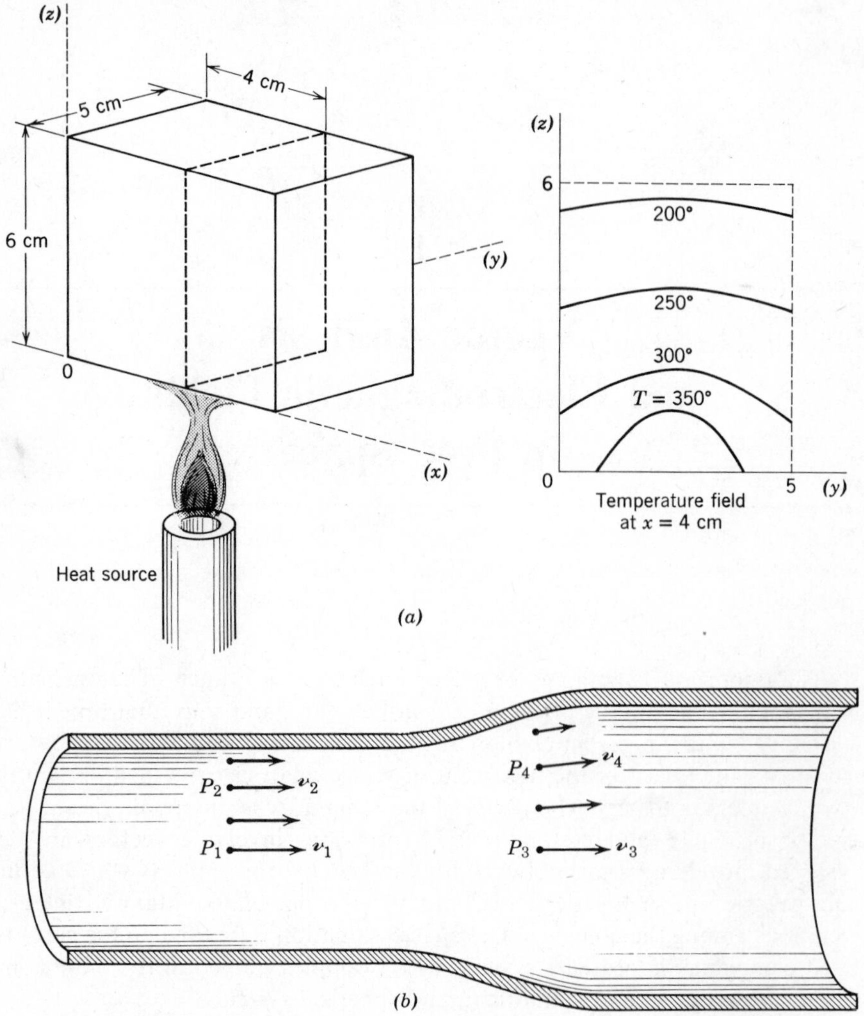

Figure 1-1. Examples of scalar and vector fields. (*a*) Temperature field inside a block of material. (*b*) Fluid velocity field inside a pipe of changing cross-section.

A variable direction, as well as magnitude, of the fluid velocity occurs in the pipe where the cross-sectional area is changing. Other examples of scalar fields are mass, density, pressure, and gravitational potential. A force field, a velocity field, and an acceleration field are examples of vector fields.

The mathematical symbol for a scalar quantity is taken to be any letter: for example, A, T, a, f. The symbol for a vector quantity is any letter set in

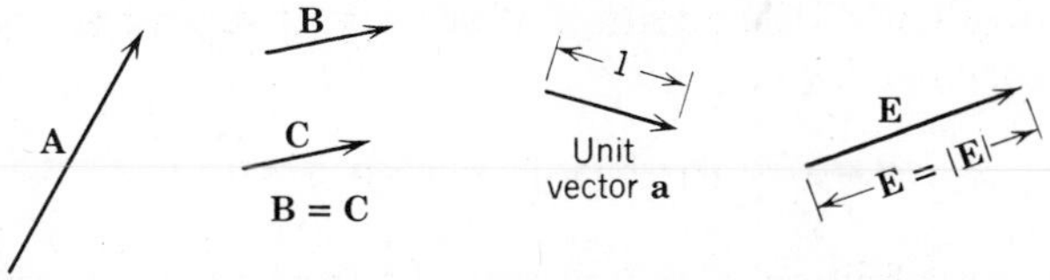

Figure 1-2. Graphic representations of a vector, equal vectors, a unit vector, and the representation of magnitude or length of a vector.

boldface roman type, for example, **A**, **H**, **a**, **g**. Vector quantities are represented graphically by means of arrows, or directed line segments, as shown in Figure 1-2. The magnitude or length of a vector **A** is written |**A**| or simply A, a positive real scalar. The *negative* of a vector is that vector taken in an opposing direction, with its arrowhead on the opposite end. A *unit vector* is any vector having a magnitude of unity. The symbol **a** is used to denote a unit vector, with a subscript employed to specify a special direction. For example, $\mathbf{a}_x$ means a unit vector having the positive-x direction. Two vectors are said to be *equal* if they have the same direction and the same magnitude. (They need not be collinear, but only parallel to each other.)

1-2 Vector Sums

The vector sum of **A** and **B** is defined in relation to the graphic sketch of the vectors, as in Figure 1-3. A physical illustration of the vector sum occurs in combining displacements in space. Thus, if a particle were displaced consecutively by the vector distance **A** and then by **B**, its final position would be denoted by the vector sum **A** + **B** = **C** shown in Figure 1-3(*a*). Reversing the order of these displacements provides the same vector sum **C**, so that

$$\mathbf{A} + \mathbf{B} = \mathbf{B} + \mathbf{A} \tag{1-1}$$

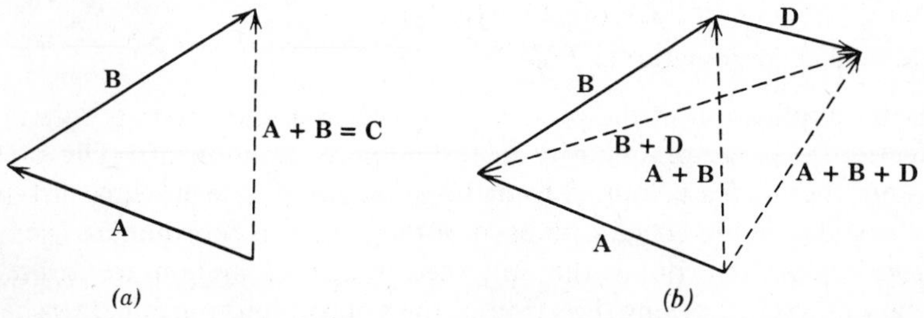

Figure 1-3. (*a*) The graphic definition of the sum of two vectors. (*b*) The associative law of addition.

the commutative law of the addition of vectors. If several vectors are to be added, an associative law

$$(\mathbf{A} + \mathbf{B}) + \mathbf{D} = \mathbf{A} + (\mathbf{B} + \mathbf{D}) \tag{1-2}$$

follows from the definition of vector sum and from Figure 1-3(*b*).

1-3 Product of a Vector and a Scalar

If a scalar quantity is denoted by u and if $\mathbf{B}$ denotes a vector quantity, their product $u\mathbf{B}$ means a vector having a magnitude u times the magnitude of $\mathbf{B}$, and having the same direction as $\mathbf{B}$ if u is a positive scalar, or the opposite direction if u is negative. The following laws hold for the products of vectors and scalars:

$$u\mathbf{B} = \mathbf{B}u \qquad \text{Commutative law} \tag{1-3}$$

$$u(v\mathbf{A}) = (uv)\mathbf{A} \qquad \text{Associative law} \tag{1-4}$$

$$(u + v)\mathbf{A} = u\mathbf{A} + v\mathbf{A} \qquad \text{Distributive law} \tag{1-5}$$

$$u(\mathbf{A} + \mathbf{B}) = u\mathbf{A} + u\mathbf{B} \qquad \text{Distributive law} \tag{1-6}$$

1-4 Coordinate Systems

The solution of physical problems often requires that the framework of a coordinate system be introduced, particularly if explicit solutions are being sought. The system most familiar to engineers and scientists is the cartesian, or *rectangular* coordinate system, although two other frames of reference often used are the *circular cylindrical* and the *spherical* coordinate systems. The symbols employed for the independent coordinate variables of these orthogonal systems are listed as follows:

1. Rectangular coordinates: (x, y, z)
2. Circular cylindrical coordinates: (ρ, ϕ, z)
3. Spherical coordinates: (r, θ, ϕ)

A useful application of the product of a vector and a scalar occurs in the specification of a vector quantity in terms of its components. The present discussion of this idea is limited to orthogonal coordinate systems. A typical point P is identified in space with each of the common coordinate systems in Figure 1-4(*a*). At that point, the unit vectors of each system are defined to lie in the positive increasing direction of the appropriate coordinate variable, as shown in Figure 1-4(*b*). The symbol $\mathbf{a}$ subscripted with the desired coordinate variable is used to denote the *unit vectors* of a particular coordinate

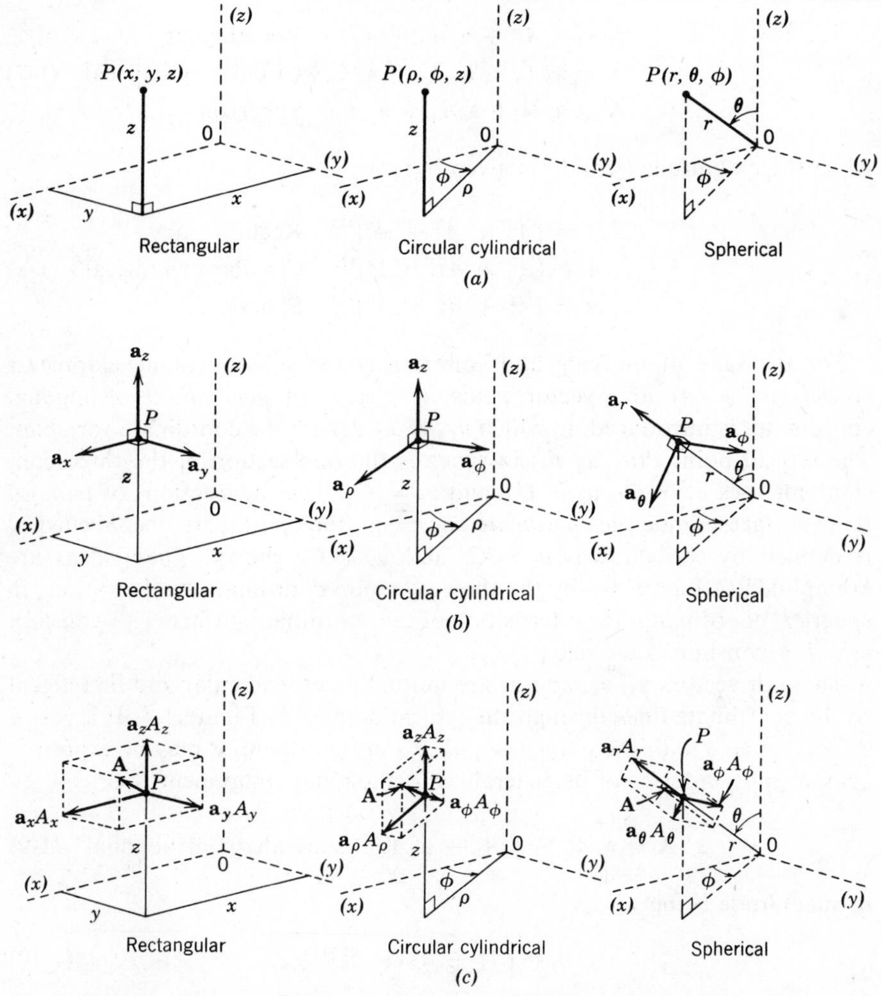

Figure 1-4. Notational conventions adopted in the three common coordinate systems. (*a*) Location of a point *P* in space. (*b*) The unit vectors at the typical point *P*. (*c*) The resolution of a vector A into its orthogonal components.

system. Thus $\mathbf{a}_x$, $\mathbf{a}_y$, $\mathbf{a}_z$ are the unit vectors in the rectangular system. In (*c*) of that figure is shown a typical vector **A** resolved into its components† in each of those coordinate systems, denoted symbolically as follows:

$$\begin{aligned} \mathbf{A} &= \mathbf{a}_x A_x + \mathbf{a}_y A_y + \mathbf{a}_z A_z && \text{Rectangular} \\ \mathbf{A} &= \mathbf{a}_\rho A_\rho + \mathbf{a}_\phi A_\phi + \mathbf{a}_z A_z && \text{Circular cylindrical} \\ \mathbf{A} &= \mathbf{a}_r A_r + \mathbf{a}_\theta A_\theta + \mathbf{a}_\phi A_\phi && \text{Spherical} \end{aligned} \tag{1-7}$$

while the magnitude (length) is given by:

$$\begin{aligned} A &= [A_x^2 + A_y^2 + A_z^2]^{1/2} && \text{Rectangular} \\ A &= [A_\rho^2 + A_\phi^2 + A_z^2]^{1/2} && \text{Circular cylindrical} \\ A &= [A_r^2 + A_\theta^2 + A_\phi^2]^{1/2} && \text{Spherical} \end{aligned} \tag{1-8}$$

For the sake of unifying and compacting the subsequent developments concerning scalar and vector fields, a system of *generalized* orthogonal coordinates is introduced, in which u_1, u_2, u_3 denote the coordinate variables. The typical point $P(u_1, u_2, u_3)$ in space is the intersection of the three constant surfaces $u_1 = C_1$, $u_2 = C_2$, and $u_3 = C_3$. The intersections of pairs of these surfaces define the *coordinate lines*; e.g., the coordinate line labelled u_1 is defined by the surfaces $u_2 = C_2$ and $u_3 = C_3$ shown. These ideas are exemplified in Figure 1-5 by the three common coordinate systems; thus, in spherical coordinates, the intersection of the coordinate surfaces r = constant and θ = constant is a circle.

The unit vectors $\mathbf{a}_1$, $\mathbf{a}_2$, and $\mathbf{a}_3$ are mutually perpendicular and lie tangent to the coordinate lines through the typical point P of Figure 1-5. If a vector **A** is associated with the point $P(u_1, u_2, u_3)$ in that figure, it may be expressed symbolically in terms of its generalized orthogonal components by

$$\mathbf{A} = \mathbf{a}_1 A_1 + \mathbf{a}_2 A_2 + \mathbf{a}_3 A_3 \quad \text{Generalized orthogonal} \tag{1-9}$$

its magnitude being

$$A = [A_1^2 + A_2^2 + A_3^2]^{1/2} \tag{1-10}$$

The scalars A_1, A_2, and A_3 are called the *components* of the vector **A**. Examples of these expressions specialized to the three common coordinate systems have already been given in (1-7) and (1-8).

† Thus, the *components* of **A** in the rectangular coordinate system are the vectors $\mathbf{a}_x A_x$, $\mathbf{a}_y A_y$, and $\mathbf{a}_z A_z$. Another common usage is to refer to only the scalar multipliers A_x, A_y, and A_z as the components of **A**, although it may be considered more proper to call these scalars the *projections* of **A** onto the respective coordinate axes.

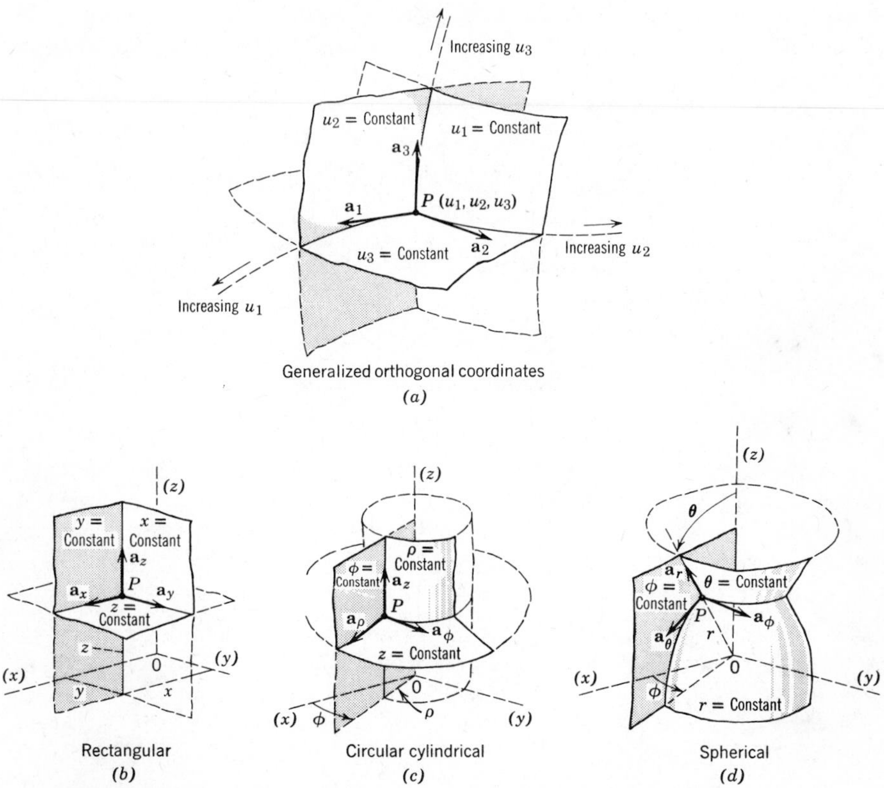

Figure 1-5. The coordinate surfaces defining the typical point *P*, and the unit vectors at *P*.

1-5 Differential Elements of Space

In the processes of integration in space to be considered shortly, the differential elements of volume, surface, and line are frequently needed. A differential element of volume dv is generated in the vicinity of a point $P(u_1, u_2, u_3)$ in space by means of the displacements $d\ell_1$, $d\ell_2$, and $d\ell_3$ on the coordinate surfaces, through the differential changes du_1, du_2, and du_3 in the coordinate variables. This situation is represented geometrically in Figure 1-6(*a*). Thus, a volume-element dv is represented in generalized orthogonal coordinates by means of the product of the differential length-elements as follows:

$$dv = d\ell_1\, d\ell_2\, d\ell_3 \tag{1-11}$$

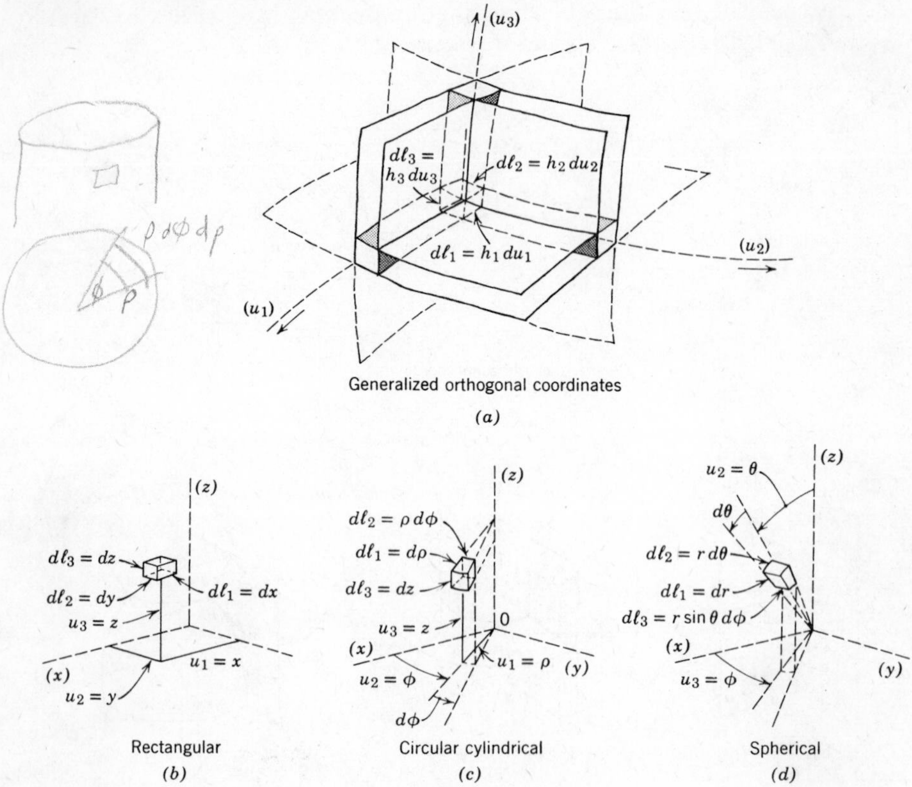

Figure 1-6. The generation of a volume-element $dv = d\ell_1\,d\ell_2\,d\ell_3$ at a typical point in space in orthogonal coordinate systems.

The relation of the length-elements to differential changes in the coordinate variables u_1, u_2, and u_3 is provided by the relations

$$d\ell_1 = h_1\,du_1 \qquad d\ell_2 = h_2\,du_2 \qquad d\ell_3 = h_3\,du_3 \tag{1-12}$$

so that (1-11) is written

$$dv = h_1 h_2 h_3\,du_1\,du_2\,du_3 \tag{1-13}$$

The coefficients h_1, h_2, and h_3 are called *metric coefficients*, needed to give the elements $d\ell$ of (1-12) their required dimension of length (meter). From a consideration of the geometry of dv in each diagram of Figure 1-6(*b*), (*c*), and (*d*), it is evident that the following length-elements and metric coefficients are applicable to the three common systems:

$$d\ell_1 = dx \qquad d\ell_2 = dy \qquad d\ell_3 = dz \qquad h_1 = h_2 = h_3 = 1 \qquad \text{Rectangular} \quad (1\text{-}14)$$

$$d\ell_1 = d\rho \qquad d\ell_2 = \rho\, d\phi \qquad d\ell_3 = dz \qquad h_1 = 1, \quad h_2 = \rho, \quad h_3 = 1 \qquad \text{Circular cylindrical} \quad (1\text{-}15)$$

$$d\ell_1 = dr \qquad d\ell_2 = r\, d\theta \qquad d\ell_3 = r \sin\theta\, d\phi \qquad h_1 = 1, \quad h_2 = r, \quad h_3 = r\sin\theta \qquad \text{Spherical} \quad (1\text{-}16)$$

The substitution of these results into (1-13) therefore, provides the volume-element dv in each system as follows:

$$\begin{aligned} dv &= dx\, dy\, dz && \text{Rectangular} \\ dv &= \rho\, d\rho\, d\phi\, dz && \text{Circular cylindrical} \quad (1\text{-}17) \\ dv &= r^2 \sin\theta\, dr\, d\theta\, d\phi && \text{Spherical} \end{aligned}$$

An element ds of a surface S in space may be left in its scalar form ds, although for some purposes it may be given a vector characterization, $d\mathbf{s}$, if desired. Suppose ds coincides with a coordinate surface u_1 = constant, as shown in Figure 1-7(*a*). Expressed as a scalar element, $ds = d\ell_2\, d\ell_3 = h_2 h_3\, du_2\, du_3$ for that example. An illustration in spherical coordinates is

cylin $ds = \rho\, d\phi\, dz$

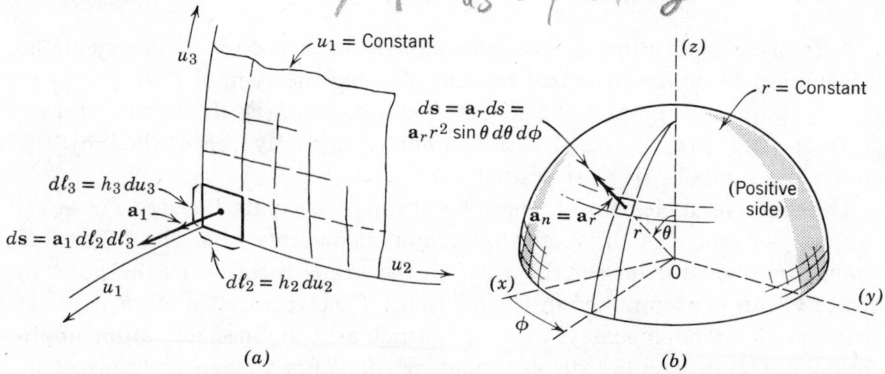

Figure 1-7. Typical surface element ds on a coordinate surface. Note the characterization of $d\mathbf{s}$ as a vector element through the multiplication with a normal unit vector. (*a*) A surface element ds on the coordinate surface u_1 constant in the generalized orthogonal system. (*b*) A surface element ds on the coordinate surface r = constant in the spherical system.

shown in Figure 1-7(b); on the r = constant coordinate surface, $ds = r^2 \sin\theta\, d\theta\, d\phi$. A vector quality is given ds through multiplying it with either the positive or the negative of the unit vector *normal* to ds. Thus, in Figure 1-7(b), the vector surface-element $d\mathbf{s} = \mathbf{a}_r\, ds$ is illustrated; $d\mathbf{s} = -\mathbf{a}_r\, ds$ is the other possible choice on the coordinate surface r = constant exemplified. These concepts are particularly useful in the flux-integration techniques discussed in Section 1-9.

Differential line-elements are frequently of interest in applications to vector integration. This subject is introduced in terms of the position vector $\mathbf{r}$ of spatial points treated in the next section.

*1-6 Position Vector

In field theory, reference may be made to a point $P(u_1, u_2, u_3)$ in space by use of the position vector, denoted by the symbol $\mathbf{r}$. The position vector of the point P in Figure 1-4, for example, is the vector $\mathbf{r}$ drawn from the origin 0 to the point P. Thus in rectangular coordinates, $\mathbf{r}$ is written

$$\mathbf{r} = \mathbf{a}_x x + \mathbf{a}_y y + \mathbf{a}_z z \tag{1-18}$$

and in circular cylindrical coordinates

$$\mathbf{r} = \mathbf{a}_\rho \rho + \mathbf{a}_z z \tag{1-19}$$

while in spherical coordinates

$$\mathbf{r} = \mathbf{a}_r r \tag{1-20}$$

A further application of the position vector $\mathbf{r}$ occurs in the symbolic designation of points in space. Instead of using the symbol $P(u_1, u_2, u_3)$ or $P(x, y, z)$, one may employ the abbreviated notation $P(\mathbf{r})$. By the same token, a scalar field $F(u_1, u_2, u_3, t)$ can be more compactly represented by the equivalent symbol $F(\mathbf{r}, t)$, if desired.

The differential element of length separating the points $P(\mathbf{r})$ and $P(\mathbf{r} + d\mathbf{r})$ in space is denoted by the vector differential displacement $d\mathbf{r}$. The differential change $d\mathbf{r}$ does not in general occur in the same direction as the position vector $\mathbf{r}$; this is exemplified in Figure 1-8(a). (The vector symbol $d\boldsymbol{\ell}$ is sometimes used interchangeably with $d\mathbf{r}$, particularly in line-integration applications). The differential displacement $d\mathbf{r}$ (or $d\boldsymbol{\ell}$) is written in terms of its generalized orthogonal components as follows:

$$d\mathbf{r} \equiv d\boldsymbol{\ell} = \mathbf{a}_1\, d\ell_1 + \mathbf{a}_2\, d\ell_2 + \mathbf{a}_3\, d\ell_3 \tag{1-21}$$

$$= \mathbf{a}_1 h_1\, du_1 + \mathbf{a}_2 h_2\, du_2 + \mathbf{a}_3 h_3\, du_3 \tag{1-22}$$

* Throughout the text sections marked with an asterisk (*) may be omitted to conserve time if desired.

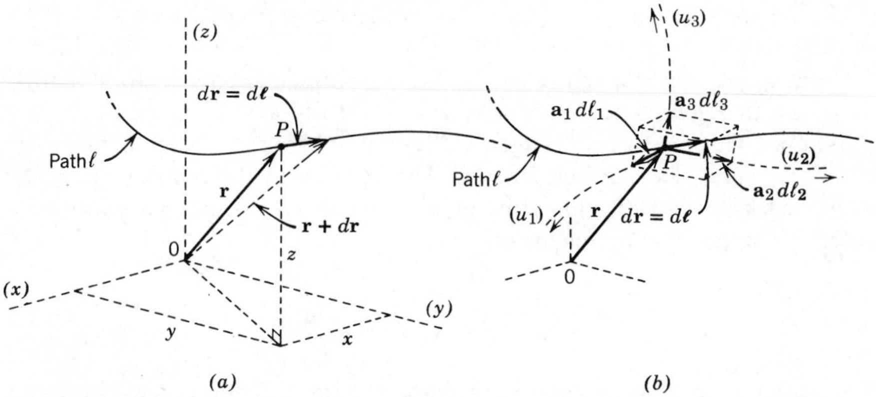

Figure 1-8. The position vector r used in defining points of space, and its differential $d\mathbf{r}$. (*a*) The position vector r and a differential position change $d\mathbf{r}$ along an arbitrary path. (*b*) Showing the components of $d\mathbf{r}$ in generalized orthogonal coordinates.

It is illustrated graphically in Figure 1-8(*b*) by means of the usual rectangular parallelepiped construction for a vector in terms of its components. Furthermore, the magnitude $d\ell$ of the vector $d\boldsymbol{\ell}$ is given by the diagonal of the rectangular parallelepiped; thus:

$$d\ell = [(h_1\,du_1)^2 + (h_2\,du_2)^2 + (h_3\,du_3)^2]^{1/2} \qquad (1\text{-}23)$$

For example, in spherical coordinates $h_1 = 1$, $h_2 = r$, and $h_3 = r \sin\theta$, so that (1-22) and (1-23) are written

$$d\mathbf{r} \equiv d\boldsymbol{\ell} = \mathbf{a}_r\,dr + \mathbf{a}_\theta r\,d\theta + \mathbf{a}_\phi r \sin\theta\,d\phi \qquad (1\text{-}24)$$

with

$$d\ell = [(dr)^2 + (r\,d\theta)^2 + (r \sin\theta\,d\phi)^2]^{1/2} \qquad (1\text{-}25)$$

The position vector **r** has useful applications in the dynamics of particles such as electrons and ions, for example. A study of Figure 1-8 reveals that if the vector displacement $d\mathbf{r}$ of a particle occurs in the time interval dt, then the ratio $d\mathbf{r}/dt$ denotes the vector velocity of the particle at $P(r)$. This particle velocity $\boldsymbol{v}$ is defined by the derivative of the position vector $\mathbf{r}(t)$

$$\boldsymbol{v} = \frac{d\mathbf{r}}{dt} = \lim_{\Delta t \to 0} \frac{\mathbf{r}(t + \Delta t) - \mathbf{r}(t)}{\Delta t} \qquad (1\text{-}26)$$

A second such derivative of $\mathbf{r}(t)$ provides the vector acceleration $\mathscr{a} = d\mathscr{v}/dt$ of the particle.

Because the vector displacement $d\mathbf{r}$ of the particle is tangent to its path ℓ as shown in Figure 1-8, the velocity $\mathscr{v} = d\mathbf{r}/dt$ will also be tangent at every point on ℓ. This property of tangency does not hold for acceleration, however, except in purely straightline motion. The velocity at the point $P(\mathbf{r})$ can be expressed systematically in terms of its generalized orthogonal coordinate velocity components by means of

$$\mathscr{v} \equiv \frac{d\mathbf{r}}{dt} = \mathbf{a}_1 \mathscr{v}_1 + \mathbf{a}_2 \mathscr{v}_2 + \mathbf{a}_3 \mathscr{v}_3 \tag{1-27}$$

For example, in a rectangular coordinate system, the notations $\mathscr{v}_1$, $\mathscr{v}_2$, and $\mathscr{v}_3$ mean $\mathscr{v}_x$, $\mathscr{v}_y$, and $\mathscr{v}_z$ respectively.

The acceleration, $\mathscr{a} = d\mathscr{v}/dt$, is a second time derivative of (1-27)

$$\mathscr{a} \equiv \frac{d\mathscr{v}}{dt} = \left[\mathbf{a}_1 \frac{d\mathscr{v}_1}{dt} + \mathscr{v}_1 \frac{d\mathbf{a}_1}{dt}\right] + \left[\mathbf{a}_2 \frac{d\mathscr{v}_2}{dt} + \mathscr{v}_2 \frac{d\mathbf{a}_2}{dt}\right] + \left[\mathbf{a}_3 \frac{d\mathscr{v}_3}{dt} + \mathscr{v}_3 \frac{d\mathbf{a}_3}{dt}\right] \tag{1-28}$$

a result involving the time derivatives of the *unit vectors* as well as the scalar components of the velocity, as the point P moves along the path ℓ. The time derivatives of the unit vectors $\mathbf{a}_1$, $\mathbf{a}_2$, and $\mathbf{a}_3$ are, moreover, expressible in terms of their derivatives with respect to the coordinate variables u_1, u_2, and u_3 as follows:

$$\frac{d\mathbf{a}_1}{dt} = \frac{\partial \mathbf{a}_1}{\partial u_1}\frac{du_1}{dt} + \frac{\partial \mathbf{a}_1}{\partial u_2}\frac{du_2}{dt} + \frac{\partial \mathbf{a}_1}{\partial u_3}\frac{du_3}{dt} \tag{1-29}$$

with similar expressions for $d\mathbf{a}_2/dt$ and $d\mathbf{a}_3/dt$.

With expressions like (1-29) substituted into (1-28), a rather lengthy expansion for the acceleration $\mathbf{a}$ is obtained. In specific coordinate systems, however, simplifications will occur, the simplest of which are in terms of rectangular components. In that case, (1-27) becomes

$$\mathscr{v} \equiv \frac{d\mathbf{r}}{dt} = \mathbf{a}_x \frac{dx}{dt} + \mathbf{a}_y \frac{dy}{dt} + \mathbf{a}_z \frac{dz}{dt} \tag{1-30}$$

whereas the acceleration, by means of (1-28) and (1-29), reduces to

$$\mathscr{a} \equiv \frac{d\mathscr{v}}{dt} = \mathbf{a}_x \frac{d^2x}{dt^2} + \mathbf{a}_y \frac{d^2y}{dt^2} + \mathbf{a}_z \frac{d^2z}{dt^2} \tag{1-31}$$

The simple form of the latter is the result of all the derivatives of the unit vectors in (1-28) being zero in the rectangular system; i.e., $\mathbf{a}_x$, $\mathbf{a}_y$, and $\mathbf{a}_z$ are invariant in a fixed rectangular system.

In other systems, some or all of the unit vectors may change direction as P moves in space. A graphical approach to obtaining the spatial derivatives of the unit vectors in an explicit coordinate system is described in the following example.

EXAMPLE 1-1. Find the following partial derivatives of the unit vector $\mathbf{a}_r$: (*a*) $\partial\mathbf{a}_r/\partial r$; (*b*) $\partial\mathbf{a}_r/\partial\theta$; (*c*) $\partial\mathbf{a}_r/\partial\phi$.

(*a*) The partial derivative $\partial\mathbf{a}_r/\partial r$ equals zero, since the unit vector $\mathbf{a}_r$ does not vary in direction with r (nor does it vary in magnitude, by the definition of a unit vector).

(*b*) The partial derivative $\partial\mathbf{a}_r/\partial\theta$ can be found graphically from the accompanying figure. If $\mathbf{a}_r$ is allowed only the differential change $d\mathbf{a}_r$ in the θ sense, then $d\mathbf{a}_r$ has the direction of the unit vector $\mathbf{a}_\theta$. The length of $d\mathbf{a}_r$ is given precisely by the angle $d\theta$, from the definition of angle (arc divided by radius, and the radius is unity), to make $d\mathbf{a}_r$ become

$$d\mathbf{a}_r\Big]_{\substack{r=\text{constant}\\ \phi=\text{constant}}} = \mathbf{a}_\theta\, d\theta$$

whence the desired result is

$$\frac{d\mathbf{a}_r}{d\theta}\Bigg]_{\substack{r=\text{constant}\\ \phi=\text{constant}}} \equiv \frac{\partial\mathbf{a}_r}{\partial\theta} = \mathbf{a}_\theta$$

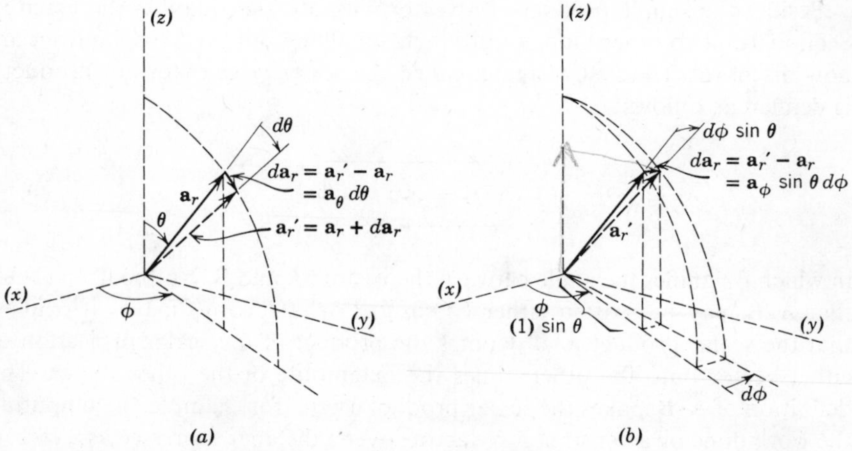

Example 1-1. (*a*) Differential $d\mathbf{a}_r$ generated by rotating $\mathbf{a}_r$ θ-wise. (*b*) Differential $d\mathbf{a}_r$ generated by rotating $\mathbf{a}_r$ ϕ-wise.

(*c*) The partial derivative $\partial \mathbf{a}_r/\partial\phi$ is found similarly from (*b*) of the figure. Allowing only the change $d\phi$ in the position of $\mathbf{a}_r$ generates the differential vector $d\mathbf{a}_r$, having a direction specified by the unit vector $\mathbf{a}_\phi$ and a magnitude given by $d\phi \sin\theta$. This makes $d\mathbf{a}_r$ (for r = constant, θ = constant) become $\mathbf{a}_\phi \sin\theta\, d\phi$ as shown, whence

$$\frac{\partial \mathbf{a}_r}{\partial \phi} = \mathbf{a}_\phi \sin\theta$$

By means of graphic techniques similar to those used in Example 1-1, one can show for spherical coordinates that all the spatial partial derivatives of the unit vectors in that system are zero except for

$$\frac{\partial \mathbf{a}_r}{\partial \theta} = \mathbf{a}_\theta \qquad \frac{\partial \mathbf{a}_r}{\partial \phi} = \mathbf{a}_\phi \sin\theta \qquad \frac{\partial \mathbf{a}_\theta}{\partial \theta} = -\mathbf{a}_r$$
$$\frac{\partial \mathbf{a}_\theta}{\partial \phi} = \mathbf{a}_\phi \cos\theta \qquad \frac{\partial \mathbf{a}_\phi}{\partial \phi} = -\mathbf{a}_r \sin\theta - \mathbf{a}_\theta \cos\theta \tag{1-32}$$

while in the circular cylindrical system, all are zero except for

$$\frac{\partial \mathbf{a}_\rho}{\partial \phi} = \mathbf{a}_\phi \qquad \frac{\partial \mathbf{a}_\phi}{\partial \phi} = -\mathbf{a}_\rho \tag{1-33}$$

1-7 Scalar and Vector Products of Vectors

Besides the simple product of a vector with a scalar quantity discussed in Section 1-3, two other kinds of products involving only vector quantities are now discussed. The first of these, called the *scalar product* (or dot product), is defined as follows:

$$\mathbf{A}\cdot\mathbf{B} \equiv AB\cos\theta \tag{1-34}$$

in which θ signifies the angle between the vectors **A** and **B**. Noting from (1-34) that $\mathbf{A}\cdot\mathbf{B}$ may be written either $(A\cos\theta)B$ or $A(B\cos\theta)$ makes it evident that the scalar product $\mathbf{A}\cdot\mathbf{B}$ denotes the product of the scalar projection of either vector onto the other, times the magnitude of the other vector. The definition of $\mathbf{A}\cdot\mathbf{B}$ makes the scalar product useful, for example, in computing the work done by a constant force acting over a distance expressed as a vector. A generalization of this idea extended to the integral expression for work is taken up in the next section.

Definition (1-34) permits the conclusion that if **A** and **B** are perpendicular, $\cos\theta$ is zero, making their scalar product zero. Again, if **A** and **B** happen to lie in the same direction, then $\mathbf{A}\cdot\mathbf{B}$ denotes the product of their lengths. These observations lead to simple results involving the scalar products of the orthogonal unit vectors $\mathbf{a}_1$, $\mathbf{a}_2$, and $\mathbf{a}_3$ of the coordinate systems illustrated in Figure 1-5. For example, $\mathbf{a}_1\cdot\mathbf{a}_2 = \mathbf{a}_2\cdot\mathbf{a}_3 = \mathbf{a}_3\cdot\mathbf{a}_1 = 0$, while $\mathbf{a}_1\cdot\mathbf{a}_1 = \mathbf{a}_2\cdot\mathbf{a}_2 = \mathbf{a}_3\cdot\mathbf{a}_3 = 1$.

From the definition (1-34), and since $\mathbf{B}\cdot\mathbf{A}$ means $BA\cos\theta$, the commutative law for the dot product follows:

$$\mathbf{A}\cdot\mathbf{B} = \mathbf{B}\cdot\mathbf{A} \tag{1-35}$$

The distributive law for the dot product of the sum of two vectors with a third vector

$$\mathbf{A}\cdot(\mathbf{B} + \mathbf{C}) = \mathbf{A}\cdot\mathbf{B} + \mathbf{A}\cdot\mathbf{C} \tag{1-36}$$

can also be proved.

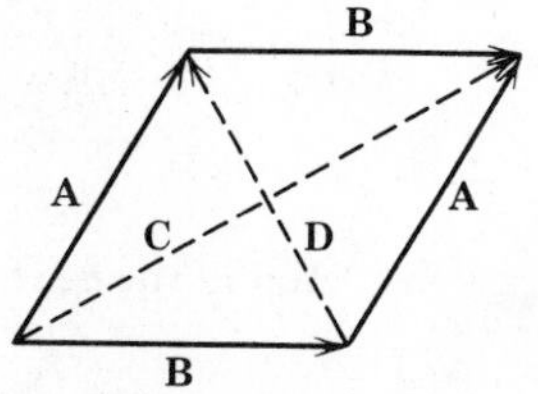

Example 1-2

EXAMPLE 1-2. Vector analysis can be used to shorten a number of proofs of geometry. Suppose one is to show that the diagonals of a rhombus are perpendicular. Represent its sides and diagonals by means of the vectors shown in the diagram. The diagonals are $\mathbf{A} + \mathbf{B} = \mathbf{C}$ and $\mathbf{A} - \mathbf{B} = \mathbf{D}$. Form the dot product of **C** and **D**.

$$(\mathbf{A} + \mathbf{B})\cdot(\mathbf{A} - \mathbf{B}) = \mathbf{A}\cdot\mathbf{A} - \mathbf{B}\cdot\mathbf{B} = A^2 - B^2$$

which must equal zero because $A = B$ for a rhombus. Therefore **C** and **D** are perpendicular.

If the vectors **A** and **B** are expressed in terms of their generalized orthogonal components in the manner of (1-9), their scalar product can be written

$$\mathbf{A}\cdot\mathbf{B} = (\mathbf{a}_1A_1 + \mathbf{a}_2A_2 + \mathbf{a}_3A_3)\cdot(\mathbf{a}_1B_1 + \mathbf{a}_2B_2 + \mathbf{a}_3B_3)$$

Upon expanding this expression by means of the distributive law (1-36) and applying the results obtained earlier for the dot products of the unit vectors, one obtains

$$\mathbf{A}\cdot\mathbf{B} = A_1B_1 + A_2B_2 + A_3B_3 \tag{1-37a}$$

For example, the expansion of the dot product of two vectors in rectangular coordinates is

$$\mathbf{A}\cdot\mathbf{B} = A_xB_x + A_yB_y + A_zB_z \tag{1-37b}$$

and in circular cylindrical coordinates

$$\mathbf{A}\cdot\mathbf{B} = A_\rho B_\rho + A_\phi B_\phi + A_zB_z \tag{1-37c}$$

EXAMPLE 1-3. Find the work done by the constant force $\mathbf{F} = 3\mathbf{a}_x + 4\mathbf{a}_y$ N in moving an object over a straight-line path defined by means of the vector $\mathbf{R} = -\mathbf{a}_x + 6\mathbf{a}_y + 3\mathbf{a}_z$ m.

The work done is $FR\cos\theta = \mathbf{F}\cdot\mathbf{R}$, found by use of (1-37b)

$$\begin{aligned}\mathbf{F}\cdot\mathbf{R} &= F_xR_x + F_yR_y + F_zR_z \\ &= (3)(-1) + (4)(6) + (0)(3) = 21 \text{ J}\end{aligned}$$

This is a special case of the integral expression for work considered in the next section.

The second kind of product of one vector with another is called the *vector product* (or cross product), defined as follows:

$$\mathbf{A}\times\mathbf{B} = \mathbf{a}_nAB\sin\theta \tag{1-38}$$

in which θ is the angle measured between $\mathbf{A}$ and $\mathbf{B}$, and $\mathbf{a}_n$ is a unit vector taken to be perpendicular to both $\mathbf{A}$ and $\mathbf{B}$ and having a direction determined from the right-hand rule provided that the rotation is taken from $\mathbf{A}$ to $\mathbf{B}$ through the angle θ. The vector product $\mathbf{A}\times\mathbf{B}$ is illustrated graphically in Figure 1-9. One may show from the diagram that

$$\mathbf{A}\times\mathbf{B} = -\mathbf{B}\times\mathbf{A} \tag{1-39}$$

which means that the vector product does not obey a commutative law. In forming the cross product, the ordering of the vectors, therefore, is an important consideration.

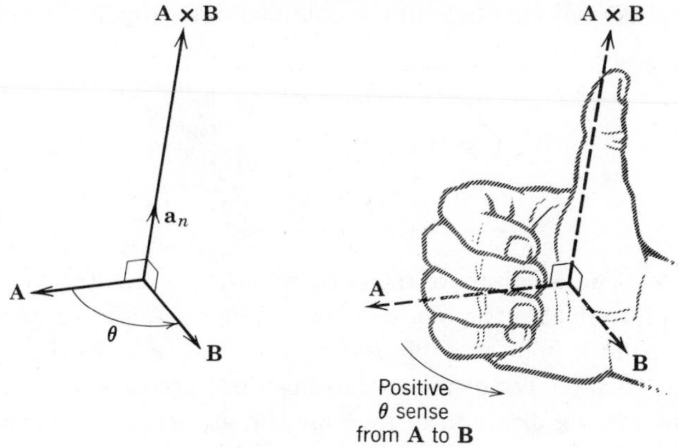

Figure 1-9. Illustrating the cross product.

If **A** and **B** are parallel vectors, sin θ is zero to make their cross product zero. If **A** and **B** happen to be perpendicular vectors, then $\mathbf{A} \times \mathbf{B}$ is a vector having a length AB and a direction perpendicular to both **A** and **B**, with the ambiguity in the direction resolved by means of the right-hand rule. These observations applied to the cross products of the orthogonal unit vectors of Figure 1-5, for example, lead to the special results: $\mathbf{a}_1 \times \mathbf{a}_1 = \mathbf{a}_2 \times \mathbf{a}_2 = \mathbf{a}_3 \times \mathbf{a}_3 = 0$; $\mathbf{a}_1 \times \mathbf{a}_2 = \mathbf{a}_3$, $\mathbf{a}_2 \times \mathbf{a}_3 = \mathbf{a}_1$, and $\mathbf{a}_3 \times \mathbf{a}_1 = \mathbf{a}_2$. However, note that $\mathbf{a}_1 \times \mathbf{a}_3 = -\mathbf{a}_2$.

A distributive law can be shown to hold for the cross product

$$\mathbf{A} \times (\mathbf{B} + \mathbf{C}) = \mathbf{A} \times \mathbf{B} + \mathbf{A} \times \mathbf{C} \tag{1-40}$$

Because of the noncommutativity of the cross product as expressed by (1-39), the order of the factors in (1-40) is important.

If the vectors **A** and **B** are given in terms of their orthogonal components in the manner of (1-9), then their vector product is written

$$\mathbf{A} \times \mathbf{B} = (\mathbf{a}_1 A_1 + \mathbf{a}_2 A_2 + \mathbf{a}_3 A_3) \times (\mathbf{a}_1 B_1 + \mathbf{a}_2 B_2 + \mathbf{a}_3 B_3)$$

The use of the distributive law (1-40) and the special results obtained for the cross products of the orthogonal unit vectors provides the following expansion:

$$\mathbf{A} \times \mathbf{B} = \mathbf{a}_1(A_2 B_3 - A_3 B_2) + \mathbf{a}_2(A_3 B_1 - A_1 B_3) + \mathbf{a}_3(A_1 B_2 - A_2 B_1)$$

which can alternatively be put into the compact determinental form

$$\mathbf{A} \times \mathbf{B} = \begin{vmatrix} \mathbf{a}_1 & \mathbf{a}_2 & \mathbf{a}_3 \\ A_1 & A_2 & A_3 \\ B_1 & B_2 & B_3 \end{vmatrix} \tag{1-41}$$

EXAMPLE 1-4. The definition of the cross product can be used to express the moment of a force $\mathbf{F}$ about a point P in space. Suppose $\mathbf{R}$ is a vector connecting the point P with the point of application Q of the force vector $\mathbf{F}$, as shown in the diagram. Then the vector moment $\mathbf{M}$ has the magnitude $M = RF \sin \theta = |\mathbf{R} \times \mathbf{F}|$. The turning direction of the moment, as well as its magnitude, are thus expressed by the vector product

$$\mathbf{M} = \mathbf{R} \times \mathbf{F} \tag{1-42}$$

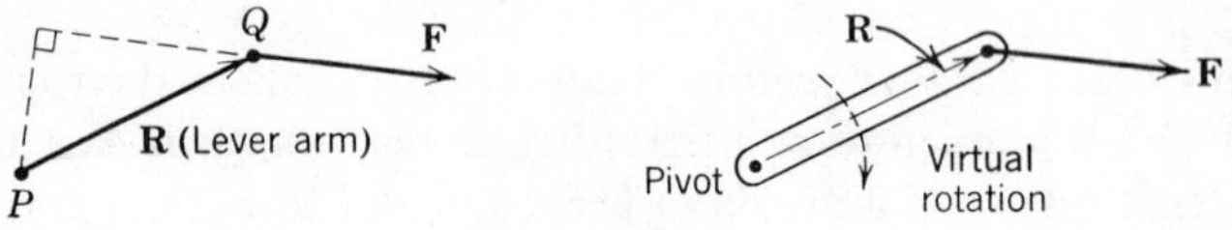

Example 1-4

EXAMPLE 1-5. A force $\mathbf{F} = 10\mathbf{a}_y$ N is applied at a point $Q(0, 3, 2)$ in space. Find the moment of $\mathbf{F}$ about the point $P(2, 0, 0)$.

The vector distance $\mathbf{R}$ between P and Q is

$$\mathbf{R} = \mathbf{a}_x(0 - 2) + \mathbf{a}_y(3 - 0) + \mathbf{a}_z(2 - 0)$$
$$= -2\mathbf{a}_x + 3\mathbf{a}_y + 2\mathbf{a}_z \text{ m}$$

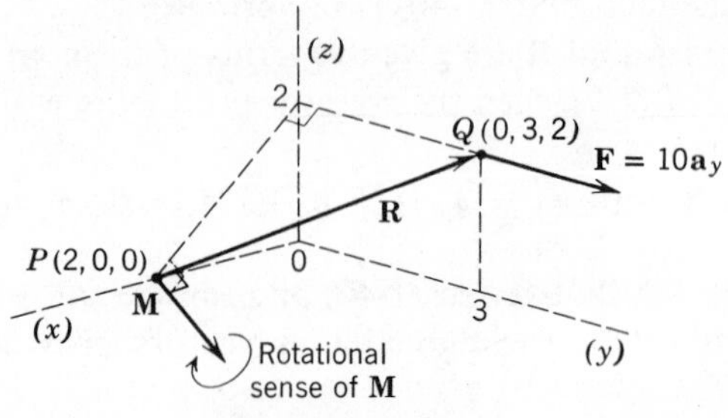

Example 1-5

The vector moment at P is found by means of (1-42) and the determinant (1-41).

$$\mathbf{M} = \mathbf{R} \times \mathbf{F} = \begin{vmatrix} \mathbf{a}_x & \mathbf{a}_y & \mathbf{a}_z \\ -2 & 3 & 2 \\ 0 & 10 & 0 \end{vmatrix} = -20\mathbf{a}_x - 20\mathbf{a}_z \text{ N-m}$$

M, shown at P in the sketch, is a vector perpendicular to the plane formed by **F** and **R**.

1-8 Vector Integration

Vector integration, for the purposes of field theory, encompasses integrals in space along lines, over surfaces, or throughout volume regions, as well as integrals in the time domain and the complex s domain. The subject of the present discussion concerns only integrations in space.

The vector notation embodies compactness as an important feature, so it is always worthwhile to examine the integrand of a vector integral carefully. The integrand may be either a scalar or a vector. Thus, the integrals

$$\int_{\ell} \mathbf{A} \cdot \mathbf{B}\, d\ell \qquad \text{Line integral}$$

$$\int_{S} (\mathbf{C} \times \mathbf{D}) \cdot d\mathbf{s} \qquad \text{Surface integral}$$

$$\int_{V} \mathbf{F} \cdot \mathbf{G}\, dv \qquad \text{Volume integral}$$

possess scalar integrands, and so produce scalar results upon integration. On the other hand, the integrals

$$\int_{\ell} \mathbf{G}\, d\ell \qquad \text{Line integral}$$

$$\int_{S} \mathbf{H} \times d\mathbf{s} \qquad \text{Surface integral}$$

$$\int_{V} \mathbf{J} \times \mathbf{K}\, dv \qquad \text{Volume integral}$$

contain vector integrands, and therefore yield vector results. In the last three examples, the integral process must take into account the different directions assumed by the integrand along the prescribed path ℓ, on the surface S, or in the volume V defined.

EXAMPLE 1-6. The different results provided by scalar and vector integrands is exemplified by simple integrals of scalar and vector displacements $d\ell$ or $d\boldsymbol{\ell}$ along some prescribed path in space. The integral

$$d = \int_{\ell} d\ell$$

summed over the path ℓ shown in (*a*) of the figure, provides its true scalar length d. On the other hand, the integral of the vector displacement $d\boldsymbol{\ell}$ on the same path

$$\mathbf{R} = \int_{\ell} d\boldsymbol{\ell}$$

produces quite a different answer, a vector result $\mathbf{R}$ determined only by the endpoints P_1 and P_2 of that path rather than by the form of the path between

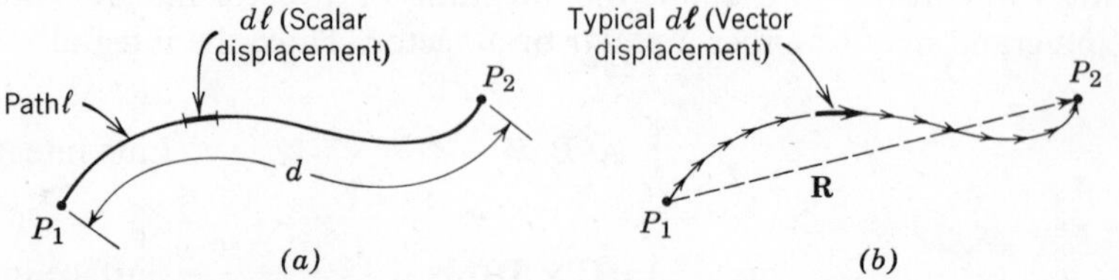

Example 1-6. (*a*) Integration of the scalar $d\ell$ over a path ℓ. (*b*) Integration of the vector $d\boldsymbol{\ell}$ over the path ℓ.

the endpoints. This vector $\mathbf{R}$ is illustrated in (*b*) of the accompanying figure. So the line integral of $d\boldsymbol{\ell}$ about a closed path is zero, whereas if $d\ell$ is the integrand, the perimeter of the closed path is the result.

An integral finding extensive utility in work or energy calculations is the scalar line integral

$$\int_{\ell} \mathbf{F} \cdot d\boldsymbol{\ell} \equiv \int_{\ell} F\, d\ell \cos\theta \tag{1-43}$$

This integral sums the scalar product $\mathbf{F} \cdot d\boldsymbol{\ell}$ over the path ℓ, as suggested by Figure 1-10. Only the projection of $\mathbf{F}$ along $d\boldsymbol{\ell}$ at each point on the path contributes to the integral result. The line integral (1-43) can be expressed in terms of the generalized orthogonal components of $\mathbf{F}$ and of $d\boldsymbol{\ell}$ in the follow-

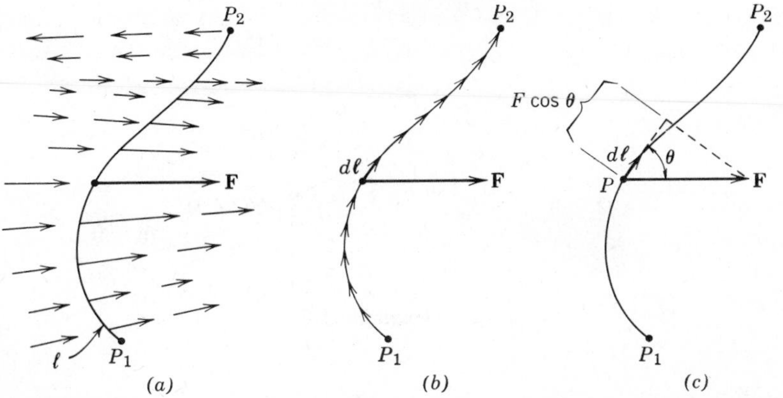

Figure 1-10. **(*a*) A path ℓ and the field F in space. (*b*) Division of ℓ into vector elements $d\ell$. (*c*) Scalar product $\mathbf{F}\cdot d\ell$ (to be summed over the path) shown at the typical point *P* on the path.**

ing way, making use of (1-9), (1-21), and (1-37a)

$$\int_\ell \mathbf{F}\cdot d\ell = \int_\ell F_1\, d\ell_1 + \int_\ell F_2\, d\ell_2 + \int_\ell F_3\, d\ell_3$$

$$= \int_\ell F_1 h_1\, du_1 + \int_\ell F_2 h_2\, du_2 + \int_\ell F_3 h_3\, du_3 \tag{1-44}$$

In the rectangular coordinate system, in which $h_1 = h_2 = h_3 = 1$, (1-44) is written

$$\int_\ell \mathbf{F}\cdot d\ell = \int_{x_1}^{x_2} F_x\, dx + \int_{y_1}^{y_2} F_y\, dy + \int_{z_1}^{z_2} F_z\, dz \tag{1-45}$$

assuming (x_1, y_1, z_1) and (x_2, y_2, z_2) are the coordinates of the endpoints P_1 and P_2 of the path ℓ.

EXAMPLE 1-7. Evaluate the line integral (1-43) between the points $P_1(0, 0, 1)$ and $P_2(2, 4, 1)$ over a path ℓ defined by the intersection of the two surfaces $y = x^2$ and $z = 1$, if **F** is the vector field

$$\mathbf{F} = \mathbf{a}_x 10x - \mathbf{a}_y 5x^2y + \mathbf{a}_z 3yz^2 \tag{1}$$

The path ℓ is illustrated in the figure.

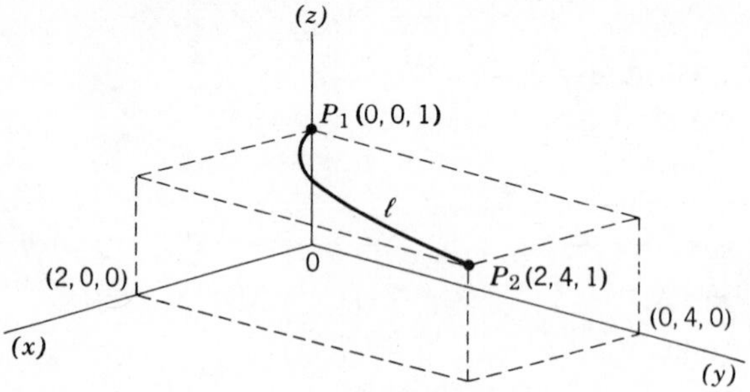

Example 1-7

Inserting $F_x = 10x$, $F_y = -5x^2y$, and $F_z = 3yz^2$ into (1-45) and since $x^2 = y$ and $dz = 0$ from the definition of ℓ, it follows that

$$\int_\ell \mathbf{F}\cdot d\boldsymbol{\ell} = \int_{x=0}^{2} 10x\,dx - \int_{y=0}^{4} 5y^2\,dy + 0$$

$$= 20 - 106.7 = -86.7$$

the desired result.

This answer can also be obtained by expressing the differential displacement dx along the path in terms of dy. From the definition of ℓ, $dy = 2x\,dx$ and $dz = 0$. Thus

$$\int_\ell \mathbf{F}\cdot d\boldsymbol{\ell} = \int_0^2 10x\,dx - \int_0^4 5y^2\,dy = \int_0^4 10\sqrt{y}\,\frac{dy}{2\sqrt{y}} - \int_0^4 5y^2\,dy = -86.7$$

EXAMPLE 1-8. A line integral such as (1-43) in general has a value depending upon the shape of the path connecting the endpoints P_1 and P_2. Evaluate the integral of Example 1-7 for the same function $\mathbf{F}$ and the same endpoints $P_1(0, 0, 1)$ and $P_2(2, 4, 1)$, but deform ℓ into the straight-line path given by the intersection of the surfaces $y = 2x$ and $z = 1$.

Integral (1-43) now becomes

$$\int_\ell \mathbf{F}\cdot d\boldsymbol{\ell} = \int_0^2 10x\,dx - \int_0^4 (5/4)y^3\,dy + 0 = -60$$

obviously different from the result obtained over the parabolic path in the last example. $\mathbf{F}$ is for this reason called a *nonconservative* field. A vector field for which the line integral (1-43) is independent of the shape of the path connecting a fixed pair of endpoints is said to be conservative. More is said later of such fields in connection with static electric charge distributions in Chapter 4.

1-9 Electric Charges, Currents, and Their Densities

The physical and the chemical properties of matter are known to be governed by the electric and magnetic forces which act among the particles comprising all material substances, whether inorganic or living cells. The fundamental electric particles of matter are of two varieties, commonly called positive and negative electric charges. Many experiments have provided the following conclusions concerning electric charges:

1. The algebraic sum of the positive and negative electric charges in a closed system never changes; that is, the total electric charge of a defined aggregate of matter is *conserved.*
2. Electric charge exists only in positive or negative integral multiples of the magnitude of the electronic charge, $e = 1.60 \times 10^{-19}$ C; this implies that electric charge is *quantized.*

From the viewpoint of classical electromagnetic theory, an electric charge aggregate will be treated as though it were capable of being indefinitely divisible, such that a volume electric-charge density, denoted by the symbol ρ_v is defined as follows:

$$\rho_v = \frac{\Delta q}{\Delta v} \text{ C/m}^3 \tag{1-46a}$$

This limit of this ratio is taken such that the volume-element in space does not become so small that it contains so few charged particles that the relatively smooth property of the density quantity ρ_v is lost, although Δv is kept small enough that the integration of the quantities containing Δv becomes a meaningful process. Figure 1-11(*a*) illustrates the meaning of these quantities relative to a volume element Δv.

Because the charge Δq residing within any element Δv may vary from point to point in a charge-bearing region, it is evident from (1-46a) that charge density is a function of space as well as possibly of time. Thus ρ_v is a *field*, written in general $\rho_v(u_1, u_2, u_3, t)$ or $\rho_v(\mathbf{r}, t)$.

In some physical problems, the charge Δq is identified with an element of *surface* or *line* instead of a volume. The limiting ratio (1-46a) should then be defined as follows:

$$\rho_s = \frac{\Delta q}{\Delta s} \text{ C/m}^2 \tag{1-46b}$$

$$\rho_\ell = \frac{\Delta q}{\Delta \ell} \text{ C/m} \tag{1-46c}$$

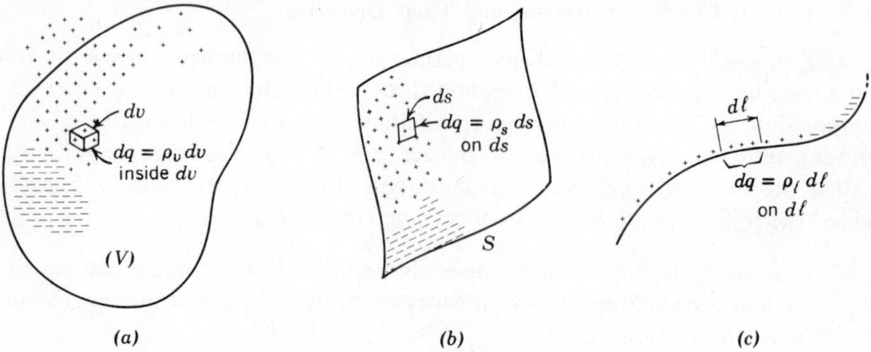

Figure 1-11. Geometries used in defining volume, surface, and line charge densities in space. (a) Quantities defining ρ_v. (*b*) Quantities defining ρ_s. (*c*) Quantities defining ρ_ℓ.

The quantities associated with these definitions of volume, surface and line charge densities are illustrated in Figure 1-11.

In some systems of charge aggregates, two species of positive and negative charge densities may be present simultaneously. A net charge density ρ_v (volume, surface, or line density) is in such an instance defined by†

$$\rho_v = \rho_v^+ + \rho_v^- \text{ C/m}^3 \tag{1-47}$$

in which ρ_v^+ and ρ_v^- denote limiting ratios defined by (1-46a) due to the positive and negative charges Δq^+ and Δq^- respectively in Δv. The occurrence of both positive metallic ions and mobile electrons in a *conductor* is an example to which (1-47) may be applied. The densities, in this case being of equal magnitudes but opposite signs ($\rho_v^+ = -\rho_v^-$), cancel, providing the net density $\rho_v = 0$ in such a compensated charge system.

The *total* amount of charge contained by a volume, surface, or line region is obtained from the integral of the appropriate density function (1-46a), (1-46b), or (1-46c). Thus in some volume region, each element dv contains the charge $dq = \rho_v\, dv$, making the total charge in v the integral

$$q = \int_v dq = \int_v \rho_v\, dv \text{ C}$$

† In some physical examples such as in a plasma discharge, electrons and several kinds of ions may be present simultaneously with different densities ρ_i. Their net density at any point in the region may then be characterized by

$$\rho_v = \sum_{i=1}^{m} \rho_i \tag{1-47a}$$

if a total of m charge species are to be found there.

Similar integral expressions may be constructed to yield the total charge on a given surface or a line in space.

A vector field $\mathbf{F}(u_1, u_2, u_3, t)$ at some given instant t, can be represented graphically by use of a myriad of vectors of appropriate lengths and directions at many points in a region of space. A vector field plotted in this way is shown in Figure 1-12(a). This is, however, a cumbersome way to graph a vector field; usually a much more satisfactory representation is by use of a *flux plot*, a

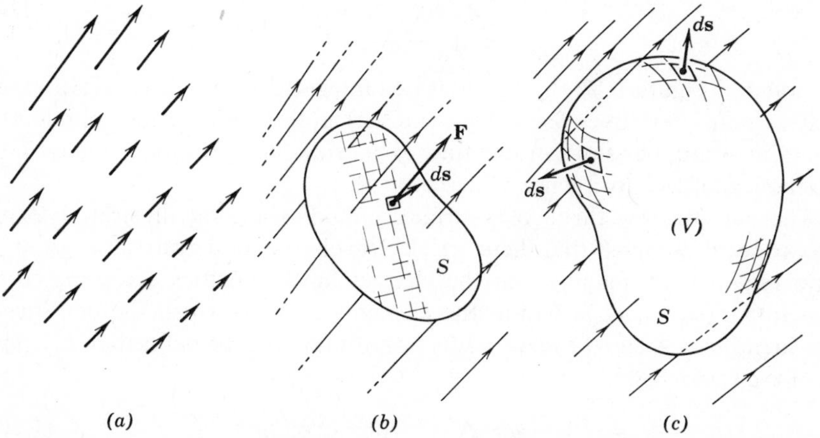

Figure 1-12. A vector field F, its flux map, and the flux through typical surfaces. (*a*) A vector field F, denoted by a system of arrows. (*b*) The flux map of the vector field F, showing an open surface *S* through which a net flux passes. (*c*) A closed surface *S*, showing zero net flux emergent from it.

method replacing the vectors with a system of lines (called *flux lines*) drawn in accordance with the following rules:

1. The directions of the flux lines agree with the directions of the field vectors.
2. The transverse densities of the flux lines are the same as the magnitudes of the field vectors.

The flux plot of the vector field of Figure 1-12(a), sketched in accordance with these rules, is noted in (b) of that figure. If a surface S is, moreover, drawn in the region of space embracing that flux, then the net lines of flux ψ passing through S can be a measure of some physical quantity (such as charge, current, power flow, and so on), depending upon the physical meaning of $\mathbf{F}$. The differential amount of flux $d\psi$ passing through any surface-element ds in space is defined by the scalar $d\psi = F\,ds\cos\theta = \mathbf{F}\cdot d\mathbf{s}$, a positive or negative result, depending on the angle between **F** and $d\mathbf{s}$. The net (positive or negative) flux of **F** through S is therefore the integral of $d\psi$ over S

$$\psi = \int_S \mathbf{F} \cdot d\mathbf{s} \tag{1-48}$$

in which $d\mathbf{s}$ is taken to emerge from that side of S assumed positive as shown in Figure 1-12(*b*). If S is a closed surface, the *net* flux through it is given by

$$\psi = \oint_S \mathbf{F} \cdot d\mathbf{s} \tag{1-49}$$

as noted in Figure 1-12(*c*). The latter will integrate to zero (an indication that just as many flux lines leave S as enter it) unless the interior volume of S contains sources or sinks of flux lines. This view will be amplified later in the discussion of the divergence of a vector field.

The current flow through a surface embodies a good illustration of the flux concept. Suppose that there are electric charges of density $\rho_v(u_1, u_2, u_3, t)$ in a region, and imagine that the charges have velocities averaging to the function $\mathcal{v}(u_1, u_2, u_3, t)$ within the elements dv with which the densities ρ_v are identified. A *current density* function **J** may then be defined at any point P in the region by

$$\mathbf{J} = \rho_v \mathcal{v} \text{ A/m}^2 \quad \text{or} \quad \text{C/sec/m}^2 \tag{1-50a}$$

This function is a measure, in the vicinity of any point P in space, of the instantaneous rate of flow of charge per unit cross-sectional area. If two species of charge density of opposite kinds, designated by ρ_v^+ and ρ_v^-, exist simultaneously in a region of space, then their total current density **J** at each point is written

$$\mathbf{J} = \rho_v^+ \mathcal{v} + \rho_v^- \mathcal{v} \text{ A/m}^2 \tag{1-50b}$$

In general, for n species with densities ρ_i and velocities $\mathcal{v}_i$ (e.g., electrons plus a mixture of ions)

$$\mathbf{J} = \sum_{i=1}^{n} \rho_i \mathcal{v}_i \text{ A/m}^2 \tag{1-50c}$$

The differential current flux di flowing through a surface element $d\mathbf{s}$ at which the current density **J** exists, is $di = \mathbf{J} \cdot d\mathbf{s}$ amperes, to make the net current i (current flux) through S

$$i = \int_S \mathbf{J} \cdot d\mathbf{s} \text{ C/sec} \quad \text{or} \quad \text{A} \tag{1-51}$$

EXAMPLE 1-9. An electron beam of circular cross-section 1 mm in diameter in a cathode ray tube (CRT) has a measured current of 1 μA, and a known average electron speed of 10^6 m/sec. Calculate the current density, charge density, and rate of mass transport in the beam.

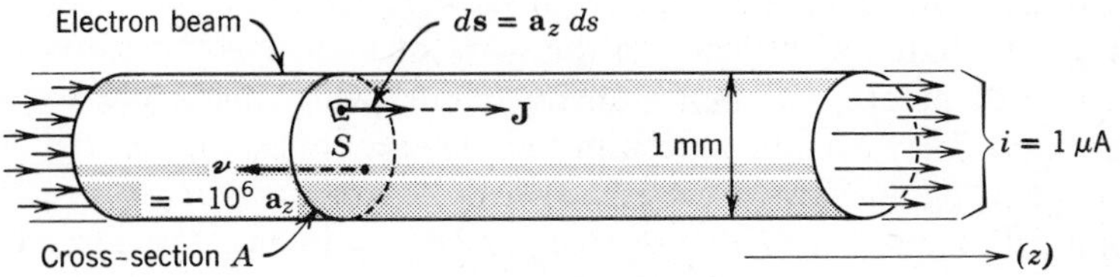

Example 1-9

Assuming a constant current density $\mathbf{J} = \mathbf{a}_z J_z$ in the cross-section, (1-51) yields the following current through any cross-section:

$$i = \int_S (\mathbf{a}_z J_z) \cdot (\mathbf{a}_z ds) = J_z \int_S ds = J_z A$$

in which A denotes the cross-sectional area of the beam. Thus, the axial current density is

$$J_z = \frac{i}{A} = \frac{10^{-6}}{\dfrac{\pi (10^{-3})^2}{4}} = \frac{4}{\pi} \text{ A/m}^2$$

The charge density in the beam, from (1-50a) in which $\mathbf{J} = \mathbf{a}_z 4/\pi$ and $v^- = -\mathbf{a}_z 10^6$, becomes

$$\rho_v^- = \frac{J_z}{v^-} = -\frac{4}{\pi} \times 10^{-6} \text{ C/m}^3$$

The rate of mass transport in the beam is the current times the electronic mass-to-charge ratio; this yields 5.7×10^{-18} kg/sec, assuming an electron mass of 9.1×10^{-31} kg.

1-10 Electric and Magnetic Fields in Terms of Their Forces

Electric and magnetic fields are fundamentally fields of force that originate from electric charges. Whether a force field may be termed *electric*, *magnetic*, or *electromagnetic* hinges upon the motional state of the electric charges relative to the point at which the field observations are made. Electric charges at rest relative to an observation point give rise to an electrostatic (time-independent) field there. The relative motion of the charges provides an additional force field called magnetic. That added field is *magnetostatic* if the charges are moving at constant velocities relative to the observation point. Accelerated motions, on the other hand, produce both time-varying electric and magnetic fields termed electromagnetic fields.

The connection of the electric and magnetic fields to their charge and current sources is provided by an elegant set of relations known as Maxwell's equations, attributed historically to the work of many scientists and mathematicians well before Maxwell's time,† but to which he made significant contributions. They are introduced in the next section. Suppose that electric and magnetic fields have been established in some region of space. The symbol for the *electric field intensity* (or just electric intensity) is the vector **E**; its units are force per unit charge (newtons per coulomb). The magnetic field is represented by means of the vector **B** called *magnetic flux density*; it has the unit weber per square meter. If the fields **E** and **B** exist at a point P in space, their presence may be detected physically by means of a charge q placed at that point. The force **F** acting on that charge is given by the *Lorentz force law*

$$\mathbf{F} = q(\mathbf{E} + \boldsymbol{v} \times \mathbf{B}) \tag{1-52a}$$

$$= \mathbf{F}_E + \mathbf{F}_B \text{ N} \tag{1-52b}$$

in which

q is the charge (coulomb) at the point P
$\boldsymbol{v}$ is the velocity (meter per second) of the charge q
E is the electric intensity (newton per coulomb) at P
B is the magnetic flux density (weber per square meter or tesla) at P
$\mathbf{F}_E = q\mathbf{E}$, the electric field force acting on q
$\mathbf{F}_B = q\boldsymbol{v} \times \mathbf{B}$, the magnetic field force acting on q

In Figure 1-13, these quantities are illustrated typically in space. The force $\mathbf{F}_E$ has the same direction as the applied field **E**, whereas the magnetic field force $\mathbf{F}_B$ is at right angles to both the applied field **B** and the velocity $\boldsymbol{v}$ of the charged particle.

The Lorentz force expression (1-52) may be used for discussing the ballistics of charged particles traveling in a region of space on which the electric and magnetic fields **E** and **B** are imposed. The deflection or the focusing of an electron beam in a cathode ray tube are common examples.

EXAMPLE 1-10. An electron at a given instant has the velocity $\boldsymbol{v} = (3)10^5\mathbf{a}_y + (4)10^5\mathbf{a}_z$ m/sec at some position in empty space. At that point the electric and magnetic fields are known to be $\mathbf{E} = 400\mathbf{a}_z$ V/m and $\mathbf{B} = 0.005\mathbf{a}_y$ Wb/m^2. Find the total force acting on the electron.

† James Clerk Maxwell (1831–1879).

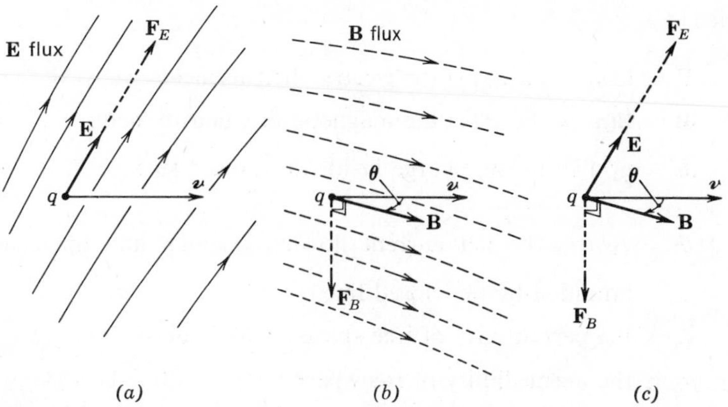

Figure 1-13. Lorentz forces acting on a moving charge q in the presence of (a) only an E field, (b) only a B field, and (c) both electric and magnetic fields.

The total force is found from the Lorentz relation (1-52a)

$$\mathbf{F} = q[\mathbf{E} + v \times \mathbf{B}] = -1.6(10^{-19})[\mathbf{a}_z 400 + (\mathbf{a}_y 3 \cdot 10^5 + \mathbf{a}_z 4 \cdot 10^5) \times \mathbf{a}_y 0.005]$$
$$= (\mathbf{a}_x 32 - \mathbf{a}_z 6.4)10^{-17}\ \text{N}$$

While this is quite a small force, the very small mass of the electron charge provides a tremendous acceleration to the particle, namely, $a = \mathbf{F}/m = (\mathbf{a}_x 3.51 - \mathbf{a}_z 0.7)10^{14}$ m/sec^2.

1-11 Maxwell's Integral Relations for Free Space

The relationships among the electric and magnetic force fields and their associated charge and current distributions in space are provided by Maxwell's equations, postulated here in integral form for the fields **E** and **B** in free space

$$\oint_S (\epsilon_0 \mathbf{E}) \cdot d\mathbf{s} = \int_V \rho_v \, dv \ \text{C} \tag{1-53}$$

$$\oint_S \mathbf{B} \cdot d\mathbf{s} = 0 \ \text{Wb} \tag{1-54}$$

$$\oint_\ell \mathbf{E} \cdot d\boldsymbol{\ell} = -\frac{d}{dt}\int_S \mathbf{B} \cdot d\mathbf{s} \ \text{V} \tag{1-55}$$

$$\oint_\ell \frac{\mathbf{B}}{\mu_0} \cdot d\boldsymbol{\ell} = \int_S \mathbf{J} \cdot d\mathbf{s} + \frac{d}{dt}\int_S (\epsilon_0 \mathbf{E}) \cdot d\mathbf{s} \ \text{A} \tag{1-56}$$

in which

$\mathbf{E} = \mathbf{E}(u_1, u_2, u_3, t)$ is the electric inensity field

$\mathbf{B} = \mathbf{B}(u_1, u_2, u_3, t)$ is the magnetic flux density field

$\int_V \rho_v \, dv = q(t)$ is the net charge inside any closed surface S

$\int_S \mathbf{J} \cdot d\mathbf{s} = i(t)$ is the net current flowing through any open surface S bounded by the closed line ℓ

ϵ_0 is the permittivity of free space ($\cong 10^{-9}/36\pi$ F/m)

μ_0 is the permeability of free space ($= 4\pi \times 10^{-7}$ H/m)

The Maxwell equations† (1-53) through (1-56) must be simultaneously satisfied by the field solutions **E** and **B** for *all* possible closed paths ℓ and surfaces S in the region of space occupied by these fields. This strict requirement might appear to limit severely the number of practical problems that can be solved by means of these integrals. Indeed, their application to the discovery of field solutions $\mathbf{E}(u_1, u_2, u_3, t)$ and $\mathbf{B}(u_1, u_2, u_3, t)$ is restricted, in the present treatment, to problems in which the charge or current distributions have particular symmetries that serve to simplify the solutions. The equivalent *differential* forms of Maxwell's equations, developed in the next chapter, have a somewhat wider range of application in problem solving at the introductory level.

A discussion of Maxwell's integral laws, along with some examples of the simpler field solutions that satisfy them, is given in the following section.

A. Gauss' Law for Electric Fields in Free Space

Maxwell's integral law (1-53)

$$\oint_S (\epsilon_0 \mathbf{E}) \cdot d\mathbf{s} = \int_V \rho_v \, dv \equiv q \tag{1-53}$$

is also known as Gauss' law for electric fields in free space. The meanings of the quantities are illustrated in Figure 1-14. Thus, given an electric field $\mathbf{E} = \mathbf{E}(u_1, u_2, u_3, t)$ in space (denoted by the flux line distribution in that figure), (1-53) means that the integration of $(\epsilon_0 \mathbf{E}) \cdot d\mathbf{s}$ over any closed surface S

† While given the collective name Maxwell's equations, historically they were in a gradual process of evolution over many years before Maxwell's time. For an enjoyable and first-rate account of the details, one is encouraged to read the historical surveys at the beginning of each chapter in Elliott, R. S. *Electromagnetics*. New York: McGraw-Hill, 1966.

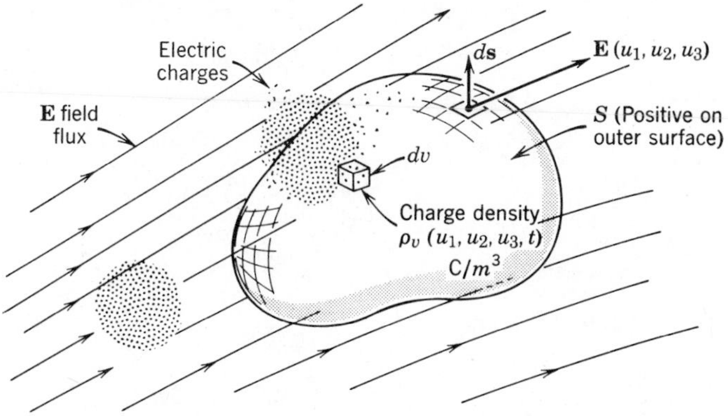

Figure 1-14. Typical closed surface S in a region containing an electric field and a related electric charge. Gauss' law must hold for all closed surfaces constructed in the region, whether charges are contained or not.

(the net, outward flux of $\epsilon_0\mathbf{E}$ emanating from S) is a measure of the amount of electric charge q contained within the volume V inside S. Since the flux of $\epsilon_0\mathbf{E}$ is to be taken positively *outward* from S, the positive sense of every surface element $d\mathbf{s}$ on S is assumed to be outward. The quantity ϵ_0 is called the *permittivity* of free space; in the mks system of units it is approximately $10^{-9}/36\pi$ F/m.†

Gauss' law (1-53) is more than just a criterion of the amount of charge contained in a closed surface. It must be satisfied for *all* possible closed surfaces that may be constructed in a region containing $\mathbf{E}(\mathbf{r}, t)$ and a related charge distribution $\rho_v(\mathbf{r}, t)$. Gauss' law may occasionally be employed to find solutions for $\mathbf{E}$ whenever the charge distribution ρ_v is known, although this is possible in only a few instances of static charge distributions having particular spatial symmetries. A few of them are given in the following examples.

EXAMPLE 1-11. Find the electric field intensity $\mathbf{E}$ of the following static charge distributions in free space: (*a*) a point charge Q; (*b*) a spherical cloud of radius r_0 containing a uniform volume density ρ_v; (*c*) a very long line charge of uniform linear density ρ_ℓ; (*d*) a very large planar (surface) charge of density ρ_s.

These charge distributions are illustrated in Figure 1-15. Closed surfaces S are shown, appropriately chosen to permit solving for $\mathbf{E}$ by the use of Gauss' law (1-53).

† The significance of the units of ϵ_0 is clarified in Chapter 4 in the discussion of capacitance. A correct interpretation of the factor ϵ_0 in (1-53) is that it is a proportionality factor accounting for the proper units (mks) of the equation.

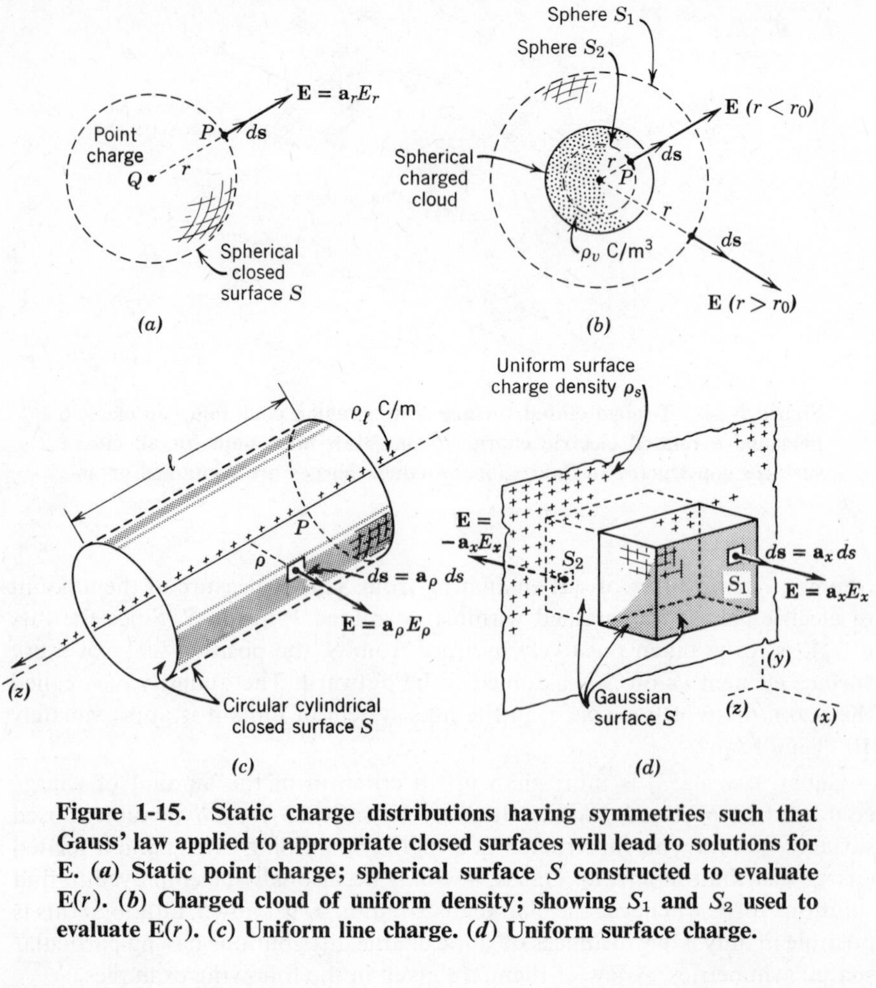

Figure 1-15. Static charge distributions having symmetries such that Gauss' law applied to appropriate closed surfaces will lead to solutions for E. (*a*) Static point charge; spherical surface *S* constructed to evaluate E(r). (*b*) Charged cloud of uniform density; showing S_1 and S_2 used to evaluate E(r). (*c*) Uniform line charge. (*d*) Uniform surface charge.

(*a*) *Field of a point charge* (symmetry about a point).—To evaluate the field **E** of the static point charge Q, choose S in Gauss' law (1-53) to be the sphere with Q at its center as in Figure 1-15(*a*). To show that **E** has only a radial component about the charge, observe that for this time-static problem ($d/dt = 0$, for all fields), (1-55) reduces to $\oint \mathbf{E}\cdot d\boldsymbol{\ell} = 0$ for all closed lines ℓ. Then integrating $\mathbf{E}\cdot d\boldsymbol{\ell}$ about any circumferential path of radius r over the sphere in Figure 1-15(*a*) yields the conclusion that E_θ and E_ϕ are zero. Furthermore, assuming Q positive, **E** must be directed radially *outward* if the integral of $\epsilon_0\mathbf{E}$ over S is to yield a positive answer.

Thus (1-53) yields

$$\oint_S \epsilon_0(\mathbf{a}_r E_r)\cdot \mathbf{a}_r\, ds = Q$$

Since $\mathbf{a}_r \cdot \mathbf{a}_r = 1$ and from the symmetry E_r is constant on S, E_r may be extracted from the integral to obtain

$$E_r = \frac{Q}{4\pi\epsilon_0 r^2}\ \text{N/C or V/m} \tag{1-57a}$$

or, in vector form

$$\mathbf{E} = \mathbf{a}_r E_r = \mathbf{a}_r \frac{Q}{4\pi\epsilon_0 r^2} \tag{1-57b}$$

Coulomb's law for the force acting on another point charge Q' in the presence of Q is deduced by combining (1-57b) with the Lorentz force relation (1-52a). In the absence of a **B** field, the force on Q' when immersed in the **E** field (1-57b) of the charge Q is

$$\mathbf{F}_E = Q'\mathbf{E} = \mathbf{a}_r \frac{Q'Q}{4\pi\epsilon_0 r^2} \tag{1-58}$$

(*b*) *Field of a charged cloud* (symmetry about a point).—For the spherical cloud containing a uniform charge density ρ_v C/m³, two cases arise. The field *outside* the cloud ($r > r_0$) can be obtained from Gauss' law (1-53) applied to a symmetrical sphere S_1 of radius r as shown in Figure 1-15(*b*). That **E** has only an E_r component is shown as in part (*a*). Then the charge q enclosed by S_1 is obtained by integrating $\rho_v\, dv$ throughout the sphere, so (1-53) becomes

$$\oint_{S_1} \epsilon_0(\mathbf{a}_r E_r)\cdot \mathbf{a}_r\, ds = \int_{V_1} \rho_v\, dv$$

Solving for E_r (constant on S_1) yields

$$E_r = \frac{\rho_v((4/3)\pi r_0^3)}{4\pi\epsilon_0 r^2} = \frac{\rho_v r_0^3}{3\epsilon_0 r^2} \qquad r > r_0 \tag{1-59}$$

an inverse-square result. It is of the form of the point-charge result (1-57a), assuming the field point outside the charged cloud ($r > r_0$).

Inside the cloud ($r < r_0$), applying (1-53) to the closed surface S_2 of Figure 1-15(*b*) yields

$$\oint_{S_2} \epsilon_0(\mathbf{a}_r E_r)\cdot \mathbf{a}_r\, ds = \int_{V_2} \rho_v\, dv$$

in which the volume integration is carried out only throughout the interior of S_2, obtaining $\rho_v(4/3)\pi r^3$. With E_r constant on S_2,

$$E_r = \frac{\rho_v r}{3\epsilon_0} \qquad r < r_0 \quad (1\text{-}60)$$

E inside the uniformly charged cloud is therefore zero at its center and varies linearly to the same value as (1-59) at the surface $r = r_0$.

(*c*) *Field of a long line charge* (symmetry about a line).—Construct a closed right circular cylinder of length ℓ and radius ρ concentric about the line charge as in Figure 1-15(*c*). From symmetry, **E** is radially directed ($\mathbf{a}_\rho E_\rho$) and of constant magnitude over the peripheral surface S_0. The left side of Gauss' law (1-53) is zero over the endcaps of S because $\mathbf{E}\cdot d\mathbf{s}$ is zero on them. Thus (1-53) becomes

$$\int_{S_0} \epsilon_0(\mathbf{a}_\rho E_\rho)\cdot\mathbf{a}_\rho \, ds = \int_\ell \rho_\ell \, d\ell$$

in which the right side reduces to a line integral over the linear charge distribution. Solving for E_ρ on S_0 yields, with $\int_\ell \rho_\ell \, d\ell = \rho_\ell \ell$ and $\int_S ds = 2\pi\rho\ell$

$$E_\rho = \frac{\rho_\ell}{2\pi\epsilon_0\rho} \qquad (1\text{-}61)$$

Thus, the electric intensity of an infinitely long, uniform line charge varies inversely with ρ.

(*d*) *Field of an infinite, planar charge* (symmetry about a plane).—A closed surface S is constructed in the form of a rectangular parallelepiped extending equally on both sides of the planar charge, as in Figure 1-15(*d*). The symmetry of the infinitely extensive charge requires that **E** be directed normally away from both sides of the charge as shown ($\mathbf{E} = \pm\, \mathbf{a}_x E_x$). Flux emanates only from the ends S_1 and S_2 of the parallelepiped, whence Gauss' law becomes

$$\int_{S_1} \epsilon_0(\mathbf{a}_x E_x)\cdot\mathbf{a}_x \, ds + \int_{S_2} \epsilon_0(-\mathbf{a}_x E_x)\cdot(-\mathbf{a}_x \, ds) = \int_S \rho_s \, ds = \rho_s A$$

A denoting the area of the ends of the parallelepiped. The two integrals over S_1 and S_2 provide exactly the same amount of outward electric flux, whence

$$E_x = \frac{\rho_s}{2\epsilon_0}$$

Writing this in vector form to include the fields on both sides of the planar charge distribution gives

$$\mathbf{E} = \mathbf{a}_x \frac{\rho_s}{2\epsilon_0} \qquad x > 0$$
$$\mathbf{E} = -\mathbf{a}_x \frac{\rho_s}{2\epsilon_0} \qquad x < 0 \tag{1-62}$$

It is evident that the electric field to either side of a uniform, infinite planar charge is everywhere constant.

Flux plots of the electric fields of the four charge distributions covered in this example are shown in Figure 1-16.

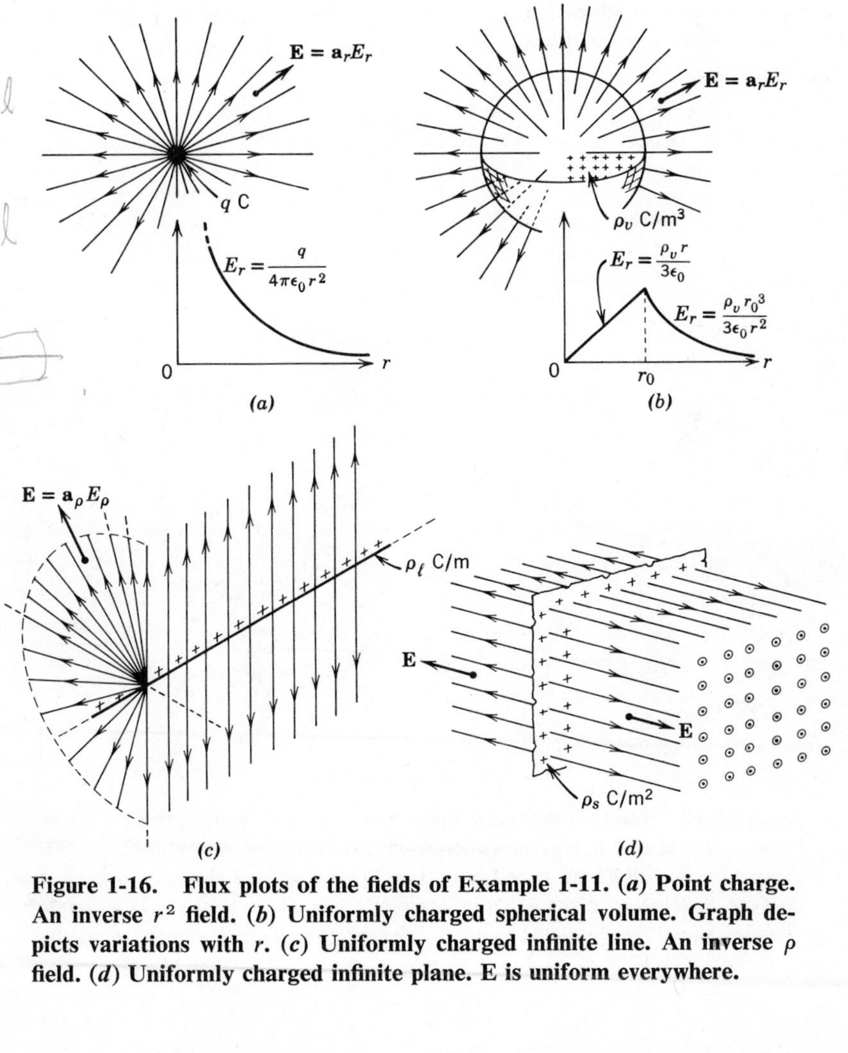

Figure 1-16. Flux plots of the fields of Example 1-11. (*a*) Point charge. An inverse r^2 field. (*b*) Uniformly charged spherical volume. Graph depicts variations with r. (*c*) Uniformly charged infinite line. An inverse ρ field. (*d*) Uniformly charged infinite plane. E is uniform everywhere.

B. Ampère's Circuital Law in Free Space

Maxwell's integral law (1-56)

$$\oint_{\ell} \frac{\mathbf{B}}{\mu_0} \cdot d\ell = \int_S \mathbf{J} \cdot d\mathbf{s} + \frac{d}{dt} \int_S (\epsilon_0 \mathbf{E}) \cdot d\mathbf{s} = i + \frac{d\psi_e}{dt} \qquad [1\text{-}56]$$

is often called Ampère's circuital law for free space. Figure 1-17 illustrates the meanings of the field quantities relative to any closed line ℓ which bounds a two-sided surface S. The positive direction of the typical element $d\mathbf{s}$ may be taken to either side of S, but the positive integration sense about ℓ must agree with the right-hand rule relative to $d\mathbf{s}$. The relation (1-56) means that the line integral of the **B**-field (modified by μ_0^{-1}) around any arbitrary closed path ℓ must, at any time t, equal the sum of the net electric current i plus the time rate of change of the net electric flux ψ_e passing through the surface S bounded by ℓ.

The two terms on the right side of (1-56) denote the two kinds of electric currents that occur physically in free space. The first, i, has already been discussed in relation to (1-50) and is given the name *convection current* when

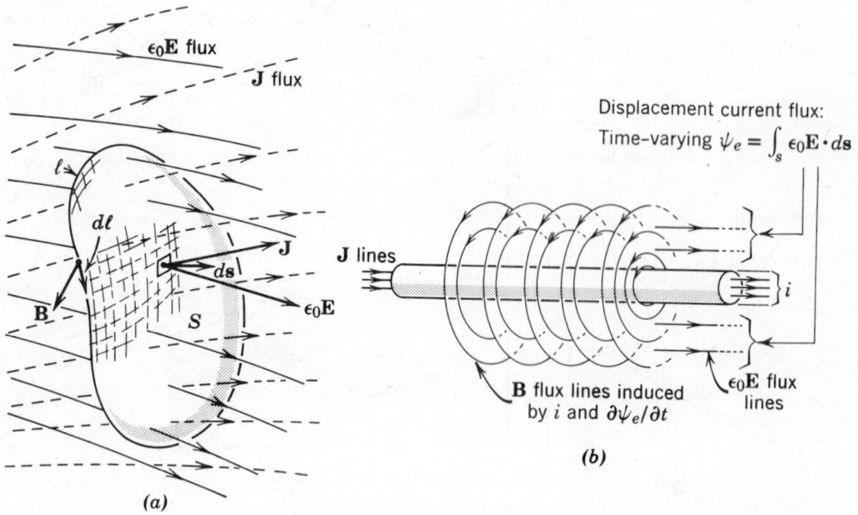

Figure 1-17. Induced magnetic fields and Ampère's law. Any closed line ℓ such as that of (*a*) may be superposed anywhere on the example of (*b*); Ampère's law must be true for it. (*a*) Typical closed line ℓ bounding a surface S, relative to the fields in Ampère's law. (*b*) A symmetric example showing the B field induced by electric currents and displacement currents.

it is comprised of one or more species of moving charges in free space; it is also called *conduction current* if it pertains to electric charges drifting or transported within a solid, liquid, or gas. The second term, $d\psi_e/dt$, is called *displacement current*, and denotes the time rate of change of the net, instantaneous electric flux ψ_e that passes through the surface S. The displacement current term is the history-making contribution of Maxwell, who provided that missing link to unify the theories of electricity and magnetism and predicted the propagation of electromagnetic waves in empty space in the absence of charges and currents. The quantity μ_0 is called the permeability of free space; it has the value $4\pi \times 10^{-7}$ H/m in the mks system of units.†

A comparison of (1-56) with Gauss' law (1-53) shows that Ampère's circuital law is more comprehensive; it involves *both* the magnetic field **B** and the time-varying electric field **E**, as well as electric currents that might be flowing in a region. Indeed, it specifies that either electric currents or time-varying electric fields in a region, or both, will give rise to a magnetic field **B** such that (1-56) must be satisfied for all possible closed lines constructed in the region.

The direct application of Ampère's circuital law (1-56) to obtaining time-varying field solutions $\mathbf{E}(\mathbf{r}, t)$ and $\mathbf{B}(\mathbf{r}, t)$ whenever, for instance, a current distribution $\mathbf{J}(\mathbf{r}, t)$ is somehow specified is not, in general, feasible. The difficulty lies in part in not knowing how to specify the current distribution without more information about the accompanying fields; the intricacies may be appreciated more fully upon recognizing that the field solutions must satisfy simultaneously *all four* of Maxwell's integral relations, (1-53) through (1-56).

One may discover, however, several simple illustrations of the application of Ampère's circuital law to the determination of the magnetic fields of *time-static* current (direct current) distributions. In the static case, (1-56) reduces to an expression from which the displacement current term is absent.

$$\oint_\ell \frac{\mathbf{B}}{\mu_0} \cdot d\boldsymbol{\ell} = \int_S \mathbf{J} \cdot d\mathbf{s} \equiv i \quad \text{Ampère's law for } \textit{static} \text{ fields} \quad (1\text{-}63)$$

EXAMPLE 1-12. Find the net, static electric current i that flows through each of the surfaces S bounded by the paths ℓ chosen for the three direct current systems of Figure 1-18.

(*a*) For the long, straight wire of Figure 1-18(*a*), the path ℓ_1 shown yields a net current $i = 0$ through S_1; while ℓ_2 embraces $i = I$ A, a positive result if $d\mathbf{s}$ on S_2 is taken to be positive in the upward direction.

† The significance of the units of μ_0 is clarified in Chapter 5, in the discussion of inductance. In (1-56), μ_0 is the factor that properly adjusts the units of the term in which it appears, to yield an equality that is dimensionally correct.

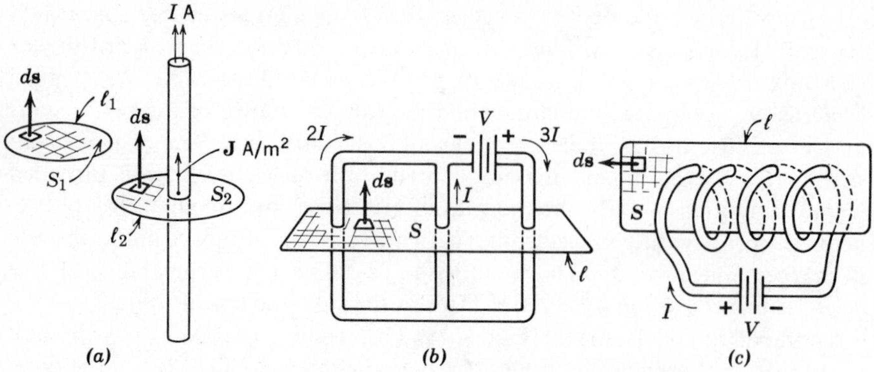

Figure 1-18. Three examples of direct current systems showing closed paths ℓ which bound surfaces S through which net currents i flow. (*a*) Infinitely long wire carrying I A. (*b*) A two-mesh dc circuit. (*c*) A four-turn coil carrying I A.

(*b*) Assume a positive $d\mathbf{s}$ in the upward direction on S. Then, by inspection of Figure 1-18(*b*), the net conduction current i through S becomes

$$i = \int_S \mathbf{J} \cdot d\mathbf{s} = 2I + I - 3I = 0$$

(*c*) For the path ℓ constructed about the coil in Figure 1-18(*c*) such that the coil pierces S four times, the net current becomes

$$\int_S \mathbf{J} \cdot d\mathbf{s} = -4I \text{ A}$$

if $d\mathbf{s}$ is assumed positive in the direction shown.

EXAMPLE 1-13. *Field of a long, round wire* (symmetry about a line).—Use Ampère's circuital law (1-63) to find the **B** field of the static current I in an infinite, straight, round wire of radius a, shown in Figure 1-19(*a*). Find **B** both inside and outside the wire.

As in Figure 1-19(*a*), assume a symmetric, closed integration path ℓ_1 having the radius ρ shown. From (1-63) the **B** field must be ϕ directed if the line integration counterclockwise (looking from above) is to yield the positive current I emerging from S_1. With $\mathbf{B} = \mathbf{a}_\phi B_\phi$ and $d\boldsymbol{\ell} = \mathbf{a}_\phi \rho \, d\phi$ on the closed contour, (1-63) yields

$$\int_{\phi=0}^{2\pi} \frac{B_\phi}{\mu_0} \rho \, d\phi = I$$

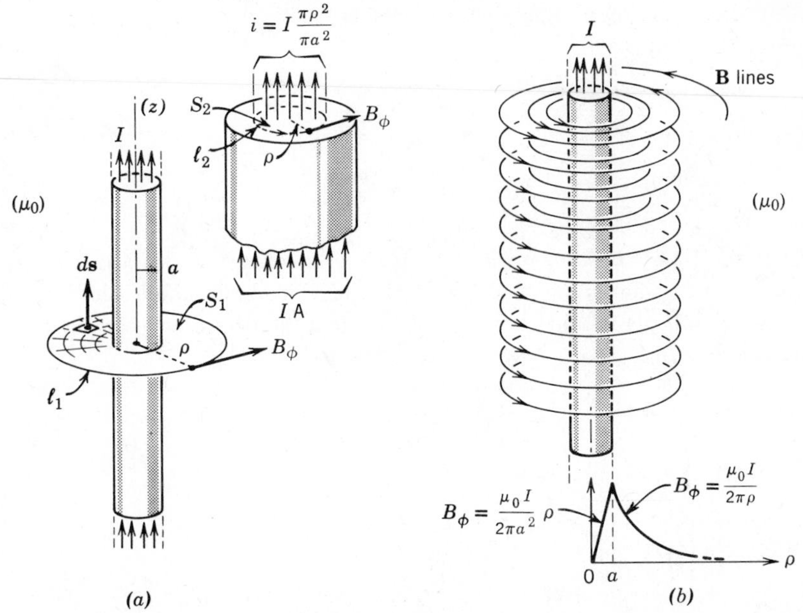

Figure 1-19. A long, straight wire carrying a static current of *I* A, and the associated magnetic field. (*a*) Portion of long, straight current-carrying wire, showing symmetric closed paths for use with Ampère's law to find B. (*b*) External magnetic flux field of the long, straight wire. Graph below depicts the flux density variations with ρ.

but B_ϕ is of constant magnitude on ℓ_1, obtaining exterior to the wire

$$B_\phi = \frac{\mu_0 I}{\int_0^{2\pi} \rho \, d\phi} = \frac{\mu_0 I}{2\pi\rho} \qquad \rho > a$$

If $\rho < a$, the closed line ℓ_2 shown in the inset of Figure 1-19(a) bounds a surface S_2 intercepting only a fraction of I, as determined by the ratio of the area of S_2 to the cross-sectional area πa^2. Then (1-63) becomes

$$\oint_{\ell_2} \frac{B_\phi}{\mu_0} \, d\ell = I \frac{\pi \rho^2}{\pi a^2}$$

yielding

$$B_\phi = \frac{\mu_0 I}{2\pi a^2} \rho$$

Thus the **B** field of the infinitely long, round wire carrying a static current I is

ϕ directed, varying inversely with ρ outside the conductor and directly with ρ inside it as follows:

$$\mathbf{B} = \mathbf{a}_\phi \frac{\mu_0 I}{2\pi\rho} \qquad \rho > a$$
$$\mathbf{B} = \mathbf{a}_\phi \frac{\mu_0 I \rho}{2\pi a^2} \qquad \rho < a \tag{1-64}$$

EXAMPLE 1-14. *Field of a flat current sheet* (symmetry about a plane).—Use Ampère's circuital law to find $\mathbf{B}$ on both sides of a thin, infinite current sheet in the $x = 0$ plane and carrying the constant, static surface current density $\mathbf{J}_s = \mathbf{a}_z J_{sz}$ A/m.

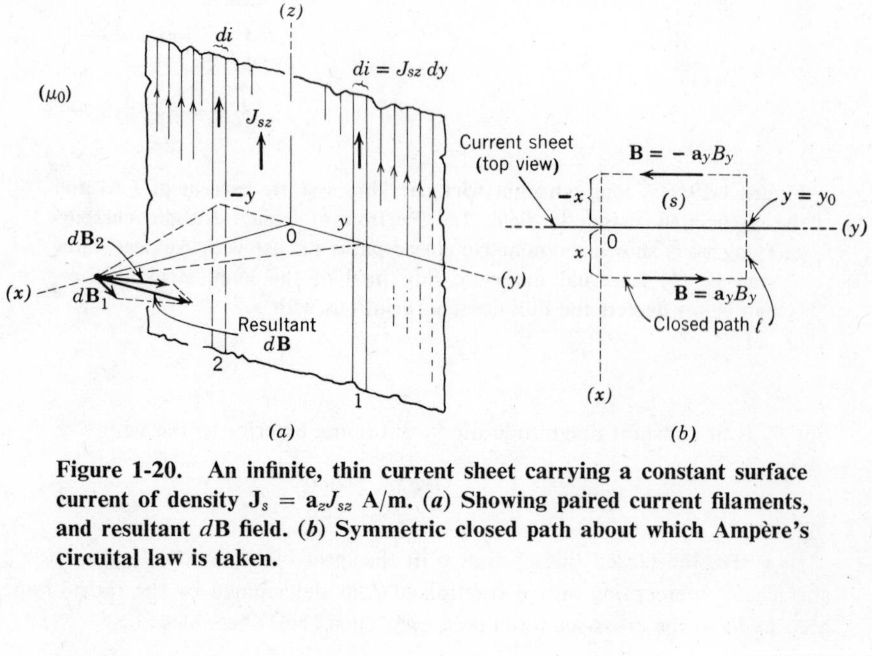

Figure 1-20. An infinite, thin current sheet carrying a constant surface current of density $\mathbf{J}_s = \mathbf{a}_z J_{sz}$ A/m. (*a*) Showing paired current filaments, and resultant $d\mathbf{B}$ field. (*b*) Symmetric closed path about which Ampère's circuital law is taken.

The infinite sheet can be viewed as pairs of thin current filaments located at y, $-y$, carrying the differential current $di = J_{sz}\,dy$ as in Figure 1-20(a). Upon recalling the exterior field result (1-64) of Example 1-13, one may conclude that the paired currents produce a net, y directed, differential field $d\mathbf{B}$ at any point on the x axis as shown. The superposed effect of the whole current sheet is therefore a y directed $\mathbf{B}$ field on the positive x side of the sheet, and a negative y directed field on the other side. Then Ampère's law (1-63) becomes, for the symmetric, rectangular path shown in Figure 1-20(*b*)

$$\int_{y=0}^{y_0} (\mathbf{a}_y B_y)\cdot\mathbf{a}_y\,dy + \int_{y=y_0}^{0} (-\mathbf{a}_y B_y)\cdot\mathbf{a}_y\,dy = \mu_0 \int_{y=0}^{y_0} J_{sz}\,dy$$

with the surface integral on the right side reducing to a line integral over any y_0 width of the sheet. Because both B_y and J_{sz} are constants over the indicated paths, this becomes $2B_y y_0 = \mu_0 J_{sz} y$, to yield $B_y = \mu_0 J_{sz}/2$. In vector form, therefore

$$\mathbf{B} = \mathbf{a}_y \frac{\mu_0 J_{sz}}{2} \qquad x > 0$$
$$\mathbf{B} = -\mathbf{a}_y \frac{\mu_0 J_{sz}}{2} \qquad x < 0 \tag{1-65}$$

EXAMPLE 1-15. Find the magnetic fields of the following *coil* configurations, each carrying a static current I: (*a*) an n turn, closely wound toroid of circular cross-section; (*b*) an infinitely long, closely wound solenoid having n turns in every length d. The coils are illustrated in Figure 1-21.

(*a*) The magnetic flux developed by I in the toroid is ϕ directed as in Figure 1-21(*a*), a result following from symmetry and the application of the right-hand rule to the positive current sense shown. Thus, inside the toroid, $\mathbf{B} = \mathbf{a}_\phi B_\phi$, exact if the winding is idealized into a *current sheet*. The application of the time-static Ampère's circuital law (1-63) to the

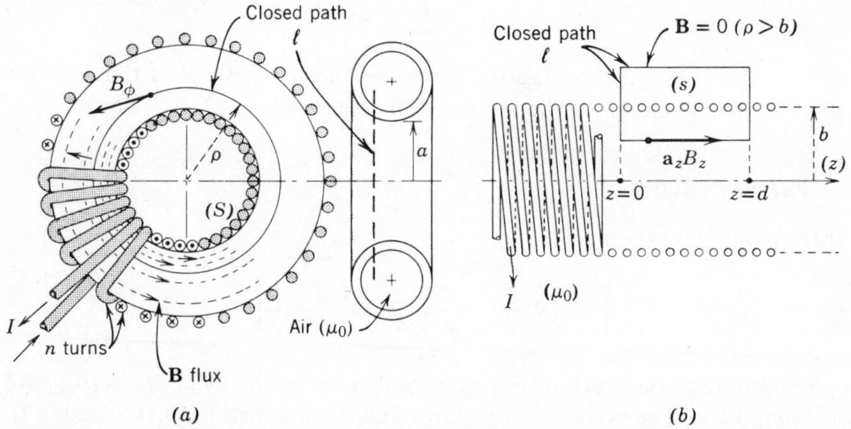

Figure 1-21. Two coil configurations, the magnetic fields of which can be found via Ampère's circuital law. (*a*) Toroidal winding of *n* turns, showing symmetric path ℓ. (*b*) Infinitely long solenoid, showing a typical rectangular closed path ℓ.

symmetric closed line ℓ of radius ρ shown therefore yields $\oint_\ell (\mathbf{a}_\phi B_\phi)\cdot\mathbf{a}_\phi\, d\ell = \mu_0 nI$, in which B_ϕ, from the symmetry, is constant around ℓ, and nI is the net current passing through S bounded by ℓ. Thus

$$B_\phi = \frac{\mu_0 nI}{2\pi\rho} \tag{1-66}$$

an inverse ρ dependent field inside the region bounded by the current sheet. If the radius ρ of ℓ in Figure 1-21(a) were chosen to cause ℓ to fall outside the torus (with $\rho < a$ for example), then S would no longer intercept any net current i. Then from the symmetry, the **B** field outside the idealized toroid must be zero.

(b) The infinitely long solenoid of Figure 1-21(b) may be regarded as a *toroid of infinite radius*; its magnetic field is thus also completely contained *within* the coil if the winding is idealized into an uninterrupted current sheet. The symmetry requires a z directed field, $\mathbf{B} = \mathbf{a}_z B_z$, independent of z. Ampère's circuital law (1-63) is applied to the rectangular closed path shown in Figure 1-21(b), two sides lying parallel to the z axis. A nonzero contribution to the line integral is obtained only over the interior path parallel to the z axis, whence

$$\int_0^d (\mathbf{a}_z B_z)\cdot\mathbf{a}_z\, dz = \mu_0 nI$$

B_z is constant over the path, whence

$$B_z = \mu_0 I \frac{n}{d} \tag{1-67}$$

the ratio n/d denoting the turns per meter length. **B** is thus constant everywhere inside the infinitely long solenoid.

C. Faraday's Law

Maxwell's integral law (1-55)

$$\oint_\ell \mathbf{E}\cdot d\ell = -\frac{d}{dt}\int_S \mathbf{B}\cdot d\mathbf{s} = -\frac{d\psi_m}{dt} \tag{1-55}$$

is attributable to the work of Faraday, and is called the *induced electromotive force (emf) law*. The essence of this law of electromagnetics is expressed in the symbolism of Figure 1-22(a). The relationship of the positive line-integration sense to the positive direction assumed for $d\mathbf{s}$ is the same as for Ampère's circuital law. Faraday's law (1-55) states that the time rate of decrease of the net magnetic flux ψ_m passing through any arbitrary surface S equals the

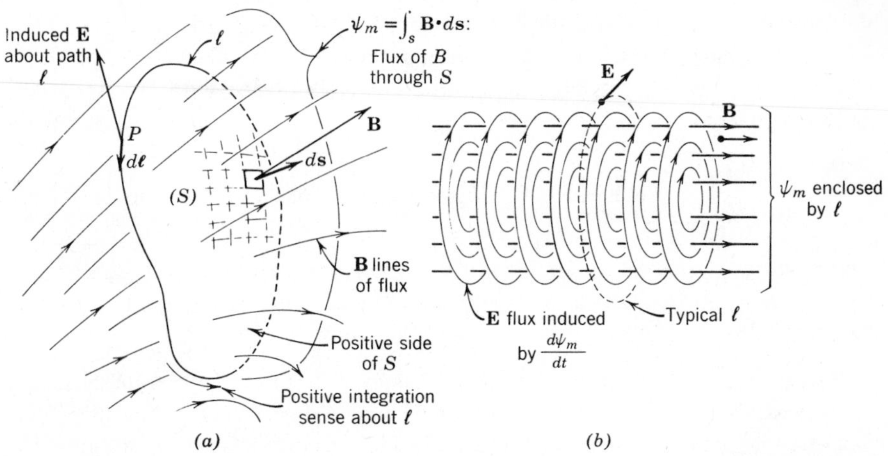

Figure 1-22. Induced electric fields and Faraday's law. Any closed line ℓ such as that of (*a*) may be superposed anywhere on the example of (*b*), such that Faraday's law must be true for it. (*a*) Typical closed line bounding a surface *S*, relative to the fields in Faraday's law. (*b*) A symmetric example showing the E field induced by a time-varying magnetic field.

integral of the **E** field around the closed line bounding *S*. This is tantamount to saying that an **E** field is generated by a time-varying magnetic flux. The **E** field, in general, must also be time-varying if (1-55) is to be satisfied at every instant.

Valid field solutions $\mathbf{E}(\mathbf{r}, t)$ and $\mathbf{B}(\mathbf{r}, t)$ satisfying Faraday's law (1-55) must also satisfy the remaining Maxwell's integral relations of (1-52) through (1-56); however, if the time variations of the fields are not too fast, in some cases a static solution for **B** satisfying the *static* form of Ampère's circuital law (1-63)

$$\oint_\ell \frac{\mathbf{B}}{\mu_0} \cdot d\ell = \int_S \mathbf{J} \cdot d\mathbf{s} = i \qquad [1\text{-}63]$$

can be assumed to be the known field (e.g., the field solutions of Examples 1-13 through 1-15). If the current densities **J** are slowly time-varying, one can assume they will give rise to a slowly time-varying **B** field. Such a static field upon which *time* variations are imposed is called *quasi-static*. Upon inserting the quasi-static field $\mathbf{B}(\mathbf{r}, t)$ into Faraday's law (1-55), a first-order approximation to the **E** field can then be obtained, assuming that the field symmetry permits the extraction of the solution for **E** from (1-55). An iterative process can sometimes then be employed to improve the accuracy of the quasi-static solution,† although if the time variations of the fields are not excessively rapid, the first-order solution will often suffice.

Faraday's law for strictly *time-static* fields is (1-55) with its right side reduced to zero

$$\oint_{\ell} \mathbf{E}\cdot d\ell = 0 \quad \text{Faraday's law for time-static fields} \tag{1-68}$$

which states that the line integral of a static **E** field about any closed path is always zero. A field obeying (1-68) is called a *conservative* field; all static electric fields are conservative.

EXAMPLE 1-16. The long solenoid of Figure 1-21(*b*) carries a suitably slowly time-varying current $i = I_0 \sin \omega t$. Determine from Ampère's law the quasi-static magnetic flux density developed inside the coil of radius a, and then use Faraday's law to find the induced electric intensity field both inside and outside the coil.

From Example 1-15(*b*), the magnetic flux density inside the long solenoid carrying a static current I was found to be (1-67). Thus, the solenoid current $I_0 \sin \omega t$ will to a first-order approximation provide the quasi-static magnetic flux density

$$\mathbf{B}(t) = \mathbf{a}_z \mu_0 I_0 \left(\frac{n}{d}\right) \sin \omega t = \mathbf{a}_z B_0 \sin \omega t \tag{1-69}$$

in which $B_0 = \mu_0 n I_0 / d$, the amplitude of **B**. This assumption is reasonably accurate for an angular frequency ω which is not too large. The electric field **E** induced by this time-varying **B** field is found by means of Faraday's law (1-53), the line integral of which is first taken around the symmetric path ℓ of radius ρ inside the coil, as shown in Figure 1-23. Faraday's law becomes

$$\oint_{\ell} (\mathbf{a}_\phi E_\phi)\cdot \mathbf{a}_\phi \, d\ell = -\frac{d}{dt}\int_S (\mathbf{a}_z B_0 \sin \omega t)\cdot \mathbf{a}_z \, ds$$

in which, from the circular symmetry, E_ϕ must be a constant on ℓ. Thus

$$E_\phi \oint_{\ell} d\ell = -\omega B_0 \cos \omega t \int_S ds$$

but $\oint_{\ell} d\ell = 2\pi\rho$ and $\int_S ds = \pi\rho^2$, so that

$$E_\phi = -\frac{\omega B_0 \rho}{2} \cos \omega t \qquad \rho < a \tag{1-70}$$

† An iterative process applied to the *differential* forms of Maxwell's equations (which are developed in Chapter 2) is described in Fano, R. M., L. J. Chu, and R. B. Adler. *Electromagnetic Fields, Energy and Forces*. New York: Wiley, 1960, Chapter 6.

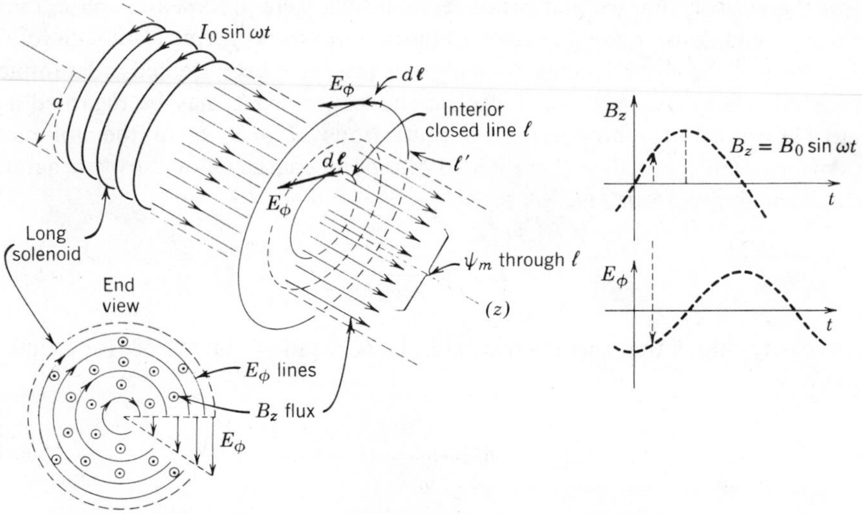

Figure 1-23. Showing the assumed integration path ℓ used for finding the induced E field of a solenoid, and the resulting E field.

is the first-order solution for the electric field intensity generated by the time-varying magnetic flux of the solenoid. Observe that E_ϕ varies in direct proportion to ρ, as shown in Figure 1-23. The negative sign in the result implies that as the net flux ψ_m through S is increasing (in the positive z direction), the sense of the induced **E** field is negative ϕ directed. This is symbolized in the time diagram of Figure 1-23. [*Note*: The induced electric field, at the position $\rho = a$ of the solenoid wire, has a direction such that it opposes the tendency for *current* to change in that wire;† this view of the self-induced electric field leads to the concept of the *self-inductance* of the coil, which will be considered in detail in Chapter 5.]

Upon applying Faraday's law to the closed line ℓ' exterior to the coil, one obtains for the electric field intensity

$$E_\phi = \frac{-\omega B_0 a^2}{2\rho} \cos \omega t \qquad \rho > a \quad (1\text{-}71)$$

The electric field generated outside the long solenoid by the time-varying magnetic flux, therefore, varies inversely with respect to ρ. Both answers are directly proportional to ω because they are governed by the time rate of change of the net magnetic flux intercepted by the surface, as noted from (1-55).

† Sometimes referred to as Lenz's law.

If the electric charges that produce an electric field are fixed in space, that electric field must obey Faraday's law in *time-static* form, (1-68). Several examples of the electric fields of charges at rest have been treated in Example 1-11. All static *distributions* of electric charges in space may be regarded as superpositions of point-charge concentrations $dq = \rho_v \, dv$ in the volume-elements dv in space. The electric field of a point charge Q, on the other hand, has been shown to be (1-57b)

$$\mathbf{E} = \mathbf{a}_r \frac{Q}{4\pi\epsilon_0 r^2} \tag{1-57b}$$

It is easily shown that this electric field obeys Faraday's law (1-68) for a time-static field

$$\oint \mathbf{E} \cdot d\ell = 0 \tag{1-68}$$

If any closed path, such as $\ell = \ell_a + \ell_b$ shown in Figure 1-24, is chosen in the space about a point charge, the integral of $\mathbf{E} \cdot d\ell$ from any point P_1 to any other point P_2 along the path ℓ_a is

$$\begin{aligned}
\int_{P_1}^{P_2} \mathbf{E} \cdot d\ell &= \int_{P_1}^{P_2} \left[\mathbf{a}_r \frac{Q}{4\pi\epsilon_0 r^2} \right] \cdot (\mathbf{a}_r \, dr + \mathbf{a}_\theta r \, d\theta + \mathbf{a}_\phi r \sin\theta \, d\phi) \\
&= \int_{r=r_1}^{r_2} \frac{Q}{4\pi\epsilon_0 r^2} \, dr \\
&= \frac{Q}{4\pi\epsilon_0} \left[\frac{1}{r_1} - \frac{1}{r_2} \right] \qquad (1\text{-}72)
\end{aligned}$$

This result† is seen to be independent of the choice of the path connecting P_1 and P_2; it is a function only of the radial distances r_1 and r_2 to the respective endpoints P_1 and P_2. Therefore, if the integration is taken around the complete path $\ell = \ell_a + \ell_b$ shown in Figure 1-24, the two integrals from P_1 to P_2 via ℓ_a and thence from P_2 back to P_1 via ℓ_b will cancel, and (1-68) follows. Static charge *distributions* like those depicted in Figure 1-24(*b*) are, in general, just collections of differential charge-elements $dq = \rho_v \, dv$; while their static electric fields are just superpositions (vector sums) of the *conservative* differential electric fields $d\mathbf{E}$ produced by each of those static charge-elements.

† The physical interpretation of the result (1-72) is of interest. It implies that the *net work* done in moving a unit test charge around a closed path is zero; such an electric field has already been termed *conservative*. Thus (1-72) forms the basis of the theory of the scalar potential field of static electric charges to be discussed in Chapter 4.

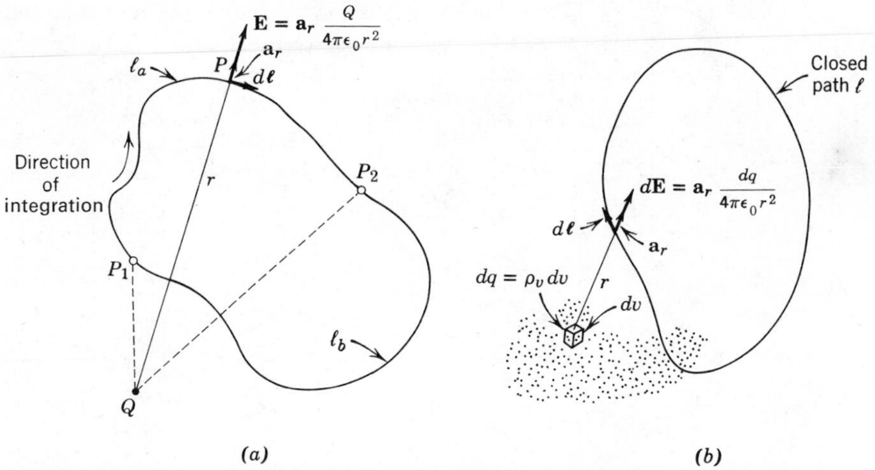

Figure 1-24. Closed paths constructed about a point charge and a charge distribution, relative to Faraday's law for static charges. (*a*) Point charge *Q*. (*b*) Charge distribution ρ_v.

One may thereby agree that Faraday's law (1-68) for static electric fields is true in general.

If an electric charge (or a group of charges) were introduced into the electric field generated by the time-varying magnetic field of a system such as the one shown in Figure 1-23, that charge would experience electric and magnetic forces predictable from the Lorentz expression (1-52). Two examples of physical devices that make use of these forces are shown in Figure 1-25. Part (*a*) of that figure shows how a time-varying magnetic field and its induced electric field are utilized, under certain conditions, in directing a beam of electrons about a circular path in the instrument known as the betatron, developed in 1941 by D. W. Kerst. The axially symmetric magnetic flux density shown between the pole pieces is generated by a low-frequency, sinusoidal current in a coil. The time variations of the magnetic flux generate an axially symmetric, ϕ directed electric field **E** predictable from Faraday's law (1-55) in the manner of Example 1-16. If electronic charges are introduced into the evacuated region between the poles, the charges will be urged into flight by two Lorentz forces: an azimuthal electric field force, $\mathbf{F}_E = -e\mathbf{E}$, which provides the electron with a velocity $\mathscr{v}$ tangent to its path of motion; and a magnetic field force $\mathbf{F}_B = -e\mathscr{v} \times \mathbf{B}$ which is perpendicular to the path as illustrated in Figure 1-25(*a*). A detailed analysis shows that a *nonuniform*, axially symmetric magnetic field is required to maintain a stable circular path

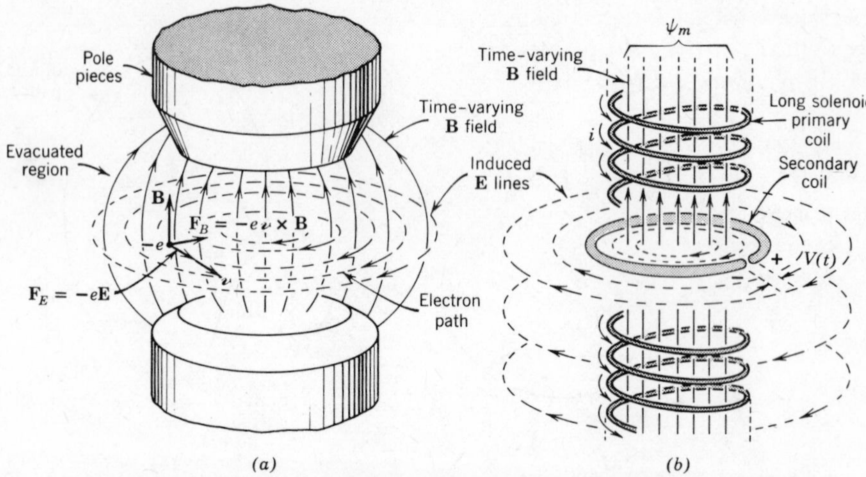

Figure 1-25. Device applications of Faraday's law (1-55). (*a*) Betatron: the magnetic field and induced electric field forces constrain a moving electron into a circular path. (*b*) Transformer principle: the motion of free electrons inside the secondary conductor provides the voltage $V(t)$ across its open ends.

for the electrons, with a greater **B** density along the axis than at the path location.

Figure 1-25(*b*) illustrates the related principle of the transformer. Supposing that the solenoid of Figure 1-23 is taken to be the transformer primary coil carrying a time-varying current i shown in Figure 1-25(*b*), its magnetic field induces the azimuthally directed **E** field shown in that figure. A secondary conductor, bent into the loop shown in Figure 1-25(*b*) and provided with a small gap between its ends, is located so that the time-varying magnetic field passes through the surface S bounded by the secondary conductor. The easily detached outer-orbit electrons of the conductor are urged by the forces of the induced **E** field to move along the secondary wire to produce an excess of negative charges toward one end of the wire, while a dearth of negative charges (tantamount to a positive charge) is established at the other end. The result is a time-varying voltage $V(t)$ at the gap. More will be said of the transformer principle in Chapter 5.

D. Gauss' Law for Magnetic Fields

Maxwell's integral law (1-54)

$$\oint_S \mathbf{B} \cdot d\mathbf{s} = 0 \qquad [1\text{-}54]$$

is also known as Gauss' law for magnetic fields. It specifies that the net magnetic flux (positive or negative) emanating from any closed surface S in space is always zero. This statement is illustrated in Figure 1-26; in (*a*) of that figure is an arbitrary closed surface S constructed in the region containing a generalized magnetic flux configuration having a density $B(\mathbf{r}, t)$ in space. Maxwell's integral law requires that a total of zero net magnetic lines emanate from every such closed surface S. This means that magnetic flux lines always form closed lines. Equivalently, it states that magnetic fields cannot terminate

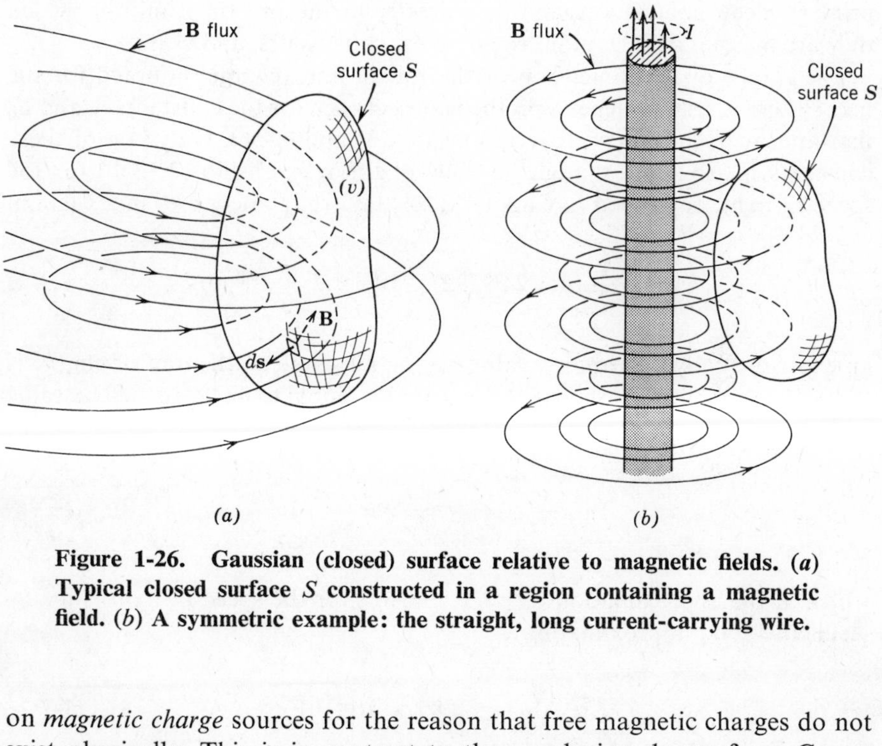

Figure 1-26. Gaussian (closed) surface relative to magnetic fields. (*a*) Typical closed surface *S* constructed in a region containing a magnetic field. (*b*) A symmetric example: the straight, long current-carrying wire.

on *magnetic charge* sources for the reason that free magnetic charges do not exist physically. This is in contrast to the conclusion drawn from Gauss' law (1-53) for electric fields; the presence of a nonzero right side involving the electric charge density function ρ_v in that relation attests to the physical existence of free electric charges.

It is easy to find physical examples that illustrate the closed nature of magnetic flux lines. The magnetic field of a long, straight, current-carrying wire of Figure 1-19 is shown once more in Figure 1-26(*b*); observe how the uninterrupted flux lines account for precisely as many magnetic flux lines entering the typical closed surface S as are emerging from it. Such is the case for all closed surfaces that one might construct in space for that field.

1-12 Units and Dimensions

The mks system of units, introduced by Giorgi in 1901, is now employed almost universally in electromagnetics. In this system, length is expressed in *meters*, mass in *kilograms*, and time in *seconds*. A fourth unit, that of either electric charge (coulomb) or electric current (coulomb per second, or ampere), is needed in the dimensional description of electromagnetic phenomena. The rationalized mks system, which eliminates a factor 4π from the Maxwell equations, has been almost universally adopted, and it is used in this text. The Giorgi mks system is especially noteworthy in that it deals with the primary electromagnetic quantities directly in the practical units in which they are measured: in coulombs, amperes, volts, watts, and ohms.

The choice of the dimension of the fourth unit (charge) adopted for the mks system is seen to depend on the values chosen for the constants ϵ_0 and μ_0 that appear in the Maxwell equations, (1-53) and (1-56). Only one of these constants is arbitrary, though, in view of the relationship (2-125b) for the speed of light, developed in Chapter 2 for uniform plane waves in a vacuum

$$c = \frac{1}{\sqrt{\mu_0 \epsilon_0}} = 2.99792 \times 10^8 \cong 3 \times 10^8 \text{ m/s} \tag{1-73}$$

an experimentally determined value. In the mks system, the unit of charge is the *coulomb*, defined by setting the constant μ_0 equal to $4\pi \times 10^{-7}$. The value of the constant ϵ_0 is then obtained from (1-73)

$$\epsilon_0 = \frac{1}{\mu_0 c^2} \tag{1-74a}$$

which, if the approximation $c \cong 3 \times 10^8$ m/s for the speed of light is made, yields the good approximation

$$\epsilon_0 \cong \frac{10^{-9}}{36\pi} \cong 8.85 \times 10^{-12} \text{ F/m} \tag{1-74b}$$

This value for ϵ_0 substituted into the Coulomb force expression (1-58) then provides the correct scale factor to obtain the force between the charges in newtons, the charges q and q' being given in coulombs and separated a distance r given in meters. One *newton* of force, that required to accelerate a 1 kg mass at the rate one meter per second per second (1 m/sec^2), is thus the product of mass (kilogram) and acceleration (meter/second^2), making 1 N = 1 km/s^2 (=10^5 dyn). The unit of energy or work is the newton-meter, or *joule* (=10^7 erg).

In Table 1-1 are listed units of the mks system by name, unit, and symbol.

Table 1-1 Physical quantities in the mks system

Physical quantity	*Symbol*	*Unit*	*Abbreviation*	*In terms of basic units*
Length	$\ell, d, \cdots$	meter	m	
Mass	m	kilogram	kg	
Time	t, T	second	sec	
Charge	q, Q	coulomb	C	
Current	i, I	ampere	A	C/sec
Frequency	f	hertz	Hz	sec^{-1}
Force	F	newton	N	$kg \cdot m/sec^2$
Energy	U	joule	J	$N \cdot m$
Power	P	watt	W	J/sec
Potential, emf	Φ, V	volt	V	$W/A = N \cdot m/C$
Electric flux	ψ_e	coulomb	C	
Capacitance	C	farad	F	$C/V = A \cdot sec/V$
Resistance	R	ohm	Ω	V/A
Conductance	G	mho	℧	A/V
Magnetic flux	ψ_m	weber	Wb	$V \cdot sec$
Magnetic flux density	B	tesla	T	$Wb/m^2 = V \cdot sec/m^2$
Inductance	L	henry	H	$Wb/A = V \cdot sec/A$
Free-space permeability	μ_0	henry/meter	H/m	$\Omega \cdot sec/m$
Free-space permittivity	ϵ_0	farad/meter	F/m	℧ $\cdot$ sec/m
Conductivity	σ	mho/meter	℧/m	

The symbolisms largely agree with the recommendations of the International Organization for Standardization (ISO).†

The numerical designation of field quantities is facilitated through the use of appropriate powers of ten. Thus 10^6 hertz $= 10^6$ Hz is written 1 MHz, in which the prefix M (mega) denotes the 10^6 factor by which the unit is multiplied. Similarly, 3×10^{-12} F is abbreviated 3 pF, with p (pico) denoting the factor 10^{-12}. Other literal prefixes to be used in this way are listed in Table 1-2.

Table 1-2 Symbols for multiplying factors

Multiplying factor	*Prefix*	*Symbol*	*Multiplying factor*	*Prefix*	*Symbol*
10^{12}	tera	T	10^{-2}	centi	c
10^{9}	giga	G	10^{-3}	milli	m
10^{6}	mega	M	10^{-6}	micro	μ
10^{3}	kilo	k	10^{-9}	nano	n
10^{2}	hecto	h	10^{-12}	pico	p
10	deka	da	10^{-15}	femto	f
10^{-1}	deci	d	10^{-18}	atto	a

† See *IEEE Spectrum*, March 1971, p. 77 for a digest of the recommendations of the IEEE Standards Committee adopted December 3, 1970.

REFERENCES

ABRAHAM, M., and R. BECKER. *The Classical Theory of Electricity and Magnetism*. Glasgow: Blackie, 1943.

CLEMENT, P. R., and W. C. JOHNSON. *Electrical Engineering Science*. New York: McGraw-Hill, 1960.

FANO, R. M., L. T. CHU, and R. P. ADLER. *Electromagnetic Fields, Energy and Forces*. New York: Wiley, 1960.

HAYT, W. H. *Engineering Electromagnetics*, 2nd ed. New York: McGraw-Hill, 1967.

LORRAIN, P., and D. R. CORSON. *Electromagnetic Fields and Waves*. San Francisco: W. H. Freeman, 1970.

PHILLIPS, H. B. *Vector Analysis*. New York: Wiley, 1944.

REITZ, R., and F. J. MILFORD. *Foundations of Electromagnetic Theory*. Reading, Mass.: Addison-Wesley, 1960.

PROBLEMS

1-1. Given the three vectors: $\mathbf{A} = 3\mathbf{a}_x + 4\mathbf{a}_y + 5\mathbf{a}_z$, $\mathbf{B} = \mathbf{a}_x - 2\mathbf{a}_y$, and $\mathbf{C} = 6\mathbf{a}_z$. Sketch these three vectors in a rectangular coordinate frame of reference, and then evaluate the following quantities:

(*a*) $\mathbf{A} + \mathbf{B}$ *Answer:* $4\mathbf{a}_x + 2\mathbf{a}_y + 5\mathbf{a}_z$
(*b*) $\mathbf{A} + \mathbf{C}$
(*c*) $\mathbf{B} \cdot \mathbf{C}$
(*d*) $\mathbf{A} - \mathbf{C}$ *Answer:* $3\mathbf{a}_x + 4\mathbf{a}_y - \mathbf{a}_z$
(*e*) $\mathbf{B} - \mathbf{A}$
(*f*) $\mathbf{C} + \mathbf{A} - \mathbf{B}$
(*g*) $\mathbf{A} \cdot \mathbf{C}$ *Answer:* 30
(*h*) $\mathbf{A} \times \mathbf{B}$
(*i*) $\mathbf{B} \times \mathbf{A}$
(*j*) $\mathbf{A} \cdot \mathbf{B} \times \mathbf{C}$ *Answer:* -60
(*k*) $\mathbf{B} \cdot \mathbf{C} \times \mathbf{A}$
(*l*) $\mathbf{A} \times \mathbf{B} \cdot \mathbf{C}$
(*m*) $(\mathbf{A} \times \mathbf{B}) \times \mathbf{C}$ *Answer:* $30\mathbf{a}_x - 60\mathbf{a}_y$
(*n*) $\mathbf{A} \times (\mathbf{B} \times \mathbf{C})$

1-2. An electric field intensity at a certain point in space is given by the function

$$\mathbf{E} = 1500\mathbf{a}_x + 1200\mathbf{a}_y - 600\mathbf{a}_z \text{ V/m}$$

Find (*a*) the magnitude of this electric field intensity at that point; and (*b*) the rectangular coordinate expression for a unit vector which has the same direction as **E** [*Hint*: Use the concept of the product of a scalar and a vector.]

1-3. Given the vectors $\mathbf{A} = 6\mathbf{a}_x - 8\mathbf{a}_y + 2\mathbf{a}_z$ and $\mathbf{B} = 4\mathbf{a}_x + 6\mathbf{a}_y + 12\mathbf{a}_z$:

(a) Find $\mathbf{A} + \mathbf{B}$, $\mathbf{A} - \mathbf{B}$, $\mathbf{A}\cdot\mathbf{B}$, $\mathbf{A} \times \mathbf{B}$.

(b) Find the unit vectors in the directions of **A** and of **B**.

(c) If **A** and **B** are taken to be the diagonals of a parallelogram, show that it is also a rhombus. Find its sides.

1-4. Suppose one is given two fields **C** and **D** in region of space such that

$$\mathbf{C}(x, y, z) = 4x\mathbf{a}_x + z\mathbf{a}_y + y^2z^2\mathbf{a}_z$$
$$\mathbf{D}(x, y, z) = y\mathbf{a}_x + 3\mathbf{a}_y - yz\mathbf{a}_z$$

(*a*) Find the vector field $\mathbf{F}(x, y, z)$, if $\mathbf{F} = \mathbf{C} \times \mathbf{D}$.

(*b*) Prove that the field $\mathbf{F}(x, y, z)$ is perpendicular to $\mathbf{C}(x, y, z)$ *everywhere* in this region of space.

1-5. Find the position vectors **r** and **r**′ of the points $P(3, 5, -2)$ and $P'(1, -4, 3)$, and show a sketch of them in a rectangular coordinate frame of reference. Find the rectangular coordinate expression for the vector $\mathbf{r} - \mathbf{r}'$, and show it in the sketch. Find its length $R = |\mathbf{r} - \mathbf{r}'|$.

1-6. Find the work done in moving an object along a straight line from the point $P_1(2, 3, -1)$ to the point $P_2(1, 4, -2)$ (expressed in meters) by the constant force $\mathbf{F} = 3\mathbf{a}_x - 2\mathbf{a}_y + \mathbf{a}_z$ N.

1-7. Find the moment of the force $\mathbf{F} = 3\mathbf{a}_y - 3\mathbf{a}_z$ about the point $P(2, 0, 0)$, if **F** is applied at the position $P'(0, 2, 3)$. Show a sketch of the quantities.

1-8. Show, for a nonzero vector **A**, that if both the conditions $\mathbf{A}\cdot\mathbf{B} = \mathbf{A}\cdot\mathbf{C}$ and $\mathbf{A} \times \mathbf{B} = \mathbf{A} \times \mathbf{C}$ are true simultaneously, then $\mathbf{B} = \mathbf{C}$, but if only one of the conditions holds, then **B** is not necessarily equal to **C**.

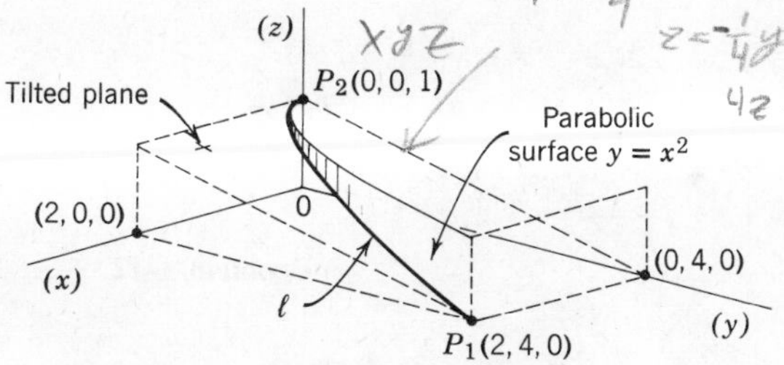

Problem 1-9

1-9. A path is defined in space as the intersection ℓ shown between the parabolic surface $y = x^2$ and a tilted plane passing through the points shown in the figure. Calculate the value of the line integral of $\mathbf{F}\cdot d\boldsymbol{\ell}$ between the points P_1 and P_2 shown, if it is assumed that the vector field **F** is given by

$$\mathbf{F}(x, y, z) = \mathbf{a}_x 5xy + \mathbf{a}_y 6xy^2 + \mathbf{a}_z 2xz$$

Assume P_1 to be the starting point of the integration.

1-10. From the following figure:

(*a*) Calculate the line integral of $\mathbf{A} \cdot d\boldsymbol{\ell}$ along the semicircular path shown in the figure, if $\mathbf{A}(\rho, \phi, z) = \mathbf{a}_\rho \rho^2 z + \mathbf{a}_\phi \sin\phi \cos\phi + \mathbf{a}_z \rho^2 \cos^2\phi$, and the starting point is at the origin.

(*b*) Evaluate the line integral along the straight-line path (the y axis) between the same endpoints, to deduce whether or not the field is conservative.

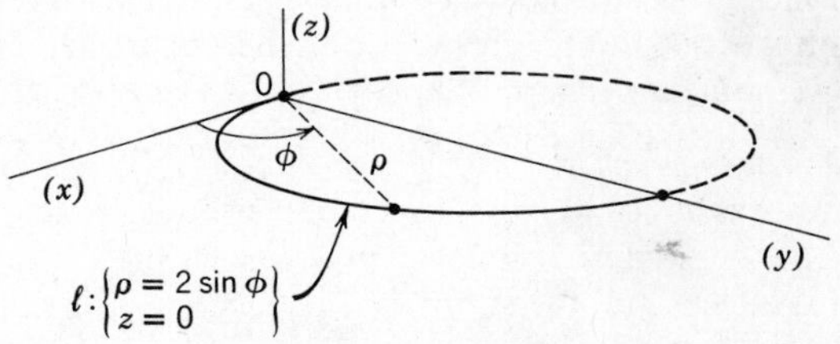

Problem 1-10

1-11. The relation among the radial unit vector $\mathbf{a}_r$ of spherical coordinates and unit vectors of the circular cylindrical coordinate system is $\mathbf{a}_r = \mathbf{a}_\rho \sin\theta + \mathbf{a}_z \cos\theta$.
Use this relation to show that $(\partial \mathbf{a}_r / \partial \phi) = \mathbf{a}_\phi \sin\theta$.

1-12. An electron is shot with an initial velocity $\mathcal{v}_0 = \mathbf{a}_x 10^5$ m/sec into a region containing a uniform magnetic field $-\mathbf{a}_z 10^{-4}$ Wb/m^2 as shown in the figure.

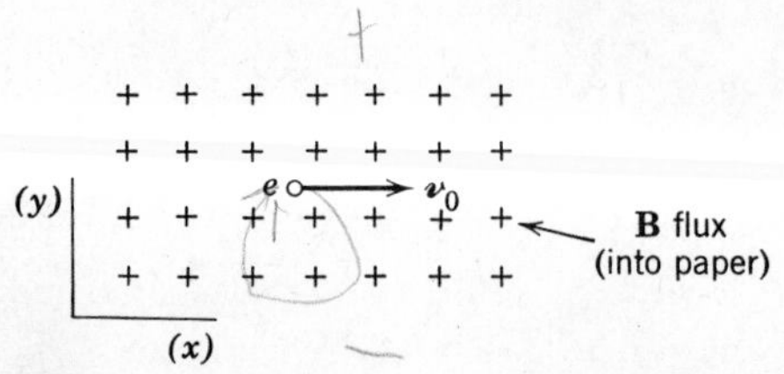

Problem 1-12

(*a*) Assuming the magnetic field to be uniform over a sufficiently large region of space, show your reasoning as to why this electron should take a circular path; then determine the location and radius of this circle.

(*b*) Determine what vector electric field intensity $\mathbf{E}$ will just overcome the effect of the given magnetic field to cause the electron to travel in a straight line along the x axis.

1-13. Determine the amount of static charge contained in free space by each of the following surfaces over which the given electric intensities are distributed:

(*a*) $\mathbf{E} = \mathbf{a}_x 1500x$ V/m over a cube 20 cm on a side, centered at the origin and with its sides coincident with coordinate surfaces.

(*b*) $\mathbf{E} = \mathbf{a}_\rho 1000/\rho$ V/m over a right circular cylinder of 10 cm radius centered on the z axis, and of a length extending from $z = -1$ to 1 m.

(*c*) $\mathbf{E} = \mathbf{a}_r 500r$ V/m over a sphere of 5 cm radius, centered at the origin.

1-14. If a total charge q is placed on a conducting sphere of radius a in free space, the mutual repulsion among the charges forces them to the surface of the sphere to produce a static surface density ρ_s C/m^2.

(*a*) What surface density exists on the sphere (in terms of q)?

(*b*) Use Gauss' law to prove that the static electric intensity field outside the sphere ($r > a$) is the same as (1-57b), that of point charge located at $r = 0$, and further prove that the electric field inside the conducting sphere is zero.

1-15. Prove the result (1-60) for the electric field inside a spherical charged cloud of density ρ_v C/m^3 in free space.

1-16. An infinitely long, cylindrical region of radius a contains a uniform charge density ρ_v C/m^3 in free space. Utilize Gauss' law and the symmetry to prove the following:

(*a*) The electric intensity outside the cylinder ($\rho > a$) is

$$E_\rho = \frac{\rho_v a^2}{2\epsilon_0 \rho}$$

an inverse ρ dependent electric field.

(*b*) The electric intensity inside the cylinder ($\rho < a$) is

$$E_\rho = \frac{\rho_v}{2\epsilon_0}\rho$$

seen to be directly ρ dependent within the cloud.

1-17. Two parallel, planar charges of infinite extent are located in free space at $x = -d/2$ and $d/2$, and they possess uniform but opposite surface charge densities $-\rho_s$ and ρ_s C/m^2.

(*a*) Show, from the superposition of the fields, that the electric field between the surface charges ($-d/2 < x < d/2$) is ρ_s/ϵ_0 V/m, while that outside the surfaces ($|x| > d/2$) is zero.

(*b*) Find the electric fields everywhere if both the surface charges are of the same sign.

1-18. Two conducting sheets of uniform thickness a and of infinite extent are placed with their nearest surfaces at $x = -d/2$ and $d/2$. If positive and negative charges are placed on the respective conductors, the mutual attraction will urge the charges to the inner surfaces. Suppose the charge densities are $-\rho_s$ and ρ_s C/m^2 on the respective conductors. Show why the electric field between the conductors $[-(d/2) < x < (d/2)]$ is the same as that obtained in Problem 1-17(*a*) and zero elsewhere.

1-19. Suppose the volume charge density inside an infinitely long, circular cylindrical region of radius a to be

$$\rho_v = \rho_0(1 + \alpha\rho^2)$$

in which ρ_0 is a scale factor and ρ is the radial distance from the cylinder axis. Determine the value of the parameter α for which the electric field everywhere outside the cylinder ($\rho > a$) is zero. For this value of α, what is the electric field inside the cylinder?

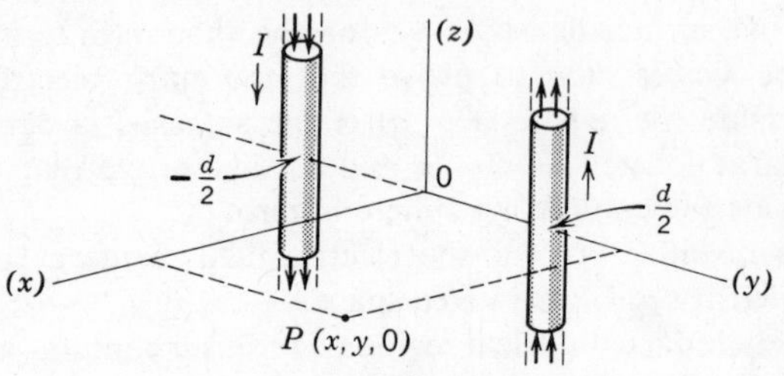

Problem 1-20

1-20. Two infinitely long, parallel, round conductors are spaced a distance d in free space and carry a static current of I A each in opposite directions as shown:

(a) Find the magnetic field $\mathbf{B}(x, y)$ at any position $P(x, y, 0)$ exterior to both conductors. (Make use of the field solution for the infinite, isolated conductor and use superposition. Express your answer in rectangular coordinates.)

(*b*) Find the value of $\mathbf{B}(0, 0)$ (magnitude and direction) at the origin, if $I = 10$ A and $d = 5$ cm.

(*c*) Sketch a rough flux plot of the resultant $\mathbf{B}$ field in the $z = 0$ plane ($\mathbf{B}$ consisting of uninterrupted flux lines).

1-21. Two parallel surface currents (current sheets) of infinite extent are located in free space at $x = -d/2$ and $x = d/2$, and they possess uniform but opposite z directed, static, surface currents of densities J_{sz} and $-J_{sz}$, respectively. Assume that the currents are charge compensated, so that electric fields are absent in this problem.

(*a*) Use the results of Example 1-14 and field superposition to show that $\mathbf{B}$ between the sheets ($-d/2 < x < d/2$) is $\mu_0 J_{sz}$, while that outside the sheets ($|x| > d/2$) is zero.

(*b*) Find $\mathbf{B}$ everywhere if both surface currents flow in the positive z direction.

(*c*) Sketch a flux plot in the $z = 0$ plane for both cases.

1-22. An infinite plane conducting slab of thickness d is situated in free space such that its two sides coincide with the $x = -d/2$ and $x = d/2$ planes. Assume a constant volume current density $\mathbf{J} = \mathbf{a}_z J_z$ A/m^2 in the conductor.

(*a*) Use Ampère's law as it is applied to Example 1-14 to prove that the magnetic flux density exterior to the conductor is

$$\mathbf{B} = \mathbf{a}_y \frac{\mu_0 J_z d}{2} \qquad x > \frac{d}{2}$$

$$\mathbf{B} = -\mathbf{a}_y \frac{\mu_0 J_z d}{2} \qquad x < \frac{d}{2}$$

(*b*) Use Ampère's law to find **B** inside the conducting slab. [*Hint*: Construct a rectangular closed line ℓ having one side parallel to the known field of part (*a*), and the other side aligned with the field to be determined.]

1-23. Extend the results of Problem 1-22 to obtain, by superposition, the magnetic field of a *pair* of infinite conducting slabs, each of thickness d, with their nearest surfaces located at $x = -D/2$ and $x = D/2$, and carrying the volume current densities J_z and $-J_z$, respectively. Sketch a graph of the magnetic flux density versus x.

1-24. Work the problem of Example 1-14 by integration, considering that the infinite current sheet consists of an infinitude of z directed current strips, each dy m wide and carrying a differential amount of current $J_{sz}\, dy$ A. [*Hint*: Use the result of Example 1-13 to show that each strip provides a differential contribution $d\mathbf{B} = \mathbf{a}_\phi \mu_0 J_{sz}\, dy/2\pi\rho$ at a normal distance ρ away. If the strips are paired as suggested by Figure 1-20, only the y component of the net $d\mathbf{B}$ remains, and the result follows from superposition (integration).]

1-25. An infinitely long, hollow, circular conductor carries a total, static current I in free space and has inner and outer radii b and c as shown. Use Ampère's law to show that the exterior field ($\rho > c$) is the same as that of a solid

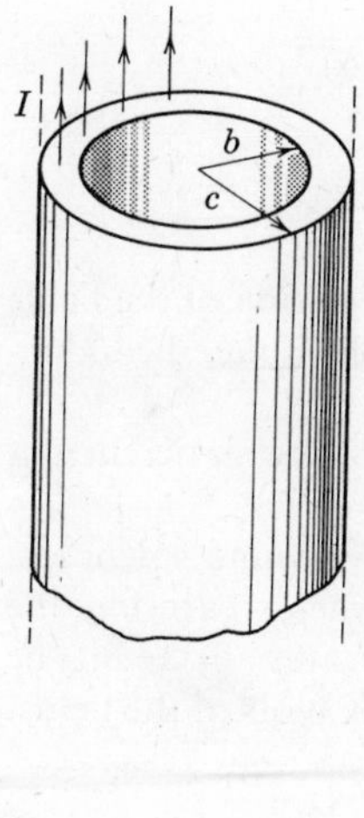

Problem 1-25

conductor carrying the same total current I. Show also that the field inside the conducting region ($b < \rho < c$) is

$$B_\phi = \frac{\mu_0 I(\rho^2 - b^2)}{2\pi(c^2 - b^2)\rho}$$

and is zero for $\rho < b$.

1-26. An infinitely long, coaxial pair of circular conductors are located in free space and have the dimensions shown. Equal and opposite total, static currents I flow in the conductors.

(*a*) Show that the magnetic fields inside the inner conductor ($\rho < a$) and between the conductors ($a < \rho < b$) are the same as for the isolated wire of Example 1-13. Show also that the field inside the outer conductor ($b < \rho < c$) is

$$B_\phi = \frac{\mu_0 I}{2\pi\rho} \frac{c^2 - \rho^2}{c^2 - b^2}$$

and is zero for $\rho > c$.

(*b*) Sketch a graph of B_ϕ versus ρ over the range $(0, c)$.

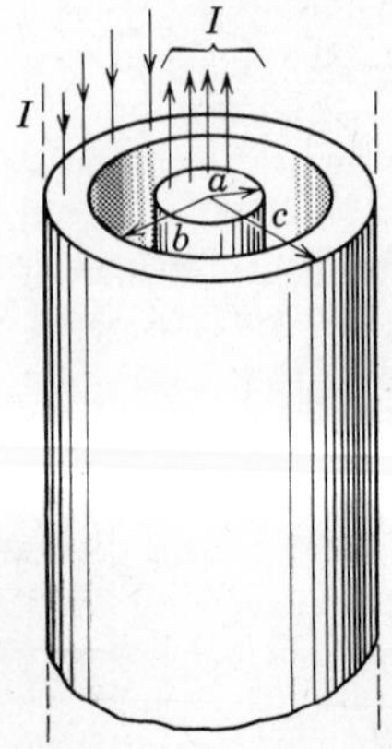

Problem 1-26

1-27. Show that the static fields of Problem 1-26 are a superposition of the fields of the hollow conductor of Problem 1-25 and those of the isolated conductor of Example 1-13.

1-28. Find, by superposition, the magnetic fields of the following static current systems (simply *add* previously obtained solutions):

(*a*) A pair of coaxial, infinitely long, ideally closely wound solenoids, each having the same number of turns per meter carrying the currents I in opposing directions. How is the answer affected if the currents travel in the same direction? Sketch flux plots of the fields. Assume the solenoids have radii a and b.

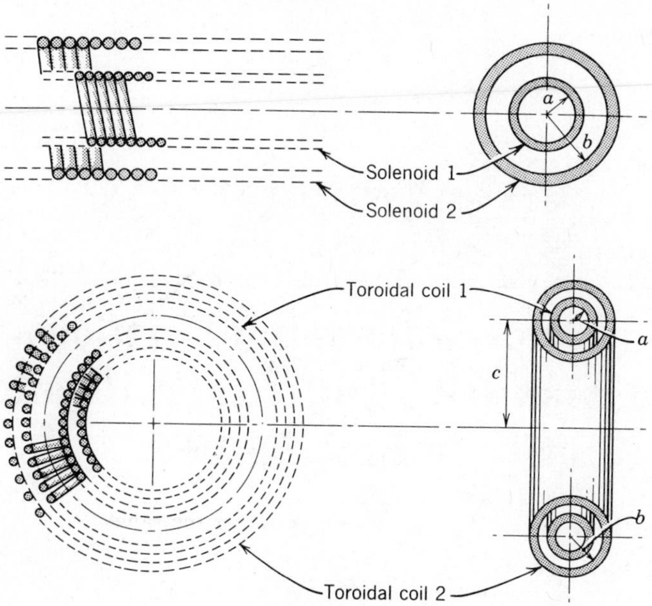

Problem 1-28

(*b*) A pair of coaxial, ideally closely wound toroids of circular cross-section, with the same number of turns and carrying currents I: first, in opposing directions; then in the same direction. Assume that the radii of the toroid cross-sections are a and b ($a < b$).

1-29. Use the time-static form (1-68) of Faraday's law to verify that the electric field of an infinitely long, uniform, static line charge is conservative; i.e., show that the line integral of (1-61) over any path connecting the arbitrary points $P_1(\rho_1, \phi_1, z_1)$ and $P_2(\rho_2, \phi_2, z_2)$ is a function of only the endpoints.

1-30. A short solenoid carrying a steady current I has the magnetic flux illustrated. Determine from Faraday's law in which direction a current is induced in the conducting ring placed at an axial position as shown and moved with a velocity v away from the solenoid.

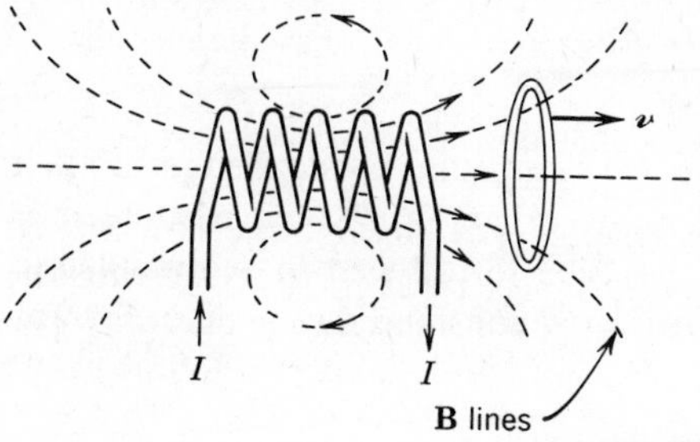

Problem 1-30

CHAPTER

2

Vector Differential Relations and Maxwell's Differential Relations in Free Space

This chapter considers the development, in generalized orthogonal coordinates, of the gradient, divergence, and curl operators of vector analysis, with forms in the common coordinate systems taken up in detail. The divergence theorem and the theorem of Stokes are utilized to derive the differential forms of Maxwell's divergence and curl equations in free space from their integral versions postulated in Chapter 1. The appropriate manipulations of Maxwell's time-varying differential equations are seen to lead to the wave equations in terms of the **B** and **E** fields, and the wavelike nature of their solutions is exemplified by considering in detail the field solutions of uniform plane waves in free space.

A pursuit of these ideas requires some background in the differentiation of vector fields, to be discussed in the following section.

2-1 Differentiation of Vector Fields

In many physical problems involving vector fields, a knowledge of their rates of change with respect to space, time, or perhaps some parameter is often of interest. This notion has already been introduced in Section 1-6 in connection with the position vector **r**. It is now considered in general for any differentiable vector field.

If $\mathbf{F}(u)$ is a vector function of a single scalar variable u, its ordinary vector derivative with respect to u is defined by the limit

$$\frac{d\mathbf{F}}{du} = \lim_{\Delta u \to 0} \frac{\Delta \mathbf{F}}{\Delta u} = \lim_{\Delta u \to 0} \frac{\mathbf{F}(u + \Delta u) - F(u)}{\Delta u} \tag{2-1}$$

provided that the limit exists (i.e., the limit is single-valued and finite). As in the instance of the derivatives of the position vector $\mathbf{r}$ considered earlier, the vector increment $\Delta\mathbf{F}$ is not necessarily aligned with the vector $\mathbf{F}$, implying that the direction of the vector $\mathbf{F}$ may change with the variable u. This circumstance is exemplified in Figure 2-1, in which the conventional triangle construction is used to define $\Delta\mathbf{F}$, the difference between $\mathbf{F}(u + \Delta u)$ and $\mathbf{F}(u)$. The derivative $d\mathbf{F}/du$ defines a function, the derivative of which in turn defines a second-order derivative function $d^2\mathbf{F}/du^2$, and so on.

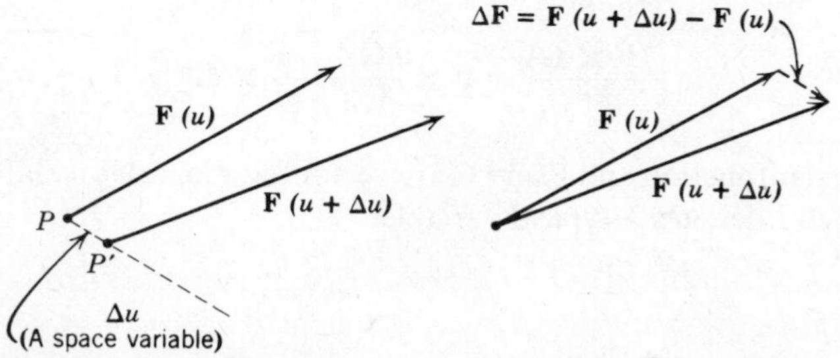

Figure 2-1. A vector function F in space, and its variation ΔF with respect to some variable *u*.

The derivatives of the sum or product combinations of scalar and vector functions are often of interest. For example, if f and $\mathbf{F}$ are respectively scalar and vector functions of the variable u, the derivative of their product is, from (2-1)

$$\frac{d(f\mathbf{F})}{du} = \lim_{\Delta u \to 0} \frac{(f + \Delta f)(\mathbf{F} + \Delta\mathbf{F}) - f\mathbf{F}}{\Delta u} = f\frac{d\mathbf{F}}{du} + \mathbf{F}\frac{df}{du} \tag{2-2}$$

Note that this result resembles in form a similar rule of the scalar calculus (in which both functions are scalars).

If $\mathbf{F}$ is a function of more than one variable, say of u_1, u_2, u_3, t, its partial derivative with respect to one of the variables (u_1) is defined

$$\frac{\partial\mathbf{F}(u_1, u_2, u_3, t)}{\partial u_1} = \lim_{\Delta u_1 \to 0} \frac{\mathbf{F}(u_1 + \Delta u_1, u_2, u_3, t) - \mathbf{F}(u_1, u_2, u_3, t)}{\Delta u_1} \tag{2-3}$$

with similar expressions for the partial derivatives with respect to the remaining variables. Successive partial differentiations yield functions such as $\partial^2\mathbf{F}/\partial u_1^2$, $\partial^2\mathbf{F}/\partial u_1\,\partial u_2$, and so on. If **F** has continuous partial derivatives of at least the second order, it is permissible to differentiate it in either order; thus

$$\frac{\partial^2\mathbf{F}}{\partial u_1\,\partial u_2} = \frac{\partial^2\mathbf{F}}{\partial u_2\,\partial u_1} \tag{2-4}$$

The partial derivative of the sum or product combinations of scalar and vector functions sometimes is useful. In particular, one can use (2-3) to prove that the following expansions are valid

$$\frac{\partial(f\mathbf{F})}{\partial t} = f\frac{\partial\mathbf{F}}{\partial t} + \mathbf{F}\frac{\partial f}{\partial t} \tag{2-5}$$

$$\frac{\partial(\mathbf{F}\cdot\mathbf{G})}{\partial t} = \mathbf{F}\cdot\frac{\partial\mathbf{G}}{\partial t} + \mathbf{G}\cdot\frac{\partial\mathbf{F}}{\partial t} \tag{2-6}$$

$$\frac{\partial(\mathbf{F}\times\mathbf{G})}{\partial t} = \mathbf{F}\times\frac{\partial\mathbf{G}}{\partial t} + \frac{\partial\mathbf{F}}{\partial t}\times\mathbf{G} \tag{2-7}$$

if f is any scalar function and **F** and **G** are vector functions of several variables, among which t denotes a typical variable.

2-2 Gradient of a Scalar Function

The space rate of change of a scalar field $f(u_1, u_2, u_3, t)$ is frequently of physical interest. For example, in the scalar temperature field $T(u_1, u_2, u_3, t)$ depicted in Figure 1-1(*a*), one can surmise from graphical considerations that the maximum space rates of temperature change occur in directions normal to the constant temperature surfaces shown. Generally, the maximum space rate of change of a scalar function, including the vector direction in which the rate of change takes place, can be characterized by means of a vector differential operator known as the gradient of that scalar function. It is developed here.

If, at any fixed time t, a single-valued, well-behaved scalar field $f(u_1, u_2, u_3, t)$ is set equal to *any constant* f_0 so that $f(u_1, u_2, u_3, t) = f_0$, a surface in space is described, as depicted by S_1 in Figure 2-2. A physical example of such a surface is any of the constant temperature surfaces of Figure 1-1(*a*). Another surface, S_2, an infinitesimal distance from S_1, is described by letting $f(u_1 + du_1, u_2 + du_2, u_3 + du_3) = f_0 + df$, in which df is taken to mean a very small, constant, scalar amount. Suppose that two nearby points,

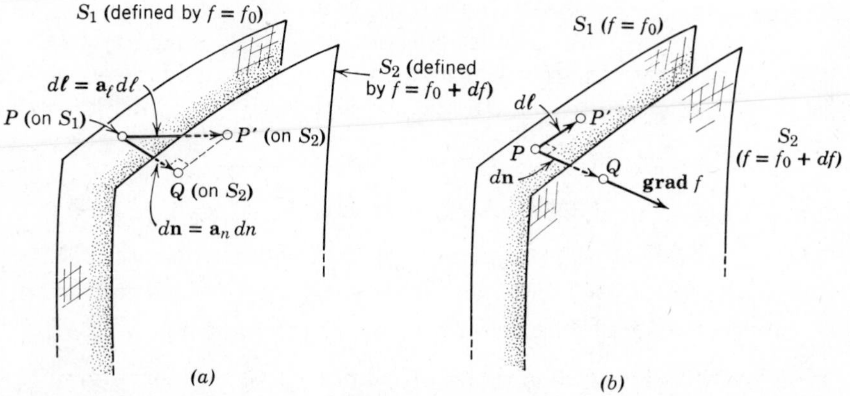

Figure 2-2. **Two nearby surfaces $f = f_0$ and $f = f_0 + df$ relative to a discussion of grad f. (*a*) Points P and P' separated by $d\ell$ and on surfaces defined by $f = f_0$ and $f = f_0 + df$. (*b*) Points P and P' on the same surface $f = f_0$, to show that grad f and $d\ell$ are perpendicular.**

P and P', are located a vector distance $d\boldsymbol{\ell}$ apart on these two surfaces as in Figure 2-2(*a*), recalling from (1-21) that one may express $d\boldsymbol{\ell} = \mathbf{a}_\ell\, d\ell$ as

$$d\boldsymbol{\ell} = \mathbf{a}_1\, d\ell_1 + \mathbf{a}_2\, d\ell_2 + \mathbf{a}_3\, d\ell_3 \tag{2-8}$$

df is the amount by which f changes in going from P to P' from the first surface to the second, written as

$$\begin{aligned} df &= \frac{\partial f}{\partial u_1}\, du_1 + \frac{\partial f}{\partial u_2}\, du_2 + \frac{\partial f}{\partial u_3}\, du_3 \\ &= \frac{\partial f}{\partial \ell_1}\, d\ell_1 + \frac{\partial f}{\partial \ell_2}\, d\ell_2 + \frac{\partial f}{\partial \ell_3}\, d\ell_3 \end{aligned} \tag{2-9}$$

The presence of the components of $d\boldsymbol{\ell}$ in (2-9) permits expressing df as the *dot product*

$$df = \left[\mathbf{a}_1 \frac{\partial f}{\partial \ell_1} + \mathbf{a}_2 \frac{\partial f}{\partial \ell_2} + \mathbf{a}_3 \frac{\partial f}{\partial \ell_3}\right] \cdot (\mathbf{a}_1\, d\ell_1 + \mathbf{a}_2\, d\ell_2 + \mathbf{a}_3\, d\ell_3)$$

Calling the bracketed quantity the *gradient of the function f*, or simply **grad** f, as follows

$$\mathbf{grad}\, f \equiv \mathbf{a}_1 \frac{\partial f}{\partial \ell_1} + \mathbf{a}_2 \frac{\partial f}{\partial \ell_2} + \mathbf{a}_3 \frac{\partial f}{\partial \ell_3}$$

or

$$\mathbf{grad}\, f \equiv \mathbf{a}_1 \frac{1}{h_1}\frac{\partial f}{\partial u_1} + \mathbf{a}_2 \frac{1}{h_2}\frac{\partial f}{\partial u_2} + \mathbf{a}_3 \frac{1}{h_3}\frac{\partial f}{\partial u_3} \tag{2-10}$$

one may write the total differential df of (2-9) in the abbreviated form

$$df = (\mathbf{grad}\, f)\cdot d\boldsymbol{\ell} \tag{2-11}$$

Two properties of **grad** f are deducible from (2-11) in the following:

1. That the vector function **grad** f defined by (2-10) is a vector perpendicular to any $f = f_0$ surface is appreciated if the points P and P', separated by a distance $d\boldsymbol{\ell}$, are placed on the same surface as in Figure 2-2(*b*). Then the amount by which f changes in going from P to P' is zero, but from (2-11), $(\mathbf{grad}\, f)\cdot d\boldsymbol{\ell} = 0$, implying that **grad** f and $d\boldsymbol{\ell}$ are perpendicular vectors. **Grad** f is therefore a vector everywhere *perpendicular* to any surface on which $f =$ constant.
2. If a displacement $d\boldsymbol{\ell}$ from the point P is assigned a constant magnitude and a variable direction, then from (2-11) and the definition (1-34) of the dot product it is seen that $df = |\mathbf{grad}\, f|\, d\ell \cos\theta$, θ denoting the direction between the **grad** f and $d\boldsymbol{\ell}$. The magnitude of **grad** f is therefore $df/(d\ell\cos\theta)$, but from Figure 2-2(*a*), $d\ell\cos\theta = dn$, the shortest (perpendicular) distance from the point P on the surface S_1 to the adjacent surface S_2 on which $f = f_0 + df$, whence

$$|\mathbf{grad}\, f| = \frac{df}{dn} \tag{2-12}$$

The vector **grad** f therefore denotes both the magnitude *and direction* of the maximal space rate of change of f, at any point in a region.

Note that the magnitude of **grad** f can also be expressed in terms of its orthogonal curvilinear components, given in the definition (2-10) by

$$|\mathbf{grad}\, f| = \left[\left(\frac{\partial f}{h_1\,\partial u_1}\right)^2 + \left(\frac{\partial f}{h_2\,\partial u_2}\right)^2 + \left(\frac{\partial f}{h_3\,\partial u_3}\right)^2\right]^{1/2} \tag{2-13}$$

The expressions for **grad** f in a specific orthogonal coordinate system are obtained from (2-10) upon substituting into it the appropriate symbols for u_1 and h_1 as discussed in Section 1-5. Thus, in the *rectangular* system

$$\mathbf{grad}\, f = \mathbf{a}_x \frac{\partial f}{\partial x} + \mathbf{a}_y \frac{\partial f}{\partial y} + \mathbf{a}_z \frac{\partial f}{\partial z} \tag{2-14a}$$

in the *circular cylindrical* system

$$\textbf{grad}\, f = \mathbf{a}_\rho \frac{\partial f}{\partial \rho} + \mathbf{a}_\phi \frac{1}{\rho}\frac{\partial f}{\partial \phi} + \mathbf{a}_z \frac{\partial f}{\partial z} \tag{2-14b}$$

and in the *spherical* coordinate system

$$\textbf{grad}\, f = \mathbf{a}_r \frac{\partial f}{\partial r} + \mathbf{a}_\theta \frac{1}{r}\frac{\partial f}{\partial \theta} + \mathbf{a}_\phi \frac{1}{r \sin\theta}\frac{\partial f}{\partial \phi} \tag{2-14c}$$

An integral property of **grad** *f*, of considerable importance in field theory, is that its line integral over any *closed* path ℓ in space is zero. Symbolically

$$\oint_\ell (\textbf{grad}\, f)\cdot d\boldsymbol{\ell} = 0 \tag{2-15}$$

holding for all well-behaved scalar functions *f*, and proved in the following manner. Consider (2-15) integrated over an *open* path between the distinct endpoints $P_0(u_1^0, u_2^0, u_3^0)$ and $P(u_1, u_2, u_3)$

$$\int_{P_0}^{P} (\textbf{grad}\, f)\cdot d\boldsymbol{\ell} \tag{2-16}$$

From (2-11) it is seen that $(\textbf{grad}\, f)\cdot d\boldsymbol{\ell}$ denotes the total differential df, so that (2-16) becomes

$$\int_{P_0}^{P} (\textbf{grad}\, f)\cdot d\boldsymbol{\ell} = \int_{P_0}^{P} df = f\Big]_{P_0}^{P}$$
$$= f(u_1, u_2, u_3) - f(u_1^0, u_2^0, u_3^0) \tag{2-17}$$

or the difference of the values of the function *f* at the endpoints *P* and P_0. Thus, *any* path connecting P_0 and *P* will provide the same result, (2-17). Carrying out (2-16) over some path *A* from P_0 to the point *P* and then back to P_0 once more over a different path *B*, the contributions of the two integrals would cancel exactly, making (2-15) the result. The integral property (2-15) of any vector field **grad** *f* is sometimes called the *conservative* property of that field, from the applications of integrals of that type to problems involving certain kinds of energy. Any field **grad** *f* is a conservative field.

2-3 The operator ∇ (Del)

Recall that the gradient of a scalar field *f* is expressed in rectangular coordinates by (2-14a)

$$\textbf{grad}\, f = \mathbf{a}_x \frac{\partial f}{\partial x} + \mathbf{a}_y \frac{\partial f}{\partial y} + \mathbf{a}_z \frac{\partial f}{\partial z} \qquad [2\text{-}14a]$$

The presence of the common function f in each term permits separating from this expression a vector partial differential operator represented by the symbol ∇ (pronounced *del*) as follows

$$\nabla \equiv \mathbf{a}_x \frac{\partial}{\partial x} + \mathbf{a}_y \frac{\partial}{\partial y} + \mathbf{a}_z \frac{\partial}{\partial z} \tag{2-18}$$

to permit writing **grad** f in an alternative symbolism, ∇f

$$\mathbf{grad}\, f \equiv \nabla f = \mathbf{a}_x \frac{\partial f}{\partial x} + \mathbf{a}_y \frac{\partial f}{\partial y} + \mathbf{a}_z \frac{\partial f}{\partial z} \tag{2-19}$$

The notations **grad** f and ∇f will henceforth be considered *interchangeable.*

It may be noted that the operator ∇ defined by (2-18) in the rectangular coordinate system can be defined in other coordinate systems as well, including the generalized orthogonal curvilinear system. This is not done here because of its lengthy form and because it serves no particular need in connection with the objectives of this text. The reader may wish to consult other sources relative to extending (2-18) to other coordinate systems.†

EXAMPLE 2-1. Suppose a scalar, time-independent temperature field in some region of space is given by

$$T(x, y) = 200x + 100y \text{ deg}$$

with x and y expressed in meters. Sketch a few isotherms (constant temperature surfaces) of this static thermal field and determine the gradient of T.

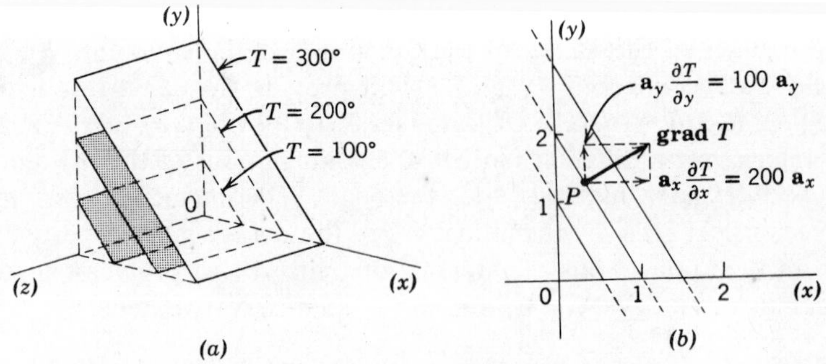

Example 2-1. (*a*) Graph of T = constant. (*b*) Side view of (*a*).

† For example, see Javid, M., and P. M. Brown. *Field Analysis and Electromagnetics.* New York: McGraw-Hill, 1963, p. 477.

The isotherms are obtained by setting T equal to specific constant temperature values. Thus, letting $T = 100$ deg yields $100 = 200x + 100y$, the equation of the tilted plane $y = -2x + 1$. This and other isothermic surfaces are shown in the accompanying figure.

The temperature gradient of $T(x, y)$ is given by (2-14a)

$$\nabla T \equiv \textbf{grad}\, T = \mathbf{a}_x \frac{\partial T}{\partial x} + \mathbf{a}_y \frac{\partial T}{\partial y} + \mathbf{a}_z \frac{\partial T}{\partial z} = 200\mathbf{a}_x + 100\mathbf{a}_y \text{ deg/m}$$

a vector everywhere perpendicular to the isotherms, as noted in (*b*) of the figure. The x and y components of the temperature gradient denote space rate of change of temperature along these coordinate axes. From (2-13), the magnitude is

$$|\textbf{grad}\, T| = \sqrt{200^2 + 100^2} = 223 \text{ deg/m}$$

denoting the maximal space rate of change of temperature at any point. One may observe that heat will flow in the direction of maximal temperature decrease; i.e., along lines perpendicular to the isotherms and thus in a direction opposite to that of the vector **grad** T at any point.

2-4 Divergence of a Vector Function

The flux representation of vector fields was described in Section 1-9. If a vector field **F** is representable by a continuous system of unbroken flux lines in a volume region as shown, for example, in Figure 2-3(*a*), the region is said to be *source free*; or equivalently, the field **F** is said to be *divergenceless*. (The divergence of **F** is zero.) On the other hand, if the flux plot of **F** consists of flux lines that are broken or discontinuous as depicted in Figure 2-3(*b*), the region contains sources of the field flux; the field **F** is then said to have a nonzero divergence in that region. The characterization of the divergence of a vector field on a mathematical basis is described here.

The *divergence of a vector field* **F**, abbreviated div **F**, is defined as the limit of the net outward flux of **F**, $\oint_S \mathbf{F}\cdot d\mathbf{s}$, per unit volume, as the volume Δv enclosed by the surface S tends toward zero. Symbolically

$$\text{div}\,\mathbf{F} \equiv \lim_{\Delta v \to 0} \frac{\oint_S \mathbf{F}\cdot d\mathbf{s}}{\Delta v} \text{ flux lines/m}^3 \tag{2-20}$$

Thus, as the closed surface S is made very small, as depicted in Figure 2-3(*c*), the limiting, net outward flux per unit volume in the neighborhood of the point P defines the divergence of the vector field **F** there. The shape of S is immaterial in this limit, as long as the dimensions of Δv tend toward zero together.

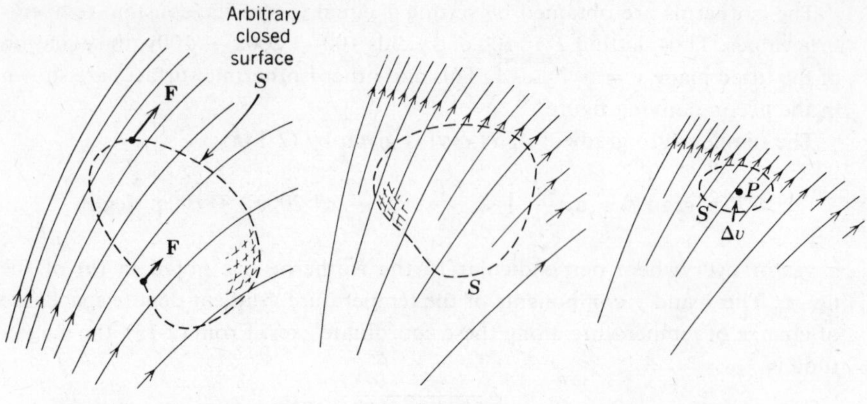

Figure 2-3. Concerning the divergence of flux fields. (*a*) A vector field F in a source free region. As many flux lines enter *S* as leave it. (*b*) A vector field F in a region containing sources (*S* possessing net, outgoing flux). (*c*) The meaning of div F: net, outward flux per unit volume as $\Delta v \rightarrow 0$.

The definition (2-20) leads to partial differential expressions for div **F** in the various coordinate systems. For example, in generalized orthogonal coordinates, div **F** is shown to become

$$\operatorname{div} \mathbf{F} = \frac{1}{h_1 h_2 h_3}\left[\frac{\partial(F_1 h_2 h_3)}{\partial u_1} + \frac{\partial(F_2 h_1 h_3)}{\partial u_2} + \frac{\partial(F_3 h_1 h_2)}{\partial u_3}\right] \tag{2-21}$$

The derivation of the differential expression (2-21) for div **F** in generalized orthogonal coordinates proceeds from the definition (2-20). Express the function **F** in terms of its generalized components as follows:

$$\mathbf{F}(u_1, u_2, u_3, t) = \mathbf{a}_1 F_1(u_1, u_2, u_3, t) + \mathbf{a}_2 F_2(u_1, u_2, u_3, t) + \mathbf{a}_3 F_3(u_1, u_2, u_3, t) \tag{2-22}$$

The definition (2-20) requires that the net efflux of **F** be found over the closed surface S bounding any limiting volume Δv, which from (1-11) or (1-13) is expressed

$$\Delta v = \Delta \ell_1 \Delta \ell_2 \Delta \ell_3 = h_1 h_2 h_3 \Delta u_1 \Delta u_2 \Delta u_3 \tag{2-23}$$

The net, outward flux of **F** is that emanating from the six sides of Δv, designa-

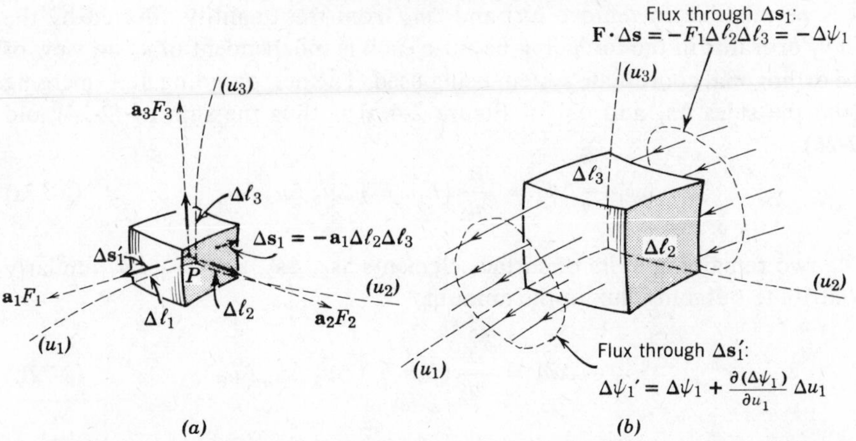

Figure 2-4. A volume-element Δv in the generalized orthogonal coordinate system, used in the development of the partial differential expression for div F. (*a*) A volume-element Δv and components of F in the neighborhood of $P(u_1, u_2, u_3)$. (*b*) Flux contributions entering and leaving opposite surfaces of Δv. Remaining four sides are similarly treated.

ted by $\Delta \mathbf{s}_1$, $\Delta \mathbf{s}_1'$, etc., in Figure 2-4(*a*). The contribution $\Delta \psi_1$ proceeding from $\Delta \mathbf{s}_1$ is just $\mathbf{F} \cdot \Delta \mathbf{s}_1 = (\mathbf{a}_1 F_1) \cdot \Delta \mathbf{s}_1$, or

$$\mathbf{F} \cdot \Delta \mathbf{s}_1 = \mathbf{F} \cdot (-\mathbf{a}_1 \Delta \ell_2 \Delta \ell_3) = -F_1 \Delta \ell_2 \Delta \ell_3 \tag{2-24}$$

$$\mathbf{F} \cdot \Delta \mathbf{s}_1 = -F_1 h_2 h_3 \Delta u_2 \Delta u_3 = -\Delta \psi_1 \tag{2-25}$$

the negative sign being the consequence of assuming a positively directed F_1 component over the outward $\Delta \mathbf{s}_1 = -\mathbf{a}_1 \Delta \ell_2 \Delta \ell_3$; i.e., the flux $\Delta \psi_1$ *enters* $\Delta \mathbf{s}_1$. In the limit, as the separation $\Delta \ell_1$ between $\Delta \mathbf{s}_1$ and $\Delta \mathbf{s}_1'$ becomes sufficiently small, the flux $\Delta \psi_1'$ leaving $\Delta \mathbf{s}_1'$ in Figure 2-4(*b*) differs from $\Delta \psi_1$ entering $\Delta \mathbf{s}_1$ by an amount given by the second term of the Taylor's expansion of $\Delta \psi_1'$ about the point P; i.e.,

$$\begin{aligned}
\Delta \psi_1'(u_1 + \Delta u_1, u_2, u_3) &= \Delta \psi_1 + \frac{\partial(\Delta \psi_1)}{\partial u_1} \Delta u_1 \\
&= F_1 \, \Delta \ell_2 \, \Delta \ell_3 + \left[\frac{\partial}{\partial u_1} (F_1 \, \Delta \ell_2 \, \Delta \ell_3) \right] \Delta u_1 \\
&= F_1 \, \Delta \ell_2 \, \Delta \ell_3 + \left[\frac{\partial}{\partial u_1} (F_1 h_2 h_3) \right] \Delta u_1 \, \Delta u_2 \, \Delta u_3
\end{aligned} \tag{2-26}$$

It is permissible to remove Δu_2 and Δu_3 from the quantity affected by the $\partial/\partial u_1$ operator in the foregoing because each is independent of u_1, in view of the orthogonal coordinate system being used. The net, outgoing flux emerging from the sides $\Delta \mathbf{s}_1$ and $\Delta \mathbf{s}_1'$ of Figure 2-4(*b*) is thus the sum of (2-24) and (2-26)

$$\Delta\psi_1' - \Delta\psi_1 = \frac{\partial}{\partial u_1}(F_1 h_2 h_3)\,\Delta u_1\,\Delta u_2\,\Delta u_3 \tag{2-27a}$$

The two remaining pairs of surface elements $\Delta \mathbf{s}_2$, $\Delta \mathbf{s}_2'$, and $\Delta \mathbf{s}_3$, $\Delta \mathbf{s}_3'$ similarly contribute outgoing flux in the amounts

$$\Delta\psi_2' - \Delta\psi_2 = \frac{\partial}{\partial u_2}(F_2 h_3 h_1)\,\Delta u_1\,\Delta u_2\,\Delta u_3 \tag{2-27b}$$

$$\Delta\psi_3' - \Delta\psi_3 = \frac{\partial}{\partial u_3}(F_3 h_1 h_2)\,\Delta u_1\,\Delta u_2\,\Delta u_3 \tag{2-27c}$$

seen to be obtainable from the symmetry and the cyclic permutation of the subscripts of (2-27a). Finally, putting (2-27a, b, c) into the numerator of (2-20) obtains the result anticipated in (2-21)

$$\operatorname{div}\mathbf{F} \equiv \lim_{\Delta v\to 0}\frac{\oint_S \mathbf{F}\cdot d\mathbf{s}}{\Delta v} = \lim_{\Delta v\to 0}\frac{\Delta\psi_1' - \Delta\psi_1 + \Delta\psi_2' - \Delta\psi_2 + \Delta\psi_3' - \Delta\psi_3}{h_1 h_2 h_3\,\Delta u_1\,\Delta u_2\,\Delta u_3}$$

whence

$$\operatorname{div}\mathbf{F} = \frac{1}{h_1 h_2 h_3}\left[\frac{\partial(F_1 h_2 h_3)}{\partial u_1} + \frac{\partial(F_2 h_1 h_3)}{\partial u_2} + \frac{\partial(F_3 h_1 h_2)}{\partial u_3}\right] \tag{2-28}$$

In rectangular coordinates, $\operatorname{div}\mathbf{F}$ is found from (2-28) by setting $h_1 = h_2 = h_3 = 1$ and $u_1 = x$, $u_2 = y$, $u_3 = z$

$$\operatorname{div}\mathbf{F} = \frac{\partial F_x}{\partial x} + \frac{\partial F_y}{\partial y} + \frac{\partial F_z}{\partial z} \qquad \text{Rectangular} \tag{2-29a}$$

while in the circular cylindrical and the spherical coordinate systems, the expressions become

$$\operatorname{div}\mathbf{F} = \frac{1}{\rho}\frac{\partial}{\partial \rho}(\rho F_\rho) + \frac{1}{\rho}\frac{\partial F_\phi}{\partial \phi} + \frac{\partial F_z}{\partial z} \qquad \text{Circular cylindrical} \tag{2-29b}$$

$$\operatorname{div}\mathbf{F} = \frac{1}{r^2}\frac{\partial}{\partial r}(r^2 F_r) + \frac{1}{r\sin\theta}\frac{\partial}{\partial \theta}(F_\theta \sin\theta) + \frac{1}{r\sin\theta}\frac{\partial F_\phi}{\partial \phi} \qquad \text{Spherical} \tag{2-29c}$$

The form (2-29a) of div **F** in the rectangular system is the basis for another notation using the del operator (∇) defined by (2-18). Upon taking the dot product of ∇ with **F** in the rectangular system of coordinates, one finds

$$\nabla \cdot \mathbf{F} = \left(\mathbf{a}_x \frac{\partial}{\partial x} + \mathbf{a}_y \frac{\partial}{\partial y} + \mathbf{a}_z \frac{\partial}{\partial z}\right) \cdot (\mathbf{a}_x F_x + \mathbf{a}_y F_y + \mathbf{a}_z F_z)$$

$$= \frac{\partial F_x}{\partial x} + \frac{\partial F_y}{\partial y} + \frac{\partial F_z}{\partial z} \tag{2-30}$$

or precisely (2-29a). This is the basis for the equivalent symbolisms

$$\text{div}\,\mathbf{F} \equiv \nabla \cdot \mathbf{F} \tag{2-31}$$

The notations div **F** and $\nabla \cdot \mathbf{F}$ will be considered *interchangeable* regardless of which coordinate system is used, even though the symbol ∇ has for our purposes been defined only in the rectangular system.

EXAMPLE 2-2. Sketch flux plots for each of the following vector fields, and find the divergence of each: (*a*) $\mathbf{F} = \mathbf{a}_x K$, $\mathbf{G} = \mathbf{a}_x K y$, $\mathbf{H} = \mathbf{a}_x K x$; (*b*) $\mathbf{J} = \mathbf{a}_\rho K$, $\mathbf{L} = \mathbf{a}_\rho (K/\rho)$.

(*a*) Applying (2-29a) to the functions **F**, **G**, and **H** in the rectangular system obtains

$$\text{div}\,\mathbf{F} = \frac{\partial(K)}{\partial x} = 0 \qquad \text{div}\,\mathbf{G} = \frac{\partial(Ky)}{\partial x} = 0 \qquad \text{div}\,\mathbf{H} = \frac{\partial(Kx)}{\partial x} = K$$

Their flux plots are shown in (*a*) through (*c*). Inspection reveals why a zero value of divergence is obtained for the fields **F** and **G**; a test closed surface S placed anywhere in the region will have zero net flux emanating from it. The nonzero div **H**, on the other hand, is evident from its flux plot because of the discontinuous flux lines, here required to possess an increasing density with x, yielding a net, nonzero outgoing flux emerging from the typical closed surface S shown.

(*b*) From (2-29b)

$$\text{div}\,\mathbf{J} = \frac{1}{\rho}\frac{\partial}{\partial \rho}(\rho K) = \frac{K}{\rho} \qquad \text{div}\,\mathbf{L} = \frac{1}{\rho}\frac{\partial}{\partial \rho}\left(\rho \frac{K}{\rho}\right) = 0$$

the flux plots of which are illustrated looking along the z axis of the cylindrical system in (*d*) and (*e*). The divergenceless character of **L** is evident from its $1/\rho$ dependence which, in this cylindrical system, provides an uninterrupted system of outgoing flux lines. The radially directed field **J**, having a constant flux density of magnitude K, on the

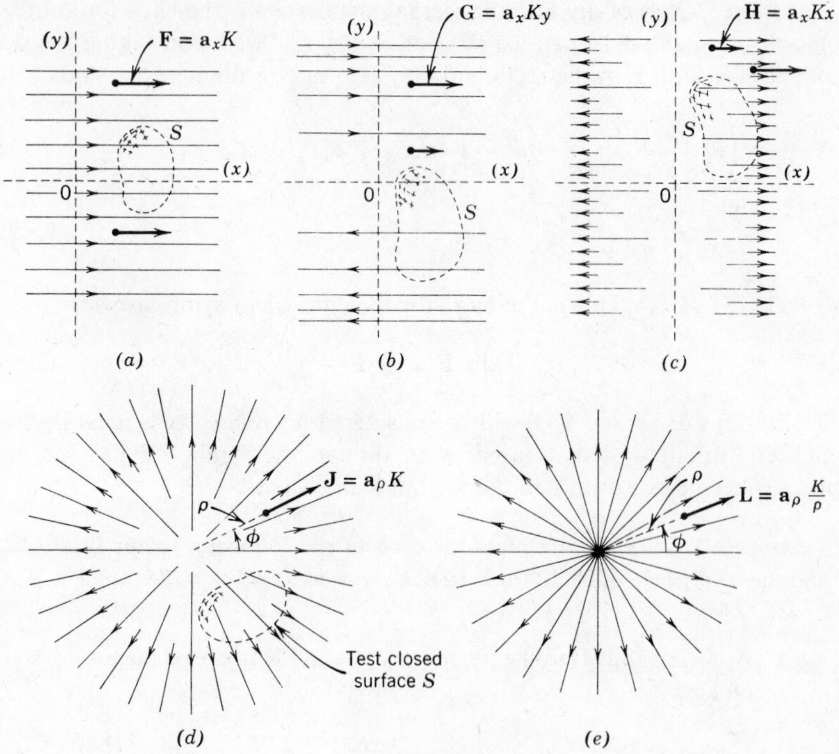

Example 2-2

other hand, clearly must pick up additional flux lines with an increase in ρ. It is therefore required to possess a divergence.

EXAMPLE 2-3. Find the divergence of the **E** field produced by the uniformly charged cloud of Figure 1-15(*b*) at any location $P(r)$ both inside and exterior to the cloud.

The field $\mathbf{E}(r)$ *outside* the cloud ($r > r_0$) is given by (1-59). Its divergence (2-29c) in spherical coordinates is

$$\operatorname{div} \mathbf{E} = \frac{1}{r^2}\frac{\partial}{\partial r}(r^2 E_r) = \frac{1}{r^2}\frac{\partial}{\partial r}\left[r^2 \frac{\rho_v r_0^3}{3\epsilon_0 r^2}\right] = 0 \qquad r > r_0 \tag{2-32}$$

This null result signifies a flux plot in the region $r > r_0$ consisting of unbroken lines, as noted in Figure 1-16(*b*). All inverse r^2 radial fields behave this way.

Inside the charged cloud ($r < r_0$), the **E** field (1-60) being proportional to r has the divergence

$$\operatorname{div} \mathbf{E} = \frac{1}{r^2}\frac{\partial}{\partial r}\left(r^2 \frac{\rho_v r}{3\epsilon_0}\right) = \frac{\rho_v}{\epsilon_0} \qquad r < r_0 \tag{2-33}$$

a nonzero, constant result, proportional to the density ρ_v of the cloud. Note that bringing ϵ_0 inside the divergence operator puts (2-33) into the form

$$\text{div}\,(\epsilon_0\mathbf{E}) = \rho_v \text{ C/m}^3 \qquad r < r_0$$

making the divergence of $(\epsilon_0\mathbf{E})$ the same as the charge density ρ_v inside the cloud. It is shown in Section 2-4-2 that this result is true *in general*, even for nonuniform charge distributions in free space.

2-4-1 Divergence Theorem

If $\mathbf{F}(u_1, u_2, u_3, t)$ is well-behaved in some region of space, then the integral identity

$$\int_V (\text{div}\,\mathbf{F})\,dv = \oint_S \mathbf{F}\cdot d\mathbf{s} \tag{2-34}$$

is true for the closed surface S bounding any volume V. Equation (2-34) implies that the volume integral of $(\text{div}\,\mathbf{F})\,dv$ taken throughout any V equals the net flux of $\mathbf{F}$ emerging from the closed surface S bounding V.

An heuristic proof of (2-34) proceeds as follows. Suppose that V is subdivided into a large number n of volume-elements, any of which is designated Δv_i with each enclosed by bounding surfaces S_i as in Figure 2-5(a). The net

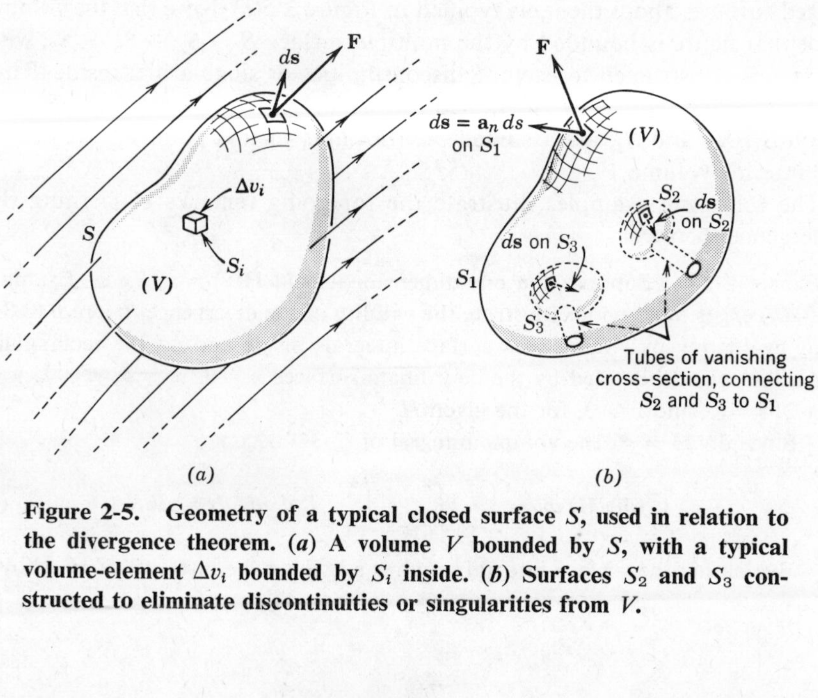

Figure 2-5. Geometry of a typical closed surface S, used in relation to the divergence theorem. (a) A volume V bounded by S, with a typical volume-element Δv_i bounded by S_i inside. (b) Surfaces S_2 and S_3 constructed to eliminate discontinuities or singularities from V.

flux emanating from Δv_i is the surface integral of $\mathbf{F}\cdot d\mathbf{s}$ over S_i, but from (2-20), this is also (div **F**) Δv_i for Δv_i sufficiently small, i.e.,

$$\oint_{S_i} \mathbf{F}\cdot d\mathbf{s} = (\text{div}\,\mathbf{F})\,\Delta v_i \tag{2-35}$$

The fluxes contributed by every S_i will sum up to yield the net flux through the exterior surface S bounding the volume V. Thus the left side of (2-35) summed over the closed surfaces Δs_i inside S yields

$$\sum_{i=1}^{n}\left[\oint_{S_i} \mathbf{F}\cdot d\mathbf{s}\right] = \oint_S \mathbf{F}\cdot d\mathbf{s} \tag{2-36}$$

Equating (2-36) to the right side of (2-35) summed over the n volume elements Δv_i yields, as the number n tends toward infinity (and as $\Delta v_i \to dv$)

$$\oint_S \mathbf{F}\cdot d\mathbf{s} = \lim_{\Delta v_i \to 0} \sum_{i=1}^{n} (\text{div}\,\mathbf{F})\,dv = \int_V (\text{div}\,\mathbf{F})\,dv \tag{2-37}$$

or just (2-34), known as the *divergence theorem.*

If the limiting process yielding (2-37) is to be valid, it is necessary that **F**, together with its first derivatives, be continuous in and on V. If **F** and its divergence $\nabla\cdot\mathbf{F}$ are not continuous, then the regions in V or on S possessing such discontinuities or possible singularities must be excluded by constructing closed surfaces about them, as typified in Figure 2-5(b). Note that the volume V of that figure is bounded by the multiple surface $S = S_1 + S_2 + S_3$, with S_2 and S_3 constructed to exclude discontinuities or singularities inside them. The normal unit vectors $\mathbf{a}_n$, identified with each vector surface element $d\mathbf{s} = \mathbf{a}_n\,ds$ on S_1, S_2, and S_3, are assumed *outward* unit vectors pointing *away* from the interior volume V.

The following examples illustrate the foregoing remarks concerning the divergence theorem.

EXAMPLE 2-4. Suppose the one-dimensional field $\mathbf{H}(x) = \mathbf{a}_x Kx$ of Example 2-2(a) exists in a region. Illustrate the validity of the divergence theorem (2-34) by evaluating its volume and surface integrals inside and on the rectangular parallelepiped bounded by the coordinate surfaces $x = 1$, $x = 4$, $y = 2$, $y = -2$, $z = 0$, and $z = 3$, for the given H.

Since div $\mathbf{H} = K$, the volume integral of (2-34) becomes

$$\int_V (\text{div}\,\mathbf{H})\,dv = \int_{z=0}^{3}\int_{y=-2}^{2}\int_{x=1}^{4} K\,dx\,dy\,dz = 36K \tag{1}$$

Evaluating the surface integral requires summing the integrals of $\mathbf{H}\cdot d\mathbf{s}$ over

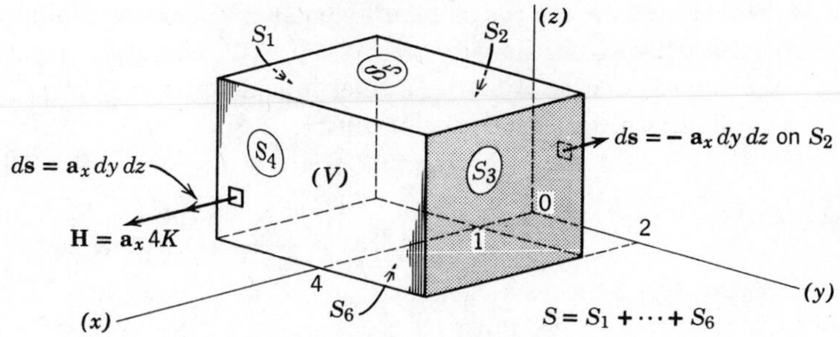

Example 2-4

the six sides of the parallelepiped. Because $\mathbf{H}$ is x directed, however, $\mathbf{H}\cdot d\mathbf{s}$ is zero over four of these sides, the surface integral reducing to the same result as (1)

$$\oint_S \mathbf{H}\cdot d\mathbf{s} = \int_{z=0}^{3}\int_{y=-2}^{2} (\mathbf{a}_x 4K)\cdot\mathbf{a}_x\,dy\,dz + \int_{z=0}^{3}\int_{y=-2}^{2} (\mathbf{a}_x K)\cdot(-\mathbf{a}_x\,dy\,dz)$$
$$= 48K - 12K = 36K \tag{2}$$

EXAMPLE 2-5. Given the ρ dependent field: $\mathbf{E} = \mathbf{a}_\rho K/\rho^{1/2}$, with K a constant, illustrate the validity of the divergence theorem by evaluating both integrals

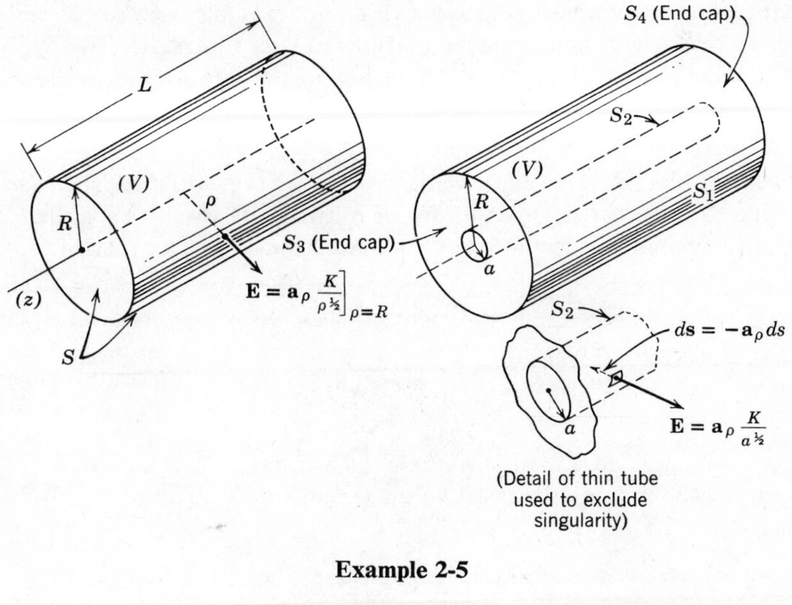

Example 2-5

of (2-34) within and on a right circular cylinder of length L, radius R, and centered about the z axis as shown.

Since **E** has a singularity at $\rho = 0$, a thin, tubular surface S_2 of radius a is constructed as shown, to exclude the singularity from the integration region, making $S = S_1 + S_2 + S_3 + S_4$. The divergence of **E**, by use of (2-29b), is

$$\nabla \cdot \mathbf{E} = \frac{1}{\rho}\frac{\partial}{\partial \rho}(\rho E_\rho) = \frac{K}{2\rho^{3/2}}$$

yielding the following *volume* integral

$$\int_V (\nabla \cdot \mathbf{E})\, dv = \frac{K}{2}\int_{\rho=a}^{R}\int_{\phi=0}^{2\pi}\int_{z=0}^{L} \frac{\rho\, d\rho\, d\phi\, dz}{\rho^{3/2}} = 2\pi K L(R^{1/2} - a^{1/2}) \qquad (1)$$

With **E** ρ directed, the surface integral of (2-34) reduces to contributions from only S_1 and S_2 (the endcaps yielding zero outward flux), whence

$$\oint_S \mathbf{E} \cdot d\mathbf{s} = \int_{\phi=0}^{2\pi}\int_{z=0}^{L}\left(\mathbf{a}_\rho \frac{K}{R^{1/2}}\right)\cdot(\mathbf{a}_\rho R\, d\phi\, dz) + \int_{\phi=0}^{2\pi}\int_{z=0}^{L}\left(\mathbf{a}_\rho \frac{K}{a^{1/2}}\right)\cdot(-\mathbf{a}_\rho a\, d\phi\, dz)$$

$$= 2\pi K L(R^{1/2} - a^{1/2}) \qquad (2)$$

agreeing with the result (1). [*Note:* Each answer has the limit $2\pi K L R^{1/2}$ as $a \to 0$.]

The usefulness of the divergence theorem embraces more generally the interchange of volume for closed-surface integrals required for establishing several theorems of electromagnetic theory. An example occurs in Poynting's theorem of electromagnetic power considered later in Chapter 7.

2-4-2 *Maxwell's Divergence Relations for Electric and Magnetic Fields* in Free Space

The definition (2-20) of the divergence of a vector field serves as a basis for deriving the differential, or point, forms of two of Maxwell's equations from their corresponding integral forms (1-53) and (1-54) for free space

$$\oint_S (\epsilon_0 \mathbf{E}) \cdot d\mathbf{s} = \int_V \rho_v\, dv \text{ C} \qquad [1\text{-}53]$$

$$\oint_S \mathbf{B} \cdot d\mathbf{s} = 0 \text{ Wb} \qquad [1\text{-}54]$$

These laws apply to closed surfaces S of arbitrary shape and size. If S is the surface bounding any *small* volume element Δv, dividing (1-53) by Δv yields

$$\frac{\oint_S (\epsilon_0 \mathbf{E}) \cdot d\mathbf{s}}{\Delta v} = \frac{\int_V \rho_v\, dv}{\Delta v} \qquad (2\text{-}38)$$

The limit of the left side, as Δv becomes sufficiently small, is div $(\epsilon_0\mathbf{E})$ from the definition (2-20). The right side denotes the ratio of the free charge Δq inside Δv to Δv itself; its limit is ρ_v. As $\Delta v \rightarrow 0$, therefore, (2-38) becomes

$$\text{div}\,(\epsilon_0\mathbf{E}) = \rho_v \ \text{C/m}^3 \tag{2-39}$$

the *differential* form of Maxwell's integral expression (1-53). Note that expressing (2-39) in rectangular coordinates using (2-29a) yields the partial differential equation

$$\frac{\partial E_x}{\partial x} + \frac{\partial E_y}{\partial y} + \frac{\partial E_z}{\partial z} = \frac{\rho_v}{\epsilon_0} \tag{2-40}$$

It is evident that the divergence of $(\epsilon_0\mathbf{E})$ at any point in a region is precisely ρ_v, the volume density of electric charge there, implying that the flux sources of **E** fields are electric charges. Equivalently, if electric field lines terminate abruptly, their termini must be electric charges.

By a similar procedure applying (1-54), one obtains the following partial differential equation in terms of **B**

$$\text{div}\,\mathbf{B} = 0 \ \text{Wb/m}^3 \tag{2-41}$$

implying that **B** fields are always divergenceless and therefore source free. The flux plot of any **B** field must, therefore, invariably consist of closed lines; free magnetic charges are thus nonexistent in the physical world. A *divergenceless* field is also called a *solenoidal* field; magnetic fields are always solenoidal.

EXAMPLE 2-6. Suppose that Maxwell's differential equation (2-39), instead of its integral form (1-53), had been postulated. Execute the reverse of the process just described, deriving (1-53) from (2-39) by integrating the latter over an arbitrary volume V and applying the divergence theorem.

Integrating (2-39) over an arbitrary volume V yields

$$\int_V \text{div}\,(\epsilon_0\mathbf{E})\,dv = \int_V \rho_v\,dv$$

Assume that **E** is well-behaved in the region in question. From a use of (2-34), the left side can be replaced by the equivalent closed-surface integral $\oint_S (\epsilon_0\mathbf{E})\cdot d\mathbf{s}$, and (1-53) follows:

$$\oint_S (\epsilon_0\mathbf{E})\cdot d\mathbf{s} = \int_V \rho_v\,dv \tag*{[1-53]}$$

2-5 Curl of a Vector Field

From (2-15) it is established that the line integral of $(\mathbf{grad}\, f)\cdot d\ell$ around any closed path is always zero. Many vector functions do not exhibit this conservative property; a physical example is the magnetic **B** field obeying Ampère's circuital law (1-56). For example, in the steady current system of Figure 1–19, the line integral of $\mathbf{B}\cdot d\ell$ taken about a circular path enclosing all or part of the wire, a nonzero current result is anticipated. Nonconservative fields such as these are said to possess a *circulation* about closed paths of integration. Whenever the closed-line integral of a field is taken about a small (vanishing) closed path and the result is expressed as a ratio to the small area enclosed, that circulation per unit area can be expressed as a vector known as the **curl** of the field in the neighborhood of a point. It follows that a conservative field has a zero value of curl everywhere; it is also called an irrotational field.

Historically, the concept of curl comes from a mathematical model of effects occurring in hydrodynamics. The early work of Helmholtz in the vortex motion of fluid fields led ultimately to the mathematical postulates by Maxwell of Faraday's conceptions of the electric fields induced by time-varying magnetic fields. A connection between curl and fluid phenomena can be established by supposing a small paddle wheel to be immersed in a stream of water, its velocity field being represented by the flux map shown in Figure 2-6. Let the paddle wheel be oriented as at A in the figure. The effect of the greater fluid velocity on one side than on the other will cause the wheel to rotate—clockwise, in the example shown. In this example, the velocity field $\boldsymbol{v}$ is said to have a vector curl directed into the paper along the axis of the paddle wheel, a sense determined by the thumb of the right hand if the fingers point in the direction of the rotation; the vector curl of $\boldsymbol{v}$ has a negative z direction at A. Similarly, physically rotating the paddle wheel axis at right angles as at

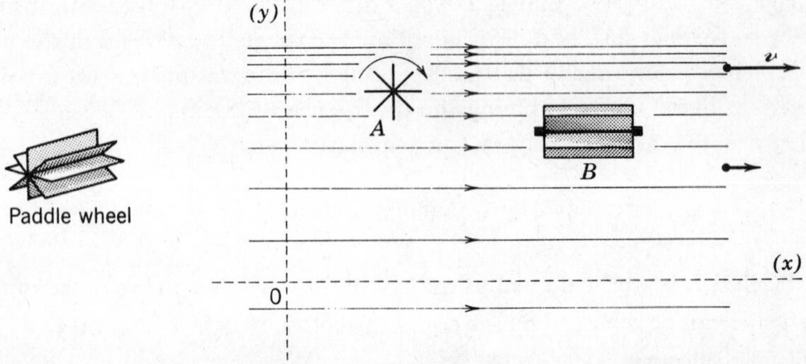

Figure 2-6. A velocity field in a fluid, with an interpretation of its curl from the rotation of a small paddle wheel.

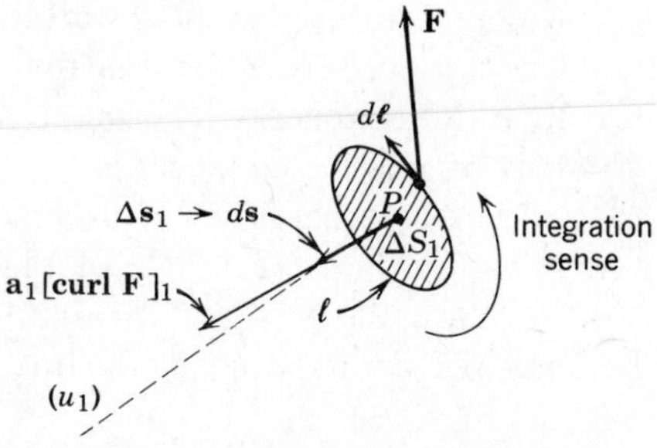

Figure 2-7. A closed line ℓ bounding the vanishing area $\Delta s_1 \to ds$, used in defining the $\mathbf{a}_1$ component of curl F at P.

B in the figure provides a way to determine the x component of the vector curl of $\mathscr{v}$, symbolized $[\mathbf{curl}\ \mathscr{v}]_x$. In rectangular coordinates, the *total* vector curl of $\mathscr{v}$ is the vector sum

$$\mathbf{curl}\ \mathscr{v} = \mathbf{a}_x[\mathbf{curl}\ \mathscr{v}]_x + \mathbf{a}_y[\mathbf{curl}\ \mathscr{v}]_y + \mathbf{a}_z[\mathbf{curl}\ \mathscr{v}]_z$$

Generally, the curl† of a vector field $\mathbf{F}(u_1, u_2, u_3, t)$, denoted **curl F**, is expressed as the vector sum of three orthogonal components as follows

$$\mathbf{curl\ F} = \mathbf{a}_1[\mathbf{curl\ F}]_1 + \mathbf{a}_2[\mathbf{curl\ F}]_2 + \mathbf{a}_3[\mathbf{curl\ F}]_3 \tag{2-42}$$

Each component is defined as a line integral of $\mathbf{F}\cdot d\ell$ about a shrinking closed line on a per-unit-area basis with the $\mathbf{a}_1$ component defined

$$\mathbf{a}_1[\mathbf{curl\ F}]_1 \equiv \mathbf{a}_1 \lim_{\Delta s_1 \to 0} \frac{\oint_\ell \mathbf{F}\cdot d\ell}{\Delta s_1} \tag{2-43}$$

The vanishing surface bounded by the closed line ℓ shown in Figure 2-7 is Δs_1, with the direction of integration around ℓ assumed to be governed by the right-hand rule.⁑ Similar definitions apply to the other two components, so the total value of **curl F** at a point is expressed

$$\mathbf{curl\ F} = \mathbf{a}_1 \lim_{\Delta s_1 \to 0} \frac{\oint_\ell \mathbf{F}\cdot d\ell}{\Delta s_1} + \mathbf{a}_2 \lim_{\Delta s_2 \to 0} \frac{\oint_\ell \mathbf{F}\cdot d\ell}{\Delta s_2} + \mathbf{a}_3 \lim_{\Delta s_3 \to 0} \frac{\oint_\ell \mathbf{F}\cdot d\ell}{\Delta s_3} \tag{2-44}$$

† In European texts, **curl F** is written *rot* **F**, and is read rotation of **F**.

⁑ The integration sense coincides with the direction in which the fingers of the right hand point if the thumb points in the direction of $\mathbf{a}_1$.

A differential expression for **curl F** in generalized coordinates is found from (2-44) by a procedure resembling that used in finding the differential expression for div **F** in Section 2.4. The shape of each closed line ℓ used in the limits of (2-44) is of no consequence, as long as the dimensions of Δs inside ℓ tend toward zero together. Thus, in finding the $\mathbf{a}_1$ component of **curl F**, ℓ is deformed into the curvilinear rectangle of Figure 2-8(*b*) with edges $\Delta\ell_2$ and $\Delta\ell_3$. The surface bounded by ℓ is $\Delta \mathbf{s}_1 = \mathbf{a}_1 \Delta\ell_2 \Delta\ell_3 = \mathbf{a}_1 h_2 h_3 \Delta u_2 \Delta u_3$, the only components of **F** contributing to the line integral in the numerator of (2-43) being

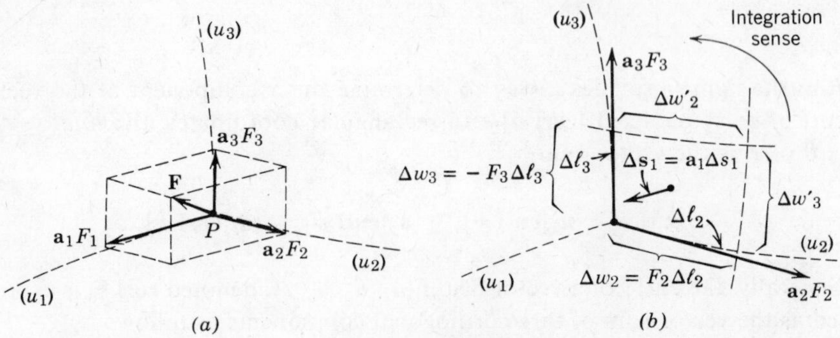

Figure 2-8. Relative to curl F in generalized orthogonal coordinates. (*a*) The components of F at a typical point *P*. (*b*) Construction of a path ℓ relative to the $\mathbf{a}_1$ component of curl F.

F_2 and F_3. Thus, along the bottom edge $\Delta\ell_2$, the contribution to $\oint_\ell \mathbf{F} \cdot d\boldsymbol{\ell}$ becomes

$$\Delta w_2 = F_2 \Delta\ell_2 \tag{2-45}$$

in which Δw_2 denotes that contribution. Along the top edge, F_2 changes an incremental amount, but in general so does the length increment, $\Delta\ell_2$, because of the curvilinear coordinate system. The line-integral contribution along the top edge is found from a Taylor's expansion of Δw_2 about P. The first two terms are sufficient if Δu_3 is suitably small; thus

$$\Delta w_2'(u_1, u_2, u_3 + \Delta u_3) = -\left[\Delta w_2 + \frac{\partial(\Delta w_2)}{\partial u_3} \Delta u_3\right] = -\left[F_2 \Delta\ell_2 + \frac{\partial(F_2 \Delta\ell_2)}{\partial u_3} \Delta u_3\right] \tag{2-46}$$

the negative sign resulting from integrating in the sense of decreasing u_2 along the top edge.

Similarly, the contribution along the left edge $\Delta\ell_3$ in Figure 2-8(*b*) is

$$\Delta w_3 = -F_3\,\Delta\ell_3 \tag{2-47}$$

while along the right edge, it is

$$\Delta w_3' = F_3\,\Delta\ell_3 + \frac{\partial(F_3\,\Delta\ell_3)}{\partial u_2}\,\Delta u_2 \tag{2-48}$$

The substitution of (2-45) through (2-48) into the definition (2-43) obtains for the $\mathbf{a}_1$ component of **curl F**

$$\mathbf{a}_1[\mathbf{curl\,F}]_1 = \mathbf{a}_1 \lim_{\Delta s_1\to 0}\frac{\oint_\ell \mathbf{F}\cdot d\ell}{\Delta s_1} = \mathbf{a}_1\lim_{\Delta s_1\to 0}\frac{\Delta w_2 + \Delta w_2' + \Delta w_3 + \Delta w_3'}{\Delta\ell_2\,\Delta\ell_3}$$

$$= \mathbf{a}_1 \lim_{\Delta s_1\to 0}\frac{F_2\,\Delta\ell_2 - \left[F_2\,\Delta\ell_2 + \dfrac{\partial(F_2\,\Delta\ell_2)}{\partial u_3}\Delta u_3\right] - F_3\,\Delta\ell_3 + \left[F_3\,\Delta\ell_3 + \dfrac{\partial(F_3\,\Delta\ell_3)}{\partial u_2}\Delta u_2\right]}{h_2 h_3\,\Delta u_2\,\Delta u_3}$$

$$= \mathbf{a}_1\frac{1}{h_2h_3}\left[\frac{\partial(F_3h_3)}{\partial u_2} - \frac{\partial(F_2h_2)}{\partial u_3}\right] \tag{2-49}$$

A similar procedure yields the two remaining vector components of **curl F** in (2-44), though from symmetry, a simple cyclic interchange of the subscripts in (2-49) leads directly to them. The expression for **curl F** in generalized orthogonal coordinates is thus

$$\mathbf{curl\,F} = \frac{\mathbf{a}_1}{h_2h_3}\left[\frac{\partial(F_3h_3)}{\partial u_2} - \frac{\partial(F_2h_2)}{\partial u_3}\right] + \frac{\mathbf{a}_2}{h_3h_1}\left[\frac{\partial(F_1h_1)}{\partial u_3} - \frac{\partial(F_3h_3)}{\partial u_1}\right] + \frac{\mathbf{a}_3}{h_1h_2}\left[\frac{\partial(F_2h_2)}{\partial u_1} - \frac{\partial(F_1h_1)}{\partial u_2}\right] \tag{2-50}$$

which is identical with the determinental form

$$\mathbf{curl\,F} = \begin{vmatrix} \dfrac{\mathbf{a}_1}{h_2h_3} & \dfrac{\mathbf{a}_2}{h_3h_1} & \dfrac{\mathbf{a}_3}{h_1h_2} \\[2ex] \dfrac{\partial}{\partial u_1} & \dfrac{\partial}{\partial u_2} & \dfrac{\partial}{\partial u_3} \\[2ex] h_1F_1 & h_2F_2 & h_3F_3 \end{vmatrix} \tag{2-51}$$

a result simplifying in the *rectangular* coordinate system to

$$\textbf{curl F} = \begin{vmatrix} \mathbf{a}_x & \mathbf{a}_y & \mathbf{a}_z \\ \dfrac{\partial}{\partial x} & \dfrac{\partial}{\partial y} & \dfrac{\partial}{\partial z} \\ F_x & F_y & F_z \end{vmatrix} \tag{2-52}$$

Upon comparing (2-52) with the cross product $\mathbf{A} \times \mathbf{B}$ of (1-41), and recalling the definition (2-18) for ∇, one is led to the equivalent symbolisms

$$\textbf{curl F} \equiv \nabla \times \mathbf{F} \tag{2-53}$$

Though ∇ has been defined only in the rectangular system, the symbolisms **curl F** and $\nabla \times \mathbf{F}$ are customarily considered interchangeable regardless of the coordinate system used.

It is seen that (2-51) also leads to the following expressions for **curl F** in the *circular cylindrical* system

$$\textbf{curl F} = \mathbf{a}_\rho\left[\frac{1}{\rho}\frac{\partial F_z}{\partial \phi} - \frac{\partial F_\phi}{\partial z}\right] + \mathbf{a}_\phi\left[\frac{\partial F_\rho}{\partial z} - \frac{\partial F_z}{\partial \rho}\right] + \mathbf{a}_z\left[\frac{1}{\rho}\frac{\partial}{\partial \rho}(\rho F_\phi) - \frac{1}{\rho}\frac{\partial F_\rho}{\partial \phi}\right] \tag{2-54}$$

and in the *spherical* coordinate system

$$\textbf{curl F} = \frac{\mathbf{a}_r}{r\sin\theta}\left[\frac{\partial}{\partial\theta}(F_\phi \sin\theta) - \frac{\partial F_\theta}{\partial \phi}\right] + \frac{\mathbf{a}_\theta}{r}\left[\frac{1}{\sin\theta}\frac{\partial F_r}{\partial \phi} - \frac{\partial}{\partial r}(rF_\phi)\right] + \frac{\mathbf{a}_\phi}{r}\left[\frac{\partial}{\partial r}(rF_\theta) - \frac{\partial F_r}{\partial \theta}\right] \tag{2-55}$$

EXAMPLE 2-7. Find the curl of $\mathbf{G} = \mathbf{a}_x Ky$, a flux plot of which is sketched in Example 2-2.

Because **G** has only a y dependent x component, from (2-52) one obtains

$$\textbf{curl G} = \begin{vmatrix} \mathbf{a}_x & \mathbf{a}_y & \mathbf{a}_z \\ 0 & \dfrac{\partial}{\partial y} & 0 \\ Ky & 0 & 0 \end{vmatrix} = \mathbf{a}_z\left[-\frac{\partial(Ky)}{\partial y}\right] = -\mathbf{a}_z K$$

a negative z directed result. So if **G** were a fluid velocity field with a paddle wheel immersed in it as in Figure 2-6, a clockwise rotation looking along the negative z direction would result, agreeing with the direction of **curl G**.

EXAMPLE 2-8. Find the curl of the **B** fields both inside and outside the long, straight wire carrying the steady current I shown in Figure 1-19.

The **B** field is given by (1-64), a ϕ directed function of ρ. The curl of **B**, obtained from (2-51), yields *inside* the wire ($\rho < a$)

$$\textbf{curl B} = \begin{vmatrix} \dfrac{\mathbf{a}_\rho}{\rho} & \mathbf{a}_\phi & \dfrac{\mathbf{a}_z}{\rho} \\ \dfrac{\partial}{\partial \rho} & \dfrac{\partial}{\partial \phi} & \dfrac{\partial}{\partial z} \\ 0 & \rho\left[\dfrac{\mu_0 I \rho}{2\pi a^2}\right] & 0 \end{vmatrix} = \frac{\mathbf{a}_z}{\rho}\frac{\partial}{\partial \rho}\left[\rho \frac{\mu_0 I \rho}{2\pi a^2}\right] = \mathbf{a}_z \mu_0 \frac{I}{\pi a^2}$$

a result proportional to the current density $J_z = I/\pi a^2$ in the wire. This special case demonstrates the validity of a Maxwell's differential relation to be developed in Section 2-5-2. The reader may further show from (2-54) that **curl B** *outside* the wire is zero, in view of the inverse ρ dependence of **B** there.

2-5-1 Theorem of Stokes

If $F(u_1, u_2, u_3, t)$ is well-behaved in some region, then the integral identity

$$\int_S (\nabla \times \mathbf{F}) \cdot d\mathbf{s} = \oint_\ell \mathbf{F} \cdot d\boldsymbol{\ell} \tag{2-56}$$

holds for every closed line ℓ in the region, if S is a surface bounded by ℓ. This is called the theorem of Stokes. An heuristic proof follows along lines resembling the proof of the divergence theorem.

Suppose the arbitrary S is subdivided into a large number n of surface elements, typical of which is Δs_i bounded by ℓ_i as in Figure 2-9(a). The line integral of $\mathbf{F} \cdot d\boldsymbol{\ell}$ around ℓ_i is inferred from the definition (2-43) of the component of the **curl F** in the direction of $\Delta \mathbf{s}_i$ to be

$$\oint_{\ell_i} \mathbf{F} \cdot d\boldsymbol{\ell} = [\textbf{curl F}] \cdot \Delta \mathbf{s}_i \tag{2-57}$$

for Δs_i sufficiently small. If the left side of (2-57) is summed over all closed contours ℓ_i on the surface S of Figure 2-9(a), the common edges of adjacent elements are traversed twice and in opposite directions to cause the integrations about ℓ_i to cancel everywhere on S except on its outer boundary ℓ. Summing the left side of (2-57) over the n interior elements Δs_i therefore obtains

$$\sum_{i=1}^{n} \left[\oint_{\ell_i} \mathbf{F} \cdot d\boldsymbol{\ell}\right] = \oint_\ell \mathbf{F} \cdot d\boldsymbol{\ell} \tag{2-58}$$

and equating to the right side of (2-57) summed over the same elements yields the result, as n approaches infinity

$$\oint_\ell \mathbf{F}\cdot d\boldsymbol{\ell} = \lim_{\Delta s_i \to 0} \sum_{i=1}^{n} [(\textbf{curl F})\cdot\Delta\mathbf{s}_i] = \int_S (\textbf{curl F})\cdot d\mathbf{s} \qquad (2\text{-}59)$$

which is Stokes' theorem (2-56).

As with the divergence theorem, it is necessary in (2-56) that $\mathbf{F}$ together with its first derivatives be continuous. If not, the discontinuities or singularities

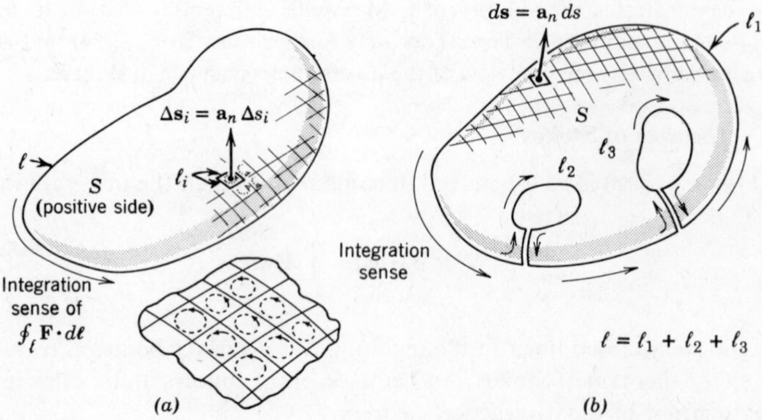

Figure 2-9. Relative to Stokes' theorem. (*a*) Showing a typical interior surface-element Δs_i bounded by ℓ_i. (*b*) Closed lines ℓ_2 and ℓ_3 constructed to eliminate discontinuities from *S*.

are excluded by constructing closed lines about them as in Figure 2-9(*b*), causing S to be bounded by the closed line $\ell = \ell_1 + \ell_2 + \ell_3$. The connective strips, of vanishing widths as shown, are however, traversed twice so their integral contributions cancel. The positive sense of $d\mathbf{s}$ should as usual agree with the integration sense around ℓ according to the right-hand rule.

EXAMPLE 2-9. Given the vector field

$$\mathbf{F}(x, y, z) = \mathbf{a}_x 5xyz + \mathbf{a}_y y^2 + \mathbf{a}_z yz \qquad (1)$$

illustrate the validity of Stokes' theorem by evaluating (2-56) over the open surface S defined by the five sides of a cube measuring 1 m on a side and about the closed line ℓ bounding S as shown.

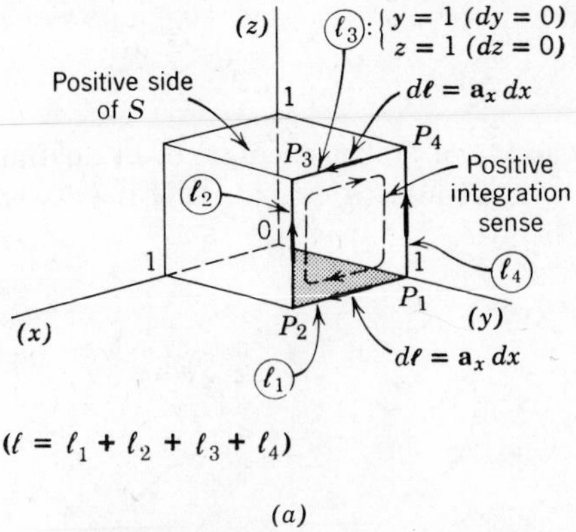

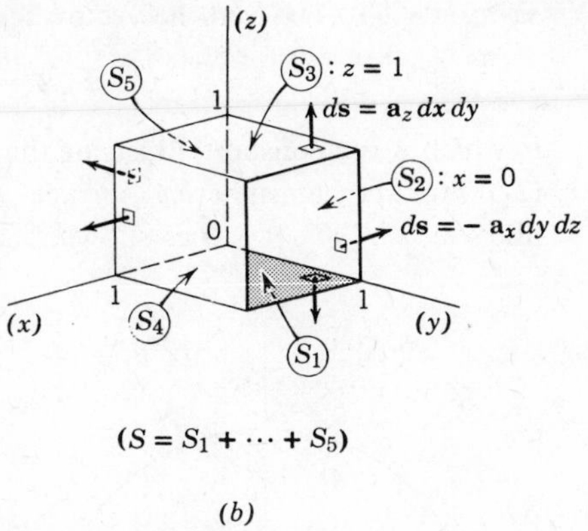

Example 2-9. **(*a*) Line elements on ℓ. (*b*) Surface elements on *S*.**

The line integral is evaluated first. The right side of (2-56) applied to ℓ becomes, making use of figure (*a*)

$$\oint_\ell \mathbf{F}\cdot d\boldsymbol{\ell} = \int_{P_1}^{P_2} \mathbf{F}\cdot\mathbf{a}_x\, dx + \int_{P_2}^{P_3} \mathbf{F}\cdot\mathbf{a}_z\, dz + \int_{P_3}^{P_4} \mathbf{F}\cdot\mathbf{a}_x\, dx + \int_{P_4}^{P_1} \mathbf{F}\cdot\mathbf{a}_z\, dz$$

$$= 0 + \int_{z=0}^{1} z\, dz + \int_{x=1}^{0} 5x\, dx + \int_{z=1}^{0} z\, dz = -5/2 \qquad (2)$$

The surface integral of (2-56) is found next. From (2-52)

$$\textbf{curl F} = \begin{vmatrix} \mathbf{a}_x & \mathbf{a}_y & \mathbf{a}_z \\ \dfrac{\partial}{\partial x} & \dfrac{\partial}{\partial y} & \dfrac{\partial}{\partial z} \\ 5xyz & y^2 & yz \end{vmatrix} = \mathbf{a}_x z + \mathbf{a}_y 5xy - \mathbf{a}_z 5xz \qquad (3)$$

whence the surface integral of (2-56) evaluated over $S_1, \ldots, S_5$ yields, using figure (*b*)

$$\int_S (\textbf{curl F})\cdot d\mathbf{s} = \int_{y=0}^{1}\int_{x=0}^{1} [5xz]_{z=0}\, dx\, dy - \int_{z=0}^{1}\int_{y=0}^{1} [z]_{x=0}\, dy\, dz$$

$$- \int_{y=0}^{1}\int_{x=0}^{1} [5xz]_{z=1}\, dx\, dy + \int_{z=0}^{1}\int_{y=0}^{1} [z]_{x=1}\, dy\, dz$$

$$- \int_{z=0}^{1}\int_{x=0}^{1} [5xy]_{y=0}\, dx\, dz = -5/2 \qquad (4)$$

which agrees with (2).

EXAMPLE 2-10. Given the vector field

$$\mathbf{F}(\theta) = \mathbf{a}_\phi K \cot\theta \tag{1}$$

in which K is a constant, illustrate the validity of Stokes' theorem by evaluating (2-55) for the hemispherical surface S with a radius a, bounded by the closed line ℓ at $\theta = 90°$, $r = a$ as shown.

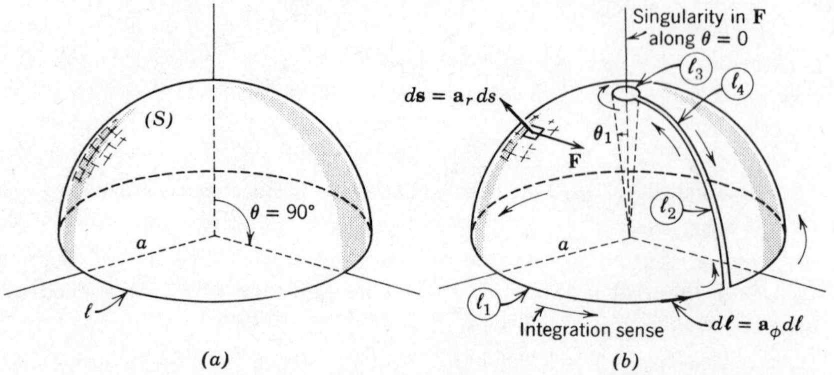

Example 2-10. (*a*) Open hemispheric surface *S*. (*b*) Exclusion of the singular point.

There is a singularity in **F** on S at $\theta = 0°$; it must be excluded to assure the validity of Stokes' theorem on the given surface. To accomplish this, a small circle ℓ_3 at $\theta = \theta_1$ and $r = a$ is constructed as in (*b*). If ds is assumed positive outward on S, then the sense of the line integration is as noted, the integrals cancelling along ℓ_2 and ℓ_4 of the connective strip as its width vanishes. The line integral around $\ell = \ell_1 + \ell_3$ thus becomes

$$\oint \mathbf{F}\cdot d\boldsymbol{\ell} = \left.\int_{\phi=0}^{2\pi} \mathbf{F}\cdot\mathbf{a}_\phi r \sin\theta\, d\phi\right]_{\substack{r=a\\ \theta=\pi/2}} + \left.\int_{\phi=0}^{-2\pi} \mathbf{F}\cdot\mathbf{a}_\phi r\sin\theta\, d\phi\right]_{\substack{r=a\\ \theta=\theta_1}}$$

$$= Ka\cos\theta_1[\phi]_0^{-2\pi} = -2\pi aK\cos\theta_1 \tag{2}$$

converging to $-2\pi aK$ as $\theta_1 \to 0$.

The surface integral is evaluated using

$$\text{curl}\,\mathbf{F} = \begin{vmatrix} \dfrac{\mathbf{a}_r}{r^2\sin\theta} & \dfrac{\mathbf{a}_\theta}{r\sin\theta} & \dfrac{\mathbf{a}_\phi}{r} \\ \dfrac{\partial}{\partial r} & \dfrac{\partial}{\partial\theta} & \dfrac{\partial}{\partial\phi} \\ 0 & 0 & (r\sin\theta)K\cot\theta \end{vmatrix} = -\mathbf{a}_r\frac{K}{r} - \mathbf{a}_\theta\frac{K}{r}\cot\theta$$

whence

$$\int_S (\textbf{curl } \mathbf{F})\cdot d\mathbf{s} = \int_{\phi=0}^{2\pi}\int_{\theta=\theta_1}^{\pi/2} (-Ka \sin\theta \, d\theta \, d\phi) = -2\pi a K \cos\theta_1 \tag{3}$$

which agrees with (2). The reader might consider how the results would have compared had one ignored the singularity.

2-5-2 *Maxwell's Curl Relations for Electric and Magnetic Fields* in Free Space

In Section 2-4-2, the divergence of a vector function was put to use in deriving the differential Maxwell equations (2-39) and (2-41) from their integral versions (1-53) and (1-54). The definition of the curl may similarly be used to obtain the differential forms of the remaining equations (1-55) and (1-56). Because the latter are correct for closed lines of arbitrary shapes and sizes, one may choose ℓ in the form of any small closed path bounding $\mathbf{a}_1 \Delta s_1$ in the vicinity of any point as in Figure 2-7. Taking the ratio of (1-55) to Δs_1 yields, with the assignment of the vector sense $\mathbf{a}_1$ to each side

$$\mathbf{a}_1 \frac{\oint_\ell \mathbf{E}\cdot d\boldsymbol{\ell}}{\Delta s_1} = \mathbf{a}_1 \frac{-\frac{d}{dt}\int_{\Delta s_1} \mathbf{B}\cdot d\mathbf{s}}{\Delta s_1} \tag{2-60}$$

From (2-43), the left side, as $\Delta s_1 \to 0$, becomes $\mathbf{a}_1[\textbf{curl } \mathbf{E}]_1$. The right side denotes the time rate of decrease of the ratio of the magnetic flux $\Delta\psi_m$ to Δs_1, but this is just the component B_1 at the point P. The limit of (2-60) therefore reduces to

$$\mathbf{a}_1[\textbf{curl } \mathbf{E}]_1 = -\frac{\partial(\mathbf{a}_1 B_1)}{\partial t} \tag{2-61}$$

relating the $\mathbf{a}_1$ component of **curl E** to the time rate of decrease of the $\mathbf{a}_1$ component of the magnetic flux density **B** at any point.† The choice of the direction assigned by $\mathbf{a}_1$ is arbitrary, implying that two similar results aligned with the directions of the unit vectors $\mathbf{a}_2$ and $\mathbf{a}_3$ and independent of (2-61) are also valid. Combining these vectorially thus obtains the *total* curl of **E** at the point

$$\mathbf{a}_1[\textbf{curl } \mathbf{E}]_1 + \mathbf{a}_2[\textbf{curl } \mathbf{E}]_2 + \mathbf{a}_3[\textbf{curl } \mathbf{E}]_3 = -\frac{\partial}{\partial t}[\mathbf{a}_1 B_1 + \mathbf{a}_2 B_2 + \mathbf{a}_3 B_3]$$

† The partial differentiation symbol $\partial/\partial t$ appearing in (2-61) replaces the total differentiation d/dt in (2-60), in view of the fact that the field **B** is a function of space as well as of time, whereas the surface integral in (2-60) is a function of t only, for a fixed Δs_1.

Making use of the notation of (2-42) yields the more compact form

$$\nabla \times \mathbf{E} = -\frac{\partial \mathbf{B}}{\partial t} \text{ V/m}^2 \tag{2-62}$$

the differential form of Faraday's law (1-55). Equation (2-62) states that the curl of the field **E** at any position is precisely the time rate of decrease of the field **B** there. This implies that the presence of a time-varying magnetic field **B** in a region is responsible for an induced time-varying **E** in that region, such that (2-62) is everywhere satisfied.

A procedure similar to that used to derive (2-62) is applicable to the Maxwell relation (1-56), yielding the differential equation

$$\nabla \times \frac{\mathbf{B}}{\mu_0} = \mathbf{J} + \frac{\partial(\epsilon_0 \mathbf{E})}{\partial t} \text{ A/m}^2 \tag{2-63}$$

It states that the curl of $\mathbf{B}/\mu_0$ at any point in a region is the sum of the electric current density **J** and the displacement current density $\partial(\epsilon_0 \mathbf{E})/\partial t$ at that point.

If the electric and magnetic fields in free space are static, the operator $\partial/\partial t$ appearing in (2-62) and (2-63) should be set to zero. This restriction provides the following curl relations for time-static fields

$$\nabla \times \mathbf{E} = 0 \tag{2-64}$$

Curl relations for static **E**, **B** fields

$$\nabla \times \frac{\mathbf{B}}{\mu_0} = \mathbf{J} \tag{2-65}$$

Equation (2-64) states that any static **E** field is irrotational (conservative), while (2-65) specifies that the curl of a static **B** field at every point in space is proportional to the current density **J** there.

2-6 *Summary of Maxwell's Equations* : Complex, Time-Harmonic Forms

One may recall that in Sections 2-4-2 and 2-5 the differential Maxwell equations for free space were obtained from their integral forms, (1-53) through (1-56). These are collected for reference in Table 2-1, columns I and III. The integral Maxwell equations were seen in Section 1-11 to be well suited for finding the field solutions of static charge or current distributions

Table 2-1 Time-dependent and complex time-harmonic forms of Maxwell's equations in free space

Integral forms		*Differential forms*	
I. Time-dependent	II. Complex, time-harmonic	III. Time-dependent	IV. Complex, time-harmonic
$\oint_S \epsilon_0 \mathbf{E} \cdot d\mathbf{s} = \int_V \rho_v \, dv$ [1-53]	$\oint_S \epsilon_0 \hat{\mathbf{E}} \cdot d\mathbf{s} = \int_V \hat{\rho}_v \, dv$	$\nabla \cdot (\epsilon_0 \mathbf{E}) = \rho_v$ [2-39]	$\nabla \cdot (\epsilon_0 \hat{\mathbf{E}}) = \hat{\rho}_v$ [2-70]
$\oint_S \mathbf{B} \cdot d\mathbf{s} = 0$ [1-54]	$\oint_S \hat{\mathbf{B}} \cdot d\mathbf{s} = 0$	$\nabla \cdot \mathbf{B} = 0$ [2-41]	$\nabla \cdot \hat{\mathbf{B}} = 0$ [2-71]
$\oint_\ell \mathbf{E} \cdot d\boldsymbol{\ell} = -\frac{d}{dt} \int_S \mathbf{B} \cdot d\mathbf{s}$ [1-55]	$\oint_\ell \hat{\mathbf{E}} \cdot d\boldsymbol{\ell} = -j\omega \int_S \hat{\mathbf{B}} \cdot d\mathbf{s}$	$\nabla \times \mathbf{E} = -\frac{\partial \mathbf{B}}{\partial t}$ [2-62]	$\nabla \times \hat{\mathbf{E}} = -j\omega \hat{\mathbf{B}}$ [2-72]
$\oint_\ell \frac{\mathbf{B}}{\mu_0} \cdot d\boldsymbol{\ell} = \int_S \mathbf{J} \cdot d\mathbf{s} + \frac{d}{dt} \int_S \epsilon_0 \mathbf{E} \cdot d\mathbf{s}$ [1-56]	$\oint_\ell \frac{\hat{\mathbf{B}}}{\mu_0} \cdot d\boldsymbol{\ell} = \int_S \hat{\mathbf{J}} \cdot d\mathbf{s} + j\omega \int_S \epsilon_0 \hat{\mathbf{E}} \cdot d\mathbf{s}$	$\nabla \times \frac{\mathbf{B}}{\mu_0} = \mathbf{J} + \frac{\partial}{\partial t}(\epsilon_0 \mathbf{E})$ [2-63]	$\nabla \times \frac{\hat{\mathbf{B}}}{\mu_0} = \hat{\mathbf{J}} + j\omega\epsilon_0 \hat{\mathbf{E}}$ [2-73]

possessing simple symmetries, though methods relying on symmetry are unfortunately limited to a few isolated problems. The differential Maxwell equations usually offer a much broader class of solutions; obtaining a number of these solutions will constitute the task of much of the remaining text.

Also of importance are the sinusoidal steady state, or *time-harmonic* solutions of Maxwell's equations. Time-harmonic fields **E** and **B** are generated whenever their charge and current sources have densities varying sinusoidally in time. Assuming the sinusoidal sources to have been active long enough that the transient field components have decayed to negligible levels permits the further assumption that **E** and **B** have reached a sinusoidal steady state. Then **E** and **B** will vary according to the factors $\cos(\omega t + \theta_e)$ and $\cos(\omega t + \theta_b)$, in which θ_e and θ_b denote arbitrary phases and ω is the angular frequency. An alternative and equivalent formulation is achieved if the fields are assumed to vary according to the complex exponential factor $e^{j\omega t}$. This assumption leads to a reduction of the field functions of space and time to functions of space only, as observed in the following.

The field quantities in the real-time forms of Maxwell's equations presented in columns I and III of Table 2-1 are symbolized

$$\begin{aligned} \mathbf{E} &\equiv \mathbf{E}(u_1, u_2, u_3, t) \qquad & \mathbf{B} &\equiv \mathbf{B}(u_1, u_2, u_3, t) \\ \mathbf{J} &\equiv \mathbf{J}(u_1, u_2, u_3, t) \qquad & \rho_v &\equiv \rho_v(u_1, u_2, u_3, t) \end{aligned} \tag{2-66}$$

The linearity of the Maxwell relations guarantees that sinusoidal time variations of charge and current sources produce **E** and **B** fields that in the steady state are also sinusoidal. Then one may replace the functions (2-66) of space and time with products of *complex functions of space* only, multiplied by the *complex factor* $e^{j\omega t}$ as follows

$$\begin{aligned} &\mathbf{E}(u_1, u_2, u_3, t) \text{ is replaced with } \hat{\mathbf{E}}(u_1, u_2, u_3)e^{j\omega t} \\ &\mathbf{B}(u_1, u_2, u_3, t) \text{ is replaced with } \hat{\mathbf{B}}(u_1, u_2, u_3)e^{j\omega t} \\ &\mathbf{J}(u_1, u_2, u_3, t) \text{ is replaced with } \hat{\mathbf{J}}(u_1, u_2, u_3)e^{j\omega t} \\ &\rho_v(u_1, u_2, u_3, t) \text{ is replaced with } \hat{\rho}_v(u_1, u_2, u_3)e^{j\omega t} \end{aligned} \tag{2-67}$$

If the complex vectors $\hat{\mathbf{E}}$, $\hat{\mathbf{B}}$, and $\hat{\mathbf{J}}$ are written in terms of the generalized coordinate system as follows: i.e.,

$$\hat{\mathbf{E}}(u_1, u_2, u_3) = \mathbf{a}_1\hat{E}_1 + \mathbf{a}_2\hat{E}_2 + \mathbf{a}_3\hat{E}_3 \tag{2-68}$$

then upon inserting (2-67) into the Maxwell equations of column III, Table 2-1, one obtains

$$\nabla\cdot(\epsilon_0\hat{\mathbf{E}}e^{j\omega t}) = \hat{\rho}_v e^{j\omega t} \qquad \nabla\times(\hat{\mathbf{E}}e^{j\omega t}) = -\frac{\partial}{\partial t}(\hat{\mathbf{B}}e^{j\omega t})$$
$$\nabla\cdot(\hat{\mathbf{B}}e^{j\omega t}) = 0 \qquad \nabla\times\left(\frac{\hat{\mathbf{B}}}{\mu_0}e^{j\omega t}\right) = \hat{\mathbf{J}}e^{j\omega t} + \frac{\partial}{\partial t}(\epsilon_0\hat{\mathbf{E}}e^{j\omega t}) \qquad \text{(2-69)}$$

The partial-derivative operators $\nabla\cdot$ and $\nabla\times$ of (2-69) affect only the space-dependent functions $\hat{\mathbf{E}}(u_1, u_2, u_3)$ and $\hat{\mathbf{B}}(u_1, u_2, u_3)$, while $\partial/\partial t$ operates only on the $e^{j\omega t}$ factors common to all the fields. Equations (2-69) therefore yield, after cancelling the $e^{j\omega t}$ factors

$$\nabla\cdot(\epsilon_0\hat{\mathbf{E}}) = \hat{\rho}_v \qquad \text{C/m}^3 \qquad \text{(2-70)}$$

$$\nabla\cdot\hat{\mathbf{B}} = 0 \qquad \text{Wb/m}^3 \qquad \text{(2-71)}$$

$$\nabla\times\hat{\mathbf{E}} = -j\omega\hat{\mathbf{B}} \qquad \text{V/m}^2 \qquad \text{(2-72)}$$

$$\nabla\times\frac{\hat{\mathbf{B}}}{\mu_0} = \hat{\mathbf{J}} + j\omega\epsilon_0\hat{\mathbf{E}} \qquad \text{A/m}^2 \qquad \text{(2-73)}$$

These are the desired complex, time-harmonic Maxwell equations for free space. They represent a simplification of the real-time forms in that the time variable t has been eliminated. Upon finding the complex solutions $\mathbf{E}(u_1, u_2, u_3)$ and $\mathbf{B}(u_1, u_2, u_3)$ that satisfy (2-70) through (2-73), the sinusoidal time dependence can be restored by multiplying each space-dependent complex solution $\hat{\mathbf{E}}$ and $\hat{\mathbf{B}}$ by $e^{j\omega t}$ and taking the real part of the result as follows

$$\mathbf{E}(u_1, u_2, u_3, t) = \text{Re}\,[\hat{\mathbf{E}}(u_1, u_2, u_3)e^{j\omega t}]$$
$$\mathbf{B}(u_1, u_2, u_3, t) = \text{Re}\,[\hat{\mathbf{B}}(u_1, u_2, u_3)e^{j\omega t}] \qquad \text{(2-74)}$$

Considerable use is to be made of (2-70) through (2-74) in subsequent discussions of the time-harmonic solutions.

One can show that a similar procedure using the replacements (2-67) leads to a complex, time-harmonic set of the *integral* forms of Maxwell's equations in free space. A comparison with their time-dependent versions is provided in Table 2-1.

Applications of the complex time-harmonic forms (2-70) through (2-73) to elementary wave solutions in free space are considered in Section 2-10. A preliminary discussion of the Laplacian operator and a development of the so-called wave equations are desirable prerequisites to finding such solutions. These are discussed next.

2-7 Laplacian and Curl Curl Operators

The gradient of a scalar field was seen in Section 2-2 to yield a vector field. Moreover, the divergence of the vector function **grad** f, denoted symbolically by $\nabla \cdot (\nabla f)$, is by the definition (2-20) a scalar measure of the flux source-per-unit-volume condition of ∇f at every point in a region. The expansions (2-10) and (2-28) for ∇f and its divergence can be combined to obtain $\nabla \cdot (\nabla f)$ in a desired coordinate system, a result to be found useful for obtaining both time-varying and time-static field solutions. Thus, in generalized coordinates, the gradient of f is expressed by (2-10)

$$\nabla f = \mathbf{a}_1 \frac{1}{h_1} \frac{\partial f}{\partial u_1} + \mathbf{a}_2 \frac{1}{h_2} \frac{\partial f}{\partial u_2} + \mathbf{a}_3 \frac{1}{h_3} \frac{\partial f}{\partial u_3} \tag{2-75}$$

To find the divergence of ∇f, the components of (2-75) become the elements F_1, F_2, and F_3 of (2-28), obtaining

$$\nabla \cdot (\nabla f) = \frac{1}{h_1 h_2 h_3} \left[\frac{\partial}{\partial u_1} \left(\frac{h_2 h_3}{h_1} \frac{\partial f}{\partial u_1} \right) + \frac{\partial}{\partial u_2} \left(\frac{h_3 h_1}{h_2} \frac{\partial f}{\partial u_2} \right) + \frac{\partial}{\partial u_3} \left(\frac{h_1 h_2}{h_3} \frac{\partial f}{\partial u_3} \right) \right] \tag{2-76}$$

This scalar result has a particularly simple form in *rectangular* coordinates, becoming

$$\nabla \cdot (\nabla f) \equiv \nabla^2 f = \frac{\partial^2 f}{\partial x^2} + \frac{\partial^2 f}{\partial y^2} + \frac{\partial^2 f}{\partial z^2} \tag{2-77}$$

The definitions of the dot product and of ∇ are seen to permit the following operator notations:

$$\nabla \cdot \nabla \equiv \frac{\partial^2}{\partial x^2} + \frac{\partial^2}{\partial y^2} + \frac{\partial^2}{\partial z^2} \equiv \nabla^2 \tag{2-78}$$

in which the notation ∇^2, called the Laplacian operator, is equivalent to $\nabla \cdot (\nabla \quad) = \nabla \cdot \nabla(\quad) =$ div (**grad** $\quad$). From (2-76), the Laplacian operator in generalized coordinates is, therefore

$$\nabla^2 \equiv \nabla \cdot \nabla \equiv \frac{1}{h_1 h_2 h_3} \left[\frac{\partial}{\partial u_1} \left(\frac{h_2 h_3}{h_1} \frac{\partial}{\partial u_1} \right) + \frac{\partial}{\partial u_2} \left(\frac{h_3 h_1}{h_2} \frac{\partial}{\partial u_2} \right) + \frac{\partial}{\partial u_3} \left(\frac{h_1 h_2}{h_3} \frac{\partial}{\partial u_3} \right) \right] \tag{2-79}$$

yielding in the *circular cylindrical* system

$$\nabla^2 f \equiv \nabla \cdot \nabla f = \frac{1}{\rho}\frac{\partial}{\partial \rho}\left(\rho \frac{\partial f}{\partial \rho}\right) + \frac{1}{\rho^2}\frac{\partial^2 f}{\partial \phi^2} + \frac{\partial^2 f}{\partial z^2} \tag{2-80}$$

while in *spherical* coordinates

$$\nabla^2 f = \frac{1}{r^2}\frac{\partial}{\partial r}\left(r^2 \frac{\partial f}{\partial r}\right) + \frac{1}{r^2 \sin\theta}\frac{\partial}{\partial \theta}\left(\sin\theta \frac{\partial f}{\partial \theta}\right) + \frac{1}{r^2 \sin^2\theta}\frac{\partial^2 f}{\partial \phi^2} \tag{2-81}$$

The Laplacian operator (2-79) is also applicable to a *vector* field $\mathbf{F}(u_1, u_2, u_3, t)$, the result of which is shown to be useful in the expansion of **curl (curl F)**. Apply (2-79) to define $\nabla^2\mathbf{F}$, the Laplacian of a vector field, as follows

$$\nabla^2 \mathbf{F} \equiv \frac{1}{h_1 h_2 h_3}\left[\frac{\partial}{\partial u_1}\left(\frac{h_2 h_3}{h_1}\frac{\partial}{\partial u_1}\right) + \frac{\partial}{\partial u_2}\left(\frac{h_3 h_1}{h_2}\frac{\partial}{\partial u_2}\right) + \frac{\partial}{\partial u_3}\left(\frac{h_1 h_2}{h_3}\frac{\partial}{\partial u_3}\right)\right] \times (\mathbf{a}_1 F_1 + \mathbf{a}_2 F_2 + \mathbf{a}_3 F_3) \tag{2-82}$$

The term-by-term expansion of the latter can be tedious, since in general $\mathbf{a}_1$, $\mathbf{a}_2$, and $\mathbf{a}_3$ are not constant unit vectors in a region; i.e., their directions depend on u_1, u_2, and u_3. In rectangular coordinates, however, (2-82) yields the relatively simple result (since $\mathbf{a}_x$, $\mathbf{a}_y$, $\mathbf{a}_z$ are constant unit vectors)

$$\nabla^2 \mathbf{F} = \mathbf{a}_x \nabla^2 F_x + \mathbf{a}_y \nabla^2 F_y + \mathbf{a}_z \nabla^2 F_z \tag{2-83}$$

in which the components $\nabla^2 F_x$, etc., are specified by (2-77). No corresponding simplicity occurs in other coordinate systems because of the spatial dependence of the unit vectors already noted. For example, if the space partial derivatives of the unit vectors are properly accounted for, as in Example 1-1 of Section 1-6, one can show from (2-82) that $\nabla^2\mathbf{F}$ in the *circular cylindrical* system becomes

$$\nabla^2 \mathbf{F} = \mathbf{a}_\rho\left[\nabla^2 F_\rho - \frac{2}{\rho^2}\frac{\partial F_\phi}{\partial \phi} - \frac{F_\rho}{\rho^2}\right] + \mathbf{a}_\phi\left[\nabla^2 F_\phi + \frac{2}{\rho^2}\frac{\partial F_\rho}{\partial \phi} - \frac{F_\phi}{\rho^2}\right] + \mathbf{a}_z \nabla^2 F_z \tag{2-84}$$

a result decidedly *not* of the form of (2-83) with F_ρ, F_ϕ, F_z merely taking the places of F_x, F_y, F_z.

Still another vector result, the curl of the vector **curl F**, designated

$\nabla \times (\nabla \times \mathbf{F})$, is of importance. The function $\nabla \times \mathbf{F}$ provides the three components given in (2-50); then performing another curl operation yields

$$
\begin{aligned}
\nabla \times (\nabla \times \mathbf{F}) = \frac{\mathbf{a}_1}{h_2 h_3} \Bigg\{ & \frac{\partial}{\partial u_2}\left[\frac{h_3}{h_1 h_2}\left(\frac{\partial(h_2 F_2)}{\partial u_1} - \frac{\partial(h_1 F_1)}{\partial u_2}\right)\right] \\
& - \frac{\partial}{\partial u_3}\left[\frac{h_2}{h_1 h_3}\left(\frac{\partial(h_1 F_1)}{\partial u_3} - \frac{\partial(h_3 F_3)}{\partial u_1}\right)\right]\Bigg\} \\
+ \frac{\mathbf{a}_2}{h_3 h_1} \Bigg\{ & \frac{\partial}{\partial u_3}\left[\frac{h_1}{h_2 h_3}\left(\frac{\partial(h_3 F_3)}{\partial u_2} - \frac{\partial(h_2 F_2)}{\partial u_3}\right)\right] \\
& - \frac{\partial}{\partial u_1}\left[\frac{h_3}{h_2 h_1}\left(\frac{\partial(h_2 F_2)}{\partial u_1} - \frac{\partial(h_1 F_1)}{\partial u_2)}\right)\right]\Bigg\} \\
+ \frac{\mathbf{a}_3}{h_1 h_2} \Bigg\{ & \frac{\partial}{\partial u_1}\left[\frac{h_2}{h_3 h_1}\left(\frac{\partial(h_1 F_1)}{\partial u_3} - \frac{\partial(h_3 F_3)}{\partial u_1}\right)\right] \\
& - \frac{\partial}{\partial u_2}\left[\frac{h_1}{h_3 h_2}\left(\frac{\partial(h_3 F_3)}{\partial u_2} - \frac{\partial(h_2 F_2)}{\partial u_3}\right)\right]\Bigg\}
\end{aligned}
\tag{2-85}
$$

Because of its complexity this result is examined only in *rectangular* coordinates, becoming

$$
\begin{aligned}
\nabla \times (\nabla \times \mathbf{F}) = \mathbf{a}_x & \left\{\frac{\partial}{\partial y}\left(\frac{\partial F_y}{\partial x} - \frac{\partial F_x}{\partial y}\right) - \frac{\partial}{\partial z}\left(\frac{\partial F_x}{\partial z} - \frac{\partial F_z}{\partial x}\right)\right\} \\
+ \mathbf{a}_y & \left\{\frac{\partial}{\partial z}\left(\frac{\partial F_z}{\partial y} - \frac{\partial F_y}{\partial z}\right) - \frac{\partial}{\partial x}\left(\frac{\partial F_y}{\partial x} - \frac{\partial F_x}{\partial y}\right)\right\} \\
+ \mathbf{a}_z & \left\{\frac{\partial}{\partial x}\left(\frac{\partial F_x}{\partial z} - \frac{\partial F_z}{\partial x}\right) - \frac{\partial}{\partial y}\left(\frac{\partial F_z}{\partial y} - \frac{\partial F_y}{\partial z}\right)\right\}
\end{aligned}
\tag{2-86}
$$

A comparison of the latter with the vector $\nabla(\nabla \cdot \mathbf{F})$ is now made. In rectangular coordinates, utilizing (2-10) and (2-28) obtains

$$
\begin{aligned}
\nabla(\nabla \cdot \mathbf{F}) = {} & \mathbf{a}_x \frac{\partial^2 F_x}{\partial x^2} + \mathbf{a}_y \frac{\partial^2 F_y}{\partial y^2} + \mathbf{a}_z \frac{\partial^2 F_z}{\partial z^2} + \mathbf{a}_x\left[\frac{\partial}{\partial x}\left(\frac{\partial F_y}{\partial y} + \frac{\partial F_z}{\partial z}\right)\right] \\
& + \mathbf{a}_y\left[\frac{y}{\partial y}\left(\frac{\partial F_z}{\partial z} + \frac{\partial F_x}{\partial x}\right)\right] + \mathbf{a}_z\left[\frac{\partial}{\partial z}\left(\frac{\partial F_x}{\partial x} + \frac{\partial F_y}{\partial y}\right)\right]
\end{aligned}
\tag{2-87}
$$

and adding and subtracting six properly chosen terms puts (2-87) into the following form

$$\nabla(\nabla\cdot\mathbf{F}) = \mathbf{a}_x\left(\frac{\partial^2 F_x}{\partial x^2} + \frac{\partial^2 F_x}{\partial y^2} + \frac{\partial^2 F_x}{\partial z^2}\right) + \mathbf{a}_y\left(\frac{\partial^2 F_y}{\partial x^2} + \frac{\partial^2 F_y}{\partial y^2} + \frac{\partial^2 F_y}{\partial z^2}\right)$$
$$+ \mathbf{a}_z\left(\frac{\partial^2 F_z}{\partial x^2} + \frac{\partial^2 F_z}{\partial y^2} + \frac{\partial^2 F_z}{\partial z^2}\right)$$
$$+ \mathbf{a}_x\left[\frac{\partial}{\partial x}\left(\frac{\partial F_y}{\partial y} + \frac{\partial F_z}{\partial z}\right) - \frac{\partial^2 F_x}{\partial y^2} - \frac{\partial^2 F_x}{\partial z^2}\right]$$
$$+ \mathbf{a}_y\left[\frac{\partial}{\partial y}\left(\frac{\partial F_z}{\partial z} + \frac{\partial F_x}{\partial x}\right) - \frac{\partial^2 F_y}{\partial z^2} - \frac{\partial^2 F_y}{\partial x^2}\right]$$
$$+ \mathbf{a}_z\left[\frac{\partial}{\partial z}\left(\frac{\partial F_x}{\partial x} + \frac{\partial F_y}{\partial y}\right) - \frac{\partial^2 F_z}{\partial x^2} - \frac{\partial^2 F_z}{\partial y^2}\right]$$

Upon comparing the terms of the latter with (2-83) and (2-86), it is seen that one has precisely $\nabla(\nabla\cdot\mathbf{F}) = \nabla^2\mathbf{F} + \nabla \times (\nabla \times \mathbf{F})$. This is a vector identity, usually written

$$\nabla \times (\nabla \times \mathbf{F}) = \nabla(\nabla\cdot\mathbf{F}) - \nabla^2\mathbf{F} \tag{2-88a}$$

Equation (2-88a) provides a useful equivalence for $\nabla \times (\nabla \times \mathbf{F})$, especially if the field $\mathbf{F}$ is divergenceless ($\nabla\cdot\mathbf{F} = 0$). Then

$$\nabla \times (\nabla \times \mathbf{F}) = -\nabla^2\mathbf{F} \qquad \text{if } \nabla\cdot\mathbf{F} = 0 \tag{2-88b}$$

While the proof of (2-88a) was carried out in the rectangular system, such differential results are independent of the coordinate system, meaning that (2-88a) and (2-88b) are true for any system.

It is worthwhile to observe that one can more easily expand $\nabla^2\mathbf{F}$ by use of the vector identity (2-88a) than by definition (2-82). Thus

$$\nabla^2\mathbf{F} = \nabla(\nabla\cdot\mathbf{F}) - \nabla \times (\nabla \times \mathbf{F}) \tag{2-89}$$

is useful for expanding $\nabla^2\mathbf{F}$ in a coordinate system other than the cartesian.

Several vector identities involving the differential operators **grad**, div, and **curl** are listed in Table 2-2 along with vector algebraic and integral identities. Proofs of the algebraic and the differential identities are achieved in the manner used to prove (2-88a), i.e., expanding both sides in rectangular coordinates leads to an identity. The integral identities (7) and (8) are recognized as those

Table 2-2 Summary of vector identities

Algebraic	*Differential*
(1) $\mathbf{F}\cdot\mathbf{G} = \mathbf{G}\cdot\mathbf{F}$	(11) $\nabla(f+g) = \nabla f + \nabla g$
(2) $\mathbf{F}\times\mathbf{G} = -\,\mathbf{G}\times\mathbf{F}$	(12) $\nabla\cdot(\mathbf{F}+\mathbf{G}) = \nabla\cdot\mathbf{F} + \nabla\cdot\mathbf{G}$
(3) $\mathbf{F}\cdot(\mathbf{G}+\mathbf{H}) = \mathbf{F}\cdot\mathbf{G} + \mathbf{F}\cdot\mathbf{H}$	(13) $\nabla\times(\mathbf{F}+\mathbf{G}) = \nabla\times\mathbf{F} + \nabla\times\mathbf{G}$
(4) $\mathbf{F}\times(\mathbf{G}+\mathbf{H}) = \mathbf{F}\times\mathbf{G} + \mathbf{F}\times\mathbf{H}$	(14) $\nabla(fg) = f\nabla g + g\nabla f$
(5) $\mathbf{F}\times(\mathbf{G}\times\mathbf{H}) = \mathbf{G}(\mathbf{H}\cdot\mathbf{F}) - \mathbf{H}(\mathbf{F}\cdot\mathbf{G})$	(15) $\nabla\cdot(f\mathbf{F}) = \mathbf{F}\cdot\nabla f + f(\nabla\cdot\mathbf{F})$
(6) $\mathbf{F}\cdot(\mathbf{G}\times\mathbf{H}) = \mathbf{G}\cdot(\mathbf{H}\times\mathbf{F}) = \mathbf{H}\cdot(\mathbf{F}\times\mathbf{G})$	(16) $\nabla\cdot(\mathbf{F}\times\mathbf{G}) = \mathbf{G}\cdot(\nabla\times\mathbf{F}) - \mathbf{F}\cdot(\nabla\times\mathbf{G})$
Integral	(17) $\nabla\times(f\mathbf{F}) = (\nabla f)\times\mathbf{F} + f(\nabla\times\mathbf{F})$
(7) $\oint_S \mathbf{F}\cdot d\mathbf{s} = \int_V \nabla\cdot\mathbf{F}\,dv$	(18) $\nabla\cdot\nabla f = \nabla^2 f$
(8) $\oint_\ell \mathbf{F}\cdot d\boldsymbol{\ell} = \int_S (\nabla\times\mathbf{F})\cdot d\mathbf{s}$	(19) $\nabla\cdot(\nabla\times\mathbf{F}) = 0$
(9) $\oint_S f(\nabla g)\cdot d\mathbf{s} = \int_V [f\nabla^2 g + (\nabla f)\cdot(\nabla g)]\,dv$	(20) $\nabla\times(\nabla f) = 0$
(10) $\oint_S [f\nabla g - g\nabla f]\cdot d\mathbf{s} = \int_V (f\nabla^2 g - g\nabla^2 f)\,dv$	(21) $\nabla\times(\nabla\times\mathbf{F}) = \nabla(\nabla\cdot\mathbf{F}) - \nabla^2\mathbf{F}$
	(22) $\nabla\times(f\nabla g) = \nabla f\times\nabla g$

of divergence and Stokes' theorem, respectively. Extensions of the divergence theorem lead to Green's integral identities (9) and (10), proved in the next section.

*2-8 Green's Integral Theorems; Uniqueness

One can specialize the divergence theorem (2-34) to a particular class of vector functions and obtain the integral identities known as Green's theorems. Suppose **F** to be a scalar field f multiplied by a conservative vector field ∇g; i.e., let $\mathbf{F} = f\nabla g$. Then (2-34) takes on the special form

$$\oint_S (f\nabla g)\cdot d\mathbf{s} = \int_V \nabla\cdot(f\nabla g)\, dv \tag{2-90}$$

assuming the functions well-behaved in and on the volume V. The integrand in the volume integral may be expanded by use of (15) in Table 2-2, whence (2-90) becomes *Green's first integral identity*

$$\oint_S (f\nabla g)\cdot d\mathbf{s} = \int_V [f\nabla^2 g + (\nabla f)\cdot(\nabla g)]\, dv \tag{2-91}$$

If one chooses to define a vector function $\mathbf{G} = g\nabla f$ instead, the same procedure leads to a result like (2-91) except for the interchange of the roles of the scalar functions f and g

$$\oint_S (g\nabla f)\cdot d\mathbf{s} = \int_V [g\nabla^2 f + (\nabla g)\cdot(\nabla f)]\, dv$$

Subtracting the latter from (2-91) obtains *Green's second integral identity*

$$\oint_S (f\nabla g - g\nabla f)\cdot d\mathbf{s} = \int_V (f\nabla^2 g - g\nabla^2 f)\, dv \tag{2-92}$$

also known as *Green's symmetric theorem.*

Green's theorems (2-91) and (2-92) are important in applications to theorems of boundary-value problems of field theory, as well as to special theorems concerning integral properties of scalar and vector functions. One such theorem concerns those differential properties of a vector field **F** that must be specified in a region to make **F** unique. One proof of this follows.

Suppose that both the divergence and curl of a well-behaved field **F** *are specified* in a volume region of space, as well as the normal component of **F** on the surface S that bounds V, designated by $\mathbf{F}\cdot\mathbf{a}_n$. Green's theorem (2-91) is utilized to show that these specifications are sufficient to make **F** a unique function. (**F** is the *only* function that will fulfill the specifications.) To prove it, suppose *another* vector function **G** exists, having exactly the same divergence and curl; i.e., $\nabla \cdot \mathbf{F} = \nabla \cdot \mathbf{G}$ and $\nabla \times \mathbf{F} = \nabla \times \mathbf{G}$; and that **G** satisfies the same boundary condition as **F**: $\mathbf{F} \cdot \mathbf{a}_n = \mathbf{G} \cdot \mathbf{a}_n$ everywhere on S. Then their *difference*, designated by $\mathbf{H} = \mathbf{F} - \mathbf{G}$, must satisfy the three conditions

$$\nabla \cdot \mathbf{H} = 0 \tag{2-93a}$$

$$\nabla \times \mathbf{H} = 0 \tag{2-93b}$$

$$\mathbf{H} \cdot \mathbf{a}_n = 0 \text{ on } S \tag{2-93c}$$

From (20) in Table 2-2 and the irrotational property (2-93b), **H** is expressible as

$$\mathbf{H} = \nabla \phi \tag{2-93d}$$

This substituted into (2-93a) provides a property of ϕ, $\nabla \cdot (\nabla \phi) = 0$, also written

$$\nabla^2 \phi = 0 \tag{2-93e}$$

Green's theorem (2-91) is true for all scalar functions f and g well-behaved in V and on S, so it must be true for $f = g = \phi$, making (2-91)

$$\oint_S (\phi \nabla \phi) \cdot d\mathbf{s} = \int_V [\phi \nabla^2 \phi + (\nabla \phi) \cdot (\nabla \phi)] \, dv$$

In view of (2-93e), this reduces to

$$\oint_S (\phi \nabla \phi) \cdot d\mathbf{s} = \int_V (\nabla \phi) \cdot (\nabla \phi) \, dv$$

From (2-93d), the left integrand is $(\phi \mathbf{H}) \cdot \mathbf{a}_n \, ds$, which from (2-93c) is just zero. The right side thus becomes

$$\int_V (\nabla \phi) \cdot (\nabla \phi) \, dv = \int_V |\mathbf{H}|^2 \, dv = 0$$

but the only way in which the squared (positive) integrand can yield a zero integral result is for **H** to be zero. Since $\mathbf{H} = \mathbf{F} - \mathbf{G} = 0$, $\mathbf{F} = \mathbf{G}$, implying that **F** is unique.

This proof shows that the specification of *both the divergence and the curl*†

† In physical terms, the specification of the divergence and the curl of **F** amounts to specifying all the *sources* of the field in a region V, a fact evident from Maxwell's equations in the case of the **E** and **H** fields.

of a vector function **F** in a region V, plus a particular boundary condition on the surface S that bounds V, are sufficient to make **F** unique. Maxwell's equations (2-39), (2-41), (2-62), and (2-63) specify the divergence and the curl of both the **E** and the **B** fields in a region (in terms of charge and current densities as well as the other **B** or **E** field), so that these relationships, together with appropriately specified boundary conditions, can similarly be expected to provide unique field solutions. Finding solutions of Maxwell's differential equations is facilitated for some problems by first manipulating them simultaneously to obtain differential equations in terms of only **B** or **E**, as is discussed next.

*2-9 Wave Equations for Electric and Magnetic Fields in Free Space

Electromagnetic field solutions **B** and **E** in free space must, by the uniqueness discussion of the previous section, satisfy the Maxwell divergence and curl relations (2-39), (2-41), (2-62), and (2-63). In a time-varying electromagnetic field problem, one is generally interested in obtaining **E** and **B** field solutions of the four Maxwell relations, a process that can often be facilitated by combining Maxwell's equations such that one of the fields (**B** or **E**) is eliminated, yielding a partial differential equation known as the *wave equation*. This is accomplished as follows.

The Maxwell differential equations for free space are repeated here for convenience:

$$\nabla \cdot (\epsilon_0 \mathbf{E}) = \rho_v \qquad [2\text{-}39]$$

$$\nabla \cdot \mathbf{B} = 0 \qquad [2\text{-}41]$$

$$\nabla \times \mathbf{E} = -\frac{\partial \mathbf{B}}{\partial t} \qquad [2\text{-}62]$$

$$\nabla \times \frac{\mathbf{B}}{\mu_0} = \mathbf{J} + \frac{\partial(\epsilon_0 \mathbf{E})}{\partial t} \qquad [2\text{-}63]$$

To eliminate **B**, taking the curl of both sides of (2-62) obtains

$$\nabla \times (\nabla \times \mathbf{E}) = -\frac{\partial}{\partial t}(\nabla \times \mathbf{B})$$

Substituting (2-63) into the right side yields, after transposing terms containing **E** to the left side

$$\nabla \times (\nabla \times \mathbf{E}) + \mu_0 \epsilon_0 \frac{\partial^2 \mathbf{E}}{\partial t^2} = -\mu_0 \frac{\partial \mathbf{J}}{\partial t} \qquad (2\text{-}94)$$

* For the purposes of the next section, Section 2-9 may be omitted if desired.

a vector partial differential equation known as the *inhomogeneous vector wave equation* for free space.

A wave equation similar to (2-94) can be obtained in terms of **B**. Thus, taking the curl of (2-63) and substituting (2-62) into the result yields the inhomogeneous vector wave equation

$$\nabla \times (\nabla \times \mathbf{B}) + \mu_0\epsilon_0 \frac{\partial^2 \mathbf{B}}{\partial t^2} = \mu_0 \nabla \times \mathbf{J} \tag{2-95}$$

Should the **E** field of (2-94) be divergenceless ($\nabla \cdot \mathbf{E} = 0$), then (2-88b) simplifies the $\nabla \times (\nabla \times \mathbf{E})$ term to $-\nabla^2\mathbf{E}$. From (2-39), **E** is divergenceless if the region is charge free ($\rho_v = 0$); furthermore, from (2-41), **B** is *always* divergenceless. Thus, in a *charge-free region*, (2-94) and (2-95) are written

$$\nabla^2 \mathbf{E} - \mu_0\epsilon_0 \frac{\partial^2 \mathbf{E}}{\partial t^2} = \mu_0 \frac{\partial \mathbf{J}}{\partial t} \tag{2-96}$$

Inhomogeneous vector wave equations for charge-free region

$$\nabla^2 \mathbf{B} - \mu_0\epsilon_0 \frac{\partial^2 \mathbf{B}}{\partial t^2} = -\mu_0 \nabla \times \mathbf{J} \tag{2-97}$$

A further simplification is possible if the region is *empty space*, i.e., it is both charge free and current free ($\rho_v = \mathbf{J} = 0$). Then the simpler *homogeneous* vector wave equations hold

$$\nabla^2 \mathbf{E} - \mu_0\epsilon_0 \frac{\partial^2 \mathbf{E}}{\partial t^2} = 0 \tag{2-98}$$

Homogeneous vector wave equations for empty space

$$\nabla^2 \mathbf{B} - \mu_0\epsilon_0 \frac{\partial^2 \mathbf{B}}{\partial t^2} = 0 \tag{2-99}$$

If in a problem the *rectangular* coordinate system is appropriate to the **E** and **B** fields governed by (2-98) and (2-99), making use of (2-83) provides the following *scalar wave equations* in terms of field components:

$$\nabla^2 E_x - \mu_0\epsilon_0 \frac{\partial^2 E_x}{\partial t^2} = 0 \tag{2-100a}$$

$$\nabla^2 E_y - \mu_0\epsilon_0 \frac{\partial^2 E_y}{\partial t^2} = 0 \tag{2-100b}$$

$$\nabla^2 E_z - \mu_0\epsilon_0 \frac{\partial^2 E_z}{\partial t^2} = 0 \tag{2-100c}$$

and

$$\nabla^2 B_x - \mu_0\epsilon_0 \frac{\partial^2 B_x}{\partial t^2} = 0 \tag{2-101a}$$

$$\nabla^2 B_y - \mu_0\epsilon_0 \frac{\partial^2 B_y}{\partial t^2} = 0 \tag{2-101b}$$

$$\nabla^2 B_z - \mu_0\epsilon_0 \frac{\partial^2 B_z}{\partial t^2} = 0 \tag{2-101c}$$

Not all six field components are necessarily present in a given problem, a fact that can provide a considerable simplification; an example is discussed in the next section.

The *complex time-harmonic* forms of the wave equations may be obtained by replacing **B** and **E** with their complex exponential forms, (2-67). If this is done for (2-98) and (2-99), one obtains after cancelling $e^{j\omega t}$

$$\nabla^2\hat{\mathbf{E}} + \omega^2\mu_0\epsilon_0\hat{\mathbf{E}} = 0 \tag{2-102}$$

$$\nabla^2\hat{\mathbf{B}} + \omega^2\mu_0\epsilon_0\hat{\mathbf{B}} = 0 \tag{2-103}$$

Homogeneous vector wave equations in complex time-harmonic form, for empty space

Since $\mathbf{E} = \mathbf{a}_x E_x + \mathbf{a}_y E_y + \mathbf{a}_z E_z$ and $\mathbf{B} = \mathbf{a}_x B_x + \mathbf{a}_y B_y + \mathbf{a}_z B_z$, (2-102) and (2-103) expand to obtain the following homogeneous, scalar wave equations in complex time-harmonic form

$$\nabla^2\hat{E}_x + \omega^2\mu_0\epsilon_0\hat{E}_x = 0 \tag{2-104a}$$

$$\nabla^2\hat{E}_y + \omega^2\mu_0\epsilon_0\hat{E}_y = 0 \tag{2-104b}$$

$$\nabla^2\hat{E}_z + \omega^2\mu_0\epsilon_0\hat{E}_z = 0 \tag{2-104c}$$

and

$$\nabla^2\hat{B}_x + \omega^2\mu_0\epsilon_0\hat{B}_x = 0 \tag{2-105a}$$

$$\nabla^2\hat{B}_y + \omega^2\mu_0\epsilon_0\hat{B}_y = 0 \tag{2-105b}$$

$$\nabla^2\hat{B}_z + \omega^2\mu_0\epsilon_0\hat{B}_z = 0 \tag{2-105c}$$

The simplest solutions of these scalar wave equations are uniform plane waves, involving as few as two field components. They are considered in the next section.

2-10 Uniform Plane Waves in Empty Space

The simplest wave solutions of Maxwell's equations are *uniform plane waves*, characterized by *uniform* fields over infinite *plane* surfaces at fixed instants. Simplifying features are that the solutions are amenable to the rectangular coordinate system, and the number of field components reduces to as few as two. These simplifications provide a background for the more complex wave structures discussed in later chapters.

Uniform plane waves have the property that, at any fixed instant, the **E** and **B** fields are uniform over plane surfaces. These planes are arbitrarily chosen; for present purposes assume that they are defined by the surfaces $z =$ constant. This is equivalent to stating that space variations of **E** and **B** are zero over $z =$ constant planes; thus assume:

1. The fields have neither x nor y dependence; i.e., $\partial/\partial x = \partial/\partial y = 0$ for all field components. It will be shown that waves propagating in the z direction result from this restriction. If the waves propagate in empty space, one requires an additional assumption.
2. Charge and current densities are everywhere zero in the region; i.e., $\rho_v = \mathbf{J} = 0$.

The complex time-harmonic forms of the Maxwell differential equations determining the wave solutions are (2-70) through (2-73). With assumption (2) they become:

$$\nabla \cdot (\epsilon_0 \hat{\mathbf{E}}) = 0 \tag{2-106}$$

$$\nabla \cdot \hat{\mathbf{B}} = 0 \tag{2-107}$$

$$\nabla \times \hat{\mathbf{E}} = -j\omega \hat{\mathbf{B}} \tag{2-108}$$

$$\nabla \times \frac{\hat{\mathbf{B}}}{\mu_0} = j\omega\epsilon_0 \hat{\mathbf{E}} \tag{2-109}$$

Combining these equations has been shown to produce the wave equations (2-102) and (2-103)

$$\nabla^2 \hat{\mathbf{E}} + \omega^2 \mu_0 \epsilon_0 \hat{\mathbf{E}} = 0 \qquad [2\text{-}102]$$

$$\nabla^2 \hat{\mathbf{B}} + \omega^2 \mu_0 \epsilon_0 \hat{\mathbf{B}} = 0 \qquad [2\text{-}103]$$

One should bear in mind that no new information is contained in the latter that is not already expressed by the preceding Maxwell's equations.

Before attempting to extract solutions from the wave equations, one may note that the curl relations, (2-108) and (2-109), furnish some interesting properties of the solutions, restricted by assumptions (1) and (2). Assuming that all six field components are present, (2-108) becomes, with $\partial/\partial x = \partial/\partial y = 0$ of assumption (1)

$$\nabla \times \hat{\mathbf{E}} \equiv \begin{vmatrix} \mathbf{a}_x & \mathbf{a}_y & \mathbf{a}_z \\ 0 & 0 & \dfrac{\partial}{\partial z} \\ \hat{E}_x & \hat{E}_y & \hat{E}_z \end{vmatrix} = -j\omega(\mathbf{a}_x\hat{B}_x + \mathbf{a}_y\hat{B}_y + \mathbf{a}_z\hat{B}_z)$$

expanding into the triplet of differential equations

$$-\frac{\partial \hat{E}_y}{\partial z} = -j\omega\hat{B}_x \tag{2-110a}$$

$$\frac{\partial \hat{E}_x}{\partial z} = -j\omega\hat{B}_y \tag{2-110b}$$

$$0 = \hat{B}_z \tag{2-110c}$$

Similarly, (2-109) provides

$$-\frac{\partial \dfrac{\hat{B}_y}{\mu_0}}{\partial z} = j\omega\epsilon_0\hat{E}_x \tag{2-111a}$$

$$\frac{\partial \dfrac{\hat{B}_x}{\mu_0}}{\partial z} = j\omega\epsilon_0\hat{E}_y \tag{2-111b}$$

$$0 = \hat{E}_z \tag{2-111c}$$

From these differential expressions, the following properties apply to the solutions about to be found:

1. *No z component* of either $\hat{\mathbf{E}}$ or $\hat{\mathbf{B}}$ is obtained, thus making the field directions entirely *transverse* to the z axis.
2. *Two independent pairs of fields*, $(\hat{E}_x, \hat{B}_y)$ and $(\hat{E}_y, \hat{B}_x)$, are yielded under the assumptions. This is seen to be the case upon setting $\hat{E}_x = 0$ in (2-110b), for example, forcing $\hat{B}_y$ to vanish while yet leaving the field pair $(\hat{E}_y, \hat{B}_x)$ intact, the latter being governed only by (2-110a) and (2-111b). When field pairs are independent of each other, they are said to be *uncoupled*.

Suppose one desires wave solutions involving only the field pair $(\hat{E}_x, \hat{B}_y)$.

Then put $\hat{E}_y = \hat{B}_x = 0$, reducing the pertinent differential equations to just (2-110b) and (2-111a)

$$\frac{\partial \hat{E}_x}{\partial z} = -j\omega \hat{B}_y \qquad [2\text{-}110b]$$

$$\frac{\partial \dfrac{\hat{B}_y}{\mu_0}}{\partial z} = -j\omega\epsilon_0 \hat{E}_x \qquad [2\text{-}111a]$$

The field solutions are obtained upon combining (2-110b) and (2-111a) to eliminate $\hat{E}_x$ or $\hat{B}_y$, yielding a scalar wave equation from which solutions can be found. Alternatively, one can make use of either vector wave equation (2-102) or (2-103), subjecting it to the same assumptions. (Only $\hat{E}_x$ and $\hat{B}_y$ are present and $\partial/\partial x = \partial/\partial y = 0$.)

Either approach obtains the following wave equation in terms of $\hat{E}_x$:

$$\frac{\partial^2 \hat{E}_x}{\partial z^2} + \omega^2 \mu_0 \epsilon_0 \hat{E}_x = 0 \qquad (2\text{-}112)$$

This is a partial differential equation in one variable (z); thus it can be written as the ordinary differential equation

$$\frac{d^2 \hat{E}_x}{dz^2} + \omega^2 \mu_0 \epsilon_0 \hat{E}_x = 0 \qquad (2\text{-}113)$$

Its solution is the familiar† superposition of two exponential solutions

$$\hat{E}_x(z) = \hat{C}_1 e^{-j\beta_0 z} + \hat{C}_2 e^{j\beta_0 z} \qquad (2\text{-}114)$$

wherein $\hat{C}_1$ and $\hat{C}_2$ are arbitrary (complex) constants and the coefficient β_0, called the *phase constant*, is given by $\beta_0 = \omega\sqrt{\mu_0\epsilon_0}$. It is to be shown that the exponential solutions $\hat{C}_1 e^{-j\beta_0 z}$ and $\hat{C}_2 e^{j\beta_0 z}$ are representations of *constant amplitude waves traveling* in the positive z and negative z directions, respectively. The complex coefficients $\hat{C}_1$ and $\hat{C}_2$ must have the units of volts per meter, denoting arbitrary *complex amplitudes* of the positive z and negative z traveling waves. Employing amplitude symbols $\hat{E}_m^+$ and $\hat{E}_m^-$ instead of $\hat{C}_1$ and $\hat{C}_2$ puts (2-114) into the form

$$\hat{E}_x(z) = \hat{E}_m^+ e^{-j\beta_0 z} + \hat{E}_m^- e^{j\beta_0 z} \quad \text{V/m} \qquad (2\text{-}115)$$

† It is assumed that the reader is familiar with the details of this solution, found in any text on ordinary differential equations.

The complex amplitudes $\hat{E}_m^+$ and $\hat{E}_m^-$ may be represented by points in the complex plane using the Argand diagram of Figure 2-10, so from their polar representations

$$\hat{E}_m^+ = E_m^+ e^{j\phi+} \quad \text{and} \quad \hat{E}_m^- = E_m^+ e^{j\phi-} \tag{2-116}$$

with ϕ^+ and ϕ^- denoting arbitrary phase angles.

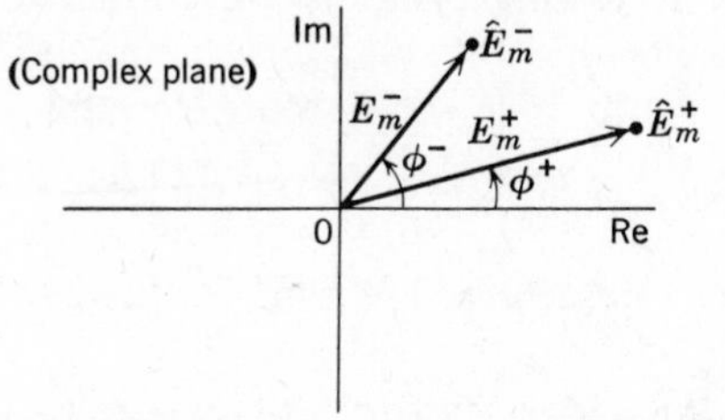

Figure 2-10. Complex amplitudes represented in the complex plane.

Once a solution of the wave equation has been obtained, the remaining field components can be found by use of Maxwell's equations. Thus, the solution (2-115) for $\hat{E}_x(z)$ inserted into the Maxwell relation (2-110b) yields

$$\hat{B}_y(z) = -\frac{1}{j\omega}\frac{\partial \hat{E}_x}{\partial z} = -\frac{\beta_0}{\omega}[-\hat{E}_m^+ e^{-j\beta_0 z} + \hat{E}_m^- e^{j\beta_0 z}]$$
$$= \sqrt{\mu_0\epsilon_0}\hat{E}_m^+ e^{-j\beta_0 z} - \sqrt{\mu_0\epsilon_0}\hat{E}_m^- e^{j\beta_0 z} \quad \text{Wb/m}^2 \tag{2-117}$$

in which β_0 once more denotes the space phase factor

$$\beta_0 \equiv \omega\sqrt{\mu_0\epsilon_0} \quad \text{rad/m} \tag{2-118}$$

The *real-time*, sinusoidal steady state expression for the electric field component is found from (2-74). Taking the real part of (2-115) after multiplying by $e^{j\omega t}$ obtains

$$E_x(z, t) = \text{Re}\,[\hat{E}_x(z)e^{j\omega t}]$$
$$= \text{Re}\,[(E_m^+ e^{j\phi+} e^{-j\beta_0 z} + E_m^- e^{j\phi-} e^{j\beta_0 z})e^{j\omega t}]$$
$$= E_m^+ \cos(\omega t - \beta_0 z + \phi^+) + E_m^- \cos(\omega t + \beta_0 z + \phi^-) \tag{2-119}$$

Note that E_m^+ and E_m^- denote the traveling wave real amplitudes, while ϕ^+ and ϕ^- are arbitrary phases relative to the instant $t = 0$ and the location $z = 0$ in space. The real-time form of $\hat{B}_y$ of (2-117) is similarly found to be

$$B_y(z, t) = [\sqrt{\mu_0\epsilon_0}E_m^+] \cos(\omega t - \beta_0 z + \phi^+) - [\sqrt{\mu_0\epsilon_0}E_m^-] \cos(\omega t + \beta_0 z + \phi^-) \tag{2-120}$$

The traveling wave nature of uniform plane waves can be grasped from a graphic interpretation of (2-119) and (2-120). Consider only the first terms of each: the positive z traveling wave. The following symbols are chosen to denote them:

$$E_x^+(z, t) = E_m^+ \cos(\omega t - \beta_0 z + \phi^+) \text{ V/m} \tag{2-121a}$$

$$B_y^+(z, t) = \sqrt{\mu_0\epsilon_0}E_m^+ \cos(\omega t - \beta_0 z + \phi^+) \text{ Wb/m}^2 \tag{2-121b}$$

Their positive z traveling nature may be observed if (2-121) is plotted as a family of cosine waves versus z, at successive instants of time t. (When observing the effects of time or space variations of a field, it is usually best to hold space *or* time fixed, while the other is allowed to vary.) At $t = 0$, (2-121a) becomes $E_x^+(z, 0) = E_m^+ \cos(-\beta_0 z + \phi^+) = E_m^+ \cos(\beta_0 z - \phi^+)$, shown plotted against the z variable as the solid line in Figure 2-11(*a*). With the period T defined by

$$T = \frac{1}{f} \text{ sec} \tag{2-122}$$

at one-eighth period later, for example, (2-121) becomes $E_x^+(z, T/8) = E_m^+ \cos(\beta_0 z - 2\pi/8 - \phi^+)$. The cosine function is thus shifted in the positive z direction by the time lapse of the eighth period as shown, yielding a positive *z motion* of the wave with increasing time. The vector field plot of Figure 2-11(*a*) shows only $\mathbf{a}_x E_x^+(z, t)$ along a typical z axis in the region. To display the field throughout a cross-section in any x-z plane, the *flux* plot of Figure 2-11(*b*) is more suitable.

The motion of the wave with increasing t is related to the phase factor $\beta_0 = \omega\sqrt{\mu_0\epsilon_0}$ appearing in the wave expressions, with $\beta_0 z$ having the units of radians (dimensionless), implying that β_0 is given in radians per meter. The z distance that the wave must travel such that 2π rad of phase shift (one complete cycle) occurs is called the *wavelength*, designated by the symbol λ and defined by

$$\beta_0 \lambda = 2\pi \text{ rad} \tag{2-123}$$

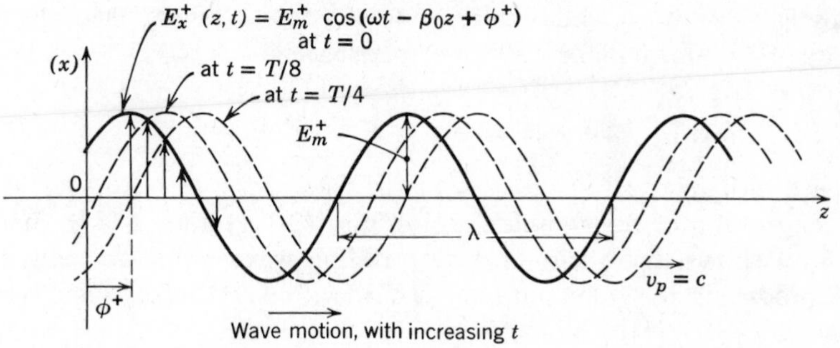

(a)

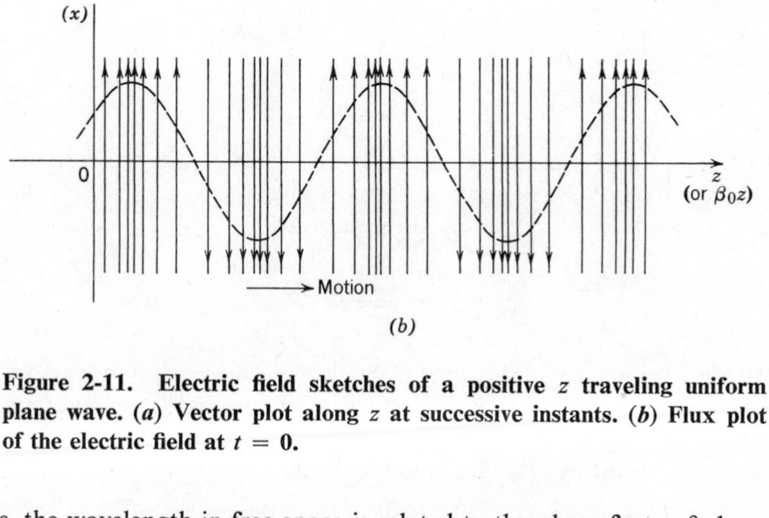

(b)

Figure 2-11. Electric field sketches of a positive z traveling uniform plane wave. (a) Vector plot along z at successive instants. (b) Flux plot of the electric field at $t = 0$.

Thus, the wavelength in free space is related to the phase factor β_0 by

$$\lambda = \frac{2\pi}{\beta_0} = \frac{2\pi}{\omega\sqrt{\mu_0\epsilon_0}} = \frac{c}{f} \text{ m} \tag{2-124}$$

An observer moving with the wave such that he experiences no phase change will move at the *phase velocity* of the wave, denoted by v_p. The equiphase surfaces of the positive z traveling wave are defined by setting the argument of (2-121) equal to a constant; i.e., $\omega t - \beta_0 z + \phi^+ = \text{constant}$, whereupon differentiating it to evaluate dz/dt yields the phase velocity

$$v_p = \frac{dz}{dt} = \frac{\omega}{\beta_0} \text{ m/sec} \tag{2-125a}$$

Because $\beta_0 = \omega\sqrt{\mu_0\epsilon_0}$, and with $\mu_0 = 4\pi \times 10^{-7}$, $\epsilon_0 \cong 10^{-9}/36\pi$, the phase velocity of a uniform plane wave in empty space is

$$v_p = \frac{1}{\sqrt{\mu_0\epsilon_0}} = c \cong 3 \times 10^8 \text{ m/sec} \tag{2-125b}$$

the speed of light.†

A comparison of the complex expressions, (2-115) and (2-117), for $\hat{E}_x(z)$ and $\hat{B}_y(z)$ shows that their separate traveling wave terms are paired into ratios producing the same constant. Thus, write (2-115) and (2-117) in the forms

$$\begin{aligned} \hat{E}_x(z) &= \hat{E}_m^+ e^{-j\beta_0 z} + \hat{E}_m^- e^{j\beta_0 z} \\ &= \hat{E}_x^+(z) + \hat{E}_x^-(z) \end{aligned} \tag{2-126}$$

and

$$\begin{aligned} \hat{B}_y(z) &= \frac{\hat{E}_m^+}{c} e^{-j\beta_0 z} - \frac{\hat{E}_m^-}{c} e^{j\beta_0 z} \\ &= \hat{B}_y^+(z) + \hat{B}_y^-(z) \end{aligned} \tag{2-127}$$

in which $\hat{E}_x^+(z)$, $\hat{E}_x^-(z)$ and $\hat{B}_y^+(x)$, $\hat{B}_y^-(z)$ symbolically denote the positive z and negative z traveling wave terms directly above them. Then the following complex ratios hold at any point in the region

$$\frac{\hat{E}_x^+(z)}{\hat{B}_y^+(z)} = -\frac{\hat{E}_x^-(z)}{\hat{B}_y^-(z)} = \frac{1}{\sqrt{\mu_0\epsilon_0}} \equiv c \cong 3 \times 10^8 \text{ m/sec} \tag{2-128}$$

to provide a means for finding one of the fields whenever the other is known. A more common variation of this technique is achieved by modifying the **B** field in empty space through a division by μ_0, defining a *magnetic intensity field* denoted by the symbol **H** for empty space as follows:

$$\frac{\mathbf{B}}{\mu_0} = \mathbf{H} \text{ A/m} \qquad \text{For empty space} \tag{2-129}$$

Thus, denoting $\hat{B}_y^+(z)/\mu_0$ by $\hat{H}_y^+(z)$, and $\hat{B}_y^-(z)/\mu_0$ by $\hat{H}_y^-(z)$ the following ratios of the traveling wave terms are valid for plane waves in empty space

$$\frac{\hat{E}_x^+(z)}{\mu_0^{-1}\hat{B}_y^+(z)} = \frac{\hat{E}_x^+(z)}{\hat{H}_y^+(z)} = \mu_0 c = \frac{\mu_0}{\sqrt{\mu_0\epsilon_0}} = \sqrt{\frac{\mu_0}{\epsilon_0}} \equiv \eta_0 \cong 120\pi\ \Omega \tag{2-130a}$$

$$-\frac{\hat{E}_x^-(z)}{\mu_0^{-1}\hat{B}_y^-(z)} = -\frac{\hat{E}_x^-(z)}{\hat{H}_y^-(z)} = \sqrt{\frac{\mu_0}{\epsilon_0}} \equiv \eta_0 \cong 120\pi\ \Omega \tag{2-130b}$$

† Experiments have shown that the speed of light, c, is more nearly 2.99792×10^8 m/sec. This value together with the assumed permeability for free space $\mu_0 = 4\pi \times 10^{-7}$ H/m, inserted into (2-125b), is seen to lead to a value for ϵ_0 which departs slightly from the approximate value $10^{-9}/36\pi$ given.

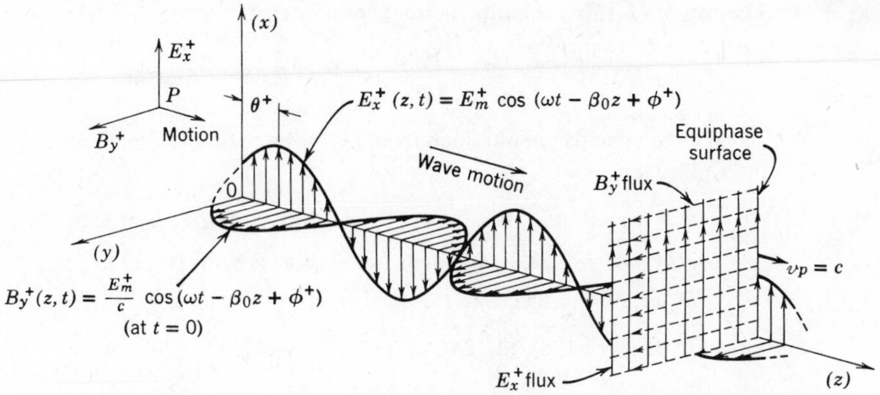

Figure 2-12. Vector plot of the fields of a uniform plane wave along the *z* axis. Note the typical plane equiphase surface, depicting fluxes of E_x^+ and B_y^+.

The real ratio, $\sqrt{\mu_0/\epsilon_0}$ (having the units volts per meter per ampere per meter, or ohms), is called the *intrinsic wave impedance for empty space*, and is denoted by the symbol η_0. The advantage of (2-130) over (2-128) is that ratio η_0 is a usefully smaller number.

The *real* impedance ratio of (2-130) shows that the electric and magnetic fields of uniform plane waves in empty space are in phase with one another, a condition evident upon comparing the positive z or the negative z traveling solutions of (2-126) and (2-127). Each contains the same phase argument in the exponential factors, ample evidence of their in-phase condition. Figure 2-12 depicts the real-time electric and magnetic fields of (2-121) plotted against z in space at $t = 0$.

EXAMPLE 2-11. Suppose a uniform plane wave in empty space has the electric field

$$\hat{\mathbf{E}}(z) = \mathbf{a}_x 1000 e^{-j\beta_0 z} \text{ V/m} \tag{1}$$

its frequency being 20 MHz. (*a*) What is its direction of travel? Its amplitude? Its vector direction in space? (*b*) Find the associated $\hat{\mathbf{B}}$ field and the equivalent $\hat{\mathbf{H}}$ field. (*c*) Express $\hat{\mathbf{E}}$, $\hat{\mathbf{B}}$, and $\hat{\mathbf{H}}$ in real-time form. (*d*) Find the phase factor β_0, the phase velocity, and the wavelength of this electromagnetic wave.

(*a*) A comparison of (1) with (2-115) or (2-126) reveals a positive z traveling wave, whence the symbolism: $\hat{\mathbf{E}}(z) = \mathbf{a}_x \hat{E}_x^+(z)$. The real amplitude is $\hat{E}_m^+ = 1000$ V/m, with the vector field x directed in space.

(*b*) Using either (2-127) or the ratio (2-128)

$$\hat{B}_y^+(z) = \frac{\hat{E}_x^+(z)}{c} = \frac{1000}{c} e^{-j\beta_0 z} = 3.33 \times 10^{-6} e^{-j\beta_0 z} \text{ Wb/m}^2$$

The use of (2-130a) obtains the magnetic intensity

$$\hat{H}_y^+(z) = \frac{\hat{E}_x^+(z)}{\eta_0} = \frac{1000}{120\pi} e^{-j\beta_0 z} = 2.65e^{-j\beta_0 z} \text{ A/m}$$

(*c*) The real-time fields are obtained from (2-74) by taking the real part after multiplication by $e^{j\omega t}$

$$E_x^+(z, t) = \text{Re}\,[1000e^{-j\beta_0 z}e^{j\omega t}] = 1000 \cos(\omega t - \beta_0 z) \text{ V/m}$$

$$B_y^+(z, t) = 3.33 \times 10^{-6} \cos(\omega t - \beta_0 z) \text{ Wb/m}^2 \text{ (or T)}$$

$$H_y^+(z, t) = 2.65 \cos(\omega t - \beta_0 z) \text{ A/m}$$

(*d*) Using (2-118), (2-125), (2-124), and (2-122) yields

$$\beta_0 \equiv \omega\sqrt{\mu_0\epsilon_0} = \frac{\omega}{c} = \frac{2\pi(20 \times 10^6)}{3 \times 10^8} = 0.42 \text{ rad/m}$$

$$v_p = c = 3 \times 10^8 \text{ m/sec}$$

$$\lambda_0 = \frac{2\pi}{\beta_0} = \frac{c}{f} = \frac{3 \times 10^8}{20 \times 10^6} = 15 \text{ m}$$

REFERENCES

ABRAHAM, M., and R. BECKER. *The Classical Theory of Electricity and Magnetism.* Glasgow: Blackie, 1943.

FANO, R. M., L. T. CHU, and R. P. ADLER. *Electromagnetic Fields, Energy and Forces.* New York: Wiley, 1960.

HAYT, W. H., Jr. *Engineering Electromagnetics*, 2nd ed. New York: McGraw-Hill, 1967.

KRAUS, J. D. *Electromagnetics.* New York: McGraw-Hill, 1950.

PHILLIPS, H. B. *Vector Analysis.* New York: Wiley, 1944.

RAMO, S., J. R. WHINNERY, and T. VAN DUZER. *Fields and Waves in Communication Electronics.* New York: Wiley, 1965.

STRATTON, J. A. *Electromagnetic Theory.* New York: McGraw-Hill, 1941.

PROBLEMS

2-1. Express as a vector function the maximum directional derivative, or gradient, of the following scalar functions:
 (*a*) $f(x) = 10x$
 (*b*) $h(x, y, z) = 10x + 15xy + 20xz^2$
 (*c*) $F(r) = 25/r$
 (*d*) $G(\rho, \phi, z) = 10 \cos\phi - 5\rho z \sin\phi$

2-2. If f and g are scalar fields, prove in rectangular coordinates that $\nabla(fg) = f\nabla g + g\nabla f$, identity (14) in Table 2-2.

2-3. Show that an inverse r dependent scalar field has a negatively r directed, inverse r^2 vector gradient; i.e., that **grad** $(1/r) = -\mathbf{a}_r/r^2$.

2-4. Given the field $\mathbf{P} = \mathbf{a}_r(K/r^n)$, in which K is a constant, show that the div $\mathbf{P}$ is $(2 - n)K/r^{n+1}$ (excluding $r = 0$). What choice of n will provide a divergenceless field? Compare this conclusion with the $\mathbf{E}$ field of a static point charge (divergenceless for $r > 0$).

2-5. Several vector fields are defined in the following. Determine for each whether or not the vector field is sourceless; i.e., find its divergence:

(a) $\mathbf{A} = \mathbf{a}_x 100$

(b) $\mathbf{B} = \mathbf{a}_r 1000/r^2$

(c) $\mathbf{C} = \mathbf{a}_x x + \mathbf{a}_y y + \mathbf{a}_z z = \mathbf{a}_r r$ (Determine div $\mathbf{C}$ in both rectangular and spherical coordinates)

(d) $\mathbf{D} = 30\mathbf{a}_x + 2xy\mathbf{a}_y + 5xz^2\mathbf{a}_z$

(e) $\mathbf{E} = 3\rho\mathbf{a}_\rho + 6\mathbf{a}_z$

(f) $\mathbf{F} = (150/r^2)\mathbf{a}_r + 10\mathbf{a}_\phi$

2-6. The following hypothetical field is given in some region of space:

$$\mathbf{F}(r, \theta) = \mathbf{a}_r r^2 + \mathbf{a}_\theta r^2 \sin\theta + \mathbf{a}_\phi r^2$$

Find div $\mathbf{F}$. What is div $\mathbf{F}$ along the vertical axis ($\theta = 0$)? On the equatorial plane ($\theta = \pi/2$)?

2-7. A particular static distribution of sources leads to the field $\mathbf{E} = \mathbf{a}_r K$ in and on a sphere of radius R centered at $r = 0$. Illustrate the truth of the divergence theorem for this field by evaluating both the surface and volume integrals for the sphere. [*Answer:* $4\pi R^2 K$]

2-8. Assume that static charges in a region produce the following ρ directed electric field $\mathbf{E} = \mathbf{a}_\rho K/\rho^n$ V/m in which K is constant and n is a parameter.

(a) What value of n is required to make $\mathbf{E}$ divergenceless? By comparison with Example 1-11(c), comment on the nature of the charge system that will produce an entirely ρ directed, divergenceless $\mathbf{E}$ field.

(b) Utilize the appropriate Maxwell's differential relation to find the volume charge density that will produce this $\mathbf{E}$ field for an arbitrary n.

2-9. Prove that $\mathbf{P}$ of Problem 2-4 is irrotational (conservative) by making use of the curl operator.

2-10. Find the curl of each field of Problem 2-5. Which are irrotational? [*Answer:* $\nabla \times \mathbf{D} = -\mathbf{a}_y 5z^2 + \mathbf{a}_z 2y$, $\nabla \times \mathbf{E} = 0$, $\nabla \times \mathbf{F} = \mathbf{a}_r(10 \cot\theta/r) - \mathbf{a}_\theta(10/r)$.]

2-11. Demonstrate that the Maxwell's equation (2-65) is satisfied by the field solutions (1-64) of a long, straight wire. What are the current densities in the two regions?

2-12. Show that the $\mathbf{B}$ fields obtained for the thin current sheet of Figure 1-20, as well as for the toroid and the long solenoid of Figure 1-21, satisfy Maxwell's curl relation (2-65).

2-13. Prove that the static $\mathbf{E}$ fields obtained for the charge distribution of Figure 1-15 are irrotational, i.e., that they satisfy Maxwell's equation (2-64).

2-14. Assume that the following time-varying $\mathbf{B}$ field is a solution of Maxwell's equations in a charge-free and current-free region ($\rho_v = \mathbf{J} = 0$): $\mathbf{B}(z, t) = \mathbf{a}_y 10^{-5} \sin \beta_0 z \sin \omega t$ Wb/m². This is a *standing wave* solution of Maxwell's equations, in which β_0 is the phase shift constant $\beta_0 = \omega\sqrt{\mu_0\epsilon_0}$, and ω is the radian frequency of the field.

(*a*) Show that (2-41) is satisfied by this field.
What does this mean physically?

(*b*) Determine the curl of $\mathbf{B}/\mu_0$ in this region.
What physical conclusion can be drawn from the answer?

2-15. Show that (2-54) and (2-55) are the consequences of expanding (2-51) in the specified coordinate systems.

2-16. Prove, by expanding in rectangular coordinates, the following vector identities:

(*a*) $\nabla \times (\nabla f) = 0$

(*b*) $\nabla \cdot (\nabla \times \mathbf{F}) = 0$

assuming f and $\mathbf{F}$ to be differentiable fields. Observe that (*a*) states that any conservative field, ∇f, is irrotational, while (*b*) reveals that any function $\nabla \times \mathbf{F}$ is solenoidal (has no flux sources).

2-17. A scalar function associated with the theory of rectangular waveguides is $f(x, y, z) = A \sin k_1 x \cos k_2 y e^{-jk_3 z}$.

(*a*) Show that its gradient is

$$\nabla f = [\mathbf{a}_x A k_1 \cos k_1 x \cos k_2 y - \mathbf{a}_y k_2 A \sin k_1 x \sin k_2 y - \mathbf{a}_z j k_3 A \sin k_1 x \cos k_2 y] e^{-jk_3 z}$$

(*b*) Show that the divergence of ∇f is

$$\nabla \cdot (\nabla f) = -A(k_1^2 + k_2^2 + k_3^2) \sin k_1 x \cos k_2 y e^{-jk_3 z}$$

(c) What is the curl of ∇f? [*Note*: A result of Problem 2-16 can be used to advantage here.]

2-18. If the surface S of Example 2-10 were deformed into a *flat plane* bounded by the same contour ℓ, evaluate the surface integral of Stokes' theorem applied to the given function $\mathbf{F}$, and compare the result with the appropriate line integral.

2-19. Given $\mathbf{F}(r) = \mathbf{a}_\phi K r$ in a region, illustrate the validity of Stokes' theorem for $\mathbf{F}$, assuming in the region a closed line ℓ defined by the equatorial perimeter at $r = a$, $\theta = 90°$, and for the two surfaces bounded by ℓ illustrated in the accompanying figure.

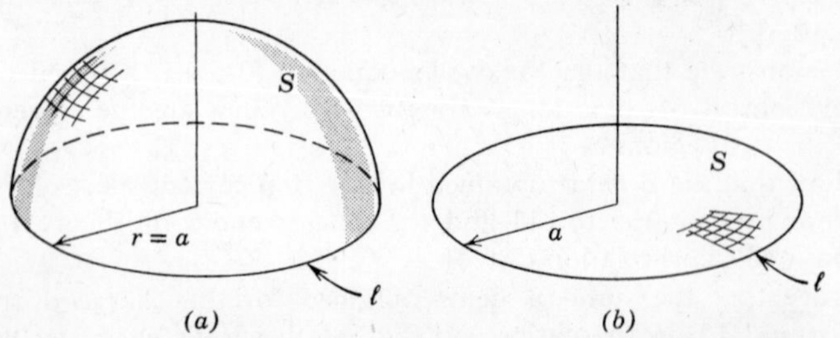

Problem 2-19. **(*a*) Hemispheric S bounded by ℓ. (*b*) Planar S bounded by ℓ.**

2-20. Convert the *integral* forms of Maxwell's equations for free space, listed in column I of Table 2-1, to their corresponding complex time-harmonic forms.

2-21. A ϕ directed electric field in some region is given by $\mathbf{E} = \mathbf{a}_\phi K\rho^2 z$ V/m, with K a constant:

(*a*) Find **curl E** at any point. Is **E** conservative?

(*b*) Evaluate the line integral of $\mathbf{E}\cdot d\boldsymbol{\ell}$ taken about a closed path $\ell = \ell_1 + \ell_2 + \ell_3 + \ell_4$ on a circular cylinder of radius 2 and height 3 as illustrated.

(*c*) Use another approach in evaluating the closed line integral of (*b*), using a surface integration via Stokes' theorem. (One such surface S is illustrated.)

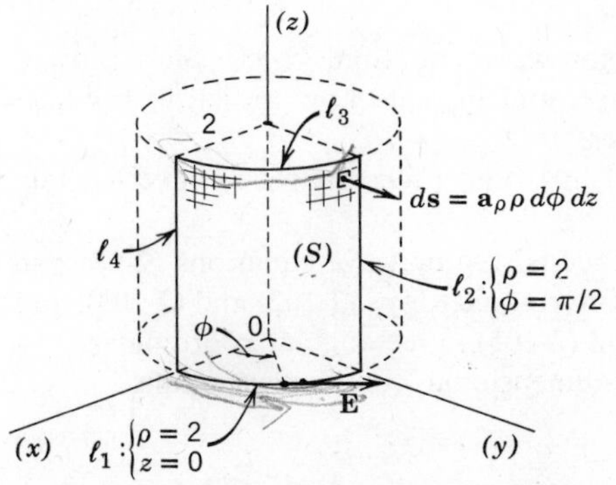

Problem 2-21

2-22. A ϕ directed field in a region is given by $\mathbf{F} = \mathbf{a}_\phi K r^2$, in which K is a constant.

(*a*) Find **curl F** at any point.

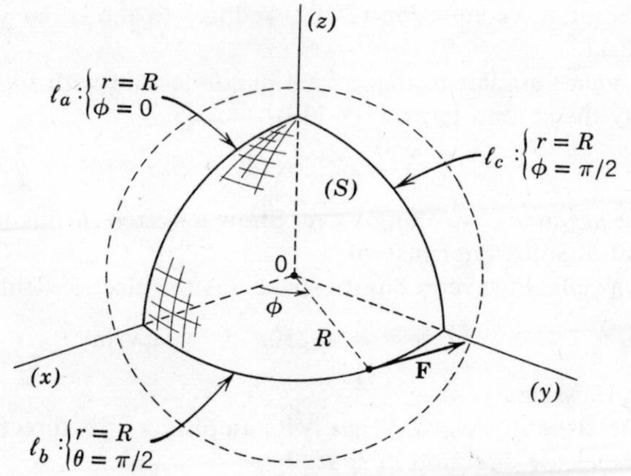

Problem 2-22

(*b*) Determine the integral of (**curl F**)$\cdot d\mathbf{s}$ over the octant surface S bounded by the closed line $\ell = \ell_a + \ell_b + \ell_c$ on the sphere of radius R shown.

(*c*) Find the answer to (*b*) another way using Stokes' theorem; from the line integral of $\mathbf{F}\cdot d\boldsymbol{\ell}$ around ℓ.

2-23. Use the general form (2-76) to show that $\nabla^2 f$ becomes (2-77), (2-80), and (2-81) in the rectangular, circular cylindrical, and spherical coordinate systems, respectively.

2-24. Use the definition (2-82) to show that the Laplacian of a vector field becomes (2-83) in rectangular coordinates.

2-25. Show from (2-82) that $\nabla^2\mathbf{F}$ becomes (2-84) in the circular cylindrical system.

2-26. Fill in the remaining steps to prove that (2-88a) is the consequence of expanding (2-87) in rectangular coordinates.

2-27. Prove the vector differential identities (15) and (16) in Table 2-2, by expansion in rectangular coordinates.

2-28. Derive the inhomogeneous, vector wave equation, (2-95). Show how it reduces to (2-99) for empty space, and indicate how the latter becomes (2-101) in rectangular coordinates.

2-29. Show how the wave equations (2-94) and (2-95) may be converted into complex time-harmonic forms.

2-30. In detail show how the homogeneous, vector wave equations (2-98) and (2-99) are converted to their time-harmonic forms (2-102) and (2-103), and how the latter become (2-104) and (2-105) in rectangular coordinates.

2-31. Suppose one is told that the one-dimensional, complex wave function

$$\hat{E}_x(z) = \hat{E}_m e^{-j\omega\sqrt{\mu_0\epsilon_0}z}$$

in which $\hat{E}_m$ is a complex amplitude (a constant), is a solution of the scalar wave equation (2-104a). By direct substitution prove that this is so.

2-32. (*a*) Show that combining Maxwell's differential equations (2-110b) and (2-111a) produces the scalar wave equation (2-112).

(*b*) Assuming that only the components $\hat{E}_x$ and $\hat{B}_y$ exist, show that the vector wave equation (2-102) reduces to the scalar wave equation (2-112).

2-33. Use arguments similar to those used in connection with Figure 2-11(*a*) to show why the second term of (2-119),

$$E_x^-(z, t) = E_m^- \cos(\omega t + \beta_0 z + \phi^-)$$

denotes a *negative* z traveling wave. Show a sketch of this field plotted at $t = 0$ and at some later instant.

2-34. A uniform, plane wave in empty space has the electric field

$$\hat{\mathbf{E}}(z)e^{j\omega t} = \mathbf{a}_x 250 e^{j(\omega t - \beta_0 z)}\ \mu\text{V/m}$$

with $f = 10$ MHz.

(*a*) Describe this wave. What is its amplitude? Its direction of travel? Its vector direction in space?

(*b*) Find the associated $\hat{\mathbf{B}}$ field (and the equivalent $\hat{\mathbf{H}}$ field) expressed in *complex*, time-harmonic form. Express $\hat{\mathbf{E}}$, $\hat{\mathbf{B}}$, and $\hat{\mathbf{H}}$ in *real-time* form.

(*c*) Find the phase velocity, period, wavelength, and phase factor.

2-35. A particular negative z traveling, uniform plane wave is described by the electric field

$$\hat{\mathbf{E}}(z) = \mathbf{a}_x \hat{E}_x^-(z) = \mathbf{a}_x 150 e^{j\beta_0 z} \text{ V/m}$$

Assume $f = 100$ MHz.

(*a*) Why, by inspection, is this wave negative z traveling?

(*b*) Determine the magnetic field $\hat{\mathbf{B}}$ associated with the given field. Find also $\hat{\mathbf{H}}$. Express $\hat{\mathbf{E}}$, $\hat{\mathbf{B}}$, and $\hat{\mathbf{H}}$ in real-time form. At $t = 0$, sketch a cycle or two of the real-time $\hat{\mathbf{E}}$ and $\hat{\mathbf{B}}$ versus z, denoting their proper phase condition and direction of travel.

(*c*) What are the phase factor, phase velocity, wavelength, and period?

2-36. Begin with the complex, time-harmonic form of Maxwell's equations subject to the same assumptions (*a*) and (*b*) of Section 2.10 for uniform plane waves, but assume that only the field pair ($\hat{E}_y$, $\hat{B}_x$) is involved.

(*a*) Manipulate the Maxwell relations (2-110a) and (2-111b) to obtain a scalar wave equation in terms of the component $\hat{E}_y$. Note that the solution is

$$\hat{E}_y(z) = \hat{E}_m^+ e^{-j\beta_0 z} + \hat{E}_m^- e^{j\beta_0 z}$$

Identify the positive z and negative z traveling wave terms.

(*b*) Show that the corresponding magnetic field solution is

$$\hat{B}_x(z) = -\frac{\hat{E}_m^+}{c} e^{-j\beta_0 z} + \frac{\hat{E}_m^-}{c} e^{j\beta_0 z}$$

Determine the corresponding expression for the magnetic intensity field $H_x(z)$.

(*c*) Use the results of (*a*) and (*b*) to establish the wave-impedance ratios $\hat{E}_y^+(z)/\hat{H}_x^+(z)$ and $\hat{E}_y^-(z)/\hat{H}_x^-(z)$. Compare these results with the impedance ratios (2-130) applicable to the field pairs $\hat{E}_x^\pm$, $\hat{H}_y^\pm$.

(*d*) Sketch a real-time wave plot similar to Figure 2-12(*a*), showing the relationship between the sinusoidal waves $E_y^+(z, t)$ and $B_x^+(z, t)$ at the instant $t = 0$. Comment on the comparison with the fields E_x^+ and B_y^+ of Figure 2-12.

2-37. Sketch a real-time wave plot, similar to Figure 2-12, showing the relationship in space (at $t = 0$) between $E_x^-(z, t)$ and $B_y^-(z, t)$, the negative z traveling wave of (2-119). Indicate the direction of the wave motion on your diagram.

2-38. Use the complex field solutions (2-126) and (2-127) to prove the complex ratios expressed by (2-128) and (2-130).

2-39. If a positive z traveling uniform plane wave with the fields $\hat{E}_x^+(z)$ and $\hat{B}_y^+(z)$ has an electric field amplitude of 1 V/m, find the amplitude of $\hat{B}_y^+$ and of $\hat{H}_y^+$ in free space.

CHAPTER

3

Maxwell's Equations and Boundary Conditions for Material Regions at Rest

Materials in nature are invariably composed of atoms or arrangements of atoms into ions or molecules, each made up of positively and negatively charged particles having various configurations in empty space and varying states of relative motion. An electric or magnetic field impressed upon a material exerts Lorentz forces upon the particles, which undergo displacements or rearrangements to modify the impressed fields accordingly. The Maxwell equations which describe the electric and magnetic field behavior in a material are thus expected to require modifications from their free-space versions to account for whatever *additional* fields the material particles produce. It is the task in this chapter to discuss these extensions of the free-space Maxwell equations.

The topic of conduction is discussed from the viewpoint of a collision model. The chapter continues with a consideration of the added effects of electric polarization within a material, providing a Maxwell divergence relation valid for materials as well as free space. Next is treated the added effect of magnetic polarization, yielding a suitably altered Maxwell curl expression for the magnetic field. The field vectors **D** and **H** are thereby defined. Boundary conditions prevailing at interfaces separating differently polarized regions are developed from the integral forms of the Maxwell equations, to compare the normal components of **D** and the tangential components of **H** at adjacent points in the regions. The discussion continues with related treatments of the Maxwell div **B** and **curl E** equations for material regions, their integral forms, and corresponding boundary conditions. The chapter concludes with a discussion of uniform plane waves in a material possessing the parameters σ, ϵ, and μ, exemplifying the use of the Maxwell equations for a linear, homogeneous, and isotropic material.

3-1 Electrical Conductivity of Metals

The electric and magnetic field behavior of material regions, solid, liquid, or gaseous, may be characterized in terms of three effects:

1. Electric charge conduction
2. Electric polarization
3. Magnetic polarization

For large classes of materials, these effects are often adequately described through use of three *parameters*: σ, the electric conductivity; ϵ, the electric permittivity; and μ, the magnetic permeability of the material. These parameters will be defined in the course of the ensuing discussions.

In terms of their charge-conduction property, materials may for some purposes be classified as *insulators* (dielectrics) which possess essentially no free electrons to provide currents under an impressed electric field; and *conductors*, in which free, outer orbit electrons are readily available to produce a conduction current when an electric field is impressed. An electrically conductive solid, commonly known as a *conductor*, is visualized in the submicroscopic world as a latticework of positive ions in which outer-orbit electrons are free to wander as *free electrons*†—negative charges not attached to any particular atoms. Upon this structure are superposed thermal agitations associated with the temperature of the conductor—the light, agile conduction electrons moving about the more massive ion lattice, imparting some of their momentum to that lattice in exchange for new random directions of flight until more interactions occur. This circumstance is depicted in Figure 3-1(a) for a typical conduction electron. The velocities of the free electrons are randomly distributed so that a mean velocity, averaged at any instant over a large number N of particles in the volume element,‡ is given by

$$\mathcal{v}_d = \frac{1}{N} \sum_{i=1}^{N} \mathcal{v}_i \text{ m/sec} \tag{3-1}$$

† In the atomic view, the free (conduction) electrons are those associated with the unfilled outer orbit, or valence band, of particular elements known as *metals*.

‡ The volume-element used in characterizing the average velocity (3-1) is chosen sufficiently large that it contains enough ions and associated conduction electrons to yield a meaningful average, and yet it is taken small enough that the averaged velocity may be characterized at a *point* in the region. That a very large number of particles are present in a small volume increment is appreciated upon noting that a typical conductor, sodium, possesses about 2.5×10^{19} atoms/mm^3 at room temperature.

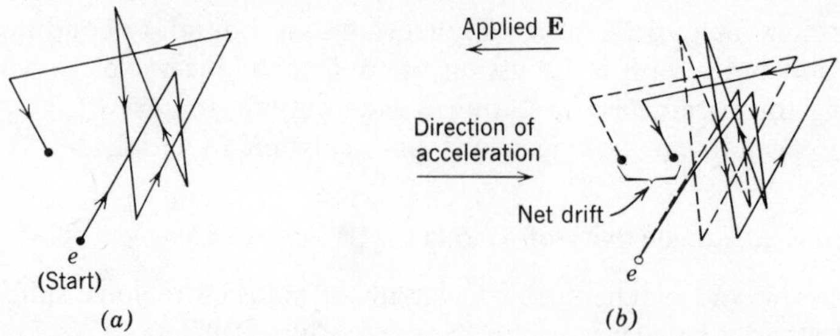

Figure 3-1. A representation of the production of a drift component of the velocity of free electrons in a metal. (*a*) A typical sequence of electron free paths resulting from collisions with the ion lattice. (*b*) Exaggerated view of the effect of drift in the direction of the acceleration due to an applied E field.

This quantity, called the drift velocity of the electrons, averages to *zero* in the absence of any externally applied electric field.

A *mean free time*, represented by the symbol τ_c, denotes the average interval between collisions in a volume element. When free electrons collide (interact) with the ion lattice they give up, on the average, a momentum $m\mathscr{v}_d$ in the mean free time τ_c between collisions, if m is the electron mass. Thus the averaged rate of momentum transfer to the ion lattice, per electron, is $m\mathscr{v}_d/\tau_c$ N of force. Upon equating this to the Lorentz electric field force applied within the conductor, one obtains

$$\frac{m\mathscr{v}_d}{\tau_c} = -e\mathbf{E} \tag{3-2}$$

and solving for $\mathscr{v}_d$ yields the steady drift velocity

$$\mathscr{v}_d = -\frac{e\tau_c}{m}\mathbf{E} \tag{3-3}$$

The expression (3-3), linearly relating the drift velocity to the applied **E** field, is of the form

$$\mathscr{v}_d = -\mu_e\mathbf{E} \tag{3-4}$$

in which the proportionality constant μ_e, taken to be a positive number, is termed the *electron mobility*, which from (3-3) is evidently

$$\mu_e = \frac{e\tau_c}{m} \text{ m}^2/\text{V-sec} \tag{3-5}$$

A high value of electron mobility is thus associated with a long mean free time τ_c.

Making use of (1-50a) and multiplying $\mathscr{v}_d$ by the volume density $\rho_v = -ne$ of the conduction electrons obtains the volume current density

$$\mathbf{J} = \rho_v \mathscr{v}_d = -ne\mathscr{v}_d = \frac{ne^2}{m} \tau_c \mathbf{E} \text{ A/m}^2 \tag{3-6}$$

with n denoting the free electron density in electrons/m^3. Equation (3-6) is an expression exhibiting a linear dependence of **J** on the applied **E** field in the conductor. Experiments show that this is an exceedingly accurate model for a wide selection of physical conductors. Equation (3-6) has the form of

$$\boxed{\mathbf{J} = \sigma \mathbf{E}} \tag{3-7}$$

sometimes given the name *point form of Ohm's law*, in which the factor σ is called the *conductivity* of the region, having the units ampere per meter squared per volt per meter, or mho per meter. For the present model to which (3-6) applies, the conductivity is expressible as the positive number

$$\sigma = \frac{ne^2}{m} \tau_c \text{ ℧/m} \tag{3-8}$$

It is thus seen that both the electron mobility and the conductivity are proportional to the mean free time, τ_c. A comparison of (3-5) and (3-8) permits expressing σ in terms of the mobility

$$\sigma = ne\mu_e = \rho_v \mu_e \tag{3-9}$$

EXAMPLE 3-1. Find the mean free time and the electron mobility for sodium, having the measured dc conductivity 2.1×10^7 V/m at room temperature.

Sodium has an atomic density of 2.3×10^{28} atom/m^3 at room temperature, and with one outer-orbit electron available, n has the same value. Thus from (3-8), the mean free time becomes

$$\tau_c = \frac{m\sigma}{ne^2} = \frac{(9.1 \times 10^{-31})(2.1 \times 10^7)}{(2.3 \times 10^{28})(1.6 \times 10^{-19})^2} = 3.3 \times 10^{-14} \text{ sec}$$

Its electron mobility is found from either of the relations

$$\mu_e = \frac{e}{m} \tau_c = \frac{\sigma}{ne} = \frac{2.1 \times 10^7}{(2.3 \times 10^{28})(1.6 \times 10^{-19})} = 5.7 \times 10^{-3} \text{ m}^2/\text{V-sec}$$

This implies from (3-4) the very slow drift velocity $v_d = 5.7$ mm/sec for an applied field of 1 V/m, emphasizing the sluggish, *viscous* nature of electron drift in a conductor.

The foregoing picture of direct current in a conductor is readily extended to the time-varying case, assuming that **E** varies slowly in comparison to the mean free time, τ_c. The force-equilibrium relation (3-2) then acquires another term due to the additional time rate of change of the average momentum of the drifting electron cloud in the conductor. Adding it to (3-2) obtains

$$m\frac{dv_d}{dt} + \frac{m}{\tau_c}v_d = -e\mathbf{E} \text{ N} \tag{3-10}$$

This differential equation has the complementary solution, assuming the initial condition $v_d = v_{d0}$ at $t = 0$, as follows:

$$v_d = v_{d0}e^{-t/\tau_c} \text{ m/sec} \tag{3-11}$$

a transient solution denoting a decay or *relaxation* in the drift velocity upon suddenly turning off the applied field **E**. Thus the *mean free time*, τ_c, introduced into force relation (3-2), has acquired the interpretation of a *relaxation time* in the event of applying or removing an electric field from a conductor. The relaxation phenomenon furthermore occurs in an exceedingly short time for typical good conductors; thus, from Example 3-1 it was shown to be of the order of 10^{-14} sec for a metal having a conductivity of about 10^7 ℧/m. The current density (3-7) is proportional to the drift velocity v_d, implying from (3-11) that *current* decays with time at the same rate upon removing the **E** field.

The differential equation (3-10) can be simplified if **E** is assumed sinusoidal. Replacing **E** and v_d with the time-harmonic forms $\hat{\mathbf{E}}e^{j\omega t}$ and $\hat{v}_d e^{j\omega t}$ obtains, after cancelling the factor $e^{j\omega t}$, the complex algebraic relation

$$j\omega m\,\hat{v}_d + \frac{m}{\tau_c}\hat{v}_d = -e\hat{\mathbf{E}} \tag{3-12}$$

yielding the time-harmonic solution for $\hat{v}_d$

$$\hat{v}_d = \frac{-\dfrac{e}{m}\hat{\mathbf{E}}}{\dfrac{1}{\tau_c} + j\omega}$$

The complex current density due to this drift velocity is therefore, from $\mathbf{J} = \rho_v \boldsymbol{\mathcal{V}}_d$

$$\hat{\mathbf{J}} = \frac{\dfrac{ne^2}{m}}{\dfrac{1}{\tau_c} + j\omega} \hat{\mathbf{E}} \text{ A/m}^2 \tag{3-13}$$

The coefficient of $\hat{\mathbf{E}}$ denotes the conductivity of the metal as in the dc result (3-7), though now a complex quantity is obtained

$$\hat{\sigma} = \frac{\dfrac{ne^2}{m}}{\dfrac{1}{\tau_c} + j\omega} \text{ ℧/m} \tag{3-14}$$

However, for typical good conductors having a mean free time, τ_c, of the order of 10^{-14} sec (Example 3-1), (3-14) reduces to the real, dc conductance result (3-8)

$$\sigma \cong \frac{ne^2}{m} \tau_c \qquad [3\text{-}8]$$

provided the angular frequency ω of the electromagnetic field is of the order of 10^{13} rad/sec or less (below the optical frequencies). Additional confidence is gained for this rather heuristic model of metallic conduction by experimental measurements made in the microwave range of frequencies, showing that the **E** and **J** fields in good conductors are in phase, implying that σ is *real* in the relationship $\mathbf{J} = \sigma\mathbf{E}$, even up to very high frequencies.

The model of electrical conductivity just described is essentially that proposed by Karl Drude in 1900. The advent of quantum mechanics since that time has provided comprehensive techniques for describing, among other things, why the conductivities of various materials behave differently with temperature and how the vast range of conductivities of physical materials comes about—of the order of 10^8 ℧/m for the best conductors at room temperature to 10^{-16} ℧/m for the best insulators—a range of some 24 orders of magnitude. The so-called band theory of solids, an outgrowth of quantum mechanics, is useful for describing the intrinsic differences among the conductors, semiconductors, and insulators.†

3-2 Electric Polarization and Div D for Materials

Insulators, or so-called dielectrics, incapable of carrying appreciable conduction currents under impressed electric fields of moderate magnitudes, are

† The reader is referred to Hutchison, T. S., and D. C. Baird. *The Physics of Engineering Solids*. New York: Wiley, 1968, for details.

the subject of this discussion. The mechanism of the dielectric polarization effects resulting from applied electric fields may be explained in terms of the microscopic displacements of the bound positive and negative charge constituents from their average equilibrium positions, produced by the Lorentz electric field forces on the charges. Such displacements are usually only a fraction of a molecular diameter in the material, but the sheer numbers of particles involved may cause a significant change in the electric field from its value in the absence of the dielectric substance.

Dielectric polarization may arise from the following causes:

1. *Electronic polarization*, in which the bound, negative electron cloud, subject to an impressed **E** field, is displaced from the equilibrium position relative to the positive nucleus.
2. *Ionic polarization*, in which the positive and negative ions of a molecule are displaced in the presence of an applied **E** field.
3. *Orientational polarization*, occurring in materials possessing permanent electric dipoles randomly oriented in the absence of an external field, but undergoing an orientation towards the applied electric field vector by amounts depending on the strength of **E**. The tendency for the so-called *polar molecules* of such a material to align parallel with the applied field is opposed by the thermal agitation effects and the mutal interaction forces among the particles. Water is a common example of a substance exhibiting orientational polarization effects.

In each type of dielectric polarization, particle displacements are inhibited by powerful restoring forces between the positive and negative charge centers. In Figure 3-2 is illustrated the polarization mechanism in a material involving two species of charge. One should imagine thermal agitations superimposed upon the average positions of the particles shown. If an external field **E** is impressed on the material, Lorentz forces $\mathbf{F}_E = q\mathbf{E}$ will be exerted on the positively charged nucleus and the negative electron cloud to produce displacements of both systems of particles. Displacement equilibrium is attained when the applied forces are balanced by the internal attractive Coulomb forces of the couplets.

The moment $\mathbf{p}_i$ of the ith displaced charge pair in a collection of polarized dipoles as in Figure 3-2(a) is defined by

$$\mathbf{p}_i = q\mathbf{d}_i \ \mathrm{C \cdot m} \tag{3-15}$$

in which q denotes the positive charge of the couplet $(q, -q)$, and $\mathbf{d}_i$ the vector separation of the couplet, directed from the negative to the positive charge.

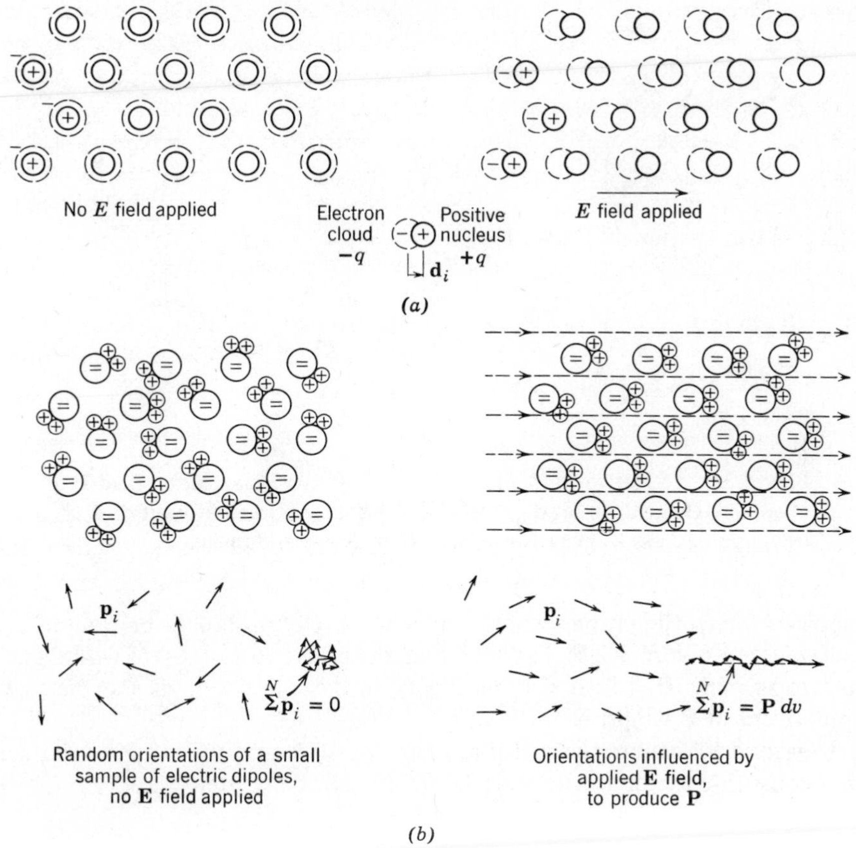

Figure 3-2. Electric polarization effects in simple models of nonpolar and polar dielectric materials. (*a*) A nonpolar substance. (*b*) A polar substance (H_2O).

The average electric dipole moment per unit volume, called the *electric polarization field* and denoted by **P**, is defined by

$$\mathbf{P} = \frac{\sum_{i=1}^{N} \mathbf{p}_i}{\Delta v} = \frac{\sum_{i=1}^{N} q_i \mathbf{d}_i}{\Delta v} \text{ C/m}^2 \tag{3-16}$$

for a volume element Δv containing N electric dipoles. If no **E** field were applied to the material, no dipoles would be induced in the case of electronic or ionic polarization; even if the material were polar (containing permanent

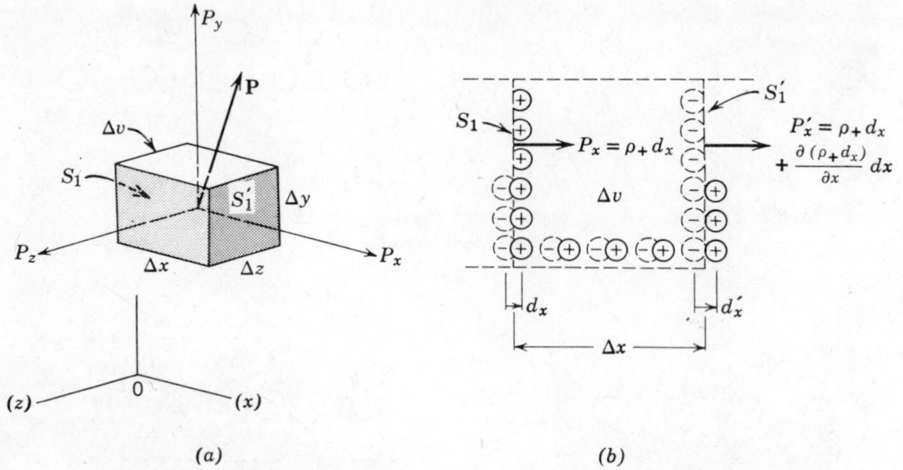

Figure 3-3. Relative to (3-21), $\rho_p = -\text{div}\,\mathbf{P}$. (*a*) Polarization field P at a volume-element in a dielectric. (*b*) Effect of nonuniformity of P_x, leaving an excess of negative polarization charge within Δv.

dipoles), their orientations would under usual circumstances be random as illustrated in Figure 3-2(*b*), in which case the numerator of (3-16) would sum to zero to make $\mathbf{P} = 0$. If $\mathbf{E}$ were applied in the x direction as shown, a net component of $\mathbf{P}$ would be induced.

If ρ_+ and ρ_- denote the densities of the positive and the negative charges that constitute the dielectric material, (3-16) can be written

$$\mathbf{P} = \frac{\sum_{i=1}^{N} \mathbf{p}_i}{\Delta v} = \frac{\sum_{i=1}^{N} q_i \mathbf{d}_i}{\Delta v} = \frac{Nq}{\Delta v} \frac{\sum_{i=1}^{N} \mathbf{d}_i}{N} = \rho_+ \mathbf{d} \tag{3-17}$$

in which $\rho_+ = Nq/\Delta v$ is the density of only the positive charges comprising the dipole-filled region, and $\mathbf{d}$ denotes $(\Sigma \mathbf{d}_i)/N$, the dipole displacement averaged over the N dipoles in Δv. An examination of the polarization field $\mathbf{P} = \rho_+ \mathbf{d}$ of (3-17), characterized as the vector in Figure 3-3(*a*), reveals the establishment of a bound charge excess within Δv, giving rise to a so-called *polarization charge density*, ρ_p, wherever $\mathbf{P}$ has a divergence. To show this, consider a typical volume element $\Delta v = \Delta x\, \Delta y\, \Delta z$ in a region containing, in general, a nonuniform polarization field $\mathbf{P}$ as shown in Figure 3-3(*b*). The x component, $P_x = \rho_+ d_x$, accounts for a net, positive bound charge passing through the left-hand face S_1 *into* Δv, amounting to

$$P_x \,\Delta y\, \Delta z = (\rho_+ d_x)\, \Delta y\, \Delta z \tag{3-18a}$$

while through the opposite side S_1', the positive bound charge coming *out of* Δv is expressed

$$P_x' \, \Delta y \, \Delta z = \left[\rho_+ d_x + \frac{\partial(\rho_+ d_x)}{\partial x} \Delta x \right] \Delta y \, \Delta z \tag{3-18b}$$

A net, negative, polarization bound charge therefore remains inside Δv, amounting to the difference of (3-18a) and (3-18b), or

$$-\frac{\partial(\rho_+ d_x)}{\partial x} \Delta x \, \Delta y \, \Delta z = -\frac{\partial P_x}{\partial x} \Delta x \, \Delta y \, \Delta z \tag{3-19}$$

With similar contributions over the other two pairs of sides, one obtains the total, negative bound charge remaining inside Δv

$$-\left(\frac{\partial P_x}{\partial x} + \frac{\partial P_y}{\partial y} + \frac{\partial P_z}{\partial z} \right) \Delta x \, \Delta y \, \Delta z \tag{3-20}$$

a measure of the net, bound charge excess $\rho_p \, \Delta v$ within Δv, if ρ_p denotes the volume density of the polarization charge excess. Equating (3-20) to $\rho_p \, \Delta v$ thus obtains, in view of (2-29a)

$$-\text{div } \mathbf{P} = \rho_p \text{ C/m}^3 \tag{3-21}$$

The divergence of $\epsilon_0 \mathbf{E}$ in a free-space region has been expressed by (2-39) to be the free charge density ρ_v. The polarization charge excess developed in a material is now seen to contribute another kind of charge density, ρ_p, so that in the presence of both free charges and a bound charge excess, the divergence of $\epsilon_0 \mathbf{E}$ in a material is in general, div $(\epsilon_0 \mathbf{E}) = \rho_v + \rho_p$. Using (3-21) in the latter thus yields

$$\nabla \cdot (\epsilon_0 \mathbf{E} + \mathbf{P}) = \rho_v \tag{3-22}$$

a divergence expression for **E** in a *material* region. A more compact version is obtained using the abbreviation **D** for $(\epsilon_0 \mathbf{E} + \mathbf{P})$ as follows

$$\mathbf{D} \equiv \epsilon_0 \mathbf{E} + \mathbf{P} \text{ C/m}^2 \tag{3-23}$$

to permit writing (3-22) in the preferred form

$$\nabla \cdot \mathbf{D} = \rho_v \text{ C/m}^3 \tag{3-24}$$

Experiments reveal that many dielectric substances are essentially linear, meaning that **P** is proportional to the **E** field applied. For such materials

$$\mathbf{P} \propto \mathbf{E}$$
$$= \chi_e \epsilon_0 \mathbf{E} \ \text{C/m}^2 \tag{3-25}$$

in which the parameter χ_e is called the *electric susceptibility* of the dielectric. The factor ϵ_0 is retained in (3-25) to make χ_e dimensionless. Then (3-22) becomes

$$\nabla \cdot [(1 + \chi_e)\epsilon_0 \mathbf{E}] = \rho_v \tag{3-26}$$

Comparing (3-26) with (3-24) shows that the bracketed quantity denotes **D**, i.e.,

$$\mathbf{D} = (1 + \chi_e)\epsilon_0 \mathbf{E} \tag{3-27}$$

It is usual to denote $1 + \chi_e$ by the dimensionless symbol

$$\epsilon_r \equiv 1 + \chi_e \tag{3-28}$$

ϵ_r is called the *relative permittivity* (or dielectric constant)† of the region. Finally, choosing the symbol ϵ, called the *permittivity* of the material, to denote $(1 + \chi_e)\epsilon_0$ as follows:

$$\epsilon \equiv (1 + \chi_e)\epsilon_0 \tag{3-29a}$$

$$\epsilon \equiv \epsilon_r \epsilon_0 \ \text{F/m} \tag{3-29b}$$

permits writing (3-27) in the following successively more compact forms

$$\mathbf{D} = (1 + \chi_e)\epsilon_0 \mathbf{E} \tag{3-30a}$$

$$\mathbf{D} = \epsilon_r \epsilon_0 \mathbf{E} \tag{3-30b}$$

$$\mathbf{D} = \epsilon \mathbf{E} \ \text{C/m}^2 \tag{3-30c}$$

In free space, $\chi_e = 0$, to reduce (3-30) properly to $\mathbf{D} = \epsilon_0 \mathbf{E}$. Also, expressing (3-29b) in the form

$$\epsilon_r = \frac{\epsilon}{\epsilon_0} \tag{3-31}$$

† Dielectric *constant* is not a preferred designation, for many materials have polarization characteristics that may vary from point to point, to make χ_e and ϵ_r functions of position. A material is said to be *inhomogeneous* if it has an electric or magnetic parameter that varies with position.

emphasizes that ϵ_r denotes a material permittivity *relative* to that of empty space.

To summarize, note that Maxwell's relation (3-22) or (3-24) is expressible in any of the equivalent forms

$$\nabla \cdot [(1 + \chi_e)\epsilon_0 \mathbf{E}] = \rho_v \tag{3-32a}$$

$$\nabla \cdot [\epsilon_r \epsilon_0 \mathbf{E}] = \rho_v \tag{3-32b}$$

$$\nabla \cdot (\epsilon \mathbf{E}) = \rho_v \tag{3-32c}$$

$$\nabla \cdot \mathbf{D} = \rho_v \text{ C/m}^3 \tag{3-32d}$$

No dielectric material is strictly linear in its electric polarization behavior, though many are very nearly so over wide ranges of applied **E** fields. If **E** is made strong enough, a material may experience polarization displacements that result in permanent dislocations of the molecular structure, or a *dielectric breakdown*, for which case (3-25) does not hold. In a *nonlinear* material the magnitude of **D** is not proportional to the applied **E** field, (though the **E** and the **P** vectors may have the same directions). Then (3-25) is written more generally

$$\mathbf{P} = \chi_e(E)\epsilon_0 \mathbf{E} \tag{3-33}$$

in which the dependence of χ_e on E is noted.

3-2-1 Dielectric Polarization Current Density

If the electric field giving rise to dielectric polarization effects is time-varying, the resulting polarization field is also time-varying. Then the displacements of the positive charge constituents in one direction, together with the negative charges moving oppositely, give rise to charge displacements through cross-sections of the material identifiable as *currents* through those cross-sections. Applying a time-derivative operator to the $\mathbf{p}_i$ terms of (3-16) thus yields a current density interpretation as follows:

$$\frac{\sum_{i=1}^{N} \frac{\partial \mathbf{p}_i}{\partial t}}{\Delta v} = \frac{\sum_{i=1}^{N} q_i \frac{\partial \mathbf{d}_i}{\partial t}}{\Delta v} = \frac{Nq}{\Delta v} \frac{\partial}{\partial t} \frac{\sum_{i=1}^{N} \mathbf{d}_i}{N} = \rho_+ \frac{\partial \mathbf{d}}{\partial t} = \frac{\partial \mathbf{P}}{\partial t} \text{ A/m}^2 \tag{3-34}$$

The resulting time derivative of the polarization field, $\partial \mathbf{P}/\partial t$, having the units of volume current density, is given the symbol $\mathbf{J}_p$ as follows:

$$\mathbf{J}_p \equiv \frac{\partial \mathbf{P}}{\partial t} \text{ A/m}^2 \tag{3-35}$$

and is called the *electric polarization current density*. The field $\mathbf{J}_p$, along with the polarization charge density field ρ_p described by (3-21), acts as an additional source of electric and magnetic fields. In particular, the special role played by $\mathbf{J}_p$ in relation to magnetic fields in a material region is discussed later in Section 3-4.

3-2-2 Integral Form of Gauss' Law for Materials

The dielectric polarization effects attributed to material regions have been seen to lead to the divergence expressions (3-21) and (3-24), relating the field quantities **P** and **D** to the polarization charge and free charge sources. The divergence theorem can be used to transform these differential equations into corresponding *integral* forms. The most important of these is (3-24) for the **D** field; i.e., $\nabla \cdot \mathbf{D} = \rho_v$. Multiplying both sides of (3-24) by dv and integrating throughout an arbitrary volume region V yields

$$\int_V \nabla \cdot \mathbf{D}\, dv = \int_V \rho_v\, dv \tag{3-36}$$

By the divergence theorem (2-34), the left side can be replaced by a closed-surface integral to yield

$$\oint_S \mathbf{D} \cdot d\mathbf{s} = \int_V \rho_v\, dv \quad \mathrm{C} \tag{3-37}$$

in which S bounds V. Equation (3-37) is the *integral form* of Maxwell's equation (3-24) for a *material* region, sometimes called *Gauss' law for material regions*. It states that the net, outward flux of **D** over any closed surface is a measure of the total, *free* charge contained by the volume V bounded by S, at any instant of time. As expected, it becomes the free-space Gauss' law (1-53) if $\chi_e = 0$, reducing **D** to $\epsilon_0 \mathbf{E}$.

Another divergence relation, $\nabla \cdot \mathbf{P} = -\rho_p$ of (3-21), has the equivalent integral form

$$\oint_S \mathbf{P} \cdot d\mathbf{s} = -\int_V \rho_p\, dv \quad \mathrm{C} \tag{3-38}$$

obtained by the method analogous to that used in converting (3-24) to Gauss' integral (3-37). Equation (3-38) states that the net outward flux of **P** emanating from the surface of V is a measure of the net polarization charge summed throughout V.

3–2–3 Spatial Boundary Conditions for Normal D and P

In many electromagnetic field problems of physical interest, it becomes necessary to discuss how the fields behave as one traverses the boundary surfaces, or *interfaces*, separating the various material regions that comprise the system. In such problems, a matching or fitting of the field solutions is required so that the *boundary conditions* at the interfaces may be satisfied. The proper boundary conditions for the fields are determined, as will be shown, from the integral forms of Maxwell's equations for material regions.

The Maxwell integral relation (3-37), $\oint_S \mathbf{D} \cdot d\mathbf{s} = \int_V \rho_v \, dv$, can be used, through an appropriately constructed closed surface, for comparing the normal components of **D** that appear just to either side of an interface separating two materials of different permittivities. Denoting the materials as region 1 and region 2 with permittivities ϵ_1 and ϵ_2, define a pillbox-shaped closed surface of small height δh and end areas Δs so that both regions to either side of the interface are penetrated as in Figure 3-4. Calling the fields

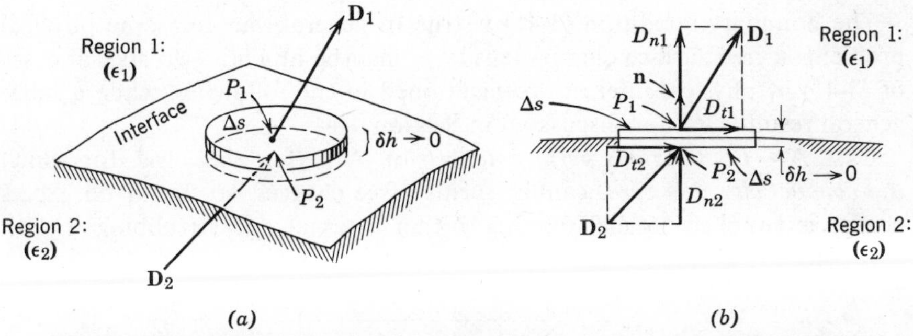

Figure 3-4. Gaussian pillbox surface constructed for deriving the boundary condition on the normal component of D. (*a*) Pillbox-shaped closed surface showing total fields at points adjacent to interface. (*b*) Edge view of (*a*), showing fields resolved into components.

$\mathbf{D}_1$ and $\mathbf{D}_2$ at points just inside regions 1 and 2, respectively, the application of the left-hand integral of (3-37) to the closed pillbox yields the net outward flux from the top and bottom surfaces Δs. At the same time, the right side is the charge enclosed by the pillbox; this is $\rho_v \, \Delta s \, \delta h$, so (3-37) becomes

$$D_{n1} \, \Delta s - D_{n2} \, \Delta s = \rho_v \, \Delta s \, \delta h \tag{3-39}$$

The right side of (3-39) vanishes as $\delta h \to 0$, assuming ρ_v denotes a volume free charge density in the region. If, however, a *surface* charge density denoted by ρ_s and defined by the limit

$$\rho_s = \lim_{\delta h \to 0} \rho_v \, \delta h \tag{3-40}$$

is present on the interface, (3-39) reduces to the general boundary condition

$$D_{n1} - D_{n2} = \rho_s \ \mathrm{C/m^2} \tag{3-41}$$

Equation (3-41) means that *the normal component of* **D** *is discontinuous to the extent of the free surface charge density present on the interface.* Since $D_{n1} = \mathbf{n}\cdot\mathbf{D}_1$ and $D_{n2} = \mathbf{n}\cdot\mathbf{D}_2$, with **n** denoting a normal unit vector directed from region 2 towards region 1 as in Figure 3-4(*b*), (3-41) is written optionally in vector notation as follows:

$$\mathbf{n}\cdot(\mathbf{D}_1 - \mathbf{D}_2) = \rho_s \ \mathrm{C/m^2} \tag{3-42}$$

The boundary condition (3-41) is true in general, but for some physical problems a free surface charge density ρ_s may be absent. Two special cases of (3-41) of physical interest are mentioned in the following, while a more general result is left for discussion in Section 4-14.

CASE A. *Both regions perfect dielectrics.* A perfect dielectric, for which the *conductivity* σ *is zero*, cannot furnish free charges, so that if no excess charge is supplied to the interface by an external agent (rubbing it with cat's fur, for example), then $\rho_s = 0$ on the interface. Then (3-41) reduces to

$$D_{n1} = D_{n2} \ \mathrm{C/m^2} \tag{3-43}$$

The normal component of **D** *is continuous at an interface separating two perfect dielectrics*, as illustrated in Figure 3-5(*a*).

CASE B. *One region is a perfect dielectric; the other is a perfect conductor.* Electric currents are limited to finite densities in the physical world. Thus from (3-7), $\mathbf{J} = \sigma\mathbf{E}$, the assumption of a *perfect conductor* in region 2 of Figure 3-4 ($\sigma_2 \to \infty$), implies that $\mathbf{E}_2$ in that region must be zero if the current densities are to have, at most, finite values. Moreover, with electromagnetic fields satisfying (2-108), $\nabla \times \hat{\mathbf{E}} = -j\omega\hat{\mathbf{B}}$, one can see that if $\mathbf{E}_2$ is zero in region 2, then $\hat{\mathbf{B}}_2$ must be zero there also. Thus for time-varying fields,

$$\sigma \to \infty \quad \text{implies} \quad \mathbf{E} = \mathbf{B} = 0 \tag{3-44}$$

in a perfect conductor. The boundary condition (3-41) or (3-42) then must reduce to $D_{n1} = \rho_s$, or in vector form

$$\mathbf{n} \cdot \mathbf{D} = \rho_s \text{ C/m}^2 \tag{3-45}$$

The surface charge density residing on a perfect conductor equals the normal component of **D** *there,* as illustrated in Figure 3-5(*b*).

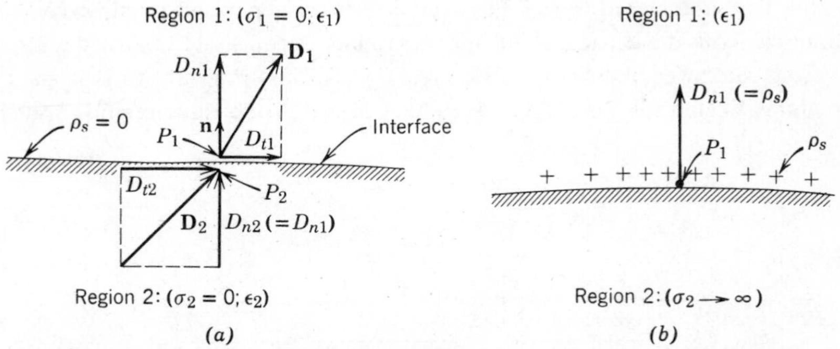

Figure 3-5. Two cases of the boundary condition for normal components of D. (*a*) Continuous D_n at an interface separating perfect dielectrics. (*b*) Equality of normal D_n to a surface charge density on a perfect conductor.

In a static field problem involving only *fixed* electric charges and no static currents, the boundary condition (3-45) holds true even though region 2 may be only *finitely conducting*, for the assumption of no static currents in the finitely conducting region 2 implies from (3-7) that $\mathbf{E}_2 = 0$ there, making $\mathbf{D}_2 = 0$ as well. Thus (3-42) reduces to (3-45).

A boundary condition similar to (3-41) can be derived comparing the normal components of the dielectric polarization vector **P**. Noting the similarity of Maxwell's integral law (3-37) and the polarization field integral (3-38) and using another pillbox construction, one can show that $P_{n1} - P_{n2} = -\rho_{sp}$, or in vector form

$$\mathbf{n} \cdot (\mathbf{P}_1 - \mathbf{P}_2) = -\rho_{sp} \text{ C/m}^2 \tag{3-46}$$

in which ρ_{sp} denotes the net, surface bound charge density lying within the pillbox. The *net* density includes the effect of both species of surface polarization charge (positive and negative) accumulated just to either side of the interface. A simpler picture is obtained if region 1 is free space, for which $\chi_{e1} = 0$ (or $\epsilon_1 = \epsilon_0$). Then $\mathbf{P}_1 = 0$, reducing to the special case

$$\mathbf{n} \cdot \mathbf{P}_2 = \rho_{sp} \text{ C/m}^2 \tag{3-47}$$

The surface polarization charge density residing at a free-space-to-dielectric interface equals the normal component of the **P** *field there.*

EXAMPLE 3.2. Two parallel, conducting plates of great extent and d m apart are statically charged with $\pm q$ C on every area A of the lower and upper plates, respectively, as noted in (*a*). The conductors are separated by air except for a homogeneous dielectric slab of thickness c and permittivity ϵ, spaced a distance b from the lower plate. (*a*) Utilize Gauss' law (3-37) to establish **D** in the three regions. Sketch the flux of **D**. (*b*) Find **E** and **P** in the three regions and show

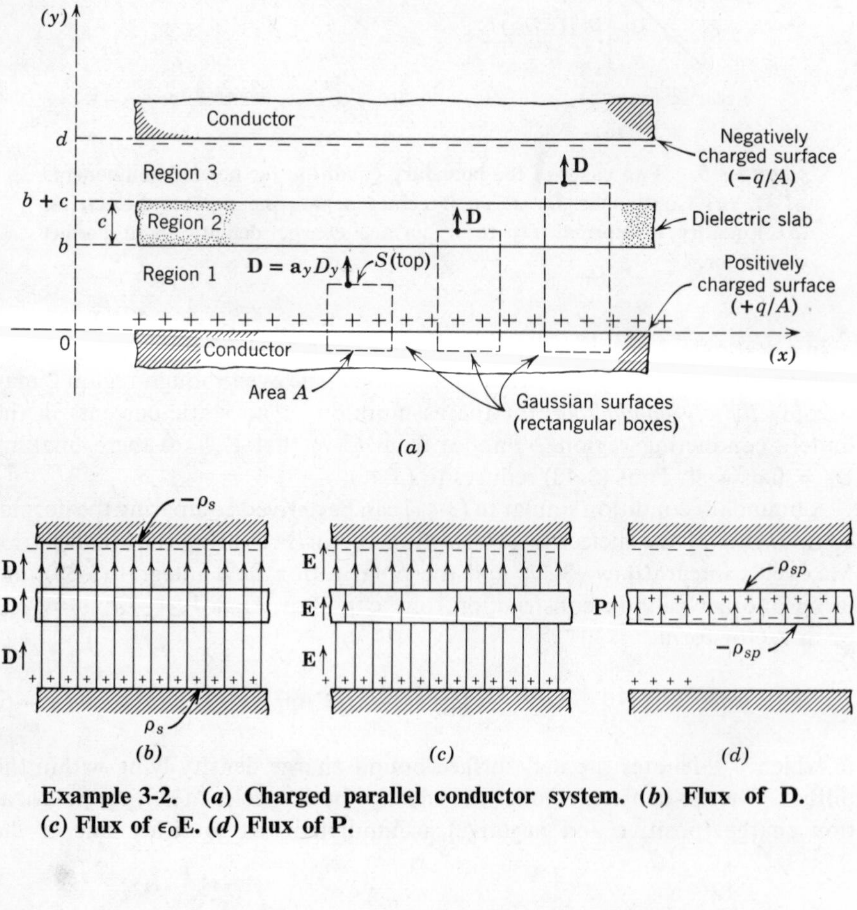

Example 3-2. (*a*) Charged parallel conductor system. (*b*) Flux of D. (*c*) Flux of ϵ_0E. (*d*) Flux of P.

their flux plots. (*c*) Determine ρ_s on the conductor surfaces, ρ_p in the dielectric, and ρ_{ps} at $y = b$ and $y = b + c$.

(*a*) **E** exists only between the conductors and by symmetry is independent of x and z. A Gaussian closed-surface S in the form of a rectangular box is placed as in Figure 1-15(*d*), to contain the free charge q. With static **E** inside the conductor zero, a **D** flux of a constant density emanates from the top of S, making the left side of Gauss' law (3-37) become

$$\oint_S \mathbf{D} \cdot d\mathbf{s} = \int_{S(top)} (\mathbf{a}_y D_y) \cdot \mathbf{a}_y \, ds = D_y \int_{S(top)} ds = D_y A$$

Equating to the right side of (3-37), the free charge $q = D_y A$, whence

$$\mathbf{D} = \mathbf{a}_y D_y = \mathbf{a}_y \frac{q}{A}, \tag{1}$$

a result correct for all three regions between the conductors because no free charge exists in or on the dielectric. The flux plot of **D** is shown in (*b*).

(*b*) **E** is obtained using (3-30c), so in the dielectric slab,

$$\mathbf{E} = \frac{\mathbf{D}}{\epsilon} = \mathbf{a}_y \frac{q}{\epsilon A} \qquad b < y < b + c \tag{2}$$

while in the air regions it is

$$\mathbf{E} = \frac{\mathbf{D}}{\epsilon_0} = \mathbf{a}_y \frac{q}{\epsilon_0 A} \qquad 0 < y < b \qquad \text{and} \qquad b + c < y < d \tag{3}$$

Since $\epsilon > \epsilon_0$ for a typical dielectric, **E** in the air regions exceeds the value in the dielectric, as shown in (*c*).

P in the dielectric is found by use of (3-23)

$$\mathbf{P} = \mathbf{D} - \epsilon_0 \mathbf{E} = \mathbf{a}_y \left(\frac{q}{A} - \epsilon_0 \frac{q}{\epsilon A} \right)$$

$$= \mathbf{a}_y \frac{q}{A} \left(\frac{\epsilon - \epsilon_0}{\epsilon} \right) = \mathbf{a}_y \frac{q}{A} \left(\frac{\epsilon_r - 1}{\epsilon_r} \right) \tag{4}$$

For $\epsilon_r > 1$, **P** in the slab is positive y directed, as shown in (*d*). In air, **P** is zero. From (3-35), *no* polarization current density $\mathbf{J}_p$ is established in the dielectric because the fields are time-static.

(*c*) The free charge densities on the conductors are obtained from (3-45), yielding $\rho_s = \pm q/A$. The polarization charge density ρ_p from (3-21) is zero because **P** is a constant vector throughout the slab. The *surface* polarization charge density ρ_{sp} is found by inserting (4) into (3-47), yielding

$$P_y = \rho_{sp} = \frac{q}{A}\left(\frac{\epsilon_r - 1}{\epsilon_r}\right) \tag{5}$$

These surface densities are noted in (*b*) and (*d*) of the figure.

3-3 Div B for Materials; Its Integral Form and a Boundary Condition for Normal B

In Section 3-2 was developed the Maxwell relation for $\nabla \cdot \mathbf{D}$ in a material by adding the effect of the electric polarization charge density ρ_p to the free-space Maxwell relation. The form of the expression for $\nabla \cdot \mathbf{B}$ in a material can be developed analogously. No additive term is required in this case, however, because *no free magnetic charges* exist physically in any known material. Thus **B** remains divergenceless in materials; i.e.,

$$\nabla \cdot \mathbf{B} = 0 \text{ Wb/m}^3 \tag{3-48}$$

Equation (3-48) is converted to its *integral form* using a technique analogous to that employed in obtaining (3-37). Multiplying both sides of (3-48) by dv, integrating it throughout an arbitrary V and applying the divergence theorem

$$\oint_S \mathbf{B} \cdot d\mathbf{s} = 0 \text{ Wb} \tag{3-49}$$

the integral form of (3-48). Equation (3-49) states that the net, outgoing flux of **B** over any closed surface S is always zero, implying that **B** flux always forms closed lines.

A boundary condition concerned with the normal components of **B** and analogous to (3-42) can be found by applying (3-49) to a vanishing Gaussian pillbox like that of Figure 3-4. The resulting boundary condition is

$$B_{n1} - B_{n2} = 0 \tag{3-50}$$

that is, *the normal component of the* **B** *field is continuous at an interface separating two adjacent regions.*

3-4 Magnetic Polarization and Curl H for Materials

The magnetic properties of a material are attributed to the tendency for the *bound currents*, circulating on an atomic scale within the substance, to align with an applied **B** field. Three types of bound currents are associated with atomic structure: those attributed to orbiting electrons, and those associated with electron spin and with nuclear spin. Each of these phenomena, represented in Figure 3-6(*a*), is equivalent to the circulation of a current I about a small closed path bounding an area $d\mathbf{s}$, the positive sense of which is related by the right-hand rule to the direction of I as in Figure 3-6(*b*). The product $I\,d\mathbf{s}$ defines the *magnetic moment* **m** contributed by those bound currents of the atomic or molecular configuration. It is shown that applying an external

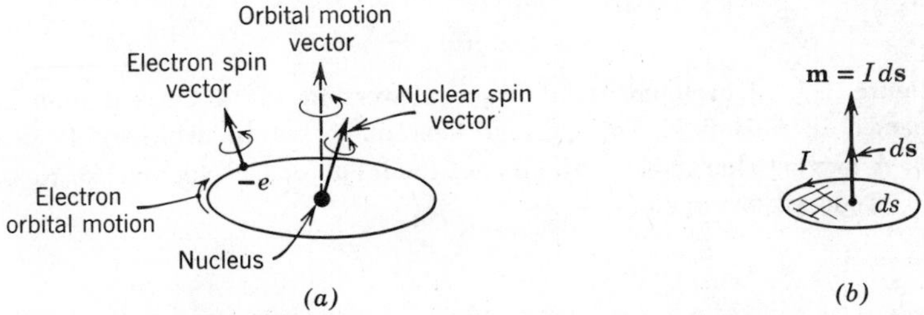

Figure 3-6. The elements of bound currents that exist in atomic structure. (*a*) Constituents of circulating currents associated with particles of a simple atom. (*b*) Magnetic moment m of a current *I* circulating about an area *d*s.

magnetic field **B** to the typical moment $\mathbf{m} = I\,d\mathbf{s}$ yields a torque exerted on **m**, tending to align **m** with the applied **B** field. One can in this manner explain the magnetic behavior of a material as though it were a collection, in empty space, of many magnetic moments **m** per unit volume. The tendency to align with the applied **B** field is shown to provide an equivalent magnetization current of density $\mathbf{J}_m$, serving to modify the magnetic field in a certain way. A description of this process, beginning with a discussion of the torque produced by the **B** field on a current element, follows.

A current loop of microscopic size has an external magnetic field behavior independent of its shape in a plane, so a square loop is assumed in lieu of the circular configuration of Figure 3-6(*b*). It is shown in Figure 3-7(*a*) in the $z = 0$ plane, immersed in the applied field $\mathbf{B} = \mathbf{a}_x B_x + \mathbf{a}_y B_y + \mathbf{a}_z B_z$. The Lorentz force acting on each of the four edges of the square current loop is obtained from (1-52)

$$d\mathbf{F}_B = dq\boldsymbol{v} \times \mathbf{B} \text{ N} \tag{3-51}$$

if the charge dq moves with a velocity $\boldsymbol{v}$ along the edges dx and dy. One may

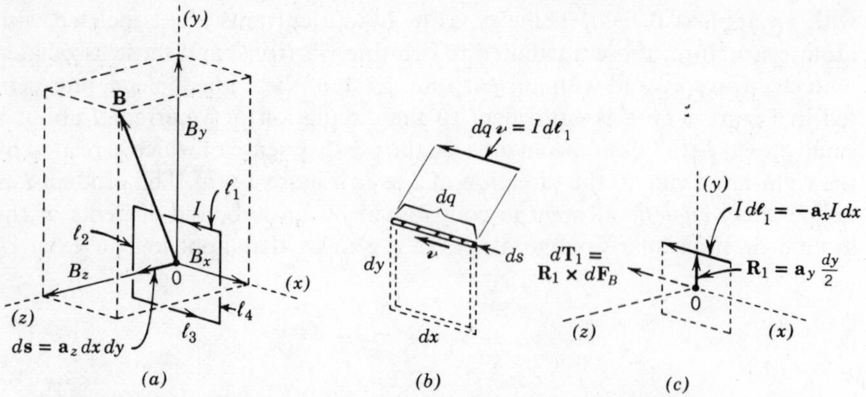

Figure 3-7. Development of torque expression for a current loop immersed in a B field. (*a*) Current loop immersed in arbitrary B field. (*b*) A moving charge element, $dq\boldsymbol{v}$, of the loop. (*c*) Development of torque $d\mathbf{T}$ produced on edge $d\ell_1$.

cast (3-51) into the following forms, noting that $dq = \rho_v\,dv = \rho_v\,d\ell\,ds$ from Figure 3-7(*b*), and utilizing (1-50a)

$$d\mathbf{F}_B = \rho_v\,d\ell\,ds\boldsymbol{v} \times \mathbf{B} = [\mathbf{J}\,d\ell\,ds] \times \mathbf{B} = I\,d\boldsymbol{\ell} \times \mathbf{B} \tag{3-52}$$

with the direction denoted by assigning a vector property to each edge length $d\boldsymbol{\ell}$. The origin of the torque arm $\mathbf{R}$ is for convenience taken at the center of the loop. Along ℓ_1, the differential torque $d\mathbf{T}_1$ is given by $\mathbf{R}_1 \times d\mathbf{F}_B$ (Example 1-4), yielding

$$d\mathbf{T}_1 = \mathbf{R}_1 \times d\mathbf{F}_B = \left[\mathbf{a}_y \frac{dy}{2}\right] \times [(-\mathbf{a}_x I\,dx) \times \mathbf{B}] = -\mathbf{a}_x I B_y \frac{dx\,dy}{2}$$

with the same result obtained for edge ℓ_3, while that acting on ℓ_2 and ℓ_4 becomes $d\mathbf{T}_2 + d\mathbf{T}_4 = \mathbf{a}_y I B_x\,dx\,dy$. Thus the torque on the complete loop becomes

$$d\mathbf{T} = (\mathbf{a}_y B_x - \mathbf{a}_x B_y) I\,dx\,dy = (\mathbf{a}_z \times \mathbf{B}) I\,ds = I(\mathbf{a}_z\,ds) \times \mathbf{B} = I\,d\mathbf{s} \times \mathbf{B}$$

and with $I\,d\mathbf{s}$ denoting the magnetic moment

$$\mathbf{m} \equiv I\,d\mathbf{s}\ \text{A}\cdot\text{m}^2 \tag{3-53}$$

one may abbreviate the result

$$d\mathbf{T} = \mathbf{m} \times \mathbf{B}\ \text{N}\cdot\text{m} \tag{3-54}$$

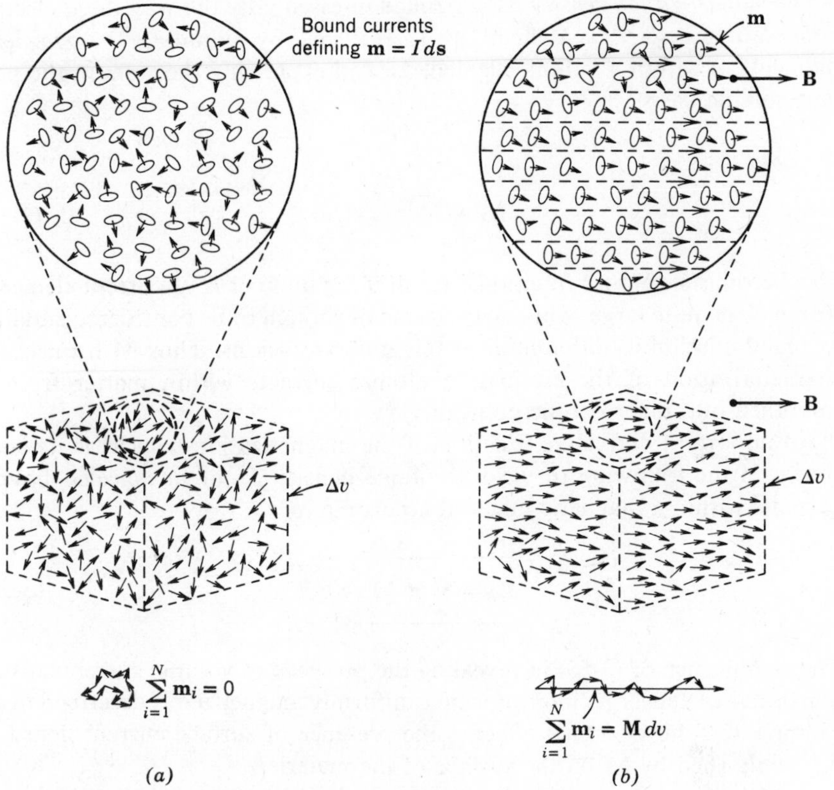

Figure 3-8. Current loop constituency of a magnetizable material, affected by an applied B field. (*a*) Random magnetic moments, in the absence of B. (*b*) Partial alignment of magnetic moments, B applied.

It is clear from (3-54) that only the components of the applied **B** field in the plane of the current element act to produce a torque on it. If **m** and **B** were parallel, $d\mathbf{T}$ would become zero; thus the torque $d\mathbf{T}$ is such that it *tends to align the current element with the applied* **B** *field.*

A very large number of current loops like those of the atomic model in Figure 3-2 comprise a *magnetic* material, susceptible to such magnetic alignment effects. In the absence of an applied **B** field, they possess random orientations accompanied by thermal agitation effects, as depicted in Figure 3-8(*a*), if one may avoid the subject of permanent magnetism occurring in some materials. Impressing a **B** field develops a torque on each current loop, as specified by (3-54), such that the loops tend to align more or less in the direction of **B** as depicted in Figure 3-8(*b*).

The *magnetization density* **M** is defined in essentially the way the dielectric polarization field **P** is defined by (3-16), i.e., by summing the *magnetic* moments **m** within a volume-element Δv and expressing the sum on the per-unit-volume basis

$$\mathbf{M} \equiv \frac{\sum_{i=1}^{N} \mathbf{m}_i}{\Delta v} \text{ A/m} \tag{3-55}$$

This becomes a smooth functional result if the number N of current elements within Δv is quite large, while Δv is yet small enough to be considered suitable for manipulation in differential or integral expressions. Thus **M** furnishes a characterization of the circulating atomic currents within matter from a smoothed-out, macroscopic point of view.

An important derivative function of the magnetization field **M** is its curl, shown in the following to yield a volume density $\mathbf{J}_m$ of uncanceled bound currents within a magnetic material according to

$$\mathbf{J}_m = \nabla \times \mathbf{M} \text{ A/m}^2 \tag{3-56}$$

The significance of (3-56) in revealing the presence of volume currents inside a material whenever its interior is nonuniformly magnetized is described in an example to follow. A side effect is the presence of surface current densities $\mathbf{J}_{sm}$ established by **M** on the surface of the material.

EXAMPLE 3-3. Suppose a **B** field is applied to a cube of magnetic material, b m on a side, such that **M** is z directed and varies *linearly* with x according to

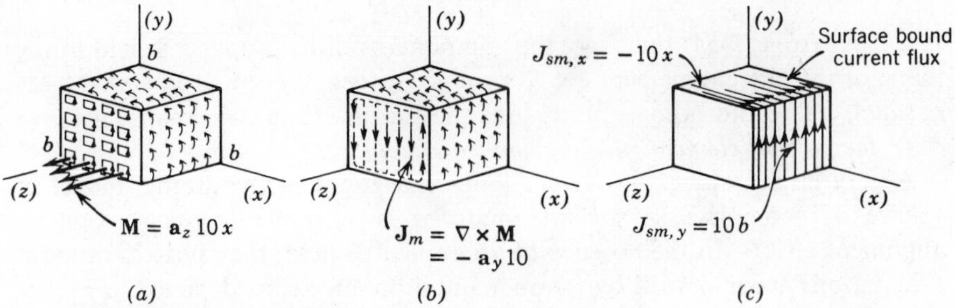

Example 3-3. **(*a*) Material sample magnetized linearly with increasing *x*. (*b*) Volume magnetization currents produced by transverse variations of M. (*c*) Surface currents produced by uncanceled segments of bound currents.**

$\mathbf{M} = \mathbf{a}_z 10x$ A/m, as shown in (*a*). Find the magnetization current density $\mathbf{J}_m$ in the material, as well as the surface magnetization current density. Sketch the bound current fields in and on the cube.

The magnetization current density $\mathbf{J}_m$ is obtained from (3-56)

$$\mathbf{J}_m = \nabla \times \mathbf{M} = \begin{vmatrix} \mathbf{a}_x & \mathbf{a}_y & \mathbf{a}_z \\ \dfrac{\partial}{\partial x} & \dfrac{\partial}{\partial y} & \dfrac{\partial}{\partial z} \\ 0 & 0 & 10x \end{vmatrix} = -\mathbf{a}_y 10 \text{ A/m}^2$$

negative *y* directed and of constant density as in (*b*).

The uncancelled segments of the bound currents at the *surface* of the block constitute a surface density of magnetization currents denoted by $\mathbf{J}_{sm}$ (A/m). On the end $x = b$, $\mathbf{J}_{sm}$ is *y* directed and has a magnitude equal to that of **M** there; i.e.,

$$\mathbf{J}_{sm}]_{x=b} = \mathbf{a}_y M_z]_{x=b} = \mathbf{a}_y 10b \text{ A/m}$$

while on the top and bottom of the block

$$\mathbf{J}_{sm}]_{y=b} = -\mathbf{a}_x M_z]_{y=b} = -\mathbf{a}_x 10x \text{ A/m}$$

$$\mathbf{J}_{sm}]_{y=0} = \mathbf{a}_x M_z]_{y=0} = \mathbf{a}_x 10x \text{ A/m}$$

No bound currents exist on the end at $x = 0$, since $M = 0$ there. These surface effects are shown as flux plots in (*c*).

A formal derivation of (3-56) proceeds with the aid of Figure 3-9. From Example 3-3 it was seen that uncanceled bound current contributions exist on the surface of a magnetized body. The examination of an incremental volume-element of such a material, depicted in Figure 3-9(*a*), similarly reveals the presence of such surface current contributions on Δv as in (*b*) of that figure, assuming for the present that only the effects of the *z* component of **M** are considered. If two such volume increments are considered side by side as in Figure 3-9(*c*), then the bound surface currents along their common sides, with densities designated by $\mathbf{J}_{sm,y}$, cancel partially to produce a net upward flow of current in the region given by

$$\Delta I_1 = (J_{sm,y} - J'_{sm,y})\,\Delta z = -\frac{\partial M_z}{\partial x}\,\Delta x\,\Delta z$$

This current passing through the cross-sectional area $\Delta x\,\Delta z$ is depicted by the bold arrow in the figure. The *y* component of the bound current density $\mathbf{J}_m$ through $\Delta x\,\Delta z$ is then $\Delta I_1/\Delta x\,\Delta z = -\partial M_z/\partial x$. Another contribution, shown in Figure 3-9(*d*) is obtained from the *x* component of **M** in the vicinity of the point; it contributes the density $\partial M_x/\partial z$ through $\Delta x\,\Delta z$. The *total* *y* component of $\mathbf{J}_m$ therefore becomes $\mathbf{J}_{m,y} = \partial M_x/\partial z - \partial M_z/\partial x$, which from (2-52) is

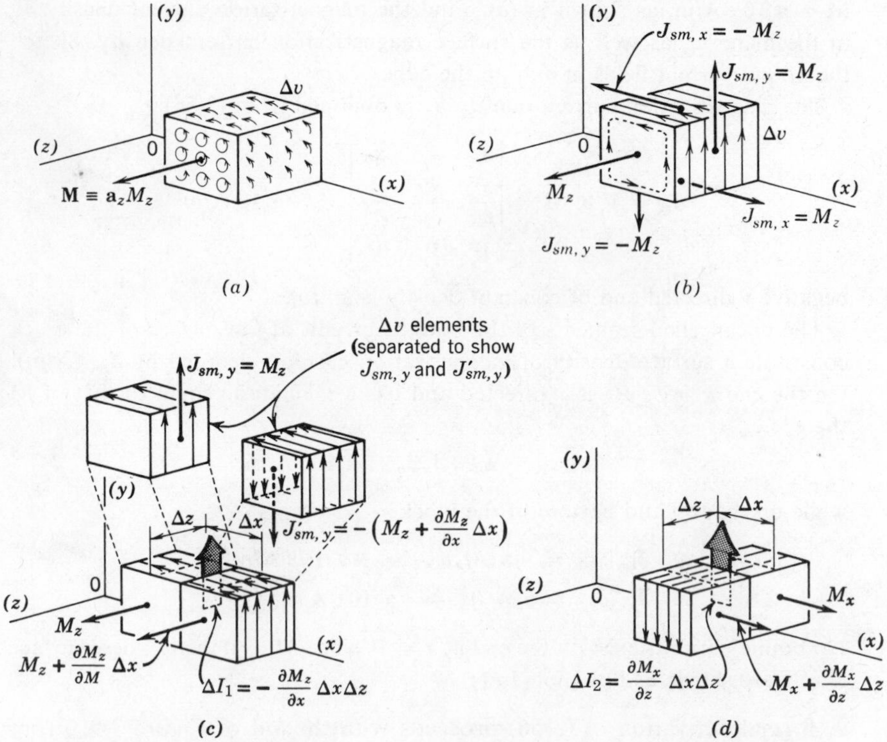

Figure 3-9. Relative to $\mathbf{J}_m = \nabla \times \mathbf{M}$. (*a*) Bound current elements producing surface currents of Δv. (*b*) Bound surface currents smoothed into rectangular components, assuming M_z only. (*c*) Net volume current ΔI_1 through $\Delta x \Delta z$: the difference of bound surface current densities. (*d*) The other contribution to the $J_{m,y}$ component.

evidently the y component of **curl M**. A similar development yields the other components $\mathbf{J}_{m,x}$ and $J_{m,z}$ of $\mathbf{J}_m$, obtaining (3-56)

$$\mathbf{J}_m = \begin{vmatrix} \mathbf{a}_x & \mathbf{a}_y & \mathbf{a}_z \\ \dfrac{\partial}{\partial x} & \dfrac{\partial}{\partial y} & \dfrac{\partial}{\partial z} \\ M_x & M_y & M_z \end{vmatrix} = \nabla \times \mathbf{M} \qquad [3\text{-}56]$$

The curl of $\mathbf{B}/\mu_0$ in *free space* has been expressed by (2-63) as the sum of a convection or a conduction current density $\mathbf{J}$ plus a displacement current density $\partial(\epsilon_0 \mathbf{E})/\partial t$ at any point. Two additional types of current densities occur generally in *materials*: $\mathbf{J}_p = \partial \mathbf{P}/\partial t$ of (3-35) and $\mathbf{J}_m = \nabla \times \mathbf{M}$ of

(3-56), arising from dielectric and magnetic polarization effects, respectively. Adding these together accounts for the total current density at any point, yielding a revision of (2-63) for a material region:

$$\nabla \times \left(\frac{\mathbf{B}}{\mu_0}\right) = \mathbf{J} + \frac{\partial(\epsilon_0 \mathbf{E})}{\partial t} + \frac{\partial \mathbf{P}}{\partial t} + \nabla \times \mathbf{M}$$

Grouping the curl terms and the time-derivative terms together obtains

$$\nabla \times \left(\frac{\mathbf{B}}{\mu_0} - \mathbf{M}\right) = \mathbf{J} + \frac{\partial(\epsilon_0 \mathbf{E} + \mathbf{P})}{\partial t} \tag{3-57}$$

Recalling from (3-23) that $\epsilon_0\mathbf{E} + \mathbf{P}$ defines $\mathbf{D}$, and further abbreviating $\mathbf{B}/\mu_0 - \mathbf{M}$ in (3-57) by use of the symbol $\mathbf{H}$, sometimes called the *magnetic intensity field*

$$\mathbf{H} \equiv \frac{\mathbf{B}}{\mu_0} - \mathbf{M} \ \text{A/m} \tag{3-58}$$

permits writing (3-57) in the compact form

$$\nabla \times \mathbf{H} = \mathbf{J} + \frac{\partial \mathbf{D}}{\partial t} \ \text{A/m}^2 \tag{3-59}$$

This is the *desired Maxwell curl expression* for the field $\mathbf{H}$ defined by (3-58), applicable to material regions. Note that it properly reduces to its free-space form (2-73) upon setting $\mathbf{P} \equiv \mathbf{M} \equiv 0$.

In a *linear* region possessing a magnetization $\mathbf{M}$, one might be inclined to express $\mathbf{M}$ proportional to the $\mathbf{B}$ field in the material (i.e., $\mathbf{M} \propto \mathbf{B}$) to provide a result analogous to (3-25) for a linear dielectric ($\mathbf{P} \propto \mathbf{E}$). Historically, however, this has not proved to be the assumption used; instead, it is customary to set $\mathbf{M}$ proportional to $\mathbf{H}$ as follows:

$$\begin{aligned} \mathbf{M} &\propto \mathbf{H} \\ &= \chi_m \mathbf{H} \end{aligned} \tag{3-60}$$

in which the dimensionless χ_m is called the *magnetic susceptibility* of a material. Inserting (3-60) into (3-58) therefore yields

$$\mathbf{H} = \frac{\mathbf{B}}{\mu_0} - \mathbf{M} = \frac{\mathbf{B}}{\mu_0} - \chi_m \mathbf{H}$$

which, upon solving for **B**, obtains

$$\mathbf{B} = (1 + \chi_m)\mu_0\mathbf{H}\ \text{Wb}/m^2 \tag{3-61}$$

The quantity $(1 + \chi_m)$, abbreviated μ_r,

$$\mu_r \equiv 1 + \chi_m \tag{3-62}$$

is called the *relative permeability* of the material. Further choosing the symbol μ, called the *permeability*, to denote the product

$$\mu \equiv (1 + \chi_m)\mu_0 \tag{3-63a}$$

$$\mu \equiv \mu_r\mu_0\ \text{H/m} \tag{3-63b}$$

permits writing (3-61) in the compact form for linear materials

$$\mathbf{B} = (1 + \chi_m)\mu_0\mathbf{H} \tag{3-64a}$$

$$\mathbf{B} = \mu_r\mu_0\mathbf{H} \tag{3-64b}$$

$$\mathbf{B} = \mu\mathbf{H}\ \text{Wb/m}^2 \tag{3-64c}$$

One should note the analogy of the steps yielding (3-64c) to those leading to (3-30), connecting **D** and **E** for linear, *electrically* polarized materials.

It is seen from (3-64b) that the relative permeability expresses the permeability of a material relative to that of free space, μ_0, if one writes

$$\mu_r \equiv \frac{\mu}{\mu_0} \tag{3-65}$$

This is evidently analogous to (3-31), the expression for the relative permittivity ϵ_r.

3-4-1 Integral Form of Ampère's Law for Materials

Maxwell's curl relation (3-59), $\nabla \times \mathbf{H} = \mathbf{J} + (\partial\mathbf{D}/\partial t)$ can be transformed into an integral relationship by using Stokes' theorem. Forming the dot product of (3-59) with $d\mathbf{s}$ and integrating over any surface S bounded by the closed line ℓ yields

$$\int_S (\nabla \times \mathbf{H})\cdot d\mathbf{s} = \int_S \mathbf{J}\cdot d\mathbf{s} + \frac{d}{dt}\int_S \mathbf{D}\cdot d\mathbf{s}$$

From Stokes' theorem (2-56), the left side can be expressed as an integral of

$\mathbf{H} \cdot d\ell$ over the closed line ℓ bounding S, assuming $\mathbf{H}$ suitably well-behaved; thus

$$\oint_\ell \mathbf{H} \cdot d\ell = \int_S \mathbf{J} \cdot ds + \frac{d}{dt} \int_S \mathbf{D} \cdot ds \text{ A} \tag{3-66}$$

the desired *integral form* of Maxwell's differential equation (3-59). Equation (3-66) is also known as *Ampère's circuital law* for materials. It states that the net circulation of $\mathbf{H}$ about any closed path ℓ is a measure of the sum of the conduction (or convection) current plus the displacement current through the surface S bounded by ℓ.

Another curl relation, (3-56), $\mathbf{J}_m = \nabla \times \mathbf{M}$ connecting the magnetization field $\mathbf{M}$ with a volume magnetization current density, was treated in the last section. It has an integral form analogously obtainable by use of Stokes' theorem, becoming

$$\oint_\ell \mathbf{M} \cdot d\ell = \int_S \mathbf{J}_m \cdot ds \text{ A} \tag{3-67}$$

This means that the circulation of the $\mathbf{M}$ field about a closed path ℓ is a measure of the net magnetization current through it. For example, a surface integration of $\mathbf{J}_m$ over a cross-section in the y-z plane of the magnetized cube in Example 3-3 is seen to yield a bound magnetization current $10b^2$ A flowing vertically through the specimen, also obtainable from a line integral of $\mathbf{M} \cdot d\ell$ around a horizontal perimeter of the cube.

3-4-2 Boundary Conditions for Tangential H and M

In a manner resembling the derivation of the boundary condition (3-41), one can compare the *tangential* components of $\mathbf{H}$ adjacent to an interface separating two materials, by applying Maxwell's integral law (3-66) to the small, rectangular closed line ℓ shown in Figure 3-10. With the magnetic fields in the adjacent media labelled $\mathbf{H}_1$ and $\mathbf{H}_2$ and resolved into normal and tangential components as in Figure 3-10, integrating the left side of (3-66) clockwise around ℓ yields $H_{t1}\,\Delta\ell - H_{t2}\,\Delta\ell$, if the height δh is taken so small that the ends do not contribute to the line integral. The right side of (3-66) involves integrations of $\mathbf{J}$ and $\mathbf{D}$ over the vanishing surface S bounded by ℓ, obtaining

$$H_{t1}\,\Delta\ell - H_{t2}\,\Delta\ell = J_n\,\Delta s + \frac{\partial D_n}{\partial t}\Delta s = J_n\,\Delta\ell\,\delta h + \frac{\partial D_n}{\partial t}\Delta\ell\,\delta h \tag{3-68}$$

if J_n and D_n denote the components normal to Δs. The last term of (3-68)

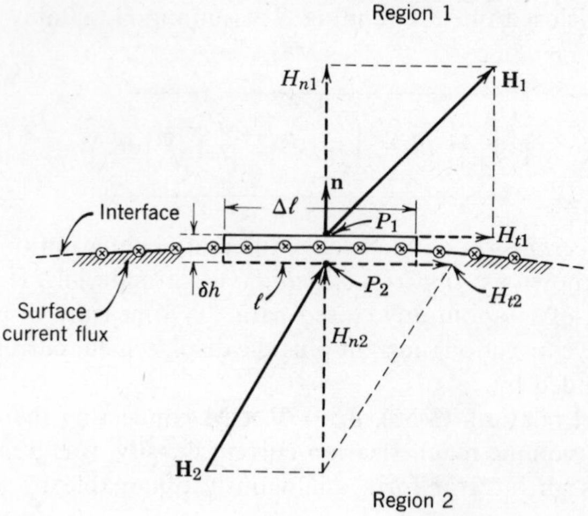

Figure 3-10. Rectangular closed line ℓ constructed to compare H_{t1} and H_{t2} using Ampère's law.

vanishes as $\delta h \to 0$; similarly, the contribution of the J_n term would also vanish if $\mathbf{J}$ were a *volume* current density. In some physical problems, however, one can assume a free *surface* current flowing solely upon the interface with a density $\mathbf{J}_s$ defined by

$$\mathbf{J}_s = \lim_{\delta h \to 0} \mathbf{J}\, \delta h \tag{3-69}$$

(It develops that $\mathbf{J}_s$ is of interest only if one of the regions is a perfect conductor, a case to be discussed shortly.) Thus, the *general boundary condition* resulting from the substitution of (3-69) into (3-68) becomes

$$H_{t1} - H_{t2} = J_{s(n)} \text{ A/m} \tag{3-70a}$$

in which the subscript (n) denotes a surface current flowing normally through the side of the rectangle, as noted in Figure 3-10. Equation (3-70a) states that the *tangential component of the* **H** *field is discontinuous at an interface to the extent of the surface current density that may be present.*

Using $\mathbf{n}$ to denote a normal unit vector directed from region 2 towards region 1 as in Figure 3-11, a vector form of (3-70a) is written

$$\mathbf{n} \times (\mathbf{H}_1 - \mathbf{H}_2) = \mathbf{J}_s \text{ A/m} \tag{3-70b}$$

to include direction as well as magnitude information.

The boundary condition (3-70) is true in general, though in its application to a boundary-value problem, it becomes two cases.

CASE A. *Both regions have finite conductivities.* In this case, free surface currents cannot exist on the interface, reducing (3-70a) to

$$H_{t1} = H_{t2} \text{ A/m} \tag{3-71}$$

Thus, the tangential component of **H** *is continuous at an interface separating two materials having, at most, finite conductivities.* This boundary condition as illustrated in Figure 3-11(*a*).

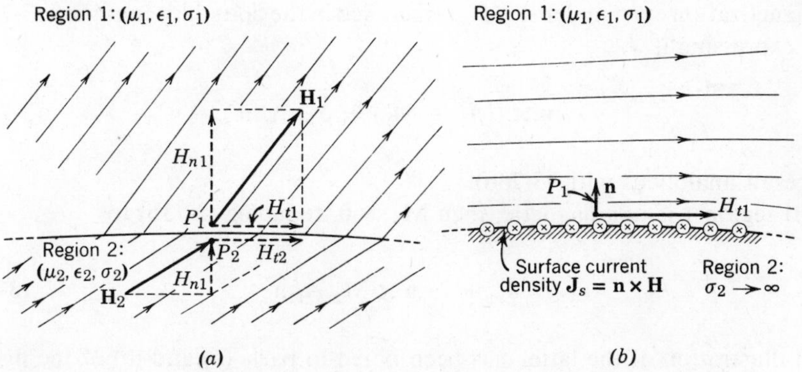

Figure 3-11. The two cases of the boundary condition (3-70b) on tangential H_t. (*a*) Continuous H_t at interface separating regions of finite conductivities. (*b*) Equality of the H_t and surface current density on a perfect conductor.

CASE B. *One region is a perfect conductor.* From (3-44) it has been noted, under time-varying conditions, that no electric or magnetic field can exist inside a perfect conductor. If region 2 were a perfect conductor, then $\mathbf{H}_2 = 0$ reducing (3-70a) to $H_{t1} = J_{s(n)}$; or in vector form, (3-70b) becomes

$$\mathbf{n} \times \mathbf{H}_1 = \mathbf{J}_s \text{ A/m} \tag{3-72}$$

the boundary condition depicted in Figure 3-11(*b*). *At the interface separating a region from a perfect conductor, the surface current density* $\mathbf{J}_s$ *has a magnitude equal to that of the tangential* **H** *there, and a direction specified by the right-hand rule.* It is shown later that no normal component of **H** or **B** may exist at the surface of a perfect conductor, implying that the tangential magnetic field is also the *total* magnetic field there.

A similarity in form is noted between Ampère's circuital law (3-66) and the relationship (3-67) for **M**. Thus, by analogy with the boundary condition (3-70a), derived by applying (3-66) to the closed rectangle as in Figure 3-10, one may establish from (3-67) the boundary condition

$$M_{t1} - M_{t2} = J_{sm(n)} \text{ A/m} \tag{3-73a}$$

This result expresses the continuity of the tangential component of **M** as one traverses an interface between two adjacent, magnetized regions. The subscript (n) denotes a surface magnetization current density *normal* to the tangential **M** components at the boundary. The vector sense of the surface magnetization current density $\mathbf{J}_{sm}$ is included in the boundary condition (3-73a) by expressing it

$$\mathbf{n} \times (\mathbf{M}_1 - \mathbf{M}_2) = \mathbf{J}_{sm} \text{ A/m} \tag{3-73b}$$

a result analogous with (3-70b).

If region 1 is nonmagnetic, then $\mathbf{M}_1 = 0$, reducing (3-73b) to

$$\mathbf{J}_{sm} = -\mathbf{n} \times \mathbf{M}_2 \text{ A/m} \tag{3-74}$$

An illustration of the latter has been noted in parts (b) and (c) of the figure accompanying Example 3-3. Applying (3-74) to the right side of the magnetized block of that example yields a surface magnetization current density $\mathbf{J}_{sm} = -\mathbf{n} \times \mathbf{M}_2 = -\mathbf{a}_x \times (\mathbf{a}_z M_z) = \mathbf{a}_y \mathbf{M}_z$, in agreement with ($c$) of that example.

Example 3-4. Suppose a very long solenoid like that of Figure 1-21(b) contains a coaxial magnetic rod of radius a, as in figure (a), the rod having a constant permeability μ. The winding is closely spaced with n turns in every d m of axial length, carrying a steady current I. (a) Determine **H** and **B** in the air and iron regions, making use of Ampère's law (3-66) and symmetry. (b) Find **M** in the rod, and determine whether any volume magnetization current density $\mathbf{J}_m$ exists in it, as well as magnetization current densities on its surface. (c) Sketch the flux of **H**, **B**, and **M** in the air and iron regions.

(a) With dc in the wire producing time-static fields, (3-66) becomes $\oint_\ell \mathbf{H} \cdot d\ell = \int_s \mathbf{J} \cdot ds$. From the axial symmetry and the implications of Ampère's law in relation to the current sense, **H** is positive z directed within the winding and essentially zero outside it. Constructing the

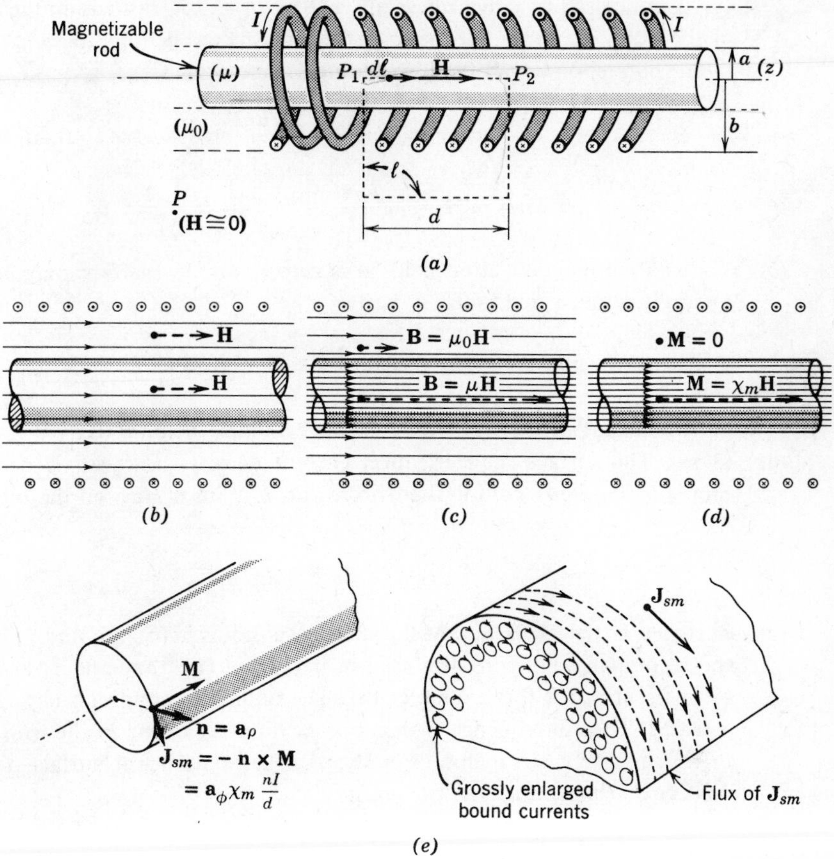

Example 3-4. **(*a*) Solenoid with magnetic core. (*b*) H field flux. (*c*) B field flux. (*d*) M field flux. (*e*) The $\mathbf{J}_{sm}$ field on the iron.**

rectangular closed path ℓ shown, $\mathbf{H}\cdot d\boldsymbol{\ell}$ integrated between P_1 and P_2 yields

$$\oint_{\ell} \mathbf{H}\cdot d\boldsymbol{\ell} = \int_{P_1}^{P_2} (\mathbf{a}_z H_z)\cdot\mathbf{a}_z\, dz = \int_0^d H_z\, dz = nI$$

in which H_z is constant over the path P_1 to P_2, yielding

$$H_z = \frac{nI}{d} \tag{1}$$

This result is correct in both the air and iron regions because nI is the current enclosed by ℓ regardless of whether P_1 and P_2 fall within the air or the iron. The turns per meter in the winding are denoted by n/d.

The corresponding **B** field is obtained from (3-64c)

$$\mathbf{B} = \mu\mathbf{H} = \mu(\mathbf{a}_z H_z) = \mathbf{a}_z \frac{\mu n I}{d} \qquad 0 < \rho < a \quad \text{Iron}$$

$$\mathbf{B} = \mu_0 \mathbf{H} = \mathbf{a}_z \frac{\mu_0 n I}{d} \qquad a < \rho < b \quad \text{Air} \tag{2}$$

(*b*) The volume magnetization field **M** is zero in air; in the ferromagnetic region it is given by (3-60)

$$\mathbf{M} = \chi_m \mathbf{H} = \mathbf{a}_z \chi_m \frac{nI}{d} = \mathbf{a}_z(\mu_r - 1)\frac{nI}{d} \tag{3}$$

M is constant in the iron rod for this example, yielding $\mathbf{J}_m = 0$ from (3-56). The surface magnetization current density, however, is determined from (3-74), calling the iron region 2. With $\mathbf{n} = \mathbf{a}_\rho$ on the interface

$$\mathbf{J}_{sm} = -\mathbf{n} \times \mathbf{M}_2 = -\mathbf{a}_\rho \times \mathbf{a}_z \chi_m \frac{nI}{d} = \mathbf{a}_\phi \chi_m \frac{nI}{d} \tag{4}$$

(*c*) Sketches of the **H**, **B**, and **M** flux-fields are shown in (*b*), (*c*), and (*d*) of the accompanying figure. It is seen from (*b*) that **H** is the same in the air as in the iron for this example; thus the boundary condition (3-71) is satisfied. The consequence is that **B** is μ_r times as strong in the iron as in the adjacent air region. Finally, $\mathbf{J}_{sm}$ has a uniform surface flux density on the iron rod as shown in (*e*).

EXAMPLE 3-5. Obtain a refractive law for the **B** field at an interface separating two isotropic materials of permeabilities μ_1 and μ_2; i.e., find the relation between the angular deviations from the normal made by $\mathbf{B}_1$ and $\mathbf{B}_2$ at points just to either side of the interface.

Assume the total **B** fields tilted from the normal by the angles θ_1 and θ_2 as in (*a*). The boundary conditions relating the tangential and the normal magnetic field components are (3-50) and (3-71); $B_{n1} = B_{n2}$ and $H_{t1} = H_{t2}$. The latter can be written

$$\frac{B_{t1}}{\mu_1} = \frac{B_{t2}}{\mu_2} \tag{3-75}$$

From the geometry of the figure, the tilt-angles obey $\tan\theta_1 = B_{t1}/B_{n1}$ and $\tan\theta_2 = B_{t2}/B_{n2}$, which combine with (3-75) to yield

$$\tan\theta_2 = \frac{\dfrac{\mu_2}{\mu_1} B_{t1}}{B_{n1}}$$

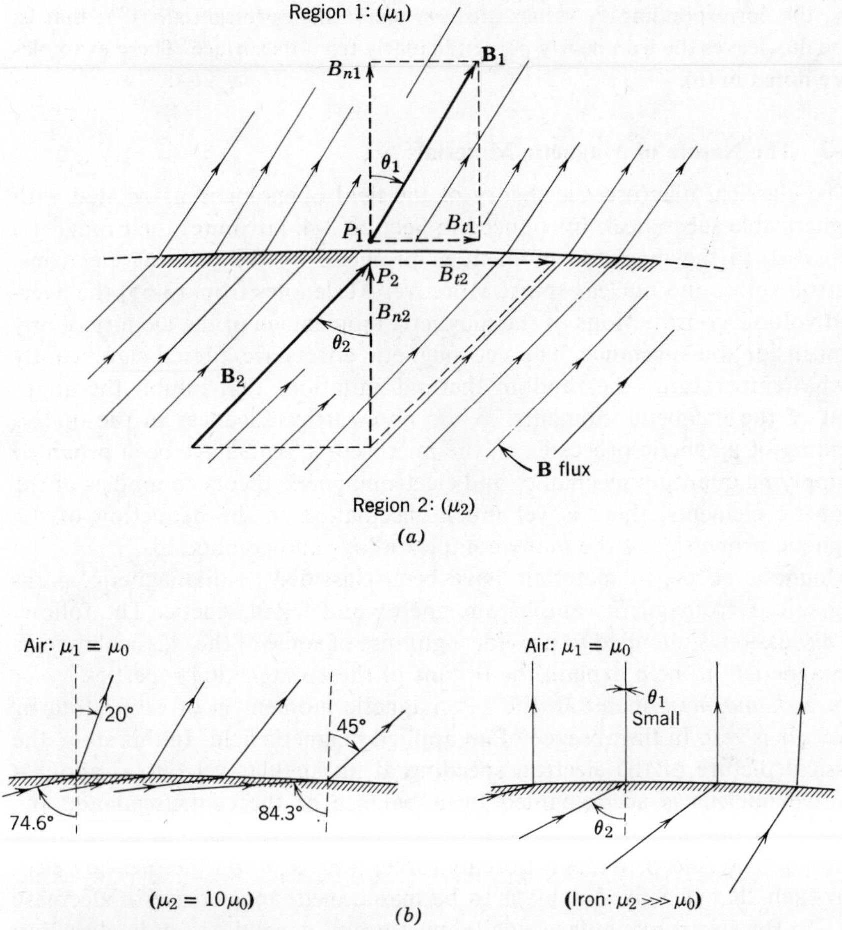

Example 3-5. (*a*) *B* flux refraction. (*b*) Refraction at air-to-magnetic-region interfaces.

Inserting the expression for tan θ_1 obtains

$$\tan \theta_2 = \frac{\mu_2}{\mu_1} \tan \theta_1 \tag{3-76}$$

As a numerical example, compare the tilt of the **B** lines at an interface separating two regions with $\mu_1 = \mu_0$ and $\mu_2 = 10\mu_0$. Assume at some point on the interface that **B** in region 1 is tilted by $\theta_1 = 45°$. From (3-76), $\theta_2 = \arctan(10 \tan 45°) = 84.3°$. Similarly, if $\theta_1 = 20°$, then $\theta_2 = 74.6°$, and so on. In the event of an air-to-iron interface ($\mu_2 \ggg \mu_1$), one may show from (3-76) that for nearly all

θ_2, the corresponding θ_1 values are *very small* angles (essentially 0°); that is, the flux leaves the iron nearly perpendicularly from its surface. These examples are noted in (*b*).

3-4-3 The Nature of Magnetic Materials

The classical macroscopic theory of the field phenomena associated with magnetizable substances, introduced in Section 3-4, attributes their magnetic properties to the magnetic moment **m** provided by the orbiting electrons, electron spins, and nuclear spins. Moreover **M** denotes from (3-55) the averaged volume contributions of the magnetic moments **m** in the vicinity of any point inside the substance. The net magnetic effects are altered significantly by the temperature—the random thermal agitations that inhibit the alignment of the magnetic moments. While noteworthy advances in the understanding of magnetic processes on the microscopic scale have been provided by applying quantum mechanics and electromagnetic theory to models of the magnetic elements, there is yet much speculation in the deduction of the magnetic properties of the many complex alloys and compounds.

Magnetic effects in materials have been classified as diamagnetic, paramagnetic, ferromagnetic, antiferromagnetic, and ferrimagnetic. The following discussion is intended to provide a glimpse of some of the classical models of magnetism to help explain the origins of these magnetic properties.†

In a *diamagnetic* material, the net magnetic moment **m** of each atom or molecule is zero in the absence of an applied magnetic field. In this state, the classical picture of the electron speeding at an angular velocity ω about a positive nucleus is accompanied by a balance of the centrifugal and the attractive Coulomb forces between those opposite charges. The application of a magnetic field provides a Lorentz force, $-e\mathbf{v} \times \mathbf{B}$, on the orbiting electron such that if a fixed orbit is to be maintained, an increase or decrease $\pm\Delta\omega$ in the electron angular velocity must occur, depending on the direction of the applied **B** field relative to the orbital plane. This amounts to a change in the electronic orbital current, thereby generating a small magnetic field, the direction of which is such as to oppose the applied field. The net, opposing magnetization field **M** thus created in any typical volume-element Δv of the material leads to a slightly negative susceptibility χ_m for such a material. Diamagnetism is presumed to exist in all materials, though in some it may be masked by other magnetic effects to be discussed. Typical small, negative values of χ_m for diamagnetic solids at room temperature are -1.66×10^{-5} for bismuth, -0.95×10^{-5} for copper, and -0.8×10^{-5} for germanium.

† An excellent digest of the theories of magnetic phenomena, including ample references, is to be found in Chapter 7 of Elliott, R. S. *Electromagnetics*. New York: McGraw-Hill, 1966.

It is to be expected that the less dense gases have even smaller diamagnetic susceptibilities, which is borne out by both calculation and experiment.

Another weak form of magnetism is known as *paramagnetism*. In a paramagnetic material, the atoms or molecules possess permanent magnetic moments due primarily to electron-spin dipole moments, randomly oriented so that the net magnetization **M** of (3-55) is zero in the absence of an applied magnetic field. The application of a **B** field to gaseous, paramagnetic nitrogen, for example, produces a tendency for the moments **m** to align with the field, a process inhibited by the collisions or interactions among the particles. In a paramagnetic solid, thermal vibrations within the molecular lattice tend to lessen the alignment effects of an applied magnetic field. The room temperature susceptibilities of typical paramagnetic salts such as $FeSO_4$, $NiSO_4$, Fe_2O_3, $CrCl_3$, etc., are of the order of 10^{-3} and inversely temperature-dependent, according to a law discovered by Pierre Curie in 1895.

The important *ferromagnetic* materials are characterized by their strong, permanent magnetic moments, even in the absence of an applied **B** field. They include iron, cobalt, nickel, the rare earths gadolinium and dysprosium, plus a number of their alloys and even some compounds not containing ferromagnetic elements. It was originally postulated by Weiss in 1907, and much later confirmed experimentally in photomicrographs by Bitter,† that a ferromagnetic material in an overall unmagnetized state in reality consists of many small, essentially *totally magnetized domains*, randomly oriented to cancel out the net magnetic field. Domain sizes have been found to range from a few microns to perhaps a millimeter across for many ferromagnetic materials. Weiss further postulated that strong intrinsic coupling or interaction forces exist between adjacent atoms to provide the fully magnetized state within a given domain. It was not until 1928 that Heisenberg of Germany and Frenkel of the U.S.S.R. independently verified, using quantum theory, that the extraordinarily strong forces holding the domain atoms in parallel alignment is attributable to the coupling forces between the net electron spins of the adjacent atoms.‡ The parallel orientation of the spin moments in a ferromagnetic domain is depicted in Figure 3-12(*a*). An idealized, perfect crystal might have a domain structure, in the absence of an applied **B** field, like that shown in Figure 3-12(*b*), although flaws such as lattice imperfections and impurities would modify this idealized picture somewhat. The walls between the domains (Bloch walls), having the appearance suggested by Figure 3-12(*c*), are transition regions between the spin alignments of the adjacent domains, and they are of the order of 100 atoms thick. The domain

† Bitter, F. "A generalization of the theory of ferromagnetism," *Phys. Rev.*, **54**, 79, 1938.

‡ Heisenberg, W. "On the theory of ferromagnetism," *Zeit. f. Phys.*, **49**, 619, 1928.

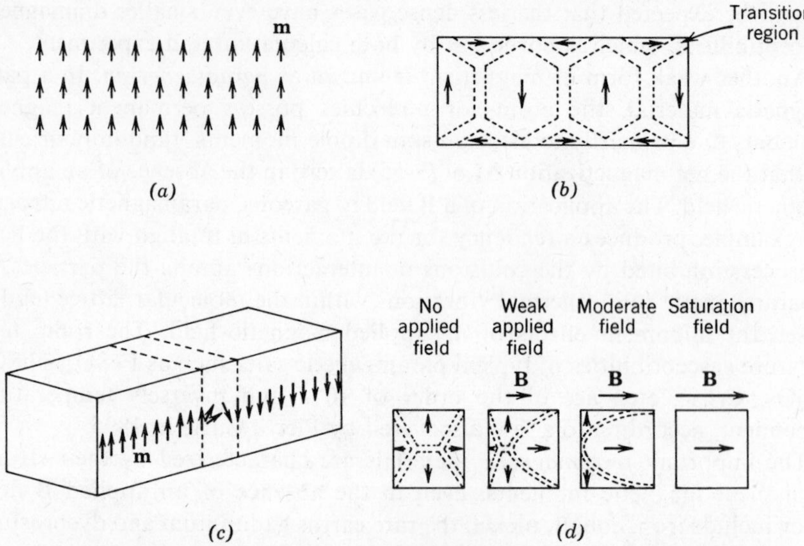

Figure 3-12. Alignment of magnetic moments in a ferromagnetic material: domain phenomena. (*a*) Magnetic moment alignments in a ferromagnetic material. (*b*) A perfect single crystal, showing domains and domain walls. (*c*) A transition region between adjacent domains. (*d*) Domain changes in a crystal with increase in the applied B field.

division by such wall structures occurs in such a way that a minimal external magnetic field is supported by the structure, to minimize the work done in forming the structure.

As an external **B** field is increasingly applied to a ferromagnetic crystal containing domains as denoted in Figure 3-12(*d*), the Bloch walls first move to favor the growth of those domains having magnetic moments aligned with the applied field, a reversible condition upon removing the field if **B** is not too large. For higher applied fields, domain-wall motion occurs which is not reversible, as noted in the third sketch of Figure 3-12(*d*). For a sufficiently large applied field, the domain magnetic moments rotate until an essentially total parallel alignment with the applied field occurs, a condition called *saturation*. The averaged effect of such changes upon the bulk magnetization **M**, in a sample volume element containing a sufficient number of domains, is shown in Figure 3-13(*a*). The arrows denote the direction of increasing or decreasing the applied *H* field.† One may note, upon decreasing the applied

† It has become customary to denote the applied magnetic field in the material as the *H* field, rather than *B*.

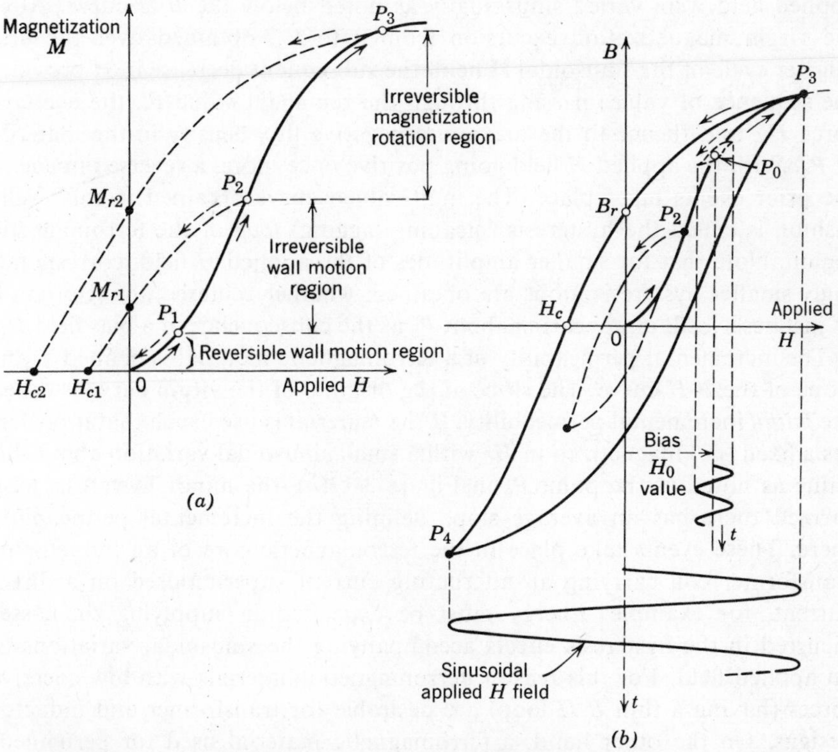

Figure 3-13. Magnetization effects due to an external magnetic field applied to a ferromagnetic material. (*a*) Magnetization process (solid line) in a virgin ferromagnetic region. Irreversible behavior shown dashed. (*b*) *B-H* hysteresis loops for a ferromagnetic material.

H field to zero from the values at P_2 or P_3, that a permanent magnetization M_{r1} (or M_{r2}) is retained in the ferromagnetic sample, signifying an irreversible and distinctly nonlinear, multivalued behavior. These M_r values are termed the remanent (remaining) magnetizations of the specimen. The applied field must be further decreased to the reverse value H_{c1} (or H_{c2}) as shown, before the permanent magnetism is removed from the material. The value H_c is rather loosely called the *coercive force*, the field required to reduce the magnetization to zero within a specimen.

If the *M-H* plot of Figure 3-13(*a*) is replotted in terms of the *B* field in the ferromagnetic material, the *B-H* curve of Figure 3-13(*b*) results, recalling from (3-58) that these quantities are related to $\mathbf{B} = \mu_0(\mathbf{H} + \mathbf{M})$. In (*b*) is depicted a complete cycle of the events of (*a*), such as might occur if the applied field were varied sinusoidally as noted below the B-H curve. After the virgin magnetization excursion from 0 to P_3, obtained over the first quarter cycle of the sinusoidal **H** field, the subsequent decrease in H provides the sequence of values passing through the remanent value B_r, the coercive force H_c, and thence to the maximum negative flux density in the material at P_4. With the applied H field going positive once more, a reversed image of the prior events takes place. The multivalued curve obtained in this cyclic fashion is called the hysteresis (meaning lagging) *loop* of the ferromagnetic region. Note that for smaller amplitudes of the applied H field, correspondingly smaller hysteresis loops are obtained, whether centered about origin 0 as just described, or appearing about P_0 as the consequence of a bias field H_0.

The incremental permeability of a ferromagnetic material is defined as the slope of the B-H curve. The slope at the origin 0 of the virgin curve is called the *initial* incremental permeability. If the material is used such that it possesses a fixed (dc) magnetization H_0 with a small sinusoidal variation about this value as noted at the point P_0 in Figure 3-13(*b*), the minor hysteresis loop formed there has an average slope defining the incremental permeability there. These events take place in the ferromagnetic core of an inductor or transformer coil carrying an alternating current superimposed on a direct current, for example. Energy must be expended in supplying the losses incurred in the hysteresis effects accompanying the sinusoidal variations of an applied field. For this reason, ferromagnetic materials with low coercive forces (having a thin B-H loop) are desirable for transformer and inductor designs. On the other hand, a ferromagnetic material used for permanent magnets should have a high coercive force H_c and a high remanent, or residual, flux density B_r (corresponding to a fat B-H loop).

Table 3-1 lists a few representative ferromagnetic alloys along with some of their magnetic properties.

An additional and usually undesirable side effect, occurring in the magnetic core of devices such as transformers, is that of the free-electron conduction currents circulating within the core material due to an electric field **E** generated inside it by a time-varying magnetic field. The densities of these currents are limited by the conductivity σ of the core material through (3-7), i.e., $\mathbf{J} = \sigma\mathbf{E}$, and are given the name *eddy currents* because of their vortex-like nature within the conductive core, resulting from their relationship to the time-varying **B** field through (2-62)

$$\nabla \times \mathbf{E} = -\frac{\partial \mathbf{B}}{\partial t} \qquad [2\text{-}62]$$

In the next section (2-62) is shown to be valid for a material region as well as for free space. Thus, with a conductive, ferromagnetic core in the solenoid

Table 3-1. Magnetic properties of ferromagnetic alloys

(A) Transformer alloys

Material	Percent composition	Permeabilities		Saturation B (Wb/m^2)	Coercive force H_c (A/m)	Conductivity ($\times 10^7$ ℧/m)
		Initial	Maximum			
Silicon iron	4 Si, 96 Fe	400	7,000	2	40	0.16
Hypersil (grain oriented)	3.5 Si, 96.5 Fe	1,500	35,000	2	16	0.2
78 Permalloy	78 Ni, 0.6 Mn, 21.4 Fe	9,000	100,000	1.07	4	0.12
Supermalloy	79 Ni, 5 Mo, 16 Fe	100,000	800,000	0.7	0.16 to 4.0	0.15

(B) Permanent magnet materials

Material	Percent composition	Coercive force (A/m)	Remanent B_r (Wb/m^2)
Carbon steel	1 Mn, 0.9 C, 98.1 Fe	4,000	1
Alnico V	8 Al, 14 Ni, 24 Co, 3 Cu, 53 Fe	44,000	1.25

as shown in Figure 3-14(*a*), a sinusoidally time-varying current in the winding produces a sinusoidal **B** field in the core material to generate an **E** field, and from (3-7) also an eddy current field therein. Its sense is thus normal to the time-varying **B** field. The losses may be reduced substantially by subdividing the conductive core into a fibrous or laminar structure as suggested by Figure 3-14(*b*), in which the subdivided conductors are insulated from one

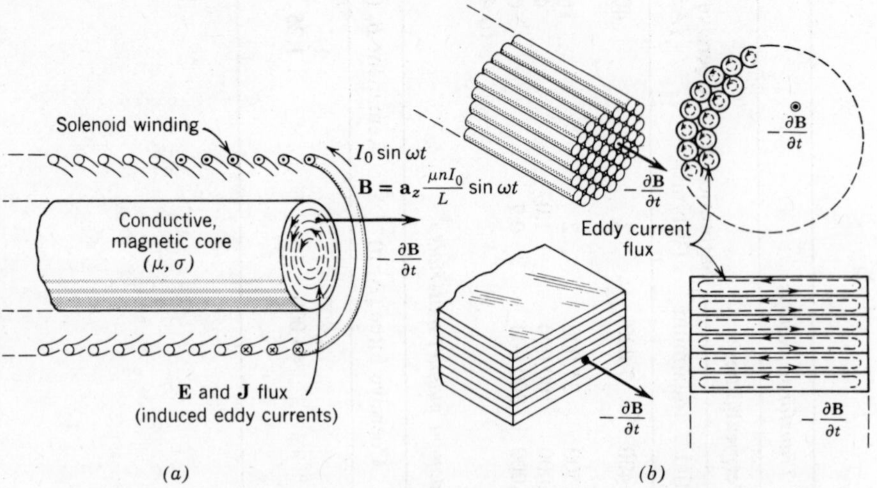

Figure 3-14. Eddy currents in conductors immersed in time-varying B fields. (*a*) Eddy currents induced in a conductive magnetic core by a time-varying B field. (*b*) Fibrous and laminar core structures used to break up eddy current paths.

another. Small, spherical magnetizable particles serve the same purpose. This constrains the eddy currents to much smaller volumes, limiting their densities substantially if the cellular substructures are made sufficiently small or thin.

In the previous discussions it was seen that paramagnetism is a characteristic of materials possessing permanent magnetic spin moments that are randomly oriented, a condition depicted in Figure 3-15(*a*). Ferromagnetic materials, due to the effects of short-range couplings between adjacent atoms, possess parallel-oriented atomic magnets within given domain boundaries that comprise the material, as suggested in Figure 3-15(*b*). If such a material is heated until the thermal energies exceed the coupling energies, the material becomes disorganized into a paramagnet, though upon cooling it reverts to a ferromagnet once more. The critical temperature at which this occurs is known as the *Curie temperature*.

Variations of the coupling phenomena responsible for ferromagnetic materials can even produce antiparallel alignments of electron spins in materials known as *antiferromagnetic*, as depicted by Figure 3-15(c). In this state, an antiferromagnet is characterized by a zero magnetic field. Manganese fluoride, for example, is paramagnetic at room temperature, but upon cooling it to $-206°C$ (called its Néel temperature, after the French physicist), it

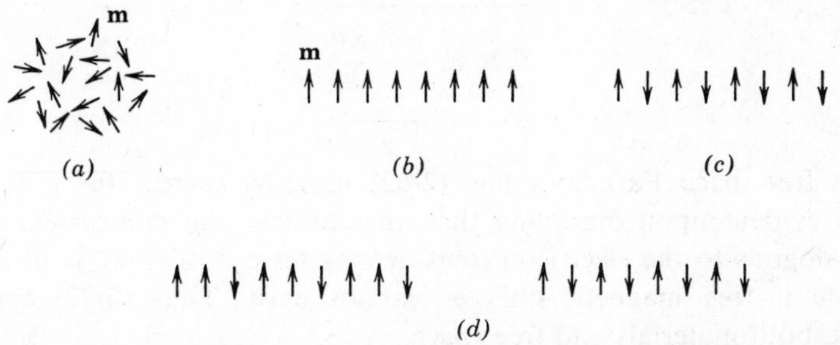

Figure 3-15. Orientations of the spin moments of various magnetic materials. (*a*) Paramagnetic. (*b*) Ferromagnetic. (*c*) Antiferromagnetic. (*d*) Ferrimagnetic materials, or ferrites.

becomes antiferromagnetic; below this temperature it exhibits no magnetic effect. An important variation of this phenomenon is *ferrimagnetism*, associated with noncancelling antiparallel arrangements of the coupled spin moments as suggested by Figure 3-15(d). Thus, in magnetite, the magnetic iron oxide $FeO \cdot Fe_2O_3$, two of the three adjacent spins are reversed such that a somewhat weaker form of ferromagnetism is produced. Magnetite is an example of the group of ferrimagnetic oxides $XO \cdot Fe_2O_3$, in which the symbol X denotes a divalent metallic ion Cd, Co, Cu, Mg, Mn, Ni, Zn, or divalent iron. When synthesized in the laboratory, these brittle, ceramiclike compounds are particularly useful for magnetic cores in high-frequency transformers and special applications ranging into the microwave frequencies because of their low conductivities comparable to those of the semiconductors, usually from 10^{-1} to 10^{-4} ℧/m. They are thus desirable because they limit eddy current losses in such applications. These conductivities may be compared with the much higher values of the typical alloys for lower-frequency applications as listed in Table 3-1, in which the values of the order of 10^6 ℧/m appear. A general account of the theory of ferro- and ferrimagnetism, together with a number of microwave applications of the latter, is found in the book by Lax and Button.†

† Lax, B., and K. J. Button. *Microwave Ferrites and Ferrimagnetics*. New York: McGraw-Hill, 1962.

3-5 Maxwell's Curl E Relation, Its Integral Form and Boundary Condition for Tangential E

In Section 3-3, the Maxwell relation (3-59) for **curl H** in a material region was developed by adding in those current densities contributed by the electric and magnetic polarization fields. The form of the **curl E** relationship for materials is obtained by analogy, but retaining its form (2-62) for free space

$$\nabla \times \mathbf{E} = -\frac{\partial \mathbf{B}}{\partial t} \tag{3-77}$$

That the free-space Faraday's law (2-62) remains correct for a material region is evident upon observing that an additive *magnetic-current-density* term, analogous to the electric-current-density term **J** of (3-7), is physically impossible if free magnetic charges cannot exist. Thus (3-77) correctly applies to both materials and free space.

Equation (3-77) is readily converted to an integral form. The scalar multiplication of (3-77) with $d\mathbf{s}$, integrating the result over any surface S bounded by a closed line ℓ, and applying Stokes' theorem yields

$$\oint_\ell \mathbf{E} \cdot d\boldsymbol{\ell} = -\frac{d}{dt} \int_S \mathbf{B} \cdot d\mathbf{s} \text{ V} \tag{3-78}$$

again unchanged from the free-space version (1-55).

The determination of (3-77) and (3-78) completes the development of Maxwell's differential and integral relations applicable to material regions, and they are summarized in the first two columns of Table 3-2.

A boundary condition, comparing the tangential components of the **E** fields to either side of an interface, may be obtained from Faraday's integral law (3-78). The details of the derivation may be avoided if one recalls that Ampère's line-integral law (3-66) leads to the boundary condition (3-70a), $H_{t1} - H_{t2} = J_{s(n)}$. The boundary condition comparing the tangential components of **E** can be analogously found by applying (3-78) to a similar thin rectangle, yielding the analog of (3-70a)

$$E_{t1} - E_{t2} = 0 \tag{3-79}$$

Thus the *tangential component of the* **E** *field is always continuous at an interface*. The right side of (3-79) is evidently zero because no magnetic currents are physically possible.

Table 3-2 Summary of Maxwell's equations and the corresponding spatial boundary conditions at an interface

Differential form	*Integral form*	*Corresponding boundary condition*
$\nabla \cdot \mathbf{D} = \rho_v$ [3-24]	$\oint_S \mathbf{D} \cdot d\mathbf{s} = \int_V \rho_v \, dv$ [3-37]	$D_{n1} - D_{n2} = \rho_s$ or $\mathbf{n} \cdot (\mathbf{D}_1 - \mathbf{D}_2) = \rho_s$ [3-42] *Case A:* σ_1, σ_2 zero: $D_{n1} = D_{n2}$ *Case B:* $\sigma_2 \to \infty$: $D_{n1} = \rho_s$ [3-45]
$\nabla \cdot \mathbf{B} = 0$ [3-48]	$\oint_S \mathbf{B} \cdot d\mathbf{s} = 0$ [3-49]	$B_{n1} = B_{n2}$ or $\mathbf{n} \cdot (\mathbf{B}_1 - \mathbf{B}_2) = 0$ [3-50]
$\nabla \times \mathbf{H} = \mathbf{J} + \dfrac{\partial \mathbf{D}}{\partial t}$ [3-59]	$\oint_\ell \mathbf{H} \cdot d\boldsymbol{\ell} = \int_S \mathbf{J} \cdot d\mathbf{s} + \dfrac{d}{dt} \int_S \mathbf{D} \cdot d\mathbf{s}$ [3-66]	$H_{t1} - H_{t2} = J_{s(n)}$ or $\mathbf{n} \times (\mathbf{H}_1 - \mathbf{H}_2) = \mathbf{J}_s$ [3-71] *Case A:* σ_1, σ_2 finite: $H_{t1} = H_{t2}$ *Case B:* $\sigma_2 \to \infty$: $\mathbf{n} \times \mathbf{H}_1 = \mathbf{J}_s$ [3-72]
$\nabla \times \mathbf{E} = -\dfrac{\partial \mathbf{B}}{\partial t}$ [3-77]	$\oint_\ell \mathbf{E} \cdot d\boldsymbol{\ell} = -\dfrac{d}{dt} \int_S \mathbf{B} \cdot d\mathbf{s}$ [3-78]	$E_{t1} = E_{t2}$ or $\mathbf{n} \times (\mathbf{E}_1 - \mathbf{E}_2) = 0$ [3-79]

A summary of the four boundary conditions derived from Maxwell's integral laws for material regions in Sections 3-2-3, 3-4-2, and in the present section, is given in Table 3-2.

EXAMPLE 3-6. (*a*) Derive a refractive law for **E** at an interface separating two nonconductive regions. (*b*) Deduce from boundary conditions the direction of **E** just outside a perfect conductor.

(*a*) The boundary conditions for the tangential and the normal components of **E** at an interface separating nonconductive regions are (3-43) and

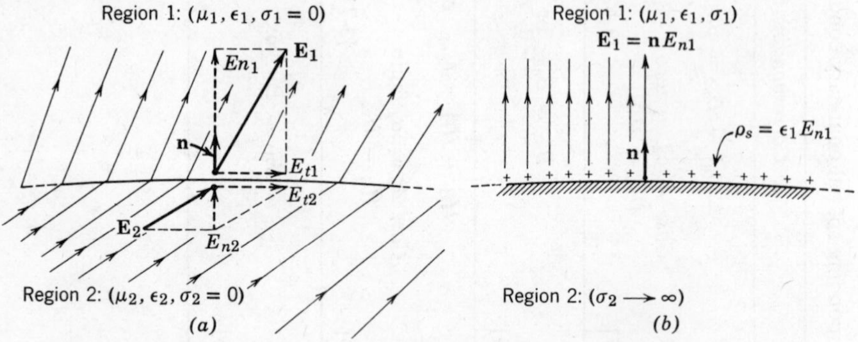

Example 3-6. (*a*) E flux refraction at an interface separating nonconductive regions. (*b*) E is everywhere normal to the surface of a perfect conductor.

(3-79); i.e., $\epsilon_1 E_{n1} = \epsilon_2 E_{n2}$ and $E_{t1} = E_{t2}$. From the latter and the geometry of (*a*), one obtains

$$\tan \theta_2 = \frac{\epsilon_2}{\epsilon_1} \tan \theta_1 \tag{3-80}$$

a result analogous with (3-76) of Example 3-5 concerned with the refraction of **B** lines.

(*b*) From (3-44), a perfectly conductive region 2 implies null fields inside it. Then by (3-79), E_{t1} in the adjacent region 1 must vanish also. The remaining normal component in region 1 is given by (3-45), $D_{n1} = \rho_s$, yielding $\rho_s = \epsilon_1 E_{n1}$ as shown in (*b*).

EXAMPLE 3-7. A uniform plane wave is described by the electric and magnetic fields

$$\mathbf{E} = \mathbf{a}_x E_x^+(z, t) = \mathbf{a}_x E_m^+ \cos(\omega t - \beta_0 z)$$

$$\mathbf{H} = \mathbf{a}_y H_y^+(z, t) = \mathbf{a}_y \frac{E_m^+}{\eta_0} \cos(\omega t - \beta_0 z)$$

and propagates in air between two perfectly conducting, parallel plates of great extent as in (*a*). The inner surfaces of the plates are located at $x = 0$ and $x = a$. Obtain expressions for (*a*) the surface charge field and (*b*) the surface currents on the two conductors.

(*a*) The given **E** is everywhere normal to the plates at $x = 0$ and $x = a$, satisfying the boundary condition of (*b*) in Example 3-6. The surface charge distributions thus become

$$\rho_s = \mathbf{n}\cdot\mathbf{D}_1 = \epsilon_0\mathbf{n}\cdot\mathbf{E}_1 = \epsilon_0\mathbf{a}_x\cdot\mathbf{a}_x E_x^+ = \epsilon_0 E_m^+ \cos(\omega t - \beta_0 z) \qquad x = 0$$

$$\rho_s = \mathbf{n}'\cdot\mathbf{D}_1 = -\epsilon_0\mathbf{a}_x\cdot\mathbf{a}_x E_x^+ = -\epsilon_0 E_m^+ \cos(\omega t - \beta_0 z) \qquad x = a$$

implying that **E** lines emerge from positive charges and terminate on negative ones.

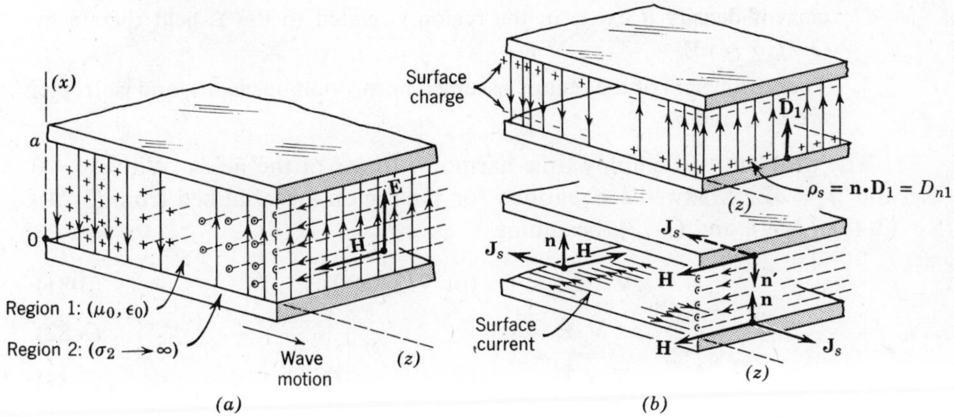

Example 3-7. (*a*) Parallel-plate system supporting a uniform plane wave field. (*b*) Charge and current distribution on conductor inner surfaces.

(*b*) The given **H**, to satisfy (3-72), must be everywhere tangential to the perfect conductors at $x = 0$ and $x = a$, yielding there

$$\mathbf{J}_s = \mathbf{n}\times\mathbf{H}_1 = \mathbf{a}_x\times\mathbf{a}_y H_y^+ = \mathbf{a}_z\frac{E_m^+}{\eta_0}\cos(\omega t - \beta_0 z) \qquad x = 0$$

$$\mathbf{J}_s = \mathbf{n}'\times\mathbf{H}_1 = -\mathbf{a}_x\times\mathbf{a}_y H_y^+ = -\mathbf{a}_z\frac{E_m^+}{\eta_0}\cos(\omega t - \beta_0 z) \qquad x = a$$

It is seen that, in any fixed *z* plane, current flows in opposite *z* directions in the two conductors.

3-6 Uniform Plane Waves in an Unbounded Conductive Region

The topic of uniform plane waves propagating in empty space was discussed in Section 2-10, in which the influence of the free-space parameters μ_0 and ϵ_0 on the various wave characteristics was observed. The study of a plane wave propagating in a material having the parameters ϵ, μ, and σ is considered in this section. It is shown that the important new effects produced by the conductivity σ is to provide wave decay in the direction of propagation, as well as a phase shift between **E** and **H**.

The assumptions made for the problem of wave propagation in an unbounded, linear, conductive region are:

1. The components of **E** and **H** have neither x nor y dependence; i.e., $\partial/\partial x = \partial/\partial y = 0$ for all field components.
2. The free-charge densities are zero in the conductive region ($\rho_v = 0$); but a current density **J** exists in the region,† related to the **E** field therein by (3-7), $\mathbf{J} = \sigma\mathbf{E}$.
3. The parameters of the region, assumed linear, homogeneous, and isotropic, are μ, ϵ, and σ.

The problem will employ time-harmonic forms of the fields. With $\hat{\rho}_v = 0$ and $\hat{\mathbf{J}} = \sigma\hat{\mathbf{E}}$, Maxwell's equations for the region are obtained from (3-24), (3-48), (3-59), and (3-77), becoming

$$\nabla\cdot(\epsilon\hat{\mathbf{E}}) = 0 \quad (\text{or } \nabla\cdot\hat{\mathbf{E}} = 0) \tag{3-81}$$

$$\nabla\cdot\hat{\mathbf{B}} = 0 \tag{3-82}$$

$$\nabla\times\hat{\mathbf{E}} = -j\omega\hat{\mathbf{B}} = -j\omega\mu\hat{\mathbf{H}} \tag{3-83}$$

$$\nabla\times\hat{\mathbf{H}} = \hat{\mathbf{J}} + j\omega\hat{\mathbf{D}} = \sigma\hat{\mathbf{E}} + j\omega\epsilon\hat{\mathbf{E}} \tag{3-84}$$

in which $\hat{\mathbf{B}} = \mu\hat{\mathbf{H}}$ and $\hat{\mathbf{D}} = \epsilon\hat{\mathbf{E}}$ of (3-30) and (3-64) are applicable.

These equations need not in fact be solved, since this has already been done analogously in Section 2.10 for plane waves in empty space. To obtain the solution by analogy, compare (3-81) through (3-84) with (2-106) through (2-109) applicable to the *empty-space* case

$$\nabla\cdot(\epsilon_0\hat{\mathbf{E}}) = 0 \quad (\text{or } \nabla\cdot\mathbf{E} = 0) \qquad [2\text{-}106]$$

$$\nabla\cdot\hat{\mathbf{B}} = 0 \qquad [2\text{-}107]$$

$$\nabla\times\hat{\mathbf{E}} = -j\omega\mu_0\hat{\mathbf{H}} \qquad [2\text{-}108]$$

$$\nabla\times\hat{\mathbf{H}} = j\omega\epsilon_0\hat{\mathbf{E}} \qquad [2\text{-}109]$$

† While this assumption refers explicitly to waves in a conductive region, the extension to wave propagation in a lossy dielectric through the use of a *loss tangent*, ϵ''/ϵ', is described in Section 3-9.

in which $\hat{\mathbf{B}} = \mu_0\hat{\mathbf{H}}$ and $\hat{\mathbf{D}} = \epsilon_0\hat{\mathbf{E}}$ apply. A comparison of the divergence relations (3-81) and (3-82) with (2-106) and (2-107) reveals that $\nabla\cdot\hat{\mathbf{E}} = 0$ and $\nabla\cdot\hat{\mathbf{B}} = 0$ for both problems. Also, the form of (3-83) is analogous with (2-108). Comparing (3-84) with (2-109), however, reveals an additional conduction-current-density term $\sigma\hat{\mathbf{E}}$ in (3-84). Upon collecting terms of the right side of (3-84) as follows

$$\nabla \times \hat{\mathbf{H}} = \sigma\hat{\mathbf{E}} + j\omega\epsilon\hat{\mathbf{E}} = (\sigma + j\omega\epsilon)\hat{\mathbf{E}} = j\omega\left(\epsilon - j\frac{\sigma}{\omega}\right)\hat{\mathbf{E}} \tag{3-85}$$

the analogy of the latter with (2-109) is evident upon replacing ϵ_0 of (2-109) with the complex permittivity, $\epsilon - j\sigma/\omega$. Thus, each of the Maxwell's equations (2-106) through (2-109) is seen to become (3-81) through (3-84) upon replacing in the former

$$\mu_0 \text{ with } \mu \qquad \text{and} \qquad \epsilon_0 \text{ with } \left(\epsilon - j\frac{\sigma}{\omega}\right) \tag{3-86}$$

These replacements applied to the wave solutions of (2-106) through (2-109) are therefore expected to yield the solutions of (3-81) through (3-84) in an unbounded conductive region. Recalling the solution (2-115) for *empty space*

$$\begin{aligned}\hat{E}_x(z) &= \hat{E}_x^+(z) + \hat{E}_x^-(z)\\ &= \hat{E}_m^+ e^{-j\omega\sqrt{\mu_0\epsilon_0}z} + \hat{E}_m^- e^{j\omega\sqrt{\mu_0\epsilon_0}z}\end{aligned} \tag*{[2-115]}$$

the replacements (3-86) in the latter yield analogous plane wave solutions for an *unbounded conductive region*

$$\begin{aligned}\hat{E}_x(z) &= \hat{E}_x^+(z) + \hat{E}_x^-(z)\\ &= \hat{E}_m^+ e^{-j\omega\sqrt{\mu[\epsilon - j(\sigma/\omega)]}z} + \hat{E}_m^- e^{j\omega\sqrt{\mu[\epsilon - j(\sigma/\omega)]}z}\end{aligned} \tag{3-87}$$

In (3-87), the pure phase factor $j\omega\sqrt{\mu_0\epsilon_0}$ of (2-115) becomes a *complex* factor abbreviated with the symbol γ, called the *propagation constant*

$$\gamma = j\omega\sqrt{\mu\left(\epsilon - j\frac{\sigma}{\omega}\right)} \tag{3-88}$$

and γ can be separated into real and imaginary parts

$$\gamma = \alpha + j\beta \text{ m}^{-1} \tag{3-89}$$

in which α, the real part of γ, is called the *attenuation constant*, and β is termed the *phase constant* of the uniform plane waves (3-87). Explicit expressions for

α and β are found by replacing γ of (3-88) with $\alpha + j\beta$, squaring both sides to remove the radical, and equating the real and imaginary parts of the result. The following positive, real solutions for α and β are obtained:

$$\alpha = \frac{\omega\sqrt{\mu\epsilon}}{\sqrt{2}}\left[\sqrt{1 + \left(\frac{\sigma}{\omega\epsilon}\right)^2} - 1\right]^{1/2} \text{ Np/m} \tag{3-90a}$$

$$\beta = \frac{\omega\sqrt{\mu\epsilon}}{\sqrt{2}}\left[\sqrt{1 + \left(\frac{\sigma}{\omega\epsilon}\right)^2} + 1\right]^{1/2} \text{ rad/m} \tag{3-90b}$$

The dimension of α and β is $(\text{m})^{-1}$, though the artificial dimensionless terms neper and radian are usually mentioned to emphasize their attenuative and phase meanings in the wave expressions.

With the substitution of (3-88) into the exponent of the wave solution (3-87), one may express it

$$\hat{E}_x(z) = \hat{E}_x^+(z) + \hat{E}_x^-(z) \tag{3-91a}$$

$$\hat{E}_x(z) = \hat{E}_m^+ e^{-\gamma z} + \hat{E}_m^- e^{\gamma z} \tag{3-91b}$$

$$\hat{E}_x(z) = (E_m^+ e^{j\phi^+})e^{-\alpha z}e^{-j\beta z} + (E_m^- e^{j\phi^-})e^{\alpha z}e^{j\beta z} \tag{3-91c}$$

in which the complex amplitudes $\hat{E}_m^\pm$ of the traveling wave terms are denoted again as in (2-116)

$$\hat{E}_m^+ \equiv E_m^+ e^{j\phi^+} \qquad \hat{E}_m^- \equiv E_m^- e^{j\phi^-} \tag{3-92}$$

A comparison of the conductive region wave solution (3-91) with the empty-space wave solution (2-115) reveals the presence of two real factors, $e^{-\alpha z}$ and $e^{\alpha z}$, accounting for *wave decay* as the positive z and the negative z traveling waves proceed in their corresponding directions of flight with increasing time. An additional view of the decay (attenuation) property of the waves is gained by converting (3-91) to its real-time form, obtained as usual by use of (2-74)

$$\begin{aligned} E_x(z, t) &= \text{Re}\,[\hat{E}_x(z)e^{j\omega t}] = \text{Re}\,[E_m^+ e^{j\phi^+} e^{-\alpha z} e^{-j\beta z} e^{j\omega t} + E_m^- e^{j\phi^-} e^{\alpha z} e^{j\beta z} e^{j\omega t}] \\ &= E_m^+ e^{-\alpha z} \cos(\omega t - \beta z + \phi^+) + E_m^- e^{\alpha z} \cos(\omega t + \beta z + \phi^-) \end{aligned} \tag{3-93}$$

These positive z and negative z attenuated traveling waves are depicted in Figure 3-16(*a*) and (*b*). A comparison of (3-93) with the real-time uniform plane wave solution (2-119) in *empty space*

$$E_x(z, t) = E_m^+ \cos(\omega t - \beta_0 z + \phi^+) + E_m^- \cos(\omega t + \beta_0 z + \phi^-) \quad [2\text{-}119]$$

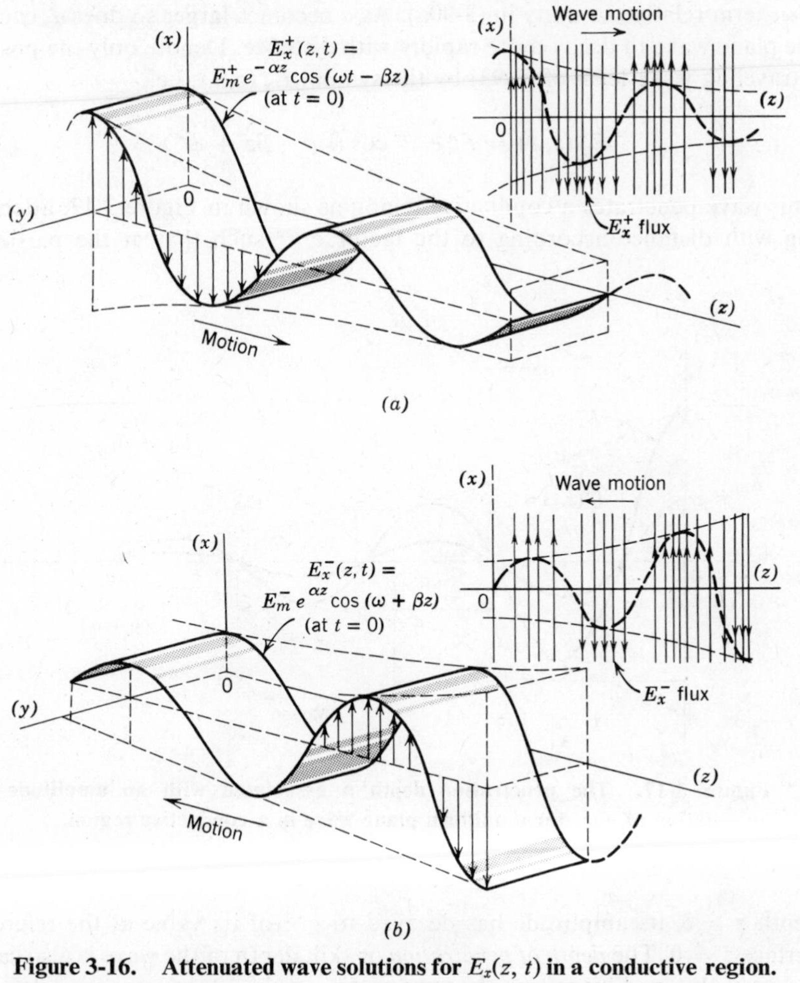

Figure 3-16. Attenuated wave solutions for $E_x(z, t)$ in a conductive region. Flux plots are shown, emphasizing field independence of x and y. (*a*) Positive z traveling, positive z attenuated wave. (*b*) Negative z traveling, negative z attenuated wave.

shows that the important new characteristic introduced by the nonzero conductivity σ is the *wave attenuation* occurring in the direction of the wave motion. Note that setting the region parameters equal to the empty-space values $\epsilon = \epsilon_0$, $\mu = \mu_0$ and $\sigma = 0$ reduces the attenuation and phase constants to $\alpha = 0$ and $\beta = \beta_0$ in (3-90a) and (3-90b).

The wave attenuation in a conductive region is governed by the size of the

$\sigma/\omega\epsilon$ term relative to unity in (3-90a). As σ becomes larger so does α, causing the plane wave to decay more rapidly with distance. Denote only the positive z traveling wave term of (3-93) by the symbol $E_x^+(z, t)$; i.e.,

$$E_x^+(z, t) = E_m^+ e^{-\alpha z} \cos(\omega t - \beta z + \phi^+) \tag{3-94}$$

This wave penetrates a conductive region as shown in Figure 3-17, attenuating with distance according to the factor $e^{-\alpha z}$ such that at the particular

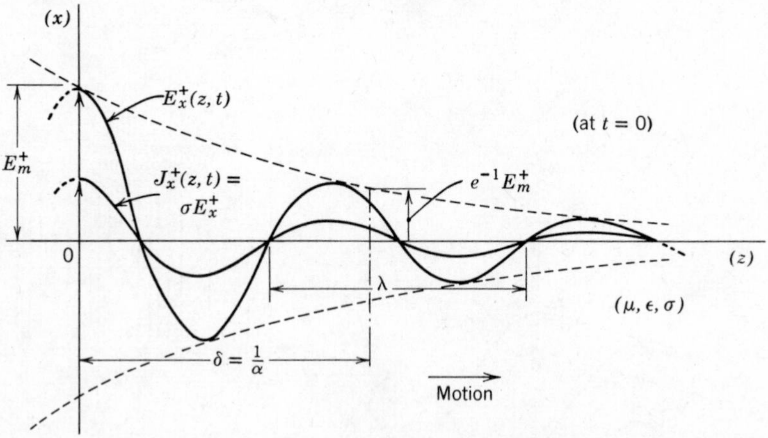

Figure 3-17. The penetration depth δ associated with an amplitude attenuation of e^{-1}, for a uniform plane wave in a conductive region.

depth $z = \delta$, its amplitude has decayed to e^{-1} of its value at the reference surface $z = 0$. The *depth of penetration* or skin depth of the wave is also called δ, being obtained by setting the exponent $-\alpha\delta = -1$, whence

$$\delta = \frac{1}{\alpha} \text{ m} \tag{3-95}$$

A current density $\mathbf{J}$ accompanies the E_x field in the conductive region as given by (3-7); i.e.,

$$J_x^+(z, t) = \sigma E_m^+ e^{-\alpha z} \cos(\omega t - \beta z + \phi^+) \text{ A/m} \tag{3-96}$$

a result *in phase* with the electric field. For a highly conductive region (with a large $\sigma/\omega\epsilon$), δ is seen from (3-95) and (3-90a) to be correspondingly small;

so in the limiting case of a *perfect conductor* ($\sigma \to \infty$), the skin depth vanishes with α indefinitely large. This provides the limiting surface current phenomenon of the boundary condition (3-72), involving the tangential **H** field at the surface of a perfect conductor.

The magnetic field accompanying $\hat{E}_x$ of (3-91) is obtained by substituting (3-91) into Maxwell's equation (3-83); or alternatively by invoking the analogy with the wave solutions in empty space, using the replacements (3-86) in (2-130). Then, for uniform plane waves in an unbounded conductive region, the following complex impedance ratios are found to apply:

$$\frac{\hat{E}_x^+(z)}{\hat{H}_y^+(z)} = \sqrt{\frac{\mu}{\epsilon - j\dfrac{\sigma}{\omega}}} \equiv \hat{\eta} \qquad -\frac{\hat{E}_x^-(z)}{\hat{H}_y^-(z)} = \sqrt{\frac{\mu}{\epsilon - j\dfrac{\sigma}{\omega}}} \equiv \hat{\eta}\ \Omega \tag{3-97}$$

with the complex ratios denoted by $\hat{\eta}$, the *intrinsic wave impedance*. The field $\hat{H}_y(z)$ is thus written in terms of the solutions $\hat{E}_x^+(z)$ and $\hat{E}_x^-(z)$ in (3-91) as follows:

$$\hat{H}_y(z) = \hat{H}_y^+(z) + \hat{H}_y^-(z) \tag{3-98a}$$

$$\hat{H}_y(z) = \frac{\hat{E}_x^+(z)}{\hat{\eta}} - \frac{\hat{E}_x^-(z)}{\hat{\eta}} \tag{3-98b}$$

$$\hat{H}_y(z) = \frac{\hat{E}_m^+}{\hat{\eta}} e^{-\gamma z} - \frac{\hat{E}_m^-}{\hat{\eta}} e^{\gamma z}\ \text{A/m} \tag{3-98c}$$

The intrinsic wave impedance defined by (3-97) can be expressed in complex polar form as follows:

$$\hat{\eta} \equiv \sqrt{\frac{\mu}{\epsilon - j\dfrac{\sigma}{\omega}}} = \frac{\sqrt{\dfrac{\mu}{\epsilon}}}{\left[1 + \left(\dfrac{\sigma}{\omega\epsilon}\right)^2\right]^{1/4}} e^{j(1/2)\arctan(\sigma/\omega\epsilon)}\ \Omega \tag{3-99a}$$

seen to be of the form

$$\hat{\eta} = \eta e^{j\theta}\ \Omega \tag{3-99b}$$

with η and θ taken to mean

$$\eta = \frac{\sqrt{\dfrac{\mu}{\epsilon}}}{\left[1 + \left(\dfrac{\sigma}{\omega\epsilon}\right)^2\right]^{1/4}} \qquad \theta = \tfrac{1}{2}\arctan\frac{\sigma}{\omega\epsilon}$$

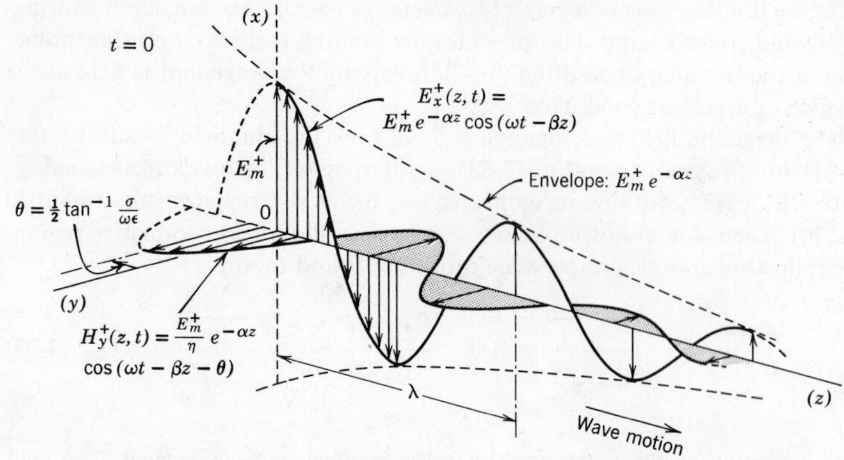

Figure 3-18. Positive *z* traveling fields of a uniform plane wave in a conductive region, shown at $t = 0$.

Evidently letting $\sigma = 0$ reduces $\hat{\eta}$ to the real result $\sqrt{\mu/\epsilon}$, applicable to uniform plane waves in a nonconductive (perfect dielectric) region. Moreover, the positive phase angle θ associated with $\hat{\eta}$ means that $\hat{H}_y^+(z)$ lags the accompanying $\hat{E}_x^+(z)$ in time phase, as shown in the real-time sketches of Figure 3-18. The *crank* method for simulating the motion of the wave with increase in the time variable t is depicted in Figure 3-19.

The additional characteristics of plane wave propagation in a conductive region are the wavelength defined in (2-123) (setting $\beta\lambda = 2\pi$ rad), yielding

$$\lambda = \frac{2\pi}{\beta} \text{ m} \tag{3-100}$$

the phase velocity v_p, obtained by putting the argument of (3-94) equal to a constant and differentiating in time, to obtain

$$v_p = \frac{\omega}{\beta} \text{ m/sec} \tag{3-101}$$

and the period

$$T = \frac{1}{f} \text{ sec} \tag{3-102}$$

The applicable value of β is of course that of (3-90b).

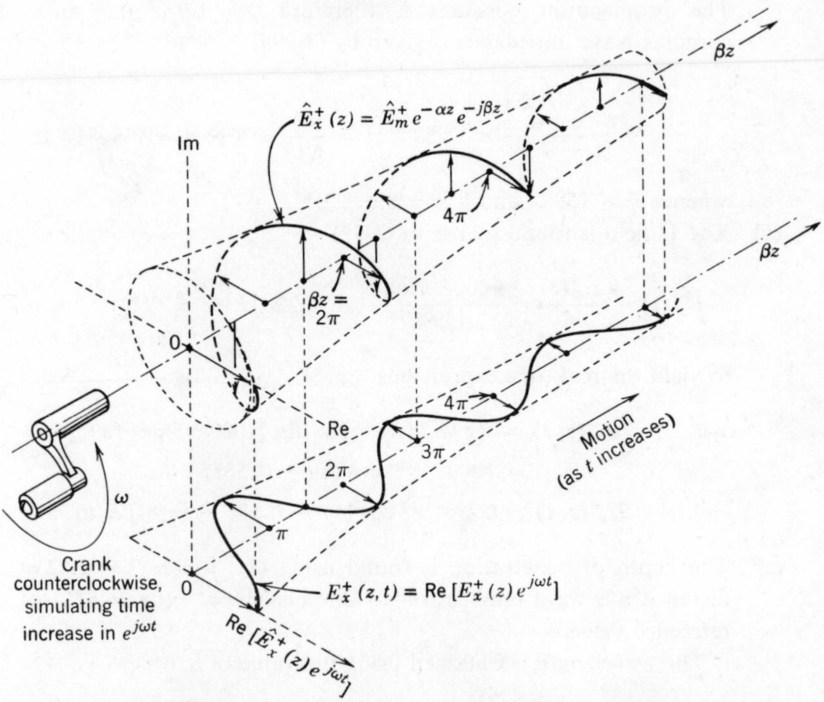

Figure 3-19. **Wave of $\hat{\mathbf{E}}(z, t)$ showing the complex phasors displayed along z at $t = 0$, and its real-time projection below.**

EXAMPLE 3-8. Suppose a uniform plane wave with the amplitude $1000e^{j0°}$ V/m propagates in the $+z$ direction at $f = 10^8$ Hz in a conductive region having the constants $\mu = \mu_0$, $\epsilon = 4\epsilon_0$, $\sigma/\omega\epsilon = 1$. (*a*) Find β, α, and $\hat{\eta}$ for the wave. (*b*) Find the associated **H** field, and sketch the wave along the z axis at $t = 0$. (*c*) Find the depth of penetration, the wavelength, and the phase velocity. Compare λ and v_p with their values in a *lossless* ($\sigma = 0$) region having the same μ and ϵ values. Assume only E_x and H_y components for the wave.

(*a*) The attenuation and phase factors are given by (3-90a) and (3-90b)

$$\alpha = \frac{\omega\sqrt{\mu\epsilon}}{\sqrt{2}}\left[\sqrt{1 + \left(\frac{\sigma}{\omega\epsilon}\right)^2} - 1\right]^{1/2} = \frac{\omega\sqrt{\mu_0 4\epsilon_0}}{\sqrt{2}}[\sqrt{1 + 1^2} - 1]^{1/2}$$

$$= \frac{2\omega}{\sqrt{2}c}[0.414]^{1/2} = 1.90 \text{ Np/m} \tag{1}$$

$$\beta = \frac{2\omega}{\sqrt{2}c}[2.414]^{1/2} = 4.58 \text{ rad/m} \tag{2}$$

The propagation constant is therefore $\gamma = 1.9 + j4.58 \text{ m}^{-1}$. The complex wave impedance is given by (3-99a)

$$\hat{\eta} = \frac{\frac{1}{2}\sqrt{\frac{\mu_0}{\epsilon_0}}}{[1 + 1^2]^{1/4}} e^{j(1/2)\arctan 1} = \frac{60\pi}{1.19} e^{j(1/2)(\pi/4)} = 159 e^{j(\pi/8)} \ \Omega \qquad (3)$$

whence $\eta = 159 \ \Omega$ and $\theta = \pi/8$ or $22.5°$.

(*b*) The $\hat{\mathbf{H}}$ field is found by use of (3-97)

$$\hat{H}_y^+(z) = \frac{\hat{E}_x^+(z)}{\hat{\eta}} = \frac{1000 e^{-\alpha z} e^{-j\beta z}}{159 e^{j\pi/8}} = 6.29 e^{-1.9z} e^{-j[4.58z + (\pi/8)]} \text{ A/m} \qquad (4)$$

to yield the real-time expressions

$$E_x^+(z, t) = \text{Re}\,[\hat{E}_x^+(z) e^{j\omega t}] = \text{Re}\,[1000 e^{-\alpha z} e^{-j\beta z} e^{j\omega t}]$$

$$= 1000 e^{-1.9z} \cos(\omega t - 4.58z) \text{ V/m} \qquad (5)$$

$$H_y^+(z, t) = 6.29 e^{-1.9z} \cos[\omega t - 4.58z - (\pi/8)] \text{ A/m} \qquad (6)$$

(*c*) The depth of penetration is found using (1): $\delta = \alpha^{-1} = 0.52$ m, the distance the wave must travel to diminish to e^{-1} (or 36.8%) of any reference value.

The wavelength is obtained using the value of β

$$\lambda = \frac{2\pi}{\beta} = \frac{2\pi}{4.58} = 1.37 \text{ m}$$

comparing with that for a lossless region (μ_0, $4\epsilon_0$) as follows

$$\lambda^{(0)} = \frac{2\pi}{\beta^{(0)}} = \frac{2\pi}{2\pi f\sqrt{\mu_0 4\epsilon_0}} = \frac{c}{2f} = \frac{3 \times 10^8}{2(10^8)} = 1.5 \text{ m}$$

The effect of finite conductivity is thus to *foreshorten* the wavelength.

The phase velocity in the conductive region is

$$v_p = \frac{\omega}{\beta} = \frac{2\pi(10^8)}{4.58} = 1.37 \times 10^8 \text{ m/sec}$$

which compares with that in the lossless region as follows:

$$v_p^{(0)} = \frac{\omega}{\beta^{(0)}} = \frac{\omega}{\omega\sqrt{\mu_0 4\epsilon_0}} = \frac{c}{2} \cong 1.5 \times 10^8 \text{ m/sec}$$

Conduction thus serves to *slow down* v_p. The foregoing numerical results may be added to Figure 3-18 to provide a picture of the wave motion in the conductive region.

3-7 Classification of Conductive Media

Conductive materials can be classified with reference to the magnitude of the conduction current density term $\sigma \mathbf{E}$ relative to the displacement current density term $j\omega\epsilon \mathbf{E}$ appearing in Maxwell's relation (3-85)

$$\nabla \times \hat{\mathbf{H}} = \sigma \hat{\mathbf{E}} + j\omega\epsilon \hat{\mathbf{E}} = j\omega\left(\epsilon - j\frac{\sigma}{\omega}\right)\hat{\mathbf{E}} \qquad \text{[3-85b]}$$

Denoting the *complex permittivity*, $\epsilon - j\sigma/\omega$, in (3-85) by the symbol

$$\hat{\epsilon} \equiv \epsilon - j\frac{\sigma}{\omega} \text{ F/m} \qquad (3\text{-}103)$$

one may represent $\hat{\epsilon}$ in the complex plane as in Figure 3-20. The angle δ_d is called the *dissipation angle*, which vanishes for a lossless region. Its tangent, defined by

$$\tan|\delta_d| = \frac{\sigma}{\omega\epsilon} \qquad (3\text{-}104)$$

is called the *loss tangent*, or dissipation factor, of the material.

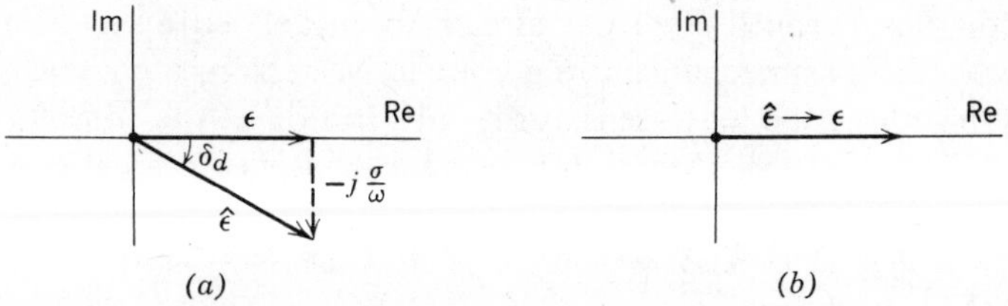

Figure 3-20. Complex permittivity for conductive and lossless regions. (*a*) Conductive region (general). (*b*) Lossless region ($\sigma \to 0$).

The importance of the loss tangent is recognized from its appearance in the expressions (3-90a) and (3-90b) for α and β, and in (3-99a) for the wave impedance of uniform plane waves propagating in a conductive material or a lossy dielectric at a given frequency.

Under impressed electric fields that are time-harmonic, the microscopic (atomic-scale) mechanisms contributing to the electric polarization $\mathbf{P}$ in a dielectric material, as discussed in Section 3-2, are often modified by damping (loss) effects. The classical model, inspired by experimental measurements on dielectric materials, assumes an oscillating system of interacting atomic or molecular particles, in which the response of the dielectric to the applied electric field involves damping mechanisms plus reasonances about certain frequencies. The damping is taken as proportional to the velocity of the particles oscillating under the impressed fields, to produce results similar in some ways to the conductivity mechanism discussed in Section 3-1 for the Drude model of a conductor when a time-harmonic field is applied. The resonances in the dielectric polarization arise from the inertia of the particles, displaced by the sinusoidal applied field and interacting with the restoring Coulomb forces. The response of the dielectric to the applied field resembles that of a three-dimensional system of masses interconnected through springs and dashpots and subjected to applied distributed vibrational forces, or analogously, a network of reactive and resistive circuit elements excited by sinusoidal voltages, with maximum losses occurring at the resonant frequencies. For typical dielectric materials, the lowest resonance is usually in or above the microwave range, with higher resonances occurring in the optical range.† The large-scale or macroscopic effect of these interaction phenomena is observed experimentally to make the permittivity of a dielectric become *complex* at a given excitation frequency, to permit writing it in terms of its real and imaginary parts

$$\hat{\epsilon} = \epsilon' - j\epsilon'' \tag{3-105}$$

Since a complex permittivity $\hat{\epsilon}$ has already been defined by (3-103) in connection with the loss mechanism in a conductive region, a comparison with (3-105) is in order. One may see that the substitution of the complex permittivity $\hat{\epsilon}$ of (3-105) into the Maxwell relation (3-85a) yields

$$\nabla \times \hat{\mathbf{H}} = j\omega\hat{\epsilon}\hat{\mathbf{E}} = j\omega(\epsilon' - j\epsilon'')\hat{\mathbf{E}} = \omega\epsilon''\hat{\mathbf{E}} + j\omega\epsilon'\hat{\mathbf{E}} \tag{3-106}$$

The conduction current density $\hat{\mathbf{J}} = \sigma\hat{\mathbf{E}}$ is omitted, since comparing (3-106) with the form of (3-85)a shows that an *equivalent conduction loss* mechanism is already accounted for by the term $\omega\epsilon''\hat{\mathbf{E}}$ in (3-106). Thus $\omega\epsilon''$ assumes the role of conductivity σ in a lossy dielectric, while ϵ' in the last term of (3-106) is identical with the real ϵ in (3-85a), corresponding to electric field energy

† Details of damping and resonance phenomena in dielectrics from the microscopic point of view and using classical or quantum-theory approaches are found in von Hippel, A. R. *Dielectric Materials and Applications*. Cambridge, Mass.: M.I.T. Press and New York: Wiley, 1954; and R. S. Elliott. *Electromagnetics*. New York: McGraw-Hill, 1966, Chapter 6. A brief digest is to be found in Ramo, S., J. R. Whinnery, and T. Van Duzer. *Fields and Waves in Communication Electronics*. New York: John Wiley, 1965, pp. 330–334.

storage in the dielectric. With the equalities

$$\epsilon = \epsilon' \qquad \sigma = \omega\epsilon'' \tag{3-107}$$

the complex permittivities expressed by (3-103) and (3-105) therefore have equivalent meanings. It is evident that ϵ'', the imaginary part of the complex $\hat{\epsilon}$, is descriptive of *all* loss mechanisms in the dielectric at a given frequency.

With the replacements (3-107), the loss tangent $\tan|\delta_d|$ of (3-104) is written in the equivalent forms

$$\tan|\delta_d| = \frac{\sigma}{\omega\epsilon} = \frac{\epsilon''}{\epsilon'} \tag{3-108}$$

Denoting the loss tangent by ϵ''/ϵ', (3-90a) for the wave attenuation in a *lossy dielectric* becomes

$$\alpha = \frac{\omega\sqrt{\mu\epsilon}}{\sqrt{2}}\left[\sqrt{1+\left(\frac{\epsilon''}{\epsilon'}\right)^2} - 1\right]^{1/2} \text{ Np/m} \tag{3-109}$$

Corresponding expressions for the phase constant β and the complex wave impedance $\hat{\eta}$ are obtained from (3-90b) and (3-99a), yielding

$$\beta = \frac{\omega\sqrt{\mu\epsilon}}{\sqrt{2}}\left[\sqrt{1+\left(\frac{\epsilon''}{\epsilon'}\right)^2} + 1\right]^{1/2} \text{ rad/m} \tag{3-110}$$

$$\hat{\eta} = \frac{\sqrt{\dfrac{\mu}{\epsilon}}}{\left[1+\left(\dfrac{\epsilon''}{\epsilon'}\right)^2\right]^{1/4}} e^{j(1/2)\arctan(\epsilon''/\epsilon')} \ \Omega \tag{3-111}$$

From the foregoing it is concluded that the characterization of the loss tangent (3-108) by $\sigma/\omega\epsilon$ is better suited to a conductor, while the form ϵ''/ϵ' is more desirable for a dielectric region.

A conductive material or a lossy dielectric supporting electromagnetic waves at a frequency ω may in general fall within one of the following three classifications: (*a*) it is a good conductor if the conductivity σ is sufficiently great that its loss tangent, $\sigma/\omega\epsilon$, becomes very large compared with unity (i.e., $\sigma/\omega\epsilon \gg 1$); (*b*) it is a good insulator if its loss tangent is sufficiently small ($\epsilon''/\epsilon' \ll 1$); and (*c*) it may be called moderately conductive (semiconducting) if it falls somewhere between these extremes, i.e., if the loss tangent is roughly of the order of unity. The expressions for the attenuation constant, phase constant, and instrinsic wave impedance associated with uniform plane wave propagation in such regions simplify to the following, for the classifications (*a*) and (*b*):

1. For a *good conductor*, assuming $\sigma/\omega\epsilon \gg 1$, (3-90a), (3-90b), and (3-99a) reduce to

$$\alpha = \sqrt{\frac{\omega\mu\sigma}{2}} \tag{3-112a}$$

$$\beta = \sqrt{\frac{\omega\mu\sigma}{2}} \tag{3-112b}$$

$$\hat{\eta} = (1 + j)\sqrt{\frac{\omega\mu}{2\sigma}} \tag{3-112c}$$

2. For a *lossy dielectric*, if $\epsilon''/\epsilon' \ll 1$, (3-109) through (3-111) become

$$\alpha = \frac{\omega\sqrt{\mu\epsilon}}{2}\left(\frac{\epsilon''}{\epsilon'}\right) \tag{3-113a}$$

$$\beta = \omega\sqrt{\mu\epsilon}\left[1 + \frac{1}{8}\left(\frac{\epsilon''}{\epsilon'}\right)^2\right] \tag{3-113b}$$

$$\hat{\eta} = \sqrt{\frac{\mu}{\epsilon}}\left[1 - \frac{3}{8}\left(\frac{\epsilon''}{\epsilon'}\right)^2 + j\frac{1}{2}\left(\frac{\epsilon''}{\epsilon'}\right)\right] \tag{3-113c}$$

The latter are obtained by including only the first two terms of the binomial expansions of the square root quantities in (3-109) through (3-111), assuming a very small loss tangent. In the limiting case of a lossless dielectric, (3-113) reduce to $\alpha = 0$, $\beta = \omega\sqrt{\mu\epsilon}$, and $\hat{\eta} = \sqrt{\mu/\epsilon}$ as expected.

Note in view of (3-95) that the inverse of (3-112a) can be used to express the depth of penetration, δ, in a *good conductor*

$$\delta = \frac{1}{\alpha} = \sqrt{\frac{2}{\omega\mu\sigma}} \text{ m} \tag{3-114}$$

a result inversely dependent on the square root of the frequency, the permeability, and the conductivity of the material. Thus, for copper having a conductivity of 5.8×10^7 ℧/m with $\mu = \mu_0$, the skin depth at 1000 Hz is about 2 mm, while at a frequency 10^6 times as large ($f = 1000$ MHz), δ is reduced by the factor 10^{-3}, becoming 0.002 mm.

3-8 Linearity, Homogeneity, and Isotropy in Materials

Electric and magnetic polarization effects in materials have been accounted for by the polarization field **P** and the magnetization field **M**, defined by (3-16) and (3-55), respectively. Their additive effects, yielding the Maxwell relations for a material region, have provided the definitions of the fields **D** and **H** by (3-23) and (3-58)

$$\mathbf{D} = \epsilon_0 \mathbf{E} + \mathbf{P} \tag{3-115}$$

$$\mathbf{B} = \mu_0 \mathbf{H} + \mu_0 \mathbf{M} \tag{3-116}$$

If the region is also capable of transporting free charges in a conduction process, the conductivity parameter σ assigned to the region expresses the proportionality of the current density **J** to the applied field **E** by (3-7)

$$\mathbf{J} = \sigma \mathbf{E} \tag{3-117}$$

Avoiding the questions of anisotropy for the moment, one may recall that the relations of the electric polarization field **P** and the magnetizations **M** to the applied fields may be expressed by (3-25) and (3-60)

$$\mathbf{P} = \chi_e \epsilon_0 \mathbf{E} \qquad [3\text{-}25]$$

$$\mathbf{M} = \chi_m \mathbf{H} \qquad [3\text{-}60]$$

With the foregoing as a background, the questions of linearity, homogeneity, and isotropy in material media are discussed. In this framework, one should assume that the temperature of the material and the sinusoidal frequency of its impressed fields are constants when defining its parameters μ, ϵ, and σ, since the dependence of the latter on temperature and frequency cannot, in general, be ignored.

A. Linearity and Nonlinearity in Materials

If the susceptibilities χ_e and χ_m are constants and thus independent of the applied fields, the material is said to be *linear* with respect to electric and magnetic polarization effects. A straight-line relationship between an applied field component E_x and the resulting polarization component P_x characterizes this linearity property. With the substitution of (3-25) and (3-60) into (3-115) and (3-116), the compact results (3-30c) and (3-64c)

$$\mathbf{D} = \epsilon \mathbf{E} \qquad [3\text{-}30c]$$

$$\mathbf{B} = \mu \mathbf{H} \qquad [3\text{-}64c]$$

have been seen to result.

Nonlinearity in a material is characterized by one or more of the parameters, μ, ϵ, and σ, being dependent on the *level* of the applied fields. Then one may choose to write (3-30c), (3-64c), and (3-117) in forms signifying this dependence

$$\mathbf{D} = \epsilon(E)\mathbf{E} \tag{3-118}$$

$$\mathbf{B} = \mu(H)\mathbf{H} \tag{3-119}$$

$$\mathbf{J} = \sigma(E)\mathbf{E} \tag{3-120}$$

An example of (3-119) is depicted by the multivalued *B-H* curve of a ferromagnetic material in Figure 3-13(*b*).

*B. Isotropy and Anisotropy in Materials

In some physical materials such as crystalline substances possessing a well-ordered atomic or molecular lattice throughout a given sample, the polarizations **P** or **M** resulting from the application of an **E** or **B** field may not necessarily have the same directions as the applied fields. Such materials are termed *anisotropic*†, meaning that different values of μ, ϵ, or σ are measurable in different directions within the substance. Differences in the polarization responses to the direction of an applied **E** field in crystals, for example, are due to the disparities in the interatomic spacings associated with the several symmetry axes of the crystalline lattice. In some crystals, in which three orthogonal principal axes may be identified, the cartesian coordinates can be chosen along the same axes. Then, for an applied field $\mathbf{E} = \mathbf{a}_x E_x + \mathbf{a}_y E_y + \mathbf{a}_z E_z$, the components of the electric polarization field **P** become

$$\begin{aligned} P_x &= \chi_{e11}(\epsilon_0 E_x) \\ P_y &= \chi_{e22}(\epsilon_0 E_y) \\ P_z &= \chi_{e33}(\epsilon_0 E_z) \end{aligned} \tag{3-121}$$

in which the susceptibility components χ_{e11}, χ_{e22}, and χ_{e33} are generally different. (The static field values for gypsum, for example, are about 8.9, 4.1, and 4.0, respectively.) These circumstances are depicted in Figure 3-21(*a*), showing the development of a polarization vector **P** having a direction different from that of the applied **E** field, a result of the unequal susceptibilities associated with the coordinate directions. It is evident that (3-121) reduces to the vector result (3-25), $\mathbf{P} = \chi_e \epsilon_0 \mathbf{E}$, whenever $\chi_{e11} \equiv \chi_{e22} \equiv \chi_{e33}$. Again, if rectangular coordinates are selected so that the applied field has

† From the Greek *an* (not), plus *iso* (same), plus *trope* (turning); hence, not having the same (property) with different directions.

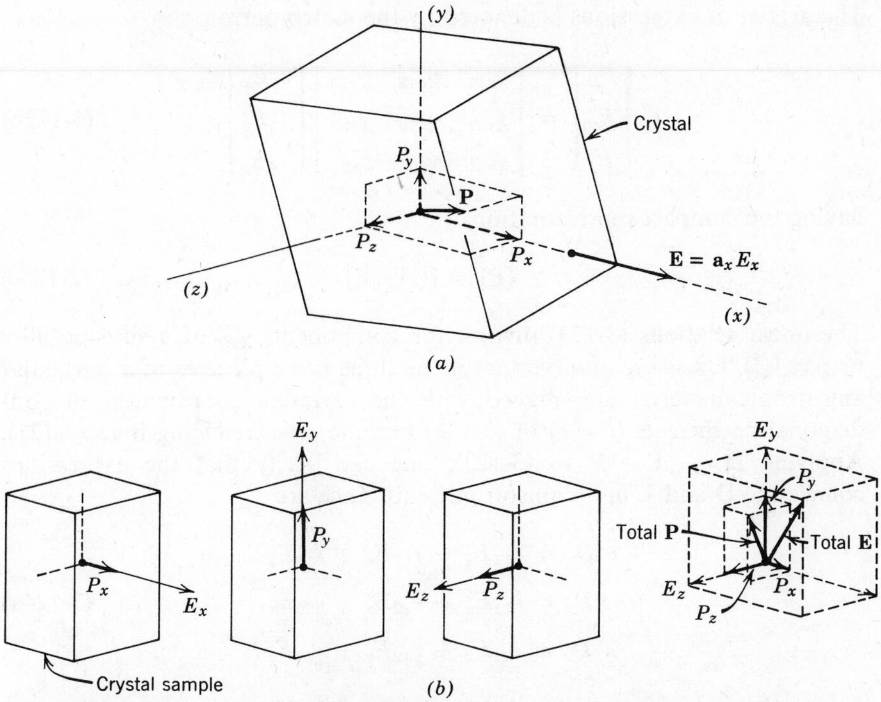

Figure 3-21. **Aspects of dielectric anisotropy in a crystal. (*a*) Polarization components resulting from an *x* directed E field component applied to an arbitrarily oriented crystal. (*b*) Polarization components resulting from applied E field components, if principal axes in a crystal are aligned with the cartesian coordinate axes.**

only an x component, i.e., $\mathbf{E} = \mathbf{a}_x \mathbf{E}_x$, applying it to an arbitrarily oriented anisotropic material yields all three components of dielectric polarization

$$\mathbf{P} = \mathbf{a}_x \chi_{e11}(\epsilon_0 E_x) + \mathbf{a}_y \chi_{e21}(\epsilon_0 E_x) + \mathbf{a}_z \chi_{e31}(\epsilon_0 E_x) \tag{3-122}$$

a result exemplified in Figure 3-21(*b*). In general, if **E** possesses all three rectangular components, applied at an arbitrary angle with respect to the crystal principle axes, one must write

$$\begin{aligned} P_x &= \chi_{e11}(\epsilon_0 E_x) + \chi_{e12}(\epsilon_0 E_y) + \chi_{e13}(\epsilon_0 E_z) \\ P_y &= \chi_{e21}(\epsilon_0 E_x) + \chi_{e22}(\epsilon_0 E_y) + \chi_{e23}(\epsilon_0 E_z) \\ P_z &= \chi_{e31}(\epsilon_0 E_x) + \chi_{e32}(\epsilon_0 E_y) + \chi_{e33}(\epsilon_0 E_z) \end{aligned} \tag{3-123a}$$

This triplet of expressions is denoted by the matrix form

$$\begin{bmatrix} P_x \\ P_y \\ P_z \end{bmatrix} = \begin{bmatrix} \chi_{e11} & \chi_{e12} & \chi_{e13} \\ \chi_{e21} & \chi_{e22} & \chi_{e23} \\ \chi_{e31} & \chi_{e32} & \chi_{e33} \end{bmatrix} \begin{bmatrix} \epsilon_0 E_x \\ \epsilon_0 E_y \\ \epsilon_0 E_z \end{bmatrix} \tag{3-123b}$$

having the compact representation

$$[\mathbf{P}] = [\chi_e][\epsilon_0 \mathbf{E}] \tag{3-123c}$$

The linear relations (3-123) involve the components χ_{eij} of a susceptibility matrix $[\chi_e]$. One may observe that if the three principle axes of a particular anisotropic material are aligned with the cartesian coordinates, the off-diagonal coefficients ($i \neq j$) of (3-123) become zero, reducing it to (3-121). Applying $\mathbf{D} = \epsilon_0 \mathbf{E} + \mathbf{P}$ to (3-123), one can verify that the expressions connecting **D** and **E** in an anisotropic substance are

$$\begin{aligned} D_x &= \epsilon_{11} E_x + \epsilon_{12} E_y + \epsilon_{13} E_z \\ D_y &= \epsilon_{21} E_x + \epsilon_{22} E_y + \epsilon_{23} E_z \\ D_z &= \epsilon_{31} E_x + \epsilon_{32} E_y + \epsilon_{33} E_z \end{aligned} \tag{3-124a}$$

having the matrix form

$$[\mathbf{D}] = [\epsilon][\mathbf{E}] \tag{3-124b}$$

It should be evident that expressions analogous to (3-123) and (3-124) can be established among the cartesian components of the vector **B** and **H** for anisotropic *magnetic* materials.

C. Homogeneity and Inhomogeneity in Materials

A material region having parameters μ, ϵ, and σ independent of the position in it is termed *homogeneous*.

Conversely if one or more of the parameters is of the space-dependent form:

$$\begin{aligned} \epsilon &= \epsilon(u_1, u_2, u_3) \\ \mu &= \mu(u_1, u_2, u_3) \\ \sigma &= \sigma(u_1, u_2, u_3) \end{aligned} \tag{3-125}$$

the material is said to be *inhomogeneous*. The mixture of earth and water occurring near the surface after a rain is an instance of an inhomogeneous

Table 3-3 Material parameters at 20°C (unless otherwise stated)

A. Nonmetals

Material	μ_r	ϵ_r, at frequency 60	ϵ_r, at frequency 10^6	ϵ_r, at frequency 10^{10}	$\frac{\epsilon''}{\epsilon'}$, at frequency 60	$\frac{\epsilon''}{\epsilon'}$, at frequency 10^6	$\frac{\epsilon''}{\epsilon'}$, at frequency 10^{10}	Dielectric strength (V/mil)
Bakelite, BM 120	1	4.87	4.74	3.68	0.080	0.028	0.0410	300
Douglas fir	1	2.05	1.93	1.80	0.004	0.026	0.030	—
Micarta 254	1	5.45	4.51	3.30	0.098	0.036	0.0400	1,020
Nylon (Dupont)	1	3.60	3.14	2.80	0.018	0.022	0.0110	400
Plexiglas	1	3.45	2.76	2.50	0.064	0.014	0.0050	990
Polyethylene	1	2.26	2.26	2.26	(<0.0002)		0.0005	1,200
Polystyrene (Dow)	1	2.55	2.55	2.54	(<0.003)		0.0003	600
Silicone fluid SC 200	1	2.78	2.78	2.74	0.0001	0.0003	0.010	—
Soil, sandy, dry	1	3.45	2.60	2.50	0.200	0.020	0.0040	—
Soil, sandy, 2.2% H_2O	1	3.25	2.50	2.50	0.700	0.025	0.0650	—
Styrofoam 103.7	1	1.03	1.03	1.03	(<0.0002)		0.0001	—
Tam Ticon B (barium titanate)	1	1240	1140	150	0.056	0.010	0.60	—
Teflon (22°C)	1	2.10	2.10	2.10	(<0.005)		0.0004	1,500
Teflon (100°C)	1	2.04	2.04	2.04	(<0.001)		0.0005	—
Water, distilled	1	81	78.2	50	—	0.040	0.200	—

B. Metals

Material	μ_r	ϵ_r	σ(℧/m)	Depth of penetration δ for plane waves (m)
Silver	1	1	6.17×10^7	$0.064/\sqrt{f}$
Copper	1	1	5.8×10^7	$0.066/\sqrt{f}$
Aluminum	1	1	3.72×10^7	$0.083/\sqrt{f}$
Sodium	1	1	2.1×10^7	$0.11/\sqrt{f}$
Brass	1	1	1.6×10^7	$0.13/\sqrt{f}$
Tin	1	1	0.87×10^7	$0.17/\sqrt{f}$
Graphite	1	1	0.01×10^7	$1.6/\sqrt{f}$

region having parameters ϵ and σ that vary with depth. The ionosphere, a gaseous mixture of positive, negative, and neutral particles, must be regarded generally as electromagnetically inhomogeneous. Artificial inhomogeneous materials are created by the variable spacing of small, metal spheres within a styrofoam or other supporting insulating material, to yield an electrically polarizable region having a variable ϵ depending on the average spatial densities of the spheres. Devices constructed in this way, using metal spheres, rods, or plates, have been used in artificial lenses for microwave applications.†

The complications of nonlinearity, inhomogeneity, and anisotropy in materials are for the most part avoided in subsequent treatments in this text. The emphasis is restricted essentially to discussions of electric and magnetic fields in materials that are linear, homogeneous, and isotropic.

3-9 Electromagnetic Parameters of Typical Materials

A tabulation of measured parameters at room temperature for typical nonmetals and nonferrous metals is given in Table 3-3. The frequency dependence of ϵ_r and the loss tangent ϵ''/ϵ' for nonmetals is evident from their values at the three widely different sinusoidal frequencies listed. Laboratory methods for measuring material parameters differ considerably depending on the frequency at which the parameters are to be determined. The permittivity and loss tangent of a nonmetal at frequencies up to several megahertz can be found using lumped-circuit methods; thus, a parallel-plate capacitor utilizing the test material as a dielectric and connected in a Q-meter arrangement might be employed. At higher frequencies, the measurement of the influence of the material on the wave transmission properties of a coaxial transmission line or a waveguide can be useful for obtaining its parameters. Several source books may be consulted for further information on such methods.‡

REFERENCES

ELLIOTT, R. S. *Electromagnetics*. New York: McGraw-Hill, 1966.

JAVID, M., and P. M. BROWN. *Field Analysis and Electromagnetics*. New York: McGraw-Hill, 1963.

JORDAN, E. C., and K. G. BALMAIN. *Electromagnetic Waves and Radiating Systems*, 2nd ed., Englewood Cliffs, N.J.: Prentice-Hall, 1968.

† For example, see Kock, W. E. "Metallic delay lenses," *Bell Syst. Tech. Jour.*, **27**, 58, January 1948.

‡ For example, see von Hippell, A. R. *Dielectrics and Waves*. New York: Wiley, 1964.

LORRAIN, P., and D. R. CORSON. *Electromagnetic Fields and Waves*. San Francisco: W. H. Freeman, 1970.
REITZ, R., and F. J. MILFORD. *Foundations of Electromagnetic Theory*. Reading, Mass.: Addison-Wesley, New York: 1960.
STRATTON, J. A. *Electromagnetic Theory*. New York: McGraw-Hill, 1941.

PROBLEMS

3-1. Suppose an electric field applied in a region has the value $\mathbf{E} = \mathbf{a}_x 1$ V/m, a relatively weak field. Find the polarization field **P**, if the material filling the region has the dielectric susceptibility:
 (*a*) Zero
 (*b*) 0.01
 (*c*) 10
 (*d*) 1000
 What is the relative permittivity of each region?

3-2. Find the polarization charge density ρ_p in a region in which the following fields exist. (K is a constant.)
 (*a*) $\mathbf{P} = \mathbf{a}_x K$
 (*b*) $\mathbf{P} = \mathbf{a}_x Kx$
 (*c*) $\mathbf{P} = \mathbf{a}_y Kz$
 (*d*) $\mathbf{P} = \mathbf{a}_r K/r^2$
 (*e*) $\mathbf{P} = \mathbf{a}_r Kr$
 (*f*) $\mathbf{P} = \mathbf{a}_x K \sin \omega t$.

3-3. Use physical reasoning to explain why a *constant* polarization field **P** in a region corresponds to a *zero* volume polarization charge in every volume-element of that region.

3-4. Using the divergence theorem, convert the differential equation (3-21) for div **P** into its integral form

$$\oint_S \mathbf{P} \cdot d\mathbf{s} = -\int_V \rho_P \, dv$$

3-5. A very long coaxial conductor pair contains a concentric, homogeneous dielectric sleeve with a permittivity ϵ as shown. Air fills the remaining regions between the conductors. Assuming positive and negative surface charges $\pm Q$ C are on every axial length ℓ of the inner and outer conductors respectively, determine the following, making use of the symmetry. In each region between the conductors, find:
 (*a*) **D**
 (*b*) **E**
 (*c*) The **P** field
 (*d*) Find ρ_s on the conductor surfaces, and the surface polarization charge density at $\rho = b$ and $\rho = c$.

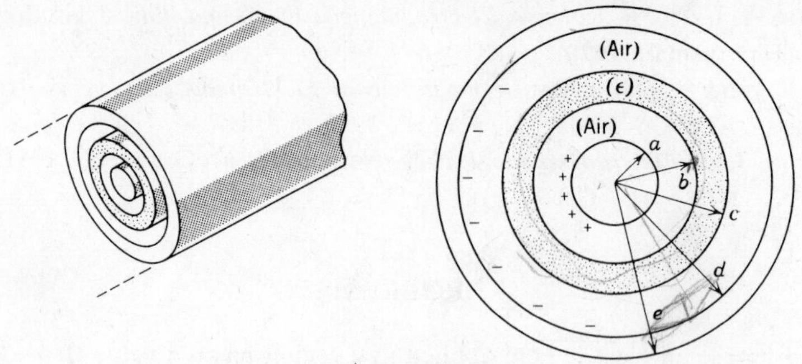

Problem 3-5

(e) If $a = 1$ cm, $b = 2$ cm, $c = 3$ cm, $d = 4$ cm, $e = 4.2$ cm, $Q/\ell = 10^{-2}\mu$ C/m, and $\epsilon = 2.1\epsilon_0$ (Teflon), find the values of **D**, **E**, and **P** at the inner surface $\rho = b$ just inside the dielectric.
[*Answer:* $\mathbf{a}_\rho$ 0.0796 μC/m², $\mathbf{a}_\rho$ 4.29 kV/m, $\mathbf{a}_\rho$ 0.042 μC/m²]

3-6. Four material substances have the following magnetic susceptibilities; (1) zero; (2) 0.01; (3) 10; (4) 10,000. What is the *relative* permeability, as well as the permeability, of each? If the field $\mathbf{H} = \mathbf{a}_x 1$ A/m exists in each material, what **B** field and magnetization field are found there?

3-7. Find the magnetization current density $\mathbf{J}_m$ within materials in which the following magnetization fields **M** exist

(a) $\mathbf{M} = \mathbf{a}_x K$
(b) $\mathbf{M} = \mathbf{a}_x Kx$
(c) $\mathbf{M} = \mathbf{a}_x Ky$
(d) $\mathbf{M} = \mathbf{a}_\rho K$
(e) $\mathbf{M} = \mathbf{a}_\rho K\rho$
(f) $\mathbf{M} = \mathbf{a}_r K$
(g) $\mathbf{M} = \mathbf{a}_r Kr$
(h) $\mathbf{M} = \mathbf{a}_r K/r$
(i) $\mathbf{M} = \mathbf{a}_r K/r^2$

3-8. A circular iron core, rectangular in cross-section, partially fills a closely wound, toroidal winding of n turns as shown in the accompanying figure.

(a) In which sense is the **H** field directed inside the toroidal region embraced by the winding?

(b) From the viewpoint of the tangential boundary conditions, comment on the continuity (or otherwise) of the **H** and **B** fields, in passing from the air region into the iron region at the radius $\rho = b$.

(c) Find **H**, **B**, and **M** in the air and iron regions. Sketch flux plots showing the relative densities of **H**, $\mathbf{B}/\mu_0$, and **M** in the two regions (using three side views of the system), assuming μ_r of the iron large compared with unity.

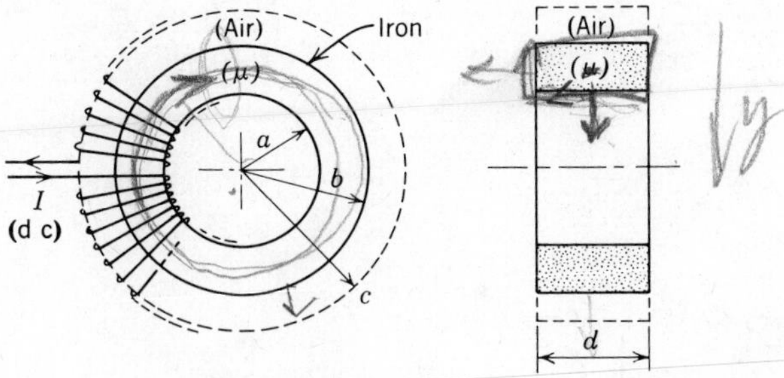

Problem 3-8

(*d*) Find $\mathbf{J}_m$ in the iron core as well as $\mathbf{J}_{sm}$ on the four sides of the core. Sketch a suitable sectional view illustrating $\mathbf{J}_m$ and $\mathbf{J}_{sm}$ fluxes in and on the core.

(*e*) Assuming $a = 2$ cm, $b = 3$ cm, $c = 4$ cm, $d = 2$ cm, $\mu_r = 100$, $n = 100$ turns, $I = 1$ A (dc), compute **H** and **B** at the positions $\rho = a$ and b (just inside the iron), and at c.

3-9. Verify that the Maxwell's differential relations $\nabla \cdot \mathbf{B} = 0$ and $\nabla \times \mathbf{H} = \mathbf{J} + \partial \mathbf{D}/\partial t$ are satisfied by the static field solutions obtained in Problem 3-8(*c*).

3-10. Beginning with the force expression (3-51), fill in the remaining details to prove (3-54).

3-11. Derive the relations for α and β given by (3-90a) and (3-90b).

3-12. Given the travelling wave solutions for plane waves in a conductive region as follows:

$$\hat{E}_x(z) = \hat{E}_m^+ e^{-\gamma z} + \hat{E}_m^- e^{\gamma z}$$

in which $\gamma = \alpha + j\beta$, find $\hat{H}_y(z)$ using Maxwell's equations. From this verify that (3-97) are valid, i.e.,

$$\frac{\hat{E}_x^+(z)}{\hat{H}_y^+(z)} = \hat{\eta} \qquad \frac{\hat{E}_x^-(z)}{\hat{H}_y^-(z)} = -\hat{\eta}$$

if $\hat{\eta}$ is the intrinsic wave impedance of (3-99a).

3-13. Show that the expressions (3-90a), (3-90b), and (3-99a) reduce to the free-space values $\alpha = 0$, $\beta = \beta_0$, and $\hat{\eta} = \eta_0$ upon putting $\mu = \mu_0$, $\epsilon = \epsilon_0$, and $\sigma = 0$.

3-14. Prove that the penetration of three skin depths by a plane wave into a conductive region produces an amplitude reduction to 5% of the reference value. Show that six skin depths yields 0.25%.

3-15. A vehicle located far above the surface of the sea transmits an electromagnetic signal at the frequency *f*. Upon striking the air-sea interface, a transmitted wave penetrates the sea as suggested by the accompanying figure. The waves at the surface are presumed to be sufficiently far from

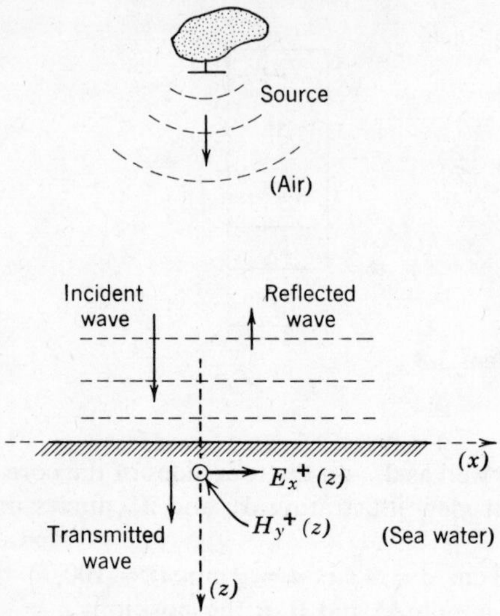

Problem 3-15

the source that they may, locally at least, be considered to be uniform plane waves.

Supposing the net transmitted electric field amplitude is $\hat{E}_m^+ = 1$ V/m, how far will the wave penetrate before reaching 5% of its surface value? Perform this calculation at two very low radio frequencies: 10 kHz (in the VLF range) and 1000 Hz (ELF), assuming sea water has the constants $\epsilon_r = 81$ and $\sigma = 4$ ℧/m at these frequencies. Comment on the effectiveness of undersea radio communication, based on your results.

3-16. By use of (3-97), derive its polar form given in (3-99a).

3-17. A uniform plane wave with the field components E_x and H_y has an electric field amplitude of $100e^{j0°}$ V/m and propagates at $f = 10^6$ Hz in a conductive region having the parameters μ_0, $9\epsilon_0$, and $(\sigma/\omega\epsilon) = 0.5$.

(*a*) Find the values of α, β, and $\hat{\eta}$ for this wave.

(*b*) Express the electric and magnetic fields in both their complex and real-time forms, with the numerical values of (*a*) inserted.

(*c*) What is the depth of penetration of this wave into the dissipative region? Determine the wavelength and the phase velocity of the wave, comparing the latter with values obtained if $\sigma = 0$ were assumed.

3-18. Use the general expressions (3-90a), (3-90b), and (3-99a) to derive (3-112), applicable to a good conductor ($\sigma/\omega\epsilon \gg 1$).

3-19. Assuming that copper has the material parameters given in Table 3-3(*B*), find the attenuation factor, phase factor, wave impedance, and depth of penetration of a plane wave traveling in copper at the frequencies (*a*) 60 Hz, (*b*) 1 MHz, and (*c*) 10 GHz.

3-20. From expression (3-114), show that at 1000 Hz the depth of penetration of a plane wave into copper is 2.1 mm, while at 1000 MHz (in the microwave range), δ becomes 2.1×10^{-3} mm. According to Table 3-3(*B*), graphite has a conductivity only 1/580 as large as that of copper. How does its plane wave depth of penetration compare with that of copper at these two frequencies?

3-21. Use the general expressions (3-109) through (3-111) as a basis for proving (3-113), the approximate expressions for α, β, and $\hat{\eta}$ associated with plane wave propagation in a dielectric, assuming $\epsilon''/\epsilon' \ll 1$.

3-22. A low-loss dielectric such as polyethylene is commonly used in flexible coaxial cables. For a uniform plane wave propagating in a sufficiently large sample of polyethylene, show that at 10^{10} Hz (in the microwave range), a thickness of about 12.7 m of the material is needed to reduce the wave amplitude to e^{-1} of the value it has at the input plane. Further show that at 60 Hz, the depth of material needed would be 5.32×10^{9} m. [*Note:* These results suggest why the dielectric losses of low-loss materials can often be neglected, particularly at lower frequencies.]

CHAPTER
4

Static and Quasi-Static Electric Fields

In this chapter, electric fields of stationary charge distributions in space are considered. Maxwell's equations, subjected to the time-static assumption, $\partial/\partial t = 0$, provide an uncoupling of the static electric fields from the static magnetic fields. Gauss' law is applied to symmetrical systems; and the scalar potential field Φ is derived to supply an intermediate, often simplifying, step useful for finding the static **E** field. Expressions for the stored energy of an electrostatic system are derived and applied to two-conductor capacitance systems. Image and flux-mapping techniques are discussed as alternative approaches to capacitance problems, and a capacitance-conductance analog is developed. Boundary-value problems of electrostatics are treated via Laplace's equation, and the chapter is concluded with a consideration of the forces of electric charge systems.

4-1 Maxwell's Equations for Static Electric Fields

In Chapter 3 the Maxwell equations and boundary conditions for time-varying electromagnetic fields in material media at rest were developed. Time-varying **B** (or **H**) and **E** (or **D**) fields are produced in a region whenever the charge and current *sources* of the fields are time-varying. For certain generic classes of field problems, it is advantageous to consider the sources non-time-varying, i.e., *time-static* (or just static). Then the charges and possibly currents responsible for the fields are stationary. The governing

Maxwell equations for time-static fields are (3-24), (3-48), (3-59), and (3-77) with the operator $\partial/\partial t$ set to zero, yielding $\nabla\cdot\mathbf{D} = \rho_v$, $\nabla\times\mathbf{E} = 0$, $\nabla\cdot\mathbf{B} = 0$, and $\nabla\times\mathbf{H} = \mathbf{J}$. The static fields are designated $\mathbf{D}(u_1, u_2, u_3)$, $\rho_v(u_1, u_2, u_3)$, and so on.

An inspection of the static Maxwell relations reveals a new property not valid for their more general, time-varying forms. Thus, the *static electric fields* **D** and **E** are governed solely by the divergence and curl properties

$$\nabla\cdot\mathbf{D} = \rho_v \tag{4-1}$$

$$\nabla\times\mathbf{E} = 0 \tag{4-2}$$

while the behavior of the *static magnetic fields* **B** and **H** is dictated by

$$\nabla\cdot\mathbf{B} = 0 \tag{4-3}$$

$$\nabla\times\mathbf{H} = \mathbf{J} \tag{4-4}$$

The coupling between the electric and magnetic field quantities, generally provided under time-varying conditions by the terms $-\partial\mathbf{B}/\partial t$ and $\partial\mathbf{D}/\partial t$ appearing in (3-77) and (3-59), is seen to be missing in these pairs of equations. The *sources* of *electrostatic fields* are, from the divergence expression (4-1), *static charges* of density ρ_v. *Magnetostatic fields*, on the other hand, have *static* (direct) *currents* for sources, as noted in (4-4). In the present chapter, solutions of the electrostatic field equations (4-1) and (4-2) are considered from several points of view, while a detailed discussion of magnetostatics by use of (4-3) and (4-4) is deferred until Chapter 5. The differential equations of electrostatics are, together with their integral forms and boundary conditions, given in Table 4-1. For a linear, homogeneous, and isotropic material, more-

Table 4-1 Maxwell's equations of electrostatics

Differential form	*Integral form*	*Boundary condition*
$\nabla\cdot\mathbf{D} = \rho_v$ [4-1]	$\oint_S \mathbf{D}\cdot d\mathbf{s} = q$ (4-5)	$D_{n1} - D_{n2} = \rho_s$ (4-7)
$\nabla\times\mathbf{E} = 0$ [4-2]	$\oint_\ell \mathbf{E}\cdot d\boldsymbol{\ell} = 0$ (4-6)	$E_{t1} - E_{t2} = 0$ (4-8)

over, (3-30c) is applicable

$$\mathbf{D} = \epsilon\mathbf{E} \tag{4-9}$$

4-2 Static Electric Fields of Fixed Charge Ensembles in Free Space

The Maxwell equations in Table 4-1 apply to fixed charges in free space, as well as to systems of dielectrics and conductors into (or onto) which charges have been introduced such that static equilibrium of the charge distribution has been reached. Examples of the applications of the Gauss law (4-5) are given in Section 1-9. One of the results, (1-58), is Coulomb's force law

$$\mathbf{F} = q'\mathbf{E} = \mathbf{a}_R \frac{qq'}{4\pi\epsilon_0 R^2} \text{ N} \tag{4-10a}$$

giving the force acting on q' in the presence of the field $\mathbf{E}$ produced by a second charge q as shown in Figure 4-1. The symbol R is used instead of the spherical coordinate variable r because the source q is not necessarily located at the origin 0. The field of q was deduced from Gauss' law in Section 1-9 to be

$$\mathbf{E} = \mathbf{a}_R \frac{q}{4\pi\epsilon_0 R^2} \text{ N/C} \quad \text{or} \quad \text{V/m} \tag{4-10b}$$

Thus (4-10a) is a special case of the Lorentz force law (1-52) in the absence of a magnetic field; i.e., $\mathbf{F} = q'\mathbf{E}$.

Maxwell's equations (4-1) and (4-2) are linear equations; therefore, any *sum* of its solutions in free space constitutes a solution. Suppose an aggregate of a point charges of arbitrary positive or negative strengths is located at fixed points P' as in Figure 4-2. The total electrostatic field at the field point P is the sum of n terms like (4-10b)

$$\mathbf{E} = \sum_{k=1}^{n} \mathbf{E}_k = \sum_{k=1}^{n} \mathbf{a}_{Rk} \frac{q_k}{4\pi\epsilon_0 R_k^2} \tag{4-10c}$$

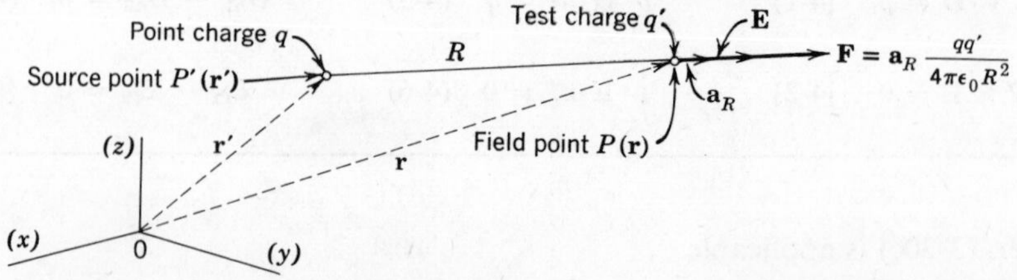

Figure 4-1. Illustrating quantities appearing in Coulomb's force law.

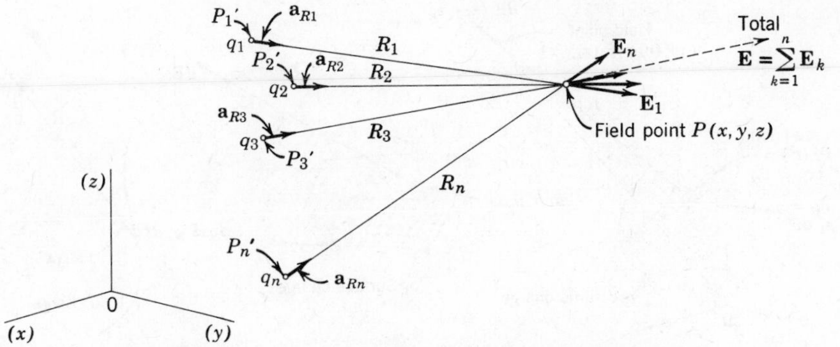

Figure 4-2. Electrostatic field of n discrete charges.

Moreover, if a charge q is placed at P, the force on it, from (1-52), becomes

$$\mathbf{F} = q \sum_{k=1}^{n} \mathbf{a}_{Rk} \frac{q_k}{4\pi\epsilon_0 R_k^2} \tag{4-10d}$$

If a system contains a large number of fixed charges, it is undesirable to use a summation like (4-10c) or (4-10d). It is preferable to replace the charge ensemble with a *function* representing the average charge density in every volume, surface, or line-element of the region. The symbols ρ_v, ρ_s, or ρ_ℓ have been used to denote these density functions as discussed in Section 1-9 relative to Figure 1-11. A continuum of charges distributed throughout some region with a density ρ_v thus possesses the charge $dq = \rho_v\, dv$ in every dv element. Generally ρ_v is a function of position and time, though for static fields, the variable t is missing.

With $dq = \rho_v\, dv'$ located at the source point $P'(x', y', z')$, the field $d\mathbf{E}$ at P due to dq is obtained from (4-10b), written

$$d\mathbf{E}(x, y, z) = \mathbf{a}_R \frac{\rho_v(x', y', z')}{4\pi\epsilon_0 R^2}\, dv' \tag{4-11}$$

The unit vector directed from the source point P' to the field point P to give the proper direction to $d\mathbf{E}$ is denoted by $\mathbf{a}_R$, while R is the scalar distance from P' to P, as in Figure 4-3. In rectangular coordinates

$$R = \sqrt{(x - x')^2 + (y - y')^2 + (z - z')^2} \tag{4-12}$$

The total static $\mathbf{E}$ field at P in Figure 4-3(a) is thus the volume integral of (4-11)

$$\mathbf{E}(x, y, z) = \int_V d\mathbf{E} = \int_V \mathbf{a}_R \frac{\rho_v(x', y', z')}{4\pi\epsilon_0 R^2}\, dv' \tag{4-13}$$

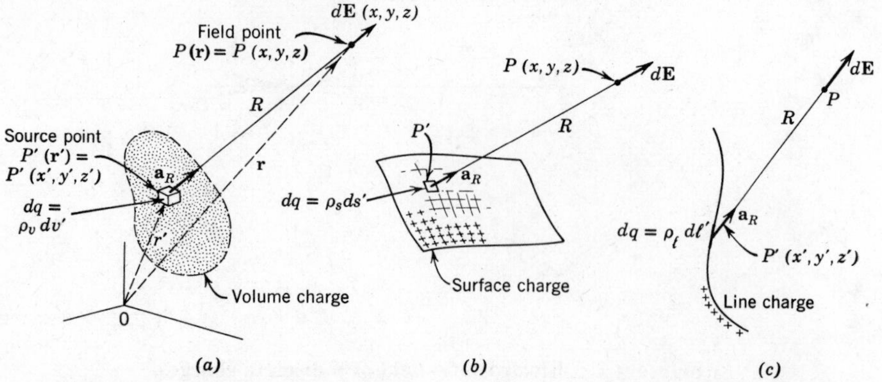

Figure 4-3. Geometries relative to the electrostatic field integrals in terms of volume, surface, and line charge distributions. (*a*) Volume charge distribution. (*b*) Surface charge distribution. (*c*) Line charge distribution.

If the static charges are distributed over a surface S or a line ℓ as in Figure 4-3(*b*) and (*c*), the following integrals apply:

$$\mathbf{E} = \int_S \mathbf{a}_R \frac{\rho_s(x', y', z')}{4\pi\epsilon_0 R^2} ds' \tag{4-14}$$

$$\mathbf{E} = \int_\ell \mathbf{a}_R \frac{\rho_\ell(x', y', z')}{4\pi\epsilon_0 R^2} d\ell' \tag{4-15}$$

The foregoing integrals are not always readily evaluated for charge distributions in space, mainly because the unit vector $\mathbf{a}_R$ changes in direction as the source point P ranges throughout the charge region. In some cases, the symmetry disposes of this problem, as in the following example.

EXAMPLE 4-1. An infinitely long, thin wire, shown in the accompanying figure, possesses a uniform charge of linear density ρ_ℓ C/m. Find $\mathbf{E}$ at the distance ρ from the charge by a direct integration.

The applicable integral is (4-15). In the circular cylindrical system, the source and fieldpoints are designated $P'(0, 0, z')$ and $P(\rho, 0, 0)$, with the latter in the $z = 0$ plane. Any charge-element along the wire is $dq = \rho_\ell\, d\ell' = \rho_\ell\, dz'$, and the distance from P' to P is $R = \sqrt{(\rho^2 + z'^2)}$. Thus (4-15) becomes

$$\mathbf{E} = \int_\ell \mathbf{a}_R\, dE = \int_{z'=-\infty}^{\infty} \mathbf{a}_R \frac{\rho_\ell\, dz'}{4\pi\epsilon_0[\rho^2 + (z')^2]}$$

The unit vector $\mathbf{a}_R$ is troublesome, for its direction changes with z'. The symmetry about 0 in (*b*) reveals, however, another contribution to $d\mathbf{E}$ due to a

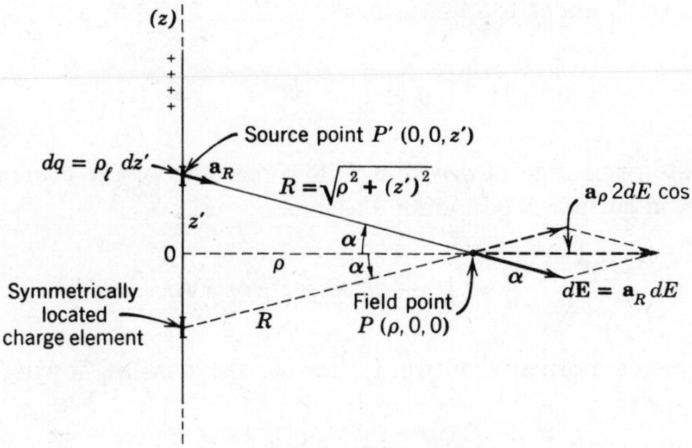

Infinitely long line charge

Example 4-1

paired charge-element at $-z'$. Thus the z component at P cancels, leaving only the ρ directed field at P given by

$$\mathbf{E} = \int_{\ell} \mathbf{a}_\rho \, dE \cos\alpha = \int_{z'=0}^{\infty} \mathbf{a}_\rho \frac{2\rho_\ell \, dz' \cos\alpha}{4\pi\epsilon_0[\rho^2 + (z')^2]}$$

$$= \int_{z'=0}^{\infty} \mathbf{a}_\rho \frac{\rho_\ell \rho \, dz'}{2\pi\epsilon_0[\rho^2 + (z')^2]^{3/2}}$$

since $\cos\alpha = \rho/R$. With ρ and $\mathbf{a}_\rho$ constants, integrating yields

$$\mathbf{E} = \mathbf{a}_\rho \frac{\rho_\ell}{2\pi\epsilon_0 \rho} \tag{4-16}$$

which checks with (1-61).

4-3 Conservation of Electric Charge

A relationship between charge and current densities is obtainable from Maxwell's equations, assuming that electric charge can neither be created nor destroyed. Let a charge density $\rho_v(u_1, u_2, u_3, t)$ occupy some volume region V. Then the net charge in V at any instant is

$$q(t) = \int_V \rho_v(u_1, u_2, u_3, t) \, dv \text{ C}$$

Note that even though ρ_v is in general a function of both space and time, the net q enclosed is a function of t only, because the definite limits on the integral dispose of the space variables. For brevity, the latter is written with the function notation understood as follows:

$$q = \int_V \rho_v \, dv \tag{4-17}$$

The time rate of change of q within V is a measure of the current flowing into the closed surface S bounding V; hence

$$\frac{\partial q}{\partial t} = \int_V \frac{\partial \rho_v}{\partial t} \, dv \text{ C/sec} \quad \text{or} \quad \text{A} \tag{4-18}$$

With $d\mathbf{s}$ directed normally outward from S, the current flowing *out of S* becomes

$$I = -\frac{\partial q}{\partial t} = \oint_S \mathbf{J} \cdot d\mathbf{s} \tag{4-19}$$

implying that the net, positive charge q inside V is *decreasing* in time. The postulate that electric charge is neither created nor destroyed permits equating the negative of (4-18) to (4-19), yielding

$$\oint_S \mathbf{J} \cdot d\mathbf{s} = -\int_V \frac{\partial \rho_v}{\partial t} \, dv \tag{4-20}$$

This means that the net outflow of current from any volume region is a measure of the time rate of decrease of electric charge inside the volume. Equation (4-20) is thus the expression of the *conservation of electric charge.*

The relation (4-20) has an equivalent differential, or point form

$$\nabla \cdot \mathbf{J} = -\frac{\partial \rho_v}{\partial t} \text{ A/m}^3 \tag{4-21a}$$

a result obtained by applying (4-20) to any limiting volume-element and using the definition (2-20) of divergence.

While (4-21a) is true for any volume-element of a current-carrying region, it is also applicable to the *surface* currents and charges at the interface between a perfect conductor and a perfect insulator, as in the system of Example 3-8. With currents and charges confined to the interface so that $\mathbf{J} \to \mathbf{J}_s$ and $\rho_v \to \rho_s$, the charge-conservation relation (4-21a) becomes

$$\nabla_T \cdot \mathbf{J}_s = -\frac{\partial \rho_s}{\partial t} \text{ A/m}^2 \tag{4-21b}$$

if $\nabla_T \cdot \mathbf{J}_s$ is taken to mean a tangential (two-dimensional) surface divergence of $\mathbf{J}_s$. For example, if the interface coincides with the y-z plane, implying $\mathbf{J}_s = \mathbf{a}_y J_{sy} + \mathbf{a}_z J_{sz}$, the two-dimensional divergence of $\mathbf{J}_s$ is written

$$\nabla_T \cdot \mathbf{J}_s = \frac{\partial J_{sy}}{\partial y} + \frac{\partial J_{sz}}{\partial z}$$

In a time-static field problem, steady current densities are divergenceless, so (4-21a) reduces in that case to

$$\nabla \cdot \mathbf{J}_s = 0 \tag{4-22}$$

Direct currents are therefore always characterized by uninterrupted, *closed* current flux lines.

EXAMPLE 4-2. Show that the surface current and surface charge fields at the conductor dielectric interfaces of Example 3-7 satisfy the two-dimensional charge-conservation relation (4-21b).

At the lower interface (at $x = 0$), the left side of (4-21b) yields

$$\nabla_T \cdot \mathbf{J}_s = \frac{\partial J_{sz}}{\partial z} = +\frac{\beta_0}{\eta_0} E_m^+ \sin(\omega t - \beta_0 z) = +\omega\epsilon_0 E_m^+ \sin(\omega t - \beta_0 z)$$

upon substituting $\beta_0 = \omega\sqrt{\mu_0\epsilon_0}$ and $\eta_0 = \sqrt{\mu_0/\epsilon_0}$. With a surface current density $\rho_s = +\epsilon_0 E_m^+ \cos(\omega t - \beta_0 z)$ on the lower conductor,

$$-\frac{\partial \rho_s}{\partial t} = +\omega\epsilon_0 E_m^+ \sin(\omega t - \beta_0 z)$$

whence (4-21b) is satisfied.

EXAMPLE 4-3. Determine the relaxation expression for the time decay of a charge distribution in a conductor, if the initial distribution at $t = 0$ is $\rho_{v0}(u_1, u_2, u_3, 0)$.

The desired result is obtained by combining (4-21a) with the expression for div $\mathbf{D}$. Replacing $\mathbf{J}$ with $\sigma\mathbf{E}$ for the conductive region obtains, from (4-21a)

$$\nabla \cdot (\sigma \mathbf{E}) + \frac{\partial \rho_v}{\partial t} = 0 \tag{4-23}$$

The region being homogeneous makes ϵ and σ constants, so (3-24) is written $\nabla \cdot \mathbf{E} = \rho_v/\epsilon$, and substituting it into the first term of (4-23) yields

$$\frac{\partial \rho_v}{\partial t} + \frac{\sigma}{\epsilon}\rho_v = 0 \tag{4-24}$$

Integrating yields the desired result

$$\rho_v(u_1, u_2, u_3, t) = \rho_{v0}(u_1, u_2, u_3, 0)e^{-(\sigma/\epsilon)t} \tag{4-25}$$

assuming the initial charge distribution at $t = 0$ to be $\rho_{v0}(u_1, u_2, u_3, 0)$. This implies that if the internal, free electric charge in a conducting region is zero, it will remain zero for all subsequent time. One may conclude that in a material having a nonzero conductivity σ, there can be no permanent *volume* distribution of free charge. Thus, the static state of a free charge supplied to a conducting body is that it must ultimately reside on the surface of the conducting body through the mutually repulsive (Coulomb) forces among the free charges.

The time constant τ of the free charge density decay process (4-25) in Example 4-3 is given by

$$\tau = \frac{\epsilon}{\sigma} \text{ sec} \tag{4-26}$$

a quantity called the *relaxation time* of the conductor. Good conductors, for which σ may be of the order of 10^7 ℧/m, have relaxation times around 10^{-18} sec, assuming a permittivity essentially that of free space. In poor conductors, τ may be of the order of microseconds, though a good insulator may have a relaxation time of hours or even days.

4-4 Gauss' Law and Static Conductor-Dielectric Systems

In Section 1-11, it was shown that Gauss' law (1-53) for free space may be used to obtain the electrostatic field of charge distributions having particular symmetries. Parallel-plane, concentric-circular-cylindrical, and concentric-spherical charge distributions in free space are particularly amenable to analysis by means of Gauss' law.

If charge distributions are combined with conducting and dielectric materials having shapes that yet preserve the symmetry, then Gauss' law in the form of (3-37) or (4-5)

$$\oint_S \mathbf{D} \cdot d\mathbf{s} = q \text{ C} \tag*{[3-37]}$$

is useful for finding $\mathbf{D}$ in the various regions, taking into account the possible polarization effects in the dielectric regions.

Example 4-4. A pair of long, coaxial, circular conducting cylinders are separated by concentric air and dielectric regions as shown, the inner ring being a dielectric with $\epsilon_r = 4$ $(a < \rho < b)$. A static charge q is assumed distributed over each length ℓ of the inner conductor, with $-q$ on the other conductor. Use Gauss' law to find $\mathbf{D}$ and $\mathbf{E}$.

From the symmetry, the $\mathbf{D}$ lines between the conductors are radial and independent of ϕ. The field is found by use of a symmetric, closed cylinder S

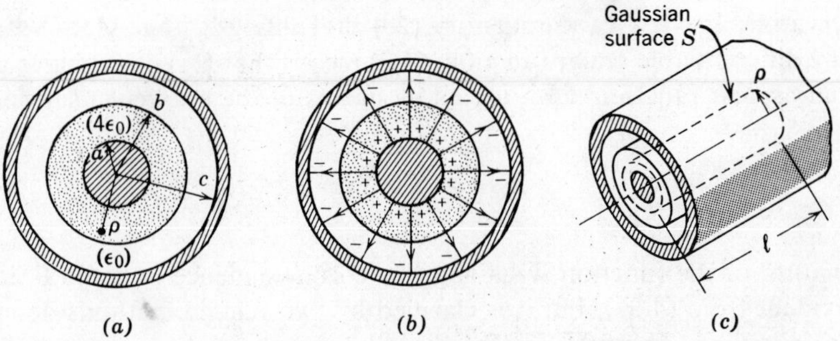

Example 4-4. **(*a*) Showing the dimensions of the coaxial pair. (*b*) Depicting the continuity of D lines. (*c*) Symmetric closed *S* for finding the fields.**

enclosing the inner conductor as in (*c*). The procedure may be compared with Example 1-11(*c*). With *S* having a length ℓ and radius ρ, and the ρ directed **D** piercing the peripheral surface S_0, Gauss' law (3-37) becomes

$$\int_{S_0} (\mathbf{a}_\rho D_\rho)\cdot\mathbf{a}_\rho\, ds = q$$

D_ρ is constant over S_0, whence

$$D_\rho = \frac{q}{2\pi\rho\ell} \tag{4-27}$$

a result applying to both the dielectric and air regions. Using (4-9), *E* in the respective regions becomes

$$E_\rho = \frac{q}{2\pi(4\epsilon_0)\rho\ell} \qquad a < \rho < b \tag{4-28}$$

$$E_\rho = \frac{q}{2\pi\epsilon_0\rho\ell} \qquad b < \rho < c \tag{4-29}$$

With $\mathbf{D} = 0$ inside the conductors, the boundary condition (4-7) reduces to

$$D_{n1} = \rho_s \tag{4-30}$$

which, applied to (4-27) at $\rho = a$ and $\rho = c$ yields the free charge densities on the conductor surfaces

$$\rho_s]_{\rho=a} = \frac{q}{2\pi a\ell} \qquad \rho_s]_{\rho=c} = \frac{-q}{2\pi c\ell}$$

4-5 Electrostatic Scalar Potential

Any electrostatic field $\mathbf{E}(u_1, u_2, u_3)$ must satisfy the curl relation (4-2), $\nabla \times \mathbf{E} = 0$, which states that any static **E** field is irrotational, and therefore conservative. In view of the identity (20) in Table 2-2, that $\nabla \times (\nabla\Phi) = 0$ for any differentiable scalar function, (4-2) means that $\mathbf{E}$ is derivable from an auxiliary scalar function $\Phi(u_1, u_2, u_3)$ by means of the gradient relation

$$\mathbf{E} = -\nabla\Phi \text{ V/m} \tag{4-31}$$

The nature of the function Φ having this correspondence to some $\mathbf{E}$ field is not evident from (4-31), but it is clarified by two related methods described in the following. The first obtains the potential Φ from the known charge distribution of density ρ_v, and once Φ has been found, the $\mathbf{E}$ field is obtained using (4-31). The second method presumes $\mathbf{E}$ known at the outset of the problem; Φ is found from an appropriate line integral of $\mathbf{E}$ over a path beginning at a designated potential reference.

A. Potential Φ Obtained from a Known Charge Density in Free Space

The relation of the electrostatic $\mathbf{E}$ to its charge sources *in free space* is (4-13)

$$\mathbf{E}(x, y, z) = \int_V \mathbf{a}_R \frac{\rho_v(x', y', z')}{4\pi\epsilon_0 R^2}\, dv' \text{ V/m} \qquad [4\text{-}13]$$

A dependence on the variables (x, y, z) and (x', y', z') is evident in this integral because, by (4-12), $R = \sqrt{(x - x')^2 + (y - y')^2 + (z - z')^2}$. One can show by direct expansion that

$$\nabla\left(\frac{1}{R}\right) = -\frac{\mathbf{a}_R}{R^2} \tag{4-32}$$

assuming that ∇ is defined in terms of derivatives with respect to the field point variables (x, y, z) as follows:

$$\nabla \equiv \mathbf{a}_x \frac{\partial}{\partial x} + \mathbf{a}_y \frac{\partial}{\partial y} + \mathbf{a}_z \frac{\partial}{\partial z} \tag{4-33}$$

This permits writing (4-13)

$$\mathbf{E}(x, y, z) = -\int_V \nabla\left(\frac{1}{R}\right) \frac{\rho_v(x', y', z')}{4\pi\epsilon_0}\, dv' = -\nabla \int_V \frac{\rho_v(x', y', z')}{4\pi\epsilon_0 R}\, dv' \tag{4-34}$$

in which an interchange between the integration and the gradient operations is permissible because the only quantity affected by ∇ is R, while the *integra-*

tion is to be carried out with respect to the *source point* variables (x', y', z'). Comparing (4-34) with (4-31) shows that the integral in (4-34) is the desired scalar function Φ; hence

$$\Phi(x, y, z) = \int_V \frac{\rho_v(x', y', z')}{4\pi\epsilon_0 R}\, dv' \text{ V} \tag{4-35a}$$

Φ is called the *scalar potential field* of the static **E** field.

It is further evident that if the charge density in (4-35a) takes the form of a surface or line charge density ρ_s or ρ_ℓ, then the integral becomes

$$\Phi(x, y, z) = \int_S \frac{\rho_s(x', y', z')}{4\pi\epsilon_0 R}\, ds' \tag{4-35b}$$

$$\Phi(x, y, z) = \int_\ell \frac{\rho_\ell(x', y', z')}{4\pi\epsilon_0 R}\, d\ell' \tag{4-35c}$$

B. Potential Φ Obtained from a Line Integral of E

The potential field Φ of a static charge distribution in space can be expressed in terms of a line integral of **E**. To show this, observe that (4-2) has the corresponding integral form (4-6)

$$\oint_\ell \mathbf{E} \cdot d\boldsymbol{\ell} = 0 \qquad [4\text{-}6]$$

true for all closed lines ℓ in space. Physically, (4-6) states that the work done on a test charge q, in moving it around any closed path ℓ in the presence of a static field **E**, is precisely zero. This is equivalent to saying that the work done on q in moving it between two fixed points P_1 and P_2 in the field is independent of the shape of the open path connecting the points. This is evident if two different paths ℓ_1 and ℓ_2 are used to connect P_1 and P_2. If the *closed* contour ℓ of (4-6) is taken to be $\ell_1 + \ell_2$, then (4-6) yields

$$\left[\int_{P_1}^{P_2} \mathbf{E} \cdot d\boldsymbol{\ell}\right]_{\ell_1} = \left[\int_{P_1}^{P_2} \mathbf{E} \cdot d\boldsymbol{\ell}\right]_{\ell_2} \tag{4-36}$$

correct for *all* paths connecting P_1 to P_2.

The property (4-36) makes it possible to derive a single-valued potential field equivalent to (4-31) as follows. Suppose $P_0(u_1^0, u_2^0, u_3^0)$ is fixed in space, called the *potential reference* and defined such that $\Phi = \Phi_0$ there. The line integral of **E**, over any path ℓ connecting P_0 and any arbitrary $P(u_1, u_2, u_3)$ as

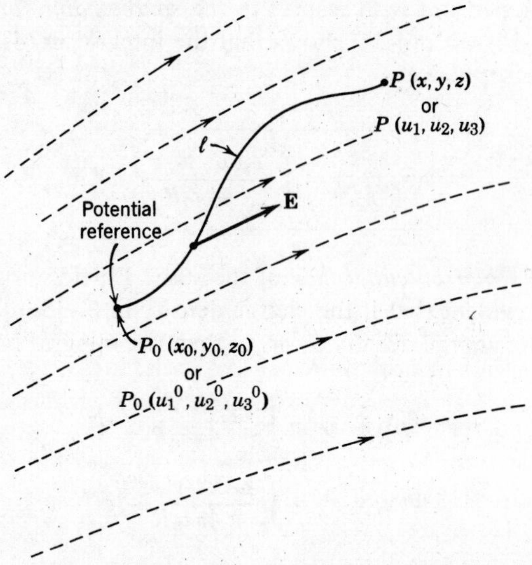

Figure 4-4. Development of the Φ field from the E field.

in Figure 4-4, is written in the forms, making use of (4-31)

$$\int_{P_0}^{P} \mathbf{E}\cdot d\boldsymbol{\ell} = -\int_{P_0}^{P} (\nabla\Phi)\cdot d\boldsymbol{\ell} = -\int_{P_0}^{P} \left(\frac{\partial\Phi}{\partial x}\,dx + \frac{\partial\Phi}{\partial y}\,dy + \frac{\partial\Phi}{\partial z}\,dz\right)$$

From (2-11), the integrand of the right side denotes the total differential $d\Phi$, whence

$$\int_{P_0}^{P} \mathbf{E}\cdot d\boldsymbol{\ell} = -\int_{P_0}^{P} d\Phi$$

the latter of which can be integrated to yield

$$\Phi(P) - \Phi_0 = -\int_{P_0(u_1^0,u_2^0,u_3^0)}^{P(u_1,u_2,u_3)} \mathbf{E}\cdot d\boldsymbol{\ell} \tag{4-37}$$

Thus, the line integral of the static **E** field over any path connecting two points in space is just the difference of the potentials at those points. For most purposes it is desirable to call $\Phi_0 = 0$ at the potential reference; then (4-37) becomes

$$\Phi(P) = -\int_{P_0(u_1^0,u_2^0,u_3^0)}^{P(u_1,u_2,u_3)} \mathbf{E}\cdot d\boldsymbol{\ell}\ \text{V} \tag{4-38a}$$

This potential field $\Phi(P)$ can be made to agree exactly with results obtained from (4-35), if one observes that (4-35) provides a zero potential at an *infinite* distance from a charge distribution grouped within a finite distance from the origin. Thus, with the reference P_0 at infinity (4-38a) provides the *absolute potential*

$$\Phi(P) = -\int_{\infty}^{P} \mathbf{E} \cdot d\boldsymbol{\ell} \text{ V} \tag{4-38b}$$

yielding the same results as (4-35). Sometimes, as in problems of academic interest involving charges that extend to infinity (e.g., the uniform line charge of infinite extent), the integrals (4-35) and (4-38b) yield infinite potentials. (The integrals then do not exist.) In such cases one should make use of (4-38a), utilizing a zero reference value at a *finite* distance from the origin.

Surfaces defined by setting $\Phi(P)$ to any constant value are called *equipotential surfaces*. Such surfaces are often of interest in static field problems because, from the gradient relation (4-31), the electrostatic field lines intersect the equipotential surfaces *normally*. This property is a useful one in some field-mapping problems considered later in this chapter. Gauss' law (4-5) assumes that electric field lines are directed away from positive charges and towards negative charges; thus, from (4-38), the potential Φ becomes more positive as one approaches positive charges; the opposite is true for negative charges.

EXAMPLE 4-5. Find the electrostatic potential at any field point located a normal distance ρ from an infinite line charge in free space having the constant density ρ_ℓ C/m as shown. Assume the zero potential reference at the position $P_0(\rho_0, \phi_0, z_0)$. Sketch a few equipotential surfaces.

The potential Φ at any location P relative to a fixed reference P_0 is (4-38a).

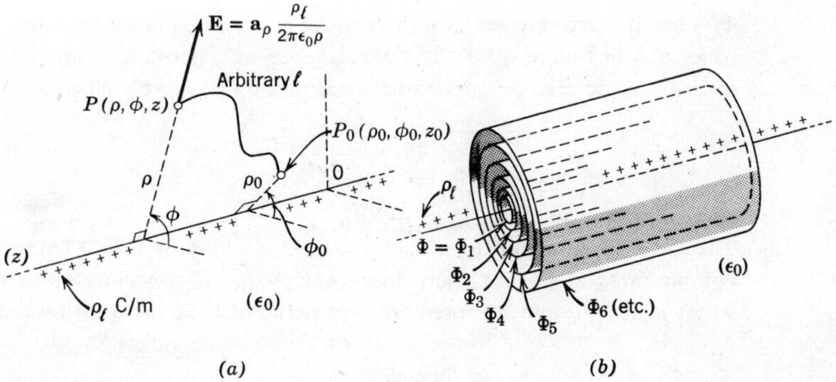

Example 4-5. Geometry of infinite line charge. (*b*) Equipotential surfaces.

The field from Example 4-1 was found to be $\mathbf{E} = \mathbf{a}_\rho \rho_\ell / 2\pi\epsilon_0\rho$. Inserting this into (4-38a) and integrating over any path connecting P_0 to P as in (*a*) obtains

$$\Phi(P) = -\int_{P_0}^{P} \mathbf{E}\cdot d\ell = -\int_{P_0}^{P}\left[\mathbf{a}_\rho \frac{\rho_\ell}{2\pi\epsilon_0\rho}\right]\cdot(\mathbf{a}_\rho\, d\rho + \mathbf{a}_\phi \rho\, d\phi + \mathbf{a}_z\, dz)$$

yielding the result independent of ϕ and z

$$\Phi(\rho) = \frac{\rho_\ell}{2\pi\epsilon_0}\, \ell n \frac{\rho_0}{\rho} \tag{4-39}$$

It is evident that putting the zero potential reference at infinity in this result ($\rho_0 \to \infty$) is not desirable, for $\Phi(P)$ then becomes infinite; a finitely located reference position is necessary. Equipotential surfaces are obtained by setting $\Phi(P)$ of (4-39) equal to the constant values $\Phi_1, \Phi_2, \Phi_3, \ldots$; yielding $\rho = \rho_1, \rho_2, \rho_3, \ldots$, the circular cylindrical surfaces shown in (*b*).

EXAMPLE 4-6. (*a*) Find the absolute potential of a point charge q located at the origin $r = 0$ in Figure 4-5(*a*), making use of the field (4-10b). Describe its equipotential surfaces. (*b*) Show that the potential field can also be obtained directly from the volume integral (4-35a) applied to concentrated charge q. Determine the potential at P if q is located at a general source point P' as in Figure 4-5(*b*).

(*a*) The $\mathbf{E}$ field in Figure 4-5(*a*) is (4-10b). Integrating $\mathbf{E}$ over any path between $P_0(r_0)$ and the arbitrary $P(r)$ yields, from (4-38a)

$$\Phi(r) = -\int_{r_0}^{r}\left[\mathbf{a}_r \frac{q}{4\pi\epsilon_0 r^2}\right]\cdot(\mathbf{a}_r\, dr + \mathbf{a}_\theta r\, d\theta + \mathbf{a}_\phi r \sin\theta\, d\phi)$$

$$= \frac{q}{4\pi\epsilon_0}\left[\frac{1}{r} - \frac{1}{r_0}\right] \tag{4-40a}$$

$\Phi(r)$ has its zero reference on the surface $r = r_0$, yielding equipotential spheres as in Figure 4-5(*c*). The *absolute potential* is found from (4-40a) by putting the zero potential surface at infinity ($r_0 \to \infty$), whence

$$\Phi(r) = \frac{q}{4\pi\epsilon_0 r} \tag{4-40b}$$

The latter is plotted in Figure 4-5(*d*).

(*b*) The absolute potential of a static charge can also be found from the volume integral (4-35a). Here the point charge is concentrated at P', so let $\rho_v\, dv \to q$ and no integration is required. Then (4-35a) becomes

$$\Phi(x, y, z) = \frac{q}{4\pi\epsilon_0 R} \tag{4-40c}$$

a result applicable to the geometry of Figure 4-5(*b*).

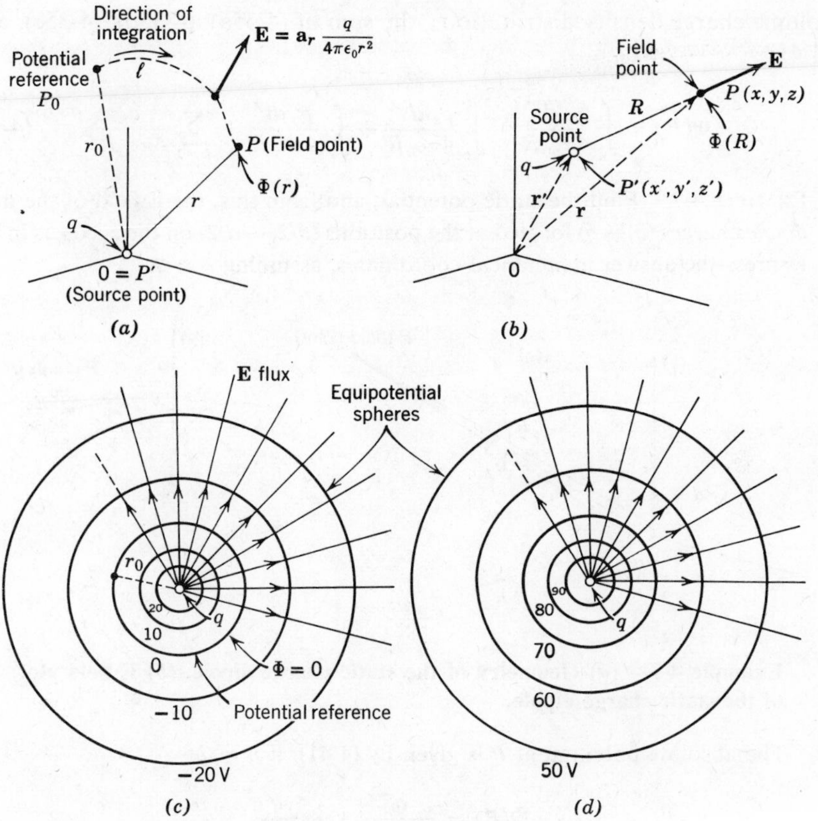

Figure 4-5. Point charge q: geometry and equipotential plots. (*a*) Geometry of a point charge at 0, showing ℓ over which E is integrated to find $\Phi(P)$ relative to P_0. (*b*) Geometry of a point charge located at $P'(r')$. P is field point at which Φ is obtained. (*c*) Equipotential surfaces of point charge, potential reference at r_0 assumed. (*d*) Equipotential surfaces of point charge: potential reference at infinity.

The result (4-40c) is useful for constructing the absolute potential of an *aggregate* of n charges in free space like that of Figure 4-2, yielding the sum of the potential contributions of each charge

$$\Phi(P) = \sum_{k=1}^{n} \frac{q_k}{4\pi\epsilon_0 R_k} \text{ V} \quad \text{Absolute potential} \quad (4\text{-}41)$$

The absolute potential of the *most general* configurations of static charges in free space is one accounting for discrete charges plus line, surface, and volume charge density distributions, the sum of (4-35a), (4-35b), (4-35c), and (4-41)

$$\Phi(P) = \int_{\ell} \frac{\rho_\ell \, d\ell'}{4\pi\epsilon_0 R} + \int_S \frac{\rho_s \, ds'}{4\pi\epsilon_0 R} + \int_V \frac{\rho_v \, dv'}{4\pi\epsilon_0 R} + \sum_{k=1}^{n} \frac{q_k}{4\pi\epsilon_0 R_k} \tag{4-42}$$

EXAMPLE 4-7. Find the static potential, and from this, the **E** field of the fixed *dipole* charges $(q, -q)$ located at the positions $(d/2, -d/2)$ on the z axis as in (*a*). Express the answer in spherical coordinates, assuming $r \gg d$.

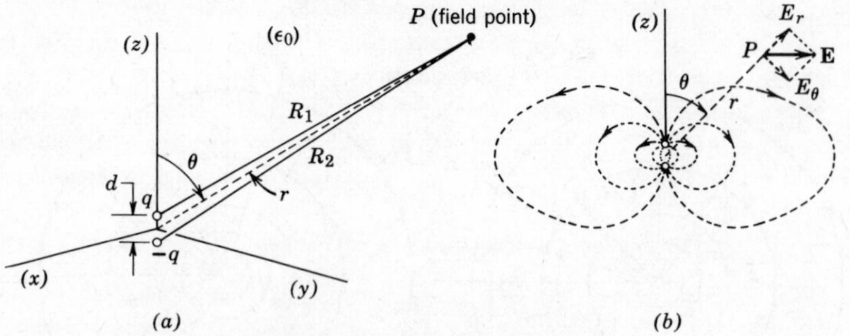

Example 4-7. (*a*) Geometry of the static charge dipole. (*b*) E field plot of the static charge dipole.

The absolute potential at P is given by (4-41), if $n = 2$

$$\Phi(P) = \frac{q}{4\pi\epsilon_0 R_1} + \frac{-q}{4\pi\epsilon_0 R_2}$$

$$= \frac{q}{4\pi\epsilon_0} \frac{R_2 - R_1}{R_1 R_2}$$

Assuming $r \gg d$, one can approximate $R_1 R_2 \cong r^2$ and $R_2 - R_1 \cong d \cos \theta$, as noted from (*a*). Then $\Phi(P)$ reduces to

$$\Phi(P) \cong \frac{qd \cos \theta}{4\pi\epsilon_0 r^2} \qquad d \ll r \tag{4-43}$$

Note that $\Phi(P)$ of a static dipole is an inverse r^2 function (for $r \gg d$), as contrasted with the inverse r potential (4-40b) of a single static charge.

One obtains **E** from (4-31) in spherical coordinates

$$\mathbf{E} = -\nabla\Phi = -\left[\mathbf{a}_r \frac{\partial \Phi}{\partial r} + \mathbf{a}_\theta \frac{\partial \Phi}{r\, \partial \theta} + \mathbf{a}_\phi \frac{\partial \Phi}{r \sin\theta \, \partial \phi}\right]$$

$$\cong \frac{qd}{4\pi\epsilon_0 r^3} [\mathbf{a}_r \, 2\cos\theta + \mathbf{a}_\theta \sin\theta] \tag{4-44}$$

an inverse r^3 function with both r and θ directed components. Its flux plot is shown in (*b*).

4-6 Capacitance

Of considerable practical utility in electrical circuits is the capacitor, commonly used to store or release electric field energy. Basically, a capacitor consists of two conductors separated by free space or suitable dielectric materials of arbitrary permittivities. Its form is generalized in Figure 4-6(*a*). A capacitor with the charges q and $-q$ can be brought to this charge state by means of a source of electric charge such as the battery shown, although it is perhaps more common to connect it to a source of sinusoidal or pulsed voltages. In this event, the charges become functions of time, $q(t)$ and $-q(t)$. The viewpoint of the present discussion is that if the time variations are sufficiently slow, a static field analysis of the system will provide results of sufficient accuracy to serve the purposes of many time-varying applications of practical interest.

A capacitor, brought to the charge state of Figure 4-6, has the properties:

1. The free charges q and $-q$ reside entirely on the conductor surfaces, accounting for a charge density ρ_s on each such that on their surfaces S_1 and S_2 reside the charges

$$q = \int_S \rho_s \, ds \qquad -q = \int_S \rho_s \, ds \tag{4-45}$$

2. From the boundary condition (4-7), the **E** field originates normally from

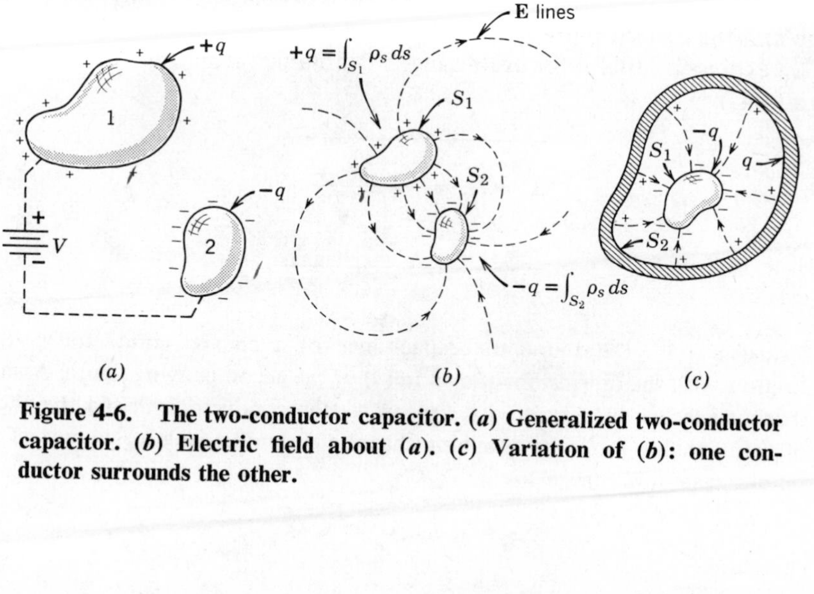

Figure 4-6. The two-conductor capacitor. (*a*) Generalized two-conductor capacitor. (*b*) Electric field about (*a*). (*c*) Variation of (*b*): one conductor surrounds the other.

the positively charged conductor and terminates normally on the negative one, with the total **D** flux equalling q (Gauss' law).

3. A consequence of the perpendicularity of **E** at the conductor surfaces is that they are *equipotential surfaces* ($\Phi = \Phi_1$ and $\Phi = \Phi_2$). Thus a single-valued potential difference $\Phi_1 - \Phi_2 = V$ exists between the conductors, obtainable from (4-38b) as follows:

$$V = \Phi_1 - \Phi_2 = \left[-\int_{\infty}^{P_1} \mathbf{E} \cdot d\boldsymbol{\ell}\right] - \left[-\int_{\infty}^{P_2} \mathbf{E} \cdot d\boldsymbol{\ell}\right] = -\int_{P_2}^{P_1} \mathbf{E} \cdot d\boldsymbol{\ell} \quad (4\text{-}46)$$

in which P_1 and P_2 can be located anywhere on the respective conductors, and the latter integral is over any path connecting P_1 and P_2. To make V positive, the *potential reference* P_2 (the lower limit) is assumed on the *negative* conductor.

If a linear dielectric medium is used in a capacitor, the effect of doubling the charges q and $-q$ on the conductors results in a doubling of the **E** field everywhere. From (4-46), then, the potential difference is also doubled. Thus in a linear capacitor, V is proportional to the charge q so that $q \propto V$, or equivalently,

$$q = CV \quad (4\text{-}47)$$

The proportionality constant C, having the units coulomb per volt or *farad*, is called the *capacitance* of the two-conductor system. It is positive whenever an increase in V (the potential of the positive conductor relative to the negative one) results in an increase in the charge q on the positive conductor (accompanied by a negative increase of $-q$ on the other conductor). For a passive element C is always positive, its value depending on physical dimension and the dielectric properties of the system.

An expression useful for evaluating C is obtained from substituting (4-46) into (4-47)

$$C = \frac{q}{V} = \frac{q}{-\int_{P_2}^{P_1} \mathbf{E} \cdot d\boldsymbol{\ell}} \text{ F} \quad (4\text{-}48)$$

EXAMPLE 4-8. Determine the capacitance of a coaxial conductor pair of length ℓ with the dimensions shown in (*a*) of the accompanying figure. Assume that a dielectric of permittivity ϵ separates the conductors. Avoid the effects of field-fringing in (*b*) by assuming that the system of length ℓ is part of the infinite system in (*c*).

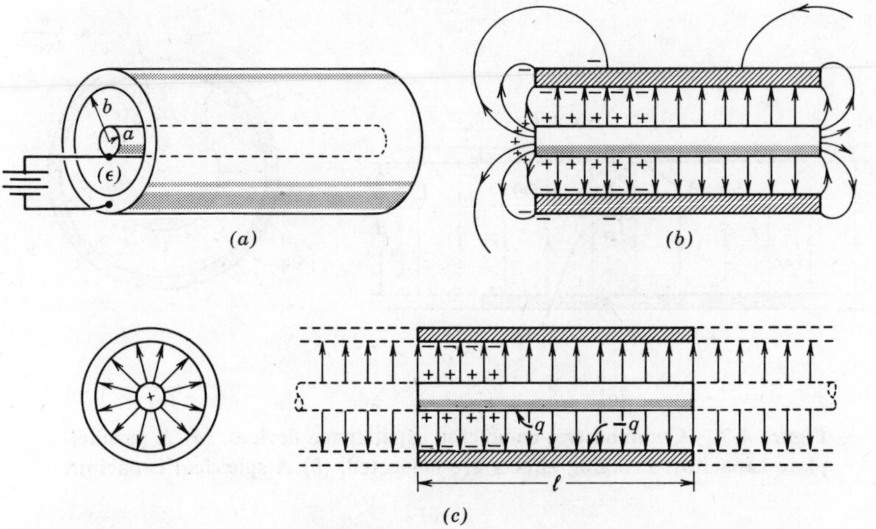

Example 4-8. ***(a)*** **Circular cylindrical coaxial capacitor and dc source. (b) Fringing of electric field at ends of finite length system.** ***(c)*** **Showing field independent of** ϕ **and** z **in a section of an infinite system.**

To find the capacitance of a length ℓ, assume the conductors charged, for every length ℓ, with $+q$ and $-q$ C on the inner and outer surfaces. The **D** field is found using Gauss' law as in Example 4-4, yielding the **E** field

$$\mathbf{E} = \mathbf{a}_\rho \frac{q}{2\pi\epsilon\rho\ell} \tag{4-49}$$

The potential difference V between a reference P_2 on the negative conductor and P_1 on the positive conductor is

$$V = -\int_{\rho=b}^{a} \left(\mathbf{a}_\rho \frac{q}{2\pi\epsilon\rho\ell}\right) \cdot (\mathbf{a}_\rho\, d\rho + \mathbf{a}_\phi \rho\, d\phi + \mathbf{a}_z\, dz) = \frac{q}{2\pi\epsilon\ell} \ln\frac{b}{a} \tag{4-50}$$

Thus, the capacitance of a length ℓ, neglecting end effects, is obtained from (4-48)

$$C = \frac{q}{V} = \frac{q}{\dfrac{q}{2\pi\epsilon\ell}\ln\dfrac{b}{a}} = \frac{2\pi\epsilon\ell}{\ln\dfrac{b}{a}}\ \text{F} \tag{4-51}$$

Note that the result is independent of q, as expected of a linear capacitor. Hence C is a function only of the dimensions and ϵ. If $a = 1$ mm, $b = 6$ mm, and $\epsilon = \epsilon_0$ (air dielectric), C/ℓ becomes 31×10^{-12} F/m (or 31 pF/m). Using a dielectric with $\epsilon = 4\epsilon_0$ yields a result four times as large.

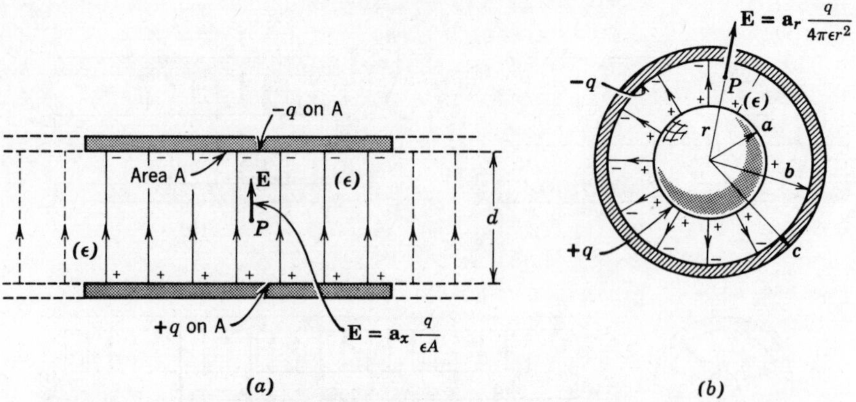

Figure 4-7. Common two-conductor capacitance devices. (*a*) A parallel-plate capacitor. Fringing effects are neglected. (*b*) A spherical capacitor.

By the techniques of the last example, one can show that the capacitance of the *parallel-plate* system of Figure 4-7(*a*), neglecting field-fringing, is

$$C = \frac{\epsilon A}{d}\ \text{F} \tag{4-52}$$

while that of the *concentric spheres* in (*b*) is

$$C = \frac{4\pi\epsilon}{\dfrac{1}{a} - \dfrac{1}{b}}\ \text{F} \tag{4-53}$$

4-7 Energy of the Electrostatic Field

The concept of stored energy in the electrostatic field has important physical interpretations and applications. As in mechanics, many problems of electrostatics can often be simplified if an energy viewpoint is adopted. While generally systems of electric charges possess both potential and kinetic energies due to their positions and motional states, in the electrostatic case only the charge positions determining the potential energy of the system need be of concern.

To establish the n charge aggregate of Figure 4-2, mechanical work must be done by some external agent in bringing the charges to their final positions. Whenever two charges q and q' are brought within a distance R of each other, work is done against the Coulomb force (4-10a) in consummating this process. Once the charges are in place, the persistence of the Coulomb force makes the stored energy potentially available whenever demanded. The discharge of a capacitor bank through a resistor exemplifies this reverse process.

The electrostatic energy stored in a system of discrete, or point, charges is found by building up the assemblage one charge at a time until all are in their intended locations. It is assumed that if they are moved slowly enough that their kinetic energies may be ignored and that radiation effects, significant if rapid charge accelerations occur, can be neglected. Assume initially that all n charges, $q_1, q_2, q_3, \ldots$, are located at infinity in their zero potential state. Upon bringing only q_1 from infinity to its final location P_1, no work is done because only q_1 is present; at least two charges are required if Coulomb forces are to exist.†

Upon next bringing q_2 from infinity to P_2 as in Figure 4-8, the work done against the field of q_1 is $U_2 = q_2\Phi_2^{(1)}$, in which $\Phi_2^{(1)}$ denotes the electrostatic potential *at* P_2 and *due to* q_1. Thus one obtains, using the absolute potential expression (4-40c)

$$U_2 = q_2\Phi_2^{(1)} = q_2 \frac{q_1}{4\pi\epsilon_0 R_{12}} = q_1 \frac{q_2}{4\pi\epsilon_0 R_{21}} \tag{4-54a}$$

$$= q_1\Phi_1^{(2)} \text{ C-V or J} \tag{4-54b}$$

in which an interchange of q_1 and q_2 is seen to yield equivalent work expressions.

Again, if a third charge q_3 is brought in to P_3 as in Figure 4-8(b), the work

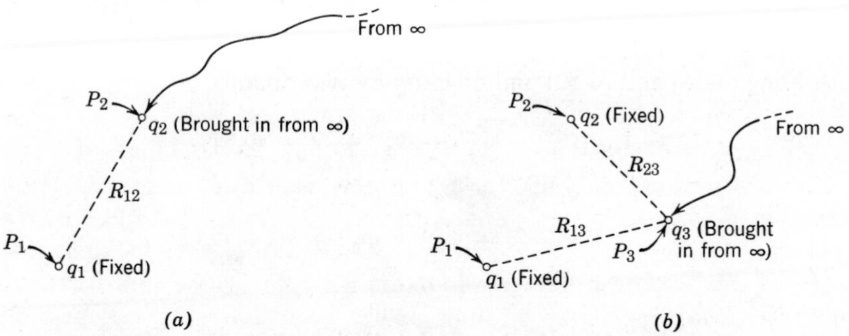

Figure 4-8. Two steps in the construction of an n charge aggregate. (*a*) From infinity q_2 is brought in in the presence of q_1. (*b*) From infinity q_3 is brought in in the presence of q_1 and q_2.

† The *self-energy* of each discrete charge, i.e., the energy required to create each diminutive electron cloud, is neglected in this development.

done against the fields of q_1 and q_2 is expressible two ways

$$U_3 = q_3\Phi_3^{(1)} + q_3\Phi_3^{(2)} = q_3\frac{q_1}{4\pi\epsilon_0 R_{13}} + q_3\frac{q_2}{4\pi\epsilon_0 R_{23}} \tag{4-55a}$$

$$= q_1\Phi_1^{(3)} + q_2\Phi_2^{(3)} \tag{4-55b}$$

and so on, for $q_4, q_5, \ldots, q_n$.

Continuing the preceding development shows that the total energy, $U_e = U_1 + U_2 + \cdots + U_n$, can be written two ways:

1. Upon adding (4-54a), (4-55a), etc.,

$$\begin{aligned} U_e = {} & q_2\Phi_2^{(1)} + q_3\Phi_3^{(1)} + q_3\Phi_3^{(2)} \\ & + q_4\Phi_4^{(1)} + q_4\Phi_4^{(2)} + q_4\Phi_4^{(3)} + \cdots \\ & + q_n\Phi_n^{(1)} + q_n\Phi_n^{(2)} + \cdots + q_n\Phi_n^{(n-1)} \end{aligned} \tag{4-56}$$

2. Adding (4-54b), (4-55b), etc. yields

$$\begin{aligned} U_e = {} & q_1\Phi_1^{(2)} + q_1\Phi_1^{(3)} + q_2\Phi_2^{(3)} \\ & + q_1\Phi_1^{(4)} + q_2\Phi_2^{(4)} + q_3\Phi_3^{(4)} + \cdots \\ & + q_1\Phi_1^{(n)} + q_2\Phi_2^{(n)} + \cdots + q_{n-1}\Phi_{n-1}^{(n)} \end{aligned}$$

which can be regrouped

$$\begin{aligned} U_e = {} & q_1(\Phi_1^{(2)} + \Phi_1^{(3)} + \cdots + \Phi_1^{(n)}) \\ & + q_2(\Phi_2^{(3)} + \Phi_2^{(4)} + \cdots + \Phi_2^{(n)}) \\ & + q_3(\Phi_3^{(4)} + \Phi_3^{(5)} + \cdots + \Phi_3^{(n)}) \\ & + \cdots \\ & + q_{n-1}\Phi_{n-1}^{(n)} \end{aligned} \tag{4-57}$$

Adding (4-56) and (4-57) and dividing by two obtains

$$\begin{aligned} U_e = {} & \tfrac{1}{2}\{q_1[\Phi_1^{(2)} + \Phi_1^{(3)} + \Phi_1^{(4)} + \cdots + \Phi_1^{(n)}] \\ & + q_2[\Phi_2^{(1)} + \Phi_2^{(3)} + \Phi_1^{(4)} + \cdots + \Phi_2^{(n)}] \\ & + q_3[\Phi_3^{(1)} + \Phi_3^{(2)} + \Phi_3^{(4)} + \cdots + \Phi_3^{(n)}] \\ & + \cdots \\ & + q_n[\Phi_n^{(1)} + \Phi_n^{(2)} + \Phi_n^{(3)} + \cdots + \Phi_n^{(n-1)}]\} \end{aligned}$$

The meaning of each bracketed sum in the latter is now assessed. In the first term, the sum $[\Phi_1^{(2)} + \Phi_1^{(3)} + \cdots + \Phi_1^{(n)}]$, abbreviated Φ_1, is the total potential of P_1 (position of q_1) due to all the charges except q_1 itself. Thus the bracketed factors signify the potential at the location of the typical charge

q_k, a potential due to all the charges except q_k. Denoting the bracketed factors by $\Phi_1, \Phi_2, \ldots, \Phi_n$ respectively, the desired result becomes $U_e = (1/2)[q_1\Phi_1 + q_2\Phi_2 + \cdots + q_n\Phi_n]$, or

$$U_e = \frac{1}{2}\sum_{k=1}^{n} q_k\Phi_k \text{ J} \tag{4-58a}$$

in which

q_k is the charge of the typical (kth) particle

Φ_k is at P_k, the absolute potential due to all the charges except the kth

If the assemblage of charges is not discrete, but rather a continuum of density ρ_v distributed throughout some *volume* region V, then (4-58a) becomes an integral upon replacing q_k with $dq = \rho_v\, dv$, obtaining

$$U_e = \frac{1}{2}\int_V \rho_v\Phi\, dv \text{ J} \tag{4-58b}$$

wherein Φ is the absolute potential at the position of ρ_v. For charge continua comprised of *surface* or *line* distributions as discussed in Sections 4-2 and 4-5, the following expressions are used in lieu of the preceding ones:

$$U_e = \frac{1}{2}\int_S \rho_s\Phi\, ds \text{ J} \tag{4-58c}$$

$$U_e = \frac{1}{2}\int_\ell \rho_\ell\Phi\, d\ell \text{ J} \tag{4-58d}$$

In computing U_e from one or a combination of these four expressions, only the charge distributions and the potentials at the charge locations need to be known.

The energy integrals (4-58), expressed in terms of the potential distribution Φ accompanying static charge distributions in space, can also be written in terms of only the **D** and **E** fields that occupy the whole of space. The result becomes

$$U_e = \tfrac{1}{2}\int_V \mathbf{D}\cdot\mathbf{E}\,dv \text{ J} \tag{4-58e}$$

To prove the latter, suppose that surface charges of density ρ_s exist on the closed conducting surface S, where S may consist of n individual conductors such that $S = S_1 + S_2 + \cdots + S_n$, with the additional possibility of a volume charge density ρ_v occupying the region V enclosed by S.† The two-conductor capacitor of Figure 4-6(*b*) or (*c*) represents such a system. The electrostatic energy of the system is the sum of (4-58b) and (4-58c)

$$U_e = \tfrac{1}{2}\oint_S \rho_s\Phi\,ds + \tfrac{1}{2}\int_V \rho_v\Phi\,dv \tag{4-59}$$

in which S denotes the simply connected closed surface of the charged conductors, and V is the region between the conductors. Using the boundary condition (3-45) but with the unit vector **n** directed away from the volume V so that $\rho_s = -\mathbf{n}\cdot\mathbf{D}$, (4-59) becomes

$$\begin{aligned} U_e &= -\tfrac{1}{2}\oint_S (\Phi\mathbf{D})\cdot\mathbf{n}\,ds + \tfrac{1}{2}\int_V \rho_v\Phi\,dv \\ &= -\tfrac{1}{2}\int_V \nabla\cdot(\Phi\mathbf{D})\,dv + \tfrac{1}{2}\int_V \rho_v\Phi\,dv \end{aligned} \tag{4-60}$$

in which the transformation of the closed-surface integral to the volume integral‡ is accomplished by use of the divergence theorem (2-34). The use of the vector identity (15) of Table 2-2, $\nabla\cdot(f\mathbf{G}) = f\nabla\cdot\mathbf{G} + \mathbf{G}\cdot\nabla f$, yields

$$U_e = -\tfrac{1}{2}\int_V \mathbf{D}\cdot(\nabla\Phi)\,dv - \tfrac{1}{2}\int_V \Phi\nabla\cdot\mathbf{D}\,dv + \tfrac{1}{2}\int_V \rho_v\Phi\,dv$$

Since $\rho_v = \nabla\cdot\mathbf{D}$, the last two integrals cancel, and with $\nabla\Phi = -\mathbf{E}$ into the

† As, for example, a volume density of free charges embedded in the dielectric region between the conductors, or a volume space charge density between the conductive electrodes of a vacuum diode.

‡ If one conductor does not enclose the other, as in the capacitor shown in Figure 4-6(*b*), the dielectric volume region V extends to infinity. The surface S enclosing V must then include a sphere at infinity, but because of the manner in which the Φ and **D** fields vanish at remote distances, it develops that the surface integral contribution over this sphere is zero.

first integral, one obtains the desired result (4-58e)

$$U_e = \tfrac{1}{2}\int_V \mathbf{D}\cdot\mathbf{E}\, dv \tag{4-61a}$$

an energy expression valid whether or not V contains a charge density ρ_v. Thus, (4-61a) provides an alternative for finding the potential energy of a static charge system, i.e., in terms of only the fields **D** and **E** in the volume region appropriate to the given system.

The integrand of (4-61a), $\mathbf{D}\cdot\mathbf{E}/2$, having the units joules per cubic meter, is called the *electrostatic energy density* at the point in the volume region. In an isotropic dielectric region, the permittivity ϵ is a scalar, yielding the energy density $\epsilon E^2/2$, whence (4-61a) becomes

$$U_e = \tfrac{1}{2}\int_V \epsilon E^2\, dv \text{ J} \tag{4-61b}$$

EXAMPLE 4-9. Find the energy stored in the electric field of the coaxial line of Example 4-8, making use of (4-58e).

In a coaxial line **D** and **E** were found in Example 4-4 to be

$$\mathbf{D} = \mathbf{a}_\rho \frac{q}{2\pi\rho\ell} \qquad \mathbf{E} = \mathbf{a}_\rho \frac{q}{2\pi\epsilon\rho\ell}$$

These substituted into (4-58e) and integrated throughout the volume of the dielectric yield

$$U_e = \tfrac{1}{2}\int_V \left(\mathbf{a}_\rho \frac{q}{2\pi\rho\ell}\right)\cdot\left(\mathbf{a}_\rho \frac{q}{2\pi\epsilon\rho\ell}\right)\rho\, d\rho\, d\phi\, dz$$

$$= \frac{q^2}{8\pi^2\epsilon\ell^2}\int_{\rho=a}^{b}\int_{\phi=0}^{2\pi}\int_{z=0}^{\ell} \frac{\rho\, d\rho\, d\phi\, dz}{\rho^2} = \ell\frac{(q/\ell)^2}{4\pi\epsilon}\,\ln\frac{b}{a} \tag{4-62}$$

A useful application of the energy integral (4-58c) is to the capacitor of Figure 4-6. The fact that the two conductors, carrying q and $-q$, are at the equipotentials $\Phi = \Phi_1$ and $\Phi = \Phi_2$ permits simplifying (4-58c) as follows:

$$U_e = \tfrac{1}{2}\int_S \rho_s \Phi\, ds = \tfrac{1}{2}\Phi_1\int_{S_1}\rho_s\, ds + \tfrac{1}{2}\Phi_2\int_{S_2}\rho_s\, ds$$

in which the surface integrals, from (4-45), denote q and $-q$ on the conductors. Thus

$$U_e = \tfrac{1}{2}q(\Phi_1 - \Phi_2) = \tfrac{1}{2}qV \text{ J} \tag{4-63a}$$

wherein V for $\Phi_1 - \Phi_2$ has been substituted from (4-46). Putting (4-47) into (4-63a) yields alternatively

$$U_e = \tfrac{1}{2}CV^2 \text{ J} \tag{4-63b}$$

$$U_e = \frac{1}{2}\frac{q^2}{C} \text{ J} \tag{4-63c}$$

which show that the stored electric field energy is proportional to the square of either V or q.

The equivalence of (4-63) to (4-61) enables finding the capacitance of a two-conductor device in terms of energy. Thus, solving for C in (4-63b) or (4-63c) and substituting for U_e with (4-61a) yields the equivalent results

$$C = \frac{2U_e}{V^2} = \frac{1}{V^2}\int_V \mathbf{D}\cdot\mathbf{E}\, dv \text{ F} \tag{4-64a}$$

$$C = \frac{q^2}{2U_e} = \frac{q^2}{\int_V \mathbf{D}\cdot\mathbf{E}\, dv} \text{ F} \tag{4-64b}$$

EXAMPLE 4-10. Determine C of the coaxial capacitor of Example 4-8 from its stored energy.

From Example 4-9, the energy of the coaxial pair of length ℓ is

$$U_e = \ell \frac{(q/\ell)^2}{4\pi\epsilon} \ln \frac{b}{a}$$

This is in terms of q, so putting it into (4-64b) yields

$$C = \frac{q^2}{2U_e} = \frac{q^2}{\dfrac{2\ell(q/\ell)^2}{4\pi\epsilon}\ln\dfrac{b}{a}} = \frac{2\pi\epsilon\ell}{\ln\dfrac{b}{a}}$$

which agrees with Example 4-8.

4-8 Poisson's and Laplace Equations

In the previous sections, the solutions of electrostatic field problems were obtained by the methods:

1. By integrating (4-13) throughout the given static charge distribution in free space to find **E**.
2. By integrating Gauss' law (4-5) with respect to certain symmetric charge and dielectric configurations to find **D**, and thus **E**.
3. By integrating (4-35) throughout a static charge distribution in free space to find the potential Φ, from which **E** is found using $-\nabla\Phi$. Conversely,

in problems for which **E** is known, the potential Φ can be obtained from (4-38), the line integral of **E**.

All three methods have the disadvantage of requiring a specification of the charge distribution producing the electrostatic field. An approach that removes this requirement by treating problems of electrostatics as *boundary-value problems* is considered in the following.

A boundary-value problem of electrostatics is concerned with finding field solutions of Maxwell's divergence and curl relations, (4-1) and (4-2), that also satisfy the boundary conditions of the problem. To this end, working with the divergence and curl relations directly requires manipulations of the components of **E** or **D**, which proves to be more cumbersome than necessary. A restatement of the problem in terms of the scalar potential field Φ is seen to be desirable.

A partial differential equation in terms of the potential $\Phi(u_1, u_2, u_3)$ can be derived by combining the Maxwell relations (4-1) and (4-2). With $\mathbf{D} = \epsilon\mathbf{E}$, (4-1) is written

$$\nabla\cdot(\epsilon\mathbf{E}) = \rho_v \tag{4-65}$$

and **E** is conservative so that (4-31) applies; thus (4-65) becomes

$$\nabla\cdot(\epsilon\nabla\Phi) = -\rho_v \tag{4-66}$$

a partial differential equation known as Poisson's equation. In this form it is correct even though the dielectric region is inhomogeneous (ϵ a function of position). If ϵ is a *constant*, (4-66) takes the more usual form: $\nabla\cdot\nabla\Phi = -\rho_v/\epsilon$ or with the notation $\nabla\cdot\nabla\Phi = \nabla^2\Phi$ of (2-79)

$$\nabla^2\Phi = -\frac{\rho_v}{\epsilon} \tag{4-67}$$

Sometimes $\nabla^2\Phi$ is called the *Laplacian of* Φ, expansions of which are given by (2-77), (2-80), and (2-81) in the common coordinate systems.

If no free charge exists in the region ($\rho_v = 0$), the generalized Poisson equation (4-66) reduces to $\nabla\cdot(\epsilon\nabla\Phi) = 0$, known as Laplace's equation, applicable to dielectric regions which may be inhomogeneous. For a region with a constant ϵ, therefore

$$\nabla^2\Phi = 0 \tag{4-68}$$

The common form of Laplace's equation (4-68), together with the particular space boundary conditions which Φ is required to satisfy, constitute a boundary-value problem in a charge-free region.

EXAMPLE 4-11. A pair of long, coaxial, circular conductors is statically charged with its inner conductor at the potential $\Phi = V$ relative to the outer conductor, assumed at zero potential. The intervening region is a homogeneous dielectric with a permittivity ϵ. Solve Laplace's equation, subject to the boundary

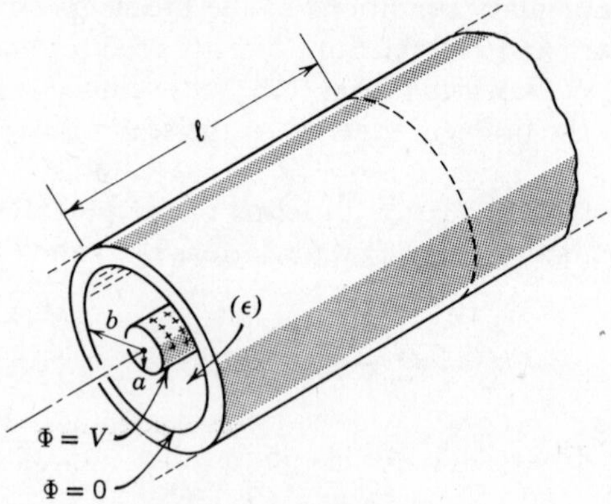

Example 4-11

conditions, for the potential anywhere between the conductors. Obtain also **E** in the dielectric, q on the conductors, and the capacitance (of a length ℓ) of the system.

From symmetry the fields are independent of ϕ and z, assuming fringing effects are neglected. Then Laplace's equation (4-68), by use of (2-80), reduces to the ordinary differential equation

$$\nabla^2\Phi \equiv \frac{1}{\rho}\frac{\partial}{\partial\rho}\left(\rho\frac{\partial\Phi}{\partial\rho}\right) = 0 \tag{4-69}$$

Integrating once obtains $\rho\,\partial\Phi/\partial\rho = C_1$, and a second integration yields the solution

$$\Phi(\rho) = C_1 \ln \rho + C_2 \tag{1}$$

The boundary conditions are applied to evaluate C_1 and C_2. At $\rho = b$, $\Phi = 0$ so that (1) yields $0 = C_1 \ln b + C_2$, to permit expressing C_2 in terms of C_1 as $C_2 = -C_1 \ln b$. Substituting this back into (1) yields

$$\Phi(\rho) = C_1(\ln \rho - \ln b) = C_1 \ln \frac{\rho}{b} \tag{2}$$

The second boundary condition, $\Phi(a) = V$, applied to (2) produces $C_1 = -V/\ln(b/a)$, whence (2) becomes

$$\Phi(\rho) = \frac{V}{\ln \dfrac{b}{a}} \ln \frac{b}{\rho} \tag{4-70}$$

the desired solution for Φ anywhere between the conductors, written in terms of V. As a check, note that setting $\rho = a$ and $\rho = b$ yields respectively the boundary values $\Phi = V$ and $\Phi = 0$.

One finds $\mathbf{E}$ from (4-70) by use of (4-31). The expansion of (2-14b) yields

$$\mathbf{E} = -\nabla\Phi = -\mathbf{a}_\rho \frac{\partial \Phi}{\partial \rho} = \mathbf{a}_\rho \frac{V}{\ln \dfrac{b}{a}} \frac{1}{\rho} \tag{4-71}$$

To find the total charge on either conductor, the charge density ρ_s is required, obtained from the boundary condition (4-30). At the inner surface $\rho = a$

$$\rho_s = D_n = D_\rho]_{\rho=a} = \epsilon E_\rho]_{\rho=a} = \frac{\epsilon V}{a \ln \dfrac{b}{a}} \tag{4-72}$$

obtaining the charge in a length ℓ

$$q = \rho_s(2\pi a\ell) = \frac{2\pi\epsilon\ell V}{\ln \dfrac{b}{a}} \tag{4-73}$$

The definition (4-47) of capacitance thus yields

$$C = \frac{q}{V} = \frac{2\pi\epsilon\ell}{\ln \dfrac{b}{a}} \tag{3}$$

which checks with (4-51) in Example 4-8.

While this example does not exhibit a great economy of effort when compared with the previous methods used to solve this one-dimensional problem, the chief merit of boundary-value methods for solving electrostatic field problems lies in their applicability to two- and three-dimensional systems lacking useful symmetries and not possessing known charge distributions. The latter is taken up in Section 4-10.

*4-9 Uniqueness of Electrostatic Field Solutions

It is of importance to know, once one has obtained (by whatever means) a solution to an electrostatic field problem that it is the only solution possible; i.e., it is a *unique* solution. The mathematical model furnished by potential theory would be of little use if it furnished several solutions to a given problem, among which the correct solution of the physical problem might have to be verified by experiment or in some other manner.

It can be shown that potential solutions of the following classes of boundary-value problems are unique solutions:

1. *The Dirichlet Problem.* A potential solution $\Phi(u_1, u_2, u_3)$ of Laplace's equation is unique if Φ satisfies a specified boundary condition

$$\Phi = \Phi_s(u_1, u_2, u_3) \tag{4-74}$$

on the closed boundary S of the region.

2. *The Neumann Problem.* A potential solution $\Phi(u_1, u_2, u_3)$ of Laplace's equation is unique within a constant value if the normal derivative of Φ satisfies a specified boundary condition

$$\frac{\partial \Phi}{\partial n} = \left.\frac{\partial \Phi}{\partial n}\right]_S \tag{4-75}$$

on the closed boundary S of the region.

3. *The Mixed Boundary-Value Problem.* A potential solution of Laplace's equation is unique if it satisfies (4-74) on a part of S, and (4-75) on the remainder.

A proof of (1) is established by supposing that there are *two* solutions, Φ and Φ', each of which satisfies Laplace's equation ($\nabla^2\Phi = 0$ and $\nabla^2\Phi' = 0$) everywhere within the volume V bounded by the closed surface S shown in Figure 4-9(*a*), and both of which satisfy the same boundary condition Φ_s as follows:

$$\Phi = \Phi' = \Phi_s(u_1, u_2, u_3) \text{ on } S \tag{4-76}$$

The specified boundary condition $\Phi_s(u_1, u_2, u_3)$ is, in general, a function of position on S. For some problems, S may consist of several (n) conductors as suggested by Figure 4-9(*b*), in which the boundary condition (4-76) is a sequence of potentials $\Phi_{s1}, \Phi_{s2}, \ldots, \Phi_{sP}$ on the respective surfaces $S_1, S_2, \ldots, S_n$. From (4-76), the difference of the two identical boundary conditions is zero, i.e.,

$$\Phi - \Phi' = 0 \text{ on } S \tag{4-77}$$

The uniqueness of Φ is established if one can also show that $\Phi - \Phi' = 0$ in V. To this end, Green's first integral identity (2-91) has the equivalent forms

$$\int_V [f\nabla^2 g + (\nabla f)\cdot(\nabla g)]\, dv = \oint_S f(\nabla g)\cdot d\mathbf{s} = \oint_S f \frac{\partial g}{\partial n}\, ds \tag{4-78}$$

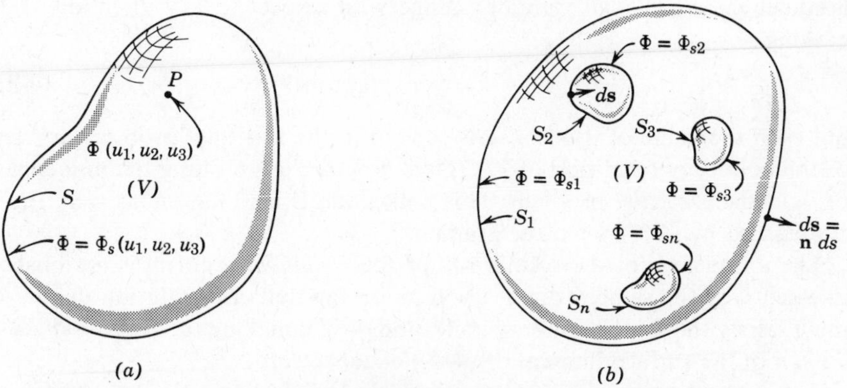

Figure 4-9. Closed surface configurations relative to boundary-value problems of electrostatics. (*a*) Volume region V bounded by a closed surface S on which the boundary condition is specified. (*b*) Variation of (*a*): V bounded by $n - 1$ interior surfaces and exterior surface S_1. A special case occurs if $S_1 = S_\infty$.

true for any pair whatsoever of well-behaved functions f and g. It must therefore hold if $f = g$, and equally well for $f = \Phi - \Phi'$, the difference of the functions being examined for uniqueness. With the latter, Green's identity takes the form

$$\int_V \{(\Phi - \Phi')\nabla^2(\Phi - \Phi') + [\nabla(\Phi - \Phi')]^2\}\, dv = \oint_S (\Phi - \Phi')\left[\frac{\partial \Phi}{\partial n} - \frac{\partial \Phi'}{\partial n}\right] ds \tag{4-79}$$

With Φ and Φ' satisfying Laplace's equation, it evidently follows that $\nabla^2(\Phi - \Phi') = 0$, causing the first term of the volume integral of (4-79) to vanish, yielding

$$\int_V [\nabla(\Phi - \Phi')]^2\, dv = \oint_S (\Phi - \Phi')\left[\frac{\partial \Phi}{\partial n} - \frac{\partial \Phi'}{\partial n}\right] ds \tag{4-80}$$

Because of (4-77), the surface integral of (4-80) is zero, obtaining

$$\int_V [\nabla(\Phi - \Phi')]^2\, dv = 0$$

The integrand is a squared quantity and is therefore everywhere positive in V, but the only way a nonnegative function can integrate to zero as indicated is if the integrand is zero everywhere in V; thus $\nabla(\Phi - \Phi') = 0$. A zero gradient means $\Phi - \Phi'$ cannot change with respect to any direction in V, making

$$\Phi - \Phi' = \text{constant in } V \tag{4-81}$$

but even the value of this constant is zero in the Dirichlet problem, in view of the boundary condition (4-77). Thus $\Phi = \Phi'$, establishing the uniqueness of Φ in the *Dirichlet* problem. This makes the **E** field unique as well, for **E** is obtained by (4-31) from the gradient of Φ.

The uniqueness of the solution Φ of the *Neumann* problem is established in essentially the same fashion, upon observing that each solution Φ and Φ' must satisfy the same boundary condition (4-75), making the factor $\partial\Phi/\partial n - \partial\Phi'/\partial n$ in the surface integral of (4-80) equal to zero.

The presence of a homogeneous, insulating dielectric with the permittivity ϵ was assumed for V in the proof given. The uniqueness of the solutions is still valid even though an inhomogeneous dielectric is present (ϵ a function of position), as well as for a dielectric partitioned into several homogeneous regions with different ϵ values. The proof follows upon subdividing V by means of surfaces lying just to either side of the interfaces, but it is not given here.†

*4-10 Laplace's Equation and Boundary-Value Problems

In Example 4-11 of Section 4-8, an instance of the direct integration of Laplace's equation (4-68b) in one dimension was described. In the present section, a method for extending the direct-integration procedure to two-dimensional conductor systems is given. The *separation of variables method* is utilized, which, via the assumption of a product-type solution, permits a conversion of the Laplace equation in two or three space variables into the same number of ordinary differential equations, solutions of which are obtained by standard methods. Laplace's equation has been found separable by this method in some 11 orthogonal coordinate systems‡. The present discussion is confined to the cartesian system.

Consider the solution of Laplace's equation in the two-dimensional cartesian system. In a charge-free, homogeneous, linear and isotropic region, (4-68) is written, by use of (2-77)

$$\frac{\partial^2\Phi}{\partial x^2} + \frac{\partial^2\Phi}{\partial y^2} = 0 \tag{4-82}$$

† A proof of this extension of the uniqueness theorem is found in Smythe. *Static and Dynamic Electricity*. New York: McGraw-Hill, 1950, p. 56.

‡ See Eisenhart, L. P. "Separable systems of Staekel," *Annals of Math.*, **35**, 1934, p. 284.

The separation of variables method begins by assuming a product solution of the form

$$\Phi(x, y) = X(x)Y(y) \tag{4-83}$$

in which $X(x)$ and $Y(y)$ respectively denote functions of x and of y only. Substitution into (4-82) yields

$$X''Y + XY'' = 0$$

in which the double primes denote differentiation with respect to x or y, whichever applies. Dividing by XY

$$\frac{X''}{X} + \frac{Y''}{Y} = 0 \tag{4-84}$$

stating that the sum of a function of x only plus a function of y only equals a constant (zero). This is possible for all values of x and y in an assigned region only if each term of (4-84) equals a constant. Denoting them by $-k_x^2$ and $-k_y^2$ yields

$$\frac{X''}{X} = -k_x^2 \qquad \frac{Y''}{Y} = -k_y^2 \tag{4-85}$$

and if (4-84) is to be satisfied, one obtains

$$k_x^2 + k_y^2 = 0 \tag{4-86}$$

This means $k_x^2 = -k_y^2$, or

$$k_x = \pm jk_y \tag{4-87}$$

implying that if one constant (k_x or k_y) is *real*, the other must be *imaginary*. Thus (4-85) are ordinary differential equations, being functions of one independent variable (x or y), and so they are written

$$\frac{d^2X}{dx^2} + k_x^2X = 0 \tag{4-88a}$$

$$\frac{d^2Y}{dy^2} + k_y^2Y = 0 \tag{4-88b}$$

If k_x is taken to be *real*, (4-88a) has exponential solutions expressible in either imaginary exponential or trigonometric form as follows:

$$X(x) = C_1 \cos k_x x + C_2 \sin k_x x \tag{4-89a}$$

or

$$X(x) = C_1' e^{jk_x x} + C_2' e^{-jk_x x} \tag{4-89b}$$

From (4-87), real k_x requires k_y to take on the *imaginary* values $\pm jk_x$, to make the solutions of (4-88b) become the real exponential or equivalent hyperbolic forms†

$$Y(y) = C_3 e^{k_x y} + C_4 e^{-k_x y} \tag{4-90a}$$

or

$$Y(y) = C_3' \cosh k_x y + C_4' \sinh k_x y \tag{4-90b}$$

Static potential field solutions of physical problems are *real* solutions, making the real trigonometric solutions (4-89a) preferable to (4-89b). Moreover, choosing the real exponential solutions (4-90a) in lieu of their hyperbolic forms yields for the product solution (4-83)

$$\Phi(x, y) = X(x)\,Y(y) = (C_1 \cos k_x x + C_2 \sin k_x x)(C_3 e^{k_x y} + C_4 e^{-k_x y}) \tag{4-91a}$$

in which $C_1, \ldots, C_4$ are real constants. If the preceding development had begun with the assumption in (4-87) of a real k_y instead of k_x, making k_x imaginary in that event, then (4-91a) would become

$$\Phi(x, y) = (C_1 e^{k_y x} + C_2 e^{-k_y x})(C_3 \cos k_y y + C_4 \sin k_y y) \tag{4-91b}$$

The choice of (4-91a) or (4-91b) depends on the boundary conditions of the given problem. Indeed, almost all boundary conditions of practical interest are such that a single solution of the form of (4-91) is insufficient to satisfy the potential conditions at the boundaries; because Laplace's equation is *linear*, an *infinite sum* of solutions like (4-91), containing different but proper values of k_x or k_y, constitutes a valid representation of $\Phi(x, y)$. It is shown in the following example that the methods of Fourier series are important in the evaluation of the cofficients of such series representations.

† The hyperbolic functions in (4-90b) are defined as the linear sums of exponential functions

$$\cosh a \equiv \frac{e^a + e^{-a}}{2} \qquad \sinh a \equiv \frac{e^a - e^{-a}}{2}$$

EXAMPLE 4-12. A two-dimensional, air-filled, infinitely long channel of semi-infinite depth in the y dimension as shown, is formed of conducting planes on three sides, insulated at the corners. The bottom plate is at V V relative to the sides at $x = 0$ and $x = a$. Find the potential anywhere inside the channel region.

A solution $\Phi(x, y)$ of Laplace's equation (4-82)

$$\frac{\partial^2\Phi}{\partial x^2} + \frac{\partial^2\Phi}{\partial y^2} = 0 \tag{1}$$

is to be found, subject to the boundary conditions

$$\Phi(0, y) = 0 \tag{2}$$

$$\Phi(a, y) = 0 \tag{3}$$

$$\Phi(x, \infty) = 0 \tag{4}$$

$$\Phi(x, 0) = V \tag{5}$$

The solution of (1) was shown to be (4-91). In view of the boundary condition (4) at $y \to \infty$, choose the form (4-91a)

$$\Phi(x, y) = (C_1 \cos k_x x + C_2 \sin k_x x)(C_3 e^{k_x y} + C_4 e^{-k_x y}) \tag{6}$$

The unknowns C_1 through C_4 and k_x are evaluated by use of the boundary conditions. Applying (2) to (6) yields

$$\Phi(0, y) = 0 = (C_1)(C_3 e^{k_x y} + C_4 e^{-k_x y})$$

to obtain $C_1 = 0$. Then (6) becomes

$$\Phi(x, y) = C_2 \sin k_x x(C_3 e^{k_x y} + C_4 e^{-k_x y}) \tag{7}$$

Applying the boundary condition (3) to the latter obtains

$$\Phi(a, y) = 0 = C_2 \sin k_x a(C_3 e^{k_x y} + C_4 e^{-k_x y})$$

satisfied only if $\sin k_x a = 0$, whence $k_x a = m\pi$, making $k_x = (m\pi/a)$ $(m = 1, 2, 3, \ldots)$. Then (7) becomes

$$\Phi(x, y) = C_2 \sin \frac{m\pi}{a} x(C_3 e^{(m\pi/a)y} + C_4 e^{-(m\pi/a)y}) \tag{8}$$

The third boundary condition (4) yields

$$\Phi(x, \infty) = 0 = \lim_{y \to \infty} \left[C_2 \sin \frac{m\pi}{a} x(C_3 e^{(m\pi/a)y} + C_4 e^{-(m\pi/a)y})\right]$$

having a zero limit as $y \to \infty$ only if $C_3 = 0$, since any nonzero C_3 would produce an infinite Φ at the remote boundary $y \to \infty$, a nonphysical result.

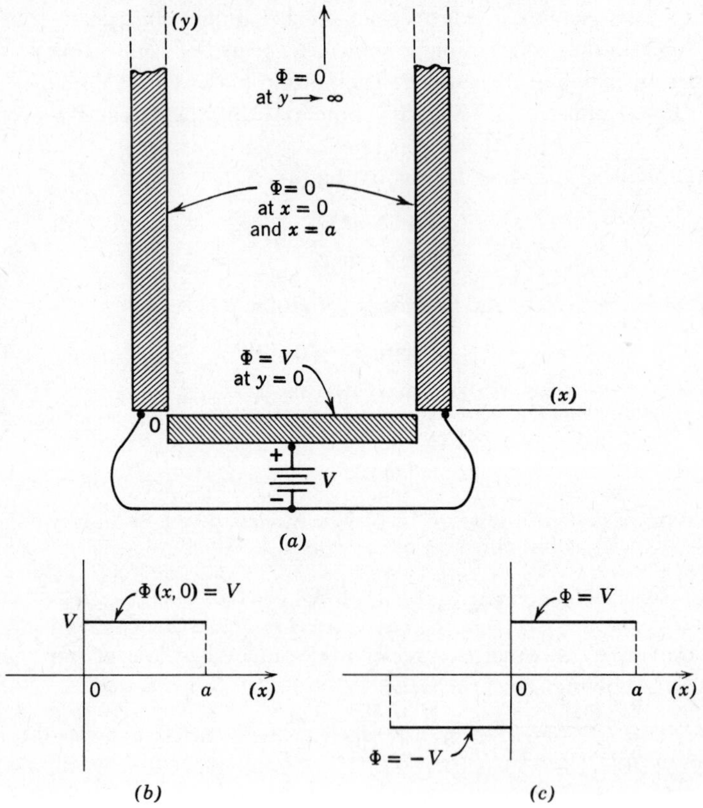

Thus (8) becomes

$$\Phi(x, y) = C_2 C_4 e^{-(m\pi/a)y} \sin \frac{m\pi}{a} x \qquad m = 1, 2, \ldots \tag{9}$$

a function exponentially decreasing in y. An attempt to apply the last boundary condition (5) to (9) yields

$$\Phi(x, 0) = V = C_2 C_4 \sin \frac{m\pi}{a} x$$

an equality impossible to satisfy for all x within the $(0, a)$ x range at the $y = 0$ boundary, but since m can assume any positive integer value $m = 1, 2, 3, \ldots$, the linearity of the differential equation (1) permits forming an infinite sum of solutions like (9) ranging over all the m integers, i.e.,

$$\Phi(x, y) = \sum_{m=1}^{\infty} A_m e^{-(m\pi/a)y} \sin \frac{m\pi}{a} x \tag{10}$$

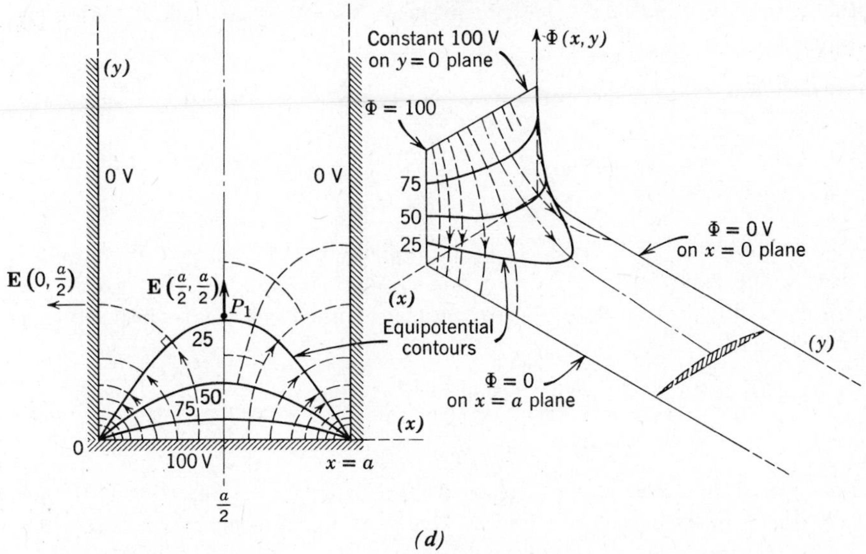

Example 4-12. **(*a*) Potential well of infinite height. (*b*) Boundary condition on the physical half range (0, *a*). (*c*) Odd function Φ assumed for the boundary condition over the complete period (−*a*, *a*). (*d*) Resulting field.**

Equation (10) is a Fourier series (trigonometric series) representation for $\Phi(x, y)$ with respect to the variable x. The unknown coefficients A_m are to be determined at $y = 0$ by applying the boundary condition (5) to the series (10), yielding

$$\Phi(x, 0) = V = \sum_{m=1}^{\infty} A_m \sin \frac{m\pi}{a} x \tag{11}$$

a *Fourier representation* of the boundary condition (5). Standard Fourier techniques yield the unknown coefficients A_m. The spatial *period* of the Fourier representation must first be defined, however. Note that the boundary condition $\Phi(x, 0) = V$ is specified over the physical x range $(0, a)$ between the channel walls, as in (*b*) of the accompanying figure.

By defining $(0, a)$ as one-half of a *total spatial period* $(-a, a)$, the rest of the range $(-a, 0)$ may be filled in with an arbitrary function Φ, as long as the Fourier expansion of $\Phi(x, 0)$ converges to $\Phi = 0$ V at the endpoints of the physical half-range $(0, a)$ as required by the boundary conditions (2) and (3). The latter is accomplished nicely by assuming $\Phi(x, 0)$ of (11) to be an odd function defined over the period $(-a, a)$, as in (*c*). The representation (11) of the boundary condition (5) then embraces the complete spatial period as follows:

$$\begin{aligned} \Phi(x, 0) &= V = \sum_{m=1}^{\infty} A_m \sin \frac{m\pi}{a} x \qquad 0 < x < a \\ \Phi(x, 0) &= -V = \sum_{m=1}^{\infty} A_m \sin \frac{m\pi}{a} x \qquad -a < x < 0 \end{aligned} \tag{12}$$

One can find the coefficients A_m by multiplying (12) by $\sin (n\pi/a)x$ and integrating the result over the orthogonality interval $(-a, a)$. The use of the orthogonality properties of the trigonometric functions yields the following coefficients:

$$A_m = \frac{2}{a}\int_0^a V \sin \frac{m\pi}{a} x\, dx = \frac{2V}{m\pi}(1 - \cos m\pi) \tag{13}$$

implying $A_1 = 2V/\pi$, $A_2 = 0$, $A_3 = 2V/3\pi$, $A_4 = 0, \ldots$. Inserting (13) into (10) thus obtains the desired Fourier representation for any position between the conductors

$$\begin{aligned} \Phi(x, y) &= \sum_{m=1}^{\infty} \left[\frac{2V}{m\pi}(1 - \cos m\pi)\right] e^{-(m\pi/a)y} \sin \frac{m\pi}{a} x \\ &= \frac{4V}{\pi}\left[e^{-(\pi/a)y} \sin \frac{\pi}{a} x + \tfrac{1}{3} e^{-(3\pi/a)y} \sin \frac{3\pi}{a} x + \cdots\right] \text{V} \end{aligned} \tag{14}$$

A sketch of $\Phi(x, y)$ is shown in (*d*). The corresponding electric field **E** is found by use of (4-31)

$$\begin{aligned} \mathbf{E} &= -\nabla\Phi = -\mathbf{a}_x \frac{\partial \Phi}{\partial x} - \mathbf{a}_y \frac{\partial \Phi}{\partial y} \\ &= -\mathbf{a}_x \sum_{m=1}^{\infty} \left[\frac{2V}{a}(1 - \cos m\pi)\right] e^{-(m\pi/a)y} \cos \frac{m\pi}{a} x \\ &\quad + \mathbf{a}_y \sum_{m=1}^{\infty} \left[\frac{2V}{a}(1 - \cos m\pi)\right] e^{-(m\pi/a)y} \sin \frac{m\pi}{a} x \\ &= -\mathbf{a}_x \frac{4V}{a}\left[e^{-(\pi/a)y} \cos \frac{\pi}{a} x + e^{-(3\pi/a)y} \cos \frac{3\pi}{a} x + \cdots\right] \\ &\quad + \mathbf{a}_y \frac{4V}{a}\left[e^{-(\pi/a)y} \sin \frac{\pi}{a} x + e^{-(3\pi/a)y} \sin \frac{3\pi}{a} x + \cdots\right] \text{V/m} \end{aligned} \tag{15}$$

The flux of **E**, orthogonal to the equipotential surfaces, is also depicted.

To illustrate the use of (14) and (15), suppose $V = 100$ V and one desires the potential at $x = a/2$, $y = a/2$, located along the central axis at P_1 in the figure.

The potential, from (14), is

$$\Phi\left(\frac{a}{2}, \frac{a}{2}\right) = \frac{400}{\pi}\left\{e^{-\pi/2}\sin\frac{\pi}{2} + \tfrac{1}{3}e^{-3\pi/2}\sin\frac{3\pi}{2} + \cdots\right\}$$

$$= 127\{0.204 - \tfrac{1}{3}(0.00848) + \cdots\} = 25.6 \text{ V}$$

The electric field there is found using (15), a result seen to depend on the a dimension. Choosing $a = 1$ m,

$$\mathbf{E}\left(\frac{a}{2}, \frac{a}{2}\right) = -\mathbf{a}_x 0 + \mathbf{a}_y \frac{400}{1}\left\{e^{-\pi/2}\sin\frac{\pi}{2} + e^{-3\pi/2}\sin\frac{3\pi}{2} + \cdots\right\}$$

$$= \mathbf{a}_y 400[0.204 - 0.00848 + \cdots] = \mathbf{a}_y 78.3 \text{ V/m}$$

From (15) it is seen that $\mathbf{E}$ is inversely dependent on a. Decreasing a to 1 cm thus increases $\mathbf{E}$ by the factor 100 to yield $\mathbf{E} = \mathbf{a}_y 7830$ V/m at the point $(a/2, a/2)$, a consequence of compressing the equipotential contours more closely together.

4-11 Image Methods

The method of images about to be described takes advantage of the uniqueness property of potential solutions. It consists of replacing a problem, involving one or more statically charged conductors, with an equivalent problem of suitably located point or line charges (so-called *image* charges) that yield precisely the same electrostatic field as the original problem. The well-known fields of point or line charges can then be utilized to obtain a solution of the original boundary-value problem. The number of charged-conductor configurations that can be solved in this manner is relatively small, but included are enough examples of physical importance to make the method worthy of treatment.

The image method is illustrated by an example in Figure 4-10. Suppose two point charges, q and $-q$, are spaced $2d$ m in free space as in (a) of that figure. The combined potential Φ at any position P is given by two terms of (4-41)

$$\Phi(x, y, z) = \frac{q}{4\pi\epsilon_0\sqrt{(d - x)^2 + y^2 + z^2}} - \frac{q}{4\pi\epsilon_0\sqrt{(d + x)^2 + y^2 + z^2}} \tag{4-92}$$

The equipotential surfaces are found by equating (4-92) to constant potentials; a family of equipotential surfaces obtained in this way is shown dashed in Figure 4-10(a). Recalling that a *conductor* immersed in an electrostatic field has its surface at a constant potential, replacing the interior of the equipotential surfaces $\Phi = \Phi_0$ and $\Phi = -\Phi_0$ with conductors as in Figure 4-10(b)

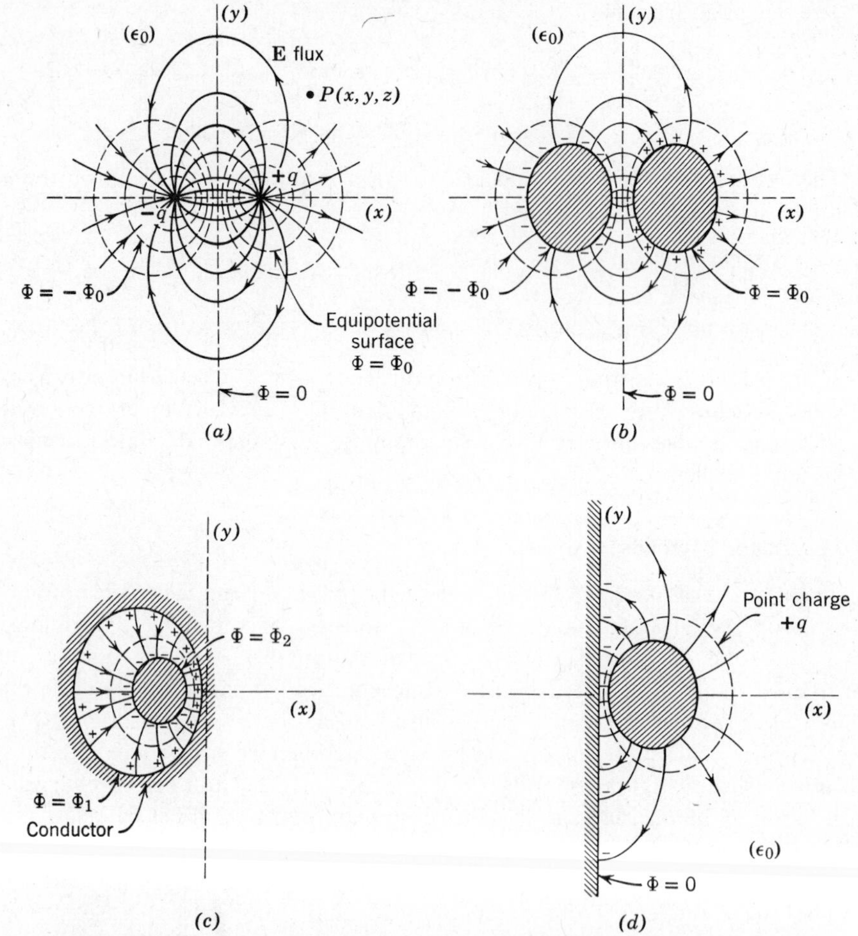

Figure 4-10. Three examples of charged conductor systems, the exact fields of which are obtained from image system (*a*). (*a*) Two electrostatic point charges and their E and Φ fields. (*b*) Replacing interior of surfaces ($\Phi = \pm\Phi_0$) with conductors. (*c*) A variation of (*b*). (*d*) Replacing region to left of $\Phi = 0$ in (*b*) with a conductor.

cannot alter the **E** field exterior to the conductors. The original image charges $\pm q$ of Figure 4-10(*a*) moreover appear as conductor surface charges totalling $\pm q$, a conclusion reached from Gauss' law (4-5) integrated over the conductor surfaces. The image charge system of Figure 4-10(*a*) therefore yields the desired fields of the two-conductor system of Figure 4-10(*b*), obtaining the same Φ and **E** solutions *outside* the conductors in the latter.

A complementary system (one conductor within another) is shown in Figure 4-10(c); its fields are also obtainable from the image system of Figure 4-10(a).

One can see that the symmetry plane $x = 0$ of Figure 4-10(a) is the equipotential surface $\Phi = 0$, evident from setting the potential expression (4-92) to zero. Thus, if a conductor having the shape of one of the equipotential surfaces is located to the right of the conducting plane at $x = 0$ as in Figure 4-10(d), the field between the conductors is once more specified by the image problem of Figure 4-10(a). The field to the left of the plane is nullified, in terms of boundary condition (4-30), by the presence on its surface of the charge density

$$\rho_s = D_n\big]_{x=0} = \epsilon_0 E_x\big]_{x=0} = -\epsilon_0 \frac{\partial \Phi}{\partial x}\bigg]_{x=0} \tag{4-93}$$

Hence, the x derivative of (4-92), with $x = 0$ in the result, yields

$$\rho_s = \frac{-qd}{2\pi[d^2 + y^2 + z^2]^{3/2}} \text{ C/m}^2 \tag{4-94}$$

Extensions of the image system of Figure 4-10(a) can be deduced from superposition as depicted in Figure 4-11. For example, a system of fixed point charges $q_1, q_2, \ldots$, placed near a large conducting plane as in Figure 4-11(a) has a static field in the right-hand space given by the sum of the fields of the original charges and their images shown. The zero potential on the median plane is maintained by that image system. A line charge of arbitrary shape placed near a plane conductor provides another image equivalence as in Figure 4-11(b), a special case of which is the straight-line charge of Figure 4-11(c). These schemes can be extended further with the image equivalent of a charge q near the perpendicular intersection of two conducting planes as in Figure 4-11(d); three image charges are needed to establish zero potential on both planes.

The *parallel-line charge* system of Figure 4-11(c), duplicated in Figure 4-12(a), is an important image system that enables finding the electrostatic fields of parallel, round conductors as developed in the following. Assume two infinitely long, parallel line charges separated $2d$ and possessing the uniform charge densities ρ_ℓ and $-\rho_\ell$. The latter are denoted by the ratios q/ℓ and $-q/\ell$, signifying the charges per length ℓ of each line. Because of the infinite extent of the system, the analysis is confined to the $z = 0$ plane, restricting it to two dimensions (x, y) as in the section view of Figure 4-12(b). The equipotential surfaces of this parallel line charge system are right circular cylinders. To show this, note that the potential $\Phi(x, y)$ at P in Figure 4-12(b) is found from the superposition of the potentials $\Phi^{(+)}$ and $\Phi^{(-)}$ due to each

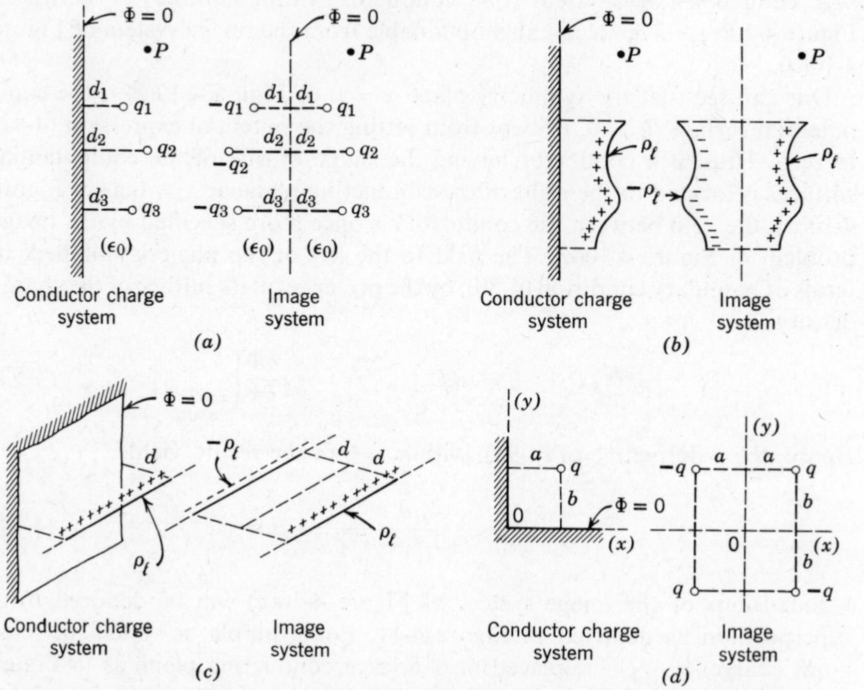

Figure 4-11. Image equivalents of static charge near infinite conducting planes. (*a*) Discrete charges near a conducting plane. (*b*) Arbitrary line charge near a conducting plane. (*c*) Line charge parallel to a conducting plane. (*d*) Point charge near intersection of two conducting planes.

line. Each produces the potential field (4-39); so with 0 chosen as the potential reference, the potentials at P due to q/ℓ and $-q/\ell$ become

$$\Phi^{(+)} = \frac{q}{2\pi\epsilon\ell}\ln\frac{d}{R_1} \qquad \Phi^{(-)} = \frac{-q}{2\pi\epsilon\ell}\ln\frac{d}{R_2} \tag{4-95}$$

Their sum is the total potential at P

$$\Phi(x, y) = \Phi^{(+)} + \Phi^{(-)} = \frac{q}{2\pi\epsilon\ell}\ln\frac{R_2}{R_1} \tag{4-96}$$

in which

$$R_1 = \sqrt{(x-d)^2 + y^2} \qquad R_2 = \sqrt{(x+d)^2 + y^2} \tag{4-97}$$

Observe from (4-96) that Φ ranges over all the real numbers, for as P approaches $-q/\ell (R_2 \to 0)$, there $\Phi \to -\infty$; while $\Phi \to \infty$ at the positive line charge.

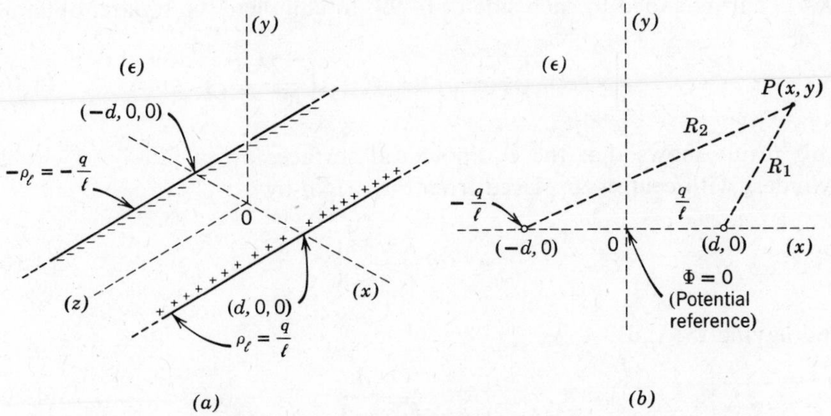

Figure 4-12. **Geometry of the parallel-line charge image system. (*a*) Parallel-line charges of uniform densities. (*b*) End view of (*a*) showing the two-dimensional geometry in the $z = 0$ plane.**

Equipotential surfaces are obtained by equating (4-96) to any desired constant potential $\Phi = \Phi_0$

$$\frac{q}{2\pi\epsilon\ell} \ell n \frac{R_2}{R_1} = \Phi_0 \tag{4-98a}$$

This means that any fixed, real ratio

$$\frac{R_2}{R_1} = K \tag{4-98b}$$

defines an equipotential surface on which $\Phi = \Phi_0$ prevails. Thus, $K = R_2/R_1 = 1$ defines the plane $x = 0$ bisecting the system. (Substituting $K = 1$ into (4-98a) reveals that $\Phi_0 = 0$ on it.) Other equipotential surfaces given by other K values are, in general, *circles* in the sectional view of Figure 4-12(*b*); if the z axis is included, they become *circular cylindrical surfaces*. This is proved by substituting (4-97) into (4-98b) as follows:

$$\frac{(x + d)^2 + y^2}{(x - d)^2 + y^2} = K^2$$

which expands into

$$x^2 - 2d\frac{K^2 + 1}{K^2 - 1}x + d^2 + y^2 = 0 \tag{4-99}$$

This reduces to the equation of a circle, $(x - h)^2 + y^2 = R^2$, if $d^2[(K^2 + 1)/$

$(K^2 - 1)]^2$ is added to each side of (4-99) to complete the square, obtaining

$$\left[x - d\frac{K^2 + 1}{K^2 - 1}\right]^2 + y^2 = \left(\frac{2Kd}{K^2 - 1}\right)^2 \tag{4-100}$$

This result shows that the equipotential surfaces are a family of circular cylinders with centers displaced from the origin by

$$h = d\frac{K^2 + 1}{K^2 - 1} \tag{4-101}$$

and having the radii

$$R = \frac{2Kd}{K^2 - 1} \tag{4-102}$$

Typical equipotential circular cylinders defined by (4-100) are illustrated in Figure 4-13. K-values less than 1 correspond to equipotential cylinders to the left of the origin, while $K > 1$ yields the cylinders on the right.

Taking the difference of the squares of (4-101) and (4-102) eliminates K to obtain $h^2 - R^2 = d^2$, whence

$$d = \sqrt{h^2 - R^2} \tag{4-103}$$

This gives the locations $\pm d$ of the image charges in Figure 4-13 in terms of R and h.

Upon now replacing the interior (or exterior) of any pair of equipotential cylinders of Figure 4-13 with *conductors* (carrying the surface charges q and

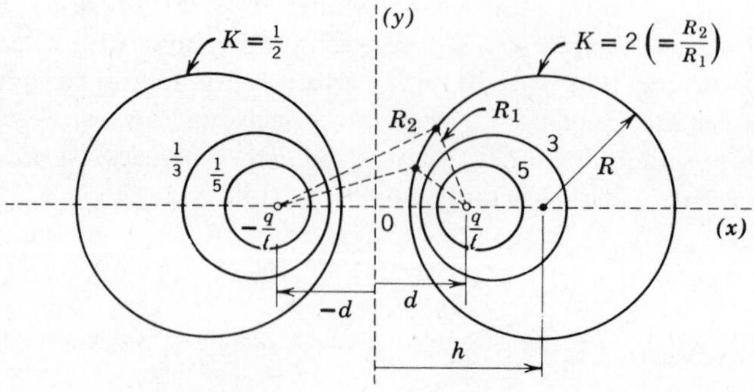

Figure 4-13. Equipotential surfaces of a parallel-line charge system, i.e., circular cylindrical surfaces.

$-q$ in every length ℓ), the electrostatic field problems such as those of Figure 4-14 can be considered to have been solved. The capacitance C of a length ℓ of the systems of Figure 4-14(a) and (b), for example, are found as follows. Dividing (4-101) by (4-102) to eliminate d obtains a quadratic expression in K, yielding

$$K = \frac{h}{R} \pm \sqrt{\left(\frac{h}{R}\right)^2 - 1} = \frac{h \pm d}{R} \tag{4-104}$$

with d given by (4-103). The positive and negative signs correspond to the positive and negative equipotential surfaces to the right and left of $x = 0$, respectively, in Figure 4-13. The potential Φ_0 of any equipotential cylinder in the *right-half* region is thus found from substituting (4-104) into (4-98a), but with the plane $x = 0$ at $\Phi = 0$ V, the potential difference V between a

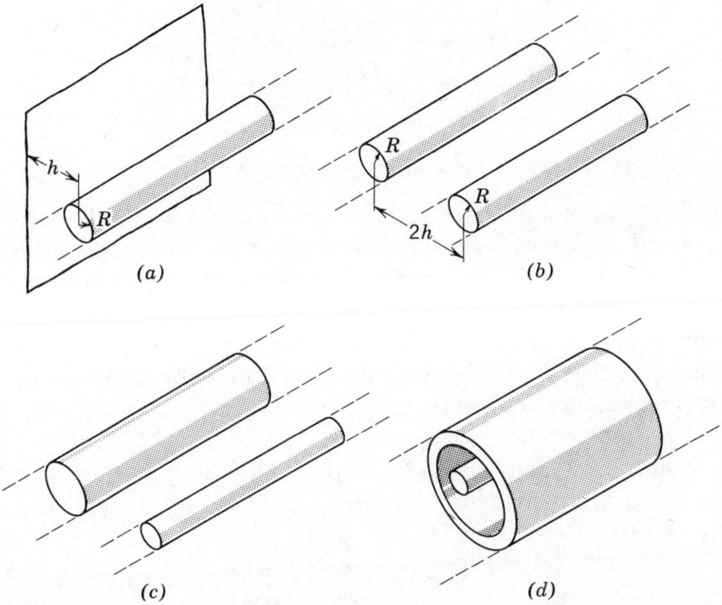

Figure 4-14. **Two-dimensional conductor systems. Solutions obtainable from image system of Figure 4-12. (*a*) Circular conductor parallel to a plane conductor. (*b*) Parallel circular conductors of equal size. (*c*) Parallel cylinders of unequal size. (*d*) Cylinders eccentrically located one inside the other.**

circular cylindrical conductor and the conducting plane of Figure 4-14(*a*) becomes $\Phi_0 - 0 = V$, yielding

$$V = \Phi_0 - 0 = \frac{q}{2\pi\epsilon\ell}\,\ell n\left[\frac{h}{R} + \sqrt{\left(\frac{h}{R}\right)^2 - 1}\right] \tag{4-105}$$

The capacitance of that system, from (4-48), is therefore

$$C = \frac{2\pi\epsilon\ell}{\ell n\left[\frac{h}{R} + \sqrt{\left(\frac{h}{R}\right)^2 - 1}\right]} \qquad \text{Wire above plane conductor} \tag{4-106}$$

In the *parallel-wire line* system of Figure 4-14(*b*), the potential (4-104) of an identical conductor in the left-half plane is just the negative of that of other conductor, yielding a potential difference between the conductors just twice that of (4-105) for the cylinder plane system. Its capacitance is therefore

$$C = \frac{\pi\epsilon\ell}{\ell n\left[\frac{h}{R} + \sqrt{\left(\frac{h}{R}\right)^2 - 1}\right]} \qquad \text{Parallel-wire line} \tag{4-107}$$

*4-12 An Approximation Method for Statically Charged Conductors

Occasionally approximation methods can be used for rapidly assessing the potentials and the capacitance of conductor systems. The technique described here depends on conductor dimensions being small compared to their separations, assuring that their surface charge distributions are not altered significantly by the proximity of the conductors. Then the electrostatic potential in the region can be obtained by simply *superposing* the potentials of the conductors taken separately.

An illustration of this concept is given in Figure 4-15. Suppose a long, circular conductor, isolated as in (*a*) of the figure, possesses for every length ℓ, a total charge q distributed uniformly over its surface. Its potential field is given by (4-39)

$$\Phi(P) = \frac{q}{2\pi\epsilon\ell}\,\ell n\,\frac{\rho_0}{\rho} \tag{4-108}$$

Two such conductors, possessing q/ℓ and q'/ℓ as in Figure 4-15(*b*) and kept reasonably apart as shown, produce a potential at P that is the sum of the potentials due to each conductor, yielding very nearly

$$\Phi(P) \cong \frac{q}{2\pi\epsilon\ell}\,\ell n\,\frac{\rho_0}{\rho} + \frac{q'}{2\pi\epsilon\ell}\,\ell n\,\frac{\rho_0'}{\rho'} \tag{4-109}$$

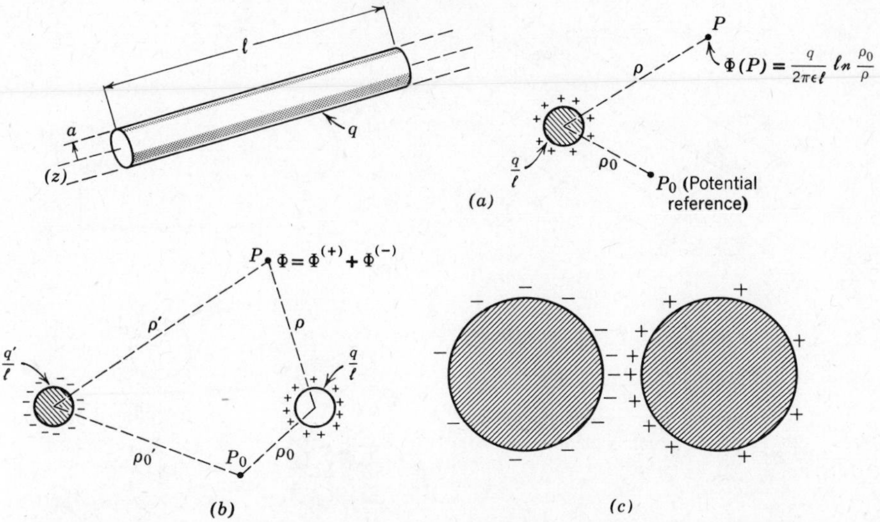

Figure 4-15. Circular cylindrical conductors, showing the effect of proximity on charge distributions and the superposed potential fields. (*a*) A round conductor and its potential field. (*b*) Wide spacing: D/a large. (*c*) Close spacing: D/a small.

This result is subject to an increasing error as the conductors are brought closer together as in Figure 4-15(*c*), in view of the charge redistribution taking place due to the attractive forces acting between the charges.

These arguments provide a basis for finding the approximate capacitance between a pair of conductors having known potential fields when taken separately. Conductors of practical interest in this class of problems are spheres and round wires. Figure 4-16 shows a few examples.

EXAMPLE 4-13. Find the approximate capacitance of the parallel-wire system of Figure 4-16(*a*), two conductors of unequal radii a_1 and a_2 separated by the center-to-center distance D.

The potential difference V between the conductors is obtained by superposing the potentials of each isolated conductor. Let the static charges on the conductors be q/ℓ and $-q/\ell$ C/m as shown, and the potential reference be at P_0 on the negative conductor. In the presence of only the conductor of radius a_1, the potential at P relative to P_0 in Figure 4-17(*a*) is obtained from (4-108), yielding

$$\Phi^{(+)}(P) = \frac{q}{2\pi\epsilon\ell} \ell n \frac{D}{a_1}$$

with the distance from the source to the reference P_0 observed to be D.

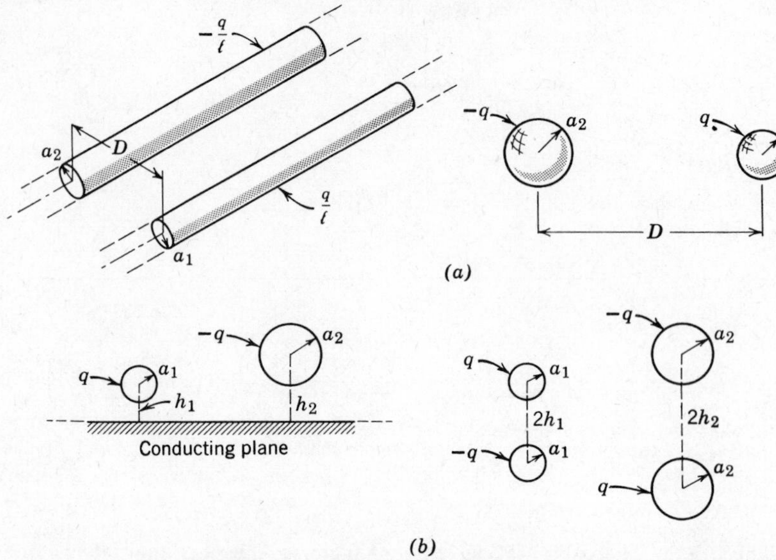

Figure 4-16. Examples of charged conductor systems amenable to approximate analysis. (*a*) Parallel round cylinders, and spheres. (*b*) Conductors (cylinders or spheres) above ground (left), and image equivalent (right).

Similarly, for the negative conductor in Figure 4-17(*b*), the potential at P relative to the same reference P_0 is

$$\Phi^{(-)}(P) = \frac{-q}{2\pi\epsilon\ell} \ln \frac{a_2}{D}$$

The sum is the total V between the conductors (neglecting charge redistribution effects); i.e.,

$$V \cong \Phi^{(+)} + \Phi^{(-)} = \frac{q}{2\pi\epsilon\ell}\left[\ln \frac{D}{a_1} - \ln \frac{a_2}{D}\right] = \frac{q}{2\pi\epsilon\ell} \ln \frac{D^2}{a_1 a_2}$$

From (4-48) the approximate capacitance becomes

$$C \cong \frac{2\pi\epsilon\ell}{\ln \dfrac{D^2}{a_1 a_2}} \tag{4-110}$$

For conductors of equal radii $a_1 = a_2 = a$, (4-110) becomes

$$C \cong \frac{\pi\epsilon\ell}{\ln \dfrac{D}{a}} \tag{4-111}$$

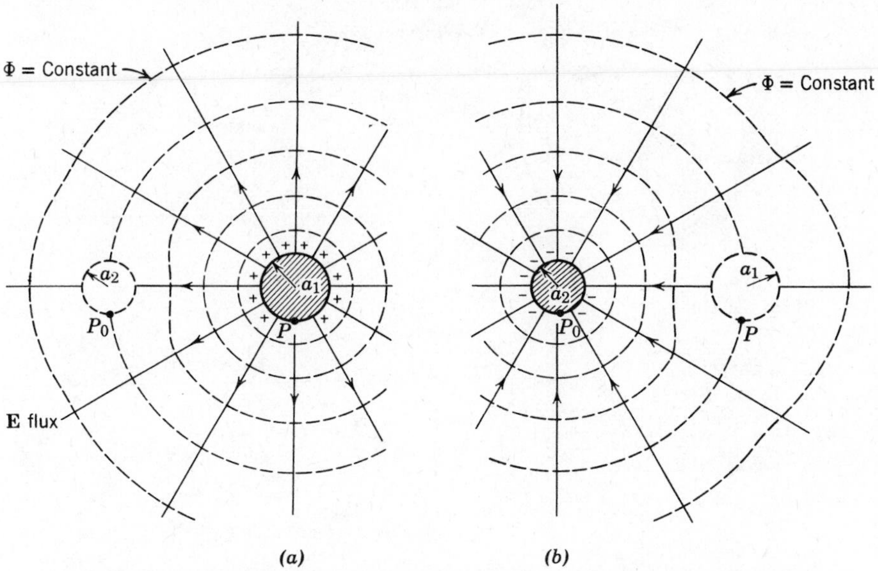

Figure 4-17. Relative to the superposition of potentials for finding approximate capacitance. (*a*) Field of positive conductor taken alone. (*b*) Field of the negative conductor only.

a result comparable to the exact expression (4-107) deduced from the image approach.

4-13 Capacitance of Two-Dimensional Systems by Field Mapping

Methods are now examined for the graphic sketching of electrostatic flux lines and equipotential surfaces of two-dimensional conductor systems. For any two-dimensional system possessing a uniform charge distribution along the z axis, the same electrostatic field sketch is seen to apply to every cross-section.

Examples of electrostatic fields between conductor pairs of arbitrary cross-sections and possessing the charges q, $-q$, in every length ℓ are shown in Figure 4-18. The sketches of the electric field flux and equipotentials of two-dimensional systems are executed in accordance with the following rules:

1. The conductors comprise equipotential surfaces between which additional equipotential surfaces may be constructed, their shapes varying gradually from that of one conductor to that of the other. Equipotential surfaces must intersect the electric flux lines orthogonally.
2. Electric flux lines form the boundaries of so-called flux tubes, as in Figure

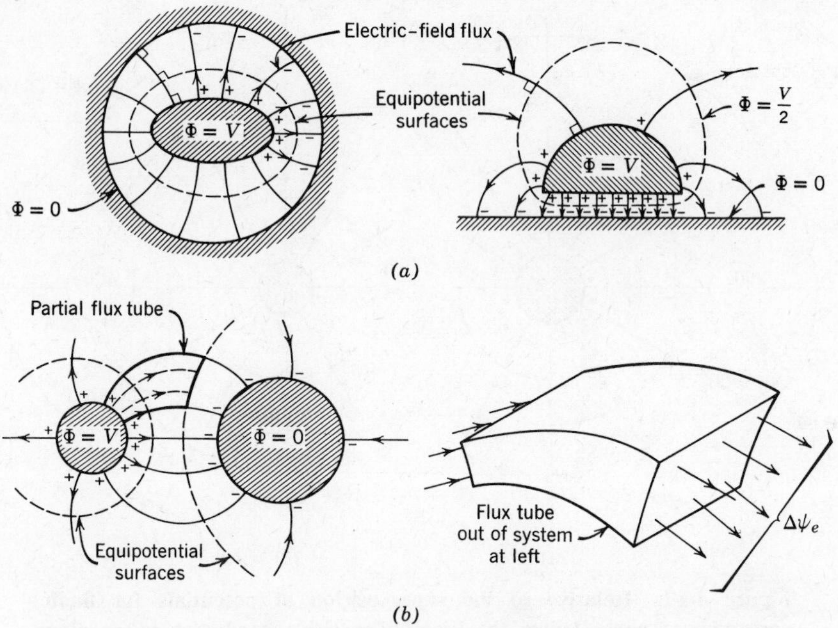

Figure 4-18. Typical two-dimensional conductor systems and flux tube interpretations. (*a*) Examples of two-dimensional conductor systems, electric flux and equipotential plots. (*b*) Flux tubes in a two-dimensional conductor system.

4-18(*b*). In a charge-free region, a flux tube contains a fixed amount of flux $\Delta\psi_e$ over any cross-section.

The capacitance per meter depth of a two-dimensional system can be found with good accuracy from a carefully executed field sketch. Given the system of Figure 4-19(*a*), flux lines originate from an assumed charge q distributed over a length ℓ of the inner conductor, terminating on $-q$ in the same length of the outer conductor. Upon replacing the equipotential surfaces with a very thin conducting foil, that system can be regarded as the *series* combination of three capacitors C_1, C_2, C_3 between the conductors. Furthermore, if the equipotentials are located such that V between the conductors is divided into three equal amounts $V_1 = V_2 = V_3 = V_0$, then the series capacitances are the same, i.e., $C_1 = C_2 = C_3 = C_0$, in view of the identical charges $\pm q$ on each. The total capacitance is therefore $C = C_0/3$ for that example. Generally, if n_s denotes the number of elements C_0 in *series*, the total capacitance is

$$C = \frac{C_0}{n_s} \tag{4-112}$$

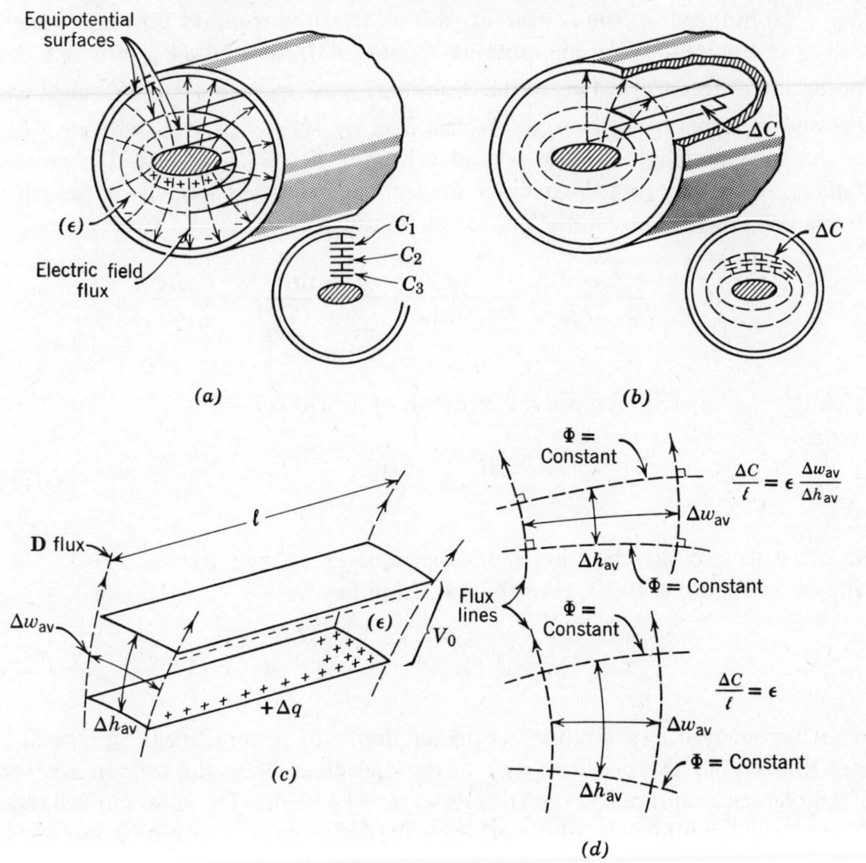

Figure 4-19. Capacitance determination from a two-dimensional electric field map. (*a*) Insertion of conducting foil at equipotential surfaces, yielding series capacitance equivalence. (*b*) Subdivision of region between equipotentials into parallel field cells. (*c*) Enlargement of field cell ΔC of (*b*). (*d*) End view of field cells. A curvilinear rectangle and square.

Each of the series capacitors of Figure 4-19(*a*) can further be subdivided into a parallel capacitance increment ΔC associated with each field cell of the system, as in Figure 4-19(*b*). With n_p *parallel* elements, $C_0 = n_p(\Delta C)$, yielding the total capacitance

$$C = \frac{n_p}{n_s} \Delta C \tag{4-113}$$

It remains to determine the field cell capacitance ΔC. Assuming charges

Δq, $-\Delta q$ induced on the *conducting-foil walls* at the top and bottom of each cell as in Figure 4-19(*c*), one obtains, from (4-48), $\Delta C = \Delta q/V_0$, in which the potential difference between the boundaries is $V_0 = -\int_{P_2}^{P_1} \mathbf{E}\cdot d\boldsymbol{\ell}$, also expressed in terms of an *average* electric field by $V_0 = E_{av}\,\Delta h_{av}$, wherein Δh_{av} is the median height of the typical cell; but $E_{av} = D_{av}/\epsilon$ and D_{av} equals $\Delta q/\Delta s_{av}$, Δs_{av} denoting the average area of the cell cross-section: its length ℓ times the average cell width Δw_{av} as in Figure 4-19(*c*). Thus ΔC becomes

$$\Delta C = \frac{\Delta q}{E_{av}\,\Delta h_{av}} = \frac{\epsilon\,\Delta q}{D_{av}\,\Delta h_{av}} = \frac{\epsilon\,\Delta q}{\dfrac{\Delta q}{\ell\,\Delta w_{av}}\,\Delta h_{av}} = \epsilon\ell\frac{\Delta w_{av}}{\Delta h_{av}}$$

yielding the capacitance per meter depth of a field cell

$$\frac{\Delta C}{\ell} = \epsilon\frac{\Delta w_{av}}{\Delta h_{av}} \tag{4-114}$$

If the cells are sketched as *curvilinear squares* defined by $\Delta w_{av} = \Delta h_{av}$ as shown in Figure 4-19(*d*), then (4-114) simplifies to

$$\frac{\Delta C}{\ell} = \epsilon \tag{4-115}$$

The incremental capacitance per meter depth of a curvilinear square flux cell thus equals the permittivity ϵ of the dielectric filling the cell. In air, for example, each square cell contributes $\epsilon_0 = 8.84$ pF/m. The total capacitance between the conductors, obtained from the series parallel combination of all cells, is found from the substitution of (4-115) into (4-113)

$$\frac{C}{\ell} = \frac{n_p}{n_s}\epsilon \text{ F/m} \tag{4-116}$$

From the development of (4-113) it is evident that n_p and n_s in (4-116) need not even be integers, as noted in the following example.

EXAMPLE 4-14. Sketch the electrostatic flux plot of the coaxial capacitor of Figure 4-20, obtaining its capacitance per meter depth. Assume air dielectric and $b/a = 2$.

Because of the symmetry, a flux plot for only one quadrant suffices. If the interval between the conductors is subdivided, by trial, into two equal potential difference intervals as in Figure 4-20(*b*), a flux map consisting of the curvilinear

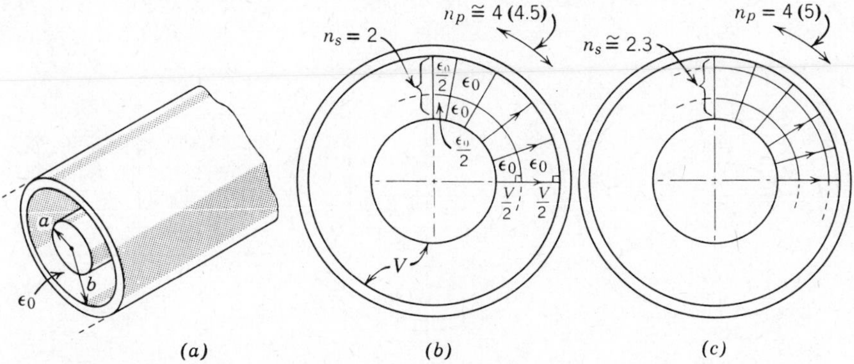

Figure 4-20. A coaxial capacitor and typical flux plots. (*a*) Coaxial capacitor: $b/a = 2$. (*b*) A flux plot using equal potential intervals. (*c*) A flux plot using five flux tubes per quadrant.

squares plus two *left-over* rectangles as shown is obtained. Then $n_s = 2$ and $n_p = 4(4.5)$, so (4-116) yields

$$\frac{C}{\ell} = \frac{n_p}{n_s}\epsilon_0 \cong \frac{4(4.5)}{2}(8.84 \times 10^{-12}) = 79.5 \text{ pF/m} \tag{1}$$

Another flux plot, dividing the quadrant into five flux tubes as shown in (*c*), yields $n_p = 4(5)$, and sketching in the equipotential surfaces to obtain the curvilinear squares as shown, $n_s = 2.3$ to yield

$$\frac{C}{\ell} \cong \frac{4(5)}{2.3}(8.84 \times 10^{-12}) = 77 \text{ pF/m} \tag{2}$$

The discrepancy between (1) and (2) is due to the unavoidable errors of estimation. It happens that this example can be checked by use of the exact (4-51), yielding

$$\frac{C}{\ell} = \frac{2\pi\epsilon_0}{\ell n \dfrac{b}{a}} = \frac{2\pi(8.84 \times 10^{-12})}{0.693} = 80.3 \text{ pF/m} \tag{3}$$

The chief merit of the flux-plotting method for two-dimensional electrostatic systems lies in its applicability to systems for which no analytical approach is feasible. In Figure 4-21 are shown such examples. Note that care must be exercised to assure the perpendicularity everywhere of the equipotential and flux lines; observe the tendency towards the compression of the flux lines at convex curves and corners because of the higher surface charge concentrations there. Advantage should always be taken of the

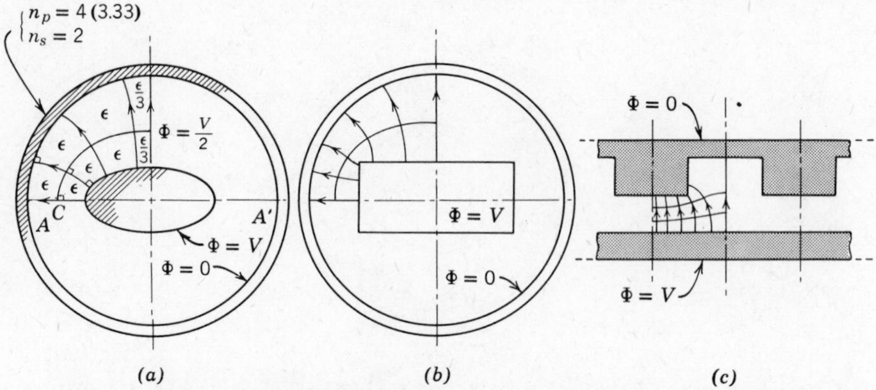

Figure 4-21. Examples of flux plots for two-dimensional conductor systems. (*a*) Elliptical cylinder inside a pipe. (*b*) Rectangular cylinder inside a pipe. (*c*) Toothed structure above a plane.

symmetry, with no more equipotentials being employed than necessary to obtain satisfactory curvilinear squares. A suitable procedure in Figure 4-21(*a*), for example, is to begin at section *A*-*A'* by placing a trial equipotential surface at point *C*, inserting appropriate orthogonal flux lines while progressing towards the right, and checking continuously for the squareness of the flux cells that develop. Needless to say, an eraser is a valuable adjunct to these trial-and-error procedures. Further suggestions and examples are found in a number of sources.†

*4-14 General Boundary Conditions for Normal D and J

In Section 3-2 was derived the boundary condition (3-41)

$$D_{n1} - D_{n2} = \rho_s \tag{4-117}$$

comparing the normal components of **D** to either side of an interface. Special cases were cited concerning (*a*) two perfect dielectrics and (*b*) a perfect dielectric and a perfect conductor. In this section are treated the remaining cases involving regions with finite conductivities, in which **E** gives rise to current densities specified by (3-7): $\mathbf{J} = \sigma\mathbf{E}$.

It is shown in general that a free charge density ρ_s accumulates at an interface, in an amount determined by the ratios of the conductivities and permittivities of the adjacent regions. To this end, the boundary condition

† For example, see Atwood, S. S. *Electric and Magnetic Fields*, 3rd ed. New York: Wiley, 1949; Ramo, S., J. Whinnery, and T. Van Duzer. *Fields and Waves in Communication Electronics*. New York: Wiley, 1965, p. 159.

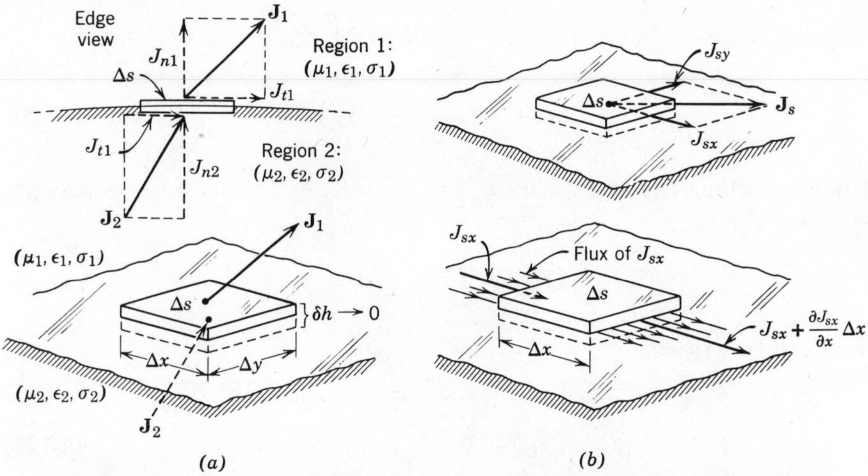

Figure 4-22. Gaussian pillbox constructed for comparing the normal components of J at an interface. (*a*) Components of $\mathbf{J}_1$ and $\mathbf{J}_2$ to either side of the interface. (*b*) Showing the variation of the *x* component of $\mathbf{J}_s$.

(4-117) cannot in itself reveal the proportions of D_{n1} and D_{n2} yielding ρ_s there. Another boundary condition is required, obtained from the current continuity relation (4-20)

$$\oint_S \mathbf{J} \cdot d\mathbf{s} = -\int_V \frac{\partial \rho_v}{\partial t} dv \tag{4-118}$$

Equation (4-118) is applied to a pillbox region of vanishing height, as used in deriving (3-41). The surface integral applied to the upper and lower surfaces of the pillbox in Figure 4-22(*a*) yields contributions $J_{n1}\,\Delta s$ and $-J_{n2}\,\Delta s$ to the net outward current flux. The tangential components J_{t1} and J_{t2} contribute only a vanishing amount of current from the sides of the pillbox, as $\delta h \to 0$. However, if a *surface* density $\mathbf{J}_s$ exists *on the interface* (permissible if region 2 is a perfect conductor), then a nonvanishing current outflow from those sides is possible, occurring if $\mathbf{J}_s$ exhibits longitudinal changes, i.e., if $\mathbf{J}_s$ has a surface divergence as shown in Figure 4-22(*b*). Then the current outflow through the four sides of the pillbox becomes

$$-J_{sx}\,\Delta y + \left(J_{sx} + \frac{\partial J_{sx}}{\partial x}\Delta x\right)\Delta y - J_{sy}\,\Delta x + \left(J_{sy} + \frac{\partial J_{sx}}{\partial y}\Delta y\right)\Delta x \tag{4-119}$$

to which is added the current flow from the upper and lower surfaces, yielding

$$J_{n1}\,\Delta s - J_{n2}\,\Delta s - J_{sx}\,\Delta y + \left(J_{sx} + \frac{\partial J_{sx}}{\partial x}\,\Delta x\right)\Delta y$$
$$-J_{sy}\,\Delta x + \left(J_{sy} + \frac{\partial J_{sy}}{\partial y}\,\Delta y\right)\Delta x = -\frac{\partial}{\partial t}(\rho_s\,\Delta s) \quad (4\text{-}120)$$

Upon cancelling terms and eliminating $\Delta s = \Delta x\,\Delta y$ factors, one obtains the *boundary condition*

$$J_{n1} - J_{n2} + \frac{\partial J_{sx}}{\partial x} + \frac{\partial J_{sy}}{\partial y} = -\frac{\partial \rho_s}{\partial t} \quad (4\text{-}121)$$

This can be written

$$J_{n1} - J_{n2} + \nabla_T \cdot \mathbf{J}_s = -\frac{\partial \rho_s}{\partial t}\ \text{A/m}^2 \quad (4\text{-}122)$$

if $\nabla_T \cdot \mathbf{J}_s$ denotes the two-dimensional *surface divergence* of the surface current density $\mathbf{J}_s$ given by

$$\nabla_T \cdot \mathbf{J}_s = \frac{\partial J_{sx}}{\partial x} + \frac{\partial J_{sy}}{\partial y} \quad (4\text{-}123)$$

The *general* boundary condition involving the continuity of the normal components of the volume current density at an interface is (4-122). It states that the normal component of $\mathbf{J}$ is discontinuous at an interface to the extent of (*a*) the time rate of decrease of the surface charge density, $-\partial\rho_s/\partial t$ and (*b*) the tangential divergence possessed by the surface current $\mathbf{J}_s$.

An alternate form of (4-122) is, with $\mathbf{J} = \sigma\mathbf{E}$

$$\sigma_1 E_{n1} - \sigma_2 E_{n2} + \nabla_T \cdot \mathbf{J}_s = -\frac{\partial \rho_s}{\partial t} \quad (4\text{-}124)$$

The general boundary condition (4-122) or (4-124) simplifies depending on the adjacent regions, three cases of which are discussed in the following:

1. *One region nonconductive; the other a perfect conductor.* Assuming region 1 lossless ($\sigma_1 = 0$) implies that $\mathbf{J}_1 = 0$, and with region 2 a perfect conductor ($\sigma_2 \to \infty$) and containing no fields, $\mathbf{J}_2 = 0$ also. Then (4-122) becomes

$$\nabla_T \cdot \mathbf{J}_s = -\frac{\partial \rho_s}{\partial t} \qquad \sigma_1 = 0, \quad \sigma_2 \to \infty \quad (4\text{-}125)$$

Thus at the surface of a perfect conductor adjoining a perfect dielectric, the time rate of decrease of ρ_s equals the surface divergence of $\mathbf{J}_s$, but (4-125) is just a restatement of the charge conservation relation (4-21b).

2. *One region has finite conductivity; the other is a perfect conductor.* With $\sigma_2 \to \infty$, $\mathbf{J}_2 = 0$, reducing (4-122) to

$$J_{n1} + \nabla_T \cdot \mathbf{J}_s = -\frac{\partial \rho_s}{\partial t} \qquad \sigma_1 \text{ finite}, \quad \sigma_2 \to \infty \tag{4-126}$$

The normal outflow of J_{n1} from a perfect conductor into an adjacent conductive region is dependent on the time rate of decrease of ρ_s and on the surface divergence of $\mathbf{J}_s$.

3. *Both regions have finite conductivities.* In the absence of a perfect conductor, $\mathbf{J}_s = 0$. Then (4-122) yields

$$J_{n1} - J_{n2} = -\frac{\partial \rho_s}{\partial t} \qquad \sigma_1, \sigma_2 \text{ finite} \tag{4-127}$$

If this is combined with (4-117), $D_{n1} - D_{n2} = \rho_s$, one can develop a relationship between the normal components of **D** (or **E**) at an interface, besides an expression for ρ_s. To avoid the use of $\partial/\partial t$ in the result, it is desirable to replace the fields with time-harmonic forms according to (2-67). Thus, after cancelling the $e^{j\omega t}$ factors and replacing $\hat{\mathbf{D}}$ with $\epsilon\hat{\mathbf{E}}$ and $\hat{\mathbf{J}}$ with $\sigma\hat{\mathbf{E}}$, (4-127) and (4-117) become

$$\sigma_1 \hat{E}_{n1} - \sigma_2 \hat{E}_{n2} = -j\omega\hat{\rho}_s \tag{4-128}$$

$$\epsilon_1 \hat{E}_{n1} - \epsilon_2 \hat{E}_{n2} = \hat{\rho}_s \tag{4-129}$$

These must be simultaneously satisfied at the interface. Eliminating $\hat{\rho}_s$ by inserting (4-129) into (4-128) obtains

$$(\sigma_1 + j\omega\epsilon_1)\hat{E}_{n1} - (\sigma_2 + j\omega\epsilon_2)\hat{E}_{n2} = 0$$

whereupon factoring $j\omega$ yields the boundary condition

$$\left(\epsilon_1 - j\frac{\sigma_1}{\omega}\right)\hat{E}_{n1} - \left(\epsilon_2 - j\frac{\sigma_2}{\omega}\right)\hat{E}_{n2} = 0 \tag{4-130a}$$

Using the *complex permittivity* notation of (3-103) obtains

$$\hat{\epsilon}_1 \hat{E}_{n1} - \hat{\epsilon}_2 \hat{E}_{n2} = 0 \qquad \sigma_1, \sigma_2 \text{ finite} \tag{4-130b}$$

The boundary condition for the normal component of **E** is, therefore, that $\hat{\epsilon}\hat{E}_n$ is *continuous* at an interface separating finitely conductive regions.

An expression for the free charge density $\hat{\rho}$ accumulated at the interface is obtained by eliminating $\hat{E}_{n1}$ or $\hat{E}_{n2}$ from (4-128) and (4-129), yielding the equivalent results

$$\hat{\rho}_s = \hat{E}_{n1} \frac{\epsilon_1\sigma_2 - \epsilon_2\sigma_1}{j\omega\hat{\epsilon}_2} = \hat{E}_{n2} \frac{\epsilon_1\sigma_2 - \epsilon_2\sigma_1}{j\omega\hat{\epsilon}_1} \tag{4-131}$$

in which $\hat{\epsilon}_1$ and $\hat{\epsilon}_2$ are given by (3-103). One concludes that a surface charge is induced upon the interface by the normal components of $\hat{E}$ if at least one region is conductive. On the other hand, *no* free surface charge exists at the interface if (*a*) both regions are nonconductive ($\sigma_1 = \sigma_2 = 0$) or (*b*) the special proportion $\epsilon_1/\epsilon_2 = \sigma_1/\sigma_2$ is true among the region parameters, presumably a rare event and of little importance. For both regions nonconductive, putting $\hat{\rho}_s = 0$ into (4-129) yields the special case $\epsilon_1 E_{n1} - \epsilon_2 E_{n2} = 0$, or just

$$D_{n1} - D_{n2} = 0 \qquad \sigma_1 = \sigma_2 = 0 \tag{4-132}$$

a result agreeing with (3-43) for the nonconductive case.

EXAMPLE 4-15. Determine the refractive law for *direct* currents at an interface separating two isotropic conductive regions. Specialize the result for one conductivity much larger than the other.

Assume the **J** vectors titled by amounts θ_1 and θ_2 as shown in (*a*). The boundary condition (4-127) for dc becomes

$$J_{n1} = J_{n2} \tag{4-133}$$

while the boundary condition involving tangential components is obtained from (3-79), with $\mathbf{J} = \sigma\mathbf{E}$

$$\frac{J_{t1}}{\sigma_1} = \frac{J_{t2}}{\sigma_2} \tag{4-134}$$

From the geometry, the tilt angles obey

$$\tan\theta_1 = \frac{J_{t1}}{J_{n1}} \qquad \tan\theta_2 = \frac{J_{t2}}{J_{n2}}$$

The latter combines with (4-133) and (4-134), whereupon inserting the expression for tan θ_1 obtains the refractive law

$$\tan\theta_1 = \frac{\sigma_2}{\sigma_2}\tan\theta_1 \tag{4-135}$$

The analogy with the refractive laws (3-76) and (3-80) for **B** and **E** might be noted. For an example in which $\sigma_2 = 10\sigma_1$, the refractive effects of direct current streamlines at an interface are shown typically in (*b*) of the accompanying figure. For $\sigma_2 \gg \sigma_1$, the near perpendicularity of the current flux occurs in

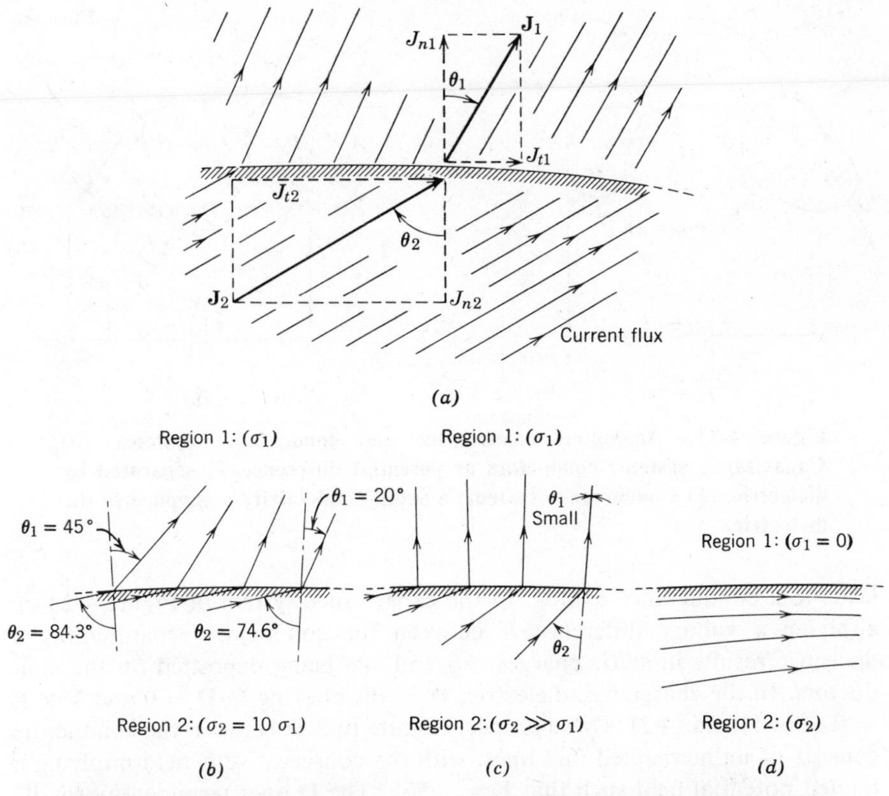

Example 4-15. (*a*) Refraction of currents. (*b*) Examples of current flux refraction if $\sigma_2 = 10\sigma_1$. (*c*) Current flux for region 2 highly conductive. (*d*) Constraint to tangential flow at interface for region 1 nonconductive.

regions 1, as noted in (*c*). If σ_1 were reduced to zero, then $\mathbf{J}_1 = 0$, constraining the current flow in region 2 to paths tangential to the conductor-insulator boundary as in (*d*), a result evident from the insertion of $J_{n1} = J_{t1} = 0$ into the boundary conditions (4-133) and (4-134).

4-15 Conductance Analog of Capacitance

A system is said to be analogous to another if a quantity in one system varies in the same way as some quantity in the other. An analogy may even exist between two quantities in the same system. If the quantities are vector fields, to be analogous they must satisfy comparable divergence and curl relationships as well as similar boundary conditions.

It is to be shown that the capacitance system of Figure 4-6 in Section 4-6

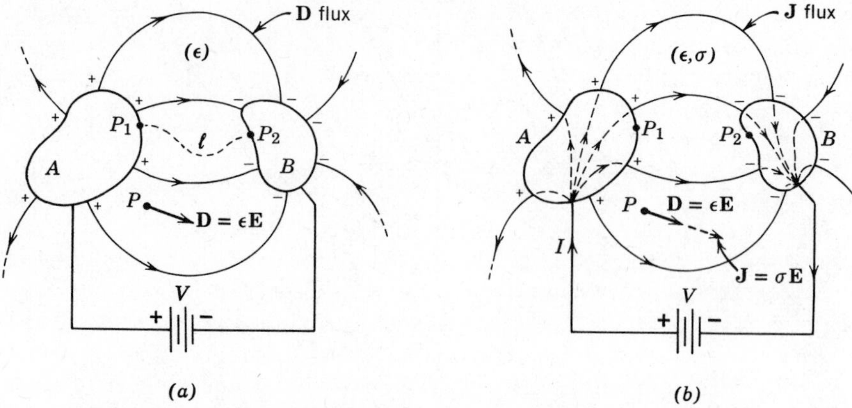

Figure 4-23. Analogous capacitance and conductance systems. (*a*) Capacitance system: conductors at potential difference *V*, separated by dielectric. (*b*) Conductance system: a small conductivity σ supplied to the dielectric.

leads to a conductance analog. In the capacitance system of Figure 4-23(*a*), applying a voltage difference V between the conductors separated by a dielectric results in static charges $+q$ and $-q$ being deposited on the conductors. In the charge free dielectric, $\mathbf{D} = \epsilon\mathbf{E}$, obeying $\nabla\cdot\mathbf{D} = 0$ and $\nabla\times\mathbf{E} = 0$ of (4-1) and (4-2). These properties state that **D** between the conductors consists of uninterrupted flux lines, with the conservative **E** field implying a related potential field such that $\mathbf{E} = -\nabla\Phi$. The **D** lines terminate normally at the conductor surfaces as required by the boundary conditions. The potential field Φ, moreover, obeys Laplace's equation (4-68), $\nabla^2\Phi = 0$, and each conductor comprises an equipotential surface with the potential difference V prevailing between them. The capacitance parameter C, moreover, applies to the system, defined by (4-48)

$$C = \frac{q}{V} = \frac{\int_S \mathbf{D}\cdot d\mathbf{s}}{-\int_{P_2}^{P_1} \mathbf{E}\cdot d\boldsymbol{\ell}} \text{ F} \tag{4-136}$$

In obtaining this ratio, $\mathbf{D}\cdot d\mathbf{s}$ is integrated over the positive conductor of Figure 4-23(*a*), while P_1 assumed on that conductor makes V positive.

A dc conductance analog of (4-136) can be established for the system if the dielectric possesses a small conductivity σ. The dielectric then carries a current of density $\mathbf{J} = \sigma\mathbf{E}$, from (3-7); **J** is the analog of **D** in the dielectric, since from (4-22), $\nabla\cdot\mathbf{J} = 0$. Thus **J** consists of uninterrupted current flux lines, supplied by V. Assuming A and B good conductors and with the dielectric a relatively poor conductor, from the refraction Example 4-15 one concludes that the current enters or leaves A and B essentially *perpendicularly*. The boundary condition (4-131), moreover, reveals what charge density ρ_s exists on each conductor surface. With $\omega = 0$ and $\sigma_1 \ll \sigma_2$, one obtains

$$\rho_s = E_{n1} \frac{\epsilon_1 \sigma_2 - \epsilon_2 \sigma_1}{\sigma_2} \cong \epsilon_1 E_{n1} \tag{4-137}$$

if the good conductor is denoted by the subscript 2 and the lossy dielectric by 1. Thus the boundary conditions of Figures 4-23(a) and (b) are essentially the same. It is thus seen that adding a small amount of conductivity to the dielectric produces virtually no change in the **E** field configuration in the dielectric. Thus besides C, an analogous positive parameter G, the *conductance* of the system, is defined by the ratio of the total current I through the dielectric to the voltage difference V between the conductors

$$G = \frac{I}{V} = \frac{\int_S \mathbf{J} \cdot d\mathbf{s}}{-\int_{P_2}^{P_1} \mathbf{E} \cdot d\boldsymbol{\ell}} \quad \mho \tag{4-138}$$

The surface S of conductor A excludes the cross-section of the connecting wire so that in (4-138) only the outflow of I into the dielectric is taken into account. The analog of C is G because $\mathbf{J}$ in (4-138) is the analog of $\mathbf{D}$ in (4-136). With σ and ϵ constants for the homogeneous, linear, and isotropic dielectric, (4-136) and (4-138) become

$$C = \frac{\epsilon \int_S \mathbf{E} \cdot d\mathbf{s}}{-\int_{P_2}^{P_1} \mathbf{E} \cdot d\boldsymbol{\ell}} \qquad G = \frac{\sigma \int_S \mathbf{E} \cdot d\mathbf{s}}{-\int_{P_2}^{P_1} \mathbf{E} \cdot d\boldsymbol{\ell}} \tag{4-139}$$

yielding the ratio

$$\boxed{\frac{G}{C} = \frac{\sigma}{\epsilon}} \tag{4-140}$$

This is also written $(RC)^{-1} = \sigma/\epsilon$ if $1/G = R$, the *resistance* between the conductors. Equation (4-140) implies that if C is known, the analogous G can be found from the applicable ratio σ/ϵ.

In view of the relaxation result (4-25) of Example 4-3, (4-140) has further implications. If V applied to the conductive capacitor system of Figure 4-23(b) were suddenly removed, the surface charges on each conductor would decay in time according to (4-25)

$$\rho_s(u_1, u_2, u_3, t) = \rho_{s0}(u_1, u_2, u_3, 0)e^{-(\sigma/\epsilon)t} \tag{4-141}$$

if σ and ϵ are the dielectric parameters. Integrating (4-141) over the positive conductor surface S yields the charge q on it at any instant

$$q(t) = \int_S \rho_s \, ds = e^{-(\sigma/\epsilon)t} \int_S \rho_{s0} \, ds = q_0 e^{-(\sigma/\epsilon)t} \tag{4-142a}$$

which, by use of (4-140) obtains a form familiar in circuit theory

$$q(t) = q_0 e^{-t/RC} \text{ C} \tag{4-142b}$$

The charge on the positive conductor thus decays exponentially with the time constant

$$\tau = RC = \frac{\epsilon}{\sigma} \text{ sec} \tag{4-143}$$

Thus τ is expressible either in terms of the derived lumped constants R and C, or the parameters ϵ and σ of the dielectric. The time-decay behavior of q on the positive conductor is depicted in Figure 4-24(*a*), with the equivalent circuit shown in (*b*).

The so-called quality factor, Q, of the capacitor is shown to be the reciprocal of the loss tangent of its dielectric. Define its Q under time-harmonic applied-voltage conditions as

$$Q = \frac{\omega \text{ (Maximum energy stored during a cycle)}}{\text{Average power loss over a cycle}} = \frac{\omega U_{\max}}{P_{\text{av},L}} \tag{4-144}$$

in which ω is the radian frequency. Assuming $V(t) = V_m \cos \omega t$ applied as in Figure 4-24(*c*), the maximum energy is stored when the voltage is V_m, to yield $U_{\max} = CV_m^2/2$ from (4-63b). Also $V(t)$ is impressed on the loss resistance

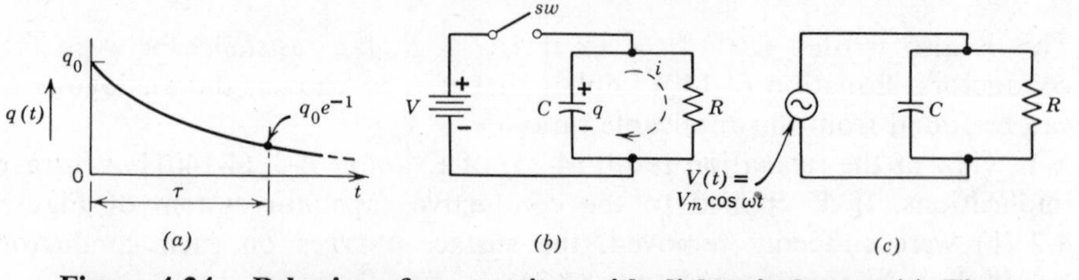

Figure 4-24. Behavior of a capacitor with dielectric losses. (*a*) Time decay of charge from initial value q_0. (*b*) Voltage removed from capacitor equivalent circuit. (*c*) Equivalent circuit, ac voltage applied.

R, yielding the time-average power loss $V_m^2/2R$. Thus, (4-144) becomes

$$Q = \frac{\omega(\frac{1}{2}CV_m^2)}{\frac{V_m^2}{2R}} = \omega RC \tag{4-145}$$

which from (4-143) is also $\omega\epsilon/\sigma$, the reciprocal of the loss tangent (3-104), which was to have been proved. By use of (4-140) and (4-143), the Q of the capacitor can be written in the various forms

$$Q = \omega RC = \frac{\omega\epsilon}{\sigma} = \frac{\epsilon'}{\epsilon''} = \omega\tau = \frac{\omega C}{G} = \frac{R}{X_c} \tag{4-146}$$

wherein $X_c \equiv (\omega C)^{-1}$ denotes the reactance of C at the frequency ω. Thus, a material with a loss tangent $\epsilon''/\epsilon' = 0.001$ at some frequency will yield a capacitor with a Q of 1000. If, moreover, its reactance X_c were 100 Ω at the given frequency, the equivalent circuit of Figure 4-24 would need to incorporate a parallel resistance $R = QX_c = 10^5$ Ω to represent the dielectric losses.

EXAMPLE 4-16. Assume that the spherical capacitor of Figure 4-7(b) contains a dielectric with the constants $\epsilon = 3\epsilon_0$ and $\sigma = 10^{-5}$ ℧/m at some frequency. Letting $a = 1$ cm and $b = 2$ cm, find C and G, and sketch the equivalent circuit.

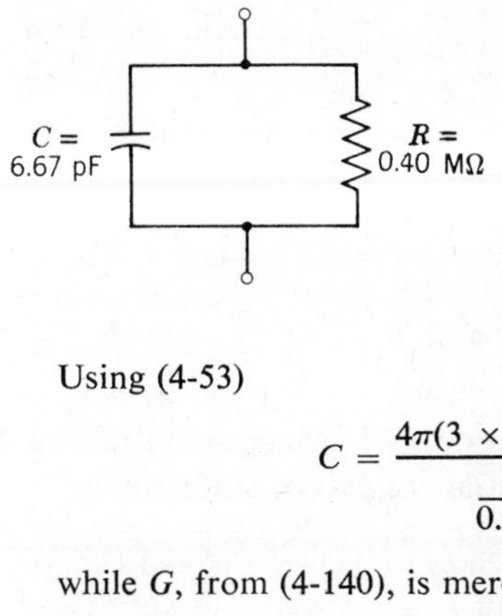

Example 4-16

Using (4-53)

$$C = \frac{4\pi(3 \times 10^{-9}/36\pi)}{\frac{1}{0.01} - \frac{1}{0.02}} = 6.67 \text{ pF}$$

while G, from (4-140), is merely C with ϵ replaced by σ

$$G = C\frac{\sigma}{\epsilon} = \frac{4\pi(10^{-5})}{\frac{1}{0.01} - \frac{1}{0.02}} = 2.51\,\mu℧$$

yielding the resistance between the spheres, $R = G^{-1} = 0.40$ MΩ. The equivalent circuit diagram consists of C in parallel with R as in the sketch.

4-15-1 Capacitance-Conductance Analog and Field Mapping

For two-dimensional capacitors, the infinite length makes it desirable to express (4-140) as the ratio

$$\frac{G}{\ell} = \frac{C}{\ell}\frac{\sigma}{\epsilon} \ \mho/\text{m} \tag{4-147}$$

Thus, from the C/ℓ ratio (4-51) for the coaxial line

$$\frac{G}{\ell} = \frac{2\pi\epsilon}{\ell n \dfrac{b}{a}}\frac{\sigma}{\epsilon} = \frac{2\pi\sigma}{\ell n \dfrac{b}{a}} \tag{4-148}$$

Such results are also applicable to the two-dimensional field-mapping techniques of Section 4-12, assuming the dielectric to possess a small conductivity σ. Then each field cell as in Figure 4-19(*c*) contributes a conductivity per meter depth obtained by putting (4-114) into (4-147)

$$\frac{\Delta G}{\ell} = \frac{\Delta C}{\ell}\frac{\sigma}{\epsilon} = \sigma\frac{\Delta w_{\text{av}}}{\Delta h_{\text{av}}} \tag{4-149a}$$

For any curvilinear *square* cell, letting $\Delta w_{\text{av}} = \Delta h_{\text{av}}$ yields

$$\frac{\Delta G}{\ell} = \sigma \qquad \text{Square cell} \tag{4-149b}$$

the analog of (4-115) as noted in Figure 4-25. The series parallel combination of all such cells between the conductors thus yields the total conductance per meter depth

$$\frac{G}{\ell} = \frac{C}{\ell}\frac{\sigma}{\epsilon} = \left[\frac{n_p}{n_s}\epsilon\right]\frac{\sigma}{\epsilon} = \frac{n_p}{n_s}\sigma \ \mho/\text{m} \tag{4-150}$$

The latter can be viewed as the conductance produced by the mesh of cell conductances connected between the equipotential conductor surfaces.

EXAMPLE 4-17. Use the flux plot of Figure 4-20(*b*) to deduce C/ℓ and G/ℓ for that coaxial system, assuming a dielectric with $\epsilon_r = 2.5$ and $\sigma = 10^{-8}$ ℧/m. What resistance is seen by a dc voltage connected to the input of 1000 m of this line, assuming an open-circuit termination?

The flux plot of Figure 4-20(*b*) yields from (4-116)

$$\frac{C}{\ell} = \frac{n_p}{n_s}\epsilon = \frac{4(4.5)}{2}(2.5 \times 8.84 \times 10^{-12}) = 198 \text{ pF/m}$$

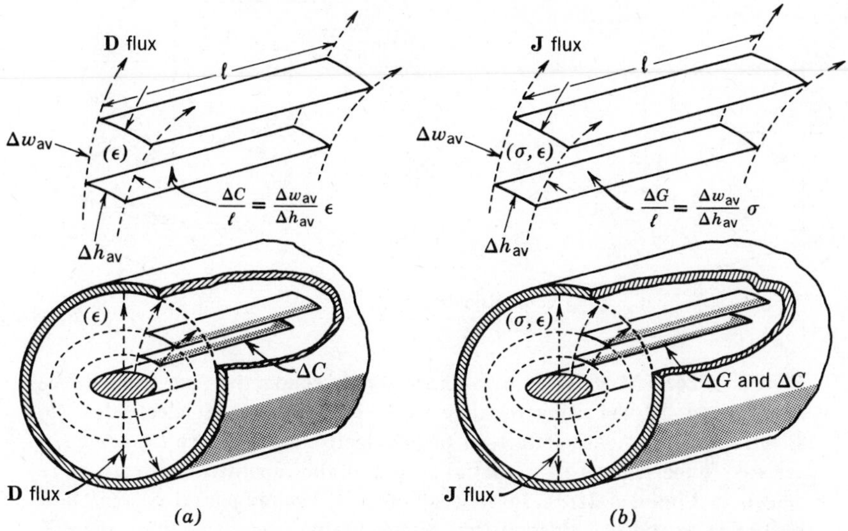

Figure 4-25. Typical two-dimensional system showing analogous quantities used in analysis of C/ℓ and G/ℓ. (*a*) Capacitive system with perfect dielectric, and flux cell. (*b*) Capacitive and conductive system, and analogous current flux cell.

Then, from (4-150)

$$\frac{G}{\ell} = \frac{n_p}{n_s}\sigma = \frac{4(4.5)}{2} 10^{-8} = 9 \times 10^{-8} \ \mho/\text{m}$$

For $\ell = 1000$ m of open-circuited line, $G = (9 \times 10^{-8})10^3 = 9 \times 10^{-5}\ \mho$, making $R = G^{-1} = 11.1\ k\Omega$, the resistance seen by the applied dc voltage.

Current-conduction *models* of two-dimensional systems as described can be constructed using commercial resistive paper, or by use of a shallow tank of electrolyte to simulate the conduction region, with electrodes of desired shapes placed in contact with the conductive medium as suggested in Figure 4-26. In (*a*) of that figure, intimate contact of the electrodes with the resistive paper is assured by using a good-conducting silver paint to produce the desired electrode shapes on the paper. A battery serves as a source of current, with a high-impedance voltmeter and pointed probe used to map equipotential contours onto the resistive paper. A low-frequency source (up to 1000 Hz or so) can be used if ac detection methods are preferred, and they are especially useful for eliminating polarization effects occurring when direct currents pass through an electrolytic liquid. The latter yields ion

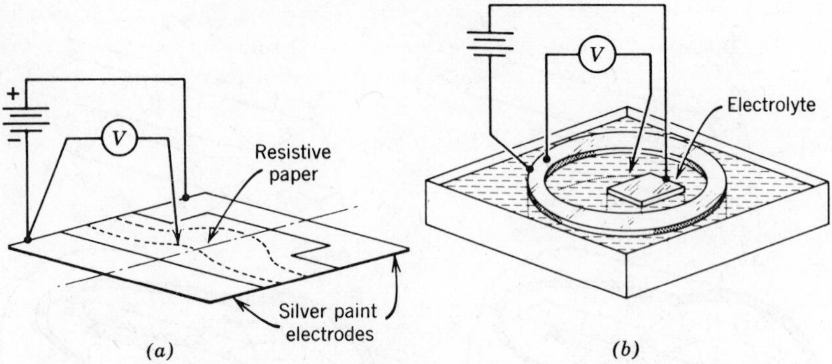

Figure 4-26. Models of two-dimensional conductive or capacitance systems. (*a*) Model using resistive paper and silver paint electrodes. (*b*) Electrolytic tank with immersed metal electrodes. (Figure (*a*) depicts a resistive paper model of a spatial period of the repetitive tooth structure shown in Figure 4-21(*c*). In view of the entirely tangential current flow occurring at the two sides of this model, the boundary condition there is $\partial\Phi/\partial n = 0$, agreeing with that of the actual system of Figure 4-21(*c*).)

accumulation near one or both electrodes, causing distortions in the equipotential distributions obtained from electrolytic tank models.†

An advantage of the current model of a flux map is that it obviates the errors of estimation incurred in the hand-plotting methods described in Section 4-12, yielding highly accurate equipotential maps when careful measurements are taken. Moreover, an ohmmeter or bridge measurement between the electrodes of a conduction model leads directly to the capacitance and conductance per meter of the system being studied, without a need for the n_p and n_s values required by hand-plotting techniques.

The electrolytic tank can be extended to axially symmetric geometries associated with electrostatic beam focusing electrodes such as used in cathode ray tubes and electron microscopes.‡ Such maps in circular cylindrical coordinates are often very difficult to obtain analytically or by hand-plotting schemes. A half-cylindrical tank containing semicylindrical electrodes and revealing their sectional views at the surface of the electrolyte permits probing the equipotential surfaces in the vicinity of the z axis. Indeed, the axial symmetry permits using merely a thin, wedge-shaped trough in which correspondingly small sectors of the cylindrical electrodes are immersed.

† The electrolytic tank was first used by C. L. Fortescue. See *Transactions of the A.I.E.E.*, **32**, 1913, p. 893.

‡ See Weber, E. *Electromagnetic Fields, Vol. 1—Mapping of Fields.* New York: Wiley, 1950, pp. 187–193.

EXAMPLE 4-18. A sheet of resistive paper measuring 1000 Ω per square (with 1000 Ω between opposite equipotential sides of a square sheet, regardless of size)† is used to model a two-conductor cable of rather unusual, though uniform, cross-sectional shape. The conductor shapes are painted on the paper with silver paint, and a measurement yields 160 Ω between those conductors. Find C/ℓ and G/ℓ of the actual cable if the dielectric has the constants $\epsilon_r = 2.5$ and $\sigma = 10^{-8}$ ℧/m.

The conductance between electrodes of the resistive sheet model is specified by (4-150)

$$G_r = \frac{n_p}{n_s}(\sigma_r \ell_r) \tag{1}$$

if G_r denotes the measured 1/160 ℧ and $\sigma_r \ell_r$ is the product of the conductivity of the resistive sheet and its thickness. For the resistance paper used, $\sigma_r \ell_r =$ 1/1000 ℧, the conductance of a curvilinear square of any flux plot applicable to this model. The usual ratio for such a plot is denoted by n_p/n_s and from (1), this is

$$\frac{n_p}{n_s} = \frac{G_r}{\sigma_r \ell_r} = \frac{(160)^{-1}}{0.001} = 6.25$$

Applying the latter to (4-116) and (4-150) obtains for the cable

$$\frac{C}{\ell} = \frac{n_p}{n_s}\epsilon = (6.25)(2.5 \times 8.84 \times 10^{-12}) = 138 \text{ pF/m}$$

$$\frac{G}{\ell} = \frac{n_p}{n_s}\sigma = 6.25 \times 10^{-2}\ \mu℧/\text{m}$$

4-15-2 Dc or Low-Frequency Resistance of Thin Conductors

Thin conductors (of small diameter compared to length) are of common occurrence in electric circuits. It is of interest to determine the resistance offered by a thin, conductive circuit to a driving source, as depicted in Figure 4-27. The circuit is immersed in a nonconductor (e.g., air). The direct current in the conductor has a density given by (3-7), $\mathbf{J} = \sigma\mathbf{E}$. Steady currents are, by (4-22), divergenceless, so the current consists of uninterrupted flux lines totalling I A through any cross-section. The static **E** field in the conductor, obeying (4-6), $\oint_\ell \mathbf{E}\cdot d\boldsymbol{\ell} = 0$, is thus conservative, so equipotential surfaces exist in the conductor, normal to the **E** and **J** field as denoted in Figure 4-27.

Equation (4-6) is equivalent to Kirchhoff's voltage law for the circuit

† From this it is inferred that any *curvilinear square* of a flux map on this paper has 1000 Ω resistance between opposite equipotential sides, or 0.001 ℧ conductance.

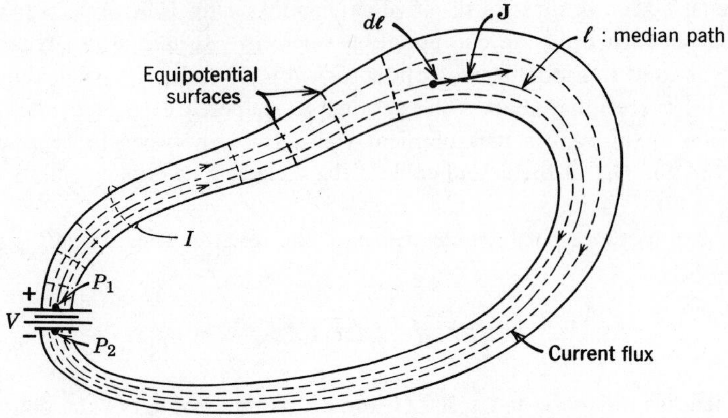

Figure 4-27. A thin dc electric circuit.

shown as follows. Taking $\mathbf{E}_g$ and $\mathbf{E}$ as the fields in the battery and the conductor respectively, integrating (4-6) clockwise over any closed path ℓ about the circuit of Figure 4-27 obtains

$$\int_{P_1}^{P_2} \mathbf{E} \cdot d\boldsymbol{\ell} + \int_{P_2}^{P_1} \mathbf{E}_g \cdot d\boldsymbol{\ell} = 0 \tag{4-151}$$

But the second integral denotes $-V$, the negative of the battery voltage;† with $\mathbf{E} = \mathbf{J}/\sigma$ in the conductor, (4-151) is written

$$\int_{P_1}^{P_2} \frac{\mathbf{J}}{\sigma} \cdot d\boldsymbol{\ell} = V \tag{4-152}$$

That (4-152) expresses the Kirchhoff law $V = IR$ is seen by noting that the current through every cross-section A is the constant value

$$I = \int_A \mathbf{J} \cdot d\mathbf{s} \tag{4-153}$$

in which $\mathbf{J}$ is not, in general, constant at each point of the cross-section. The need for knowing $\mathbf{J}$ at every point is disposed of, if I is expressed in terms of an *average* density $\mathbf{J}_{\text{av}}$; i.e.

$$I = J_{\text{av}} A \tag{4-154}$$

† The negative sign of the $-V$ term in (4-151) is justified from the direction of the generated $\mathbf{E}_g$ fields in the source, which go from + to −.

wherein J_{av} is tangential to a properly chosen median line ℓ as denoted in Figure 4-21. For a thin wire, ℓ in (4-152) may be taken as the wire axis, and with $\mathbf{J}_{av} = \mathbf{a}_\ell I/A$ into (4-154), (4-152) becomes

$$\int_{P_1}^{P_2} \left(\frac{I}{\sigma A}\,\mathbf{a}_\ell\right)\cdot \mathbf{a}_\ell\, d\ell = V$$

yielding

$$I = \frac{V}{\displaystyle\int_{P_1}^{P_2} \frac{d\ell}{\sigma A(\ell)}} \tag{4-155}$$

This is of the form $I = V/R$, the Kirchhoff voltage law for the circuit, in which the *dc resistance* of the conducting path is

$$R = \int_{P_1}^{P_2} \frac{d\ell}{\sigma A(\ell)} \ \text{V/A} \quad \text{or} \quad \Omega \tag{4-156}$$

Its reciprocal, $R^{-1} = G$, is its *conductance*. The notation $A(\ell)$ emphasizes that the conductor cross-sectional area might not be uniform, depending generally on the position along ℓ. For a conductor of constant cross-section, (4-156) reduces to

$$R = \frac{\ell}{\sigma A}\ \Omega \tag{4-157}$$

if ℓ denotes the conductor length. These resistance expressions, correct for direct currents, are reasonable approximations at sufficiently low frequencies for which the skin effect, associated with reduced field penetration into a conductor with increasing frequency, is neglected.

As an example, the dc resistance of 10 m of 0.1 in. (0.00254 m) diameter copper wire (obtaining σ from Table 3-3) is

$$R = \frac{10}{(5.8 \times 10^7)(0.00254^2\ \pi/4)} = 0.034\ \Omega$$

A wire this size made of aluminum, for which $\sigma = 3.72 \times 10^7$ ℧/m, will have a resistance about 56% greater than the copper one.

*4-16 Electrostatic Forces and Torques

In Section 4-7 were developed expressions for the work done by an external source in establishing a system of electrostatic charges in a region. Such charges reside physically on the conducting bodies of the system, which may also include dielectric regions. The force on any of the conductors or dielectric bodies can be deduced from an assumed differential displacement $d\ell$ of that body, upon computing the change in energy dU_e accompanying the displacement. It is shown that the electrostatic force can be found from the *gradient* of the electrostatic energy of the system, if the energy is expressed in terms of the coordinate location of the body being displaced. Forces obtained in this way are said to be found by the method of *virtual work*. This method is developed for two cases.

CASE A. *System of conductors with fixed charges.* Suppose one is concerned with a system of dielectric and conducting bodies of Figure 4-28(*a*), the conductors being assumed isolated from one another so that they possess fixed amounts of free charge. (Batteries or other sources used to bring them to their charge states have been removed.) Let one element (conductor or dielectric) be displaced by a differential distance $d\ell$, due to electric field forces acting on it. The mechanical work done by the system is

$$dU = \mathbf{F} \cdot d\boldsymbol{\ell} = F_x\,dx + F_y\,dy + F_z\,dz \text{ J} \tag{4-158}$$

Since no additional energy is being supplied (sources are disconnected), the work (4-158) is done at the expense of the stored electrostatic energy of the system, energy being conserved, such that

$$\underbrace{dU_e}_{\text{Electrostatic energy change}} + \underbrace{dU}_{\text{Mechanical work done}} = 0 \tag{4-159}$$

implying an energy decrease in the amount

$$-dU_e = dU = \mathbf{F} \cdot d\boldsymbol{\ell} = F_x\,dx + F_y\,dy + F_z\,dz \tag{4-160}$$

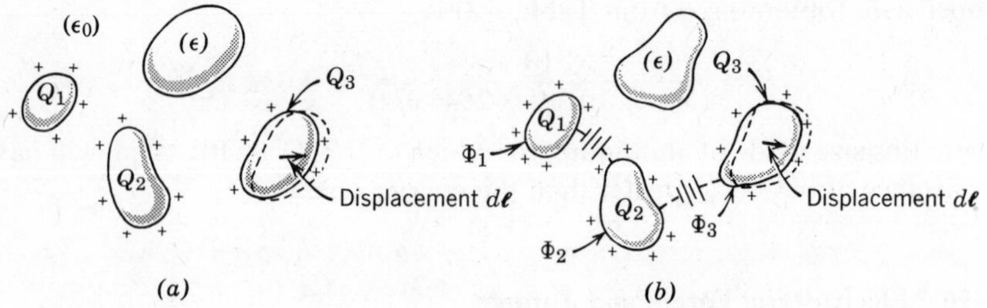

Figure 4-28. Two electrostatic systems of conducting and dielectric bodies. A virtual displacement $d\ell$ of a body is assumed for the purpose of calculating the force on it. (*a*) System with fixed charges. (*b*) Conductors at fixed potentials.

but dU_e can be written also in terms of the x, y, and z variations in U_e as the body moves $d\boldsymbol{\ell} = \mathbf{a}_x\,dx + \mathbf{a}_y\,dy + \mathbf{a}_z\,dz$

$$dU_e = \frac{\partial U_e}{\partial x}\,dx + \frac{\partial U_e}{\partial y}\,dy + \frac{\partial U_e}{\partial z}\,dz = (\nabla U_e)\cdot d\boldsymbol{\ell} \qquad (4\text{-}161)$$

the latter being evident from the vector representation (2-11) for a total differential. A comparison of (4-160) and (4-161) reveals that

$$\boxed{\mathbf{F} = -\nabla U_e \text{ N}} \qquad (4\text{-}162\text{a})$$

implying the cartesian components of $\mathbf{F}$ given by

$$F_x = -\frac{\partial U_e}{\partial x} \qquad F_y = -\frac{\partial U_e}{\partial y} \qquad F_z = -\frac{\partial U_e}{\partial z} \qquad (4\text{-}162\text{b})$$

It is seen from (4-162) that knowing how the total electrostatic field energy U_e of the system changes with dx, dy, and dz displacements of one of its elements is sufficient to determine the force on that element. This is called the *virtual work* method for finding the force, since no actual physical displacements are required.

If, instead of being subjected to a translation, the desired body is rotated about an axis, assuming constant charges on the conductors, then (4-158) is written

$$dU = \mathbf{T}\cdot d\boldsymbol{\theta} = T_1\,d\theta_1 + T_2\,d\theta_2 + T_3\,d\theta_3 \qquad (4\text{-}163)$$

in which $\mathbf{T} = \mathbf{a}_1 T_1 + \mathbf{a}_2 \mathrm{T}_2 + \mathbf{a}_3 \mathrm{T}_3$ is the torque developed, and $d\boldsymbol{\theta}$ is the vector differential angular displacement. One can analogously show that the components of the vector torque $\mathbf{T}$ become

$$T_1 = -\frac{\partial U_e}{\partial \theta_1} \qquad T_2 = -\frac{\partial U_e}{\partial \theta_2} \qquad T_3 = -\frac{\partial U_e}{\partial \theta_3} \qquad (4\text{-}164)$$

CASE B. *System conductors at fixed potentials.* The system consists of n charged conductors held at the fixed potentials $\Phi_1, \Phi_2, \ldots, \Phi_n$ by charge sources (such as batteries). Dielectric bodies may also be included, as in Figure 4-28(*b*). The displacement $d\boldsymbol{\ell}$ of an element is in this case accompanied by changes in the charges on each conductor. For example, if two parallel conducting plates connected to a battery were moved apart, the positive and negative charges on the plates would both decrease to maintain the constant, impressed voltage difference. This means that the total electrostatic energy on the system changes with the displacement, but also it means that work is done by the sources in producing the changes in the charge states of the conductors, to maintain their fixed potentials. The work done by the *sources* (batteries) during the displacement of the desired element is

$$dU_s = \sum_{k=1}^{n} \Phi_k \, dq_k \text{ J} \tag{4-165}$$

in which the potentials Φ_k on the n conductors are constants. The energy conservation relation now becomes

$$\underbrace{dU_e}_{\substack{\text{Electrostatic}\\ \text{energy change}}} + \underbrace{dU}_{\substack{\text{Mechanical}\\ \text{work done}}} = \underbrace{dU_s}_{\substack{\text{Work done by sources}\\ \text{to maintain fixed potentials}}} \tag{4-166}$$

Since each conductor charge undergoes a change dq_k while being maintained at the potential Φ_k, from (4-58a) the total electrostatic energy changes by

$$dU_e = \tfrac{1}{2} \sum_{k=1}^{n} \Phi_k \, dq_k \tag{4-167}$$

or just one-half the work (4-165) done by the sources. Thus

$$dU_s = 2 \, dU_e \tag{4-168}$$

stating that the work done by the sources is twice the change in the total electrostatic energy; the remainder is the mechanical work dU done in moving the element in question by the distance $d\ell$. Putting (4-168) into (4-166) therefore yields

$$dU_e = dU = \mathbf{F} \cdot d\boldsymbol{\ell} = F_x \, dx + F_y \, dy + F_z \, dz \tag{4-169}$$

which means

$$\mathbf{F} = \nabla U_e \text{ N} \tag{4-170}$$

EXAMPLE 4-19. Find the force between two point charges $\pm q$ separated a distance x in free space, using the concept of virtual displacement.

The electrostatic energy is obtained using (4-58a), with $n = 2$. From (4-40), the potential Φ_1 due to $+q$ at the location of $-q$ is $q/4\pi\epsilon_0 x$, while that due to

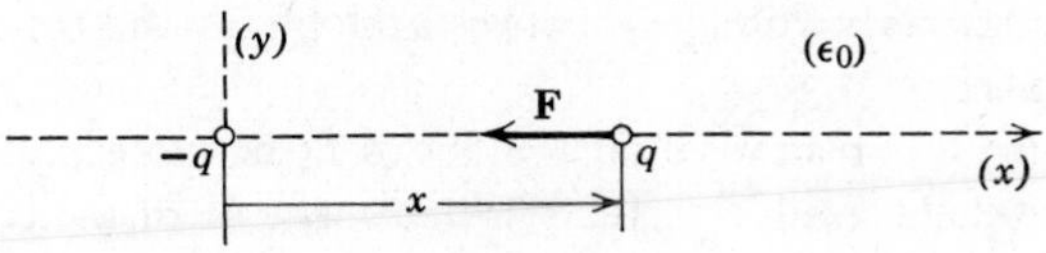

Example 4-19

$-q$ and the location of $+q$ is $-q/4\pi\epsilon_0 x$. The total energy is therefore

$$U_e = \tfrac{1}{2}\sum_{k=1}^{2} q_k\Phi_k = \frac{1}{2}\left[(-q)\frac{q}{4\pi\epsilon_0 x} + (q)\frac{-q}{4\pi\epsilon_0 x}\right] = -\frac{q^2}{4\pi\epsilon_0 x}$$

In this isolated system, the force of $+q$ is found from (4-162)

$$F_x = -\frac{\partial U_e}{\partial x} = -\frac{-q^2}{4\pi\epsilon_0}\frac{\partial}{\partial x}\left(\frac{1}{x}\right) = -\frac{q^2}{4\pi\epsilon_0 x^2}$$

to the left (attractive) as noted in the accompanying figure. This answer agrees with that obtained from (1-52), making use of $\mathbf{E}$ (due to $-q$) at the location of $+q$

$$\mathbf{F} = q\mathbf{E} = q\left(\mathbf{a}_x\frac{-q}{4\pi\epsilon_0 x^2}\right) = \mathbf{a}_x\frac{-q^2}{4\pi\epsilon_0 x^2}$$

EXAMPLE 4-20. Two parallel conducting plates are separated by an air dielectric. Each has an area A, separated a distance x as shown. Neglecting fringing at the edges, obtain the force on either plate from the field energy,

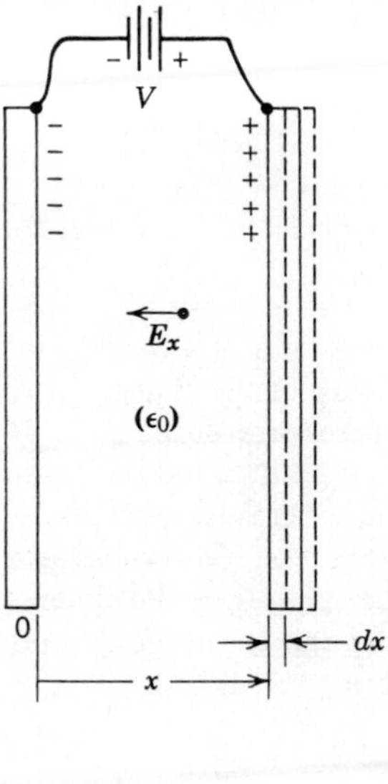

Example 4-20

assuming (*a*) a constant voltage V between the plates and (*b*) constant charges $\pm Q$ on the plates.

(*a*) Assuming a constant voltage V between the plates, and with the plate at $x = 0$ held fixed, a virtual displacement dx of the other yields $F_x = \partial U_e/\partial x$ from (4-170). With $C = \epsilon A/x$, one obtains

$$F_x = \frac{\partial U_e}{\partial x} = \frac{\partial}{\partial x}(\tfrac{1}{2}CV^2) = \frac{V^2}{2}\frac{\partial}{\partial x}\left(\frac{\epsilon_0 A}{x}\right) = -\frac{\epsilon_0 A V^2}{2x^2}$$

The negative result denotes an attractive force, since the stored energy increases with a decrease in the plate separation x.

(*b*) With V disconnected, fixed charges $\pm Q$ reside on the plates. Then a constant $E_x = V/x$ exists between the plates regardless of their separation (neglecting fringing). The electrostatic energy is conveniently expressed by $U_e = (1/2)QV = (1/2)QE_x x$, and with Q and E_x both independent of x, (4-162) obtains

$$F_x = -\frac{\partial U_e}{\partial x} = -\frac{\partial}{\partial x}(\tfrac{1}{2}QE_x x) = -\tfrac{1}{2}QE_x = -\frac{\epsilon_0 A V^2}{2x^2}$$

REFERENCES

ELLIOTT, R. S. *Electromagnetics.* New York: McGraw-Hill, 1966.

HAYT, W. H. *Engineering Electromagnetics*, 2nd ed. New York: McGraw-Hill, 1967

LORRAIN, P., and D. R. CORSON. *Electromagnetic Fields and Waves.* San Francisco: W. H. Freeman, 1970.

REITZ, R., and F. J. MILFORD. *Foundations of Electromagnetic Theory.* Reading, Mass.: Addison-Wesley, 1960.

PROBLEMS

4-1. Show that both the divergence and the curl of the **E** field of a fixed point charge q in free space are zero, except at the location of the charge.

4-2. Two small, electrically charged, conducting spheres are suspended by weightless strings of length ℓ attached to a common point. Find the angle θ by which each is deflected from the vertical if the spheres have a mass m and are charged by amounts q_1 and q_2, as shown. Assume the charges to act as though concentrated at the centers of the spheres. If the spheres weigh 1 g each and the strings are 10 cm long, what charge on the spheres causes the strings to deflect 45° from the vertical, if $q_1 = q_2$?

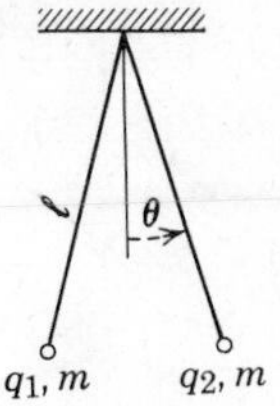

Problem 4-2

4-3. (*a*) A thin conductor bent into a circle of radius a as shown is charged with a uniform density ρ_ℓ. Use a direct integration to show that the field at any point $P(0, 0, z)$ along the z axis is

$$\mathbf{E}(0, 0, z) = \mathbf{a}_z \frac{\rho_\ell a z}{2\epsilon_0(a^2 + z^2)^{3/2}}$$

Sketch roughly $|\mathbf{E}|$ versus z to either side of the wire.

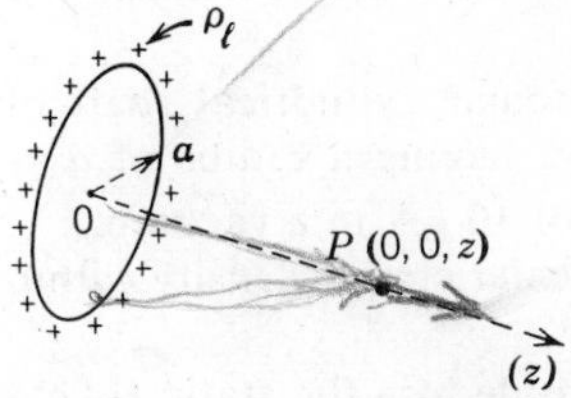

Problem 4-3

(*b*) Show that the answer to (*a*) at great distances from the charge converges to the field of a point charge, if the answer is expressed in terms of the total charge

4-4. Assume that a very thin, circular, insulating disk of radius a is charged uniformly with a density ρ_s. Find the **E** field along the disk axis by integrating directly for **E**. Show that this answer reduces to the field of a point charge as $z \to \infty$, if it is expressed in terms of the total charge on the disk.

4-5. Find the electrostatic field at any point $P(0, 0, z)$ of a uniformly charged, infinite sheet lying in the $z = 0$ plane. Use a direct integration of (4-14) to obtain $\mathbf{E}(0, 0, z)$.

4-6. A spherical conductor of radius a is statically charged with a total charge Q. What charge distribution exists in the static state? Prove, by a direct integration for **E**, that the electrostatic field outside the sphere is the same as though Q were concentrated at the location of the center of the sphere, while inside the sphere $\mathbf{E} = 0$.

4-7. Utilize Gauss' law and symmetry to deduce the answers to Problem 4-6.

4-8. Assume that the following charges are placed respectively on the inner and outer conductors of the concentric spherical pair shown:

(*a*) $Q, 0$
(*b*) $-Q, 0$

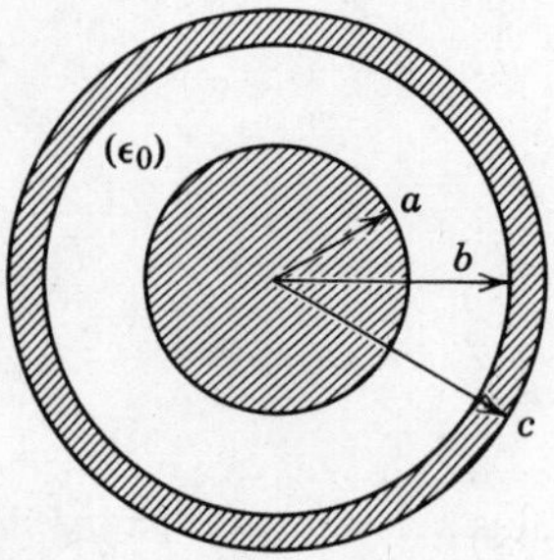

Problem 4-8

(c) $Q, -Q$
(d) $-Q, Q$
(e) $Q, -3Q$
(f) $-Q, 3Q$

Sketch the static E fields for each case. Use arrows to denote the field directions, indicating how much charge appears on the three conductor surfaces in each instance.

4-9. An electron beam consists of a z directed, circular cylindrical shaft of electrons with an average velocity of 10^5 m/sec, having a radius of $a =$ 2 mm and carrying a measured steady current of 10 μA in a vacuum.

(a) Find the vector current density and the scalar charge density within the beam.

(b) Neglecting end effects, use Gauss' law to deduce the static **E** field both inside and outside the beam. What is the value of **E** at the surface of the beam?

4-10. Use Gauss' law to deduce an approximate expression for the electric field *very close* to the ring charge of Problem 4-3.

4-11. A circular coaxial line of great length has dielectric materials of permittivities ϵ_1 and ϵ_2 filling the concentric regions as shown. The interface at $\rho = b$ is assumed midway between the conductor radii a and c.

(a) Assuming the charges Q, $-Q$ reside on every length ℓ of the inner and outer conductors respectively, deduce from Gauss' law and

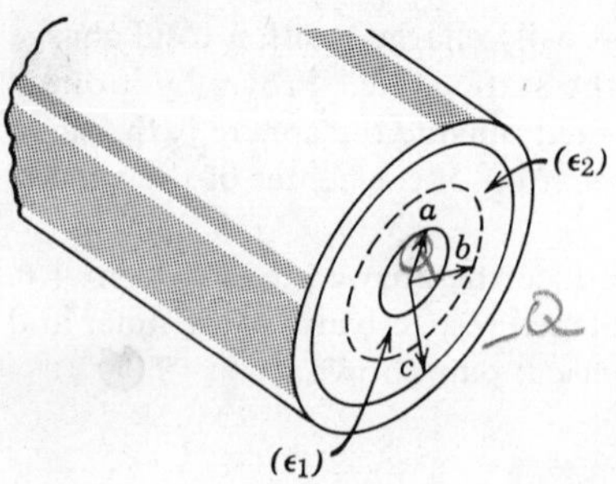

Problem 4-11

boundary conditions the expressions for **D** and **E** in the regions between the conductors.

(*b*) Show, by use of graphs drawn to a relative scale, the variations of D_ρ and E_ρ with ρ for the cases (1) $\epsilon_1 = 2\epsilon_2$ and (2) $\epsilon_1 = \epsilon_2/2$. Which provides the least opportunity for dielectric breakdown, assuming each material will break down at the same value E_0? Explain.

4-12. Assume for the system of Problem 4-11 that $\epsilon_1 = 4\epsilon_0$. What value of ϵ_2 is required if the maximum **E** fields in both regions are to be the same?

4-13. Problem 4-12 suggests a way that a series of dielectric layers with different permittivities can be used in a coaxial system to maintain a very nearly constant **E** field between the conductors. Determine the radial dependence required of the permittivity of an *inhomogeneous* dielectric, if a constant magnitude of **E** is to be maintained everywhere between the conductors. (Use Gauss' law.)

4-14. Prove the relation (4-32) concerning $\nabla(1/R)$. [*Hint:* Expand in rectangular coordinates.]

4-15. Using the potential expression (4-43) for the static electric dipole, derive (4-44). Compare the **E** fields along the $\theta = 0°$ and $\theta = 90°$ directions, and comment.

4-16. Rework Problem 4-3 by a direct integration for the potential; from this result find **E**.

4-17. (*a*) A hollow conducting sphere in free space is charged with Q as shown. Use the line integral of **E** to find Φ at any point P outside the sphere, as well as inside it. Sketch Φ versus r.

(*b*) Repeat (*a*), but assume that Q is somehow distributed uniformly

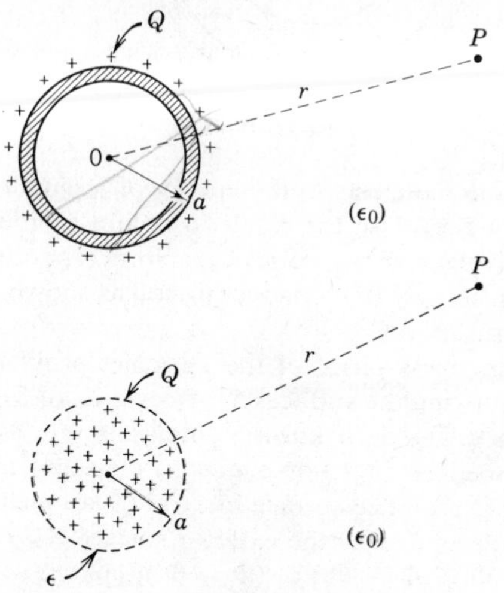

Problem 4-17

throughout the volume of a dielectric sphere of permittivity ϵ, as in the lower diagram. Sketch Φ versus r for this case.

4-18. Rework Problem 4-4, this time integrating directly for Φ at any point on the disk axis, and use the result to derive **E**.

4-19. (*a*) Make use of Gauss' law and symmetry to obtain **D** and **E** of the parallel-plate and spherical capacitors of Figure 4-7, assuming the two conductors are separated by a dielectric of permittivity ϵ. (In the parallel-plate case, neglect field-fringing.)

(*b*) Use the line integral of **E** to deduce Φ at any point between the conductors, assuming the potential reference at the negative conductor.)

(*c*) Verify (4-52) and (4-53) for the capacitance of the two systems, making use of the results of (*b*).

4-20. (*a*) Determine, by use of (4-61b), the stored electrostatic energy of the parallel-plate and spherical capacitor of Problem 4-19.

(*b*) From part (*a*), find the capacitance of each system.

4-21. Find the capacitance of a length ℓ of the coaxial system given in Problem 4-11, using the methods:

(*a*) Find the potential difference between the conductors from the line integral of the **E** field, using the result to find C.

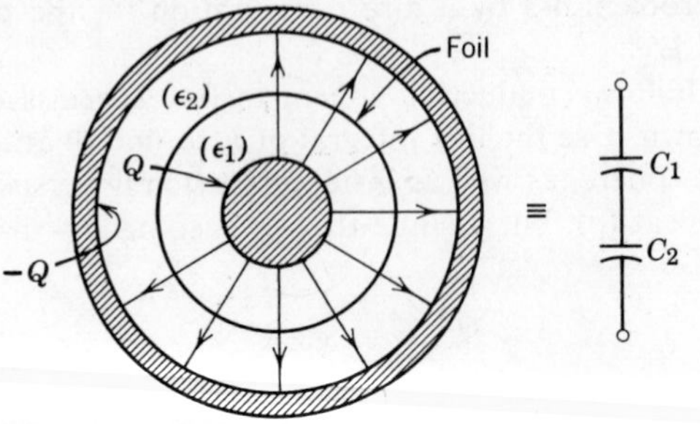

Problem 4-21

(*b*) Determine the electrostatic energy of a length ℓ of the system by use of (4-58e). Use the result to obtain C. Identify the answer as the equivalent of two series capacitors separated at their interface by a vanishingly thin conducting foil as shown in the accompanying figure.

4-22. Repeat the uniqueness proof of the Dirichlet problem for the case of n conductors occupying the surfaces $S_1, S_2, \ldots, S_n$ of Figure 4-9(*b*). Assume the conductors charged to known potentials $\Phi_1, \Phi_2, \ldots, \Phi_n$. The conductors are embedded in a homogeneous dielectric of permittivity ϵ and infinite extent. [*Note:* The surface integral (4-80) yields integrals over S_1, $S_2, \ldots, S_n$, while that over the exterior surface S_∞ vanishes because the quantity $(\Phi - \Phi')\nabla(\Phi - \Phi')$ or $(\Phi - \Phi')[(\partial\Phi/\partial n) - (\partial\Phi'/\partial)]$ decreases as $1/r^3$, whereas ds on the sphere increases only as r^2.]

4-23. A parallel-plate structure like that of Figure 4-7 has its lower plate (at $x = 0$) at zero potential, while its upper plate is at the potential V. Assuming no fringing effects, solve Laplace's equation for the potential at any point between the plates, subject to the proper boundary conditions. Show that **E** is everywhere constant between the plates.

4-24. Assuming for the spherical capacitor of Figure 4-7 that $\Phi(a) = V$ and $\Phi(b) = 0$, obtain by direct integration of Laplace's equation the expression for the Φ and **E** fields between the spheres.

4-25. Using the definitions of the hyperbolic functions, show that the solution (4-91a) of the Laplace equation (4-82) may equivalently be expressed

$$\Phi(x, y) = (C_1 \cos k_x x + C_2 \sin k_x x)(C_3' \cosh k_x y + C_4' \sinh k_x y)$$

in which C_1, C_2, C_3', C_4' are arbitrary constants. The corresponding form of the alternate solution (4-91b) is seen to be

$$\Phi(x, y) = (C_1' \cosh k_y x + C_2' \sinh k_y x)(C_3 \cos k_y y + C_4 \sin k_y y)$$

4-26. (*a*) An infinitely long, rectangular U shaped conducting channel is insulated at the corners from a conducting plate forming the fourth side with interior dimensions as shown. Use an appropriate solution of the two-dimensional Laplace equation subject to appropriate boundary conditions to find the $\Phi(x, y)$ at any interior point, showing that

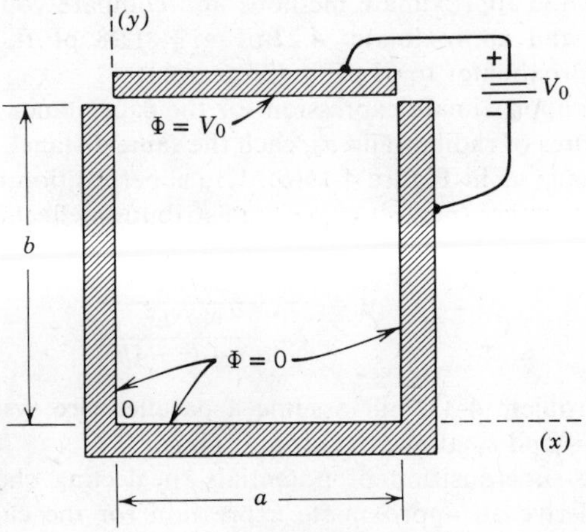

Problem 4-26

$$\Phi(x, y) = \frac{4V_0}{\pi}\left[\frac{\sinh \dfrac{\pi y}{b}}{\sinh \dfrac{\pi a}{b}} \sin \frac{\pi x}{b} + \frac{1}{3}\frac{\sinh \dfrac{3\pi y}{b}}{\sinh \dfrac{3\pi a}{b}} \sin \frac{\pi x}{b} + \cdots\right]$$

[*Hint:* Noting that *both* solutions $e^{k_x y}$ and $e^{-k_x y}$ of (4-91a) are

needed to satisfy the boundary condition $\Phi(x, 0) = 0$, it is desirable to use the equivalent hyperbolic solution of Problem 4-25, observing that $\sinh u \to 0$ and $\cosh u \to 1$ as $u \to 0$.]

(*b*) Find the potential at $(a/2, b/2)$ if $V_0 = 100$ V and $a \doteq b$.

4-27. Use the solution obtained in Problem 4-26 to find **E** in the channel region. Assuming $a = b = 1$ m and $V_0 = 100$ V, find **E** at the three points (1/2, 0), (1/2, 1/2), and (1/2, 1) along the center line.

4-28. Prove (4-94) for the charge density on the infinite, planar conductor of Figure 4-10(*d*).

4-29. If the two line charges in Figure 4-13 are in air and have the densities $\pm 1\ \mu$C/m, find the potentials on the equipotential surfaces shown (for which R_2/R_1 are 5, 3, 2, 1, 1/2, 1/3, and 1/5.)

4-30. Assuming air dielectric, find C/ℓ for the cylindrical conductor above ground of Figure 4-14(*a*), as well as for the parallel-wire system of (*b*) in that figure, if h/R is 2, 5, 10, 100, 1000.

4-31. Compare the C/ℓ obtained for a parallel-wire line by use of (4-107) with that obtained from the approximation (4-111), if $h/R = 2$, 5, and 10, assuming air dielectric.

4-32. Given two parallel-wire lines with air dielectric as follows:

(*a*) Two 0.1 in. (No. 10 AWG) wires spaced 3 ft

(*b*) Two 3/4 in. diameter brass tubes spaced 2 in. center-to-center

Calculate the capacitance per meter for each system. Use expressions from *both* exact and approximate methods and compare your results. [*Answer:* (*a*) Exact and approximate: 4.22 pF/m = 1.28 pF/ft. (*b*) Exact: 16.96 pF/m; approximate: 16.63 pF/m (2% error)]

4-33. Derive an approximate expression for the capacitance between two long, parallel wires of radii a_1 and a_2, each the same distance h above an infinite ground plane as in Figure 4-16(*b*). Use superposition combined with the method of images (neglect charge redistribution effects). [*Answer:*]

$$\frac{C}{\ell} = \frac{2\pi\epsilon}{\ln \dfrac{4d^2h^2}{a_1a_2(d^2+4h^2)}}$$

4-34. Repeat Problem 4-33, but assume a parallel-wire system with different spacings h_1 and h_2 above the ground plane.

4-35. Using the superposition of potentials (neglecting charge redistribution effects), derive an approximate expression for the capacitance between conducting *spheres* of radii a and b, spaced $2d$ between centers. Assume the radii small compared to $2d$. [*Answer*]:

$$C = \frac{4\pi\epsilon}{\dfrac{1}{a} + \dfrac{1}{b} - \dfrac{1}{d}}$$

4-36. Repeat Problem 4-35, assuming in addition that both spheres are a distance h from the center of each sphere to a ground plane.

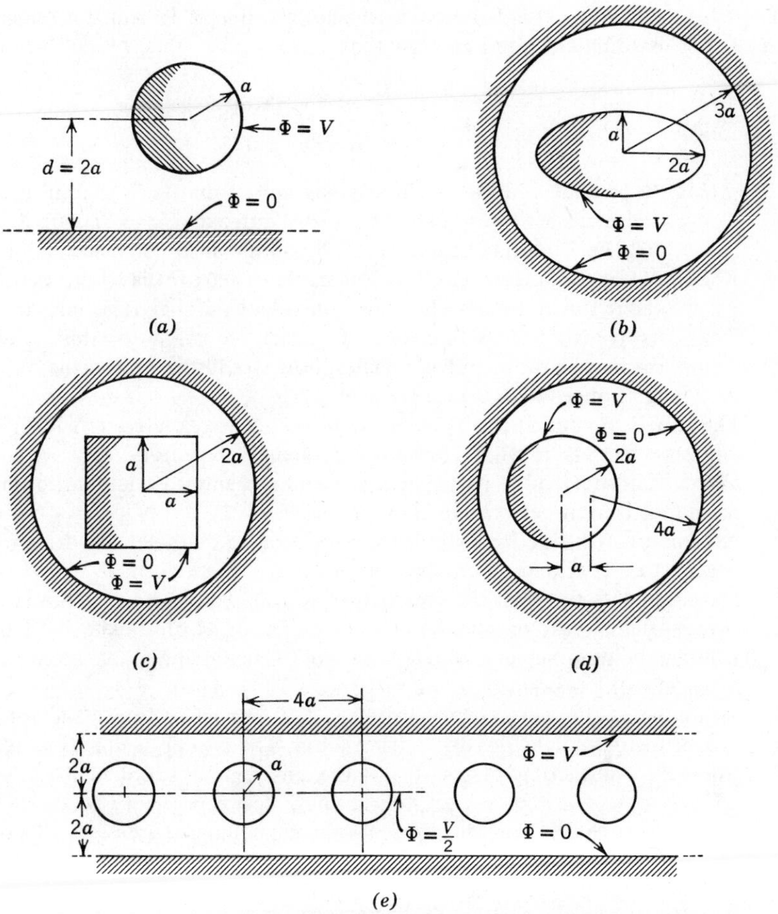

Problem 4-37. (*a*) Rod above plane conductor (compare with analytical answer). (*b*) Elliptic cylinder inside round pipe. (*c*) Square conducting bar inside round pipe. (*d*) An eccentric line. (*e*) Potential difference is impressed between planes: equispaced grid rods are neutral (no net charge).

4-37. Use field-mapping techniques to find C/ℓ of the accompanying two-dimensional configurations (assume air dielectric). Use curvilinear squares as a basis, taking advantage of symmetry wherever possible.

4-38. Using the given flux plots, estimate the capacitance per meter depth of the two-dimensional systems of Figure 4-21, assuming air dielectric.

4-39. Supposing the systems of Figure 4-21 employ a dielectric with $\epsilon_r = 2.25$ and $\sigma = 10^{-7}$ ℧/m at some frequency, find the capacitance and conductance per meter depth for each system.

4-40. (*a*) Assuming the dielectric in the coaxial line of Example 4-8 has the parameters ϵ and σ, show that

$$\frac{G}{\ell} = \frac{2\pi\sigma}{\ln \frac{b}{a}} \ \mho/\text{m}$$

(*b*) A coaxial line with dimensions $a = 1$ mm, $b = 5$ mm uses a dielectric with $\epsilon_r = 2.26$ and a loss tangent $\epsilon''/\epsilon' = 0.0002$ at $f = 10^6$ Hz. Find its capacitance and conductance per meter length.

(*c*) A ten-centimeter length of the cable of part (*b*) is to be used as a capacitor at 10^6 Hz. Find its equivalent parallel *R*, *C* circuit, and its *Q*. If the loss tangent were nearly constant for this material over some range of frequencies, how would the *Q* and the parallel resistance change with frequency?

4-41. The spherical capacitor of Figure 4-7(*b*) uses a lossy dielectric with $\epsilon_r = 4$ and $\sigma = 10^{-4}$ $\mho$/m at a particular frequency. Assuming $a = 5$ cm and $b = 5.5$ cm, determine its capacitance and the shunt conductance. What resistance is seen between the conductors?

4-42. Assume two conducting cylindrical rods 2 cm in diameter are driven into wet earth, separated 12 cm between centers, to a depth of 1 m. Bridge measurements between the conductors at 1000 Hz reveal that the system is equivalent to the parallel combination of $C = 47$ pF and $R = 1870$ Ω. Deduce the parameters ϵ_r and σ of the soil, if field fringing at the rod ends is neglected. [*Answer:* $\epsilon_r = 4.2$, $\sigma = 4.2 \times 10^{-4}$ $\mho$/m]

4-43. A sample of the wet earth of Problem 4-42 is packed into a rectangular box of inside dimensions $10 \times 10 \times 2$ cm. The two opposite 10×10 cm sides are conducting plates. From the answers to Problem 4-42, what resistance would one expect to measure between the two conducting plates through the soil medium? What capacitance? [*Answer:* 2000 Ω, 44.2 pF]

4-44. A conductive paper model, similar to that of Figure 4-25, is made of the system of Figure 4-21(*b*) by applying silver-paint electrodes in the shapes shown onto resistive paper measuring 1000 Ω per square. From the field sketch in Figure 4-21(*b*), what resistance between the electrodes would one expect to measure? What conductance?

4-45. A conventional variable air capacitor consists of an equispaced, multiplate stator and a multiplate rotor meshing into it, to provide a variable capacitance with rotation. Measurements reveal that the capacitance changes nearly linearly from 40 to 460 pF with rotor rotation from 0 to 270°. Find the electrostatic torque on the rotor when set at an arbitrary angle ϕ, with 1000V applied.

CHAPTER

5

Static and Quasi-Static Magnetic Fields

The static magnetic fields of steady currents and the electromagnetic fields of relatively slowly time-varying currents are considered in this chapter. Ampère's law is applied to symmetrical current configurations and to magnetic circuits containing high permeability cores, for the purpose of obtaining their magnetic fields. The static magnetic potential, a vector function, is inferred, and from this, the Biot–Savart law. Faraday's law then leads into the concepts of self and mutual inductance, and the energy and forces of the magnetic field.

5-1 Maxwell's Equations and Boundary Conditions for Static Magnetic Fields

In Section 4-1 it was pointed out that static magnetic fields are required to satisfy only the Maxwell equations (4-3) and (4-4)

$$\nabla \cdot \mathbf{B} = 0 \tag{5-1}$$

$$\nabla \times \mathbf{H} = \mathbf{J} \tag{5-2}$$

The divergenceless property (5-1) specifies that **B** flux lines are always closed, while (5-2) states that the sources of static magnetic fields are steady currents of density **J**. The divergenceless property of any direct current distribution in space is moreover assured by (4-22)

$$\nabla \cdot \mathbf{J} = 0 \tag{5-3}$$

although this property of direct current is not independent of Maxwell's equations, in view of the fact that (5-3) is a consequence of taking the divergence of (5-2). The three foregoing differential equations have integral counterparts given by the static versions of (3-49), (3-66), and (4-20) as follows:

$$\oint_S \mathbf{B} \cdot d\mathbf{s} = 0 \tag{5-4}$$

$$\oint_\ell \mathbf{H} \cdot d\boldsymbol{\ell} = i \tag{5-5}$$

$$\oint_S \mathbf{J} \cdot d\mathbf{s} = 0 \tag{5-6}$$

while the constitutive relationship between **B** and **H** at any point, for the linear, homogeneous, and isotropic materials considered in this chapter, is given by (3-64c)

$$\mathbf{B} = \mu \mathbf{H} \tag{5-7}$$

The boundary conditions for magnetic fields have already been derived in Chapter 3 under the general assumption of time variations for the fields, though they remain unaltered under static conditions. These are given by (3-50), (3-71), and (4-127) as follows:

$$B_{n1} - B_{n2} = 0 \tag{5-8}$$

$$H_{t1} - H_{t2} = 0 \tag{5-9}$$

$$J_{n1} - J_{n2} = 0 \tag{5-10}$$

assuring the continuity of the normal components of the static **B** and **J** fields at any interface, as well as the tangential components of **H**.

The presence of a current in a finitely conductive region implies the presence of an **E** field, in view of relation (3-7) that $\mathbf{J} = \sigma \mathbf{E}$, yielding the possibility of coupling the static magnetic field with an electrostatic field.

5-2 Ampère's Circuital Law

Ampère's circuital law for the static magnetic field in *free space* was initially discussed in Section 1-11. The presence of a magnetic material with a permeability μ in the region of interest was taken into account by the definition (3-58) of the field **H**, the law in this event becoming (5-5)

$$\oint_\ell \mathbf{H} \cdot d\boldsymbol{\ell} = i \qquad [5\text{-}5]$$

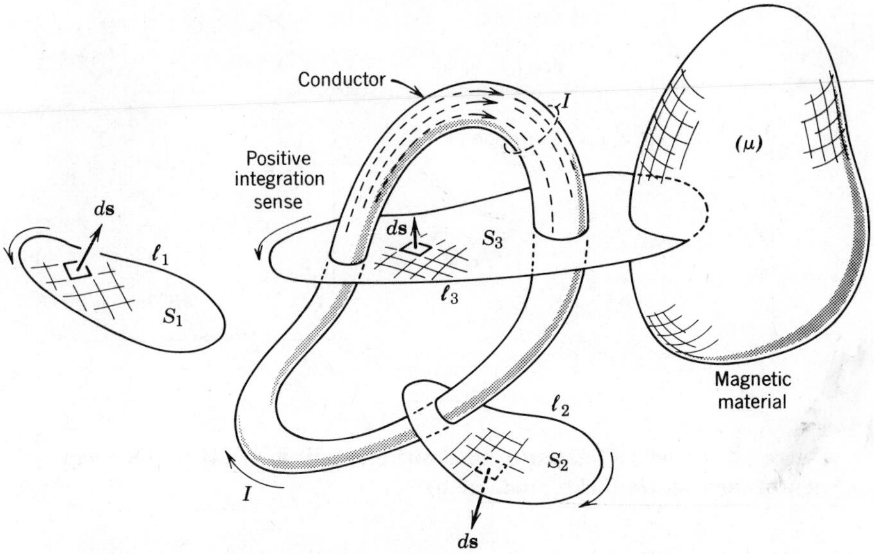

Figure 5-1. Showing typical closed paths ℓ_1, ℓ_2, and ℓ_3 chosen to illustrate Ampère's law and the interpretation of the current *i*.

Figure 5-1 illustrates (5-5) relative to a conductor compelled to carry a steady current *I*. Thus, the line integral of **H**, around the closed path ℓ_1 shown, yields the value zero because the current *i* enclosed by that particular choice of path is zero. On the other hand, the current piercing S_2 is precisely the current *I* carried by the conductor, whereas $i = 0$ for the assumed path ℓ_3 because the current *I* flows both into and out of S_3 to provide canceling contributions to *i*.

Two important interpretations of Ampère's circuital law are the following:

1. Steady current sources possess magnetic flux line distributions which, at positions in space near the sources, are directed in accordance with the right-hand rule.
2. Ampère's circuital law may be used as the basis for finding the **H** field (and thus **B**) of a steady current if the physical symmetry of the problem permits extricating the desired field from the integral.

Applications of 2 to finding the static magnetic fields of systems exhibiting simple symmetries have been given in Examples 1-13, 1-15, and 3-4. Additional examples involving conductors wound about symmetrically shaped magnetic materials are given here.

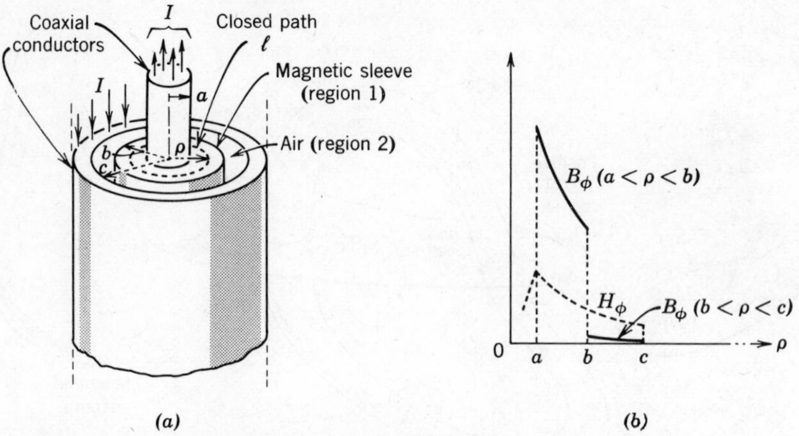

Figure 5-2. Coaxial line partly filled with magnetic material. (*a*) Cutaway view of the line. (*b*) Fields produced by *I*.

EXAMPLE 5-1. Two long, coaxial, circular conductors carry the steady current I as shown in Figure 5-2. Assume constant current densities over each conductor cross-section. The region $a < \rho < b$ is filled with a magnetic material of constant permeability μ; the region $b < \rho < c$ is air. Find $\mathbf{B}$ and $\mathbf{H}$ in the two regions. Sketch their graphs versus ρ, assuming $\mu = 100\mu_0$.

From symmetry and the application of the right-hand rule, the magnetic field is everywhere ϕ directed, i.e., $\mathbf{H} = \mathbf{a}_\phi H_\phi$. Ampère's law (5-5) applied to a symmetric path ℓ of radius ρ (shown in region 1) yields

$$\oint_\ell (\mathbf{a}_\phi H_\phi)\cdot\mathbf{a}_\phi\, d\ell = H_\phi \oint_\ell d\ell = I$$

and because $\oint d\ell = 2\pi\rho$, solving for H_ϕ obtains

$$\mathbf{H} = \mathbf{a}_\phi H_\phi = \mathbf{a}_\phi \frac{I}{2\pi\rho} \tag{5-11}$$

This result is independent of μ, which means that it applies to both the magnetic region 1 and the air region 2. Thus the $\mathbf{B}$ field in each region is found by inserting (5-11) into (5-7)

$$\begin{aligned} \mathbf{B} &= \mathbf{a}_\phi \frac{\mu I}{2\pi\rho} \qquad a < \rho < b \\ \mathbf{B} &= \mathbf{a}_\phi \frac{\mu_0 I}{2\pi\rho} \qquad b < \rho < c \end{aligned} \tag{5-12}$$

These results show that if region 1 had a permeability $\mu = 100\,\mu_0$, B_ϕ just inside the magnetic region (at $\rho = b-$) would be 100 times as dense as on the air side. This is illustrated by the solid curve of Figure 5-2(*b*). Thus nearly all of the flux of **B** resides within the magnetic material, if $\mu \gg \mu_0$.

EXAMPLE 5-2. Suppose an n turn, closely wound toroidal coil with a rectangular cross-section is filled with a magnetic material of constant permeability μ from a to b as in Figure 5-3(*a*). With a current I in the coil, find **B** and **H** in the two regions; sketch their relative magnitudes if $\mu = 100\,\mu_0$ for the magnetic material. Compare the total magnetic flux ψ_m in the core if it is all air with that obtained if it is all magnetic material, assuming $\mu = 100\,\mu_0$.

From the symmetry it is evident that Ampère's law is useful for finding **H**; choose ℓ as a circle with the radius ρ shown in Figure 5-3. From the symmetry and the right-hand rule, **H** must be ϕ directed and of constant magnitude on ℓ. Equation (5-5) then yields

$$\oint_\ell (\mathbf{a}_\phi H_\phi)\cdot \mathbf{a}_\phi\, d\ell = H_\phi \oint_\ell d\ell = nI$$

whence

$$H_\phi = \frac{nI}{2\pi\rho} \tag{5-13}$$

Using (5-7), **B** in the magnetic and air regions of the core becomes

$$\begin{aligned} \mathbf{B} &= \mathbf{a}_\phi B_\phi = \mathbf{a}_\phi \frac{\mu n I}{2\pi\rho} & a < \rho < b \\ \mathbf{B} &= \mathbf{a}_\phi B_\phi = \mathbf{a}_\phi \frac{\mu_0 n I}{2\pi\rho} & b < \rho < c \end{aligned} \tag{5-14}$$

The graphs of these quantities are shown in Figure 5-3(*b*).

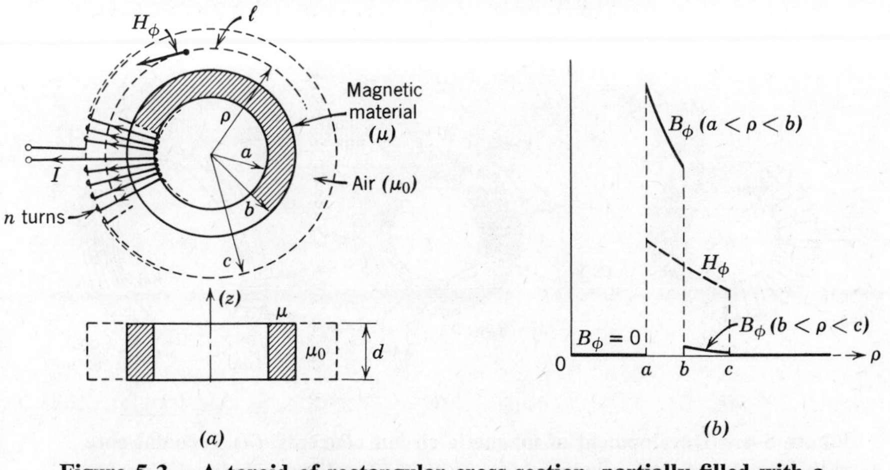

Figure 5-3. A toroid of rectangular cross-section, partially filled with a magnetic material. (*a*) Dimensions of toroid. (*b*) Interior fields.

If the core is *all air*, the total **B** flux in it becomes

$$\psi_m = \int_{S(core)} \mathbf{B} \cdot d\mathbf{s} = \int_{z=0}^{d} \int_{\rho=a}^{c} \frac{\mu_0 n I}{2\pi\rho} \, d\rho \, dz = \frac{\mu_0 n I d}{2\pi} \ln \frac{c}{a}$$

For a completely magnetic core, $\mu = 100\mu_0$ would appear in the foregoing answer in lieu of μ_0, demonstrating the considerable increase in magnetic flux possible if an iron core is utilized.

5-3 Magnetic Circuits

It has been noted that a magnetic material of large permeability can aid in producing large magnetic flux densities compared to what would exist without its use. From (5-1) it is evident that physical magnetic fields must always consist of closed flux lines. By constraining the **B** flux to occupy the interior of closed (or nearly closed) paths of magnetic material, one may speak of *magnetic circuits* with reference to those closed paths.

Figure 5-4(*a*) shows an idealized magnetic circuit: a closely spaced toroidal winding establishing a magnetic field within it, with essentially no magnetic flux outside the core, whether or not the core material is magnetic. If the winding is localized on the core as in (*b*), the effect of a high-permeability core material ($\mu \gg \mu_0$) is such that the magnetic flux ψ_m generated by the current I in the coil still appears almost wholly within the boundaries of the core. The magnetic flux must consist of closed lines as required by the divergence property (5-1), and because of the constraint supplied by the refractive law (3-76) (requiring that **B** flux leave the surface of the high-permeability

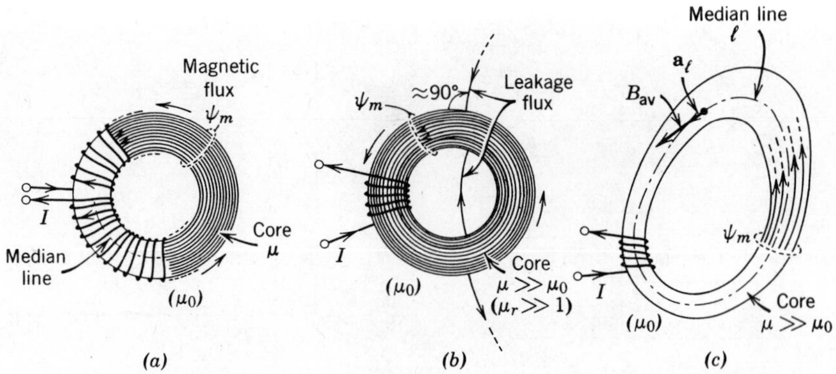

Figure 5-4. Development of magnetic circuit concepts. (*a*) Torodial core with closely spaced winding. (*b*) With a localized winding, showing leakage flux. (*c*) A generalized magnetic circuit: leakage flux neglected.

magnetic core very nearly perpendicularly), one concludes that very little can appear outside the core as *leakage flux* if the permeability of the core is sufficiently large. In ferromagnetic cores having relative permeabilities of 10^2 to 10^4 or more, the leakage flux developed external to a core may therefore ordinarily be neglected. The analytical determination of the leakage flux usually requires a rigorous solution of the boundary-value problem of the magnetic system; in general, this is a difficult process.

For present purposes concerned with magnetic circuits as in Figure 5-4, the magnetic core is assumed linear, homogeneous, and isotropic; furthermore, the leakage flux is ignored, implying a *constant flux* ψ_m through any cross-section of a single-mesh core. This flux is

$$\psi_m = \int_S \mathbf{B} \cdot d\mathbf{s} \text{ Wb} \tag{5-15}$$

if S is any cross-section. The need for knowing $\mathbf{B}$ at each point in the cross-section is obviated if ψ_m is expressed in terms of an *average* flux density B_{av} over S; i.e.,

$$\psi_m = B_{\text{av}} A \tag{5-16}$$

assuming B_{av} lies tangent to a median line ℓ, as in Figure 5-4(c). Even for a toroid of constant cross-section, the median line will not lie precisely at the core center, for one may recall from the solutions obtained in Example 5-2 the inverse ρ dependence of B_ϕ in the core. In the following, the B_{av} is assumed at the *center* of the core cross-section; i.e., the median line ℓ is taken as the core center line. Then (5-16) becomes a good approximation if the core is thin.

To find the flux ψ_m developed by the current I in the core of the single magnetic circuit of Figure 5-4(c), apply Ampère's law to the median path ℓ; i.e.,

$$\oint_\ell \mathbf{H} \cdot d\boldsymbol{\ell} = nI \text{ A}$$

in which $d\boldsymbol{\ell} = \mathbf{a}_\ell \, d\ell$, and $\mathbf{H} = \mathbf{B}/\mu = \mathbf{a}_\ell B_{\text{av}}/\mu$. Making use of (5-16)

$$\oint_\ell \left(\mathbf{a}_\ell \frac{\psi_m}{\mu A(\ell)} \right) \cdot \mathbf{a}_\ell \, d\ell = nI$$

In the generalized case, the cross-sectional core area A can be a variable depending upon the location along the median path ℓ, and it is designated $A(\ell)$. The core flux ψ_m through any cross-section along ℓ is constant if the

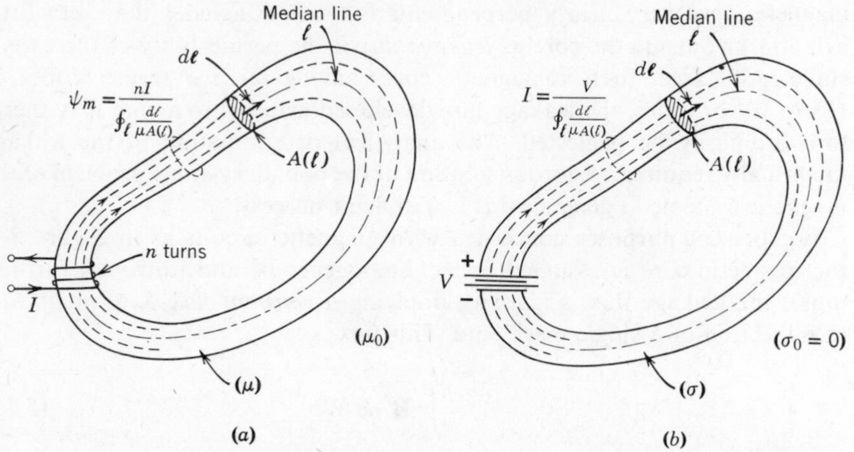

Figure 5-5. Dc magnetic and electric circuit analogs. Leakage flux is neglected in the magnetic circuit. (*a*) Magnetic circuit. Magnetic flux is generated by the source nI. (*b*) Electric circuit. Current flux is generated by the source V.

leakage flux is neglected, obtaining

$$\psi_m = \frac{nI}{\displaystyle\oint_\ell \frac{d\ell}{\mu A(\ell)}} \text{ Wb} \tag{5-17}$$

Thus (5-17) is seen to be analogous with Ohm's law (4-155), applicable to the thin, dc circuit of Figure 4-27 and reproduced in Figure 5-5(*b*). The quantity $\oint_\ell \mathbf{H}\cdot d\ell = nI$ in the numerator of (5-17) is sometimes called *magnetomotive force*, analogous to the applied voltage V (electromotive force) in Figure 5-5(*b*). The denominator, called the *reluctance* of the magnetic circuit, is given the symbol $\mathscr{R}$:

$$\mathscr{R} = \oint_\ell \frac{d\ell}{\mu A(\ell)} \text{ A/Wb} \quad \text{or} \quad \text{H}^{-1} \tag{5-18}$$

a quantity analogous to (4-156), the resistance R of the dc electric circuit. Its reciprocal (analogous to conductance) is called the *permeance* of the magnetic circuit.

If the magnetic core has a *constant cross-section* A and a *constant permeability* μ, the reluctance (5-18) reduces to $\mathscr{R} = \ell/\mu A$, whence (5-17) yields the special result for the core flux

$$\psi_m = \frac{nI}{\mathscr{R}} = \frac{nI}{\dfrac{\ell}{\mu A}} \quad \text{Single-mesh; constant } A, \mu \tag{5-19}$$

This result applies to the magnetic circuit of Figure 5-4(b) neglecting leakage flux and assuming a reasonably thin core.

More general magnetic circuits might consist of series arrangements of magnetic materials as in Figure 5-6. A narrow air gap (of length ℓ_g) can also be included, of interest in the design of relays and in the linearization of iron core inductors, or a gap might be a mechanical necessity as in a motor or generator. For the *series* system of Figure 5-6(*a*), applying the reluctance integral (5-18) to the successive portions ℓ_1, ℓ_2, ℓ_3 and ℓ_g over which the permeabilities and cross-sections are constants, obtains

$$\mathscr{R} = \frac{\ell_1}{\mu_1 A_1} + \frac{\ell_2}{\mu_2 A_2} + \frac{\ell_3}{\mu_3 A_3} + \frac{\ell_g}{\mu_0 A_g} \equiv \mathscr{R}_1 + \mathscr{R}_2 + \mathscr{R}_3 + \mathscr{R}_g \tag{5-20}$$

analogous to the resistance of the series electric circuit shown. The field-fringing effects near the edges of a small gap are neglected in the air gap reluctance term $\mathscr{R}_g$. The magnetic flux in the circuit can thus be found by use of (5-17).

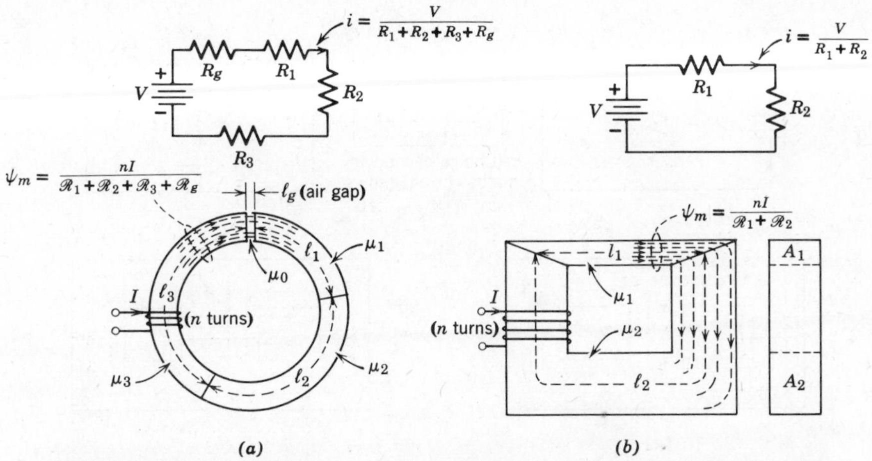

Figure 5-6. Examples of series magnetic circuits, and their electric circuit analogs. (*a*) A series magnetic circuit and its electric circuit analog. (*b*) A rectangular configuration of high permeability materials.

In (5-20) the air gap permeability μ_0 ordinarily is much smaller than μ_1, μ_2, and μ_3 of the magnetic materials in a bonafide magnetic circuit. This means that for even a small air gap, the gap reluctance term can often be orders of magnitude larger than the reluctance of the rest of the circuit. A good approximation in such cases is that the core flux is determined essentially by the air gap reluctance only; i.e., $\mathscr{R} \cong \mathscr{R}_g$.

For practical reasons concerned with fabrication problems, magnetic cores of rectangular shape, like that of Figure 5-6(*b*), are in common use in devices such as relays, inductors, and transformers. The approximations of the magnetic circuit concept become greater in such configurations because of the difficulty in assigning correct median lengths to the various legs of the rectangle, particularly if the cross-sections are large compared with the overall core dimensions.

An extension of the theory of magnetic circuits to systems having more than one magnetic path is possible again through the use of the electric circuit analogy, as illustrated in Figure 5-7. Because the fluxes divide among the branches of the magnetic circuit in just the way the currents do in a dc electric circuit, it is seen that writing Ampère's law around the *two* magnetic meshes of Figure 5-7(*a*), for example, yields the following equations:

$$nI = \mathscr{R}_1\psi_{m1} + \mathscr{R}_2\psi_{m2} \tag{5-21}$$

$$0 = -\mathscr{R}_3(\psi_{m1} - \psi_{m2}) + \mathscr{R}_2\psi_{m2}$$

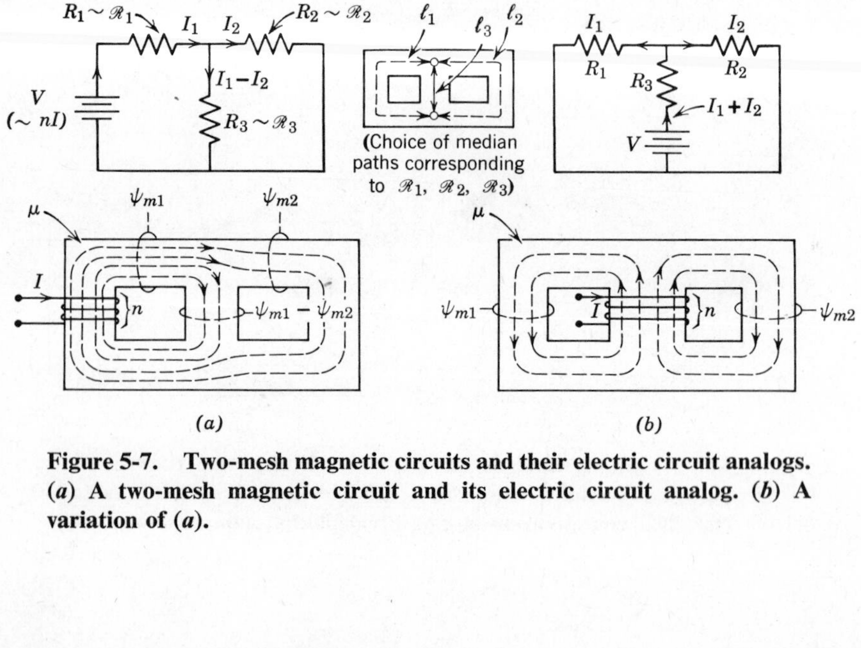

Figure 5-7. Two-mesh magnetic circuits and their electric circuit analogs. (*a*) A two-mesh magnetic circuit and its electric circuit analog. (*b*) A variation of (*a*).

in which $\mathscr{R}_1$, $\mathscr{R}_2$, $\mathscr{R}_3$ are found from the median paths ℓ_1, ℓ_2, ℓ_3 in Figure 5-7(*a*). For linear core materials, (5-21) can be solved simultaneously to find the magnetic fluxes ψ_{m1} and ψ_{m2}.

The accuracy of the analysis of magnetic circuits through reluctance methods is affected not only by the leakage flux problem and the assignment of median paths, but also by the nonlinear *B-H* curves of ferromagnetic materials. Nonlinearity, as exhibited in Figure 3-13, requires that the permeability be expressed as a *function* of the **H** field in the core, or $\mu(H)$. One cannot find **H**, on the other hand, until a value of μ has been assigned to the circuit (or values of μ to its branches). Iterative processes are frequently successful in such problems. Thus, if a trial value of magnetic flux is assumed for the circuit, the value of μ may be obtained; this result can then be used to determine a new value of magnetic flux. This process is repeated until the desired accuracy of the answer is obtained.

EXAMPLE 5-3. A toroidal iron core of square cross-section, with a 2 mm air gap and wound with 100 turns, has the dimensions shown. Assume the iron has the constant $\mu = 1000\mu_0$. Find (*a*) the reluctances of the iron path and the air gap and (*b*) the total flux in the circuit if $I = 100$ mA.

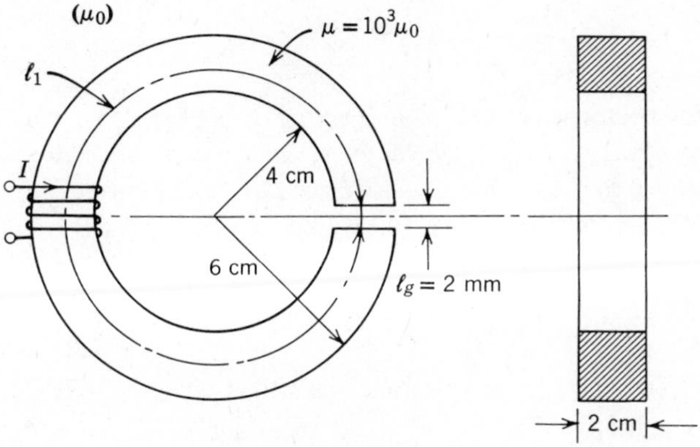

Example 5-3

(*a*) The reluctance of the iron path, having a median length $\ell_1 \cong 2\pi(0.05) = 0.314$ m and cross-sectional area $A_1 = 4 \times 10^{-4}$ m^2, is

$$\mathscr{R}_1 \cong \frac{\ell_1}{\mu_1 A_1} = \frac{0.314 - 0.002}{10^3(4\pi \times 10^{-7})4 \times 10^{-4}} = 0.621 \times 10^6 \text{ H}^{-1}$$

The air gap reluctance, assuming no fringing, becomes

$$\mathscr{R}_g = \frac{\ell_g}{\mu_0 A_1} = \frac{0.002}{4\pi \times 10^{-7}(4 \times 10^{-4})} = 3.98 \times 10^6 \text{ H}^{-1}$$

(*b*) The magnetic flux is given by (5-17), i.e., the magnetomotive force nI of the coil divided by the reluctance of the series circuit

$$\psi_m = \frac{nI}{\mathscr{R}_1 + \mathscr{R}_g} = \frac{10^2(0.1)}{4.6 \times 10^6} = 2.18 \times 10^{-6}\ \text{Wb}$$

With the air gap absent, ψ_m is limited only by the reluctance $\mathscr{R}_1$ of the iron path, becoming $\psi_m = 15.97 \times 10^{-6}$ Wb.

5-4 Vector Magnetic Potential

In Section 4-5 was shown how the irrotational property (4-2) of the static **E** field permits expressing **E** as the gradient of some auxiliary scalar potential function Φ through (4-31). It was also shown how Φ can be found by use of (4-35a) integrated over the free charge sources ρ_v.

An equivalent approach for determining static *magnetic* fields is also by use of an auxiliary potential field, in this case a vector. Noting that any **B** field has the solenoidal property (5-1), i.e., $\nabla \cdot \mathbf{B} = 0$, permits expressing **B** in terms of an auxiliary vector function **A** by means of the curl relation

$$\mathbf{B} = \nabla \times \mathbf{A} \tag{5-22}$$

in view of the vector identity (19) in Table 2-2, $\nabla \cdot (\nabla \times \mathbf{A}) = 0$. The function **A** defined by (5-22) is called the *vector magnetic potential* field.

The vector magnetic potential **A** is related to steady current density sources **J** responsible for the field **B** as follows. In a static magnetic field problem, the relation (5-2), $\nabla \times \mathbf{H} = \mathbf{J}$, is satisfied by the **H** field. It is also written

$$\nabla \times \mathbf{B} = \mu \mathbf{J} \tag{5-23}$$

for a region in which μ is constant; substituting (5-22) for **B** into (5-23) yields

$$\nabla \times (\nabla \times \mathbf{A}) = \mu \mathbf{J} \tag{5-24}$$

This vector differential equation is simplified by use of the vector identity (2-88a)

$$\nabla \times (\nabla \times \mathbf{A}) \equiv \nabla(\nabla \cdot \mathbf{A}) - \nabla^2 \mathbf{A} \tag{5-25}$$

To assure the uniqueness of the potential **A**, both its curl and divergence must be specified. The curl is given by (5-22), and div **A** appearing in (5-25) has not yet been assigned. Assuming $\nabla \cdot \mathbf{A} = 0$ does not conflict with any prior assumption, permitting $\nabla \times (\nabla \times \mathbf{A})$ in (5-24) to be replaced with

$-\nabla^2\mathbf{A}$ to yield

$$\nabla^2\mathbf{A} = -\mu\mathbf{J} \tag{5-26}$$

This result, sometimes called the *vector Poisson equation* because of its similarity to (4-67), is an inhomogeneous, linear differential equation relating **A** to its sources **J**, with μ a constant in the region in question. The virtue of (5-26) lies in the availability of several methods for finding its solutions, among which are the method of separation of variables, and an integration approach described in the next section.

5-5 An Integral Solution for A in Free Space; Biot–Savart Law

An integral solution of (5-26) can be inferred as follows, assuming an unbounded region of free space ($\mu = \mu_0$). In cartesian coordinates, the left side of (5-26) is written, using (2-83)

$$\nabla^2\mathbf{A} = \mathbf{a}_x\nabla^2 A_x + \mathbf{a}_y\nabla^2 A_y + \mathbf{a}_z\nabla^2 A_z$$

whence (5-26) becomes the three scalar differential equations

$$\nabla^2 A_x = -\mu_0 J_x \qquad \nabla^2 A_y = -\mu_0 J_y \qquad \nabla^2 A_z = -\mu_0 J_z \tag{5-27}$$

Each of the latter is *analogous* to the Poisson equation (4-67)

$$\nabla^2\Phi = -\frac{\rho_v}{\epsilon} \qquad [4\text{-}67]$$

the integral solution of which, in unbounded free space ($\epsilon = \epsilon_0$) containing a static charge of density ρ_v, has been shown to be (4-35a)

$$\Phi(u_1, u_2, u_3) = \int_V \frac{\rho_v(u_1', u_2', u_3')}{4\pi\epsilon_0 R}\, dv' \qquad [4\text{-}35\text{a}]$$

Therefore, the analogous solutions of the three scalar differential equations (5-27) in free space must be

$$A_x(u_1, u_2, u_3) = \int_V \frac{\mu_0 J_x(u_1', u_2', u_3')}{4\pi R}\, dv'$$

$$A_y(u_1, u_2, u_3) = \int_V \frac{\mu_0 J_y(u_1', u_2', u_3')}{4\pi R}\, dv'$$

$$A_z(u_1, u_2, u_3) = \int_V \frac{\mu_0 J_z(u_1', u_2', u_3')}{4\pi R}\, dv'$$

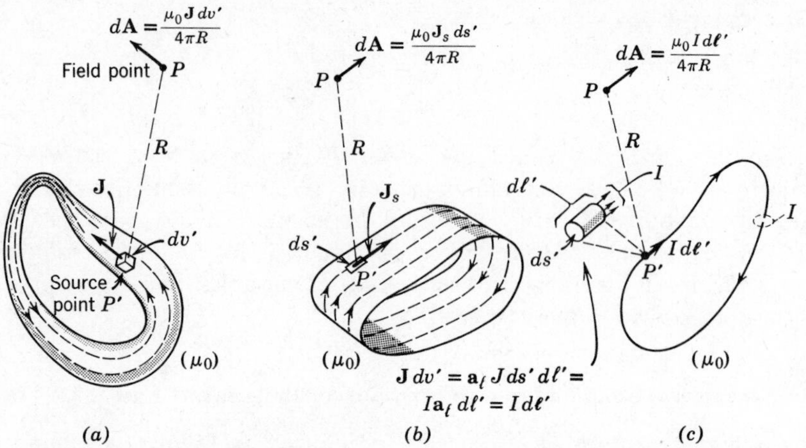

Figure 5-8. Three types of steady current distributions in space. (*a*) Volume distribution of elements $\mathbf{J}\, dv'$. (*b*) Surface distribution of elements $\mathbf{J}_s\, ds'$. (*c*) Line distribution of elements $\mathbf{J}\, dv' \rightarrow I\, d\boldsymbol{\ell}'$.

Adding these three integrals vectorially yields the desired integral solution of (5-26)

$$\mathbf{A}(u_1, u_2, u_3) = \int_V \frac{\mu_0 \mathbf{J}(u_1', u_2', u_3')}{4\pi R}\, dv' \text{ Wb/m} \tag{5-28a}$$

The meaning of R in (5-28a) is the same as in (4-35a); it denotes the distance from the source point P' to the field point P at which $\mathbf{A}$ is to be found. Once the $\mathbf{A}$ has been obtained by means of (5-28a), the corresponding $\mathbf{B}$ field is obtained from the curl of $\mathbf{A}$, using (5-22).

The geometry of a system with current sources of density $\mathbf{J}$ producing the vector magnetic potential $\mathbf{A}$ given by (5-28a) is shown in Figure 5-8. Note that the integrand of (5-28a) is a differential $d\mathbf{A}$ given by

$$d\mathbf{A}(u_1, u_2, u_3) = \frac{\mu_0 \mathbf{J}(u_1', u_2', u_3')}{4\pi R}\, dv'$$

from which it is seen that the current source $\mathbf{J}\, dv'$ at the typical source point $P'(u_1', u_2', u_3')$ produces, at any fixed field point P, a vector contribution $d\mathbf{A}$ *parallel* to the element $\mathbf{J}\, dv'$. Moreover, the magnitude of its influence at P is inversely proportional to the distance R. These relationships are depicted in Figure 5-8(*a*).

In case of either a surface current (a current sheet) or a line current,† as noted in Figures 5-8(*b*) and (*c*), (5-28a) reduces to the following surface and line integral:

$$\mathbf{A}(u_1, u_2, u_3) = \int_S \frac{\mu_0 \mathbf{J}_s(u_1', u_2', u_3')}{4\pi R}\, ds' \tag{5-28b}$$

$$\mathbf{A}(u_1, u_2, u_3) = \int_\ell \frac{\mu_0 I\, d\boldsymbol{\ell}'}{4\pi R} \tag{5-28c}$$

In practice, steady surface and line currents are approximated by physical currents flowing in thin, sheet conductors or thin wires. The vector magnetic potential results (5-28a, b, c) deserve comparison with the analogous results (4-35a, b, c) for the scalar electric potential fields of static charge distributions.

EXAMPLE 5-4. Find the vector magnetic potential in the plane bisecting a straight piece of thin wire of finite length $2L$ in free space, assuming a direct current I as in Figure 5-9. Find **B** from **A**.

The fixed field point is on the plane $z = 0$ at $P(\rho, 0, 0)$. The typical current source element at $P'(0, 0, z')$ is $I\, d\boldsymbol{\ell}' = \mathbf{a}_z I\, dz'$, and R from P' to P is $R = \sqrt{\rho^2 + (z')^2}$, putting the line integral (5-28c) in the form

$$\mathbf{A}(\rho, 0, 0) = \int_{z'=-L}^{L} \frac{\mu_0 \mathbf{a}_z I\, dz'}{4\pi\sqrt{\rho^2 + (z')^2}}$$

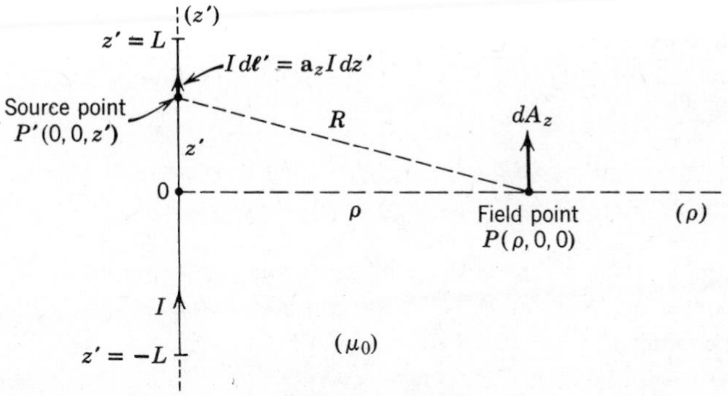

Figure 5-9. Geometry of a thin wire carrying a steady current *I*.

† The line element of current is shown in Figure 5-8(*c*), enlarged into the volume element $\mathbf{J}\, dv' = \mathbf{a}_\ell J\, d\ell'\, ds' = (J\, ds')(\mathbf{a}_\ell\, d\ell')$, which becomes just $I\, d\boldsymbol{\ell}'$ if the product $J\, ds'$ denotes the finite current I in the line source.

The unit vector $\mathbf{a}_z$ has the same direction at all P', yielding at P

$$\mathbf{A} = \mathbf{a}_z \frac{\mu_0 I}{4\pi} \int_{-L}^{L} \frac{dz'}{\sqrt{\rho^2 + (z')^2}} = \mathbf{a}_z \frac{\mu_0 I}{4\pi} [\ell n\,(z' + \sqrt{\rho^2 + (z')^2})]_{-L}^{L}$$

$$= \mathbf{a}_z \frac{\mu_0 I}{4\pi} \ell n \frac{\sqrt{L^2 + \rho^2} + L}{\sqrt{L^2 + \rho^2} - L} \tag{5-29}$$

One finds $\mathbf{B}$ at P using (5-22) in circular cylindrical coordinates

$$\mathbf{B} = \nabla \times \mathbf{A} = \begin{vmatrix} \dfrac{\mathbf{a}_\rho}{\rho} & \mathbf{a}_\phi & \dfrac{\mathbf{a}_z}{\rho} \\ \dfrac{\partial}{\partial \rho} & 0 & 0 \\ 0 & 0 & A_z \end{vmatrix} = -\mathbf{a}_\phi \frac{\partial A_z}{\partial \rho} = \mathbf{a}_\phi \frac{\mu_0 I}{2\pi\rho} \frac{L}{\sqrt{L^2 + \rho^2}} \tag{5-30}$$

For $\rho \ll L$, (5-30) simplifies to

$$\mathbf{B} = \frac{\mu_0 I}{2\pi\rho} \mathbf{a}_\phi \tag{5-31}$$

a result very nearly correct when near a finite-length wire, or correct at any ρ distance for an infinitely long wire. In Example 1-13 (5-31) agrees with (1-64).

EXAMPLE 5-5. Find the $\mathbf{A}$ and $\mathbf{B}$ fields of a thin wire loop of radius a and carrying a steady current I, as in Figure 5-10(a). Make approximations to provide valid answers at large distances from the loop (assume $a \ll r$).

Without detracting from the generality, the field point P can be located directly above the y axis as shown in Figure 5-10(b). The $\mathbf{A}$ field at P is given by (5-28c), in which $I\,d\boldsymbol{\ell}' = \mathbf{a}_\phi Ia\,d\phi'$. The variable direction of $\mathbf{a}_\phi$ in the integrand is handled by pairing the effects of the current elements $I\,d\boldsymbol{\ell}_1'$ and $I\,d\boldsymbol{\ell}_2'$ at the symmetrical locations about the y axis in Figure 5-10(b). From the geometry,

$$\begin{aligned} I\,d\boldsymbol{\ell}_1' &= \mathbf{a}_\phi Ia\,d\phi' = (-\mathbf{a}_x \sin\phi' + \mathbf{a}_y \cos\phi')Ia\,d\phi' \\ I\,d\boldsymbol{\ell}_2' &= (-\mathbf{a}_x \sin\phi' - \mathbf{a}_y \cos\phi')Ia\,d\phi' \end{aligned} \tag{1}$$

to provide a cancellation of the y components of the potential contributions of the pair of elements at P, leaving a net $d\mathbf{A}$ at P that is $-x$ directed. Thus (5-28c) becomes

$$-A_x = A_\phi = 2\int_{\phi' = -\pi/2}^{\pi/2} \frac{\mu_0 Ia \sin\phi'\,d\phi'}{4\pi R} \tag{2}$$

From the law of cosines applied to the triangle POP' in the figure, $R^2 = a^2 + r^2 - 2ar\cos\alpha = a^2 + r^2 - 2ar\sin\theta\sin\phi'$. If $r \gg a$, one can approxi-

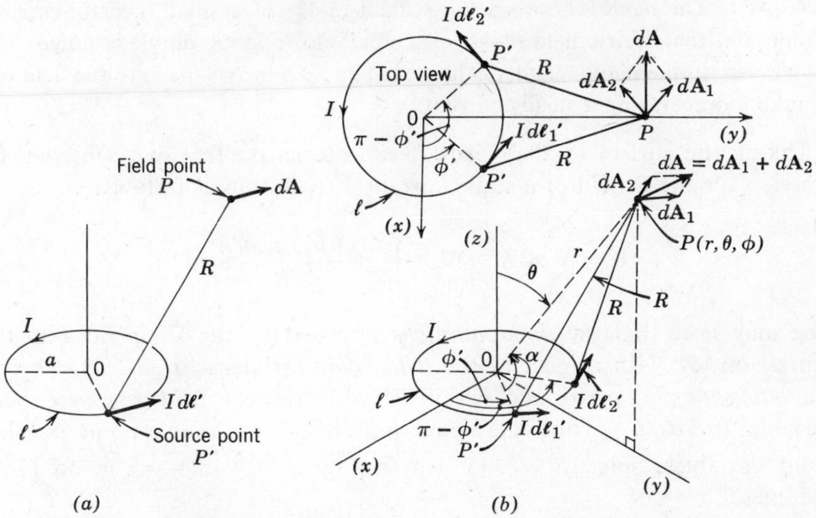

Figure 5-10. Circular loop, showing the spherical coordinate geometry adopted for finding the static magnetic field at *P*. (*a*) Circular loop carrying a current *I*. (*b*) Making use of symmetry to obtain fields at *P*.

mate, making use of the binomial theorem,†

$$R \cong r\left[1 - 2\frac{a}{r}\sin\theta\sin\phi'\right]^{1/2} \cong r\left[1 - \frac{a}{r}\sin\theta\sin\phi' + \cdots\right]$$

The reciprocal, for small a, is similarly approximated

$$\frac{1}{R} \cong \frac{1}{r} + \frac{a}{r^2}\sin\theta\sin\phi'$$

Thus puts (2) into the form

$$A_\phi \cong \frac{2\mu_0 Ia}{4\pi}\int_{\phi'=-\pi/2}^{\pi/2}\left[\frac{1}{r} + \frac{a}{r^2}\sin\theta\sin\phi'\right]\sin\phi'\,d\phi' \tag{3}$$

The integral of the $(\sin\phi')/r$ term is zero, so integrating the second term yields the answer

$$A_\phi \cong \frac{\mu_0 a^2 I\sin\theta}{4r^2} \tag{5-32}$$

Taking $\mathbf{B} = \nabla \times \mathbf{A}$ in spherical coordinates therefore yields

$$\mathbf{B} \cong \frac{\mu_0 a^2 I}{4r^3}[\mathbf{a}_r 2\cos\theta + \mathbf{a}_\theta\sin\theta] \tag{5-33}$$

† From the binomial theorem one may see that, in the expansion of $(1 \pm b)^n$, if $b \ll 1$ then $(1 \pm b)^n \cong 1 \pm nb$.

if $a \ll r$. The *duality* between the **B** field (5-33) of a small, current-carrying loop and the electric field (4-44) of a small electrostatic dipole is noted. This gives rise to the name magnetic dipole, when reference is made to the field of a small loop carrying a steady current.

Taking the curl of (5-28a) leads to an alternative free-space integral expression for the **B** field of a static current distribution as follows:

$$\mathbf{B} = \nabla \times \mathbf{A} = \nabla \times \int_V \frac{\mu_0 \mathbf{J}(u_1', u_2', u_3')}{4\pi R}\, dv' \tag{5-34}$$

One may note that the *differentiations* imposed by the ∇ operator in this expression are with respect to the *field point* variables (u_1, u_2, u_3), whereas the *integration* is performed within V with respect to the *source point* variables (u_1', u_2', u_3'). Thus R is a function of both the source point and field point variables, since $R = \sqrt{(x - x')^2 + (y - y')^2 + (z - z')^2}$, so (5-34) becomes

$$\mathbf{B} = \int_V \frac{\mu_0}{4\pi} \nabla \times \left[\frac{\mathbf{J}}{R}\right] dv'$$

One can write $\nabla \times [\mathbf{J}/R]$ from the vector identity (17) in Table 2-2

$$\nabla \times \left[\frac{\mathbf{J}}{R}\right] = \nabla\left(\frac{1}{R}\right) \times \mathbf{J} + \frac{1}{R} \nabla \times \mathbf{J}$$

The last term is zero because **J** is a function of only the source point variables; furthermore, the factor $\nabla(1/R)$ can be expressed

$$\nabla\left(\frac{1}{R}\right) = -\frac{\mathbf{a}_R}{R^2}$$

if $\mathbf{a}_R$ is a unit vector pointing from P' to P. Thus

$$\nabla \times \left[\frac{\mathbf{J}}{R}\right] = \mathbf{J} \times \left(\frac{\mathbf{a}_R}{R^2}\right)$$

obtaining

$$\mathbf{B}(u_1, u_2, u_3) = \int_V \frac{\mu_0 \mathbf{J} \times \mathbf{a}_R}{4\pi R^2}\, dv' \tag{5-35a}$$

This integral for **B**, expressed directly in terms of the static current distribution **J** in free space, is known as the *Biot-Savart law*. It provides an alternative approach for obtaining the magnetic fields of static current distributions in free space. Figure 5-11 shows the geometry relative to (5-35a), depicting a

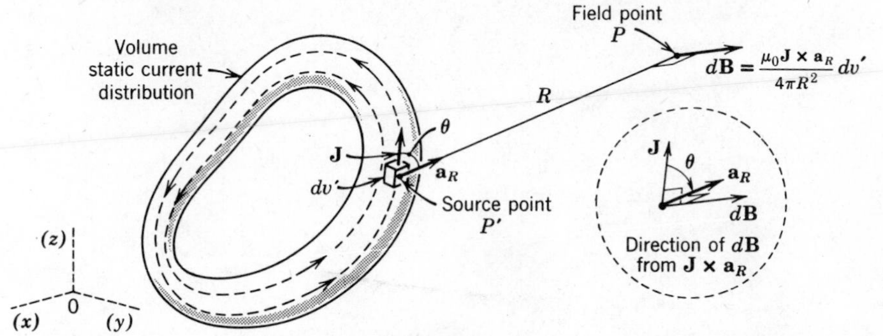

Figure 5-11. A volume distribution of currents, showing the $d\mathbf{B}$ contribution of a typical current element $\mathbf{J}\,dv'$ from the Biot–Savart law.

system of steady currents with densities $\mathbf{J}$, and a typical field point P at which $\mathbf{B}$ is found by means of (5-35a). The differential contribution $d\mathbf{B}$ is given by the integrand of (5-35a)

$$d\mathbf{B} = \frac{\mu_0 \mathbf{J} \times \mathbf{a}_R}{4\pi R^2} dv'$$

meaning that $d\mathbf{B}$ contributed at P by $\mathbf{J}\,dv'$ is mutually perpendicular to both the current element vector $\mathbf{J}$ and the unit vector $\mathbf{a}_R$, as depicted in Figure 5-11.

Specializations of the Biot–Savart law to surface or to line currents are readily obtained. Thus, if the volume current of Figure 5-11 is contracted to a *thin filament* of negligible cross-section, putting $\mathbf{J}\,dv' \to I\,d\ell'$ into (5-35a) obtains

$$\mathbf{B}(u_1, u_2, u_3) = \int_\ell \frac{\mu_0 I\, d\ell' \times \mathbf{a}_R}{4\pi R^2} \text{ T} \tag{5-35b}$$

EXAMPLE 5-6. Use the Biot–Savart law to find the $\mathbf{B}$ field of the thin wire of length $2L$ and carrying a steady current, as given in Example 5-4.

The form (5-35b) of the law is applicable. In the circular cylindrical system as shown in Figure 5-12, $I\,d\ell' = \mathbf{a}_z I\,dz'$, while $\mathbf{a}_R$ is resolved into components as follows: $\mathbf{a}_R = \mathbf{a}_\rho \sin\alpha - \mathbf{a}_z \cos\alpha = R^{-1}(\mathbf{a}_\rho \rho - \mathbf{a}_z z')$. With $R = \sqrt{\rho^2 + (z')^2}$, (5-35b) becomes

$$\mathbf{B} = \int_{z'=-L}^{L} \frac{\mu_0 I}{4\pi} \frac{(\mathbf{a}_z\, dz') \times (\mathbf{a}_\rho \rho - \mathbf{a}_z z')}{[\rho^2 + (z')^2]^{3/2}} = \mathbf{a}_\phi \frac{\mu_0 I \rho}{4\pi} \int_{-L}^{L} \frac{dz'}{[\rho^2 + (z')^2]^{3/2}}$$

and integrating obtains

$$\mathbf{B} = \mathbf{a}_\phi \frac{\mu_0 I}{4\pi\rho} \left[\frac{z'}{\sqrt{\rho^2 + (z')^2}} \right]_{-L}^{L} = \mathbf{a}_\phi \frac{\mu_0 I}{2\pi\rho} \frac{L}{\sqrt{\rho^2 + L^2}} \tag{5-36}$$

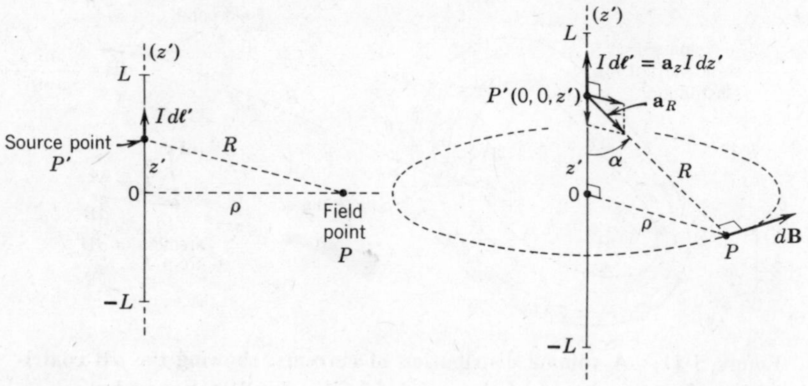

Figure 5-12. Geometry of the straight wire of length $2L$, using the Biot–Savart law to find B.

Close to a wire of finite length ($\rho \ll L$), or for an infinitely long wire, (5-36) becomes

$$\mathbf{B} = \mathbf{a}_\phi \frac{\mu_0 I}{2\pi\rho} \tag{5-37}$$

results that agree with those of Example 5-4.

5-6 Quasi-Static Electromagnetic Fields

In previous sections of this chapter, only purely *static* magnetic fields, associated with steady current distributions, were considered. Such fields are required to satisfy the Maxwell integral laws (5-4) and (5-5) for all closed surfaces or lines in the regions in question, or equivalently the differential laws (5-1) and (5-2) for all points in the regions. The boundary conditions, also to be satisfied at all interfaces, are (5-8) and (5-9). If the current sources are generalized to the time-varying case, their fields are then no longer purely magnetic but become electromagnetic, governed by all four of the Maxwell's equations (3-24), (3-48), (3-59), and (3-77), with the boundary conditions embracing the relations (3-42), (3-50), (3-70), and (3-79). For current sources that vary slowly in time, however, approximate methods, termed *quasi-static*, may sometimes be employed to advantage. An instance has already been given in Example 1-16.

Quasi-static field solutions can be termed first-order solutions, because they do not satisfy Maxwell's equations exactly except in the zero frequency limit. Another view, better appreciated in Chapter 11 on radiation and antennas, is that the dimensions of the current-carrying system must be small compared with the wavelength λ_0 in free space† if the system is to be amenable to a quasi-static method of attack. This constraint is equivalent to ignoring the finite velocity of propagation of the field from the sources to the nearby field points of interest, amounting to ignoring field radiation effects. A more sophisticated approach to quasi-static field solutions, utilizing an appropriate power series representation of the fields, is described elsewhere.‡ The quasi-static approach to field problems is sometimes the only method that provides ready solutions to an otherwise difficult boundary-value problem. It has applications in the discussion of the voltages induced in stationary or moving coils immersed in magnetic fields that may or may not be varying in time, as well as in the development of circuit theory, particularly regarding concepts of self and mutual inductance, to be discussed in subsequent sections.

EXAMPLE 5-7. Demonstrate that the approximate quasi-static fields of the long solenoid of Example 1-16 obey the Maxwell's equations (3-59) and (3-77) exactly only in the static field limit $\omega \to 0$.

The quasi-static **B** and **E** fields inside the solenoid were found to be

$$\mathbf{B}(t) = \mathbf{a}_z B_0 \sin \omega t \qquad \mathbf{E}(\rho, t) = -\mathbf{a}_\phi \frac{\omega \rho B_0}{2} \cos \omega t \tag{1}$$

Testing whether these fields satisfy (3-77), $\nabla \times \mathbf{E} = -\partial \mathbf{B}/\partial t$, one finds

$$\nabla \times \mathbf{E} = \begin{vmatrix} \dfrac{\mathbf{a}_\rho}{\rho} & \mathbf{a}_\phi & \dfrac{\mathbf{a}_z}{\rho} \\ \dfrac{\partial}{\partial \rho} & 0 & 0 \\ 0 & \rho E_\phi & 0 \end{vmatrix} = -\mathbf{a}_z \omega B_0 \cos \omega t \tag{2}$$

revealing that **B** and **E** of (1) do indeed satisfy (3-77). This is to be expected, because **E** was originally obtained using the integral form of (3-77), but Maxwell's equation (3-59), reducing to $\nabla \times \mathbf{H} = \partial \mathbf{D}/\partial t$ within the solenoid, is *not* satisfied by (1). This is evident upon obtaining $\nabla \times \mathbf{H} = \nabla \times (\mathbf{B}/\mu_0) = 0$, since **B** of (1) is independent of position inside the solenoid; whereas $\partial \mathbf{D}/\partial t$ becomes

$$\frac{\partial \mathbf{D}}{\partial t} = \epsilon_0 \frac{\partial \mathbf{E}}{\partial t} = \mathbf{a}_\phi \frac{\omega^2 \epsilon_0 B_0 \rho}{2} \sin \omega t$$

† Suppose one assumes that a device such as a coil or capacitor should not exceed $0.01\lambda_0$ in its maximum dimension, adopted as a criterion for sufficient smallness to enable employing quasi-static analysis in the description of its fields. Operation of the device at a frequency of 100 MHz implies that its size should then not exceed 0.03 m (3 cm), since $\lambda_0 = 3$ m at this frequency.

‡ See Fano, R. M., L. J. Chu, and R. B. Adler. *Electromagnetic Fields, Energy and Forces*. New York: Wiley, 1960, p. 221 ff.

a vanishing result only if $\omega \rightarrow 0$. Thus (3-59) is satisfied only in the static field limit, though an approximate equality prevails if ω is sufficiently small.

5-7 Open-Circuit Induced Voltage

The transformer makes use of Faraday's Law (3-77) to couple electromagnetic energy from one electric circuit to another through the time-varying magnetic field. Typical physical arrangements are diagramed in Figure 5-13. In (*a*) is shown the configuration of Figure 1-25(*b*): a primary coil consisting of a long solenoid, encircled by a secondary coil. Single-turn secondary coils are shown for simplicity; many turns are commonly used to enhance the induced voltage $V(t)$. A ferromagnetic or a ferrite core can also be used in a magnetic circuit arrangement as in Figure 5-13(*c*), to augment substantially the magnetic flux intercepted by the secondary coil.

The voltage $V(t)$ developed at an open-circuit gap in the secondary coil† of a transformer is shown to be

$$V(t) = -\frac{d\psi_m}{dt} \text{ V} \tag{5-38}$$

in which ψ_m denotes the magnetic flux intercepted by the surface S bounded

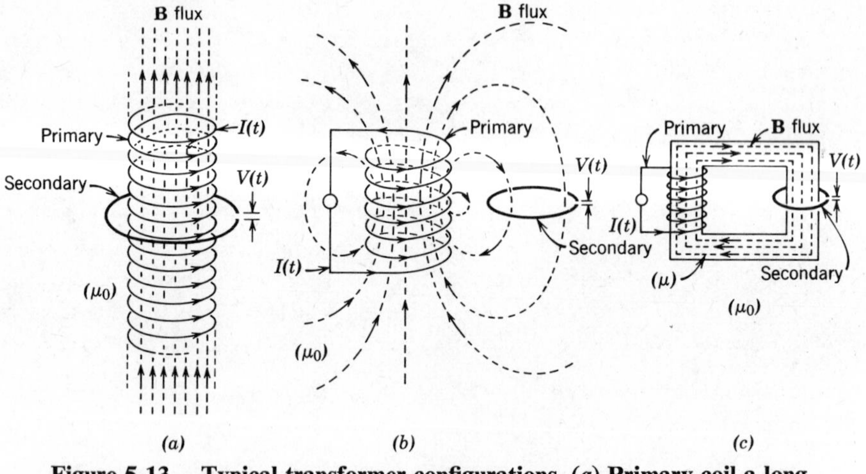

Figure 5-13. Typical transformer configurations. (*a*) Primary coil a long solenoid. (*b*) Short solenoid primary, secondary laterally displaced. (*c*) Configuration of (*b*) with ferromagnetic core.

† It should be borne in mind that the designation secondary coil is arbitrary; either coil of a transformer may be designated as the primary coil, with the other coil taking the name secondary.

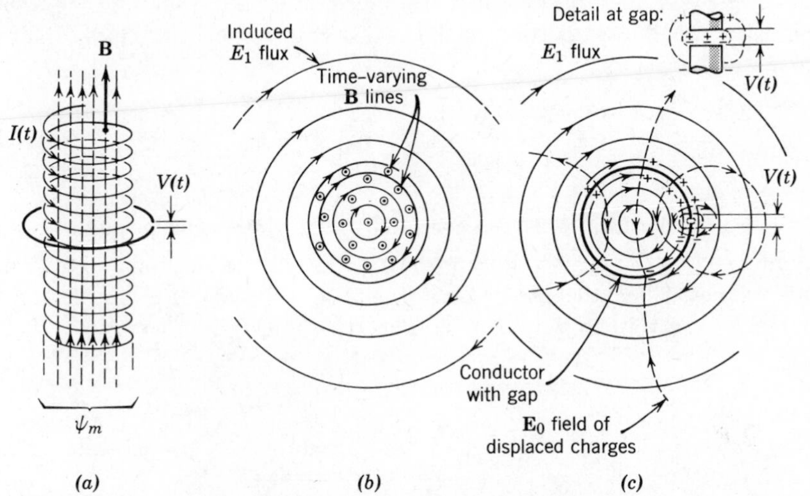

Figure 5-14. Development of the open-circuit voltage $V(t)$ of a transformer. (*a*) Transformer configuration used to prove (5-38). (*b*) Sectional view of $\mathbf{E}_1$ induced by the time-varying B of (*a*). (*c*) Showing charges displaced by $\mathbf{E}_1$ to produce $\mathbf{E}_0$, canceling total E along wire.

by the secondary winding. Suppose the coil shown in Figure 5-14(*a*) carries a time-varying current $I(t)$. In the surrounding region, the accompanying magnetic field $\mathbf{B}(u_1, u_2, u_3, t)$ induces an azimuthally directed, time-varying **E** field as described in Example 1-16 and depicted in the cross-sectional view of Figure 5-14(*b*). The secondary coil is shaped such that flux of **B** passes through the surface S bounded by the coil. This assures the alignment of the conductor with the induced **E** field such that the free electrons in the conductor are urged by the **E** field forces to move along the conductor as noted in Figure 5-14(*c*). Thus an excess of electron charge accumulates at one end of the wire, while a dearth of electrons (a positive charge) is established at the other, producing about the gap another electric field denoted by $\mathbf{E}_0$. Then the total **E** field about the system becomes $\mathbf{E} = \mathbf{E}_1 + \mathbf{E}_0$. Faraday's Law (3-78) written about the closed path including the secondary coil and its gap thus becomes

$$\oint_{\ell} \mathbf{E} \cdot d\boldsymbol{\ell} \equiv \int_{\text{conductor}} (\mathbf{E}_1 + \mathbf{E}_0) \cdot d\boldsymbol{\ell} + \int_{\text{gap}} (\mathbf{E}_1 + \mathbf{E}_0) \cdot d\boldsymbol{\ell} = -\frac{d\psi_m}{dt} \qquad (5\text{-}39)$$

The relationship between the total electric field $\mathbf{E}_1 + \mathbf{E}_0$ along the conductor and the current density **J** within it is given by (3-7), becoming $\mathbf{J} = \sigma(\mathbf{E}_1 + \mathbf{E}_0)$ along the coil. Assuming for the moment a *perfect* conductor, **J** must tend toward zero if $\mathbf{E}_1 + \mathbf{E}_0$ is kept to a necessary finite value, making $\mathbf{E}_1 + \mathbf{E}_0 = 0$ along the conductor portion of the closed path ℓ. This simplifies (5-39) to obtain

$$\oint_\ell \mathbf{E}\cdot d\ell = \int_{\text{gap}} (\mathbf{E}_1 + \mathbf{E}_0)\cdot d\ell = -\frac{d\psi_m}{dt} \tag{5-40}$$

implying that the total $\oint_\ell \mathbf{E}\cdot d\ell$ generated by the time-varying magnetic flux ψ_m embraced by ℓ appears wholly at the gap. The closed-line integral of (5-40) is sometimes called the *induced electromotive force* (emf) about ℓ, and is denoted by the voltage symbol $V(t)$. Then (5-40) is written

$$V(t) \equiv \oint \mathbf{E}\cdot d\ell = -\frac{d\psi_m}{dt}\ \text{V} \tag{5-41}$$

Thus the induced emf, or equivalently the gap voltage $V(t)$, depends only on the time rate of change of magnetic flux through the surface S bounded by the closed line ℓ described by the wire. The explicit values of $\mathbf{E}_1$ and $\mathbf{E}_0$ are *not required to be known* on the path. Furthermore, the wire path ℓ may be distorted, if desired, into any arbitrary shape; e.g., a square, a helix, and so on, in which cases (5-41) is still valid. A helix-shaped (many turn) conductor is useful for increasing the induced voltage across the gap, and it is commonly used in practical transformer and inductor designs.

If in the foregoing discussions a *finitely* conducting wire had been assumed, the result (5-41) would have been modified only trivially if the conductivity σ were sufficiently large (of the order of 10^7 ℧/m, as for most good conductors).

EXAMPLE 5-8. A thin wire is bent into a circle of radius b and placed with its axis concentric with that of the solenoid in Example 1-16. Find $V(t)$ induced across a small gap left in the conductor, for the two cases of Figure 5-15: (*a*) $b > a$ and (*b*) $b < a$. Include the *polarity* of $V(t)$ in the answers.

(*a*) If $b > a$, (5-41) combined with (1-67) yields, for the solenoid current $I_0 \sin \omega t$

$$V(t) = -\frac{d}{dt}\int_S \mathbf{B}\cdot d\mathbf{s} = -\frac{d}{dt}\int_S \left[\mathbf{a}_z \frac{\mu_0 n I_0 \sin \omega t}{d}\right]\cdot(\mathbf{a}_z\, ds)$$

$$= -\frac{\omega\mu_0 n\pi a^2 I_0}{d}\cos \omega t \qquad b > a \tag{5-42}$$

since $\int_S ds = \pi a^2$.

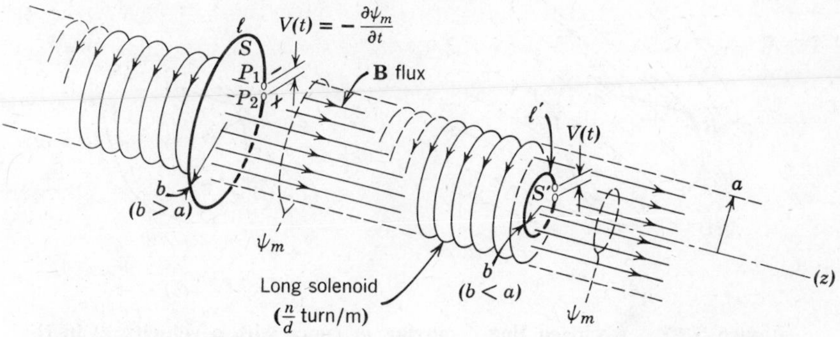

Figure 5-15. Showing open-circuit coils ℓ and ℓ' and the induced voltages $V(t)$ obtained from time-varying ψ_m.

The polarity of $V(t)$ is found by use of a right-hand-rule interpretation of the induced voltage law (5-41). Assuming, at a given t, that ψ_m through ℓ is increasing in the positive z sense in Figure 5-15, aligning the thumb of the right hand in that direction points the fingers towards the terminal P_2 at the gap, which at that moment is the positive terminal. The presence of the *negative* sign in the answer (5-42), however, requires that the true polarity of $V(t)$ becomes the opposite of the indicated polarity in Figure 5-15, at that instant.

(*b*) If $b < a$, the surface S' bounded by the wire ℓ' is smaller than the solenoid cross-section; (5-41) then becomes

$$V(t) = -\frac{d}{dt}\int_{S'} \mathbf{B}\cdot d\mathbf{s} = -\frac{\omega\mu_0 n\pi b^2 I_0}{d}\cos\omega t \qquad b < a \qquad (5\text{-}43)$$

5-8 Motional Electromotive Force and Voltage

The Faraday law (3-78) provides the connection between the time-varying magnetic flux ψ_m passing through a surface S and its induced **E** field. It states that the closed-line integral of **E** over the closed path ℓ bounding S exactly equals the time rate of change of the magnetic flux through S, or

$$\oint_\ell \mathbf{E}\cdot d\boldsymbol{\ell} = -\frac{d\psi_m}{dt} \qquad [3\text{-}78]$$

In (3-78), the closed-line integral of $\mathbf{E}\cdot d\boldsymbol{\ell}$ about ℓ, termed the circulation of **E** about the closed path, has also come to be known as the induced electromotive force, or just emf, produced by the time-varying magnetic field

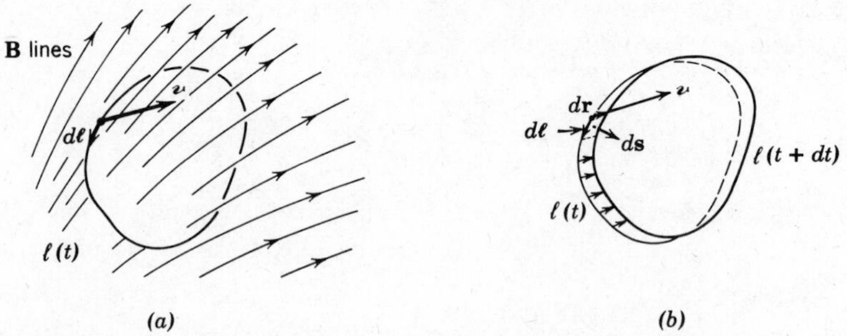

Figure 5-16. A closed line ℓ moving in space with a velocity v in the presence of a time-varying B field. (*a*) A moving contour ℓ in the presence of a time-varying B field. (*b*) The contour ℓ shown at the times t and $t + dt$.

through S.† In the present section, the emf induced by a motion of the path ℓ relative to the frame of reference of the magnetic field is discussed.

Faraday's law (3-78) can conveniently be resolved into two terms on its right side, accounting for the induced emf's about ℓ produced (*a*) by the time variation of the **B** field over the surface bounded by ℓ and (*b*) by the relative motion of the closed path ℓ with respect to the coordinate frame of reference of the **B** field, as shown in Figure 5-16(*a*). This form of the law, useful in the analysis of moving-coil devices such as generators, motors, d'Arsonval type instruments, etc. is developed in the following.

If the path ℓ, about which Faraday's law (3-78) is integrated, changes its position in time, this phenomenon may be accounted for by expanding the time derivative of the magnetic flux enclosed by ℓ into two terms as follows:

$$\oint_\ell \mathbf{E}\cdot d\boldsymbol{\ell} = -\frac{d\psi_m}{dt} = -\frac{d}{dt}\int_S \mathbf{B}\cdot d\mathbf{s} = -\int_S \frac{d}{dt}[\mathbf{B}\cdot d\mathbf{s}]$$

$$= -\int_S \left[\frac{\partial \mathbf{B}}{\partial t}\cdot d\mathbf{s} + \mathbf{B}\cdot\frac{\partial(d\mathbf{s})}{\partial t}\right] \tag{5-44a}$$

The significance of the last term in (5-44a) is as follows. Suppose the surface S, and therefore its bounding closed line ℓ, is moving in a region as depicted in Figure 5-16(*b*). Each element $d\boldsymbol{\ell}$ of ℓ is imagined to move the distance $d\mathbf{r}$ in a time dt, ascribing a velocity $v = d\mathbf{r}/dt$ to $d\boldsymbol{\ell}$, where in general v is not

† Electromotive force (emf), $\oint \mathbf{E}\cdot d\boldsymbol{\ell}$, obviously does not have the units of force, but rather volts, or joules per coulomb (work per unit charge about ℓ). For this reason the word electromotance has been suggested as an improved term for emf. Because it is widely used, however, emf is the term employed in this text.

necessarily perpendicular to $d\ell$ and not necessarily constant about ℓ. The change in area incurred by this motion is $d\mathbf{s} = d\mathbf{r} \times d\ell$. Then

$$\frac{\partial(d\mathbf{s})}{\partial t} = \frac{\partial}{\partial t}(d\mathbf{r} \times d\ell) = \boldsymbol{v} \times d\ell$$

since $d\mathbf{r}$ occurs in the time dt, while $d\ell$ is a fixed quantity. Thus the last term of (5-44a), by use of the identity (6) in Table 2-2, is written to obtain

$$\oint_{\ell} \mathbf{E} \cdot d\ell = -\int_{S} \frac{\partial \mathbf{B}}{\partial t} \cdot d\mathbf{s} + \oint_{\ell} (\boldsymbol{v} \times \mathbf{B}) \cdot d\ell \text{ V} \qquad (5\text{-}44\text{b})$$

accounting for two contributions to the induced emf as follows:

1. The first term of the right side of (5-44b) accounts for the induced emf about ℓ provided by the time rate of change of the $\mathbf{B}$ field integrated over the surface S bounded by ℓ.
2. The second term yields the additional induced emf arising from the motion of the path ℓ relative to the coordinate frame of reference in which $\mathbf{B}$ is specified.

A comparison of $\boldsymbol{v} \times \mathbf{B}$ appearing in (5-44b) with the Lorentz magnetic force term $\mathbf{F}_B$ of (1-52) reveals that $\boldsymbol{v} \times \mathbf{B}$ constitutes an equivalent electric field $\mathbf{E}'$, capable of influencing an electric charge q if present on ℓ. A wire conductor, bent into a closed path ℓ, furnishes a convenient source of conduction charges subject to this Lorentz force effect whenever relative motion between the charges and $\mathbf{B}$ occurs.

If ℓ is *stationary* in space ($\boldsymbol{v} = 0$), (5-44b) reduces to

$$\oint_{\ell} \mathbf{E} \cdot d\ell = -\int_{S} \frac{\partial \mathbf{B}}{\partial t} \cdot d\mathbf{s} \text{ V} \qquad \ell \text{ stationary} \qquad (5\text{-}44\text{c})$$

Suppose that a wire loop ℓ is immersed in a *steady* magnetic field, but ℓ is *moving in space*. Then (5-44b) becomes

$$\oint_{\ell} \mathbf{E} \cdot d\ell = \oint_{\ell} (\boldsymbol{v} \times \mathbf{B}) \cdot d\ell \text{ V} \qquad \text{Static } \mathbf{B} \qquad (5\text{-}44\text{d})$$

The *sense* of the emf (hence, a gap voltage) generated about ℓ by the motional electric field $\boldsymbol{v} \times \mathbf{B}$ is obtained by comparing the integrands of (5-44d). If a wire contour were shrinking in size while immersed in a steady $\mathbf{B}$ field as in

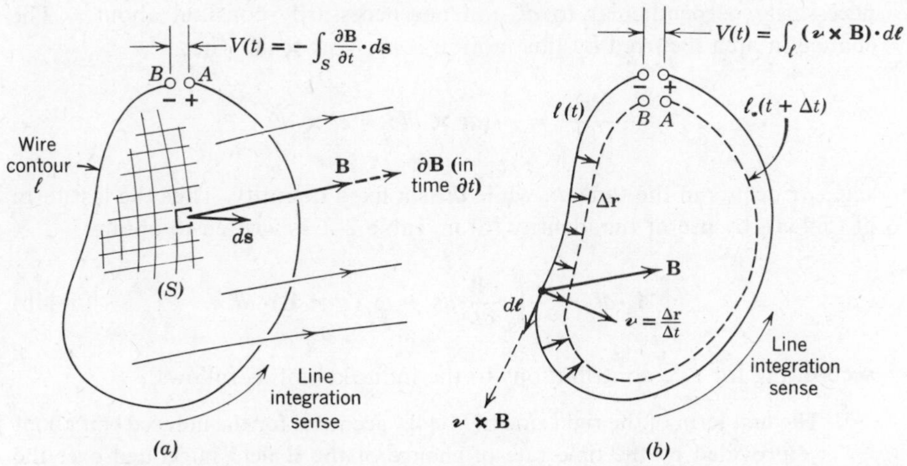

Figure 5-17. Conventions relative to Faraday's law. (*a*) Polarity of $V(t)$ induced by time-varying B, with ℓ stationary. (*b*) A shrinking wire contour, showing sense of induced field $\boldsymbol{v} \times \mathbf{B}$, with B static.

Figure 5-17(*b*), for example, an equivalent electric field $\boldsymbol{v} \times \mathbf{B}$ is induced along ℓ having the same direction as the positive line-integration sense assumed as shown. A gap left in the wires of Figures 5-17(*a*) or (*b*) would thus develop gap voltages with polarities as shown.

It is evident that if *both* a time variation of the **B** field and a motion of the contour ℓ prevailed simultaneously, the superposition of the two effects would require the use of (5-44b) in the analysis. It is emphasized that the forms (3-78) and (5-44b) of Faraday's law are equivalent and provide the same answer to a given induced emf problem.

EXAMPLE 5-9. A rigid, rectangular conducting loop with the dimensions a and b is located between the poles of a permanent magnet as shown. Let $\mathbf{B} = \mathbf{a}_z B_0$, constant as shown over the left portion of the loop, and assume the loop is pulled to the right at a constant velocity $\boldsymbol{v} = \mathbf{a}_x v_0$. Find (*a*) the emf induced around the loop, (*b*) the direction of the current caused to flow in the loop, (*c*) the force on the wire resulting from the current flow, and (*d*) the magnitude and polarity of the open-circuit voltage $V(t)$ appearing at a gap in the wire at P shown.

(*a*) The sense of the line integration is assumed counterclockwise looking from the front, as in (*b*) of the accompanying figure. The emf induced about the loop is found by use of (5-44d),

$$\oint_{\ell(t)} \mathbf{E} \cdot d\boldsymbol{\ell} = \oint_{\ell(t)} (\boldsymbol{v} \times \mathbf{B}) \cdot d\boldsymbol{\ell}$$

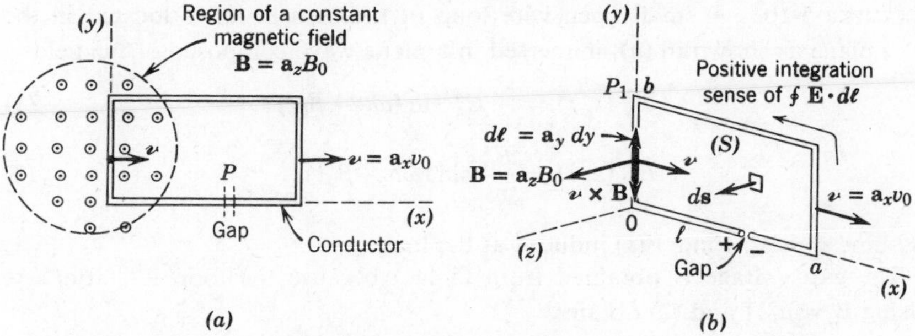

Example 5-9. ***(a)* Moving wire loop in a constant magnetic field. *(b)* Geometry showing assumed line-integration sense.**

On P_1 $(\boldsymbol{v} \times \mathbf{B})\cdot d\boldsymbol{\ell} = [\mathbf{a}_x v_0) \times (\mathbf{a}_z B_0)]\cdot\mathbf{a}_y\, dy = -v_0 B_0\, dy$, obtaining

$$\oint_{\ell(t)} \mathbf{E}\cdot d\boldsymbol{\ell} = \int_{y=b}^{0} (-v_0 B_0)\, dy = v_0 B_0 b \tag{1}$$

(*b*) The positive sign of (1) denotes that the induced emf about ℓ is in the same sense (counterclockwise) as the direction of integration. The result (1) therefore causes a current to flow in the same direction.

(*c*) The force acting on the wire carrying the current I immersed in the field $\mathbf{B}$ is obtainable from (1-52a), $\mathbf{F}_B = q(\boldsymbol{v} \times \mathbf{B})$. The force on a differential charge $dq = \rho_v\, dv$ being $d\mathbf{F}_B = dq(\boldsymbol{v} \times \mathbf{B})$, and with $\rho_v \boldsymbol{v}$ of (1-50a) the current density $\mathbf{J}$ in the wire, one obtains

$$d\mathbf{F}_B = \mathbf{J} \times \mathbf{B}\, dv \tag{5-45a}$$

The product $\mathbf{J}\, dv$ defines a volume current element $I\, d\boldsymbol{\ell}$ in a thin wire, so (5-45a) becomes

$$d\mathbf{F}_B = I\, d\boldsymbol{\ell} \times \mathbf{B} \tag{5-45b}$$

Integrating (5-45b) over the length $0P_1$ of the wire obtains the total force

$$\mathbf{F}_B = \int_\ell I\, d\boldsymbol{\ell} \times \mathbf{B} = \int_{y=b}^{0} I(\mathbf{a}_y\, dy) \times (\mathbf{a}_z B_0) = -\mathbf{a}_x B_0 I b \tag{2}$$

in which the integration in the direction of the current produces the proper vector sense of the force. $\mathbf{F}_B$ is a force to the left in the figure, opposing the motion of the wire.

(*d*) A small gap at P in the wire renders the loop open circuited, reducing I to zero and yielding $V(t) = \oint_\ell \mathbf{E}\cdot d\boldsymbol{\ell} = v_0 B_0 b$ across the gap. The polarity is determined by the direction of $\boldsymbol{v} \times \mathbf{B}$, directed around the loop towards the positive terminal of the gap as in (*b*) of the figure.

EXAMPLE 5-10. A small open wire loop of radius a in air is located in the y - z plane as shown in (a), immersed in a plane wave composed of the fields

$$E_y^+(z, t) = -E_m^+ \sin(\omega t - \beta_0 z) \tag{1}$$

$$H_x^+(z, t) = \frac{E_m^+}{\eta_0} \sin(\omega t - \beta_0 z) \tag{2}$$

Assume $a \ll \lambda_0$. Find $V(t)$ induced at the loop gap.

The gap voltage is obtained from (5-44c) because the loop is stationary. Using $\mathbf{B} = \mu_0\mathbf{H}$ and (2) obtains

$$\frac{\partial \mathbf{B}}{\partial t} = \frac{\partial}{\partial t}\left[\frac{\mathbf{a}_x\mu_0 E_m^+}{\eta_0}\sin(\omega t - \beta_0 z)\right] = \mathbf{a}_x \frac{\mu_0\omega E_m^+}{\eta_0}\cos(\omega t - \beta_0 z)$$

With $a \ll \lambda_0$, the coil occupies essentially the position $z = 0$ on the plane wave

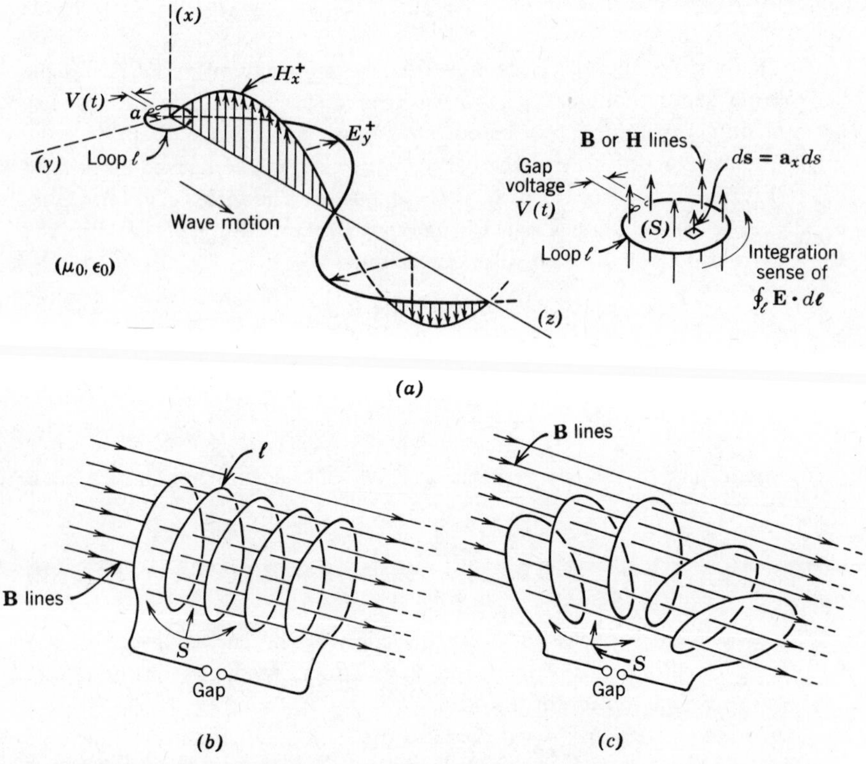

Example 5-10. (*a*) Loop immersed in a field. (*b*) Five-turn coil immersed in a time-varying field. (*c*) Effect of spreading the turns; fewer B lines intercept *S*.

moving past, so (5-44c) yields

$$V(t) = -\int_S \left[\mathbf{a}_x \frac{\mu_0 \omega E_m^+}{\eta_0} \cos(\omega t - \beta_0 z)\right]_{z=0} \cdot \mathbf{a}_x \, ds = \frac{-\omega E_m^+ \pi a^2}{c} \cos \omega t \qquad (3)$$

If $E_m^+ = 100 \, \mu V$, $f = 1$ MHz, and $a = 1$ m (satisfying the criterion $a \ll \lambda$), the induced voltage becomes $V(t) = 396 \cos \omega t \, \mu$V. The five-turn coil shown in (*b*) provides a gap voltage five times that of (*a*), in view of the structure of the surface S bounded by the wire contour ℓ, while from (*c*), the effect of opening the structure is to reduce ψ_m intercepting S, reducing $V(t)$ accordingly.

5-9 Induced Emf from Time-Varying Vector Magnetic Potential

The emf induced about a closed path ℓ linking a time-varying magnetic flux ψ_m can also be expressed in terms of the *vector magnetic potential* **A** developed in Section 5-4. This is accomplished by use of (5-22), $\mathbf{B} = \nabla \times \mathbf{A}$, to permit writing Faraday's law (3-78) as follows:

$$\oint_\ell \mathbf{E} \cdot d\boldsymbol{\ell} = -\frac{d\psi_m}{dt} = -\frac{d}{dt}\int_S \mathbf{B} \cdot d\mathbf{s} = -\frac{d}{dt}\int_S (\nabla \times \mathbf{A}) \cdot d\mathbf{s}$$

$$= -\frac{d}{dt}\oint_\ell \mathbf{A} \cdot d\boldsymbol{\ell} \text{ V} \qquad (5\text{-}46a)$$

upon applying Stokes' theorem (2-56) to obtain the last result. If ℓ is stationary (motional emf is absent), (5-46a) is written simply

$$\oint_\ell \mathbf{E} \cdot d\boldsymbol{\ell} = -\oint_\ell \frac{\partial \mathbf{A}}{\partial t} \cdot d\boldsymbol{\ell} \text{ V} \quad \text{Stationary path } \ell \qquad (5\text{-}46b)$$

One can see from (5-46a) that the flux through the path ℓ is expressed two ways

$$\psi_m = \int_S \mathbf{B} \cdot d\mathbf{s} = \oint_\ell \mathbf{A} \cdot d\boldsymbol{\ell} \text{ Wb} \qquad (5\text{-}47)$$

Thus it follows that a knowledge of the vector magnetic potential **A** on the closed ℓ determines the magnetic flux ψ_m passing through S bounded by ℓ.

EXAMPLE 5-11. Find the emf induced about a rectangular stationary path ℓ in free space, in the plane of two long, parallel wires carrying the currents $I(t)$ and $-I(t)$ as shown. Find the emf two ways using the Faraday law expressed (*a*) in terms of the **B** field and (*b*) in terms of **A**.

(*a*) From (1-64), the quasi-static **B** field exterior to a long, *single* wire carrying $I(t)$ is $\mathbf{B} = \mathbf{a}_\phi \mu_0 I(t)/2\pi\rho$, if ρ is the normal distance from the wire to the field point. In the present example involving two wires, a

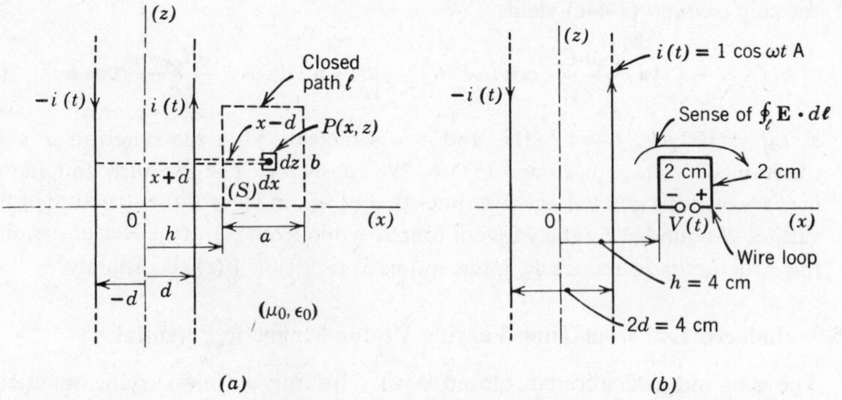

Example 5-11. ***(a)*** **Geometry of a parallel-wire system and a rectangular closed path** ℓ**.** ***(b)*** **Showing polarity of gap voltage** $V(t)$**, corresponding to sense of** $\oint \mathbf{E}\cdot d\ell$ **integrated about wire loop.**

cartesian coordinate system is adopted as in (*a*). At any x on S bounded by ℓ, the quasi-static **B** field due to both wires is the vector sum

$$\mathbf{B} = \mathbf{a}_y\left[\frac{\mu_0 I(t)}{2\pi(x-d)} - \frac{\mu_0 I(t)}{2\pi(x+d)}\right]$$

The latter into (5-44c) obtains the induced emf about ℓ

$$\oint_\ell \mathbf{E}\cdot d\ell = -\int_S \frac{\partial \mathbf{B}}{\partial t}\cdot d\mathbf{s} = -\frac{\partial I}{\partial t}\frac{\mu_0 b}{2\pi}\,\ell n\,\frac{(h+a-d)(h+d)}{(h+a+d)(h-d)} \tag{1}$$

the desired result.

(*b*) To find the induced emf using **A**, note from Example 5-4 for a *single wire* of length $2L$ that

$$\mathbf{A} = \mathbf{a}_z\frac{\mu_0 I(t)}{4\pi}\,\ell n\,\frac{\sqrt{L^2+\rho^2}+L}{\sqrt{L^2+\rho^2}-L}$$

The latter is improved by noting that **A** for $\rho \ll L$ is desired. The first two terms of the binomial expansion for the square root quantities obtains

$$A_z = \frac{\mu_0 I}{4\pi}\,\ell n\,\frac{\sqrt{1+\left(\frac{\rho}{L}\right)^2}+1}{\sqrt{1+\left(\frac{\rho}{L}\right)^2}-1} = \frac{\mu_0 I}{4\pi}\,\ell n\,\frac{\left[1-\frac{1}{2}\left(\frac{\rho}{L}\right)^2\right]+1}{\left[1+\frac{1}{2}\left(\frac{\rho}{L}\right)^2\right]-1}$$

$$\cong \frac{\mu_0 I(t)}{4\pi}\,\ell n\left[1+4\left(\frac{L}{\rho}\right)^2\right] \cong \frac{\mu_0 I(t)}{2\pi}\,\ell n\,\frac{2L}{\rho}$$

valid for $\rho \ll L$. For parallel wires, **A** is the vector sum

$$\mathbf{A} = \mathbf{a}_z \frac{\mu_0 I}{2\pi}\left[\ell n \frac{2L}{x-d} - \ell n \frac{2L}{x+d}\right] = \mathbf{a}_z \frac{\mu_0 I(t)}{2\pi} \ell n \frac{x+d}{x-d}$$

With **A** z directed, the induced emf by use of (5-46b) becomes

$$\oint_\ell \mathbf{E}\cdot d\boldsymbol{\ell} = -\oint_\ell \frac{\partial \mathbf{A}}{\partial t}\cdot d\boldsymbol{\ell}$$

$$= -\frac{\partial I}{\partial t}\frac{\mu_0}{2\pi}\left\{\left[\ell n \frac{x+d}{x-d}\right]_{x=h}\int_{z=0}^{b} dz + \left[\ell n \frac{x+d}{x-d}\right]_{x=h+a}\int_{z=b}^{0} dz\right\}$$

$$= -\frac{\partial I(t)}{\partial t}\frac{\mu_0 b}{2\pi}\ell n \frac{(h+d)(h+a-d)}{(h-d)(h+a+d)} \qquad (2)$$

which agrees with (1). Note that the integration has been taken clockwise about ℓ, to conform to the assumption in (*a*) of a positive y directed $d\mathbf{s}$ on S.

If the current $I(t) = 1 \sin \omega t$ A flows in the wires, with $f = 1000$ Hz and $d = h/2 = a = b = 2$ cm, the induced emf (2) becomes

$$\oint_\ell \mathbf{E}\cdot d\boldsymbol{\ell} = -(2\pi \times 10^3)\cos \omega t \frac{(4\pi \times 10^{-7})(2 \times 10^{-2})}{2\pi} \ell n \frac{3}{2}$$

$$= -17.4 \cos \omega t \ \mu\text{V}$$

This is also the gap voltage $V(t)$ developed if an open-circuited *wire loop* replaces ℓ, with a polarity as in (*b*).

From (5-46b) for the induced emf about a fixed closed path,

$$\oint_\ell \mathbf{E}\cdot d\boldsymbol{\ell} = -\oint_\ell \frac{\partial \mathbf{A}}{\partial t}\cdot d\boldsymbol{\ell} \qquad [5\text{-}46\text{b}]$$

one might be inclined to argue, because (5-46b) is true for *all* closed paths ℓ, that the electric field can be expressed at any point in the region from equating the integrands; i.e.,

$$\mathbf{E} = -\frac{\partial \mathbf{A}}{\partial t}$$

It is, however, noted that adding an arbitrary function $-\nabla\Phi$ to the latter, obtaining

$$\mathbf{E} = -\nabla\Phi - \frac{\partial \mathbf{A}}{\partial t} \qquad (5\text{-}48)$$

provides an **E** field that still satisfies (5-46b), in view of the property (2-15), $\oint_\ell (\nabla\Phi)\cdot d\boldsymbol{\ell} = 0$, true for any scalar function Φ. Thus (5-48) is in general the correct expression for **E** in terms of its potential fields **A** and Φ. The physical meanings of each contribution to **E** in (5-48) is appreciated upon noting, in the *time-static* limit, that (5-48) reduces to

$$\mathbf{E} = -\nabla\Phi \qquad \partial/\partial t = 0 \quad (5\text{-}49)$$

Comparison of (5-49) with (4-31) identifies Φ as the scalar potential field established by the free-charges of the system, whether they be volume, surface, or line charges. The potential integral (4-35a) provides this relationship, extended in (5-48) to time-varying charge distributions. Secondly, the **A** field in (5-48) is connected with the current distributions of the system through the integral (5-28). Summarizing, the total electric field (5-48) is written†

$$\mathbf{E} = \underbrace{\mathbf{E}_0}_{\text{Due to charges}} + \underbrace{\mathbf{E}_1}_{\text{Due to time-varying currents}} \qquad (5\text{-}50)$$

$$\mathbf{E} = \overbrace{-\nabla\Phi} - \overbrace{\frac{\partial \mathbf{A}}{\partial t}}$$

in which the quasi-static potentials Φ and **A** in free space are given by (4-35a) and (5-28)

$$\Phi = \int_V \frac{\rho_v \, dv'}{4\pi\epsilon_0 R} \text{ V} \qquad (5\text{-}51)$$

$$\mathbf{A} = \int_V \frac{\mu_0 \mathbf{J} \, dv'}{4\pi R} \text{ Wb/m} \qquad (5\text{-}52)$$

A system exemplifying these processes is shown in Figure 5-18, in which a current is urged by the generator through a conducting path connected to a

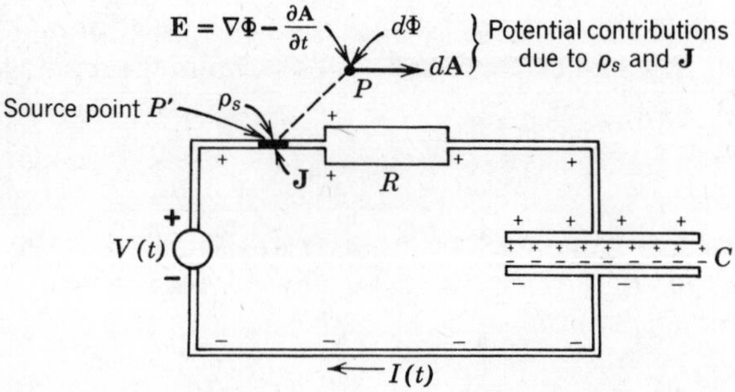

Figure 5-18. An *R-C* circuit driven by *V*(*t*), showing field point *P* at which E is established by the charge and current density distributions.

† From Section 5-7, one may observe that the notations $\mathbf{E}_0$ and $\mathbf{E}_1$ with the meanings defined in (5-50) were used in that discussion.

resistive device R and a charge storage device C. Charges in this system are the low-density surface charges (ρ_s) found on the conducting wires (required to maintain boundary conditions there), plus heavy concentrations on the closely spaced capacitor plates; together they contribute to the total quasi-static potential Φ at every point P of the system. Current densities **J** inside the conductors contribute to the vector magnetic potential **A** at P. If electrically or magnetically polarizable materials were, moreover, included in the system, the integrals (5-51) and (5-52) would have to be modified by adding in the effects of the dielectric polarization density **P** to (5-51) and of the magnetization current density **M** to (5-52).†

The quasi-static formalisms expressed by (5-50), (5-51), and (5-52) are not very useful as a problem-solving tool when applied directly to a circuit like that of Figure 5-18, because the charge and current distributions are seldom known exactly. With some simplifying assumptions, however, these expressions combined with Faraday's law can be used to derive the Kirchhoff voltage law of circuit theory, as described in the next section. A more general case, including the time retardation effects associated with the finite velocity of propagation of the electromagnetic fields in systems large compared to a wavelength, is considered in Chapter 11, leading to a discussion of antennas and radiating systems.

5-10 Voltage Generators and Kirchhoff's Laws

The electromagnetic forces urging charges around closed conducting paths can be obtained from energy-conversion devices known as *generators*. Generators of electricity take a variety of forms. Chemical (voltaic) cells, or batteries of them, are examples of *electrochemical* generators. Direct and alternating current rotating machines are *electromechanical* energy convertors, with voltages derived from the time-varying magnetic flux intercepted by conductors in relative motion with respect to that field. A vacuum tube or a transistor may serve as the active (negative resistance) element of an *electronic* (or quantum electronic) generator. Other energy-conversion devices such as the photocell and the thermocouple are classified respectively as *photoelectric* and *thermoelectric*, because their voltages are derived from the energy of a light or heat source. To help clarify the meaning of the term generator in developing the connection of field theory with circuit theory, two classes of generators, electrochemical and electromechanical, are discussed.

† The latter integral extensions are not considered here, though the direct effects of the polarization densities **P** and **M** on Maxwell's equations have already been treated in Sections 3-2 and 3-3. For a treatment of the added effects of **P** and **M** on the potential integrals, see Lorrain, P. and D. R. Corson. *Electromagnetic Fields and Waves*, 2nd ed. San Francisco: Freeman, 1970, pp. 92 and 384.

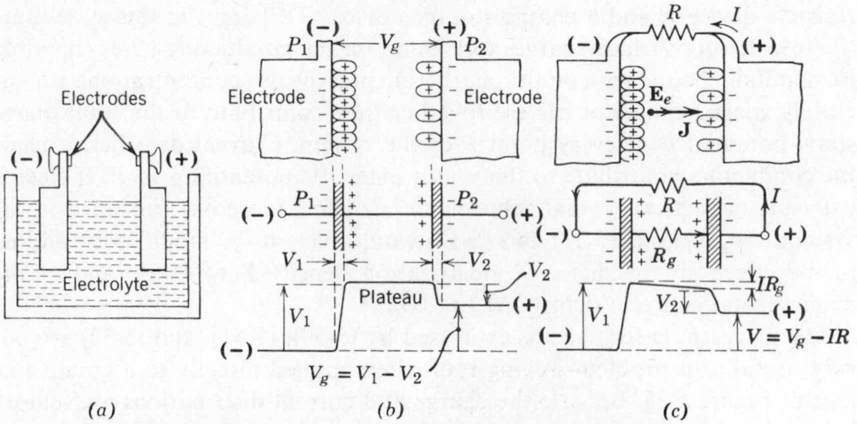

Figure 5-19. The electrochemical cell and its emf behavior. (*a*) Components of an electrochemical cell. (*b*) Half cell voltages establishing open circuit $V_g = V_1 - V_2$. (*c*) Effect of internal resistance when delivering current.

A. The Electrochemical Generator

An *electrochemical* cell (voltaic cell) consists of two dissimilar electrodes in contact with an electrolyte, usually an aqueous solution of inorganic salts as depicted in Figure 5-19(*a*). Electrodes are used to conduct an electronic current to an external circuit, and they can consist of a metal or carbon rod enveloped by a metallic oxide. In the common *dry* (Leclanché) cell, for example, an electrolyte consisting of an aqueous paste of ammonium and zinc chlorides is used between a zinc electrode (negative) and magnesium dioxide in contact with an inert carbon rod (positive). Conduction takes place in the electrolyte by the motion of ions, with a chemical reaction occurring just outside the electrodes to maintain a potential difference between each electrode and the electrolyte. An electrochemical cell can be thought of as two half-cells associated with each electrode-electrolyte interface, at which potential double layers of oppositely charged particles (electrons and ions) are established until equilibrium potentials are reached, as suggested by Figure 5-19(*b*). The open-circuit generated voltage V_g of the cell is thus the difference of the two barrier potentials V_1 and V_2. If no current is delivered, the electrolyte is an equipotential region noted by the flat central plateau in the potential diagram, with no **E** field inside it. The behavior of the electrochemical system is thus equivalent to the lower diagram of Figure 5-19(*b*), a series pair of charge double layers maintained by the chemical reactions at the electrode-electrolyte interfaces.† To maintain V_g, energy is supplied at the expense of one or more of the materials comprising the cell. When they are used up, the cell might be restored by replacing the materials, or in some instances by applying energy externally to reconvert them to their original forms. A cell that must be restored by adding new materials is called a *primary* cell; it is not rechargeable. A cell is called a *secondary* or a storage cell if it can be rejuvenated by externally driving the current backwards through the cell to reverse the electrolytic action that took place during discharge. The reactants as well as the products of the electrochemical reactions are in general gaseous, liquid, or solid. They can be stored in one or both electrodes or in the electrolyte as the reaction proceeds, or, as in the case of fuel cells, they may be removed continuously.

When a cell is connected to a resistive loop as in Figure 5-19(*c*), the resulting current is predictable from a Kirchhoff voltage law, derived from field theory as described in Figure 5-20(*a*) to emphasize the role of the external conductor.

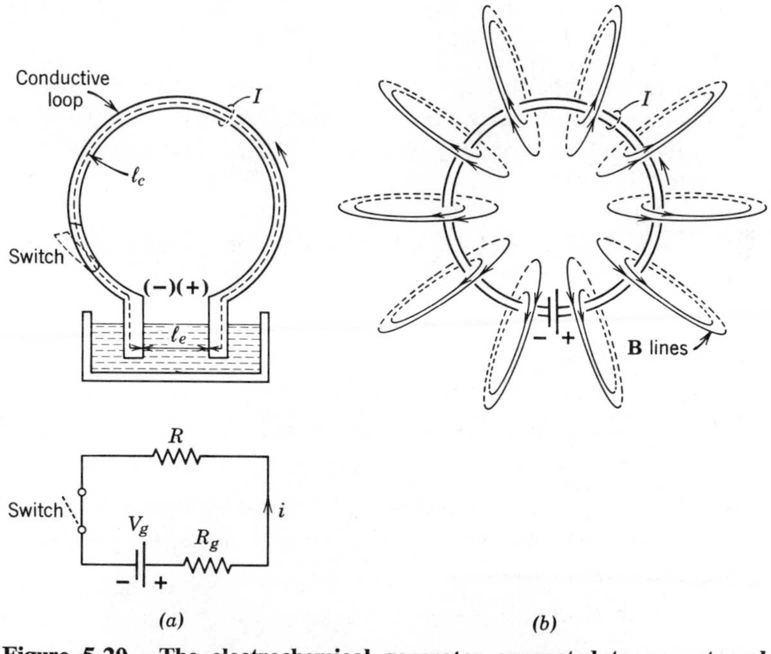

Figure 5-20. The electrochemical generator connected to an external resistive circuit. (*a*) Actual circuit and equivalent symbolism. (*b*) Magnetic flux ψ_m generated by *I*.

† For details of the chemical reactions, the reader is referred to *Encyclopedia of Chemical Technology*, 2nd ed., Vol. 3. New York: Interscience, 1964.

The total field **E**, at any field point P either inside or outside the conductor or the cell, is expressed by the sum

$$\mathbf{E} = \mathbf{E}_g + \mathbf{E}' \tag{5-53}$$

in which

$\mathbf{E}'$ = the electric field *induced* at any field point by the charges and time-varying currents, effects already described by the field integrals (5-51) and (5-52) yielding the total

$$\mathbf{E}' = -\nabla\Phi - \frac{\partial \mathbf{A}}{\partial t} \tag{5-54}$$

and

$\mathbf{E}_g$ = the *generated* electric field (generated electrochemically within the cell in this example, $\mathbf{E}_g = 0$ elsewhere

Inserting (5-54) into (5-53) permits writing the total **E** at any P

$$\mathbf{E} = \mathbf{E}_g - \nabla\Phi - \frac{\partial \mathbf{A}}{\partial t} \tag{5-55}$$

If (5-55) is integrated about a closed path ℓ coincident with the circuit, the Kirchhoff voltage law of circuit theory is obtained. Choose ℓ to coincide with the circuit as in Figure 5-20(a). Rearranging (5-55) and integrating over ℓ yields

$$\oint_\ell \mathbf{E}_g \cdot d\boldsymbol{\ell} = \oint_\ell \mathbf{E} \cdot d\boldsymbol{\ell} + \oint_\ell \nabla\Phi \cdot d\boldsymbol{\ell} + \oint_\ell \frac{\partial \mathbf{A}}{\partial t} \cdot d\boldsymbol{\ell} \tag{5-56}$$

The generated field $\mathbf{E}_g$ exists only inside the cell, so the left side reduces to just the generator open-circuit voltage, V_g

$$\oint_\ell \mathbf{E}_g \cdot d\boldsymbol{\ell} = \int_{P_1}^{P_2} \mathbf{E}_g \cdot d\boldsymbol{\ell} \equiv V_g \tag{5-57a}$$

From (3-7), the fields in the conductor and electrolyte become $\mathbf{J}/\sigma_c$ and $\mathbf{J}/\sigma_e$ respectively, so the second integral of (5-56) yields, at low frequencies

$$\oint_\ell \mathbf{E} \cdot d\boldsymbol{\ell} = \int_{\ell_c} \frac{\mathbf{J}}{\sigma_c} \cdot d\boldsymbol{\ell} + \int_{\ell_e} \frac{\mathbf{J}}{\sigma_e} \cdot d\boldsymbol{\ell} = Ri + R_g i \tag{5-57b}$$

in which R and R_g denote, from Section 4-15-2, the resistances of the external circuit and the generator, respectively. Both R and R_g are generally temperature dependent. Also R_g is affected by the available ion concentration in the electrolyte, a condition influenced by the age of the cell.

The third integral of (5-56) *vanishes*, in view of (2-15), since the gradient of a scalar potential integrated about a closed path is always zero.

The last term of (5-56) accounts for the circuit *self-inductance*. A magnetic flux ψ_m is generated by the current I in the circuit as depicted in Figure 5-20(*b*), expressed by (5-47) as follows:

$$\psi_m = \int_S \mathbf{B} \cdot d\mathbf{s} = \oint_\ell \mathbf{A} \cdot d\boldsymbol{\ell} \qquad [5\text{-}47]$$

The last term of (5-56) is thus equivalent to the various forms

$$\oint_\ell \frac{\partial \mathbf{A}}{\partial t} \cdot d\boldsymbol{\ell} = \frac{d}{dt} \oint_\ell \mathbf{A} \cdot d\boldsymbol{\ell} = \frac{d}{dt} \int_S \mathbf{B} \cdot d\mathbf{s} = \frac{d\psi_m}{dt} \qquad (5\text{-}57\text{c})$$

expressing the emf induced about the circuit ℓ by the time rate of change of the magnetic flux ψ_m linking that path. With the three results (5-57) into (5-56), one obtains the Kirchhoff voltage expression for the circuit of Figure 5-20

$$V_g = (R_g + R)I + \frac{d\psi_m}{dt} \qquad (5\text{-}58)$$

The term $d\psi_m/dt$ is the *self-induced voltage* generated about the circuit ℓ by the time-varying ψ_m linking ℓ (also called the self-flux of the circuit). If the surrounding region is magnetically linear, the self-flux is proportional to the current producing it; thus, $\psi_m \propto I$. The proportionality constant, called the *self-inductance* of the circuit, is designated by L as follows:

$$\psi_m = LI \text{ Wb} \qquad (5\text{-}59)$$

to yield a definition for the self-inductance

$$L = \frac{\psi_m}{I} \text{ Wb/A} \quad \text{or} \quad \text{H} \qquad (5\text{-}60)$$

in which ψ_m denotes the flux linked by the circuit.

With (5-59) inserted into (5-58), the Kirchhoff voltage expression for the circuit of Figure 5-20 is written $V_g = (R + R_g)I + d(LI)/dt$, and if L is a constant (independent of time), one obtains the Kirchhoff voltage law

$$V_g = (R + R_g)I + L\frac{dI}{dt} \text{ V} \qquad (5\text{-}61)$$

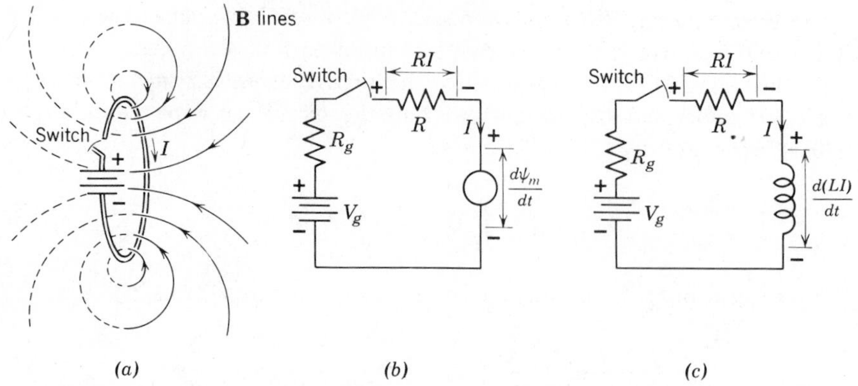

Figure 5-21. Series electric circuit and models. (*a*) The physical dc circuit. (*b*) Circuit model depicting voltage terms of (5-61). (*c*) Circuit model using inductance symbol *L*.

The transient and dc (steady state) solutions of this circuit differential equation are well-known and are omitted here. A further discussion of the self-inductance parameter L, from the energy point of view, is discussed in the next section.

The Kirchhoff equation (5-61) leads to the circuit model shown in Figure 5-21. The effects of a time-changing magnetic flux linking the circuit, as noted in (*a*) of the figure, is to produce a back voltage term, $d\psi_m/dt$, seen from its polarity markings in (*b*) to oppose any tendency for the current to change. The circuit convention representing this phenomenon is the lumped-inductance element L of Figure 5-21(*c*), across which the back voltage $L\,dI/dt$ is imagined to be generated.

B. The Electromechanical Generator

Another example of a generator is the electromechanical energy converter, or rotating machine. Its emf is derived from a magnetic flux linked by the machine windings, a flux that, in one version of such machines, becomes time-varying by virtue of the motion of the conductor windings relative to a static magnetic field. A generic model is diagramed in Figure 5-22(*a*). The magnetic flux is obtained from a permanent magnet or a field winding as shown. A cylindrical iron armature forming part of the magnetic circuit carries a winding that intercepts magnetic flux when the armature is rotated. The purpose of the armature is to provide physical support for the winding and to decrease the reluctance of the magnetic circuit by leaving only a small air gap, thereby enhancing the magnetic flux intercepted by the armature winding. A single-loop winding is illustrated for simplicity, although practical machines use many turns distributed about the armature to increase the induced emf. The wires are usually in slots as noted in Figure 5-22(*b*) to lessen the effective air gap even more and reduce the mechanical forces on the armature conductors through a transference of the forces to the core material. The armature iron is also laminated to reduce eddy current losses (see Figure 3-14).

If the armature of the generator is left open-circuited and rotated with an angular speed ω rad/sec, the gap voltage $V(t)$ is obtained from (5-44d)

$$V(t) = \oint_{\ell} (\boldsymbol{v} \times \mathbf{B}) \cdot d\ell \qquad [5\text{-}44d]$$

as seen from details in Figure 5-22(*c*). Thus, a radially directed magnetic field of constant value B_0 imposed upon a single-turn coil of radius a and length d produces an open-circuit voltage $V = 2B_0\, da\omega$ V, as long as the rotating coil is immersed in a constant magnetic field. The polarity is shown in Figure 5-22(*c*), determining the direction of the current in an externally connected load. If the voltage were taken off slip rings, the waveform of V would approximate a square wave as the sides of the coil are moved from the **B** field of one pole of the stator into the reversed magnetic flux lines of the other pole. A proper shaping of the poles, to make the air gap width variable

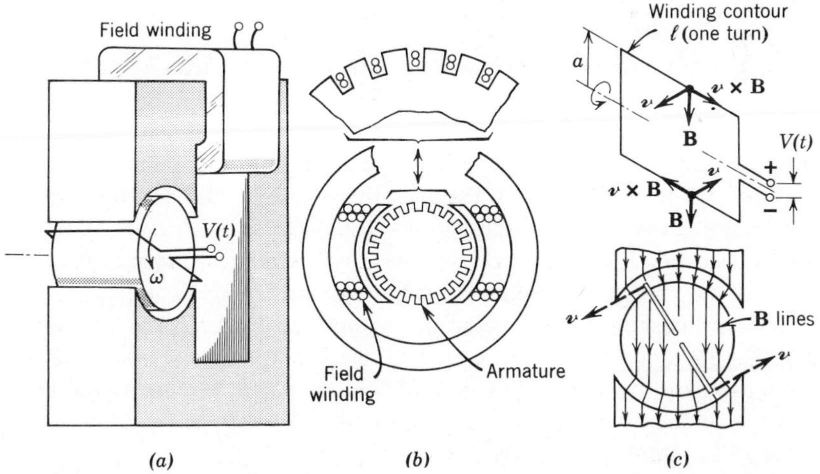

Figure 5-22. The simple electromechanical energy convertor (generator). (*a*) Simple generator showing field and armature windings. (*b*) Enhancement of air gap using armature slots. (*c*) Voltage-inducing effect of armature-winding motion.

with the angular position of the armature winding, could produce an essentially sinusoidal voltage $V(t)$, making a sinusoidal alternator of the machine. Finally, the use of an interrupted contactor (commutator) instead of the slip ring arrangement produces a rectified or unilateral output voltage polarity, to yield a direct current machine. An analysis of the induced emf of such machines is left to appropriate books on the subject.†

If the output terminals of an electromechanical energy convertor are connected to an external circuit, the resulting current is influenced not only by the external circuit, but also by the reactions of the armature winding itself. One of these reactions is the back torque that must be supplied by the motor driving the generator to keep the latter at the desired speed. Because of the presence of iron in both the field and the armature structures, the forces and torques developed between the armature and the stator are best expressed in terms of the changes taking place in the system magnetic energies with rotation. An interpretation is developed in Section 5-13 dealing with virtual forces.

Another important reaction to current flow in the generator is the effect of the armature-winding inductance. The linking of the winding current with the self-flux produced by that current yields an opposition to changes in current with time resulting from the self-voltage generated by the changing self-flux, a phenomenon already observed relative to the circuit of Figure 5-20. In this way, an armature self-inductance can be defined as in (5-60)

$$L_a = \frac{\psi_m}{I} \text{ H} \tag{5-62}$$

expressed as the ratio of the self-flux produced by the armature current, to the current itself. An equivalent circuit of the armature winding with a connected load is depicted in Figure 5-23, showing the generated voltage $V(t)$ resulting from the rotation of the armature winding in the impressed static **B** field, the self-inductance L_a of the armature winding, and a series resistance R_a representing the Ohmic winding losses. (Other losses such as iron hysteresis and eddy current losses, as well as rotational wind resistance and bearing friction losses, may be represented in more elaborate equivalent diagrams, and they are omitted here.) If the externally connected load has the resistance R and self-inductance L as developed in relation to the external circuit of Figures 5-20 and 5-21, one can deduce the equivalent circuit of the loaded rotating machine as in Figure 5-23. The Kirchhoff voltage differential equation for this system is evidently

$$V(t) = (R + R_a)I + (L + L_a)\frac{dI}{dt} \tag{5-63}$$

† For example, see Thaler, G. J., and M. L. Wilcox. *Electric Machines: Dynamics and Steady State*. New York: Wiley, 1966.

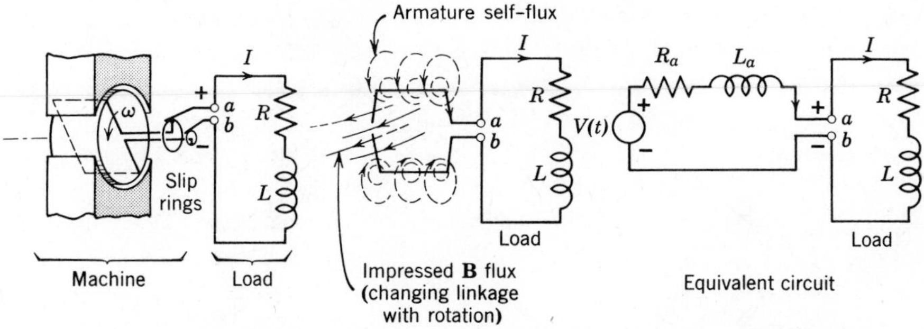

Figure 5-23. Development of an equivalent circuit of a rotating machine connected to a load.

with $V(t)$ denoting the machine generated voltage deduced from the basic expression (5-44d).

5-11 Magnetic Energy and Self-Inductance

In this section, the glib assertions of the last section concerning the inductance of a current-carrying circuit are examined from the viewpoint of the energy required by the circuit to supply its heat losses and to build up the magnetic field. The generalized definitions of the self-inductance of a single circuit, and in the next section, the mutual inductance between pairs of circuits, are established in this way. This point of view regards the inductance parameter as the basic criterion of the magnetic field energy, or work done in establishing the magnetic field.

A. Self-Inductance in Terms of A and J

Consider the series circuit of Figure 5-24. An external energy source of terminal voltage $V(t)$ is connected to a conductive circuit of arbitrary shape, carrying a current I. It is assumed that the currents form closed paths, i.e., the current-continuity relation is (4-22), $\nabla \cdot \mathbf{J} = 0$. Strictly speaking, the latter requires that the current be dc, although it is very nearly satisfied up to fairly high frequencies as long as the *overall circuit dimensions are not an appreciable fraction of a free-space wavelength.* At the higher frequencies, however, the current penetration into the conductor is severely limited by the skin effect, with negligible electromagnetic field penetration occurring at very high frequencies.† The work done by the source $V(t)$ in bringing the current up to the value I, expressed in terms of the electric and magnetic fields developed

† See Section 9-3, part A for a discussion of the skin effect.

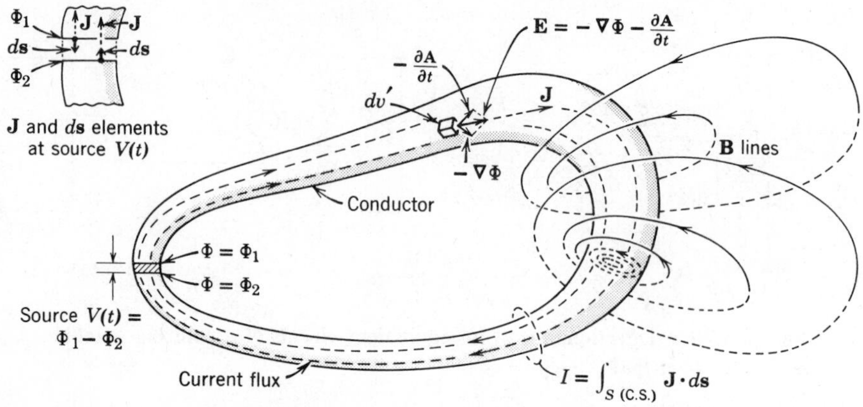

Figure 5-24. Electric and magnetic field quantities associated with a current-carrying circuit.

in and around a conductive circuit, leads to the circuit parameters (resistance and inductance) as shown in the following.

Observe in Figure 5-24 that the conductive circuit, the interior denoted by V_c, is bounded by S (conductor surface), with endcaps at the gap where the voltage V is impressed. At the gap V is specified by the quasistatic equipotentials $\Phi = \Phi_1$ and $\Phi = \Phi_2$ at the endcaps such that $V = \Phi_1 - \Phi_2$. With the current $I = -\int \mathbf{J}\cdot d\mathbf{s}$ delivered by V *into* the endcap at the positive terminal, and $\int \mathbf{J}\cdot d\mathbf{s}$ coming *out of* the negative side, the energy supplied by $V(t)$ in the amount $V\,dq = VI\,dt$ is written

$$VI\,dt = \left[(\Phi_1 - \Phi_2)\int_S \mathbf{J}\cdot d\mathbf{s}\right] dt \text{ J} \tag{5-64}$$

The electric field anywhere in the conductor is (5-48)

$$\mathbf{E} = -\nabla\Phi - \frac{\partial \mathbf{A}}{\partial t} \tag{5-65}$$

The latter is given an energy rate interpretation by dotting (5-65) with $\mathbf{J}\,dv$ and integrating the result throughout the volume V_c of the conductor; thus

$$\int_{V_c} \mathbf{E}\cdot\mathbf{J}\,dv = -\int_{V_c} (\nabla\Phi)\cdot\mathbf{J}\,dv - \int_{V_c} \frac{\partial \mathbf{A}}{\partial t}\cdot\mathbf{J}\,dv \text{ W}$$

By the identity (15) in Table 2-2, $\mathbf{J}\cdot(\nabla\Phi) = \text{div}\,(\Phi\mathbf{J})$, since $\text{div}\,\mathbf{J} = 0$. With this into the second volume integral and applying the divergence theorem (2-34), one obtains

$$-\oint_S (\Phi\mathbf{J})\cdot d\mathbf{s} = \int_{V_c} \mathbf{E}\cdot\mathbf{J}\, dv + \int_{V_c} \mathbf{J}\cdot\frac{\partial \mathbf{A}}{\partial t}\, dv \tag{5-66}$$

From the continuity of the current flux, only tangential currents appear at the conductor walls in Figure 5-24, except at the gap endcaps. There, $\Phi = \Phi_1$ on one endcap and $\Phi = \Phi_2$ on the other, reducing the surface integral of (5-66) to just

$$-\oint_{S(\text{gap})} (\Phi\mathbf{J})\cdot d\mathbf{s} \equiv (\Phi_1 - \Phi_2)\int_{S(\text{gap})} \mathbf{J}\cdot d\mathbf{s} = VI$$

the power delivered by V to the circuit at any instant. The second term of (5-66) is a measure of the irreversible heat energy expended in the volume; its value is I^2R, defining the low-frequency conductor resistance† R by

$$R = \frac{1}{I^2}\int_{V_c} \sigma E^2\, dv \tag{5-67}$$

Inserting the last two expressions into (5-66) obtains

$$VI = RI^2 + \int_{V_c} \mathbf{J}\cdot\frac{\partial \mathbf{A}}{\partial t}\, dv$$

but the energy expended by V in the time dt is

$$VI\, dt = RI^2\, dt + \int_{V_c} \mathbf{J}\cdot(d\mathbf{A})\, dv \tag{5-68a}$$

symbolized

$$dU_s = dU_h + dU_m \tag{5-68b}$$

By integrating (5-68a) with respect to time, the result

$$U_s = \int_0^t VI\, dt = R\int_0^t I^2\, dt + \int_{V_c} \left[\int_0^A \mathbf{J}\cdot(d\mathbf{A})\right] dv \tag{5-69}$$

is obtained, yielding the work done by V in bringing the circuit to its final

† The question of conductor resistance, defined in terms of the heat generated by it, is examined in Example 7-1.

state. The last term is interpreted as the *energy* U_m *expended in establishing the magnetic field* (the energy *stored* in the field)

$$U_m = \int_{V_c} \left[\int_0^A \mathbf{J} \cdot d\mathbf{A} \right] dv \text{ J} \qquad \text{In general} \quad (5\text{-}70)$$

The interpretation of (5-70) is straightforward. The current density at any point in the conductor is **J**, with **A** the vector magnetic potential there. Both **J** and **A** are fields, so they are generally dependent on position in V_c. Equation (5-70) states that the energy stored in the magnetic field is the integral of $[\int_0^A \mathbf{J} \cdot d\mathbf{A}]\, dv$ throughout the conductor volume, in which $\int_0^A \mathbf{J} \cdot d\mathbf{A}$ denotes, at any dv, the integral of $J\, dA \cos\theta$ as the potential **A** there is built up from zero to its final magnitude A. Note that the integrand has the units of joules per cubic meter.

For a linear circuit (a linear magnetic environment), **A** anywhere in the conductor is proportional to the current density **J** (hence, to the total current I). If the circuit were nonlinear, the relationship between **A** and the value of **J** at each volume-element in the conductor would not be a straight line, but for a linear circuit, the energy expression (5-70) simplifies as in the following.

The integration within the brackets of (5-70) entails a buildup in time of the vector magnetic potential from zero to its final value **A**. For a *linear* magnetic environment, the vector potential anywhere in V_c is proportional to the densities **J** therein. Suppose **J** is built up in a straight-line fashion from zero to its final (quasi-static) value $\mathbf{J}^{(f)}$ in the time t_0 as well, as depicted in Figure 5-25. Put $\mathbf{J} = \tau \mathbf{J}^{(f)}$, in which $\tau = t/t_0$, a normalized time variable. The linearity implies that **A** at the same point becomes $\mathbf{A} = \tau \mathbf{A}^{(f)}$, making $d\mathbf{A} = \mathbf{A}^{(f)}\, d\tau$. With these substitutions, the magnetic energy (5-70) becomes

$$U_m = \tfrac{1}{2} \int_{V_c} \mathbf{A} \cdot \mathbf{J}\, dv \text{ J} \qquad \text{Linear circuit} \quad (5\text{-}71)$$

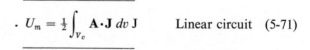

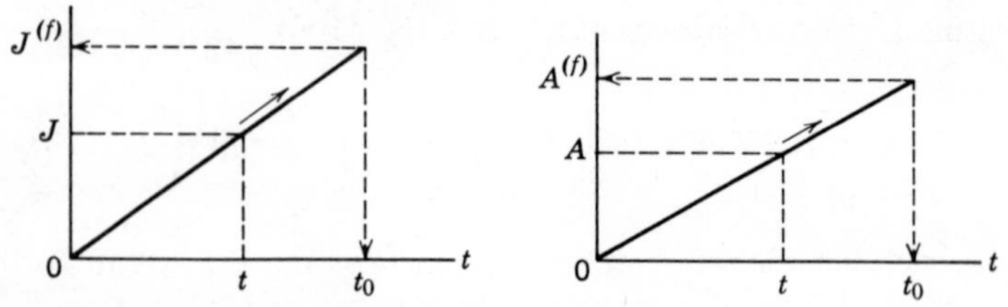

Figure 5-25. Simultaneous buildup of J and A at a typical volume-element dv in a conductor, as current is brought from zero to final value I.

if the final value (f) superscript notation is dropped. Note that (5-71) is applicable to a *linear* system only. Like its more general version, (5-70), it expresses the energy expended in establishing the magnetic field, through an integration required to be taken only throughout the conductor volume region possessing the current densities **J**.

The self-inductance of a linear circuit can be defined in terms of the energy (5-71). It contains a product of **J** and **A** and is thus proportional to I^2, whence

$$U_m = \tfrac{1}{2}\int_{V_c} \mathbf{A}\cdot\mathbf{J}\, dv \propto I^2 \qquad (5\text{-}72)$$

$$= \tfrac{1}{2}LI^2 \text{ J}$$

in which the proportionality constant L is termed the *self-inductance* of the circuit, expressed in joule per square ampere, or henry. Solving for L thus permits expressing the self-inductance in terms of the magnetic energy as follows:

$$L = \frac{2U_m}{I^2} = \frac{1}{I^2}\int_{V_c} \mathbf{A}\cdot\mathbf{J}\, dv \text{ H} \qquad (5\text{-}73)$$

assuming the circuit is linear (i.e., immersed in a linear magnetic environment).

*B. Self-Inductance of a Circuit in Free Space

For a linear circuit devoid of magnetic materials (e.g., an air core coil, parallel-wire line, etc.), (5-72) and (5-73) can be simplified by use of the free-space integral (5-28a) for **A**

$$\mathbf{A}(u_1, u_2, u_3) = \int_{V_c} \frac{\mu_0 \mathbf{J}(u_1', u_2', u_3')}{4\pi R}\, dv' \qquad [5\text{-}28a]$$

The circuit in Figure 5-26 depicts the quantities needed in the evaluation of **A** at a typical field point P by use of (5-28a). Substituting it into (5-72), the magnetic energy integral (5-71) becomes

$$U_m = \tfrac{1}{2}\int_{V_c} \left[\int_{V_c} \frac{\mu_0 \mathbf{J}'\, dv'}{4\pi R}\right]\cdot\mathbf{J}\, dv$$

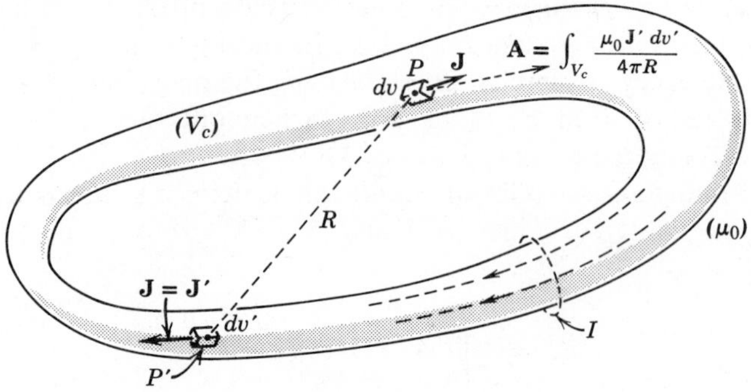

Figure 5-26. Circuit in free space, showing source point P' and field point P relative to energy and self-inductance integrals.

which can also be written

$$U_m = \frac{1}{2}\int_{V_c}\int_{V_c} \frac{\mu_0 \mathbf{J}' \cdot \mathbf{J} \, dv' \, dv}{4\pi R} \text{ J} \quad \text{Free space} \quad (5\text{-}74)$$

The result (5-74) is independent of the order of integration, but note the use of the primed current density $\mathbf{J}'$ at the source point P' to avoid confusion with $\mathbf{J}$ at the field point P. The corresponding self-inductance expression becomes, using (5-72)

$$L = \frac{2U_m}{I^2} = \frac{1}{I^2}\int_{V_c}\int_{V_c} \frac{\mu_0 \mathbf{J}' \cdot \mathbf{J} \, dv' \, dv}{4\pi R} \text{ H} \quad \text{Free space} \quad (5\text{-}75)$$

No explicit use is made here of (5-75) in self-inductance calculations. The reader interested in applications of (5-75) might consult other sources on this subject.†

C. Self-Inductance from an Integration throughout All Space

Another expression for the magnetic energy of a circuit can be obtained from (5-70) in terms of the **B** and **H** fields of the system. The current densities **J** in the conductor are related to **H** therein by (5-2) for quasistatic fields: $\mathbf{J} = \nabla \times \mathbf{H}$. Making use of the vector identity (16) in Table 2-2, $\nabla \cdot (\mathbf{F} \times \mathbf{G}) =$

† For example, see Elliott, R. S. *Electromagnetics*. New York: McGraw-Hill, 1966, p. 309.

$\mathbf{G}\cdot(\nabla\times\mathbf{F}) - \mathbf{F}\cdot(\nabla\times\mathbf{G})$, $\mathbf{J}\cdot d\mathbf{A}$ in (5-70) can be written

$$\mathbf{J}\cdot(d\mathbf{A}) = (d\mathbf{A})\cdot(\nabla\times\mathbf{H}) = \nabla\cdot[\mathbf{H}\times(d\mathbf{A})] + \mathbf{H}\cdot\nabla\times(d\mathbf{A})$$
$$= \nabla\cdot[\mathbf{H}\times(d\mathbf{A})] + \mathbf{H}\cdot d\mathbf{B}$$

Inserting this into (5-70) and applying the divergence theorem (2-34) to the first volume integral yields

$$U_m = \int_V \left[\int_0^A \nabla\cdot(\mathbf{H}\times d\mathbf{A})\right] dv + \int_V \left[\int_0^B \mathbf{H}\cdot d\mathbf{B}\right] dv$$
$$= \oint_S \left[\int_0^A (\mathbf{H}\times d\mathbf{A})\right]\cdot d\mathbf{s} + \int_V \left[\int_0^B \mathbf{H}\cdot d\mathbf{B}\right] dv$$

but the surface integral in the latter vanishes as S is expanded to include all of space, because **H** and **A** decrease at least as r^{-2} and r^{-1} respectively in remote regions, while surface area is picked up only as r^2. Thus the magnetic energy expended in establishing the fields of a quasistatic circuit becomes

$$U_m = \int_V \left[\int_0^B \mathbf{H}\cdot d\mathbf{B}\right] dv \ \text{J} \qquad \text{In general} \quad (5\text{-}76)$$

As with (5-70), the energy (5-76) is correct whether or not the circuit is linear, although (5-70) requires integration only throughout the conductor volume, whereas (5-76) must be integrated throughout *all of space* to obtain the same result.

One can simplify (5-76) if the system is linear, by making use of the fact that (5-76) is analogous in form with (5-70). Since the latter becomes (5-71) for a linear system, one should thus expect (5-76) to yield

$$U_m = \frac{1}{2}\int_V \mathbf{B}\cdot\mathbf{H}\, dv \ \text{J} \qquad \text{Linear circuit} \quad (5\text{-}77)$$

The integrand $\mathbf{B}\cdot\mathbf{H}/2$, seen to have the units of joules per cubic meter, is called the magnetic energy density in the volume region V.

Another expression for the self-inductance of the circuit of Figure 5-24 is obtained by equating (5-77) to the definition (5-72) for L, whence

$$L = \frac{2U_m}{I^2} = \frac{1}{I^2}\int_V \mathbf{B}\cdot\mathbf{H}\, dv \tag{5-78}$$

One can separate (5-77), if desired, into two volume integrations as follows:

$$U_m = \tfrac{1}{2}\int_{V_i} \mathbf{B}\cdot\mathbf{H}\, dv + \tfrac{1}{2}\int_{V_e} \mathbf{B}\cdot\mathbf{H}\, dv \tag{5-79}$$

attributing the total energy U_m to two contributions: one associated with the volume V_i *in* the conductor, plus another *outside* it. With (5-79) substituted into (5-78), the total inductance is expressed

$$L = \frac{2U_m}{I^2} = \frac{1}{I^2}\int_{V_i} \mathbf{B}\cdot\mathbf{H}\, dv + \frac{1}{I^2}\int_{V_e} \mathbf{B}\cdot\mathbf{H}\, dv = L_i + L_e \tag{5-80}$$

The first term, L_i, is called the *internal self-inductance*; while the remaining integration taken outside the conductor yields the *external self-inductance*, L_e.

EXAMPLE 5-12. Find only the *internal* self-inductance associated with every length ℓ of a very long, straight wire carrying a low-frequency current I.

For any length of the single, infinitely long wire shown, the energy in the *external* magnetic field is infinite, a fact revealed upon integrating (5-77) for the energy associated with the exterior fields $\mathbf{B}$ and $\mathbf{H}$; however, the energy stored *within* a length ℓ of the conductor is finite. The associated internal

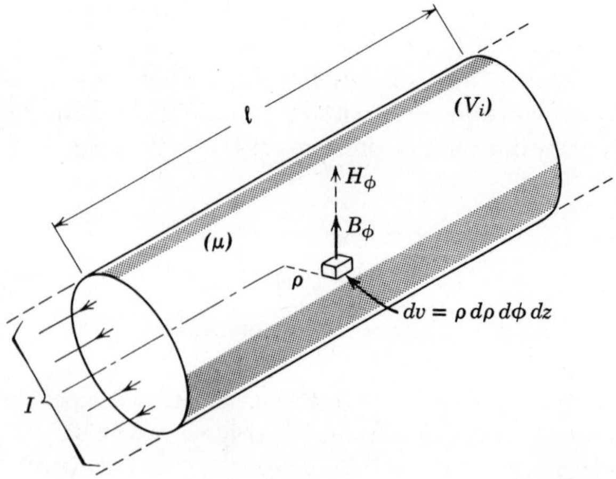

Example 5-12

inductance is obtained from (5-80)

$$L_i = \frac{2U_{m,\text{in}}}{I^2} = \frac{1}{I^2}\int_{V_i} \mathbf{B}\cdot\mathbf{H}\, dv \tag{5-81}$$

By use of (1-64) for B_ϕ inside the wire (the factor μ used in the event of a magnetic conductor), one obtains from (5-81)

$$L_i = \frac{1}{I^2}\int_{V_i} \mu H_\phi^2\, dv = \frac{1}{I^2}\int_{z=0}^{\ell}\int_{\phi=0}^{2\pi}\int_{\rho=0}^{a} \frac{\mu(I\rho)^2}{4\pi^2 a^4}\,\rho\, d\rho\, d\phi\, dz = \frac{\mu\ell}{8\pi} \tag{5-82}$$

a result independent of the wire radius. A nonmagnetic wire therefore has the internal inductance per unit length, $L_i/\ell = \mu_0/8\pi = 0.05\ \mu\text{H/m}$.

EXAMPLE 5-13. Find the total self-inductance of every length ℓ of a long coaxial line with the dimensions shown. Assume uniform current densities in the conductors.

The total self-inductance is obtained using (5-78). The magnetic fields within and between the conductors, obtained by the methods of Example 5-1, are

$$H_\phi = \frac{I\rho}{2\pi a^2} \qquad 0 < \rho < a$$

$$H_\phi = \frac{I}{2\pi\rho} \qquad a < \rho < b$$

$$H_\phi = \frac{I}{2\pi(c^2 - b^2)}\left(\frac{c^2}{\rho} - \rho\right) \qquad b < \rho < c$$

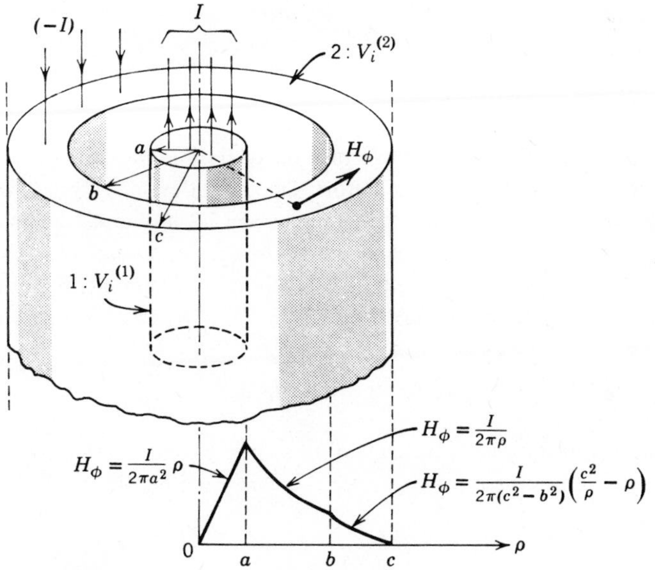

Example 5-13

with $H_\phi = 0$ outside the system. The *internal* self-inductance of the inner conductor (1) is

$$L_i^{(1)} = \frac{2U_{m,\text{in}}^{(1)}}{I^2} = \frac{1}{I^2}\int_{V_i^{(1)}} \mathbf{B}\cdot\mathbf{H}\,dv = \frac{\mu\ell}{8\pi} \tag{1}$$

a result the same as (5-82) for the isolated wire.

An *external* self-inductance, attributed to the field between the conductors, is

$$\begin{aligned} L_e &= \frac{2U_{m,\text{ex}}}{I^2} = \frac{1}{I^2}\int_{V_e} \mathbf{B}\cdot\mathbf{H}\,dv \\ &= \frac{\mu_0}{4\pi^2}\int_{z=0}^{\ell}\int_{\phi=0}^{2\pi}\int_{\rho=a}^{b} \frac{d\rho}{\rho}\,d\phi\,dz = \frac{\mu_0\ell}{2\pi}\ln\frac{b}{a} \end{aligned} \tag{2}$$

Another *internal* contribution by the field in conductor (2) yields

$$\begin{aligned} L_i^{(2)} &= \frac{2U_{m,\text{in}}^{(2)}}{I^2} = \int_0^{\ell}\int_0^{2\pi}\int_b^{c} \frac{\mu}{4\pi^2(c^2-b^2)^2}\left(\frac{c^2}{\rho}-\rho\right)^2 \rho\,d\rho\,d\phi\,dz \\ &= \frac{\mu\ell}{2\pi(c^2-b^2)^2}\left[c^4\ln\frac{c}{b} - c^2(c^2-b^2) + \tfrac{1}{4}(c^4-b^4)\right] \end{aligned} \tag{3}$$

The *total* self-inductance L of the coaxial line is therefore the sum of (1), (2), and (3). If at high frequencies the two conductors are assumed perfectly conducting to prevent field penetration into them, the self-inductance reduces to (2)

$$L_{hf} \cong L_e = \frac{\mu_0\ell}{2\pi}\ln\frac{b}{a} \tag{5-83}$$

EXAMPLE 5-14. Determine the low-frequency self-inductance of length ℓ of the long, parallel-wire line in free space shown, by use of (5-73).

The integration of (5-73) is performed inside the conductors where **J** exists, so **A** need be found only in the conductors. One might employ (5-28a) to evaluate **A**, but for a *single* wire, applying symmetry to (5-22) obtains the answer more quickly. Thus, with $\partial/\partial z = \partial/\partial\phi = 0$ and only a B_ϕ component, $\mathbf{B} = \mathbf{a}_\phi B_\phi = \nabla\times\mathbf{A} = \mathbf{a}_\phi(-\partial A_z/\partial\rho)$, implying that **A** has only a z component as noted in part (*b*) of the accompanying figure. From $B_\phi = -\partial A_z/\partial z$, integrating yields

$$A_z = -\int B_\phi\,d\rho + C \tag{1}$$

with C an arbitrary constant, depending on the potential reference chosen for A_z. Thus A_z values inside and outside the wire become, from (1)

$$A_z = -\int \frac{\mu_0 I\rho}{2\pi a^2}\,d\rho + C_1 = -\frac{\mu_0 I\rho^2}{4\pi a^2} + C_1 \qquad \rho < a \tag{2}$$

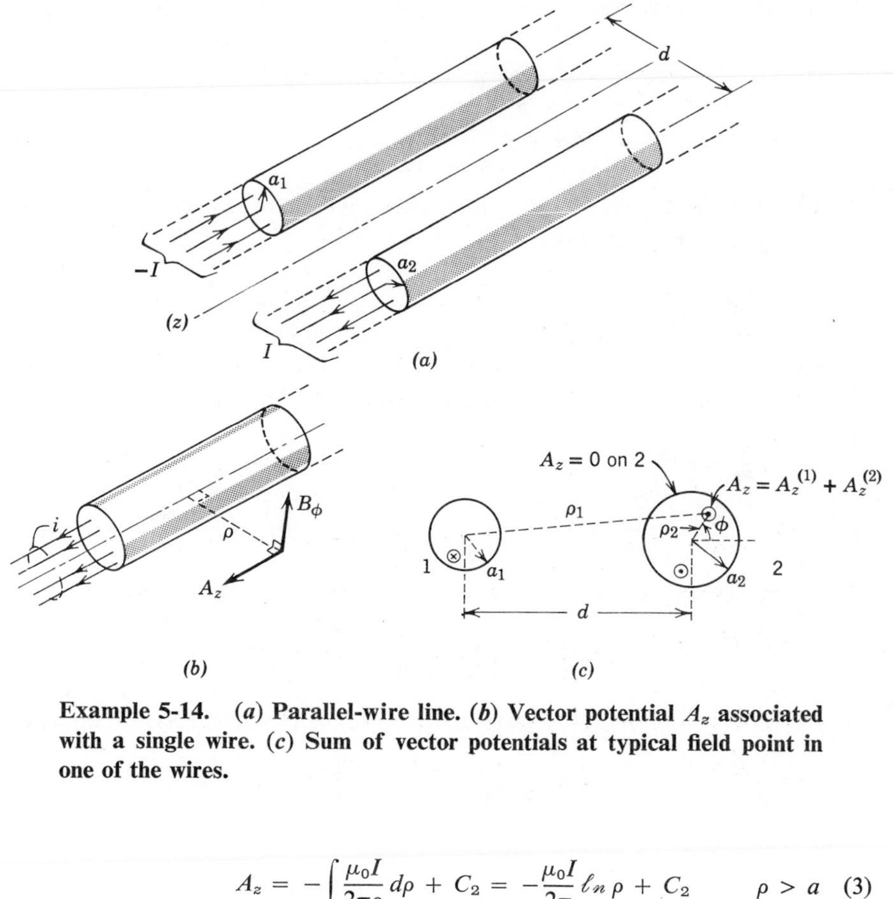

Example 5-14. ***(a)*** **Parallel-wire line.** ***(b)*** **Vector potential A_z associated with a single wire.** ***(c)*** **Sum of vector potentials at typical field point in one of the wires.**

$$A_z = -\int \frac{\mu_0 I}{2\pi\rho}\,d\rho + C_2 = -\frac{\mu_0 I}{2\pi}\ln\rho + C_2 \qquad \rho > a \quad (3)$$

In the presence of *both* conductors, carrying I and $-I$ as in (c) of the figure, the total A_z is obtained by adding the contribution of each conductor, called $A_z^{(1)}$ and $A_z^{(2)}$. For conductor 2, choosing zero potential at $\rho = a_2$ such that from (2), $A_z^{(2)} = 0 = -\mu_0 I/4\pi + C_1$. From (3), with $A_z^{(2)} = 0 = -(\mu_0 I/2\pi)\ln a_2 + C_2$, one obtains $C_1 = \mu_0 I/4\pi$ and $C_2 = (\mu_0 I \ln a_2)/4\pi$. Thus the contribution of wire 2, satisfying $A_z^{(2)} = 0$ on its surface, is

$$\begin{aligned} A_z^{(2)} &= \frac{\mu_0 I}{4\pi}\left(1 - \frac{\rho_2^2}{a^2}\right) \qquad \rho_2 < a_2 \\ A_z^{(2)} &= -\frac{\mu_0 I}{2\pi}\ln\frac{a_2}{\rho_2} \qquad \rho_2 > a_2 \end{aligned} \quad (4)$$

Similarly, the potential of wire 1 satisfying $A_z^{(1)} = A_0$ (an arbitrary value) at its surface $\rho_1 = a_1$ becomes

$$A_z^{(1)} = -\frac{\mu_0 I}{4\pi}\left(1 - \frac{\rho_1^2}{a_1^2}\right) - A_0 \qquad \rho_1 < a_1$$
$$A_z^{(1)} = -\frac{\mu_0 I}{2\pi}\ell n\frac{a_1}{\rho_1} - A_0 \qquad \rho_1 > a_1 \tag{5}$$

The total potential $A_z = A_z^{(1)} + A_z^{(2)}$ inside conductor 2, using $\rho_1 = a^2 + \rho_2^2 + 2\rho_2 a \cos\phi$ from part (c), is

$$A_z = \frac{\mu I}{4\pi}\left[1 - \frac{\rho_2^2}{a_2^2} - 2\,\ell n\frac{a_1}{\sqrt{d^2 + \rho_2^2 + 2\rho_2 d\cos\phi}}\right] - A_0$$

In conductor 2, $J_z = I/\pi a_2^2$, and with $dv = \rho_2\,d\rho_2\,d\phi\,dz$, the inductance contribution of a length ℓ of conductor 2 only, from (5-73), is

$$L^{(2)} = \frac{1}{I^2}\int_V \mathbf{A}\cdot\mathbf{J}\,dv$$
$$= \int_{\rho=0}^{a_2}\int_{\phi=0}^{2\pi}\left\{\frac{\mu_0\ell}{4\pi a_2^2}\left[1 - \frac{\rho_2^2}{a_2^2} - 2\,\ell n\frac{a_1}{[d^2+\rho_2^2+2\rho_2 d\cos\phi]^{1/2}}\right] - \frac{A_0}{\pi a_2^2 I}\right\}\rho_2\,d\rho_2\,d\phi$$

The integral contribution of the third term in the integrand can be written

$$\int_{\rho=0}^{a_2}\int_{\phi=0}^{2\pi}\left[2\,\ell n\frac{d}{a_1} + \ell n\left(1 + \frac{\rho_2^2}{d^2} + \frac{2\rho_2}{d}\cos\phi\right)\right]\rho_2\,d\rho_2\,d\phi$$

in which the second term integrates to zero (Peirce's† integral 523), obtaining

$$L^{(2)} = \mu_0\ell\left[\frac{1}{8\pi} + \frac{1}{2\pi}\ell n\frac{d}{a_2}\right] - \frac{A_0}{I}$$

A similar consideration of conductor 1 yields by analogy

$$L^{(1)} = \mu_0\ell\left[\frac{1}{8\pi} + \frac{1}{2\pi}\ell n\frac{d}{a_1}\right] + \frac{A_0}{I}$$

making the total inductance $L^{(1)} + L^{(2)}$ of the two-wire system

$$L = \frac{\mu_0\ell}{4\pi} + \frac{\mu_0\ell}{2\pi}\ell n\frac{d^2}{a_1 a_2} \tag{5-84}$$

Comparison with (5-82) shows that the leading term of (5-84) is the internal inductance, making the last term the external inductance.

D. Self-Inductance by the Method of Flux Linkages

The resolution of the self-inductance of a circuit into the sum of internal and external self-inductances provided by (5-80) is closely related to another technique known as the *method of flux linkages*. This approach is based on

† Peirce, B. O. *A Short Table of Integrals*. Boston: Ginn, 1910.

the use of the energy definition, (5-78), but with the integration in all of space replaced by a surface integral intercepting all the magnetic flux of the system, the self-inductance being thereby characterized by the linkage of that flux with the circuit current. The method is described here.

For most circuits, the total magnetic flux generated by the current can be partitioned (exactly or approximately) into two amounts: that lying entirely outside the conductor, plus that flux wholly internal to the conductor. Such a flux division occurs precisely for the single, round wire noted in Figure 5-27(*a*), and very nearly so for the parallel two-wire line shown in the same figure,† especially for wires with diameters small compared to their separation. Another example is the loop shown in Figure 5-27(*c*); for thin wire, the flux tubes can be separated into those wholly inside or outside the wire as shown. The volume occupied by the magnetic field (all of space) is thus divisible into closed flux tubes that surround or are embedded in the current.

The magnetic energy contained in all of space has been given by (5-77)

$$U_m = \frac{1}{2}\int_V \mathbf{B}\cdot\mathbf{H}\, dv \qquad [5\text{-}77]$$

Suppose the volume of the typical flux tube in Figure 5-27(*b*) is subdivided into elements $dv = ds\, d\ell'$, in which $d\ell'$ is aligned with the tube wall (and therefore with the **B** field) and ds denotes the cross-sectional area of the tube. Then $\mathbf{B}\cdot\mathbf{H}\, dv = (\mathbf{a}_\ell B)\cdot\mathbf{H}\, ds\, d\ell' = (\mathbf{a}_\ell\, d\ell')\cdot\mathbf{H}B\, ds = \mathbf{H}\cdot d\boldsymbol{\ell}'\, d\psi_m$. Thus, if the integration (5-77) of the latter is to include all elements dv where **B** and **H** prevail, $\mathbf{H}\cdot d\boldsymbol{\ell}'$ should be integrated about the closed median line ℓ' of the flux tube shown, with the remaining surface integration taken over an open surface S chosen to intercept *all* the flux tubes of the circuit. For the single-turn circuit of Figure 5-27(*b*) or (*c*), the appropriate S intercepting all flux tubes is that bounded by a closed line ℓ essentially coincident with the wire axis. Thus the energy integral (5-77) can be written‡

$$U_m = \frac{1}{2}\int_V \mathbf{B}\cdot\mathbf{H}\, dv = \frac{1}{2}\int_S \left[\oint_{\ell'} \mathbf{H}\cdot d\boldsymbol{\ell}'\right] d\psi_m = \frac{1}{2}\int_S i(\ell')\, d\psi_m \text{ J} \qquad (5\text{-}85)$$

† Due to the proximity effects of the low-frequency currents, the magnetic flux in the interior of parallel wires is not concentric about the centers of the wires, but about points moved slightly apart from the centers. This effect is responsible for some of the magnetic flux being partly inside and partly outside the wire, as noted in Figure 5-27(*a*).

‡ An energy expression analogous to (5-85), but derived from (5-71), is

$$U_m = \frac{1}{2}\int_S \psi_m(\ell')\, di$$

in which di is a typical elementary current filament passing through the cross-section S of the conductor, and $\psi_m(\ell')$ denotes the magnetic flux linked by that filament. Problem 5-25 concerns a derivation.

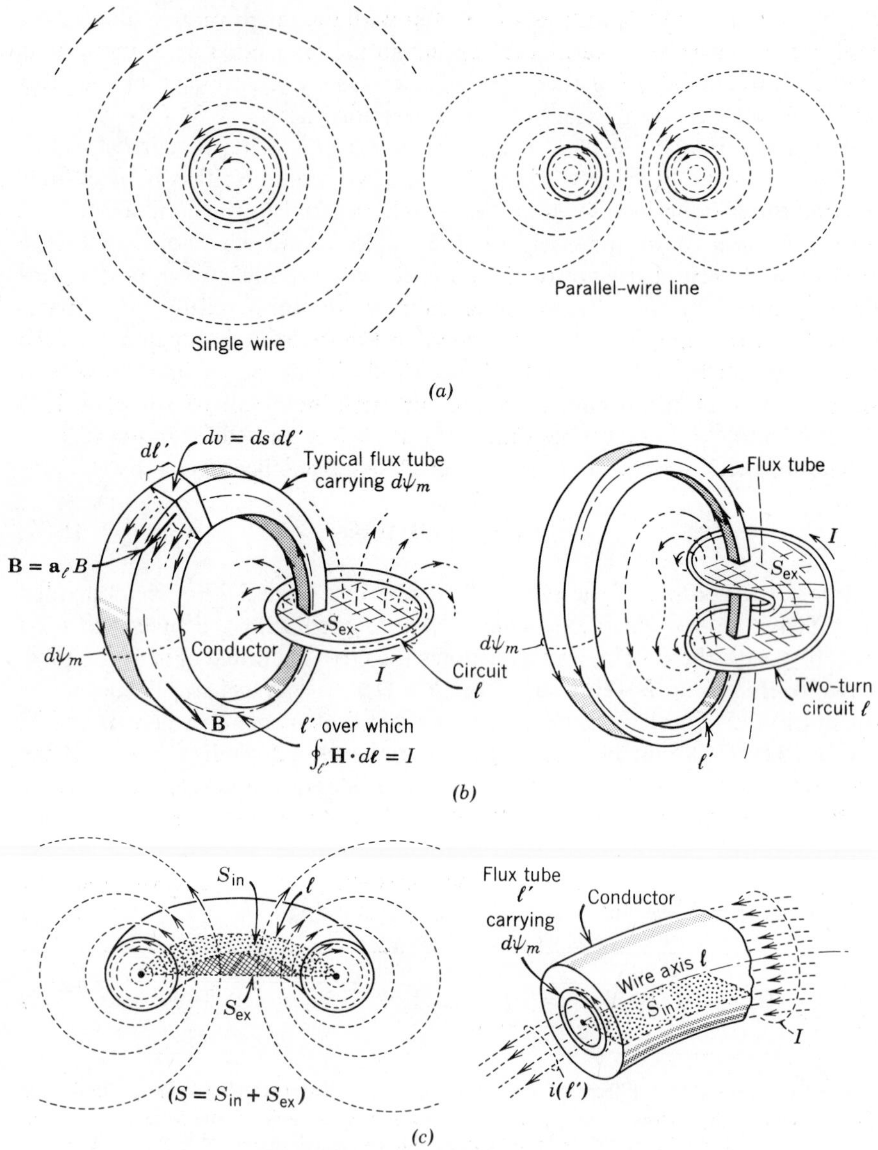

Figure 5-27. Concerning the method of flux linkages. (*a*) Examples of internal and external flux-partitioning. (*b*) Single-turn circuit (left) showing external flux tube linking I once, and a two-turn circuit (right) with a flux tube linking I twice (passing through S_{ex} twice). (*c*) Wire loop, showing internal and external flux (left), and a typical internal flux tube (right) linking $i(\ell')$, a fraction of I.

with S bounded by the circuit ℓ. [*Note:* The last integral is the consequence of $\oint \mathbf{H} \cdot d\ell'$, integrated about any closed flux tube ℓ', being just the current $i(\ell')$ enclosed by ℓ'.] For all *exterior* flux tubes, passing through S_{ex} as shown in Figure 5-27(*b*), the *total* current I is linked by ℓ', whereas a variable fraction $i(\ell')$ of I is linked by flux tubes ℓ' located inside the conductor and passing through S_{in}, as shown in Figure 5-27(*c*). In the event of a circuit ℓ having more than one turn as in Figure 5-27(*b*), an exterior flux tube ℓ' may even encompass I *more than once* (in general, as many as n times for an n turn coil). It is thus evident in such cases that the same flux tube ℓ' can contribute to (5-85) over the surface S_{ex} several times, thereby increasing the magnetic energy and the self-inductance correspondingly.

Whenever the magnetic flux of a circuit is separable into internal and external linkages passing through S_{in} and S_{ex} as depicted in Figure 5-27(*c*), it is convenient to separate (5-85) into the contributions

$$U_m = \tfrac{1}{2}\int_{S_{in}} i(\ell')\, d\psi_m + \tfrac{1}{2}I\int_{S_{ex}} d\psi_m = \tfrac{1}{2}\int_{S_{in}} i(\ell')\, d\psi_m + \frac{I\psi_{m,ex}}{2} \tag{5-86}$$

In the latter, one is cautioned to observe that the quantity $\psi_{m,ex} = \int_{S_{ex}} \mathbf{B} \cdot d\mathbf{s}$ appearing in the external energy term denotes a *total* flux through S_{ex}, which can be the result of some or all the flux tubes passing through that surface more than once, e.g., as in Figure 5-27(*b*), or for the many turn coils illustrated in Figure 5-28.

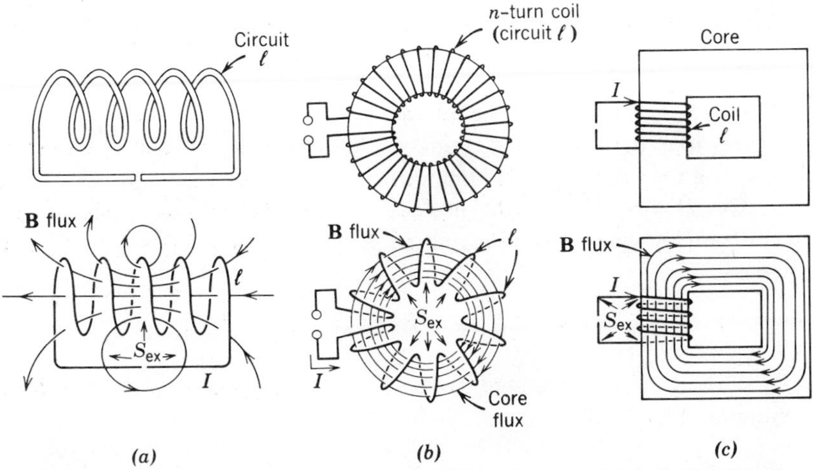

Figure 5-28. Examples of many-turn coils having negligible internal self-inductance. (*a*) Air core solenoid. (*b*) Toroidal winding. (*c*) Coil and iron core.

By use of (5-86), the self-inductance of the circuit ℓ is expressed in terms of internal and external contributions as follows:

$$L = \frac{2U_m}{I^2} = \frac{1}{I^2}\int_S i(\ell')\,d\psi_m = \frac{1}{I^2}\int_{S_{\text{in}}} i(\ell')\,d\psi_m + \frac{\psi_{m,\text{ex}}}{I} \tag{5-87}$$

such that the external inductance is given by

$$L_e = \frac{\psi_{m,\text{ex}}}{I} = \frac{1}{I}\int_{S_{\text{ex}}} \mathbf{B}\cdot d\mathbf{s} \tag{5-88a}$$

or just the total magnetic flux penetrating S_{ex} divided by the current I. The internal inductance is

$$L_i = \frac{1}{I^2}\int_{S_{\text{in}}} i(\ell')\,d\psi_m \tag{5-88b}$$

a consequence of $i(\ell')\,d\psi_m$ integrated over the appropriate internal strip S_{in} connecting with the wire axis, as depicted in Figure 5-27(*c*). An illustration of the use of the latter for a long straight wire is taken up in Example 5-15. While the internal conductor volume of a circuit may be small, the magnetic fields may be relatively large there; individual circumstances will dictate whether or not the internal inductance is negligible. For circuits having large external fluxes, such as those with iron cores, the total self-inductance is generally well approximated by (5-88a), the external self-inductance.

EXAMPLE 5-15. Determine the internal self-inductance of every length ℓ of the infinitely long wire shown, carrying the low-frequency current I. Use (5-88b), employing the method of flux linkages.

A typical flux tube ℓ' carrying $d\psi_m = \mathbf{B}\cdot d\mathbf{s}$ through S_{in} is shown in the accompanying figure. With the internal **B** obtained from (1-64), the flux in the tube is

$$d\psi_m = \mathbf{B}\cdot d\mathbf{s} = B_\phi\,d\rho\,dz = \frac{\mu I\rho}{2\pi a^2}\,d\rho\,dz$$

The current $i(\ell')$ intercepted by $d\psi_m$ is the fraction $I(\pi\rho^2/\pi a^2) = I(\rho^2/a^2)$, obtaining from (5-88b)

$$L_i = \frac{1}{I^2}\int_{S_{\text{in}}} i(\ell')\,d\psi_m = \frac{1}{I^2}\int_{z=0}^{\ell}\int_{\rho=0}^{a}\left(I\frac{\rho^2}{a^2}\right)\left(\frac{\mu I\rho\,d\rho\,dz}{2\pi a^2}\right) = \frac{\mu\ell}{8\pi} \tag{5-89}$$

which agrees with (5-82).

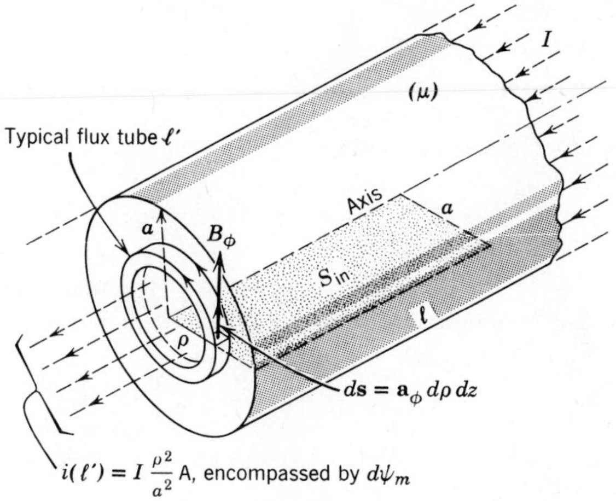

Example 5-15

EXAMPLE 5-16. Determine the approximate self-inductance of a length ℓ of a long, parallel-wire line shown in (*a*), using the flux linkage method. Assume the radii small compared to the spacing *d*.

For conductors well separated as in (*b*) the internal field is essentially that of an isolated conductor, making the internal self-inductance for both conductors just twice (5-89), i.e., $L_i = \mu\ell/4\pi$.

The external inductance is found by using (5-88a), the ratio of the magnetic flux through S_{ex} of (*a*), divided by *I*, but the total flux is just twice that through S_{ex} due to one wire, given by

$$\int_{S_{ex}} \mathbf{B} \cdot d\mathbf{s} = \int_{z=0}^{\ell} \int_{\rho=a}^{d-a} \left(\frac{\mu_0 I}{2\pi\rho} \mathbf{a}_\phi \right) \cdot \mathbf{a}_\phi \, d\rho \, dz \cong \frac{\mu_0 I \ell}{2\pi} \ln \frac{d}{a} \qquad (1)$$

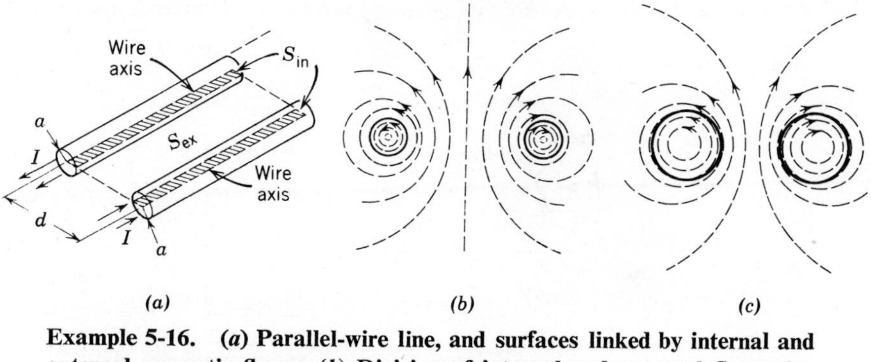

Example 5-16. (*a*) Parallel-wire line, and surfaces linked by internal and external magnetic fluxes. (*b*) Division of internal and external fluxes for thin wires. (*c*) Proximity effect for thick wires.

yielding for both wires

$$L_e = \frac{\psi_{m,ex}}{I} = \frac{\mu_0 \ell}{\pi} \ln \frac{d}{a} \tag{2}$$

The total self-inductance is the sum

$$L = L_i + L_e = \frac{\mu \ell}{4\pi} + \frac{\mu_0 \ell}{\pi} \ln \frac{d}{a} \text{ H} \tag{5-90}$$

a result seen to agree with the exact expression (5-84) upon putting $a_1 = a_2$ into the latter and assuming nonmagnetic wire. For a nonmagnetic parallel-wire line with $d = 12$ in. and $a = 0.1$ in., one obtains

$$L/\ell = 10^{-7} + (4 \times 10^{-7}) \ln 120 = 2.02 \ \mu\text{H/m}$$

Neglecting the internal inductance would incur about 5% error in this example, not a negligible amount.

In calculating self-inductance at low frequencies, the internal-inductance contribution is in some cases quite small; in others it should not be neglected. The internal inductance of a single-layer air core coil of several turns, illustrated in Figure 5-28(a), contributes little to the self-inductance if the volume of the wire is small compared to the region where the significant fields are located. In the closely spaced toroid as in Figure 5-28(b), with *every* turn intercepting *all* the core flux, the self-inductance is proportional to the square of the turns, as seen in Example 5-17. The addition of an iron core in the form of the low-reluctance magnetic circuit of Figure 5-28(c) increases the self-inductance substantially more. In these cases, the added effect of the internal inductance is insignificant.

EXAMPLE 5-17. Find the self-inductance of an n turn toroid with a rectangular cross-section as shown, for two cases: (a) with an air core, assuming closely spaced turns, and (b) with the core a linear ferromagnetic material (constant μ).

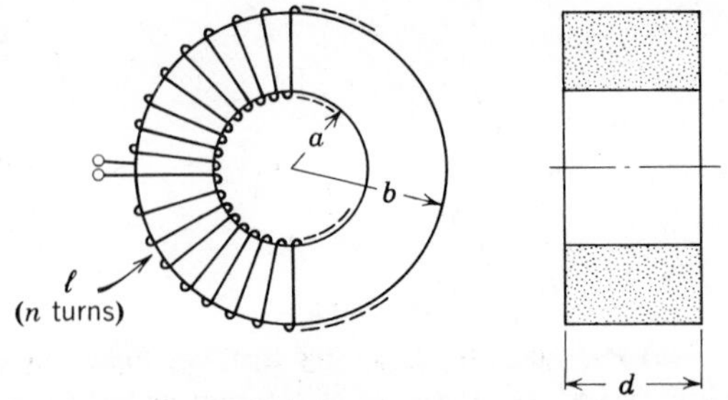

Example 5-17

(*a*) The magnetic flux in the air core, from Example 5-2, is

$$\psi_{m,\text{core}} = \frac{\mu_0 n I d}{2\pi} \ln \frac{b}{a}$$

An inspection of the surface S_{ex} bounded by the circuit ℓ, as given in Figure 5-28(*b*), reveals that S_{ex} intercepts the core flux n times,† yielding $\psi_{m,\text{ex}} = n\psi_{m,\text{core}}$ through S_{ex}. Thus the self-inductance from (5-88a) becomes

$$L \cong L_e = \frac{\psi_{m,\text{ex}}}{I} = \frac{n\psi_{m,\text{core}}}{I} = \frac{\mu_0 n^2 d}{2\pi} \ln \frac{b}{a} \tag{5-91a}$$

with the internal inductance neglected. Thus, a 100-turn air core toroid with dimensions $a = 1$ cm, $b = 3$ cm, $d = 0.5$ cm has the inductance

$$L = (4\pi \times 10^{-7} \times 100^2 \times 0.005 \ln 3)/2\pi = 11.0\ \mu\text{H}$$

Doubling the turns to 200 is seen to quadruple the inductance.

(*b*) Inserting an iron core with the permeability μ, (5-91a) becomes

$$L = \frac{\mu n^2 d}{2\pi} \ln \frac{b}{a} \tag{5-91b}$$

Using a linear ferromagnetic material with $\mu_r = 1000$ makes the inductance of the 100-turn toroid just 1000 times as large, yielding $L = 11.0$ mH.

*E. Neumann's Formula for External Inductance in Free Space

An extension of the flux linkage expression (5-87) leads to Neumann's formula, applicable to circuits in free space. Equation (5-87) consists of internal and external self-inductance terms as follows:

$$L = \frac{2U_m}{I^2} = \frac{1}{I^2}\int_{S_{\text{in}}} i(\ell')\, d\psi_m + \frac{\psi_{m,\text{ex}}}{I} = L_i + L_e \qquad [5\text{-}87]$$

Consider first only the external inductance term (5-88a) of (5-87), involving the flux $\psi_{m,\text{ex}}$ linked by the external surface S_{ex} bounded by the circuit ℓ. From (5-47), this is expressed

$$\psi_{m,\text{ex}} = \int_{S_{\text{ex}}} \mathbf{B}\cdot d\mathbf{s} = \int_{S_{\text{ex}}} (\nabla \times \mathbf{A})\cdot d\mathbf{s} = \oint_{\ell} \mathbf{A}\cdot d\boldsymbol{\ell}\ \text{Wb} \tag{5-92}$$

† Or equivalently, every flux tube $d\psi_m$ encompasses the current In times in this example.

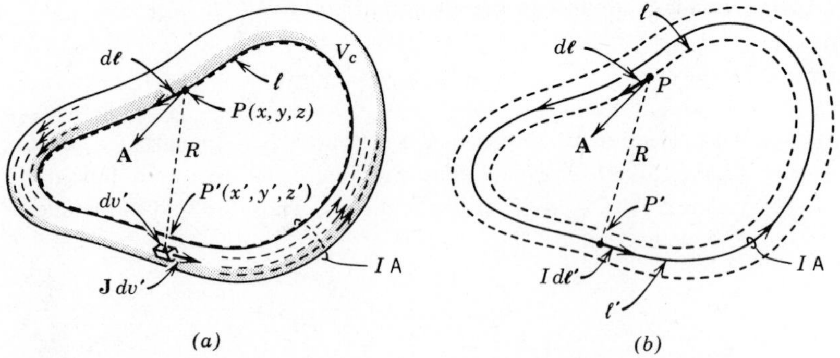

Figure 5-29. A closed circuit in free space, relative to external self-inductance calculations. (*a*) Wire circuit, showing source and field points *P'* and *P*. (*b*) Simplification of (*a*), with sources *I dℓ'* concentrated on the wire axis.

With (5-92), (5-88a) becomes

$$L_e = \frac{\psi_{m,\mathrm{ex}}}{I} = \frac{1}{I}\int_{S_{\mathrm{ex}}} \mathbf{B}\cdot d\mathbf{s} = \frac{1}{I}\oint_{\ell} \mathbf{A}\cdot d\boldsymbol{\ell} \tag{5-93}$$

In free space, the vector magnetic potential **A** can be found by use of (5-28a)

$$\mathbf{A}(u_1, u_2, u_3) = \int_{V_c} \frac{\mu_0 \mathbf{J}(u_1', u_2', u_3')}{4\pi R}\, dv' \qquad [5\text{-}28\text{a}]$$

Applied to the circuit of Figure 5-29(*a*), (5-28a) obtains **A** at the typical field point *P* located on ℓ bounding S_{ex}. Another integration of $\mathbf{A}\cdot d\boldsymbol{\ell}$ about ℓ in accordance with (5-93) then obtains the external self-inductance of the circuit. These steps are combined by inserting (5-28a) into (5-93), yielding

$$L_e = \frac{1}{I}\oint_{\ell}\left[\int_{V_c} \frac{\mu_0 \mathbf{J}\, dv'}{4\pi R}\right]\cdot d\boldsymbol{\ell} \ \mathrm{H} \tag{5-94a}$$

This quadruple integration is simplified for a thin-wire circuit if I is considered concentrated on the wire axis as in Figure 5-29(*b*). Then $\mathbf{J}\, dv'$ becomes $I\, d\boldsymbol{\ell}'$, reducing (5-94a) to

$$L_e = \frac{1}{I}\oint_{\ell}\left[\oint_{\ell'} \frac{\mu_0 I\, d\boldsymbol{\ell}'}{4\pi R}\right]\cdot d\boldsymbol{\ell} = \oint_{\ell}\oint_{\ell'} \frac{\mu_0\, d\boldsymbol{\ell}'\cdot d\boldsymbol{\ell}}{4\pi R} \tag{5-94b}$$

a result known as Neumann's formula for the external inductance of a thin circuit in free space. The order of the integrations relative to $d\ell'$ and $d\ell$, and hence, relative to the source point and field point coordinates, is immaterial.

From (5-87), the *total* self-inductance L is obtained by adding (5-94b) to the internal inductance term L_i. Since the latter is a measure of the internal stored magnetic energy, L_i is expressible using either (5-81) or (5-88b); thus $L = L_i + L_e$ becomes, in *free space*

$$L = \frac{1}{I^2}\int_{V_{in}} \mathbf{B}\cdot\mathbf{H}\, dv + \oint_\ell \oint_{\ell'} \frac{\mu_0\, d\ell'\cdot d\ell}{4\pi R}$$
$$= \frac{1}{I^2}\int_{S_{in}} i(\ell')\, d\psi_m + \oint_\ell \oint_{\ell'} \frac{\mu_0\, d\ell'\cdot d\ell}{4\pi R} \tag{5-95}$$

EXAMPLE 5-18. Find the self-inductance of a thin, circular loop of wire in free space, with dimensions as in (*a*). Use the Neumann formula (5-94b).

The current assumed concentrated on the wire axis as in (*b*) allows the use of (5-94b). In cylindrical coordinates, $d\ell' = \mathbf{a}_\phi b\, d\phi'$ at the source point $P'(b, \phi', 0)$

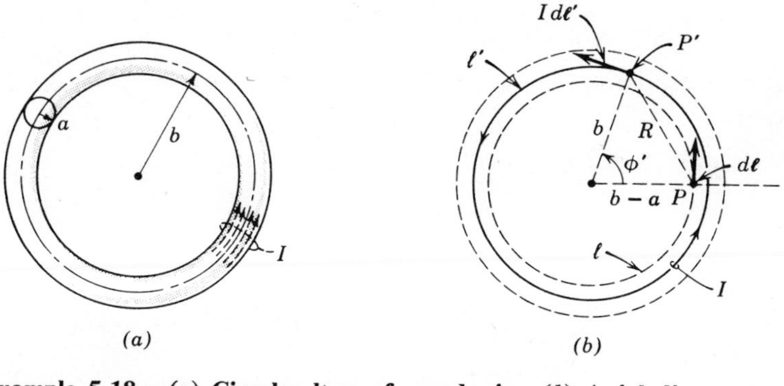

Example 5-18. (*a*) Circular loop of round wire. (*b*) Axial, line current approximation of (*a*).

on the axis ℓ'. From the circular symmetry, the location of the field point P on ℓ is immaterial, so put P at $\phi = 0$; i.e., at $P(b - a, 0, 0)$. The distance from P' to P is given by the law of cosines

$$R = \sqrt{b^2 + (b-a)^2 - 2b(b-a)\cos\phi'} \tag{1}$$

while $d\ell'\cdot d\ell$ in (5-94b), from part (*b*), means $d\ell' d\ell \cos\phi'$ (implying that only the component of $\mathbf{A}$ parallel to $d\ell$ at P is required in the integration.) Then

(5-94b) becomes

$$L_e = \int_{\phi=0}^{2\pi} \int_{\phi'=0}^{2\pi} \frac{\mu_0 b(b-a)\cos\phi'\,d\phi'\,d\phi}{4\pi\sqrt{b^2+(b-a)^2-2b(b-a)\cos\phi'}}$$

$$= \frac{\mu_0 b(b-a)}{2} \int_{\phi'=0}^{2\pi} \frac{\cos\phi'\,d\phi'}{\sqrt{b^2+(b-a)^2-2b(b-a)\cos\phi'}} \tag{5-96}$$

This result is not integrable in closed form, though with numerical values of a and b it yields to computer solution. An alternative makes use of tabulated values of the complete elliptic integrals $K(k)$ and $E(k)$. A conversion of (5-96) in terms of such integrals is accomplished as follows. Change the variable ϕ' to 2α, making $d\phi' = 2d\alpha$ and $\cos\phi' = \cos 2\alpha = 2\cos^2\alpha - 1$, with the limits on α going from 0 to π. Then R in (5-96) becomes

$$R = \sqrt{b^2+(b-a)^2-2b(b-a)(2\cos^2\alpha-1)}$$

$$= \sqrt{(2b-a)^2-4b(b-a)\cos^2\alpha}$$

$$= \sqrt{(2b-a)^2(1-k^2\cos^2\alpha)} \tag{2}$$

if $k^2 = 4b(b-a)/(2b-a)^2$. The complete elliptic integrals, defined by

$$K(k) \equiv \int_0^{\pi/2} \frac{d\theta}{\sqrt{1-k^2\sin\theta}} \qquad E(k) = \int_0^{\pi/2} \sqrt{1-k^2\sin^2\theta}\,d\theta \tag{5-97}$$

are incorporated into (5-96) as follows. The integral in (5-96), making use of (2), becomes

$$\int_{\phi'=0}^{2\pi} = \int_0^{\pi} \frac{2\cos 2\alpha\,d\alpha}{(2b-a)\sqrt{1-k^2\cos^2\alpha}} = \frac{1}{\sqrt{b(b-a)}} \int_0^{\pi} \frac{k\cos 2\alpha\,d\alpha}{\sqrt{1-k^2\cos^2\alpha}} \tag{3}$$

but the numerator of (3) is written

$$k\cos 2\alpha = k(2\cos^2\alpha - 1) = 2k\cos^2\alpha - k + \frac{2}{k} - \frac{2}{k}$$

$$= \frac{2}{k}(k^2\cos^2\alpha - 1) + \left(\frac{2}{k} - k\right)$$

to yield a further conversion of (3)

$$\frac{1}{\sqrt{b(b-a)}}\int_0^{\pi} = \frac{1}{\sqrt{b(b-a)}}\left\{\int_0^{\pi} \frac{\left(\frac{2}{k}-k\right)d\alpha}{\sqrt{1-k^2\cos^2\alpha}} - \frac{2}{k}\int_0^{\pi}\sqrt{1-k^2\cos^2\alpha}\,d\alpha\right\} \tag{4}$$

An inspection of the last integral shows that

$$\int_0^{\pi}\sqrt{1-k^2\cos^2\alpha}\,d\alpha = 2\int_0^{\pi/2}\sqrt{1-k^2\cos^2\alpha}\,d\alpha$$

$$= 2\int_0^{\pi/2}\sqrt{1-k^2\sin^2\alpha}\,d\alpha$$

which from (5-97) is just $2E(k)$. A similar consideration of the preceding integral in (4) reveals that it is just $2K(k)$, so (5-96) becomes

$$L_e = \mu_0 \sqrt{b(b-a)}\left[\left(\frac{2}{k} - k\right)K(k) - \frac{2}{k}E(k)\right] \tag{5-98}$$

The tabulated values† of $K(k)$ and $E(k)$ can be used in (5-97) to evaluate L_e

Table 5-1 Summary of magnetic energy and self-inductance relations

Magnetic energy	*Self-inductance*
In terms of **A** and **J** integrated throughout conductor volume	
In general:	
$U_m = \int_{V_c}\left[\int_0^A \mathbf{J}\cdot d\mathbf{A}\right] dv$ (5-70)	
Linear circuit:	
$U_m = \frac{1}{2}\int_{V_c} \mathbf{A}\cdot\mathbf{J}\, dv$ (5-71)	$L = \frac{2U_m}{I^2} = \frac{1}{I^2}\int_{V_c} \mathbf{A}\cdot\mathbf{J}\, dv$ (5-73)
In free space:	In free space:
$U_m = \frac{1}{2}\int_{V_c}\int_{V_c} \frac{\mu_0 \mathbf{J}'\cdot\mathbf{J}}{4\pi R}\, dv'\, dv$ (5-74)	$L = \frac{1}{I^2}\int_{V_c}\int_{V_c} \frac{\mu_0 \mathbf{J}'\cdot\mathbf{J}}{4\pi R}\, dv'\, dv$ (5-75)
In terms of **B** and **H** integrated throughout all space	
In general:	
$U_m = \int_V\left[\int_0^B \mathbf{H}\cdot d\mathbf{B}\right] dv$ (5-76)	
Linear circuit:	
$U_m = \frac{1}{2}\int_V \mathbf{B}\cdot\mathbf{H}\, dv$ (5-77)	$L = \frac{2U_m}{I^2} = \frac{1}{I^2}\int_V \mathbf{B}\cdot\mathbf{H}\, dv$ (5-78)
$= U_{m,\text{in}} + U_{m,\text{ex}}$	$= L_i + L_e$
$= \frac{1}{2}\int_{V_{\text{in}}} \mathbf{B}\cdot\mathbf{H}\, dv + \frac{1}{2}\int_{V_{\text{ex}}} \mathbf{B}\cdot\mathbf{H}\, dv$ (5-79)	$= \frac{1}{I^2}\int_{V_{\text{in}}} \mathbf{B}\cdot\mathbf{H}\, dv + \frac{1}{I^2}\int_{V_{\text{ex}}} \mathbf{B}\cdot\mathbf{H}\, dv$
Extension to method of flux linkages:	
$U_m = \frac{1}{2}\int_S i(\ell')\, d\psi_m$ (5-85)	$L = \frac{1}{I^2}\int_S i(\ell')\, d\psi_m$
$= \frac{1}{2}\int_{S_{\text{in}}} i(\ell')\, d\psi_m + \frac{I\psi_{m,\text{ex}}}{2}$ (5-86)	$= \frac{1}{I^2}\int_{S_{\text{in}}} i(\ell')\, d\psi_m + \frac{\psi_{m,\text{ex}}}{I}$ (5-87)
	In free space:
	$L = \frac{1}{I^2}\int_{S_{\text{in}}} i(\ell')\, d\psi_m + \oint_\ell \oint_{\ell'} \frac{\mu_0 d\boldsymbol{\ell}'\cdot d\boldsymbol{\ell}}{4\pi R}$ (5-95)

† For example, see Jahnke, E., and F. Emde. *Tables of Functions*, 4th ed. New York: Dover, 1945.

of a circular loop with desired dimensions. For thin wires ($a \ll b$), the elliptic integrals are approximated by

$$K(k) \cong \ell n \left[\frac{8b}{a} - 4\right] \qquad E(k) \cong 1 \qquad a \ll b \quad (5\text{-}99)$$

yielding the simplification

$$L_e \cong \mu_0 b\left(\ell n \frac{8b}{a} - 2\right) \qquad a \ll b \quad (5\text{-}100)$$

For example, a 2 mm diameter wire bent into a circle of 10 cm radius has the external inductance $L_e = (4\pi \times 10^{-7})(0.1)(\ell n\, 800 - 2) = 0.588\ \mu\text{H}$. The internal magnetic field of the loop is virtually that of a straight, isolated wire, making their internal inductances nearly the same. Applying the results of Example 5-12, the approximate internal inductance of the loop becomes

$$L_i \cong \frac{\mu(2\pi b)}{8\pi} = \frac{\mu b}{4} \qquad (5\text{-}101)$$

With $b = 10$ cm and assuming nonmagnetic wire, $L_i = 0.031\ \mu\text{H}$. Thus the self-inductance expressed by (5-93) becomes $L = L_e + L_i = 0.619\ \mu\text{H}$, in which L_e is seen to be the predominant term.

A summary of expressions for magnetic energy described in the foregoing discussion, together with expressions for the circuit inductance when the system is linear, is given in Table 5-1.

*F. Kirchhoff Voltage Relation from Energy Considerations

In concluding the remarks about the circuit of Figure 5-30(a), a Kirchhoff type voltage equation resembling (5-63) can be obtained for it from the energy expression (5-68a)

$$VI\,dt = RI^2\,dt + \int_{V_c} \mathbf{J}\cdot d\mathbf{A}\,dv \qquad [5\text{-}68a]$$

abbreviated

$$dU_s = dU_h + dU_m \qquad [5\text{-}68b]$$

Dividing by dt obtains

$$VI = RI^2 + \frac{dU_m}{dt} \qquad (5\text{-}102)$$

signifying the instantaneous power delivered by V: the sum of the instan-

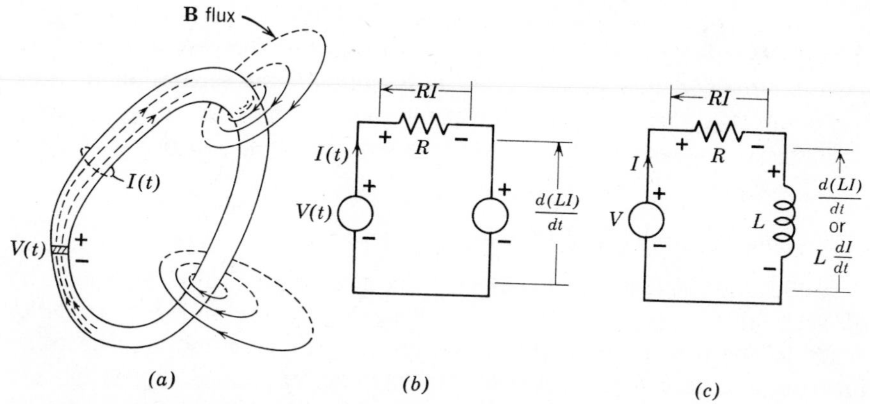

Figure 5-30. Development of circuit models of the circuit of Figure 5-24. (*a*) Physical circuit driven by $V(t)$. (*b*) Circuit model depicting terms of (5-106). (*c*) Circuit model using lumped elements.

taneous heat loss plus dU_m/dt, the power delivered to the magnetic field (*rate of magnetic energy storage or release*). Dividing by I produces a *voltage relation*

$$V = RI + \frac{1}{I}\frac{dU_m}{dt} \tag{5-103}$$

in which U_m, the instantaneous magnetic stored energy, is specified by any of the expressions listed in Table 5-1, depending on whether the system is magnetically linear. For a *linear* circuit, a self-inductance L is attributable to the circuit energy by (5-78)

$$U_m = \tfrac{1}{2}LI^2 \tag{5-104}$$

With L constant, the last term of (5-103) becomes

$$\frac{1}{I}\frac{dU_m}{dt} = \frac{1}{I}\frac{d}{dt}(\tfrac{1}{2}LI^2) = \frac{d(LI)}{dt} \tag{5-105}$$

making (5-103) a voltage relation comparable to the Kirchhoff expression (5-61); i.e.

$$V = RI + \frac{d(LI)}{dt} \tag{5-106}$$

Equation (5-106) states that the applied voltage $V(t)$ supports two effects:

(*a*) a voltage drop RI associated with the circuit resistance R and (*b*) a back voltage $d(LI)/dt$ or $L\,dI/dt$ produced by the time-varying magnetic flux linking the circuit, a flux produced by I. Because of the separation of these effects into two terms, one may properly lump the resistive voltage and the self-induced voltage to yield the series circuit model shown in Figure 5-30.

5-12 Coupled Circuits and Mutual Inductance

Besides the single circuit of Figure 5-24, also of physical interest is a *pair* of such circuits, coupled electromagnetically by the time-varying fields generated by their currents. Examples are the iron core and air core transformers of Figure 5-31(*a*), which may have active sources in one or both windings. A generalization is illustrated in (*b*).

The analysis of coupled circuits from the magnetic energy point of view closely parallels that for the single circuit. Consider the circuit pair of Figure 5-31(*b*) with one driving source $V(t)$ in circuit 1, producing the primary current $I_1(t)$. The latter generates a field $\mathbf{B}_1$, the flux of which links not only circuit 1 but some fraction of that flux (governed by the geometry and the presence of ferromagnetic bodies) also links circuit 2, generating an emf about each circuit in accordance with the Faraday law, (3-78). The ensuing current I_2 produces a field B_2 reacting similarly upon circuit 2 while also partly linking circuit 1, thereby establishing an additional back emf in each

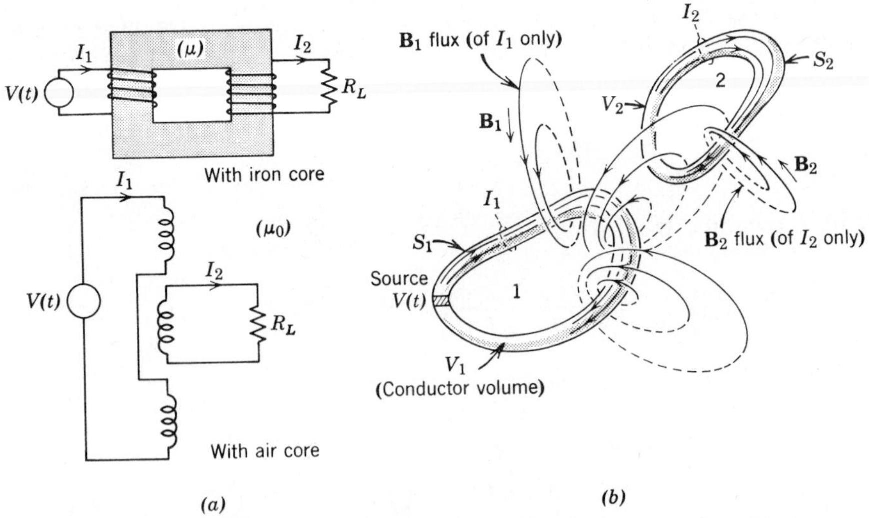

Figure 5-31. Magnetically coupled circuits. (*a*) Typical coupled circuits. (*b*) Generalized coupled circuits.

to modify I_2 and I_1 accordingly. The influence of these mutual coupling effects on current flow can conveniently be treated by use of Kirchhoff voltage equations, developed later in this section. The mutual magnetic coupling between the circuits leads to their *mutual inductance* parameters, developed in the following.

A simple extension of the power integral (5-66) to the pair of circuits of Figure 5-31(*b*) yields

$$-\oint_{S_1} (\Phi\mathbf{J})\cdot d\mathbf{s} = \int_{V_1} \mathbf{E}\cdot\mathbf{J}\, dv + \int_{V_2} \mathbf{E}\cdot\mathbf{J}\, dv + \int_{V_1} \frac{\partial \mathbf{A}}{\partial t}\cdot\mathbf{J}\, dv + \int_{V_2} \frac{\partial \mathbf{A}}{\partial t}\cdot\mathbf{J}\, dv \tag{5-107}$$

in which V_1, V_2 denote the volumes inside the two conductors, with S_1 taken as the surface enclosing V_1 exclusive of the driving source $V(t)$. The left side of (5-107) denotes the instantaneous power VI_1 delivered, while the two-volume integrals of $\mathbf{E}\cdot\mathbf{J}$ are the Ohmic losses $R_1 I_1^2$ and $R_2 I_2^2$ within the conductors. Multiplying (5-107) by dt yields

$$VI_1\, dt = R_1 I_1^2\, dt + R_2 I_2^2\, dt + \int_{V_1} \mathbf{J}\cdot d\mathbf{A}\, dv + \int_{V_2} \mathbf{J}\cdot d\mathbf{A}\, dv \tag{5-108a}$$

abbreviated

$$dU_s = dU_{h1} + dU_{h2} + dU_{m1} + dU_{m2} \tag{5-108b}$$

Integrating (5-108b) obtains

$$U_s = \int_0^t R_1 I_1^2\, dt + \int_0^t R_2 I_2^2\, dt + \int_{V_1} \left[\int_0^A \mathbf{J}\cdot d\mathbf{A}\right] dv + \int_{V_2} \left[\int_0^A \mathbf{J}\cdot d\mathbf{A}\right] dv \tag{5-109}$$

denoting the work done by $V(t)$ in bringing the system up to the levels I_1 and I_2 at the instant t. The volume integrals in (5-109) represent the energy expended by V in establishing the magnetic fields of the coupled circuits; i.e., the energy *stored* in the magnetic fields in the amount

$$U_m = \int_{V_1} \left[\int_0^A \mathbf{J}\cdot d\mathbf{A}\right] dv + \int_{V_2} \left[\int_0^A \mathbf{J}\cdot d\mathbf{A}\right] dv \text{ J} \quad \text{In general} \tag{5-110}$$

The integrations are required only within the conductors, since no densities $\mathbf{J}$ exist outside them. Equation (5-110) is correct whether or not the system is linear.

If the system of Figure 5-31(*b*) is *linear*, one can assert that the contributions to the total $\mathbf{A}$ at any point in the region are proportional to the current densities **J** in the circuits. Then

$$U_m = \tfrac{1}{2}\int_{V_1} \mathbf{A}\cdot\mathbf{J}\,dv + \tfrac{1}{2}\int_{V_2} \mathbf{A}\cdot\mathbf{J}\,dv \quad \text{J} \qquad \text{Linear system} \tag{5-111}$$

obtained analogously from (5-110) in the manner that (5-70) led to (5-71).

It is advantageous to reexpress (5-111) in terms of the vector potential contributions of each current. Let the total vector potential at any field point P in either conductor be written

$$\mathbf{A} = \mathbf{A}_1 + \mathbf{A}_2 \tag{5-112}$$

with $\mathbf{A}_1$ and $\mathbf{A}_2$ denoting the potentials at P due to the currents in circuits 1 and 2, respectively. Then (5-111) splinters into the four contributions

$$U_m = \tfrac{1}{2}\int_{V_1} \mathbf{A}_1\cdot\mathbf{J}\,dv + \tfrac{1}{2}\int_{V_2} \mathbf{A}_2\cdot\mathbf{J}\,dv + \tfrac{1}{2}\int_{V_2} \mathbf{A}_1\cdot\mathbf{J}\,dv + \tfrac{1}{2}\int_{V_1} \mathbf{A}_2\cdot\mathbf{J}\,dv \tag{5-113a}$$

abbreviated as follows

$$U_m = U_{m1} + U_{m2} + U_{m12} + U_{m21} \tag{5-113b}$$

Note that U_{m1}, for example, denotes the magnetic self-energy of circuit 1 taken alone (with circuit 2 open-circuited), with (5-71) revealing that U_{m1} is the energy associated with the *self-inductance* of circuit 1, called L_1. A similar remark applies to U_{m2}, leading to the self-inductance L_2 of circuit 2. With **J** in conductor 1 as well as $\mathbf{A}_1$ both proportional to I_1, U_{m1} becomes proportional to I_1^2. Similarly, U_{m2}, U_{m12}, and U_{m21} are proportional to I_2^2, I_1I_2, and I_1I_2 respectively, yielding from (5-113a)

$$U_{m1} = \tfrac{1}{2}\int_{V_1} \mathbf{A}_1\cdot\mathbf{J}\,dv = \tfrac{1}{2}L_1I_1^2 \tag{5-114a}$$

$$U_{m2} = \tfrac{1}{2}\int_{V_2} \mathbf{A}_2\cdot\mathbf{J}\,dv = \tfrac{1}{2}L_2I_2^2 \tag{5-114b}$$

$$U_{m12} = \tfrac{1}{2}\int_{V_2} \mathbf{A}_1\cdot\mathbf{J}\,dv = \tfrac{1}{2}M_{12}I_1I_2 \tag{5-114c}$$

$$U_{m21} = \tfrac{1}{2}\int_{V_1} \mathbf{A}_2\cdot\mathbf{J}\,dv = \tfrac{1}{2}M_{21}I_1I_2 \quad \text{J} \tag{5-114d}$$

The constants M_{12} and M_{21} appearing in (5-114c) and (5-114d) are known as the *mutual inductances* of the pair of circuits, related to the additional mutual magnetic energies associated with the magnetic coupling of the circuits. It is now shown that the mutual inductances M_{12} and M_{21} are *identical* for linear systems, namely

$$M_{12} = M_{21} = M \tag{5-115}$$

with the symbol M chosen to denote either parameter.

That (5-115) is true for a linear system is demonstrated upon reexpressing (5-113a) in terms of the volume integral of $\mathbf{B} \cdot \mathbf{H}$ by use of (5-77)

$$U_m = \tfrac{1}{2} \int_V \mathbf{B} \cdot \mathbf{H} \, dv \qquad [5\text{-}77]$$

This result, derived for the single circuit of Figure 5-24, is equally valid for the coupled circuits of Figure 5-31. Suppose **B** and **H** of the coupled system are expressed as the sums

$$\begin{aligned} \mathbf{B} &= \mathbf{B}_1 + \mathbf{B}_2 \\ \mathbf{H} &= \mathbf{H}_1 + \mathbf{H}_2 \end{aligned} \tag{5-116}$$

in which $\mathbf{B}_1 = \nabla \times \mathbf{A}_1 = \mu \mathbf{H}_1$, $\mathbf{B}_2 = \nabla \times \mathbf{A}_2 = \mu \mathbf{H}_2$, whence $\mathbf{B}_1$, $\mathbf{H}_1$ are taken to be due to I_1 in circuit 1, while $\mathbf{B}_2$, $\mathbf{H}_2$ are proportional to I_2 in circuit 2. Then (5-77) expands into the four terms

$$U_m = \tfrac{1}{2} \int_V \mathbf{B}_1 \cdot \mathbf{H}_1 \, dv + \tfrac{1}{2} \int_V \mathbf{B}_2 \cdot \mathbf{H}_2 \, dv + \tfrac{1}{2} \int_V \mathbf{B}_1 \cdot \mathbf{H}_2 \, dv + \tfrac{1}{2} \int_V \mathbf{B}_2 \cdot \mathbf{H}_1 \, dv \tag{5-117}$$

in which the integrations are to be taken throughout all of the space where the fields **B** and **H** exist. A comparison of the four integrals in (5-117) with those of (5-113a) reveals a one-to-one energy correspondence, implying that the self and mutual inductances defined in (5-114) can also be written

$$L_1 = \frac{2U_{m1}}{I_1^2} = \frac{1}{I_1^2} \int_{V_1} \mathbf{A}_1 \cdot \mathbf{J} \, dv = \frac{1}{I_1^2} \int_V \mathbf{B}_1 \cdot \mathbf{H}_1 \, dv \tag{5-118a}$$

$$L_2 = \frac{2U_{m2}}{I_2^2} = \frac{1}{I_2^2} \int_{V_2} \mathbf{A}_2 \cdot \mathbf{J} \, dv = \frac{1}{I_2^2} \int_V \mathbf{B}_2 \cdot \mathbf{H}_2 \, dv \tag{5-118b}$$

$$M_{12} = \frac{2U_{m12}}{I_1 I_2} = \frac{1}{I_1 I_2} \int_{V_2} \mathbf{A}_1 \cdot \mathbf{J} \, dv = \frac{1}{I_1 I_2} \int_V \mathbf{B}_1 \cdot \mathbf{H}_2 \, dv \tag{5-118c}$$

$$M_{21} = \frac{2U_{m21}}{I_1 I_2} = \frac{1}{I_1 I_2} \int_{V_1} \mathbf{A}_2 \cdot \mathbf{J} \, dv = \frac{1}{I_1 I_2} \int_V \mathbf{B}_2 \cdot \mathbf{H}_1 \, dv \ \mathrm{H} \tag{5-118d}$$

but in the latter, the product $\mathbf{B}_1 \cdot \mathbf{H}_2$ equals $\mathbf{B}_2 \cdot \mathbf{H}_1$ because

$$\mathbf{B}_1 \cdot \mathbf{H}_2 = \mu \mathbf{H}_1 \cdot \mathbf{H}_2 = \mathbf{H}_1 \cdot \mathbf{B}_2 \tag{5-119}$$

Thus (5-118c) and (5-118d) are identical, proving (5-115), that $M_{12} = M_{21}$. Combining (5-114) and (5-115) into (5-113b) permits writing the total magnetic energy in the form

$$U_m = \tfrac{1}{2} L_1 I_1^2 + \tfrac{1}{2} L_2 I_2^2 + M I_1 I_2 \text{ J} \tag{5-120}$$

Hence, a knowledge of the inductance parameters and instantaneous currents determines the magnetic energy state of coupled circuits at any instant. Since the self-inductance expressions (5-118a, b) have already been considered in detail, the expressions (5-118c, d) concerning the mutual inductance M will occupy the attention of the remainder of this section.

For coupled circuits in *free space*, M can be expressed by a volume integral in terms of the current sources, yielding a result resembling (5-75) for self-inductance. Hence, substituting (5-28a) for $\mathbf{A}$ into (5-118c) or (5-118d) obtains

$$M_{12} = M_{21} = M = \frac{1}{I_1 I_2} \int_{V_1} \int_{V_2} \frac{\mu_0 \mathbf{J}' \cdot \mathbf{J}}{4\pi R} \, dv' \, dv \text{ H} \quad \text{Free space} \tag{5-121}$$

with primes again used to distinguish the source point current element $\mathbf{J}' \, dv'$ from the unprimed field point element as in (5-75). In Figure 5-32(*a*) is shown the geometry relative to the integrations. The Neumann integral (5-121) is not discussed further here; the reader is referred to other sources† for applications.

More general expressions for M can be derived from magnetic flux and current linkage interpretations of (5-114c) and (5-114d), to include the effects of magnetic materials. Subdivide circuit 2 into closed current filaments ℓ_2' carrying the differential current di as in Figure 5-32(*b*), each linking a portion $\psi_{12}(\ell_2')$ of the flux of circuit 1. Equation (5-114c) for the mutual energy U_{m12} then becomes

$$U_{m12} = \tfrac{1}{2} \int_{V_2} \mathbf{A}_1 \cdot \mathbf{J} \, dv = \tfrac{1}{2} \int_{S_2} \oint_{\ell_2'} \mathbf{A}_1 \cdot d\ell' J \, ds = \tfrac{1}{2} \int_{S_2} \psi_{12}(\ell_2') \, di \tag{5-122a}$$

† See Elliott, R. S., *loc. cit.*

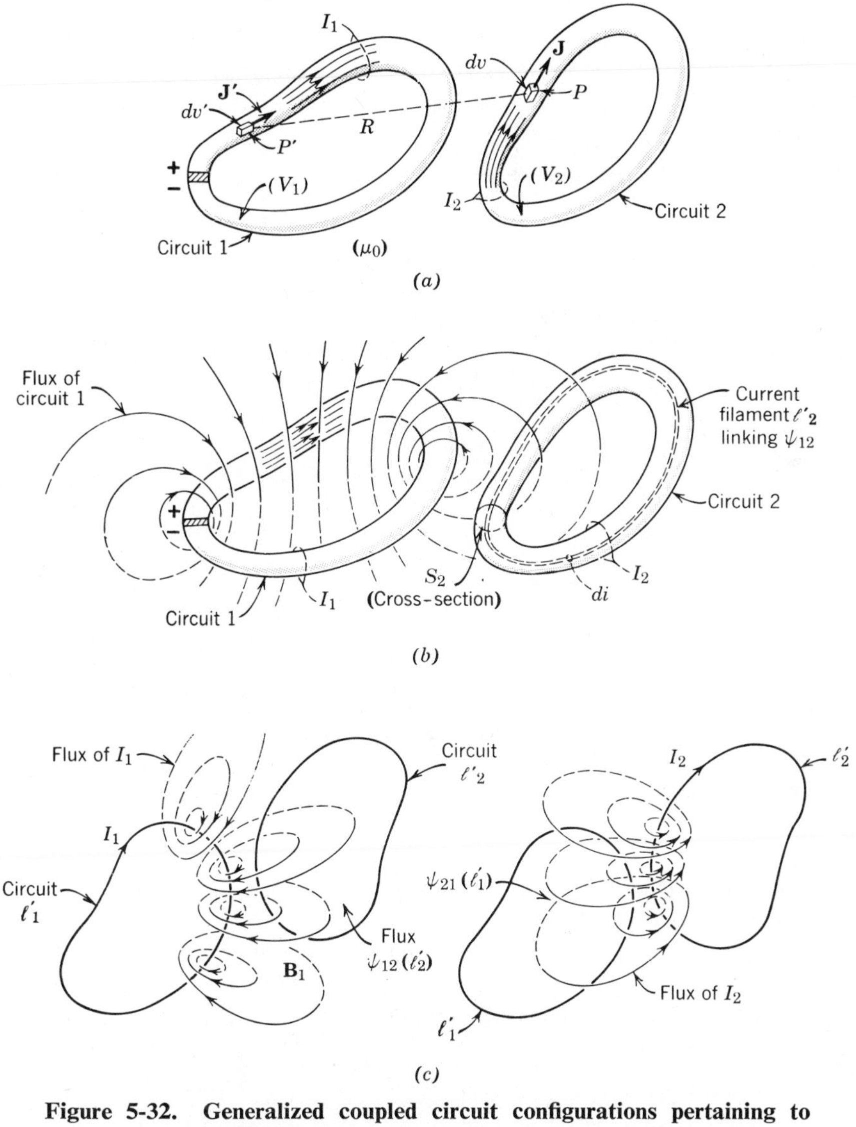

Figure 5-32. Generalized coupled circuit configurations pertaining to mutual energy and inductance calculations. (*a*) Linear coupled circuits in free space. (*b*) Linear coupled circuits in general (iron present or not), showing the portion $\psi_{12}(\ell'_2)$ of the flux of I_1 linking current filament ℓ'_2. (*c*) Special case of (*b*): thin circuits. Depicting portions ψ_{12} (left) and ψ_{21} (right) of the fluxes I_1 and I_2.

a result that follows upon noting that $\mathbf{A}_1 \cdot \mathbf{J}\, dv = \mathbf{A}_1 \cdot (\mathbf{a}_\ell J)\, d\ell'\, ds = \mathbf{A}_1 \cdot d\boldsymbol{\ell}'\, di$, and observing from (5-47) that $\oint \mathbf{A}_1 \cdot d\boldsymbol{\ell}'$ denotes $\psi_{12}(\ell_2')$, the portion of the flux of I_1 linking ℓ_2'. Thus U_{m12} is found by integrating $\psi_{12}(\ell_2')\, di$ over the *cross-section* S_2 of wire 2, as depicted in Figure 5-32(*b*). Similarly, (5-114d) becomes

$$U_{m21} = \tfrac{1}{2} \int_{V_1} \mathbf{A}_2 \cdot \mathbf{J}\, dv = \tfrac{1}{2} \int_{S_1} \psi_{21}(\ell_1')\, di \qquad \text{(5-122b)}$$

The use of the flux linkage expressions (5-122a, b) is facilitated by assuming I_1 and I_2 to be concentrated along the wire axes. Then $\psi_{12}(\ell_2')$ and $\psi_{21}(\ell_1')$ in (5-122a) and (5-122b) become constants, yielding the simpler results

$$U_{m12} \cong \tfrac{1}{2}\psi_{12} \int_{S_2} di = \tfrac{1}{2}\psi_{12} I_2 \qquad \text{(5-122c)}$$

$$U_{m21} \cong \tfrac{1}{2}\psi_{21} I_1 \text{ J} \qquad \text{(5-122d)}$$

in which

ψ_{12} = the portion of the flux of I_1 linked by circuit 2

ψ_{21} = the portion of the flux of I_2 linked by circuit 1

The simplifications (5-122c) and (5-122d) are excellent approximations if the circuits are *thin*, as depicted in Figure 5-32(*c*).

The mutual inductance M is finally obtained by substituting the energies (5-122) into the definitions (5-118c) and (5-118d), making use of $M_{12} = M_{21} = M$ of (5-115); thus

$$M = \frac{2U_{m12}}{I_1 I_2} = \frac{2U_{m21}}{I_1 I_2} = \frac{1}{I_1 I_2} \int_{S_2} \psi_{12}(\ell_2')\, di = \frac{1}{I_1 I_2} \int_{S_1} \psi_{21}(\ell_1')\, di \qquad \text{Exact} \quad \text{(5-123a)}$$

$$M = \frac{2U_{m12}}{I_1 I_2} = \frac{2U_{m21}}{I_1 I_2} \cong \frac{\psi_{12}}{I_1} = \frac{\psi_{21}}{I_2} \qquad \text{For thin circuits} \quad \text{(5-123b)}$$

The latter approximations (5-123b) are usually acceptable in practical mutual inductance calculations.

EXAMPLE 5-19. Find M for the iron core toroidal transformer illustrated, the

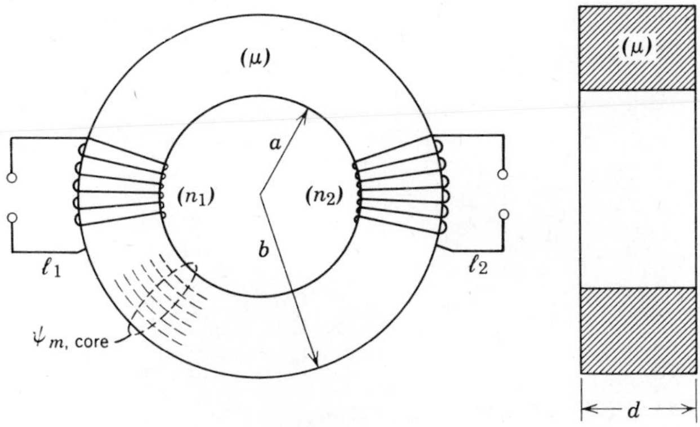

Example 5-19

windings having n_1 and n_2 turns and assuming no leakage flux. Compare M^2 with the product L_1L_2.

M for thin coils is conveniently found by use of (5-123b). For I_1 in ℓ_1, the *core* flux obtained in Example 5-2 is

$$\psi_{m,\text{core}} = \frac{\mu n_1 I_1 d}{2\pi} \ln \frac{b}{a} \tag{1}$$

but ψ_{12} linked by ℓ_2 (i.e., passing through $S_{\text{ex},2}$ bounded by ℓ_2) is n_2 times $\psi_{m,\text{core}}$, obtaining from (5-123b)

$$M = \frac{\psi_{12}}{I_1} = \frac{n_2 \psi_{m,\text{core}}}{I_1} = \frac{\mu n_1 n_2 d}{2\pi} \ln \frac{b}{a} \tag{2}$$

The same answer is obtained using $M = \psi_{21}/I_2$.

The self-inductances of the coils, from Example 5-17, are

$$L_1 = \frac{\mu n_1^2 d}{2\pi} \ln \frac{b}{a} \qquad L_2 = \frac{\mu n_2^2 d}{2\pi} \ln \frac{b}{a} \tag{3}$$

Thus the product L_1L_2 equals the square of M given by (2). This is expected for coupled circuits whenever *all* the magnetic flux links each turn of the windings.

The idealization that all the magnetic flux produced by one circuit completely links the other, as in Example 5-19, is never quite attained in practice, even when high-permeability cores are utilized to minimize flux leakage. There is invariably some leakage, as depicted in Figure 5-33(a), causing M^2

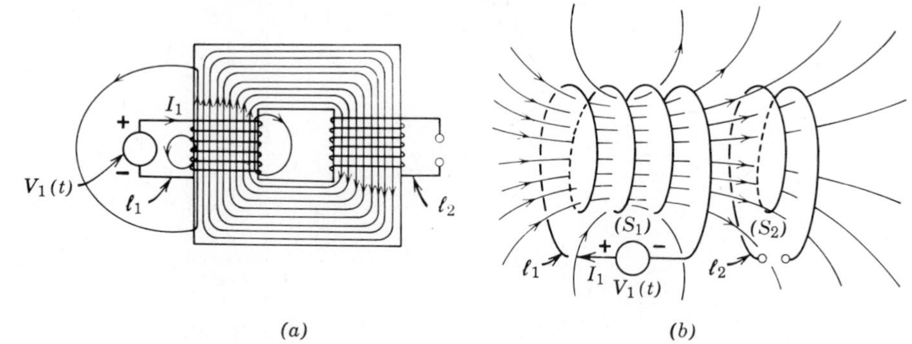

Figure 5-33. Magnetic coupling between circuits yielding high and low coupling coefficients. (*a*) Iron core transformer with small leakage ($k \to 1$). (*b*) Circuits coupled in air, for high-frequency applications.

to be *less* than L_1L_2. This circumstance is expressed by the so-called coefficient of coupling between circuits, symbolized by k and defined

$$k = \frac{M}{\sqrt{L_1L_2}} \tag{5-124}$$

The latter permits expressing M as a function of the self-inductance of each circuit whenever k is known; i.e.,

$$M = k\sqrt{L_1L_2}\ \text{H} \tag{5-125}$$

The maximum value attainable by k is unity, while for circuits totally uncoupled, $k = 0$. If coils are coupled using high-permeability cores, k may have a value as high as 0.99 or better, though with air as the coupling medium as in Figure 5-33(*b*), a much smaller k is usual, in view of one circuit linking a correspondingly smaller fraction of the total self-flux of the other.

The *circuit model* of coupled circuits can be deduced in the same manner as for single circuits. Since a pair of circuits is involved, two Kirchhoff voltage relations are desired. Three interrelated methods can be employed to obtain the Kirchhoff voltage equations: (*a*) a method based on the scalar and vector potentials Φ and $\mathbf{A}$ of the electromagnetic fields, described in Section 5-10; (*b*) a technique based on energy considerations, treated in Section 5-11, part F; and (*c*) an approach making use of the Faraday law, (3-78).

The Kirchhoff voltage equations of coupled circuits are derived from

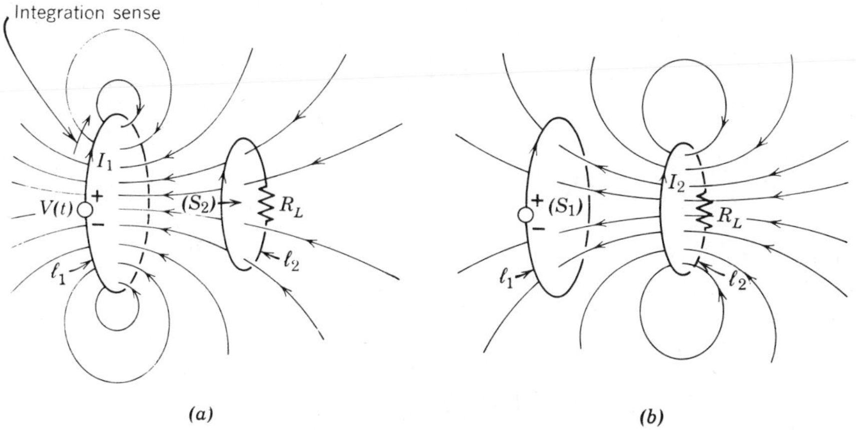

Figure 5-34. Self and mutual fluxes produced by I_1 and I_2 in coupled circuits, and emf's induced. (*a*) Flux of I_1 only. The self-flux ψ_1 links ℓ_1, inducing

$$\oint_{\ell_1} \mathbf{E}_1 \cdot d\boldsymbol{\ell} = -\frac{d\psi_1}{dt} \qquad (1)$$

The mutual flux $\psi_{12} = \int_{S_2} \mathbf{B}_1 \cdot d\mathbf{s}$ links ℓ_2, inducing

$$\oint_{\ell_2} \mathbf{E}_{12} \cdot d\boldsymbol{\ell} = -\frac{d\psi_{12}}{dt} \qquad (2)$$

(*b*) Flux of I_2 only. The self-flux ψ_2 links ℓ_2, inducing

$$\oint_{\ell_2} \mathbf{E}_2 \cdot d\boldsymbol{\ell} = -\frac{d\psi_2}{dt} \qquad (3)$$

The mutual flux $\psi_{21} = \int_{S_1} \mathbf{B}_2 \cdot d\mathbf{s}$ links ℓ_1, inducing

$$\oint_{\ell_1} \mathbf{E}_{21} \cdot d\boldsymbol{\ell} = -\frac{d\psi_{21}}{dt} \qquad (4)$$

application of the Faraday law, (3-78)

$$\oint_{\ell} \mathbf{E} \cdot d\boldsymbol{\ell} = -\frac{d\psi_m}{dt} \qquad [3\text{-}78]$$

to the closed paths ℓ_1 and ℓ_2 defining the circuits. In (3-78) **E** denotes the *total* field existing at the elements $d\boldsymbol{\ell}$ of the paths ℓ_1 and ℓ_2, with ψ_m the total flux intercepted by each circuit—flux generated by both I_1 and I_2. To help visualize this process, in Figure 5-34 are shown the separate fluxes of I_1 and I_2. Only one independent voltage source $V(t)$ is used. The senses of I_1 and I_2 are arbitrary, being assumed as shown. Figure 5-34 shows that the total **E** generated along ℓ_1 consists of two contributions: that induced by $-d\psi_1/dt$, in which ψ_1 is the self-flux linking ℓ_1 and due to I_1; and that induced by $-d\psi_{21}/dt$, in which ψ_{21} is the mutual flux linking ℓ_1 and produced by I_2; plus an $\mathbf{E}_g$ generated only within the source $V(t)$. Thus $\mathbf{E}\cdot d\boldsymbol{\ell}$ anywhere along conductor ℓ_1 is

$$\mathbf{E}\cdot d\boldsymbol{\ell} = \frac{\mathbf{J}}{\sigma}\cdot d\boldsymbol{\ell} = (\mathbf{E}_1 + \mathbf{E}_{21} + \mathbf{E}_g)\cdot d\boldsymbol{\ell} \tag{5-126}$$

in which $\mathbf{J}/\sigma = \mathbf{E}$ is substituted in view of the continuity (3-79) of the tangential $\mathbf{E}$ field at the conductor surface. Integrating (5-126) about ℓ_1 yields

$$\oint_{\ell_1} \frac{\mathbf{J}}{\sigma}\cdot d\boldsymbol{\ell} = \oint_{\ell_1} \mathbf{E}_g\cdot d\boldsymbol{\ell} + \oint_{\ell_1} \mathbf{E}_1\cdot d\boldsymbol{\ell} + \oint_{\ell_1} \mathbf{E}_{21}\cdot d\boldsymbol{\ell} \tag{5-127}$$

with the arbitrary choice of the integration sense about ℓ_1 indicated in Figure 5-34. At low frequencies the left integral becomes $R_1 I_1$, in which R_1 is the resistance of ℓ_1 by the arguments of Section 4-15-2. The second term reduces to $V(t)$ because $\mathbf{E}_g$ is zero everywhere except in the generator, while the last two integrals denote the negative time rates of flux (1) and (4) given in Figure 5-34. Thus (5-127) yields

$$R_1 I_1 = V(t) - \frac{d\psi_1}{dt} - \frac{d\psi_{21}}{dt} \text{ V}$$

The fluxes $\psi_1 = \int_{S_1} \mathbf{B}_1\cdot d\mathbf{s}$ and $\psi_{21} = \int_{S_1} \mathbf{B}_2\cdot d\mathbf{s}$ linked by ℓ_1 are the *positive* quantities $\psi_1 = L_1 I_1$ and $\psi_{21} = M I_2$, since those fluxes emerge from the positive side of S_1 bounded by ℓ_1 in Figure 5-34. With these substitutions one obtains

$$V(t) = R_1 I_1 + L_1 \frac{dI_1}{dt} + M\frac{dI_2}{dt} \tag{5-128a}$$

the desired Kirchhoff voltage relation for the circuit ℓ_1.

Applying a similar line of reasoning to the other circuit, one obtains the desired Kirchhoff voltage relation for ℓ_2

$$0 = (R_2 + R_L)I_2 + L_2\frac{dI_2}{dt} + M\frac{dI_1}{dt} \tag{5-128b}$$

These coupled differential equations correspond to the circuit model in Figure 5.35. The use of this model makes it evident, without recourse to field theory, that upon removing R_L, for example, the open-circuit voltage obtained across gap terminals at *c-d* is just $M\, dI_1/dt$. Other features of coupled circuits

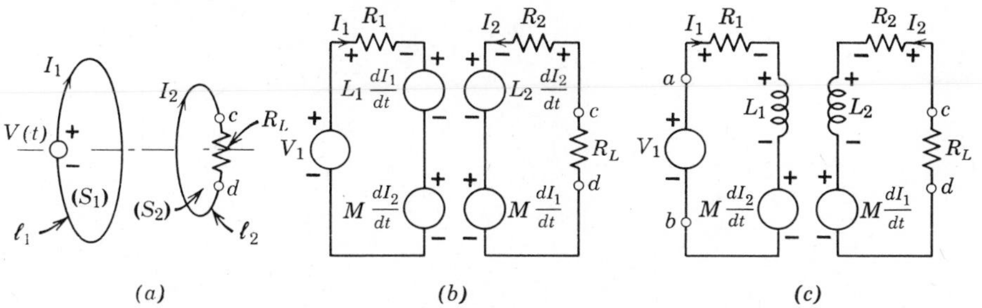

Figure 5-35. Magnetically coupled circuits and circuit models. (*a*) The physical coupled circuits, with assumed current directions. (*b*) Circuit model showing elements corresponding to terms of (5-128). (*c*) Circuit model using symbolic convention to denote circuit self-inductances.

from the point of view of this model are treated in standard texts on circuit theory.†

5-13 Magnetic Forces and Torques

While the force acting on a current-carrying circuit in the presence of an external magnetic field can often be obtained by use of the Ampère force law, (5-45a), frequently it is more expedient to obtain it from the stored magnetic field energy. It is shown how the force or torque acting on a current-carrying circuit or a nearby magnetic material region is deduced from an application of the conservation of energy principle to a virtual displacement or rotation of the desired body. This process is analogous to the determination of forces or torques exerted on charged conductors or dielectrics in the presence of an electrostatic field, discussed in Section 4-16.

Suppose the magnetic circuit of Figure 5-36(*a*), having an air gap of variable width x, derives its energy from the source V supplying a direct current to the winding. If the armature were displaced a distance $d\ell$ at the gap due to the magnetic field force $\mathbf{F}$ acting on it, the mechanical work done would be

$$dU = \mathbf{F} \cdot d\boldsymbol{\ell} = F_x\,dx + F_y\,dy + F_z\,dz \text{ J} \tag{5-129}$$

This work is done by V at the expense of the energy in the magnetic field

† See, for example, Pearson, S. I., and G. J. Maler. *Introductory Circuit Analysis*. New York: Wiley, 1965, pp. 54–63.

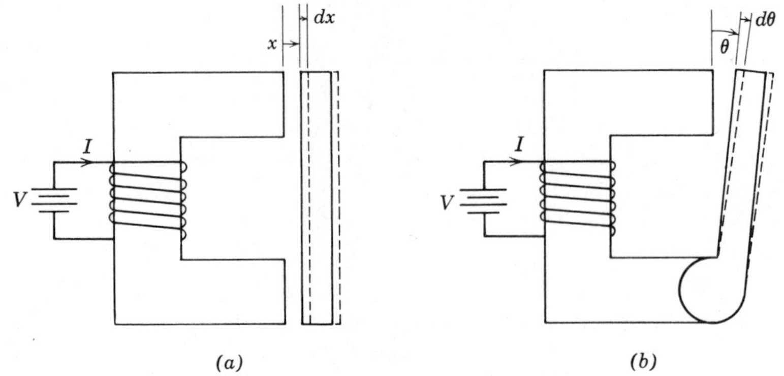

Figure 5-36. Single circuits utilizing magnetic cores subject to relative translation or rotation. (*a*) Armature translates. (*b*) Armature rotates.

such that the following energy balance is maintained:

$$\underbrace{dU_m}_{\text{Magnetostatic energy change}} + \underbrace{dU}_{\text{Mechanical work done}} = \underbrace{dU_s}_{\text{Work done by source } V} \tag{5-130}$$

The change in the magnetic energy, upon changing the air gap in Figure 5-36(*a*), produces a corresponding inductance change. The magnetostatic energy U_m is $1/2LI^2$ from (5-72), so the energy change occurring with I held constant becomes

$$dU_m = \tfrac{1}{2}I^2\,dL \tag{5-131}$$

Omitting the I^2R heat losses associated with the coil resistance in the equivalent circuit of this system depicted in Figure 5-30(*c*), the work dU_s exerted by V to maintain (5-130) is done against the voltage induced by the flux change $d\psi_m$ in the time dt such that $V = -d\psi_m/dt$. With $\psi_m = LI$ from (5-88a), and with I maintained at a constant value, the induced voltage becomes $V = -d\psi_m/dt = -I\,dL/dt$. The work dU_s done by the source in the time dt to overcome this voltage is therefore

$$dU_s = -VI\,dt = I^2\,dL \tag{5-132}$$

which is just twice (5-131), the change in the stored energy. Combining (5-129), (5-131), and (5-132) into the energy balance, (5-130) thus yields

$(1/2)I^2\, dL + \mathbf{F}\cdot d\ell = I^2\, dL$, reducing to $\mathbf{F}\cdot d\ell = (1/2)I^2\, dL$, or

$$dU = dU_m \tag{5-133}$$

The latter shows that the mechanical work just equals the change in the magnetostatic field energy. Thus, of the electrical energy supplied by V, *one-half goes to increasing the magnetic energy of the system, while the other half is used up as mechanical work done by the magnetic force.*

The differential magnetostatic energy change dU_m can be written in terms of the coordinate variations of U_m as the armature moves the distance $d\ell = \mathbf{a}_x\, dx + \mathbf{a}_y\, dy + \mathbf{a}_z\, dz$ if desired; i.e.,

$$dU_m = \frac{\partial U_m}{\partial x}\, dx + \frac{\partial U_m}{\partial y}\, dy + \frac{\partial U_m}{\partial z}\, dz = (\nabla U_m)\cdot d\ell \tag{5-134}$$

a gradient form allowable in view of (2-11). A comparison of (5-134) with (5-129), making use of (5-133), leads to the cartesian components of $\mathbf{F}$

$$F_x = \frac{\partial U_m}{\partial x} \qquad F_y = \frac{\partial U_m}{\partial y} \qquad F_z = \frac{\partial U_m}{\partial z}\ \text{N} \tag{5-135a}$$

Since $U_m = (1/2)LI^2$ from (5-72), the force components with I constant can also be written in terms of the derivations of the self-inductance L as follows:

$$F_x = \frac{I^2}{2}\frac{\partial L}{\partial x} \qquad F_y = \frac{I^2}{2}\frac{\partial L}{\partial y} \qquad F_z = \frac{I^2}{2}\frac{\partial L}{\partial z}\ \text{N} \tag{5-135b}$$

To evaluate $\mathbf{F}$, the magnetostatic energy U_m (or the self-inductance L) should be given in terms of the coordinates of the displaced element of the system. In Figure 5-36(*a*), for example, U_m would be expressed in terms of the single coordinate x denoting the air gap width.

Suppose a portion of the iron core, instead of being translated, is constrained to a rotation about an axis as in Figure 5-36(*b*). Then the differential work (with $dU = dU_m$) done by the magnetic force in the angular displacement $d\boldsymbol{\theta} = \mathbf{a}_1\, d\theta_1 + \mathbf{a}_2\, d\theta_2 + \mathbf{a}_3\, d\theta_3$ becomes

$$dU_m = \mathbf{T}\cdot d\boldsymbol{\theta} = T_1\, d\theta_1 + T_2\, d\theta_2 + T_3\, d\theta_3 \tag{5-136}$$

wherein $\mathbf{T} = \mathbf{a}_1 T_1 + \mathbf{a}_2 T_2 + \mathbf{a}_3 T_3$ denotes the vector torque due to the magnetic force. Then results analogous with (5-135a, b), in terms of the variations of the magnetic energy with respect to angular changes, obtain as follows:

$$T_1 = \frac{\partial U_m}{\partial \theta_1} \qquad T_2 = \frac{\partial U_m}{\partial \theta_2} \qquad T_3 = \frac{\partial U_m}{\partial \theta_3}\ \text{N}\cdot\text{m} \tag{5-137a}$$

and in terms of the variations in the circuit self-inductance with respect to the angular motions, one obtains

$$T_1 = \frac{I^2}{2}\frac{\partial L}{\partial \theta_1} \qquad T_2 = \frac{I^2}{2}\frac{\partial L}{\partial \theta_2} \qquad T_3 = \frac{I^2}{2}\frac{\partial L}{\partial \theta_3} \ \text{N·m} \tag{5-137b}$$

EXAMPLE 5-20. A magnetic relay has a movable armature with two air gaps of width x as shown in the accompanying figure. The n turn coil carries a current I derived from the source V. The core and armature, both of permeability μ, have the median lengths and cross-sectional areas ℓ_1, A_1; ℓ_2, A_2,

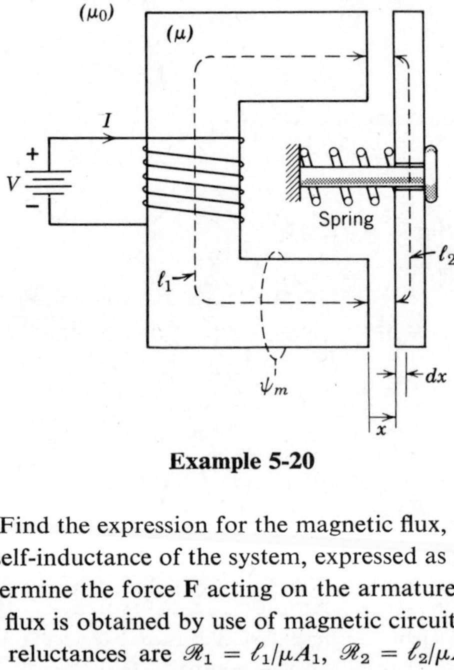

Example 5-20

respectively. (*a*) Find the expression for the magnetic flux, the magnetic energy stored, and the self-inductance of the system, expressed as functions of the gap width x. (*b*) Determine the force **F** acting on the armature.

(*a*) The core flux is obtained by use of magnetic circuit methods in Section 5-3. The reluctances are $\mathscr{R}_1 = \ell_1/\mu A_1$, $\mathscr{R}_2 = \ell_2/\mu A_2$, and that of the two air gaps in series is $2x/\mu_0 A_1$; whence

$$\psi_{m,\text{core}} = \frac{nI}{\mathscr{R}_1 + \mathscr{R}_2 + \dfrac{2x}{\mu_0 A_1}} \tag{1}$$

L is well approximated by the external self-inductance (5-88a). The core flux passes n times through the surface S_{ex} bounded by the coil, so that

$$L = \frac{n\psi_{m,\text{core}}}{I} = \frac{n^2}{\mathscr{R}_1 + \mathscr{R}_2 + \dfrac{2x}{\mu_0 A_1}} \tag{2}$$

The magnetic energy of the system is therefore

$$U_m = \tfrac{1}{2}LI^2 = \frac{1}{2}\frac{(nI)^2}{\mathscr{R}_1 + \mathscr{R}_2 + \dfrac{2x}{\mu_0 A}} \tag{3}$$

It is evident that increasing the air gap results in a decrease in the core flux, the self-inductance, and the stored energy.

(*b*) The force on the armature is obtained from (5-135a) or (5-135b); **F** has only an x component, as expected from the physical layout; thus

$$F_x = \frac{\partial U_m}{\partial x} = \frac{I^2}{2}\frac{\partial L}{\partial x} = \frac{-(nI)^2}{\mu_0 A\left[\mathscr{R}_1 + \mathscr{R}_2 + \dfrac{2x}{\mu_0 A}\right]^2} \tag{4}$$

The negative sign means F_x is in the direction of decreasing gap width x, corresponding to an increase in magnetic energy.

REFERENCES

ELLIOTT, R. S. *Electromagnetics*. New York: McGraw-Hill, 1966.

LORRAIN, P., and D. R. CORSON. *Electromagnetic Fields and Waves*, 2nd ed. San Francisco: W. H. Freeman, 1970.

REITZ, R., and F. J. MILFORD. *Foundations of Electromagnetic Theory*. Reading, Mass.: Addison-Wesley, 1960.

PROBLEMS

5-1. For the coaxial line of Figure 5-2, use Ampère's law to find **B** and **H** within the inner and outer conductors. Also, prove that the exterior fields are zero. Show that the boundary conditions are satisfied at all interfaces.

5-2. A ferromagnetic toroid of square cross-section surrounds a long round wire carrying the steady current I as shown. Assume the material has a permeability $\mu = 5000\mu_0$ and dimensions $a = 2$ cm, $b = 3$ cm, $h = 1$ cm.

(*a*) Use boundary conditions and symmetry to establish that the **H** field is the same in the iron as in the surrounding air.

(*b*) Compare **B** in the iron with that in the air, and compute the core flux ψ_m.

5-3. Show that the flux in the high-permeability core of Problem 5-2 is the same as that obtained from a one-turn coil carrying the current I and wrapped tightly or loosely about the core, provided that leakage is neglected. Show that the same is true for an n turn coil carrying the current I/n.

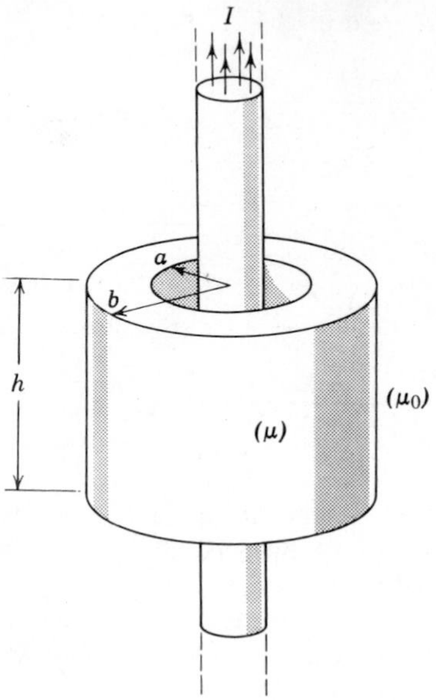

Problem 5-2

5-4. Show that the forces on two long, parallel wires separated a distance d and carrying the steady currents I_1 and I_2 are attractive forces for the currents in the same direction, repulsive, if in opposite directions. Use the Lorentz force law to prove that the force acting on every length ℓ of one wire in the presence of the **B** field of the other has the magnitude, for thin wires, $F = \mu_0 I_1 I_2 \ell / 2\pi d$.

5-5. The iron ring illustrated has the permeability $\mu = 6000\mu_0$, and dimensions $R = 10$ cm, $b = 2$ cm. The air gap is 0.5 cm wide and the 100 turn winding carries 6 A dc.

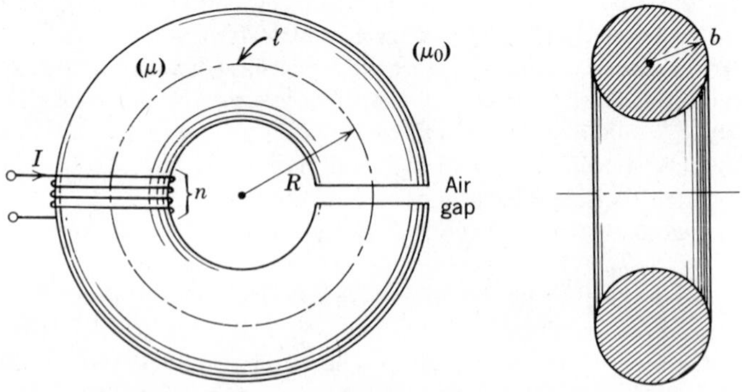

Problem 5-5

(*a*) Sketch a labeled analogous electric circuit, indicating analogous quantities ($nI \sim V$, etc.). Find the core flux, assuming R defines the mean path ℓ.

(*b*) Determine **B** and **H** along ℓ in the iron and the air gap. Verify that the boundary conditions are satisfied at the iron air interface in that gap.

(*c*) To what value will the flux increase if the gap is absent?

5-6. A magnetic circuit has a core like that of Figure 5-6(*b*). If $\ell_1 = 10$ cm, $A_1 = 2\text{ cm}^2$, $\ell_2 = 25$ cm, $A_2 = 5\text{ cm}^2$, $\mu_1 = \mu_2 = 8000\mu_0$ and the coil of 500 turns carries 0.1 A, find the magnetic flux in the core and B_{av} in each cross-section. (Sketch the analogous electric circuit.)

5-7. The magnetic circuit of Figure 5-7(*a*) has a 200-turn coil carrying 0.1 A dc. The core $\mu = 8000\mu_0$. The overall core dimensions are 4×7 cm by 1 cm thick, with apertures measuring 2×2 cm each, giving each leg a cross-sectional area of 1 cm^2. Neglecting leakage and using median paths centered in the core, compute the fluxes in the branches by use of the analogous dc electric circuit. Find B_{av} in each branch.

5-8. Repeat Problem 5-7 for the magnetic circuit of Figure 5-7(*b*), assuming the identical coil, current, and core dimensions used in that problem and making use of the analogous dc electric circuit.

5-9. Prove that $\nabla(1/R) = \mathbf{a}_R/R^2$, used in the derivation of (5-35). (Expand in rectangular coordinates, obtaining an identity.)

5-10. (*a*) By direct integration of the Biot–Savart law, show that the static **B** along the z axis of a thin, circular loop of wire of radius a in free space and carrying the current I is

$$\mathbf{B}(0, 0, z) = \mathbf{a}_z \frac{\mu_0 I a^2}{2(a^2 + z^2)^{3/2}}$$

if $z = 0$ is at the loop center.

(*b*) What is **B** at the center of the loop? Show that $\mathbf{A} = 0$ there. Why are these results not a contradiction of $\mathbf{B} = \nabla \times \mathbf{A}$?

(*c*) Show that the answer to part (*a*) checks with that obtained in Example 5-5, if the latter is specialized to the z axis.

5-11. (*a*) Show that **B** along the z axis of a thin, square loop of length a on a side and carrying I A dc in free space is

$$\mathbf{B}(0, 0, z) = \mathbf{a}_z \frac{\mu_0 I a^2}{2\pi\left[z^2 + 2\left(\frac{a}{2}\right)^2\right]^{1/2}} \frac{1}{\left[z^2 + \left(\frac{a}{2}\right)^2\right]}$$

if the z axis is normal to the plane of the loop, with $z = 0$ at its center. [*Hint:* Use results obtained in Example 5-4.]

(*b*) What is **B** at the center of the loop? Demonstrate that $\mathbf{A} = 0$ there. Why are these results not in contradiction to $\mathbf{B} = \nabla \times \mathbf{A}$?

5-12. A rectangular wire loop is placed a distance d from a long, straight wire

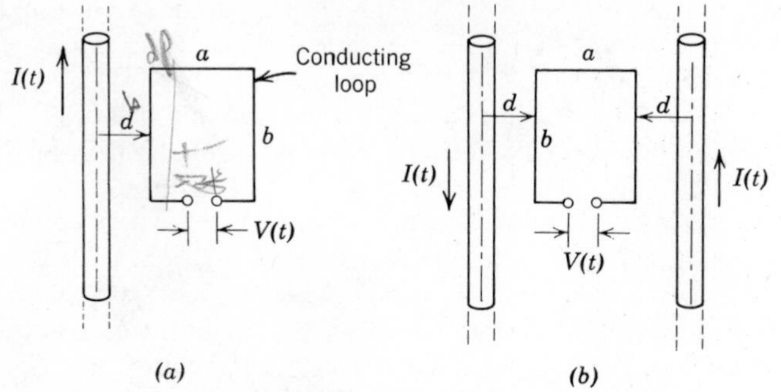

Problem 5-12

in air as seen in figure (*a*). The current in the wire is $I(t) = I_0 \sin \omega t$. What quasi-static **B** field is produced by $I(t)$?

(*a*) Find an expression for $V(t)$ across the gap.

(*b*) If $I_0 = 10$ A, $f = 100$ Hz, $d = 1$ cm, $a = 5$ cm, $b = 10$ cm, find $V(t)$, including the polarity.

(*c*) Repeat part (*a*) for the two-wire system of drawing (*b*). Utilize symmetry to permit a rapid assessment of the answer.

5-13. The current $I = I_0 \sin \omega t$ flows in the n turn toroid in free space, with the rectangular cross-section shown. The magnetic flux is assumed quasi-static.

(*a*) Use Ampère's law to find $\mathbf{B}(\rho, t)$. Evaluate the core flux at any instant. Sketch the flux (side view), showing its direction in relation to an assumed current sense.

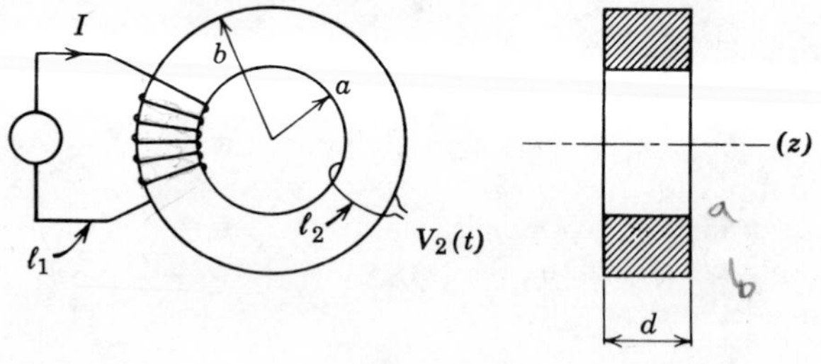

Problem 5-13

(*b*) Find an expression for $V_2(t)$ at the gap of the single-turn secondary coil shown.

(*c*) If $a = 5$ cm, $b = 8$ cm, $d = 3$ cm, $n_1 = 200$, $f = 100$ Hz, $\mu = 5000\mu_0$, compute the core flux and $V_2(t)$ for $I_0 = 0.1A$.

(*d*) Evaluate $V_2(t)$ for part (*c*), but with ten turns in the secondary coil.

5-14. A thin, rectangular wire loop with a small gap is located as illustrated near a long, straight wire.

(*a*) If $I = I_0 \cos \omega t$ A, determine the open-circuit gap voltage, including polarity. Sketch I and V versus t. (Assume loop dimensions small compared with a free-space wavelength of the associated **B** field at the frequency ω.)

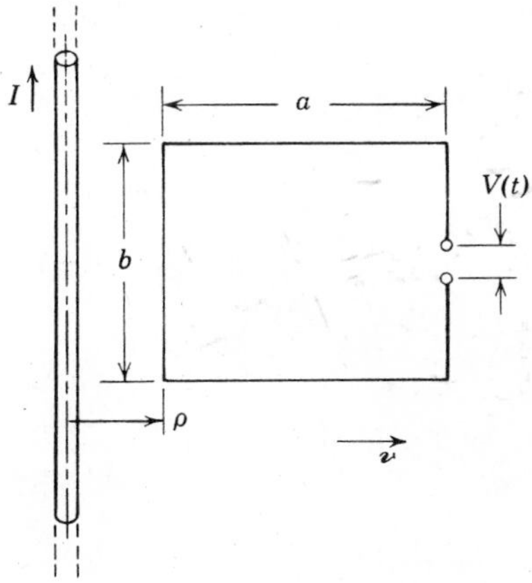

Problem 5-14

(*b*) With dc in the wire and the loop moved away from the wire with a velocity $\mathcal{v} = \mathbf{a}_\rho v_\rho$, what is $V(t)$ induced across the gap? (Include polarity.)
(*c*) Find $V(t)$ if the conditions of part (*a*) and the velocity $\mathcal{v}$ given in part (*b*) prevail simultaneously; i.e., assume $I = I_0 \cos \omega t$ in the straight wire.

5-15. For the generator in Figure 5-22, assume a radial **B** field of constant strength B_0 over a gap region extending $\pm 75°$ from the vertical, and zero over the remainder. Derive the generated voltage expression $V = 2da\omega B_0$ (for the time that the coil remains in the field B_0). Sketch the generated voltage versus the angle ϕ measured from the vertical.

5-16. (*a*) Using (5-77), show that the stored energy of the toroid in Problem 5-13 is $U_m = (\mu n^2 dI^2/4\pi) \ln (b/a)$.
(*b*) Use (*a*) to show that the self-inductance is

$$L = \frac{\mu n^2 d}{2\pi} \ln \frac{b}{a}$$

Compare this result with (5-91b).
(*c*) Find the magnetic energy of the toroid with dimensions as in Problem 5-13(*c*), assuming $I = 0.1$ A. What is the self-inductance?

5-17. (*a*) Find the approximate magnetic energy of the toroid with gap in Example 5-3, using the calculated flux to determine B_{av} over the cross-section. What percentage of the energy is stored in the air gap?
(*b*) Repeat the energy calculation of (*a*), but for no air gap.

(*c*) Use the results of (*a*) and (*b*) to determine the self-inductance with and without an air gap. Comment on the results.

5-18. Calculate, from the external flux linkage, the self-inductance of the toroidal coil with gap shown in Example 5-3. With no air gap, to what value (and by what factor) does the self-inductance increase?

5-19. Use results of Example 1-15 to find the magnetic energy stored in any length d of an infinitely long solenoid having n/d closely spaced turns per meter length in air. Show that the self-inductance per unit length is given by

$$\frac{L}{d} = \mu_0 \pi b^2 \left(\frac{n}{d}\right)^2$$

5-20. (*a*) Use the flux-linkage expression (5-88a) to determine the approximate self-inductance of an n turn winding about the toroidal core with the dimensions as specified. Neglect internal inductance and assume zero leakage flux and constant μ for the iron. In the flux

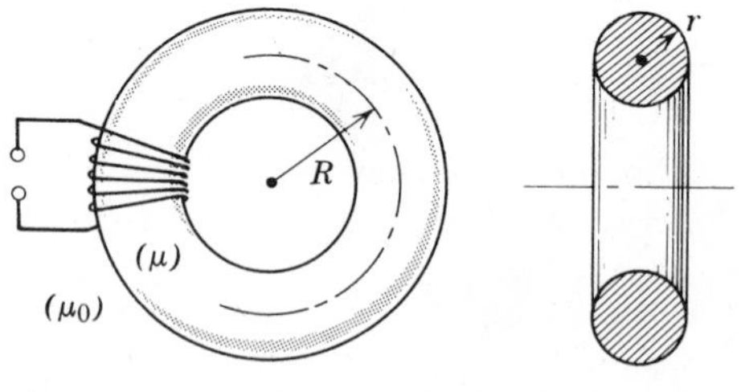

Problem 5-20

calculation, assume an average flux density B_{av} to be that along the core center line at R. Show that $L \cong \mu n^2 r^2 / 2R$.

(*b*) Rework part (*a*), except calculate L from the exact integration for the flux over the cross-section.

5-21. By use of the approximations (5-99) for elliptic integrals, prove (5-100) for the external inductance of a circular loop.

5-22. (*a*) Use the external and internal inductances (5-100) and (5-101) to obtain the approximate self-inductance of the circular loop in air

$$L = \frac{\mu b}{4} + \mu_0 b \left(\ell n \frac{8b}{a} - 2 \right)$$

in which μ denotes the conductor permeability.

(*b*) Use (*a*) to compute the low-frequency and high-frequency inductances of a loop of nonmagnetic wire of 2 mm diameter bent into a 2 cm diameter circle. Is the internal inductance negligible in the low-frequency case?

5-23. (*a*) Compute the self-inductance of a round loop of wire of 0.1 mm thickness bent into a circle of 0.5 cm radius.

(*b*) The loop is wound about a toroidal core of mean radius $R = 1.5$ cm,

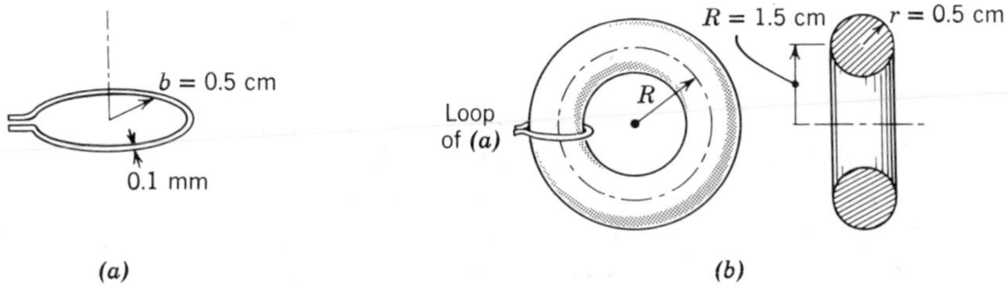

Problem 5-23

a cross-sectional radius $r = 0.5$ cm. Assuming zero leakage flux and $\mu = 1000\mu_0$ for the core, determine the factor by which L increases over its free-space value in (*a*).

5-24. If a small, toroidal, low-loss ferrite bead with a constant permeability μ is threaded into a circuit as shown, approximately how much self-inductance is added to the circuit? [*Hint:* Assume the bead dimensions

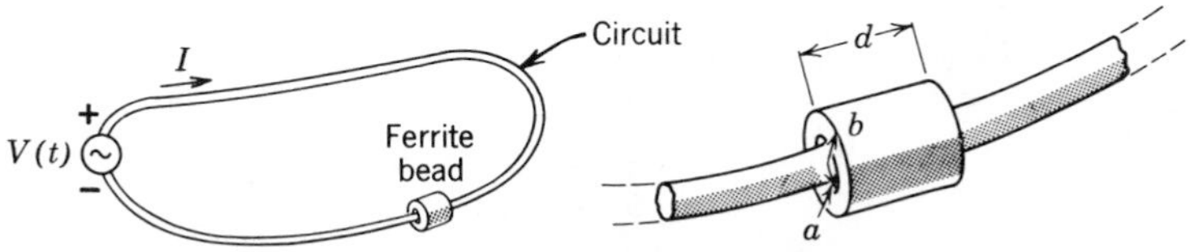

Problem 5-24

small compared to the overall circuit dimensions, such that the **H** field with and without the bead are essentially the same. To find the fields inside the bead, assume the wire in the neighborhood to be essentially straight, i.e., see Problem 5-2.]

5-25. (*a*) Use (5-71) to obtain the energy of the circuit shown

$$U_m = \tfrac{1}{2}\int_S \psi_m(\ell')\, dI \tag{1}$$

in which dI is the current in a typical filament ℓ' within the conductor, S is the conductor cross-section, and $\psi_m(\ell')$ is the magnetic flux linked by the filament. Show that the self-inductance is

$$L = \frac{1}{I^2}\int_S \psi_m(\ell')\, dI \tag{2}$$

if I is the circuit current. [*Hint:* With the circuit divided into filaments ℓ' carrying currents $dI = J\,ds$, the integrand of (5-71) is written $\mathbf{J}\cdot\mathbf{A}\, d\ell'\, ds = J\, ds(\mathbf{A}\cdot d\boldsymbol{\ell}')$, the integration being taken over all such filaments.]

(*b*) Compare the energy expression (1) with (5-85). In particular, note how the integration surfaces S differ, and explain why one of these

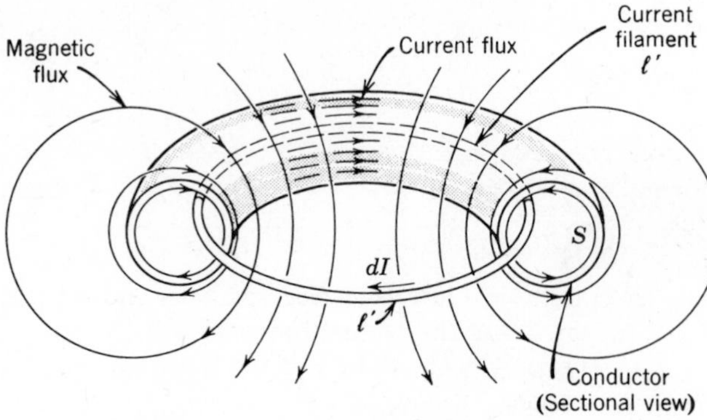

Problem 5-25

results leads to a decomposition into internal and external inductances, while the other does not.

(*c*) Show in the high-frequency limit that the self-inductance expression (2) reduces to $L = \psi_m/I$. Compare it with (5-88a).

5-26. Prove (5-107) for the power delivered to coupled circuits, beginning with (5-65).

5-27. Starting with (5-110) for the magnetic energy of coupled circuits, show that it reduces to (5-111) for linear circuits.

5-28. Use (5-121) to deduce the Neumann formula for two thin circuits in free space

$$M = \int_{\ell_1}\int_{\ell_2} \frac{\mu_0\, d\boldsymbol{\ell}' \cdot d\boldsymbol{\ell}}{4\pi R}$$

Sketch a pair of circuits with labeling appropriate to the use of this integral.

5-29. Use the Neumann formula for thin circuits given in Problem 5-28 to derive the mutual inductance between two coaxial, circular loops with radii a and b, and separated by the distance d in free space as shown, obtaining

$$M = \mu_0 \sqrt{ab}\left[\left(\frac{2}{k} - k\right)K(k) - \frac{2}{k}E(k)\right]$$

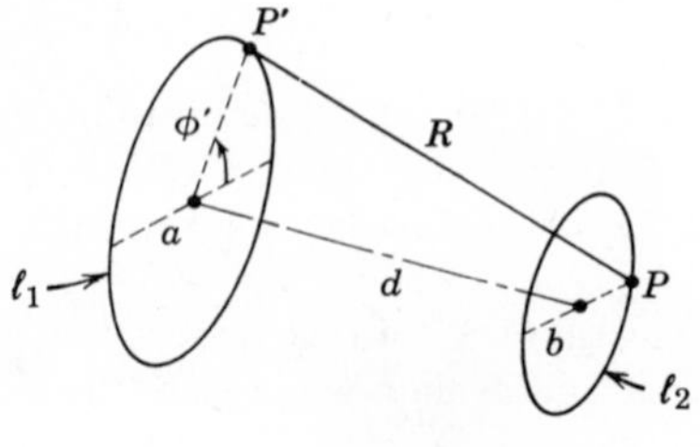

Problem 5-29

in which $K(k)$ and $E(k)$ are the complete elliptic integrals (5-97), and

$$k = 2\sqrt{\frac{ab}{d^2 + (a + b)^2}}$$

[*Hint:* Proceed along lines suggested by Example 5-13, noting that the distance between a source point P' and a field point P is

$$R = \sqrt{d^2 + a^2 + b^2 - 2\,ab\cos\phi'}.]$$

5-30. Given a fixed circular loop ℓ_1 in free space as shown, suggest how to maximize and minimize the mutual inductance with respect to a second loop ℓ_2, by the appropriate rotation of the latter for the three illustrated cases. Briefly explain.

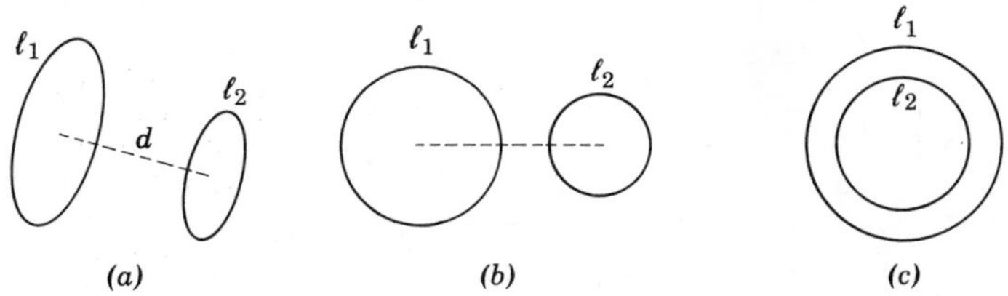

Problem 5-30. **(*a*) Coaxial circuits. (*b*) Coplanar circuits. (*c*) Coaxial and coplanar circuits.**

Which system provides the maximum mutual inductance? Explain.

5-31. (*a*) Use (5-123b) to derive an approximate expression for M between the coaxial loops shown in Problem 5-29, assuming the radius b small compared to a, permitting the assumption of an essentially uniform **B** over the smaller coil. Show that

$$M \cong \frac{\mu_0 \pi (ab)^2}{2(a^2 + d^2)^{3/2}}$$

[*Hint:* Use the solution to Problem 5-10 for **B** along the loop axis.]

(*b*) Use the result of (*a*) to evaluate M between two loops in air, if $a = 5$ cm, $b = 1$ cm, and for the two cases: $d = 0$ (coaxial and coplanar loops) and $d = 5$ cm. Assume wire diameter $= 1$ mm.

(*c*) Evaluate the coupling coefficient k for the two results of part (*b*).

5-32. A second coil with 200 turns is wound on the iron core in Example 5-3. Use flux linkage methods to determine the self-inductance of each winding, and assuming zero leakage, find M. [*Note:* M can be found by flux linkage techniques or from the assumption of unity coupling coefficient between the windings.]

5-33. Use the flux linkage method to find M between the long wire and rectangular loop in Problem 5-12. Explain why this answer is nearly correct if the long wire is deformed into a closed circuit, assuming the rectangular loop small and in close proximity to the larger circuit.

5-34. (*a*) Use the results of Example 5-11 to find M between a rectangular loop of wire in air and a parallel-wire line carrying opposing currents $I(t)$. The loop is assumed to have the dimensions of the closed path ℓ shown in (*a*). (Call the parallel-wire line the circuit ℓ_1.)

(*b*) Find M if $a = d = b = h/2 = 2$ cm. Use M to deduce the magnetic flux intercepted by the loop, if the parallel-wire line carries 1 A dc. Use M to find the open-circuit gap voltage developed by the loop if $I(t) = 1 \sin \omega t$ A flows in the parallel-wire line, assuming $f = 1000$ Hz.

5-35. A split toroid with a permeability μ and the dimensions shown has n turns wound about it. It is clamped firmly (with negligible air gap) about a more or less straight-wire region of a much larger circuit carrying the sinusoidal current $I_1(t) = I_m \sin \omega t$.

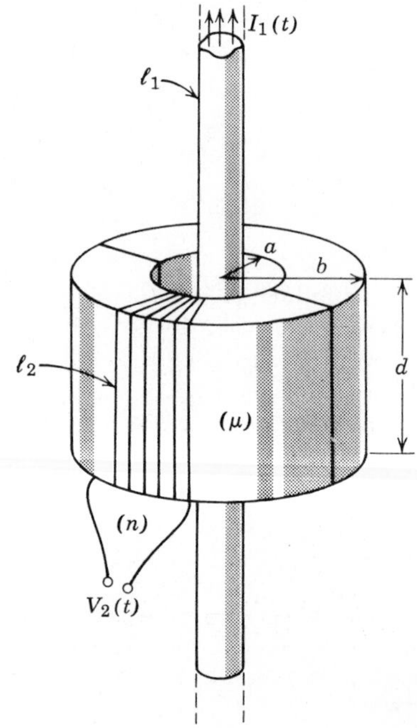

Problem 5-35

(*a*) Derive an expression for M between ℓ_1 and the n-turn coil ℓ_2.

(*b*) If the core permeability is $5{,}000\mu_0$ and $a = 1$ cm, $b = 2$ cm, $d = 2$ cm, $n = 100$, $I_1(t) = 10 \sin \omega t$ A, $f = 60$ Hz, what is M, and how much open-circuit voltage is developed by the winding? What is the effect of increasing the frequency by a factor of ten? [*Hint:* Use either $V_2(t) = -d\psi_{12}/dt$ from Section 5-12 or make use of $\psi_{12} = MI_1$ from (5-123b), yielding the equivalent result $V_2(t) = -M\, dI_1/dt$.]

5-36. Use the Faraday law to prove the Kirchhoff voltage expression (5-128b) in detail, for the circuit ℓ_2 of Figure 5-34(a).

5-37. (a) Use the coupled equations (5-128) converted to time-harmonic form to prove that the impedance seen by a generator V in Figure 5-35 is

$$\hat{Z}_{ab} = \hat{Z}_1 + \frac{\omega^2 M^2}{\hat{Z}_2} = \hat{Z}_1 + \frac{\omega^2 k^2 L_1 L_2}{\hat{Z}_2}$$

if $\hat{Z}_1$ and $\hat{Z}_2$ denote the series impedances of the circuits ℓ_1 and ℓ_2 taken alone.

(b) If a coil of inductance L_1 and negligible resistance (high Q) is coupled to another high Q, shorted coil of inductance L_2 as shown in (a), show that the input impedance becomes the inductive result: $\hat{Z}_{ab} \cong j\omega L_1 (1 - k^2)$. What happens to the input impedance when the second coil is coupled very closely?

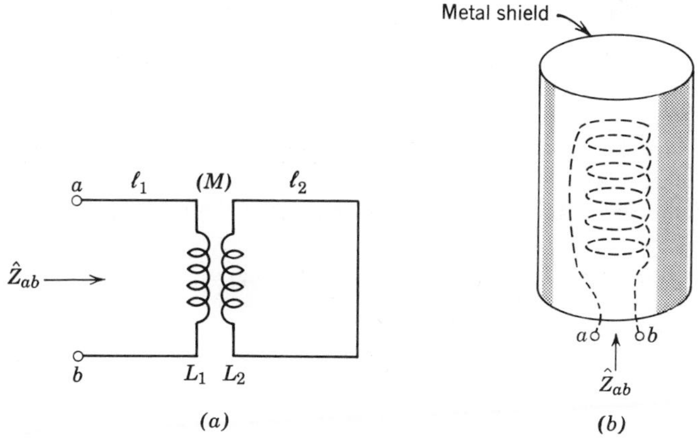

Problem 5-37

(c) A coil inside a metal shield as shown in (b) can be regarded as coupled into a single-turn, shorted secondary coil. (Describe the induced current in the shield can.) How does a closely fitted shield affect $\hat{Z}_{ab}$, compared to a somewhat larger shield?

5-38. The split-toroid arrangement of Problem 5-35 is modified into the clamp-on ammeter shown, the coil ℓ_2 consisting of a few turns of low-resistance wire connected to an ac ammeter of low or negligible resistance. This system is used to meter the amplitude of $\hat{I}_1$ in a circuit ℓ_1 without breaking into that circuit. Analyze this arrangement, expressing the ammeter reading as a function of $\hat{I}_1$. Under what conditions is the scale calibration of the ammeter relatively frequency independent?

5-39. In Example 5-20, let $n = 100$ turns, $I = 100$ mA, $\ell_1 = 8$ cm, $\ell_2 = 4$ cm,

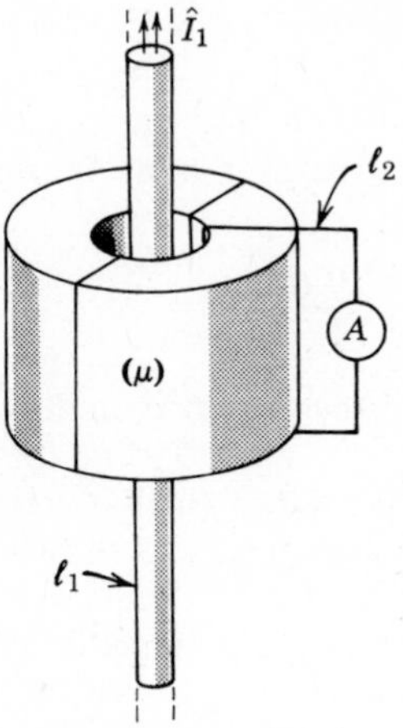

Problem 5-38

$A_1 = 4\text{ cm}^2$, $A_2 = 1\text{ cm}^2$, gap width $x = 1$ mm, and $\mu = 5000\mu_0$. Find the core flux and the force on the armature at the specified gap width; repeat with the gap closed.

5-40. A magnetic relay with a rotating armature like that of Figure 5-36(*b*) has mean paths ℓ_1, ℓ_2 and cross-sectional areas A_1, A_2 in the iron of permeability μ, as specified for the relay of Example 5-20. With an air gap established by the small angle $\theta = x/\ell_2$ (with x the mean air gap), find expressions for the magnetic flux, stored energy, self-inductance, and torque acting on the armature, all in terms of the small angle θ.

CHAPTER

6

Normal-Incidence Wave Reflection and Transmission at Plane Boundaries

This chapter is concerned with boundary-value problems in one space dimension. The reflection from a perfectly conducting plane upon which a uniform plane wave is incident is considered first. Replacing the perfect conductor with a lossy dielectric extends the problem into a two-region system, for which the wave transmitted into the dielectric is also of interest. The definition of wave impedance and reflection coefficient permits a systematic analysis of the multiple-layer problem, dealing with the reflected and transmitted waves excited by a normally incident wave. Next a development of the Smith chart is discussed, with applications to the foregoing problems. The chapter concludes with the concept of standing waves and standing-wave ratio for a lossless region.

6-1 Boundary-Value Problems

A boundary-value problem in electromagnetics is one involving two or more regions (separated by one or more interfaces) for which solutions are desired such that (*a*) Maxwell's equations are satisfied by those field solutions in each of the regions, and (*b*) the boundary conditions discussed in Chapter 3 are satisfied at the interfaces. Examples are illustrated in Figure 6-1. In Figure 6-1(*a*) is shown a rudimentary boundary-value problem: a plane wave normally incident upon a perfect conductor, yielding a reflected wave.

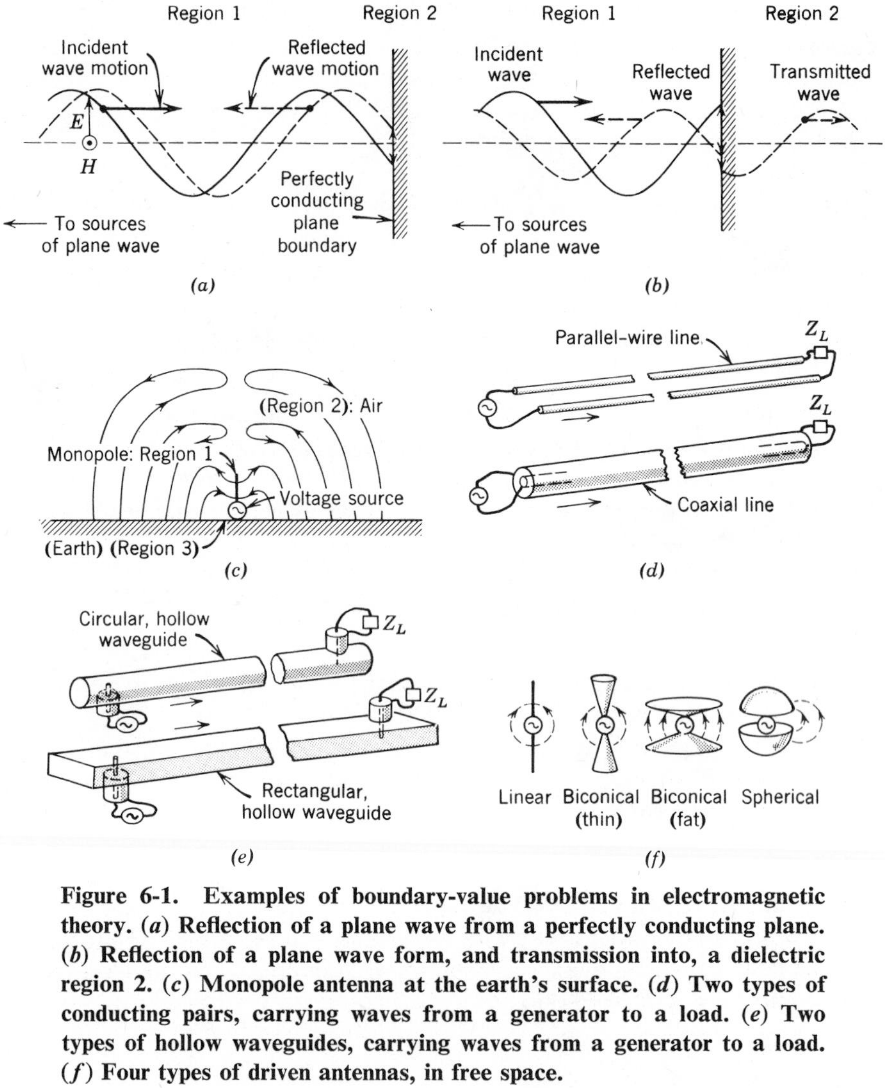

Figure 6-1. Examples of boundary-value problems in electromagnetic theory. (*a*) Reflection of a plane wave from a perfectly conducting plane. (*b*) Reflection of a plane wave form, and transmission into, a dielectric region 2. (*c*) Monopole antenna at the earth's surface. (*d*) Two types of conducting pairs, carrying waves from a generator to a load. (*e*) Two types of hollow waveguides, carrying waves from a generator to a load. (*f*) Four types of driven antennas, in free space.

In (*b*) is a two-region system separated by a plane interface. A given plane wave traveling in region 1 leads to the additional waves shown, such that the boundary conditions at the interface are satisfied. In these problems, the given incident wave is presumed to originate from an appropriate electromagnetic source (a generator) at the far left.

Whenever the source of electromagnetic energy is included in a boundary-value problem, one may say he is discussing the *complete* boundary-value problem. If the reflected wave does not couple significantly with the generator, a discussion of the complete problem may not be necessary. In Figure 6-1(c) is shown a three-region problem consisting of a driven monopole antenna source transmitting electromagnetic energy into the surrounding space (region 2) and into the earth (region 3). In Figure 6-1(d) and (e) are shown other complete boundary-value problems involving generators (sources) driving waves down one- or two-conductor systems (waveguides or transmission lines) to a load at the far end. Systems such as these are considered in Chapters 8–10.

6-2 Reflection from a Plane Conductor at Normal Incidence

A fundamental boundary-value problem of electromagnetics involves the reflection of a normally incident uniform plane wave from a plane perfect conductor. Assuming a plane of infinite extent avoids edge (diffraction) effects, and with the simplification of normal incidence, the problem is reduced to two dimensions (t and z). The geometry is shown in Figure 6-2. The sources of the incident wave are assumed at the far left in lossless region 1. Assuming x polarization, the incident wave is given in the real-time domain by (2-121)

$$E_x^+(z, t) = E_m^+ \cos(\omega t - \beta z) \text{ V/m} \tag{6-1}$$

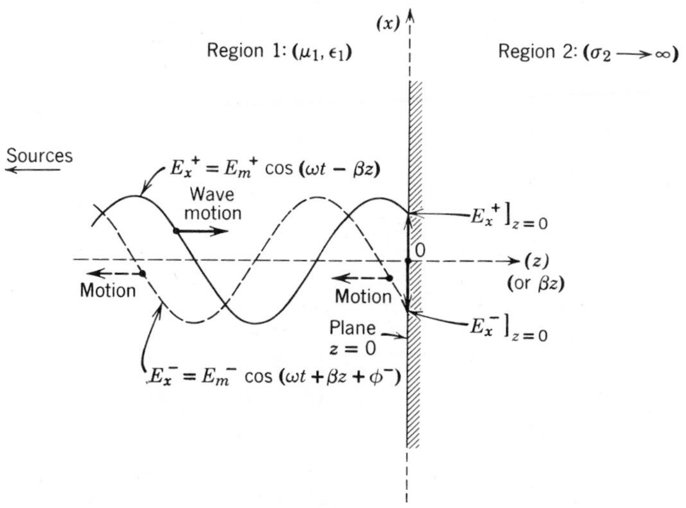

Figure 6-2. Reflection of normally incident plane wave from perfect conducting plane.

letting the phase angle $\phi^+ = 0$ for convenience, but the incident wave (6-1) alone cannot satisfy the tangential field boundary conditions (3-72) and (3-79) at the interface. One must add a reflected wave solution, its effect being such as to cancel the incident field everywhere on the perfect conductor at every instant t. This occurs only if the second solution has the same frequency and if its equiphase surfaces are parallel to the wall. The only other independent solution of Maxwell's equations which meets these requirements is the negative z traveling wave solution of (2-119)

$$E_x^-(z, t) = E_m^- \cos(\omega t + \beta z + \phi^-) \tag{6-2}$$

The unknown amplitude E_m^- and phase ϕ^- are found by applying the boundary condition (3-79). The details are more readily carried out in the complex time-harmonic form; hence, the sum of (6-1) and (6-2), in complex notation, takes the form of (2-115)

$$\begin{aligned} \hat{E}_x(z) &= \hat{E}_x^+(z) + \hat{E}_x^-(z) \\ &= \hat{E}_m^+ e^{-j\beta z} + \hat{E}_m^- e^{j\beta z} \text{ V/m} \end{aligned} \tag{6-3}$$

The boundary condition (3-79), that the total tangential electric field must vanish at the surface of the perfect conductor, is written $\hat{E}_x(0) = 0$; so (6-3) becomes $0 = \hat{E}_m^+ + \hat{E}_m^-$, whence

$$\hat{E}_m^+ = -\hat{E}_m^- \tag{6-4}$$

Thus total reflection occurs, with the reflected wave amplitude equalling the negative of the incident wave. Inserting (6-4) into (6-3), the total electric field at *any* location to the left of the conducting plane becomes

$$\hat{E}_x(z) = \hat{E}_m^+[e^{-j\beta z} - e^{j\beta z}] = -j2\hat{E}_m^+ \sin \beta z \tag{6-5}$$

a result with a wave amplitude $2\hat{E}_m^+$, just twice that of the incident wave. The dependence of (6-5) on z is unlike the traveling wave nature of either wave constituent in (6-3). It has instead a standing wave character, in view of the factor $\sin \beta z$. A graphical space-time sketch of this standing wave is facilitated upon converting (6-5) to its *real-time* form by use of (2-74). Assuming the *real* amplitude E_m^+, one obtains

$$\begin{aligned} E_x(z, t) &= \text{Re}\,[\hat{E}_x(z)e^{j\omega t}] = \text{Re}\,[-j2E_m^+ \sin \beta z\, e^{j\omega t}] \\ &= \text{Re}\,[e^{-j90^\circ}2E_m^+ \sin \beta z\, e^{j\omega t}] = 2E_m^+ \sin \beta z \sin \omega t \end{aligned} \tag{6-6}$$

A sketch depicting the dependence on z at successive t is shown in Figure 6-3(a).

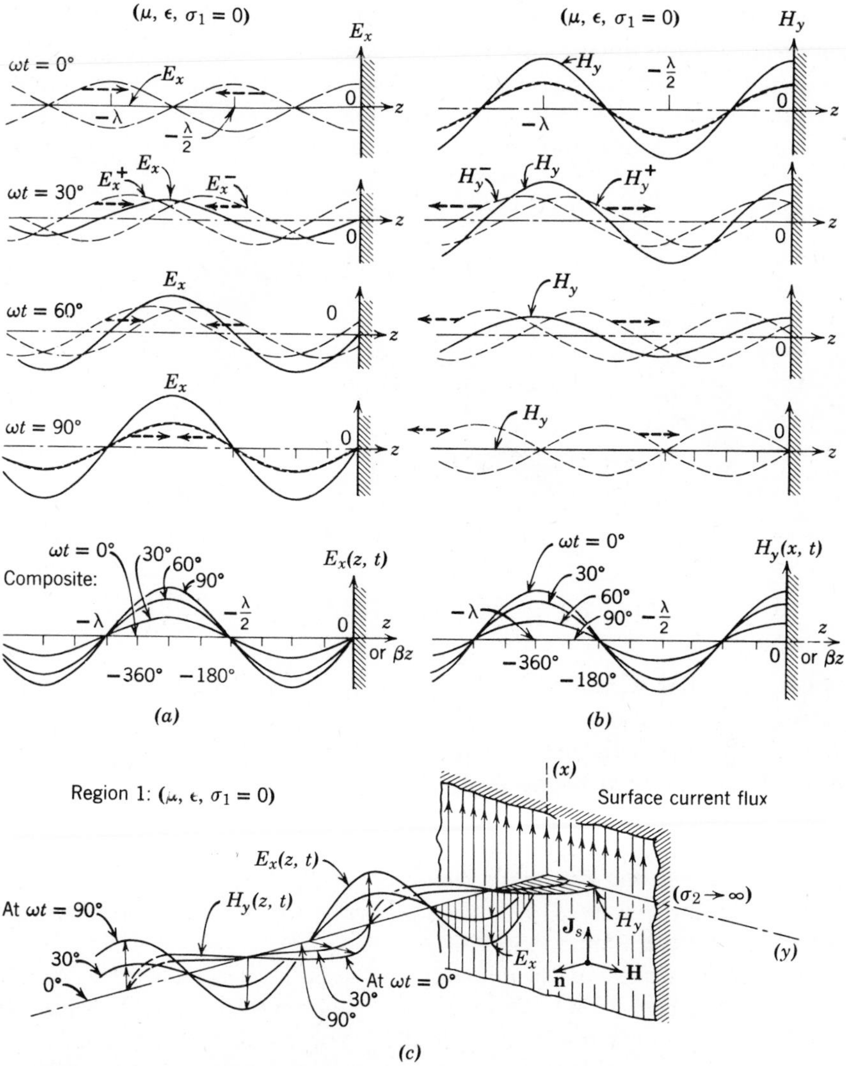

Figure 6-3. Standing waves resulting from a plane wave normally incident upon a perfect conductor. (*a*) Incident, reflected, and total electric fields. (*b*) Incident, reflected, and total magnetic fields. (*c*) Showing the vector electric and magnetic fields of (*a*) and (*b*).

The total magnetic field accompanying the electric field (6-5) is obtained directly by substituting (6-5) into Maxwell's curl relation (2-108). This was, in effect, already done in Section 3-6, however, in which it was shown in (3-98b) that magnetic field traveling waves are related to corresponding electric fields by the intrinsic wave impedance. Hence, to (6-3) correspond the two terms of the magnetic field

$$\begin{aligned}\hat{H}_y(z) &= \hat{H}_y^+(z) + \hat{H}_y^-(z)\\ &= \frac{\hat{E}_m^+}{\eta} e^{-j\beta z} - \frac{\hat{E}_m^-}{\eta} e^{j\beta z} \text{ A/m}\end{aligned} \tag{6-7}$$

in which $\eta \equiv (\mu/\epsilon)^{1/2}$ is, from (3-99a), the intrinsic wave impedance of the lossless region. If (6-4) is inserted into (6-7), the complex *magnetic field* reduces to

$$\hat{H}_y(z) = \frac{\hat{E}_m^+}{\eta}[e^{-j\beta z} + e^{j\beta z}] = \frac{2\hat{E}_m^+}{\eta}\cos\beta z \tag{6-8}$$

The real-time form of (6-8) (with $\hat{E}_m^+$ taken to be the pure real E_m^+) becomes

$$H_y(z, t) = \text{Re}\,[\hat{H}_y(z)e^{j\omega t}] = \frac{2E_m^+}{\eta}\cos\beta z\cos\omega t \tag{6-9}$$

another standing wave. It is plotted in Figure 6-3(*b*) for comparison with the electric field. A *space phase shift* of 90° occurs between the peaks of the electric and magnetic field standing waves, with the maximum magnetic intensity appearing at the perfectly conducting surface $z = 0$.

The magnetic field (6-9) cannot fall abruptly to zero upon passing into the interior of the perfect conductor without inducing an electric *surface current*, predictable from the boundary condition (3-72). Observe that the induced surface current density $\mathbf{J}_s$ is x directed and cophasal over the conducting plane as shown in Figure 6-3(*c*).

One can see a close physical analogy between the electromagnetic standing waves of Figure 6-3 and the mechanical standing waves of displacements and tensions along a transversely oscillating string anchored at one end† as shown in Figure 6-4(*a*). In (*b*) is shown another example of standing waves resulting from the reflection of electromagnetic waves from a conducting plane. Although the waves emanating from the horn are essentially spherical in the vicinity of the horn, at suitable distances away and over a limited transverse region they are very nearly plane waves, so that the solutions (6-5) and (6-8) are applicable in the vicinity of the plane reflector. If sufficient

† For example, see Halliday, D., and R. Resnick. *Physics for Students of Science and Engineering*. New York: Wiley, 1962, p. 412.

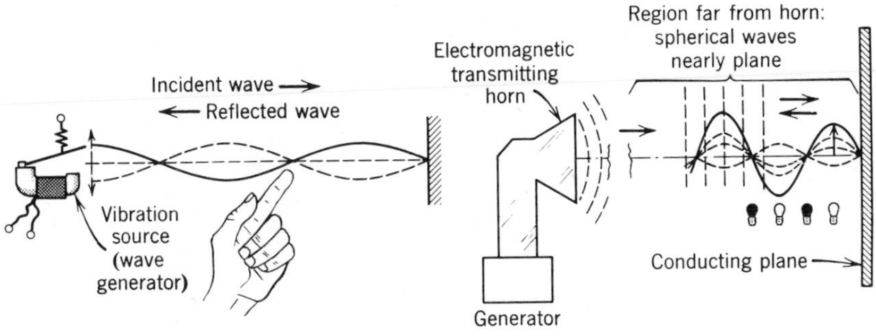

(a) (b)

Figure 6-4. Experiments involving standing waves. (*a*) Standing waves on a string connected to a rigid body and a wave generator. Null locations are checked visually. (*b*) Electromagnetic standing waves near conducting plane. Waves may originate from a distant source as shown. A neon bulb reveals maxima and nulls.

power is available, a small neon bulb might be used for detecting the nulls in the standing waves, yielding a rough measure of wavelength.

6-3 Wave Solutions Using Real-Time and Complex Fields

The space-time behavior of electromagnetic fields is best represented graphically in the real-time domain. Their representation by complex functions possesses important advantages, however, in view of the simplifications occurring in the mathematics. A comparison of the solution details of the wave problem of the preceding section worked out in both the complex and real-time domains reveals some of the advantages of the complex notation.

A comparison of the two methods is tabulated in the following.

In complex notation	In real notation
(1) Assume the forward and backward traveling plane wave fields:	
$\hat{E}_x^+(z) = \hat{E}_m^+ e^{-j\beta z}$ with $\hat{E}_m^+ = E_m^+$	$E_x^+(z, t) = E_m^+ \cos(\omega t - \beta z)$
$\hat{E}_x^-(z) = \hat{E}_m^- e^{j\beta z}$ where $\hat{E}_m^- = E_m^- e^{j\phi^-}$	$E_x^-(z, t) = E_m^- \cos(\omega t + \beta z + \phi^-)$

(2) The corresponding magnetic fields are:

$$\hat{H}_y^+(z) = \frac{E_m^+}{\eta} e^{-j\beta z} \qquad H_y^+(z, t) = \frac{E_m^+}{\eta} \cos(\omega t - \beta z)$$

$$\hat{H}_y^-(z) = -\frac{\hat{E}_m^-}{\eta} e^{j\beta z} \qquad H_y^-(z, t) = -\frac{E_m^-}{\eta} \cos(\omega t + \beta z + \phi^-)$$

(3) The total fields are the sums:

$$\hat{E}_x(z) = E_m^+ e^{-j\beta z} + \hat{E}_m^- e^{j\beta z} \qquad E_x(z, t) = E_m^+ \cos(\omega t - \beta z) + E_m^- \cos(\omega t + \beta z + \phi^-)$$

$$\hat{H}_y(z) = \frac{E_m^+}{\eta} e^{-j\beta z} - \frac{\hat{E}_m^-}{\eta} e^{j\beta z} \qquad H_y(z, t) = \frac{E_m^+}{\eta} \cos(\omega t - \beta z) - \frac{E_m^-}{\eta} \cos(\omega t + \beta z + \phi^-)$$

(4) Apply the boundary condition ($E_{t1} = 0$) from (3-79), yielding:

$$0 = E_m^+ + \hat{E}_m^- \qquad 0 = E_m^+ \cos \omega t + E_m^- \cos(\omega t + \phi^-)$$

whence:

$$\hat{E}_m^- = -E_m^+$$

(to make $\hat{E}_m^-$ pure real, the negative of E_m^+)

(true for all t only if $\phi^- = 0$ and $E_m^- = -E_m^+$)

(5) Substituting into the total field expressions obtains:

$$\hat{E}_x(z) = E_m^+[e^{-j\beta z} - e^{j\beta z}] = -j2E_m^+ \sin \beta z$$

$$E_x(z, t) = E_m^+[\cos(\omega t - \beta z) - \cos(\omega t + \beta z)] = 2E_m^+ \sin \beta z \sin \omega t$$

$$\hat{H}_y(z) = \frac{E_m^+}{\eta}[e^{-j\beta z} + e^{j\beta z}] = \frac{2E_m^+}{\eta} \cos \beta z$$

$$H_y(z, t) = \frac{E_m^+}{\eta}[\cos(\omega t - \beta z) + \cos(\omega t + \beta z)] = \frac{2E_m^+}{\eta} \cos \beta z \cos \omega t$$

These solutions correspond to (6-5) and (6-8).

These solutions correspond to (6-6) and (6-9)

The advantages of working with exponential solutions rather than the more cumbersome trigonometric expressions are evident upon noting that the variable t is absent when the complex notation is adopted, whereas ωt terms must be retained in the real forms of the fields.

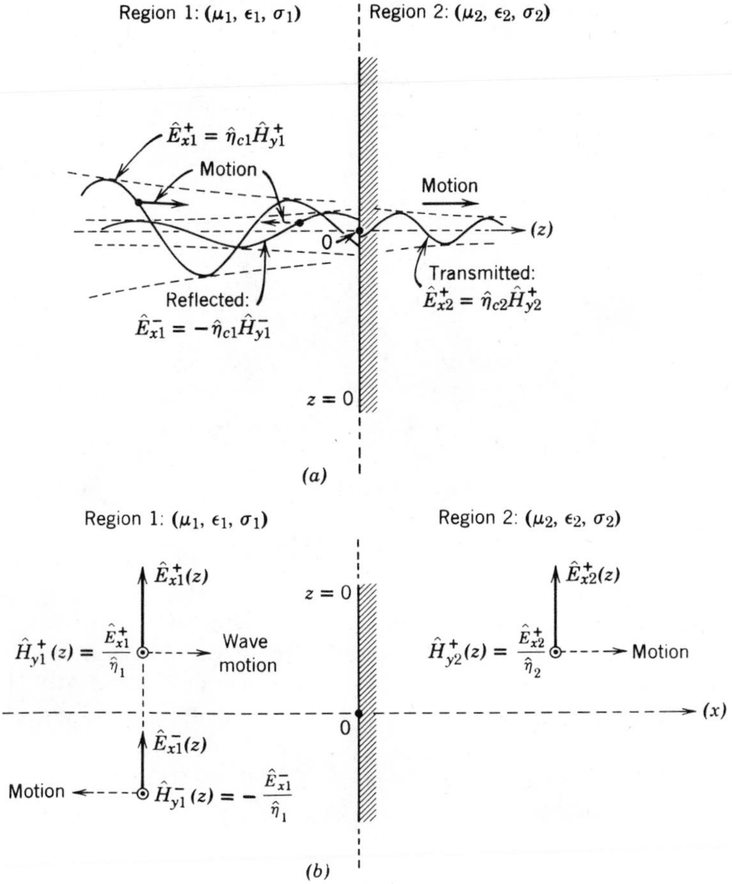

Figure 6-5. Plane wave normally incident upon an interface separating two lossy regions. (*a*) Incident and reflected waves in region 1, transmitted wave in region 2. (*b*) Vector representations denoting the fields of (*a*).

6-4 Two-Region Reflection and Transmission

The wave problem of Figure 6-2 can be generalized by assuming region 1 conductive ($\sigma_1 \neq 0$) instead of lossless, and region 2 with a finite conductivity instead of being a perfect reflector. The system is shown in Figure 6-5. An *incident*, plane wave originating from the far left is given by the positive z traveling wave terms of (3-91b) and (3-98c)

$$\hat{E}_{x1}^{+}(z) = \hat{E}_{m1}^{+}e^{-\gamma_1 z} \qquad \hat{H}_{y1}^{+}(z) = \frac{\hat{E}_{m1}^{+}}{\hat{\eta}_1}e^{-\gamma_1 z} \tag{6-10}$$

wherein $\hat{\eta}_1$ is specified by (3-99a) for conductive region 1

$$\hat{\eta} = \sqrt{\frac{\mu}{\hat{\epsilon}}} = \frac{\sqrt{\frac{\mu}{\epsilon}}}{\left[1 + \left(\frac{\sigma}{\omega\epsilon}\right)^2\right]^{1/4}} e^{j(1/2)\,\mathrm{arc\,tan}\,(\sigma/\omega\epsilon)}\ \Omega \tag{6-11}$$

or equivalently by (3-111)

$$\hat{\eta} = \frac{\sqrt{\frac{\mu}{\epsilon}}}{\left[1 + \left(\frac{\epsilon''}{\epsilon'}\right)^2\right]^{1/4}} e^{j(1/2)\,\mathrm{arc\,tan}\,(\epsilon''/\epsilon')}\ \Omega \tag{6-12}$$

The propagation constant of region 1 is γ_1, given by (3-89)

$$\gamma = \alpha + j\beta\ \mathrm{m}^{-1} \tag{6-13}$$

in which α and β are obtained from (3-90a, b), or equivalently from (3-109) and (3-110).

The continuity of the tangential fields across the interface in Figure 6-5(*a*) gives rise to another plane wave at the same frequency in region 2. This wave is not sufficient to satisfy the boundary conditions (3-71) and (3-79) at the interface, however. One more wave, *reflected* in region 1, is required if the boundary conditions are to be met. The three waves are shown in Figure 6-5(*a*) in real-time, and as complex vectors in Figure 6-5(*b*). Thus, in region 1, the *reflected* wave is required as follows:

$$\hat{E}_{x1}^{-}(z) = \hat{E}_{m1}^{-} e^{\gamma_1 z} \qquad \hat{H}_{y1}^{-}(z) = -\frac{\hat{E}_{m1}^{-}}{\hat{\eta}} e^{\gamma_1 z} \tag{6-14}$$

in which $\hat{\eta}_1$ and γ_1 are given by (6-11) and (6-13). In region 2, the transmitted wave is

$$\hat{E}_{x2}^{+}(z) = \hat{E}_{m2}^{+} e^{-\gamma_2 z} \qquad \hat{H}_{y2}^{+}(z) = \frac{\hat{E}_{m2}^{+}}{\hat{\eta}_2} e^{-\gamma_2 z} \tag{6-15}$$

with expressions for $\hat{\eta}_2$ and γ_2 similar to (6-11) and (6-13) being applicable. No reflected wave can exist in region 2, because that region is infinite in extent towards the right in Figure 6-5, while the only sources of the fields are to the far left in region 1.

Satisfying the boundary conditions at the interface in Figure 6-5(*b*) requires setting the total tangential fields equal to each other at $z = 0$. In region 1, the total electric and magnetic fields are given by the sums of (6-10) and (6-14)

$$\hat{E}_{x1}(z) = \hat{E}_{m1}^{+} e^{-\gamma_1 z} + \hat{E}_{m1}^{-} e^{\gamma_1 z} \qquad \hat{H}_{y1}(z) = \frac{\hat{E}_{m1}^{+}}{\hat{\eta}_1} e^{-\gamma_1 z} - \frac{\hat{E}_{m1}^{-}}{\hat{\eta}_1} e^{\gamma_1 z} \tag{6-16}$$

In region 2, they are simply (6-15). The boundary condition (3-79) requires the equality of the electric fields of (6-15) and (6-16) at $z = 0$; i.e.,

$$[\hat{E}_{m1}^{+} e^{-\gamma_1 z} + \hat{E}_{m1}^{-} e^{\gamma_1 z} = \hat{E}_{m2}^{+} e^{-\gamma_2 z}]_{z=0} \tag{6-17}$$

obtaining

$$\hat{E}_{m1}^{+} + \hat{E}_{m1}^{-} = \hat{E}_{m2}^{+} \tag{6-18}$$

The other boundary condition (3-71) requires the continuity of the magnetic fields there, obtaining

$$\frac{\hat{E}_{m1}^{+}}{\hat{\eta}_1} - \frac{\hat{E}_{m1}^{-}}{\hat{\eta}_1} = \frac{\hat{E}_{m2}^{+}}{\hat{\eta}_2} \tag{6-19}$$

The linear results (6-18) and (6-19) involve the known impedances $\hat{\eta}_1$ and $\hat{\eta}_2$ of the regions, as well as the complex amplitudes of the incident, the reflected, and the transmitted waves. Assuming the incident wave to be *given* ($\hat{E}_{m1}^{+}$ is known), the other amplitudes are obtained from the simultaneous solution of (6-18) and (6-19). Rearranging them with $\hat{E}_{m1}^{+}$ on the right yields

$$\hat{E}_{m1}^{-} - \hat{E}_{m2}^{+} = -\hat{E}_{m1}^{+}$$

$$\frac{\hat{E}_{m1}^{-}}{\hat{\eta}_1} + \frac{\hat{E}_{m2}^{+}}{\hat{\eta}_2} = \frac{\hat{E}_{m1}^{+}}{\hat{\eta}_1}$$

Their simultaneous solution obtains the complex amplitude of the *reflected* wave

$$\hat{E}_{m1}^{-} = \hat{E}_{m1}^{+} \frac{\hat{\eta}_2 - \hat{\eta}_1}{\hat{\eta}_2 + \hat{\eta}_1} \tag{6-20}$$

Similarly, the transmitted wave has the amplitude

$$\hat{E}_{m2}^{+} = \hat{E}_{m1}^{+} \frac{2\hat{\eta}_2}{\hat{\eta}_2 + \hat{\eta}_1} \tag{6-21}$$

Additional confidence is gained in the results (6-20) and (6-21) upon considering two special cases: (*a*) for which region 2 is a perfect conductor and (*b*) for which regions 1 and 2 have identical parameters (no interface exists). In case (*a*), with $\hat{\eta}_2 = 0$, (6-21) yields $\hat{E}_{m2}^+ = 0$, a result expected from the null fields within a perfect conductor; while (6-20) obtains $\hat{E}_{m1}^- = -\hat{E}_{m1}^+$, agreeable with (6-4) as one should expect. In case (*b*), identical regions means $\hat{\eta}_1 = \hat{\eta}_2$, whence from (6-20) and (6-21), $\hat{E}_{m1}^- = 0$ and $\hat{E}_{m2}^+ = \hat{E}_{m1}^+$, implying the reasonable conclusion that no reflection occurs if the region has no discontinuity.

EXAMPLE 6-1. A uniform plane wave with the amplitude $\hat{E}_{m1}^+ = 100e^{j0°}$ V/m in air is normally incident on the plane surface of a lossless dielectric with the parameters $\mu_2 = \mu_0$, $\epsilon_2 = 4\epsilon_0$, and $\sigma_2 = 0$. Find the amplitudes of the reflected and transmitted fields.

The geometry is shown in Figure 6-5. Region 1 is air, so $\hat{\eta}_1 \doteq \eta_0 = \sqrt{\mu_0/\epsilon_0} = 120\pi\ \Omega$. For region 2, $\hat{\eta}_2 = \sqrt{\mu_0/4\epsilon_0} = 60\pi\ \Omega$. The complex amplitudes of the reflected and transmitted waves are given by (6-20) and (6-21)

$$\hat{E}_{m1}^- = 100\frac{60\pi - 120\pi}{60\pi + 120\pi} = -33.3 \text{ V/m}$$

$$\hat{E}_{m2}^+ = 100\frac{2(60\pi)}{60\pi + 120\pi} = 66.7 \text{ V/m}$$

These amplitudes into (6-15) and (6-16) provide the total fields in each region

$$\hat{E}_{x1}(z) = 100e^{-j\beta_1 z} - 33.3e^{j\beta_1 z} \qquad \hat{E}_{x2}(z) = 66.7e^{-j\beta_2 z}$$

$$\hat{H}_{y1}(z) = \frac{100}{120\pi}e^{-j\beta_1 z} - \frac{(-33.3)}{120\pi}e^{j\beta_1 z} \qquad \hat{H}_{y2}(z) = \frac{66.7}{60\pi}e^{-j\beta_2 z}$$

in which $\gamma_1 = j\beta_1 = j\omega\sqrt{\mu_0\epsilon_0}$ and $\gamma_2 = j\beta_2 = j\omega\sqrt{\mu_0(4\epsilon_0)}$, the values of which can be inserted into the wave expressions once ω is specified. Observe that setting $z = 0$ produces continuous tangential electric and magnetic fields across the interface, as expected.

6-5 Normal Incidence for More Than Two Regions

An extension of the two-region problem of the last section to three or more regions leads to a multiplicity of reflected and transmitted wave terms which, in the sinusoidal steady state, yield single forward- and backward-traveling plane waves in each region. Suppose the three-region system of Figure 6-6(*a*) has the wave $\hat{E}_{xA}^+(z) = \hat{E}_{mA}^+e^{-\gamma z}$ impinging normally upon it as shown. A study of the subsequent phenomena in the time domain, after the arrival of the incident wave labelled *A* in Figure 6-6(*b*), reveals the generation of an infinite sequence of forward and backward waves in the system. Thus, two time-harmonic waves designated *B* and *C* are established successively in regions 1 and 2, the forward wave *C* in region 2 striking the second interface

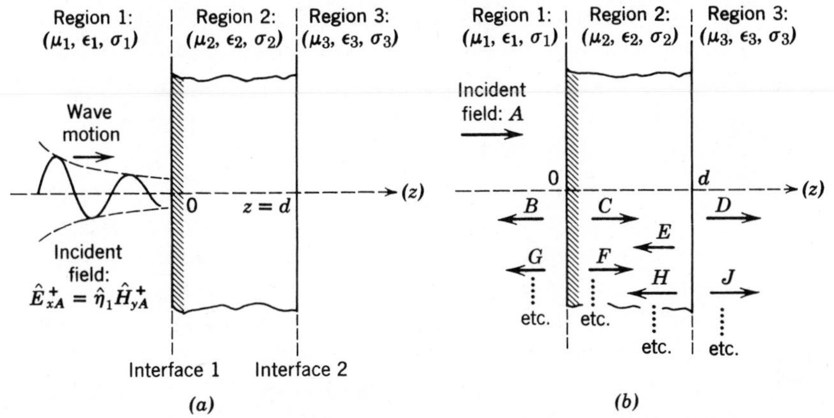

Figure 6-6. Three-region system upon which a uniform plane wave is normally incident. (*a*) Three-region system, showing the plane wave field incident upon a thickness *d* of region 2. (*b*) Depicting the effects of the incident field on reflected and transmitted waves, with increasing time.

to produce a transmitted wave D, plus another reflected wave E returning to interface 1. A continuation of this process, as time increases, produces an infinite sequence of reflected and transmitted waves, the linear sum of which obtains *sinusoidal steady state* forward- and backward-traveling waves in the respective regions. Thus, in region 1, the net positive z traveling electric field will consist only of the postulated x polarized, incident wave A, denoted by

$$\hat{E}^+_{xA}(z) = \hat{E}^+_{mA} e^{-\gamma_1 z}$$

while the reflected wave in that region consists of an infinite sequence of contributions of the waves $B, G, \ldots$; i.e.,

$$\hat{E}^-_{xB}(z) = \hat{E}^-_{mB} e^{\gamma_1 z} \qquad \hat{E}^-_{xG}(z) = \hat{E}^-_{mG} e^{\gamma_1 z} \cdots$$

Each wave term of the latter has a common factor $e^{\gamma_1 z}$ so that the infinite sum, in the sinusoidal steady state, becomes

$$\hat{E}^-_{mB} e^{\gamma_1 z} + \hat{E}^-_{mG} e^{\gamma_1 z} + \cdots = \left[\sum \hat{E}^-_{mi}\right] e^{\gamma_1 z} = \hat{E}^-_{m1} e^{\gamma_1 z} \tag{6-22}$$

reducing to a net reflected wave in region 1 designated by

$$\hat{E}^-_{x1}(z) = \hat{E}^-_{m1} e^{\gamma_1 z} \tag{6-23}$$

in which $\hat{E}_{m1}^-$ denotes its complex amplitude. Every term of (6-22) has an associated magnetic field related by the intrinsic wave impedance of region 1, yielding

$$\hat{H}_{y1}^-(z) = -\frac{\hat{E}_{m1}^-}{\hat{\eta}} e^{\gamma_1 z} \tag{6-24}$$

The net, sinusoidal steady state forward and backward waves in region 1 are depicted in Figure 6-7. Similar arguments applied to the infinite sequences of waves in regions 2 and 3 lead to the net field vectors shown.

The sinusoidal steady state wave solution of the three-region problem of Figure 6-7, with a known incident field amplitude $\hat{E}_{m1}^+$, involves finding the amplitudes $\hat{E}_{m1}^-$, $\hat{E}_{m2}^+$, $\hat{E}_{m2}^-$, and $\hat{E}_{m3}^+$, a total of four unknowns. The four boundary conditions, involving the continuity of the tangential E and H fields at the interfaces, are sufficient to generate four linear equations in terms of these amplitudes. To illustrate the procedure for Figure 6-7, the material parameters (μ, ϵ, σ) of each region are given, permitting γ and $\hat{\eta}$ of each to be calculated by use of (6-13) and (6-11). The depth d of region 2 is also specified. The total fields in the three regions are

$$\hat{E}_{x1}(z) = \hat{E}_{m1}^+ e^{-\gamma_1 z} + \hat{E}_{m1}^- e^{\gamma_1 z} \tag{6-25a}$$

Region 1

$$\hat{H}_{y1}(z) = \frac{\hat{E}_{m1}^+}{\hat{\eta}_1} e^{-\gamma_1 z} - \frac{\hat{E}_{m1}^-}{\hat{\eta}_1} e^{\gamma_1 z} \tag{6-25b}$$

$$\hat{E}_{x2}(z) = \hat{E}_{m2}^+ e^{-\gamma_2 z} + \hat{E}_{m2}^- e^{\gamma_2 z} \tag{6-26a}$$

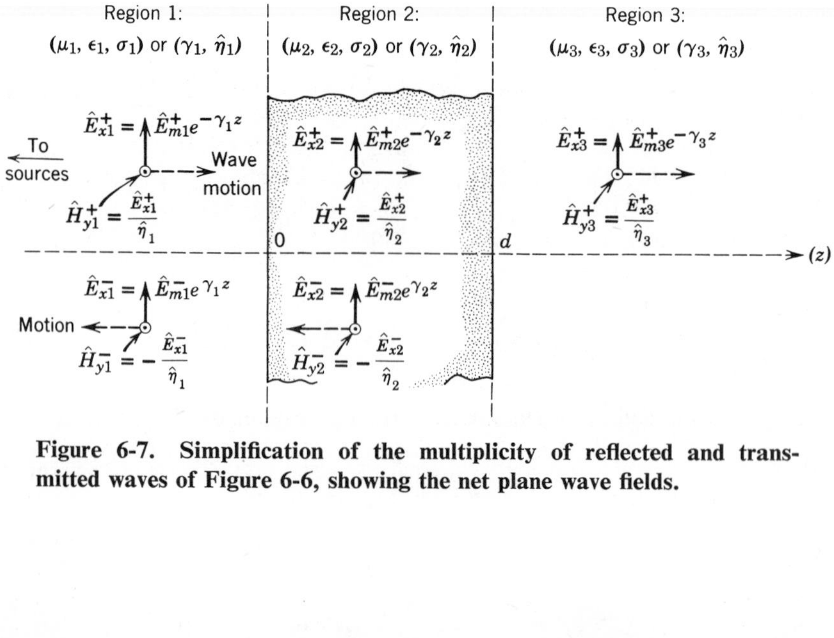

Figure 6-7. Simplification of the multiplicity of reflected and transmitted waves of Figure 6-6, showing the net plane wave fields.

$$\hat{H}_{y2}(z) = \frac{\hat{E}^+_{m2}}{\hat{\eta}_2} e^{-\gamma_2 z} - \frac{\hat{E}^-_{m2}}{\hat{\eta}_2} e^{\gamma_2 z} \qquad \text{Region 2} \qquad (6\text{-}26b)$$

$$\hat{E}_{x3}(z) = \hat{E}^+_{m3} e^{-\gamma_3 z} \qquad (6\text{-}27a)$$

$$\hat{H}_{y3}(z) = \frac{\hat{E}^+_{m3}}{\hat{\eta}_3} e^{-\gamma_3 z} \qquad \text{Region 3} \qquad (6\text{-}27b)$$

The boundary conditions (3-71) and (3-79) are satisfied by equating (6-25a) to (6-26a) and (6-25b) to (6-26b) at $z = 0$, and equating (6-26a) to (6-27a) and (6-26b) to (6-27b) at $z = d$. Then the rearrangement of the resulting four simultaneous equations, placing the known $\hat{E}^+_{m1}$ on the right, yields

$$\hat{E}^-_{m1} - \hat{E}^+_{m2} - \hat{E}^-_{m2} = -\hat{E}^+_{m1} \qquad (6\text{-}28a)$$

$$\frac{\hat{E}^-_{m1}}{\hat{\eta}_1} + \frac{\hat{E}^+_{m2}}{\hat{\eta}_2} - \frac{\hat{E}^-_{m2}}{\hat{\eta}_2} = \frac{\hat{E}^+_{m1}}{\hat{\eta}_1} \qquad (6\text{-}28b)$$

$$\hat{E}^+_{m2} e^{-\gamma_2 d} + \hat{E}^-_{m2} e^{\gamma_2 d} - \hat{E}^+_{m3} e^{-\gamma_3 d} = 0 \qquad (6\text{-}28c)$$

$$\frac{\hat{E}^+_{m2}}{\hat{\eta}_2} e^{-\gamma_2 d} - \frac{\hat{E}^-_{m2}}{\hat{\eta}_2} e^{-\gamma_2 d} - \frac{\hat{E}^+_{m3}}{\hat{\eta}_3} e^{-\gamma_3 d} = 0 \qquad (6\text{-}28d)$$

This is suitable for solution by fourth-order determinants, but is a tedious process, to say nothing of the higher-order results obtained when three or more interfaces are present. An alternative procedure is described in the next section.

6-6 Solution using Reflection Coefficient and Wave Impedance

The system of Figure 6-7 is generalized into n regions in Figure 6-8. Excited by the normally incident, time-harmonic wave ($\hat{E}^+_{x1}$, $\hat{H}^+_{y1}$) in region 1, each region acquires, in the sinusoidal steady state, the forward- and backward-traveling fields ($\hat{E}^+_x$, $\hat{H}^+_y$) and ($\hat{E}^-_x$, $\hat{H}^-_y$) except for the last ($k = n$) region, in which only the forward-traveling components $\hat{E}^+_{xn}$, $\hat{H}^+_{yn}$ appear. The total electric field for each region† becomes

$$\hat{E}_x(z) = \hat{E}^+_m e^{-\gamma z} + \hat{E}^-_m e^{\gamma z} = \hat{E}^+_m e^{-\gamma z}\left[1 + \frac{\hat{E}^-_m}{\hat{E}^+_m} e^{2\gamma z}\right] = \hat{E}^+_m e^{-\gamma z}[1 + \hat{\Gamma}(z)] \qquad (6\text{-}29)$$

in which $\hat{\Gamma}(z)$ is called the *reflection coefficient* at any location z in the region,

† Since these results apply to any (kth) region, an additional k subscript should be applied to all quantities. For simplicity such subscripts have been dropped.

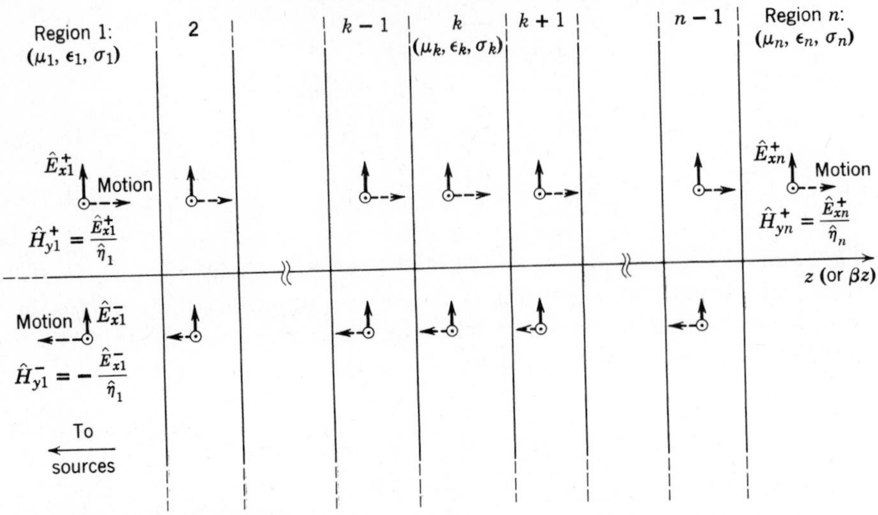

Figure 6-8. A multilayer system of n layers, upon which a uniform plane wave is normally incident from the left.

defined as the complex ratio of the reflected wave to the incident wave as follows:

$$\hat{\Gamma}(z) \equiv \frac{\hat{E}_m^-}{\hat{E}_m^+} e^{2\gamma z} \tag{6-30}$$

The corresponding total magnetic field is

$$\hat{H}_y(z) = \hat{H}_m^+ e^{-\gamma z} + \hat{H}_m^- e^{\gamma z} = \frac{\hat{E}_m^+}{\hat{\eta}} e^{-\gamma z}\left[1 - \frac{\hat{E}_m^-}{\hat{E}_m^+} e^{2\gamma z}\right] = \frac{\hat{E}_m^+}{\hat{\eta}} e^{-\gamma z}[1 - \hat{\Gamma}(z)] \tag{6-31}$$

A *total-field impedance* $\hat{Z}(z)$ is defined at any z location by the ratio of the total electric field (6-29) to the total magnetic field (6-31)

$$\hat{Z}(z) = \frac{\hat{E}_x(z)}{\hat{H}_y(z)} = \hat{\eta}\,\frac{1 + \hat{\Gamma}(z)}{1 - \hat{\Gamma}(z)}\ \Omega \tag{6-32}$$

A converse expression for $\hat{\Gamma}(z)$ in terms of $\hat{Z}(z)$ is obtained from (6-32) by solving for $\hat{\Gamma}(z)$

$$\hat{\Gamma}(z) = \frac{\hat{Z}(z) - \hat{\eta}}{\hat{Z}(z) + \hat{\eta}} \tag{6-33}$$

a form convenient for finding $\hat{\Gamma}(z)$ whenever $\hat{Z}(z)$ is known.

Another useful expression is one that enables finding $\hat{\Gamma}$ at any location z' in a region in terms of that at another position z. At z', the reflection coefficient is expressed by use of (6-30): $\hat{\Gamma}(z') = (\hat{E}_m^-/\hat{E}_m^+)e^{2\gamma z'}$. Dividing the latter by (6-30) eliminates the wave amplitudes, yielding the desired result

$$\hat{\Gamma}(z') = \hat{\Gamma}(z)e^{2\gamma(z'-z)} \tag{6-34}$$

In the application of (6-29) through (6-34) to the wave system of Figure 6-8, one should note the following properties of $\hat{\Gamma}(z)$ and $\hat{Z}(z)$ at any interface separating two regions:

1. The total field impedance $\hat{Z}(z)$ is *continuous* across the interface; i.e., at an interface defined by $z = a$

$$\hat{Z}(a^+) = \hat{Z}(a^-) \tag{6-35}$$

evident from the continuity of the tangential electric and magnetic fields appearing in the definition (6-32).

2. The reflection coefficient $\hat{\Gamma}(z)$ is *discontinuous* across the interface. This follows from (6-33), for, because $\hat{Z}(z)$ must be continuous across the interface, $\hat{\Gamma}(z)$ cannot be if the wave impedance $\hat{\eta}$ is different in the adjacent regions.

The procedure for finding the complex amplitudes of the forward- and backward-traveling waves in a multilayer system like that of Figure 6-8 is illustrated in two examples.

EXAMPLE 6-2. A uniform plane wave is normally incident in air upon a slab of plastic with the parameters shown, a quarter wave thick at the operating frequency $f = 1$ MHz. The x polarized wave has the amplitude $\hat{E}_{m1}^+ = 100e^{j0°}$ V/m. Utilize the concepts of reflection coefficient and total field impedance to find the remaining wave amplitudes.

To obviate carrying cumbersome phase terms across the interfaces, assume separate z origins 0_1, 0_2, and 0_3 shown in (b) of the figure. The wave amplitudes are referred to these origins. First, values of $\hat{\eta}$ for each region are found by

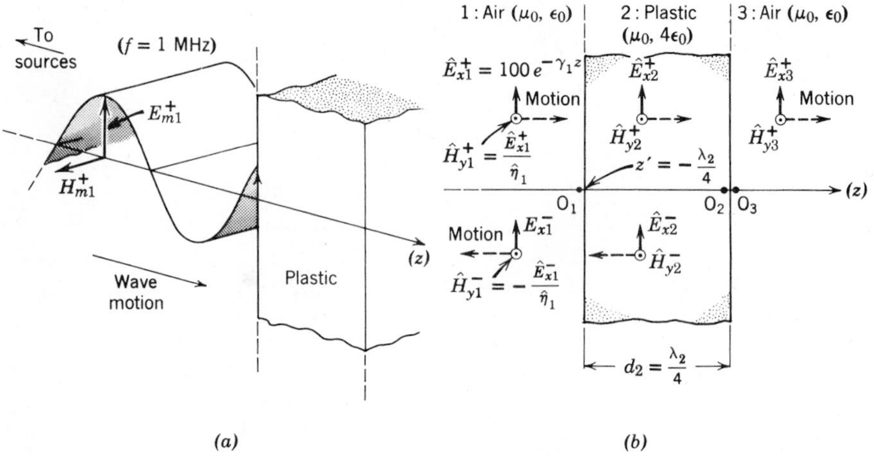

Example 6-2. (*a*) Uniform plane wave normally incident on a plastic slab. (*b*) Side view showing wave components in the regions.

using (6-12); thus, $\hat{\eta}_1 = \hat{\eta}_3 = \sqrt{\mu_0/\epsilon_0} = 120\pi\ \Omega$; in the plastic slab, $\hat{\eta}_2 = \sqrt{\mu_0/4\epsilon_0} = 60\pi\ \Omega$. The propagation constants $\gamma = \alpha + j\beta$ are computed from (3-90a, b) or (3-109) and (3-110); thus, in lossless region 2,

$$\gamma_2 = j\beta_2 = j\omega\sqrt{\mu_0(4\epsilon_0)} \qquad (= j2\pi/\lambda_2)$$

Then finding the complex wave amplitudes proceeds as follows:

(*a*) One begins in region 3, containing no reflected wave. $\hat{\Gamma}_3(z)$, from (6-30), is therein zero, yielding the total field impedance from (6-32): $\hat{Z}_3(z) = \hat{\eta}_3(1 + 0)/(1 - 0) = \hat{\eta}_3 = 120\pi\ \Omega$. By (6-35), the total field impedance $\hat{Z}_2(0)$ just inside region 2 has the same value, i.e., $\hat{Z}_2(0) = \hat{Z}_3(0) = 120\pi\ \Omega$.

(*b*) By use of (6-33), $\hat{\Gamma}_2$ at 0_2 in region 2 becomes

$$\hat{\Gamma}_2(0) = \frac{\hat{Z}_2(0) - \hat{\eta}_2}{\hat{Z}_2(0) + \hat{\eta}_2} = \frac{120\pi - 60\pi}{120\pi + 60\pi} = \tfrac{1}{3}$$

Equation (6-34) is employed† to translate $\hat{\Gamma}_2(0)$ to the value $\hat{\Gamma}_2(-d_2)$ at the input plane of region 2. With $z' = -d_2 = -\lambda_2/4$ and $\gamma_2 = j\beta_2 = j2\pi/\lambda_2$

$$\hat{\Gamma}_2\left(-\frac{\lambda_2}{4}\right) = \hat{\Gamma}_2(0)e^{2\gamma_2[(-\lambda_2/4)-0]}$$
$$= \hat{\Gamma}_2(0)e^{2(j2\pi/\lambda_2)(-\lambda_2/4)} = \tfrac{1}{3}e^{-j\pi} = -\tfrac{1}{3}$$

† The advantage of specifying the thickness of the lossless region in terms of *wavelength* ($d_2 = \lambda_2/4$) is evident from the determination of $\hat{\Gamma}_2(-d_2)$. Note, in view of $\gamma = j\beta = j(2\pi/\lambda)$ for a lossless region, that the product $\gamma(z' - z)$ appearing in the exponential factor does not require an explicit numerical value for β.

(*c*) Steps (*a*) and (*b*) are repeated to find $\hat{Z}$ and $\hat{\Gamma}$ in the next region to the left. First, the use of (6-32) at $z = -d_2$ in region 2 obtains

$$\hat{Z}_2(-d_2) = \hat{\eta}_2 \frac{1 + \hat{\Gamma}_2(-d_2)}{1 - \hat{\Gamma}_2(-d_2)} = 60\pi \frac{1 + (-\frac{1}{3})}{1 - (-\frac{1}{3})} = 30\pi\ \Omega$$

which from the continuity relation (6-35) yields $\hat{Z}_2(-d_2) = 30\pi\ \Omega = \hat{Z}_1(0)$. The reflection coefficient at the output plane of region 1, from (6-33), is

$$\hat{\Gamma}_1(0) = \frac{\hat{Z}_1(0) - \hat{\eta}_1}{\hat{Z}_1(0) + \hat{\eta}_1} = \frac{30\pi - 120\pi}{30\pi + 120\pi} = -\tfrac{3}{5}$$

The reflected wave amplitude $\hat{E}_{m1}^-$ is now obtained, using the definition (6-30) of reflection coefficient. Applied at $z = 0$ in region 1, given $\hat{E}_{m1}^+ = 100e^{j0°}$, it yields

$$\hat{E}_{m1}^- = \hat{E}_{m1}^+\hat{\Gamma}_1(z)e^{-2\gamma_1 z}\big]_{z=0} = (100e^{j0°})(-\tfrac{3}{5}) = -60\ \text{V/m}$$

Then the total electric field in air region 1, from (6-29), is

$$\hat{E}_{x1}(z) = \hat{E}_{m1}^+ e^{-\gamma_1 z} + \hat{E}_{m1}^- e^{\gamma_1 z} = 100e^{-j\beta_0 z} - 60e^{j\beta_0 z}\ \text{V/m}$$

The total *magnetic field* is obtained by use of (6-31)

$$\hat{H}_{y1}(z) = \frac{100}{120\pi} e^{-j\beta_0 z} - \frac{(-60)}{120\pi} e^{j\beta_0 z} = 0.266e^{-j\beta_0 z} + 0.159e^{j\beta_0 z}\ \text{A/m}$$

(*d*) The rest of the problem concerns finding $\hat{E}_{m2}^+$, $\hat{E}_{m2}^-$, and $\hat{E}_{m3}^+$. For example, $\hat{E}_{m2}^+$ is obtained by specializing $\hat{E}_{x1}(z)$ to $z = 0$ at the interface, whence

$$\hat{E}_{x1}(0) = 100 - 60 = 40\ \text{V/m} = \hat{E}_{x2}(-d_2)$$

in which the last equality is evident from the continuity condition (3-79). The total electric field in region 2 is $\hat{E}_{x2}(z) = \hat{E}_{m2}^+ e^{-\gamma z}[1 + \hat{\Gamma}_2(z)]$, from (6-29), but at $z = -d_2$, all quantities in (6-29) are known except $\hat{E}_{m2}^+$; solving for it obtains

$$\hat{E}_{m2}^+ = \hat{E}_{x2}(z)\frac{e^{\gamma_2 z}}{1 + \hat{\Gamma}_2(z)}\bigg]_{z=-d_2} = 40\frac{e^{(j2\pi/\lambda_2)(-\lambda_2/4)}}{1 + (-\frac{1}{3})}$$
$$= 60e^{-j(\pi/2)} = -j60\ \text{V/m}$$

Then applying (6-30) to region 2, $\hat{E}_{m2}^-$ is obtained from $\hat{E}_{m2}^+$ just found. A similar procedure applied at the second interface yields $\hat{E}_{m3}^+$, completing the problem.

EXAMPLE 6-3. An x polarized wave arrives from the left at $f = 1$ MHz with an amplitude $\hat{E}_{m1}^+ = 100e^{j0°}$ V/m. It is incident upon a lossless slab an eighth

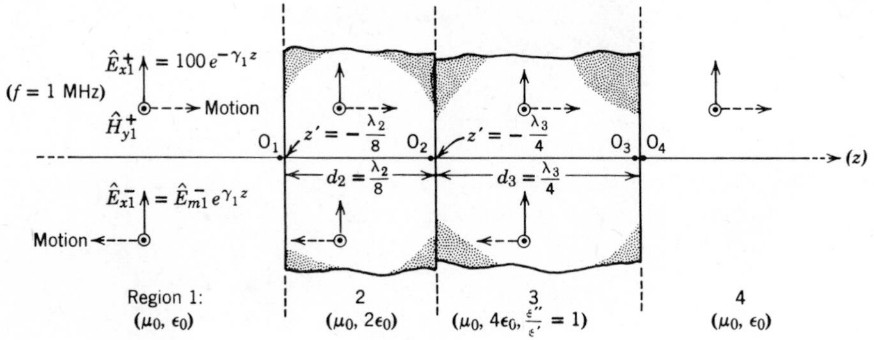

Example 6-3

of a wavelength thick, backed with a quarter wave lossy slab, with parameters as shown in the diagram. Find the remaining wave amplitudes.

The origins assumed for the four regions are noted in the diagram. A tabulation of α, β, λ, and $\hat{\eta}$ obtained for the regions using (3-109), (3-110), and (3-111) is given here:

Region	μ_r	ϵ_r	$\frac{\epsilon''}{\epsilon'}$	α (m^{-1})	β (m^{-1})	λ (m)	$\hat{\eta}$ (Ω)
1	1	1	0	0	0.0209	300	377
2	1	2	0	0	0.0296	212	266
3	1	4	1	0.0191	0.0461	136	$159e^{j22.5°}$
4	1	1	0	0	0.0209	300	377

(*a*) Beginning in region 4, which contains no reflection, $\hat{\Gamma}_4(0)$ is by (6-30) zero, yielding, from (6-32), $\hat{Z}_4(0) = \hat{\eta}_4 = 120\pi\ \Omega = \hat{Z}_3(0)$.

(*b*) Inserting the latter into (6-33) obtains

$$\hat{\Gamma}_3(0) = \frac{377 - 159e^{j22.5°}}{377 + 159e^{j22.5°}} = 0.451e^{-j21.4°}$$

to yield from (6-34) at the input plane $z' = -d_3 = -\lambda_3/4$

$$\hat{\Gamma}_3(-d_3) = \hat{\Gamma}_3(0)e^{2\gamma_3(-d_3-0)} = \hat{\Gamma}_3(0)e^{-2\alpha_3 d_3}e^{-j2\beta_3 d_3}$$
$$= (0.451e^{-j21.4°})(e^{-2(0.0191)34})e^{-j180°} = 0.1233e^{-j201.4°}$$

(*c*) Steps (*a*) and (*b*) are repeated to find $\hat{Z}$ and $\hat{\Gamma}$ at the output plane of region 2. Thus (6-32) yields

$$\hat{Z}_3(-d_3) = \hat{\eta}_3 \frac{1 + \hat{\Gamma}_3(-d_3)}{1 - \hat{\Gamma}_3(-d_3)} = 159e^{j22.5°}\frac{1 + 0.1233e^{-j201.4°}}{1 - 0.1233e^{-j201.4°}}$$
$$= 126.2e^{j27.9°}\ \Omega$$

and from (6-35), $\hat{Z}_3(-d_3) = \hat{Z}_2(0)$, yielding from (6-33)

$$\hat{\Gamma}_2(0) = \frac{\hat{Z}_2(0) - \hat{\eta}_2}{\hat{Z}_2(0) + \hat{\eta}_2} = \frac{126.2e^{j27.9°} - 266}{126.2e^{j27.9°} + 266} = 0.434e^{j150.3°}$$

The latter transforms, by use of (6-34) at the input plane $z = -d_2$

$$\hat{\Gamma}_2(-d_2) = \hat{\Gamma}_2(0)e^{2\gamma_2(-d_2-0)} = (0.434e^{j150.3°})e^{2(j2\pi/\lambda_2)(-\lambda_2/8)} = 0.434e^{j60.3°}$$

(*d*) The total field impedance there, from (6-32), is

$$\hat{Z}_2(-d_2) = \hat{\eta}_2 \frac{1 + \hat{\Gamma}_2(-d_2)}{1 - \hat{\Gamma}_2(-d_2)} = 266 \frac{1 + 0.434e^{j60.3°}}{1 - 0.434e^{j60.3°}} = 390e^{j42.9°}\ \Omega$$

which, by continuity across the interface, yields $\hat{Z}_1(0) = 390e^{j42.9°}\ \Omega$. From (6-33)

$$\hat{\Gamma}_1(0) = \frac{\hat{Z}_1(0) - \hat{\eta}_1}{\hat{Z}_1(0) + \hat{\eta}_1} = \frac{390e^{j42.9°} - 377}{390e^{j42.9°} + 377} = 0.393e^{j87.1°}$$

The reflected wave amplitude is obtained using (6-30); applying it at $z = 0$ yields $\hat{E}_{m1}^- = \hat{E}_{m1}^+\hat{\Gamma}_1(0) = (100)(0.393e^{j87.1°}) = 39.3e^{j87.1°}$, whence the total fields in region 1 become, from (6-29) and (6-31)

$$\hat{E}_{x1}(z) = 100e^{-j\beta_1 z} + 39.3e^{j(\beta_1 z + 87.1°)}\ \text{V/m}$$

$$\hat{H}_{y1}(z) = \frac{100}{377} e^{-j\beta_1 z} - \frac{39.3}{377} e^{j(\beta_1 z + 87.1°)}\ \text{A/m}$$

(*e*) The remaining task concerns finding $\hat{E}_{m2}^+$, $\hat{E}_{m2}^-$, $\hat{E}_{m3}^+$, $\hat{E}_{m3}^-$, and $\hat{E}_{m4}^+$. The procedure has already been outlined in part (*d*) of Example 6-2.

6-7 Graphical Solutions Using the Smith Chart

A convenient way to attack multiregion wave problems is by use of the Smith chart, named for its originator.† This chart enables finding graphically the total field impedance $\hat{Z}(z)$ from a given reflection coefficient $\hat{\Gamma}(z)$ or vice versa, while from a rotation about the chart, the reflection coefficient $\hat{\Gamma}(z')$ and the corresponding impedance $\hat{Z}(z')$ may be obtained at any other position z' within the region. The theoretical basis for this tool is developed in the following.

Beginning with (6-32), normalize it through division by $\hat{\eta}$ as follows:

$$\frac{\hat{Z}(z)}{\hat{\eta}} = \hat{z}(z) = \frac{1 + \hat{\Gamma}(z)}{1 - \hat{\Gamma}(z)} \tag{6-36}$$

† See articles by Smith, P. H. "Transmission-line calculator," *Electronics*. January 1939; and "An improved transmission-line calculator," *Electronics*. January 1944.

The dimensionless ratio $\hat{Z}(z)/\hat{\eta}$, symbolized $\hat{z}(z)$, is called the *normalized total field impedance*. The latter†, solved for $\hat{\Gamma}(z)$, yields the converse of (6-36)

$$\hat{\Gamma}(z) = \frac{\hat{z}(z) - 1}{\hat{z}(z) + 1} \tag{6-37}$$

Writing $\hat{\Gamma}$ and $\hat{z}$ in the rectangular complex forms

$$\hat{z} = r + jx \qquad \hat{\Gamma} = \Gamma_r + j\Gamma_i \tag{6-38}$$

one obtains from (6-36) the following:

$$r + jx = \frac{1 + \Gamma_r + j\Gamma_i}{1 - \Gamma_r - j\Gamma_i} \tag{6-39}$$

Rationalizing the denominator and equating the real and imaginary parts of each side yields

$$r = \frac{1 - \Gamma_r^2 - \Gamma_i^2}{(1 - \Gamma_r)^2 + \Gamma_i^2} \tag{6-40a}$$

$$x = \frac{2\Gamma_i}{(1 - \Gamma_r)^2 + \Gamma_i^2} \tag{6-40b}$$

Setting r and x equal to any real constants in (6-40) yields two mappings (transformations) from coordinate straight lines in the complex $\hat{z}$ plane into circles in the complex $\hat{\Gamma}$ plane, as depicted in (*a*), (*b*), and (*c*) of Figure 6-9.‡ This is proved as follows. With r = constant in (6-40a) and manipulating it into the form

$$\left(\Gamma_r - \frac{r}{r + 1}\right)^2 + \Gamma_i^2 = \left(\frac{1}{r + 1}\right)^2 \tag{6-41a}$$

one obtains in the $\hat{\Gamma}$ plane a family of circles with radii $1/(r + 1)$ and centers displaced horizontally to the positions $P_0(r/(r + 1), 0)$. For example, the $r = 2$ line in Figure 6-9(*a*) transforms, by use of (6-41a), into the circle

† From the theory of complex variables, (6-36) or (6-37) are bilinear transformations, having the property of transforming circles (or straight lines) in the plane of one of its complex variables into circles in the plane of the other complex variable. See also Churchill, R. V. *Complex Variables and Applications*. New York: McGraw-Hill, 1948.

‡ Since the r = constant and x = constant lines of Figure 6-9(*a*) intersect orthogonally, their maps as circles onto the $\hat{\Gamma}$ plane of Figure 6-9(*d*) also intersect at right angles. This is the conformal property of a bilinear transformation.

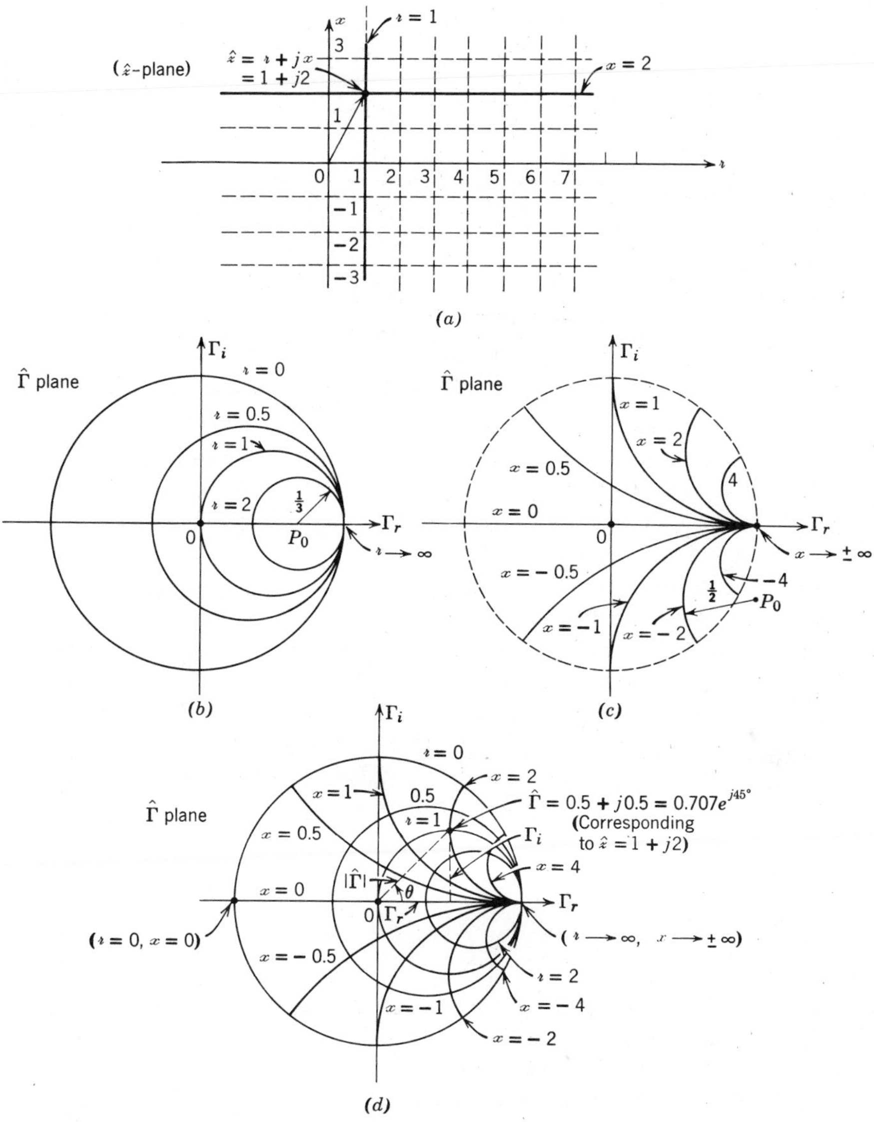

Figure 6-9. **Development of the Smith Chart. (*a*) Typical r-constant and x-constant coordinate lines in the plane. (*b*) Circles of constant r mapped by (6-41a) onto the $\hat{\Gamma}$ plane. (*c*) Circles of constant x mapped by (6-41b) onto the $\hat{\Gamma}$ plane. (*d*) Complete mapping of r-constant and x-constant lines onto the $\hat{\Gamma}$ plane: the Smith chart.**

labelled $r = 2$ in Figure 6-9(b), with a radius of 1/3 and its center at $P_0(2/3, 0)$.

Similarly, any $x =$ constant line in Figure 6-9(a) maps into the circles in (c) of that figure, evident from manipulating (6-40b) into

$$(\Gamma_r - 1)^2 + \left(\Gamma_i - \frac{1}{x}\right)^2 = \frac{1}{x^2} \qquad \text{(6-41b)}$$

This is a family of circles with radii $1/x$ and centers at the locations $P_0(1, 1/x)$. Typical circles corresponding to $x = 0, \pm 0.5, \pm 1, \pm 2$, and ± 4 are shown in Figure 6-9(c).

The superposition of the circles of Figures 6-9(b) and (c) yields (d), the so-called Smith chart. Any point (Γ_r, Γ_i) on this chart is a solution to (6-36) or, conversely, to (6-37). Thus, if some reflection coefficient $\hat{\Gamma}(z) = \Gamma_r + j\Gamma_i = |\hat{\Gamma}(z)|e^{j\theta}$ is located at $P(\Gamma_r, \Gamma_i)$ on the chart, that point also provides the coordinates (r, x), read off the $r =$ constant and $x =$ constant circles, yielding the solution $\hat{z} = r + jx$ of (6-36) that was sought. For example, given that $\hat{\Gamma}(z) = 0.5 + j0.5 = 0.707e^{j45°}$, $(\Gamma_r, \Gamma_i) = (0.5, 0.5)$ on the Smith chart produces $(r, x) = (1, 2)$ at that same point, implying the solution $\hat{z} = r + jx = 1 + j2 = 2.23e^{j63.4°}$. Thus $\hat{\Gamma}(z) = 0.707e^{j45°}$ and $\hat{z}(z) = 2.23e^{j62.4°}$ are a solution pair satisfying (6-36) and its converse, (6-37).

More accurate solution pairs to (6-36) are available from the enlarged Smith chart of Figure 6-10. Besides the usual overlay of circles of constant r and constant x onto the $\hat{\Gamma}$ plane, on the chart rim are three scales, the inner of which denotes the angle θ of the reflection coefficient, a quantity usually expressed in the polar form, $\hat{\Gamma} = |\hat{\Gamma}|e^{j\theta}$. The radial, dashed line overlay on the chart gives the angle θ. The concentric, dashed line circles denote the magnitude $|\hat{\Gamma}|$, with a range from zero to one.

The two *outer* rim scales on the chart pertain to the angular rotation associated with the exponential factor of (6-34)

$$\hat{\Gamma}(z') = \hat{\Gamma}(z)e^{2\gamma(z'-z)} \qquad \text{[6-34]}$$

in going from a z location to a new position z' in a region. For the case of a *lossless* region, wherein the propagation constant γ is the pure imaginary $\gamma = j\beta = j2\pi/\lambda$, (6-34) simplifies to

$$\hat{\Gamma}(z') = \hat{\Gamma}(z)e^{2(j2\pi/\lambda)(z'-z)} \qquad \text{(6-42a)}$$

stating that the reflection coefficient at the location z' is found by multiplying $\Gamma(z)$, at some other position z, times the pure phase factor $\exp[2(j2\pi/\lambda)(z'-z)]$. The factor 2 denotes that the phase shift that Γ undergoes is *twice* that associated with the wave motion in going from z to z'. Both outer rim scales are

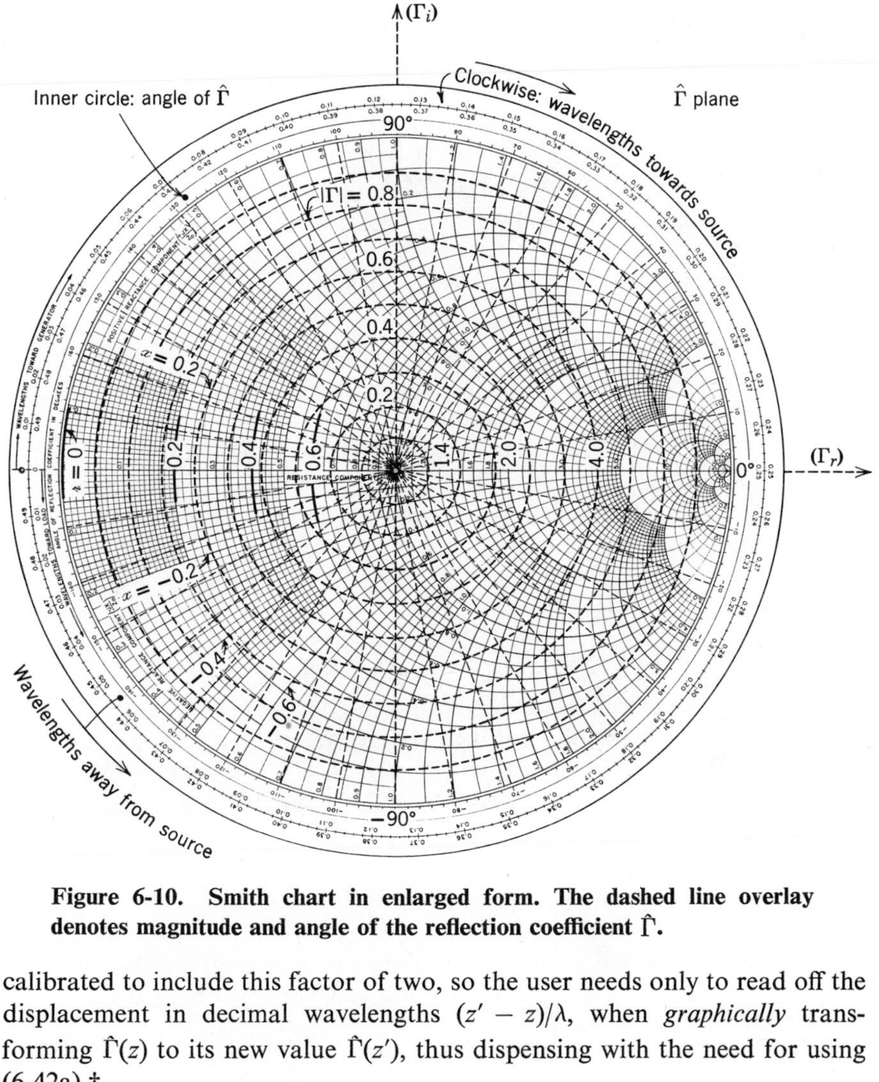

Figure 6-10. Smith chart in enlarged form. The dashed line overlay denotes magnitude and angle of the reflection coefficient $\hat{\Gamma}$.

calibrated to include this factor of two, so the user needs only to read off the displacement in decimal wavelengths $(z' - z)/\lambda$, when *graphically* transforming $\hat{\Gamma}(z)$ to its new value $\hat{\Gamma}(z')$, thus dispensing with the need for using (6-42a).†

† One may note that if the direction of motion in going from z to z' is *towards the sources* of the waves in a region (to the *left* in Figure 6-8), the quantity z' is then more negative than z, making the corresponding phase shift produced by the exponential factor of (6-42a) a *negative*, or *clockwise* angular shift around the $\hat{\Gamma}$-plane of the Smith chart. The *outer* rim scale of Figure 6-10 is calibrated to denote that phase shift (in decimal wavelengths) in moving *towards* the source (generator). The middle scale is used when moving away from the source of the waves.

Whenever the region possesses *losses*, the propagation constant γ becomes $\gamma = \alpha + j\beta$, yielding from (6-34) the general result

$$\hat{\Gamma}(z') = \hat{\Gamma}(z)e^{2\alpha(z'-z)}e^{2(j2\pi/\lambda)(z'-z)} \tag{6-42b}$$

In this case, besides the phase factor there appears a double attenuation factor $\exp[2\alpha(z' - z)]$, implying a decrease in the magnitude of the reflection coefficient as one proceeds towards the source. The application of the Smith chart to wave problems involving either lossless or attenuative regions is illustrated by the following examples.

EXAMPLE 6-4. Rework Example 6-2 by making use of the Smith chart. This problem concerns a plane wave of amplitude 100 V/m, normally incident in air upon a quarter wave lossless slab.

(*a*) In region 3 of (*a*), containing no reflection, the total field impedance from (6-32) is $\hat{Z}_3(z) = \hat{\eta}_3 = 120\pi\ \Omega$. From (6-35), the impedance just across the interface is $\hat{Z}_2(0) = \hat{Z}_3(0) = 120\pi\ \Omega$. Normalizing the latter using $\hat{\eta}_2 = 60\pi\ \Omega$ obtains

$$\hat{z}_2(0) = \frac{\hat{Z}_2(0)}{\hat{\eta}_2} = \frac{120\pi}{60\pi} = 2 \quad (= r + jx)$$

Thus $r = 2$ and $x = 0$ at $z = 0$ in region 2, entered onto the Smith chart as in part (*b*) of the accompanying figure. (Although the reflection coefficient $\hat{\Gamma}_2(0)$ can be found at the location of $\hat{z}_2(0)$, it may be ignored if desired.) The normalized impedance at $z' = -\lambda_2/4$ in region 2 is obtained from a phase rotation of $\hat{\Gamma}_2(0)$ according to (6-42a), in which $z' = -\lambda_2/4$, $z = 0$, and $\lambda = \lambda_2$. The use of (6-42a) is unnecessary because the *rim scales are calibrated in terms of the phase rotation* given by (6-42a), a negative rotation (towards the source) by the amount $(z' - z) = -0.25\ \lambda_2$ in this example. In (*c*), the rotation to the new value $\hat{\Gamma}(z') = \hat{\Gamma}_2(-\lambda_2/4)$ is depicted. At the same point, $\hat{z}_2(-\lambda_2/4)$ is found, becoming $\hat{z}_2(-\lambda_2/4) = 0.5 + j0$. Denormalizing obtains $\hat{Z}_2(-\lambda_2/4) = \hat{\eta}_2\hat{z}_2(-\lambda_2/4) = 60\pi(0.5) = 30\pi\ \Omega$.

(*b*) From (6-35), the impedance just inside region 1 has the same value: $\hat{Z}_1(0) = 30\pi\ \Omega$. To obtain the reflection coefficient there, normalize $\hat{Z}_1(0)$ obtaining

$$\hat{z}_1(0) = \frac{\hat{Z}_1(0)}{\hat{\eta}_1} = \frac{30\pi}{120\pi} = 0.25 + j0$$

The latter, entered onto the chart as in (*d*), yields

$$\hat{\Gamma}_1(0) = 0.6e^{-j180^\circ} = -0.6$$

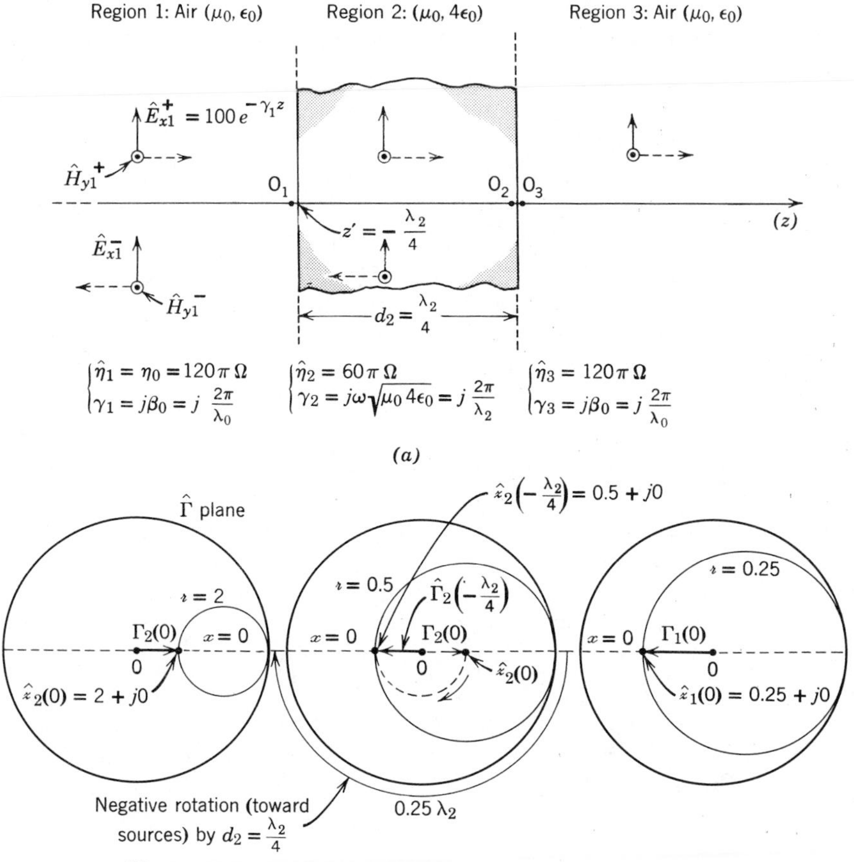

Example 6-4

in agreement with that obtained analytically in Example 6-2. Remaining details proceed exactly as given in Example 6-2.

EXAMPLE 6-5. Rework Example 6-3, making use of the Smith chart. With region 3 in (*a*) a *lossy* material, an attenuative as well as a phase shift effect is associated with its waves.

Beginning in region 4, because of no reflected wave, $\hat{Z}_4(z) = \hat{\eta}_4 = 377\ \Omega$, and from (6-35), $\hat{Z}_3(0) = \hat{Z}_4(0) = 120\pi\ \Omega$. $\hat{Z}_3(0)$ is normalized using $\hat{\eta}_3$ yielding

$$\hat{z}_3(0) = \frac{\hat{Z}_3(0)}{\hat{\eta}_3} = \frac{377}{159e^{j22.5^\circ}} = 2.19 - j0.907$$

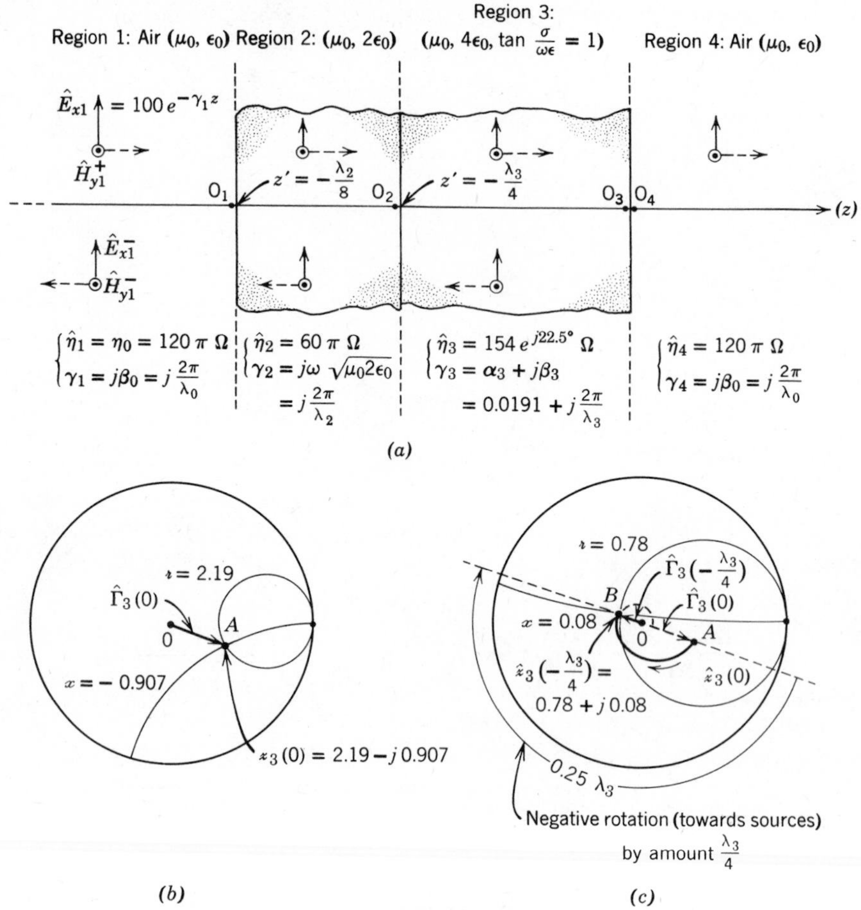

Example 6-5

labeled A in figure (b). (The value of $\hat{\Gamma}_3(0)$ available at A is ignored if it is not desired.) To find the normalized impedance at the input plane of region 3 using the Smith chart, one must utilize (6-42b)

$$\hat{\Gamma}(z') = \hat{\Gamma}(z)e^{2\alpha(z'-z)}e^{2(j2\pi/\lambda)(z'-z)} \qquad [6\text{-}42b]$$

noting that in moving from z to z' in a region, $\hat{\Gamma}$ undergoes a decrease in magnitude due to exp $[2\alpha(z' - z)]$, besides changing its phase according to the imaginary exp $[2(j2\pi/\lambda)(z' - z)]$. The latter, in moving from $z = 0$ to $z' = -\lambda_3/4$, entails a *phase rotation* of 0.25 λ_3 clockwise around the chart, read off the outer *rim scale* as shown in figure (c). The effect of the double attenuation factor in (6-42b) is determined using $\alpha_3 = 0.0191$ Np/m and $z - z' = d_3 = \lambda_3/4 = 34$ m, obtaining

$$e^{2\alpha_3(z'-z)} = e^{-2(0.0191)(34)} = 0.274$$

Thus $\hat{\Gamma}_3(0)$ in (*b*) is also diminished in *magnitude* by the factor 0.274, yielding $\hat{\Gamma}_3(-\lambda_3/4)$ at B on figure (*c*). The normalized impedance there is $\hat{z}_3(-\lambda_3/4) = 0.78 + j0.08 = 0.79e^{j5.4^\circ}$. Denormalizing yields

$$\hat{Z}_3\left(-\frac{\lambda_3}{4}\right) = \hat{\eta}_3\hat{z}_3\left(-\frac{\lambda_3}{4}\right) = 159e^{j22.5^\circ}(0.79e^{j5.4^\circ}) = 126e^{j27.9^\circ}\ \Omega$$

The remainder of the problem involving the lossless regions 2 and 1 proceeds in the manner already detailed in the previous example.

*6-8 Standing Waves

In Figure 6-3 was observed the standing wave produced by the total reflection of a plane wave normally incident upon a perfect conductor. From the composite diagram is seen the basis for the term standing wave; the total field magnitudes have a stationary appearance in space, similar to standing waves on a vibrating string as in Figure 6-4(*a*). The undulations, from maximum to null amplitudes every quarter wave, occur in accordance with the $\sin \beta z$ or $\cos \beta z$ factors in the total field expressions (6-6) and (6-9).

The example of Figure 6-3 represents a special case of standing waves produced whenever plane waves of equal amplitudes move in opposite directions through a lossless region. In general, an arbitrary percentage of the incident wave is reflected, determined by the reflection coefficient amplitude at the interface. The region may, moreover, be lossy. An analysis of standing-wave behavior requires the total electric and magnetic field expressions, given in time-harmonic form by (6-29) and (6-31)

$$\hat{E}_x(z) = \hat{E}_m^+ e^{-\gamma z} + \hat{E}_m^- e^{\gamma z} = \hat{E}_m^+ e^{-\gamma z}[1 + \hat{\Gamma}(z)] \tag{6-43}$$

$$\hat{H}_y(z) = \frac{\hat{E}_m^+}{\hat{\eta}} e^{-\gamma z} - \frac{\hat{E}_m^-}{\hat{\eta}} e^{\gamma z} = \frac{\hat{E}_m^+}{\hat{\eta}} e^{-\gamma z}[1 - \hat{\Gamma}(z)] \tag{6-44}$$

in which $\gamma = \alpha + j\beta$ given by (3-90a, b), and $\hat{\eta}$ is specified by (3-99a). In some standing-wave discussions, only the wave magnitudes are of interest. The magnitudes of (6-43) and (6-44) are written

$$|\hat{E}_x(z)| = |\hat{E}_m^+|e^{-\alpha z}|1 + \hat{\Gamma}(z)| \tag{6-45a}$$

$$|\hat{H}_y(z)| = \frac{|\hat{E}_m^+|}{\eta} e^{-\alpha z}|1 - \hat{\Gamma}(z)| \tag{6-45b}$$

noting that the magnitudes of the phase quantity, $|e^{-j\beta z}|$, and of the angular factor of the wave impedance, $|e^{j\theta}|$, are both unity.

Of particular simplicity are the fields of a *lossless* region for which $\gamma = j\beta$ and $\hat{\eta} = \eta$, reducing (6-43) and (6-44) to

$$\hat{E}_x(z) = \hat{E}_m^+ e^{-j\beta z} + \hat{E}_m^- e^{j\beta z} = \hat{E}_m^+ e^{-j\beta z}[1 + \hat{\Gamma}(z)] \tag{6-46}$$

$$\hat{H}_y(z) = \frac{\hat{E}_m^+}{\eta} e^{j-\beta z} - \frac{\hat{E}_m^-}{\eta} e^{j\beta z} = \frac{\hat{E}_m^+}{\eta} e^{-j\beta z}[1 - \hat{\Gamma}(z)] \tag{6-47}$$

Figure 6-11 displays a real-time plot of $E_x(z, t)$ and $H_y(z, t)$, showing the incident and reflected wave terms of (6-46) and (6-47) at successive instants along a portion of the z axis. An inspection of the *total* fields with varying t and z in the lower diagrams shows how a standing wave is developed in the region; there the total fields appear to be traveling waves with a changing amplitude as they move in the z direction. Thus $|E_x(z, t)|$ and $|H_y(z, t)|$ change from a maximum to a minimum, and vice versa, every quarter wave (90°) along the z axis, a consequence of the forward and backward wave terms becoming phase aiding and then phase-opposing at that spacing as the waves move in their respective directions. The maximum of the total electric field envelope, $E_{max} = |E_x(z, t)|_{max}$, is observed to coincide in space with the minimum $H_{min} = |H_y(z, t)|_{min}$, and vice versa, a result of the sign reversal in the reflected magnetic field term in (6-47).

The so-called *standing-wave ratio* (SWR), associated with incident and reflected uniform waves in a lossless region as exemplified in Figure 6-11, is defined as the ratio of the maximum amplitude, E_{max}, of the electric field envelope, to the minimum amplitude, E_{min}, occurring a quarter wave away; i.e.,

$$\text{SWR} = \frac{|E_x(z, t)|_{max}}{|E_x(z, t)|_{min}} \equiv \frac{E_{max}}{E_{min}} \tag{6-48}$$

It is seen from Figure 6-11 that the envelope maximum E_{max} occurs where the amplitudes E_m^+ and E_m^- are aiding, while E_{min} is produced a quarter wave away where they are in opposition, such that

$$\begin{aligned} E_{max} &= E_m^+ + E_m^- = |\hat{E}_m^+| + |\hat{E}_m^-| \\ E_{min} &= E_m^+ - E_m^- = |\hat{E}_m^+| - |\hat{E}_m^-| \end{aligned} \tag{6-49}$$

Similar conclusions can be reached concerning H_{max} and H_{min} along the

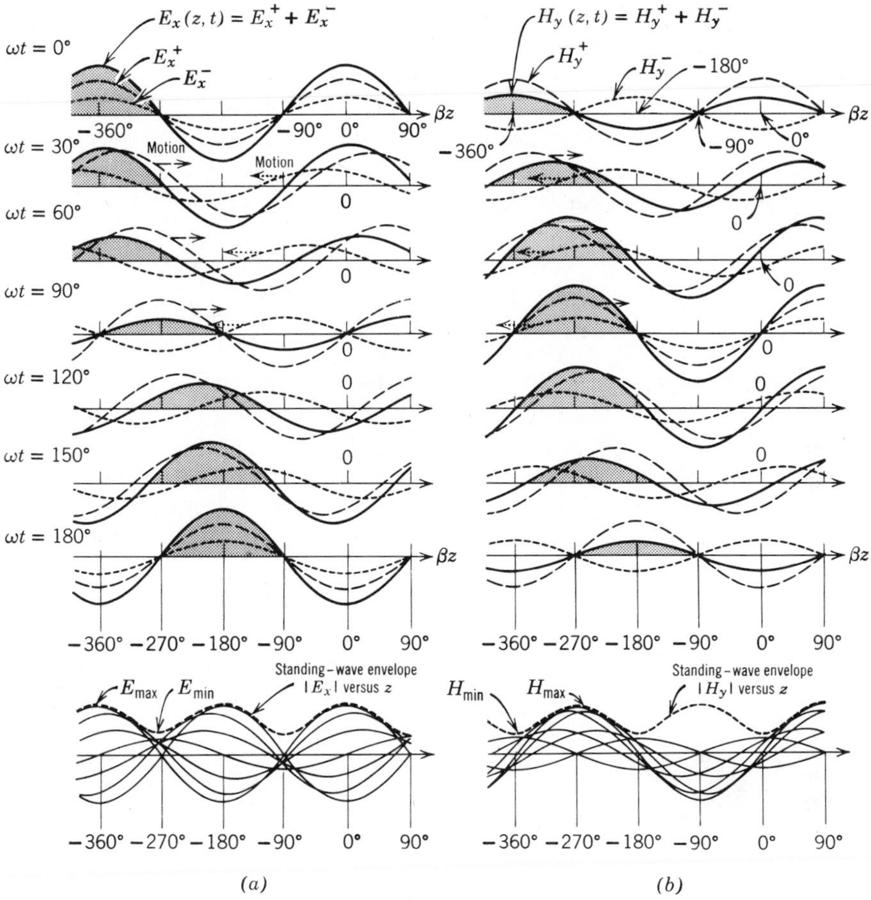

Figure 6-11. Real-time diagrams of forward- and backward-traveling waves of E_x and H_y at successive instants, in a region where reflection occurs. The composite standing wave pattern below is the result. (*a*) Electric field. (*b*) Magnetic field.

magnetic field standing-wave envelope; thus SWR, from (6-48), becomes

$$\text{SWR} = \frac{E_{\max}}{E_{\min}} = \frac{|\hat{E}_m^+| + |\hat{E}_m^-|}{|\hat{E}_m^+| - |\hat{E}_m^-|} = \frac{|\hat{H}_m^+| + |\hat{H}_m^-|}{|\hat{H}_m^+| - |\hat{H}_m^-|} = \frac{H_{\max}}{H_{\min}} \tag{6-50}$$

For example, with the launching of a forward traveling plane wave field with $|\hat{E}_m^+| = 100$ V/m in some lossless region, and a reflection occurring such that

$|\hat{E}_m^-| = 20$ V/m, the standing-wave ratio, from (6-50), becomes SWR $= 120/80 = 1.5$. A reflectionless region (with $|\hat{E}_m^-| = 0$) will have the minimum possible SWR of unity.

There are advantages in analyzing standing-wave phenomena by use of the complex forms of the fields. Since the standing-wave diagrams of Figure 6-11 are total field *magnitudes* plotted against z, it behooves one to reexamine the wave magnitude expressions (6-45). For a lossless region ($\alpha = 0$), they become

$$|\hat{E}_x(z)| = |\hat{E}_m^+||1 + \hat{\Gamma}(z)|, \qquad |\hat{H}_y(z)| = \frac{|\hat{E}_m^+|}{\eta}|1 - \hat{\Gamma}(z)| \tag{6-51}$$

results of special interest in that they involve the reflection coefficient $\hat{\Gamma}(z)$, a quantity readily available from the Smith chart. It is evident from (6-51) that the maximum wave magnitude, $|\hat{E}_x(z)|_{\max}$, occurs in the lossless region where $|1 + \hat{\Gamma}(z)|$ is maximal; i.e., where it has the value $1 + |\hat{\Gamma}(z)|$. Thus, $E_{\max} = |\hat{E}_m^+|(1 + |\hat{\Gamma}(z)|)$. Similarly, $E_{\min} = |\hat{E}_m^+|(1 - |\hat{\Gamma}(z)|)$. Hence, the SWR defined by (6-50) becomes

$$\text{SWR} = \frac{E_{\max}}{E_{\min}} = \frac{1 + |\hat{\Gamma}(z)|}{1 - |\hat{\Gamma}(z)|} \tag{6-52a}$$

For example, the reflection-coefficient magnitude $|\hat{\Gamma}(z)| = 0.2$ (20% reflection) yields from (6-52a) the SWR $= (1 + 0.2)/(1 - 0.2) = 1.5$, as before. Since the reflection coefficient magnitude has the range $0 \le |\hat{\Gamma}| \le 1$, from (6-52a) the SWR is limited to the range $1 \le \text{SWR} < \infty$.

The Smith chart, from which the reflection coefficient $\hat{\Gamma}(z)$ is readily found, is also convenient for finding the SWR graphically. For a lossless region containing the total fields (6-46) and (6-47), the locus of $\hat{\Gamma}(z)$ versus z is a circle as shown typically on the chart in Figure 6-12(a). This locus is sometimes called the SWR circle. The complex quantities $1 + \hat{\Gamma}(z)$ and $1 - \hat{\Gamma}(z)$ occur on the SWR circle at the points A and B, as in (b) of the figure. The quantities $1 + |\hat{\Gamma}(z)|$ and $1 - |\hat{\Gamma}(z)|$ are evidently the distances $0'C$ and $0'D$ in Figure 6-12(c), yielding from (6-52a)

$$\text{SWR} = \frac{0'C}{0'D} \tag{6-52b}$$

The use of (6-52b) can be avoided, however, since the normalized impedance $\mathscr{z}$, at the point C in the figure, has a value equal to the SWR in question, a fact easily proved by applying (6-39) at that point. Since $\mathscr{x} = 0$ and $\Gamma_i = 0$

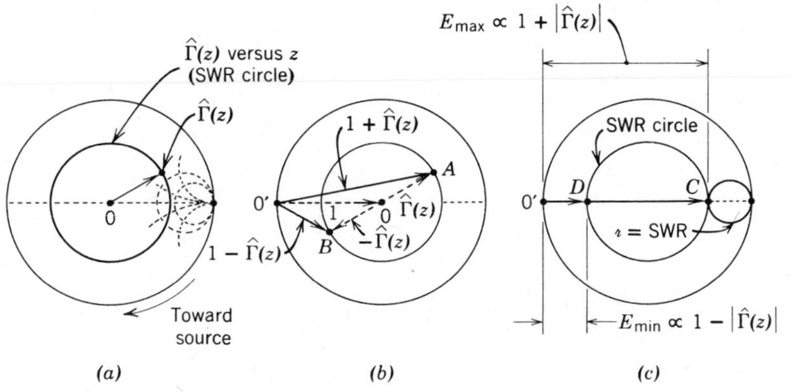

Figure 6-12. Smith chart field interpretations in a lossless region. (*a*) Locus of $\hat{\Gamma}(z)$ versus z: the SWR circle. (*b*) The quantities $1 + \hat{\Gamma}(z)$ and $1 - \hat{\Gamma}(z)$. (*c*) Showing z locations where $\hat{E}(z)$ is maximum and minimum.

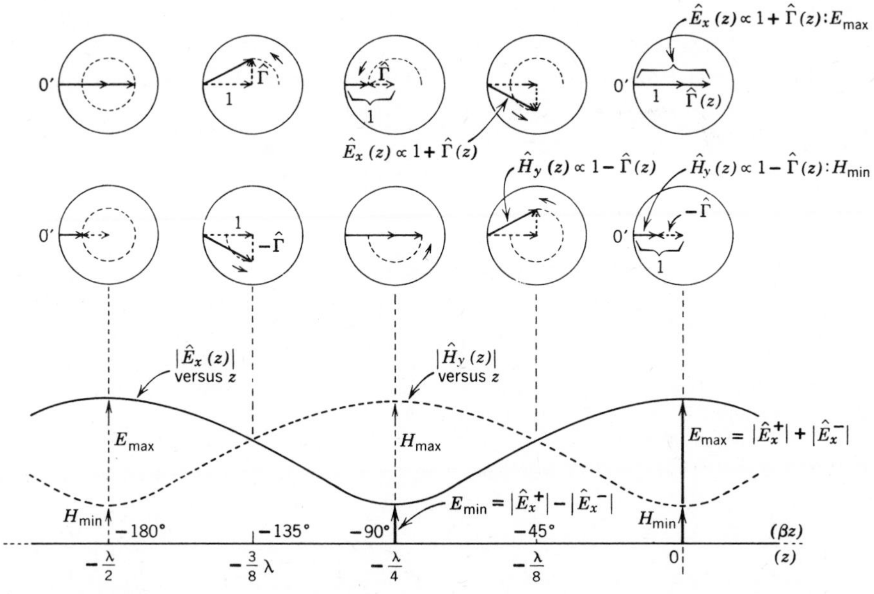

Figure 6-13. Forward z and backward z traveling field magnitudes deduced from the Smith chart (above) and the corresponding standing-wave field magnitude graphs (below).

there, (6-39) yields

$$\imath = \frac{1 + \Gamma_r}{1 - \Gamma_r} = \frac{1 + |\hat{\Gamma}(z)|}{1 - |\hat{\Gamma}(z)|} \tag{6-53}$$

or just the SWR given by (6-52a). Thus the SWR circle can be drawn on the Smith chart by noting it must pass through the point $\imath$ = SWR on the real axis.

The SWR circle on the Smith chart can be used to obtain the z variations of the electric and magnetic field magnitudes in a lossless region. From (6-51), $|\hat{E}_x(z)|$ and $|\hat{H}_y(z)|$ are proportional to the quantities $|1 + \hat{\Gamma}(z)|$ and $|1 - \hat{\Gamma}(z)|$, respectively, but these are just $0'A$ and $0'B$ in Figure 6-12(*b*), whence the relative field magnitudes versus z yield the Smith diagrams of Figure 6-13. The lower graph shows the standing waves of the two field magnitudes obtained therefrom. The occurrence of H_{min} at the position of E_{max}, and vice versa, is noted as mentioned before.

REFERENCES

LORRAIN, P., and D. R. Corson. *Electromagnetic Fields and Waves.* San Francisco: W. H. Freeman, 1970.

FANO, R. M., L. T. CHU, and R. P. ADLER. *Electromagnetic Fields, Energy and Forces.* New York: Wiley, 1960.

RAMO, S., J. R. WHINNERY, and T. VAN DUZER. *Fields and Waves in Communication Electronics.* New York: Wiley, 1965.

PROBLEMS

6-1. Make use of the appropriate boundary condition to find the surface current density $\mathbf{J}_S$ induced onto the surface $z = 0$ of Figure 6-3, expressed in both real-time and complex forms. What is the depth of penetration into region 2?

6-2. The result (6-8) was found by applying (6-4) to the magnetic field expression (6-7). Show that (6-8) is also obtained by inserting (6-5) into the appropriate Maxwell curl equation.

6-3. (*a*) For the system of Figure 6-5, prove for the case in which regions 1 and 2 have identical material constants that (6-20) and (6-21) reduce to the results expected.

(*b*) Use (6-20) and (6-21) to show that assuming region 2 perfectly conducting yields results agreeing with those obtained in Section 6-2.

6-4. Use the solutions for $\hat{E}_{x1}(z)$, $\hat{H}_{y1}(z)$, $\hat{E}_{x2}(z)$, and $\hat{H}_{y2}(z)$ of Example 6-1 to show that the tangential field boundary conditions at the interface are satisfied.

6-5. Assume a two-region system like that of Figure 6-5, a plane wave arriving at normal incidence in air with the amplitude $100e^{j0°}$ V/m and a frequency of 100 MHz. The slab is Teflon ($\epsilon_r = 2.1$, $\epsilon''/\epsilon' = 0$).

(*a*) What are the intrinsic wave impedance and propagation constant of each region?

(*b*) Use (6-20) and (6-21) to find the complex amplitudes of the reflected and transmitted waves.

(*c*) Write the expressions for $\hat{E}_{x1}(z)$, $\hat{H}_{y1}(z)$, $\hat{E}_{x2}(z)$, and $\hat{H}_{y2}(z)$ in the two regions. Show from these that the boundary conditions at the interface are satisfied.

6-6. A uniform plane wave in air at $f = 100$ MHz is normally incident on a wall of plastic of great extent and of thickness $d_2 = \lambda_2/8$, having the material parameters μ_0, $9\epsilon_0$, and $\sigma_2 = 0$. Choosing z origins in the regions as in the figure accompanying Example 6-2, determine:

(*a*) The total field impedance at the *output plane* of region 2. Find the reflection coefficient at that location.

(*b*) Determine the reflection coefficient at the input plane of the plastic, and evaluate the total field impedance there.

(*c*) Find $\hat{\Gamma}_1(0)$, and if $\hat{E}_{m1}^+ = 100$ V/m, compute $\hat{E}_{m1}^-$, writing the expression for the total electric field in region 1 using (6-29). Find also the total magnetic field in region 1.

6-7. Complete the calculations of Example 6-2 by determining $\hat{E}_{m2}^-$ and $\hat{E}_{m3}^+$ to obtain the total fields:

$$\hat{E}_{x2}(z) = -j60e^{-j\beta_2 z} - j20e^{j\beta_2 z} = 60e^{-j(\beta_2 z + 90°)} + 20e^{j(\beta_2 z - 90°)} \text{ V/m}$$
$$\hat{H}_{y2}(z) = 0.318e^{-j(\beta_2 z + 90°)} + 0.106e^{j(\beta_2 z + 90°)} \text{ A/m}$$
$$\hat{E}_{x3}(z) = 80e^{-j(\beta_3 z + 90°)} \text{ V/m} \qquad \hat{H}_{y3}(z) = 0.212e^{-j(\beta_3 z + 90°)} \text{ A/m}$$

6-8. Finish Problem 6-6 by finding $\hat{E}_{m2}^+$, $\hat{E}_{m2}^-$, and $\hat{E}_{m3}^+$, and obtain the total electric and magnetic fields in regions 2 and 3.

6-9. A plane wave in air is normally incident, at $f = 10$ MHz, upon a plane

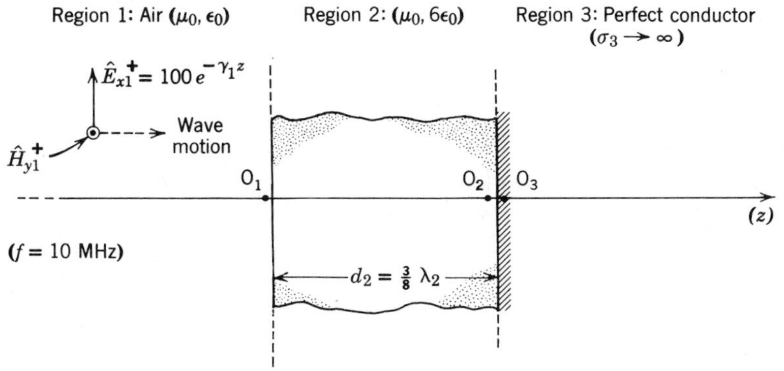

Problem 6-9

slab backed by a perfect conductor as shown in the accompanying figure. The slab is $(3/8)\lambda_2$ thick at this frequency. Assume the arriving wave amplitude to be $100e^{j0°}$ V/m.

(*a*) Assuming z origins in the regions as noted, find the total field impedance at the output plane of the plastic slab. What is the reflection coefficient there? [*Answer:* 0 Ω, −1]

(*b*) Determine the total field impedance at 0_1 (the output plane of region 1) and the reflection coefficient there. [*Answer:* $Z_1(0) = -j154\ \Omega$, $\hat{\Gamma}_1(0) = e^{-j135°}$]

(*c*) Find the reflected wave amplitude in region 1, and the total electric field there. [*Answer:* $\hat{E}_{m1}^- = 100e^{-j135°}$, $\hat{E}_{x1}(z) = 100e^{-j\beta_0 z} + 100e^{j(\beta_0 z - 135°)}$ V/m]

(*d*) How thick (in meters) is the slab? What is its electrical thickness (βz) in degrees? [*Answer:* 4.6 m, 135°]

6-10. Determine the total field impedance at the output plane of region 1 in Problem 6-9, assuming a quarter wave slab thickness ($d_2 = \lambda_2/4$). [*Answer:* ∞ Ω]

6-11. Repeat Problem 6-9, assuming region 2 lossy; i.e., assume a loss tangent of 1.2 at $f = 10$ MHz. [*Answer:* (*a*) 0 Ω, −1; (*b*) $\hat{Z}_1(0) = 123e^{j12°}\ \Omega$, $\hat{\Gamma}_1(0) = 0.515e^{j172°}$; (*c*) $\hat{E}_{m1}^- = 51.5e^{j172°}$, $\hat{E}_{x1}(z) = 100e^{-j\beta_0 z} + 51.5e^{j(\beta_0 z + 172°)}$; (*d*) 4.06 m, $\beta_2 d = 135°$]

6-12. A plane wave is normally incident upon region 2 of thickness d, separating region 1 from region 3 as shown, with medium parameters symbolized as shown. Use (6-32) to derive the following expressions (*a*) and (*b*) for the output plane impedance of region 1, in terms of the terminal region impedance $\hat{\eta}_3$ and the parameters $\hat{\gamma}_2$, $\hat{\eta}_2$, and d of region 2 as follows:

$$\hat{Z}_1(0) = \hat{Z}_2(-d) = \hat{\eta}_2 \frac{1 + \hat{\Gamma}_2(-d)}{1 - \hat{\Gamma}_2(-d)}$$

$$= \hat{\eta}_2 \frac{(\hat{\eta}_3 + \hat{\eta}_2)e^{\gamma_2 d} + (\hat{\eta}_3 - \hat{\eta}_2)e^{-\gamma_2 d}}{(\hat{\eta}_3 + \hat{\eta}_2)e^{\gamma_2 d} - (\hat{\eta}_3 - \hat{\eta}_2)e^{-\gamma_2 d}} \quad (a)$$

$$= \hat{\eta}_2 \frac{\hat{\eta}_3 \cosh \gamma_2 d + \hat{\eta}_2 \sinh \gamma_2 d}{\hat{\eta}_3 \sinh \gamma_2 d + \hat{\eta}_2 \cosh \gamma_2 d} \quad (b)$$

Note that (*b*) makes use of the definitions of the hyperbolic cosine and sine functions: $\cosh \theta = (1/2)(e^{\theta} + e^{-\theta})$, $\sinh \theta = (1/2)(e^{\theta} - e^{-\theta})$.

6-13. (*a*) Assuming for the system of Problem 6-12 that region 2 is lossless and a quarter wave thick ($d = \lambda_2/4$), show that the total field impedance at 0_1 is $\hat{Z}_1(0) = \eta_2^2/\hat{\eta}_3$.

(*b*) Assuming region 2 lossless and a half wave thick ($d = \lambda_2/2$), show that $\hat{Z}_1(0) = \hat{\eta}_3$.

6-14. The result of Problem 6-13 can be used to find the relative permittivity of a quarter wave matching plate, used in the low-reflection coating of lenses and prisms. If a lossless glass region has $\epsilon_{r3} = 3$ (region 3 in Problem 6-12), what must be the relative permittivity of a quarter wave thick

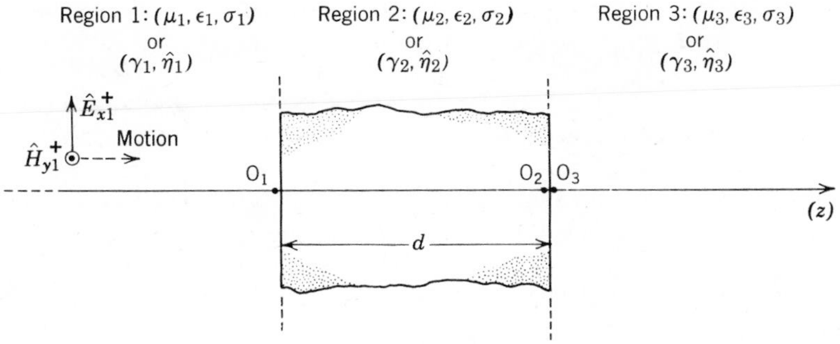

Problem 6-12

lossless dielectric coating (region 2) if a normally incident plane wave in air (region 1) is not to be reflected? Assume nonmagnetic materials. [*Answer:* $\epsilon_{r2} = \sqrt{3}$]

6-15. If a plane wave is normally incident on a half wave lossless slab in air, explain why no reflected wave exists in region 1, regardless of the parameters μ_2 and ϵ_2 of the slab (see Problem 6-13).

6-16. In detail prove that equating the real and imaginary parts of (6-39) leads to (6-40). Show that these results can be manipulated into (6-41), circles that map into the $\hat{\Gamma}$ plane for constant r and x.

6-17. From a Smith chart, find the values of $\hat{\Gamma}$ or $\hat{z}$ corresponding to the following values of $\hat{z}$ or $\hat{\Gamma}$. Check answers using (6-36) or (6-37) where requested.

(*a*) $\hat{\Gamma} = 0.707e^{j45°}$ [*Answer:* $\hat{z} = 1 + j2 = 2.23e^{j63.4°}$]

(*b*) $\hat{\Gamma} = 0.5$ [*Answer:* $\hat{z} = 3$] Check.

(*c*) $\hat{\Gamma} = 0.5e^{-j143.1°}$ [*Answer:* $\hat{z} = 0.366 - j0.293$] Check.

(*d*) $\hat{z} = 1 + j2$ [*Answer:* see (*a*) above]

(*e*) $\hat{z} = 0.4 - j0.2$ [*Answer:* $\hat{\Gamma} = 0.448e^{-j153°}$] Check.

(*f*) $\hat{z} \rightarrow \infty$ [*Answer:* $\hat{\Gamma} = 1$]

(*g*) $\hat{z} = 0$ [*Answer:* $\hat{\Gamma} = -1$] Check.

6-18. For the two-region (air–Teflon) system with a normally incident plane wave described in Problem 6-5, use the Smith chart to assist, where applicable, in finding the following:

(*a*) Determine the normalized total field impedance $\hat{z}_1(0)$ at the output plane of region 1.

(*b*) What is the complex amplitude $\hat{E}_{m1}^-$ of the reflected wave in region 1? Obtain the total electric field in region 1.

6-19. Work Problem 6-6, making full use of the Smith chart. In particular, find the following:

(*a*) Determine the normalized total field impedance at 0_2. Use the Smith chart to obtain $\hat{\Gamma}_2(0)$ there.

(*b*) From an appropriate rotation about the chart, find the reflection coefficient at the input plane of region 2. Determine also $\hat{Z}_2(-d_2)$ there.

(*c*) By use of the normalized impedance in the air at 0_1, use the Smith chart to determine $\hat{\Gamma}_1(0)$ there. Find the reflected wave amplitude $\hat{E}_{m1}^-$, and obtain the total electric field in region 1.

6-20. Work Problem 6-9, making use of the Smith chart. [*Note:* The total reflection at the perfect conductor yields, throughout *both* lossless regions, reflection coefficients of unity magnitude.]

6-21. Work Problem 6-10 using the Smith chart. [*Note:* For a lossless region, a half wave rotation about the chart reproduces the impedance.]

6-22. Work Problem 6-11, making use of the Smith chart. [*Note:* In this case involving region 2 with losses, one may see how the reflection coefficient magnitude is reduced, from its unity value at the terminal plane, to successively smaller values due to the spiralling towards the chart center.]

6-23. Suppose a plane wave is normally incident on the system illustrated, with region 2 assumed lossy. Demonstrate by use of the Smith chart, for the

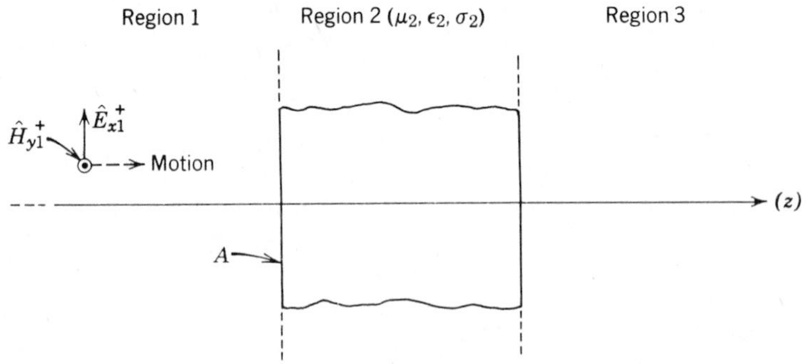

Problem 6-23

lossy region *sufficiently thick*, that the total impedance at the interface A approaches the intrinsic wave impedance $\hat{\eta}_2$, a result independent of the properties of the regions to the right of region 2.

6-24. Find the standing-wave ratio in region 1 of Example 6-2, by the following methods:

(*a*) Using (6-50), employing the forward- and backward-traveling wave magnitudes.

(*b*) Using (6-52a), employing the reflection coefficient magnitude.

(*c*) From the Smith chart results of Example 6-4, utilizing the intersection of the SWR circle with C as in Figure 6-12(*c*).

6-25. Utilize results of Problem 6-7 to find the SWR in regions 2 and 3 of Example 6-2.

6-26. Find the SWR in regions 1, 2, and 4 of Example 6-3, making use of any of the methods suggested in Problem 6-24. (Why is it improper to speak of the SWR in a lossy region?)

CHAPTER
7

The Poynting Theorem and Electromagnetic Power

Energy can be transported through empty space and within or along conductive or dielectric wave transmission devices by means of electromagnetic waves. The power flow through a closed surface in the region occupied by such waves may be interpreted from the surface integration of a power-flux density vector $\mathscr{P} \equiv \mathbf{E} \times \mathbf{H}$, known as the Poynting vector. The validity of this procedure is justified from the point of view of a theorem due to J. H. Poynting. Applications to the power flow associated with a wire carrying a direct current and with plane waves in lossless or conductive regions are considered. The related questions of time-instantaneous and time-average power-flux density and total power-flux through surfaces are treated using the real-time form of the fields. Simpler expressions for time-average power-flux density are then shown to arise from the employment of complex, time-harmonic forms of the fields.

7-1 The Theorem of Poynting

It is shown that the flow of electromagnetic power through a closed surface is obtained from a surface integral of the time-instantaneous quantity

$$\mathscr{P} \equiv \mathbf{E} \times \mathbf{H} \text{ VA/m}^2\text{, or W/m}^2 \tag{7-1}$$

known as the Poynting vector.† The units of (7-1) suggest a power flux density interpretation of $\mathscr{P}$. Taking the divergence of $\mathscr{P}$ obtains the two-term expansion

$$\nabla \cdot \mathscr{P} = \nabla \cdot (\mathbf{E} \times \mathbf{H}) = \mathbf{H} \cdot \nabla \times \mathbf{E} - \mathbf{E} \cdot \nabla \times \mathbf{H} \tag{7-2}$$

in view of the identity (16) in Table 2-2. The appearance of $\nabla \times \mathbf{E}$ and $\nabla \times \mathbf{H}$ in (7-2) prompts the substitution of Maxwell's equations (3-59) and (3-77), $\nabla \times \mathbf{E} = -\partial \mathbf{B}/\partial t$ and $\nabla \times \mathbf{H} = \mathbf{J} + \partial \mathbf{D}/\partial t$, yielding

$$\nabla \cdot \mathscr{P} = -\mathbf{H} \cdot \frac{\partial \mathbf{B}}{\partial t} - \mathbf{E} \cdot \frac{\partial \mathbf{D}}{\partial t} - \mathbf{J} \cdot \mathbf{E} \tag{7-3}$$

Utilizing the rules of differentiation (2-6) and (2-7), and with $\mathbf{B} = \mu \mathbf{H}$, one can write

$$\frac{\partial}{\partial t}\left(\frac{\mathbf{H} \cdot \mathbf{B}}{2}\right) = \frac{1}{2}\left[\mathbf{H} \cdot \frac{\partial \mathbf{B}}{\partial t} + \mathbf{B} \cdot \frac{\partial \mathbf{H}}{\partial t}\right] = \frac{1}{2}\left[\mathbf{H} \cdot \frac{\partial (\mu \mathbf{H})}{\partial t} + \mu \mathbf{H} \cdot \frac{\partial \mathbf{H}}{\partial t}\right]$$

$$= \mathbf{H} \cdot \frac{\partial \mathbf{B}}{\partial t} \tag{7-4}$$

assuming μ is not a function of time. Similarly, with $\mathbf{D} = \epsilon \mathbf{E}$, and ϵ not a function of time,

$$\frac{\partial}{\partial t}\left(\frac{\mathbf{E} \cdot \mathbf{D}}{2}\right) = \mathbf{E} \cdot \frac{\partial \mathbf{D}}{\partial t} \tag{7-5}$$

Substituting (7-4) and (7-5) into (7-3) yields

$$\nabla \cdot \mathscr{P} = -\frac{\partial}{\partial t}\left[\frac{\mathbf{H} \cdot \mathbf{B}}{2} + \frac{\mathbf{E} \cdot \mathbf{D}}{2}\right] - \mathbf{J} \cdot \mathbf{E} \tag{7-6}$$

This result shows that the power-flux-density vector $\mathscr{P}$ has a divergence in a region if at least one term on the right side of (7-6) is nonzero. Integrating (7-6) throughout an arbitrary volume region V obtains

$$\int_V \nabla \cdot \mathscr{P}\, dv = -\frac{\partial}{\partial t}\int_V \left[\frac{\mathbf{H} \cdot \mathbf{B}}{2} + \frac{\mathbf{E} \cdot \mathbf{D}}{2}\right] dv - \int_V \mathbf{J} \cdot \mathbf{E}\, dv \tag{7-7}$$

Assuming $\mathscr{P}$ in (7-7) meets the conditions of the divergence theorem discussed

† First defined in Poynting, J. H., "On the transfer of energy in the electromagnetic field," *Phil. Trans. Royal Society*, **175**, 343, 1884.

in Section 2-4-1, it can be reexpressed

$$-\oint_S \mathscr{P}\cdot d\mathbf{s} = \frac{\partial}{\partial t}\int_V \left[\frac{\mathbf{H}\cdot\mathbf{B}}{2} + \frac{\mathbf{E}\cdot\mathbf{D}}{2}\right] dv + \int_V \mathbf{J}\cdot\mathbf{E}\, dv \text{ W} \tag{7-8}$$

This is the integral form of the *theorem of Poynting*, interpreted physically in relation to Figure 7-1 as follows:

1. The left side of (7-8) denotes the *ingoing power-flux over S*, assuming $d\mathbf{s}$ outward-directed. In subsequent discussions, the symbol $P(t)$ is chosen to denote the time-instantaneous, net, ingoing power-flux as follows:

$$P(t) \equiv -\oint_S \mathscr{P}\cdot d\mathbf{s} \text{ W} \quad \text{Ingoing power-flux} \tag{7-9}$$

2. The first term of the right side of (7-8) denotes, at any instant, the *time rate of increase of total electromagnetic energy within the volume V enclosed by S*, in view of (4-61a) and (5-77) for electric field and magnetic field energies defined under static conditions

$$U_e \equiv \int_V \frac{\mathbf{E}\cdot\mathbf{D}}{2}\, dv \qquad U_m \equiv \int_V \frac{\mathbf{H}\cdot\mathbf{B}}{2}\, dv \text{ J} \tag{7-10}$$

3. The last term of (7-8) represents the *total dissipated or generated power within V* at any instant. If the projection of the current density vector $\mathbf{J}$ along $\mathbf{E}$ lies in the direction of $\mathbf{E}$, the power is dissipated in the region. An example occurs in a conductive region to which (3-7) applies; the substitution of $\mathbf{J} = \sigma\mathbf{E}$ into (7-8) then identifies the last term as an Ohmic *power-loss* term. In the event of a negative $\mathbf{E}$ directed projection of $\mathbf{J}$ along $\mathbf{E}$ in the region, the power obtained from the last term of (7-8) is

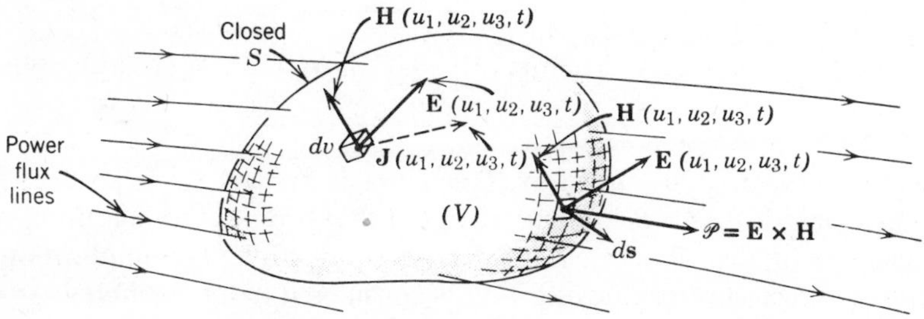

Figure 7-1. A typical volume in a region, depicting quantities associated with Poynting's theorem.

interpreted as *generated* power, in view of the reversal in the sign of the integrated result.

To summarize the observations (1) through (3) just given, (7-8) states that the net, inward power-flux $P(t) = -\oint_S \mathscr{P}\cdot d\mathbf{s}$, supplied by the field over a closed surface S, must equal the sum of the time rate of increase of electromagnetic energy inside V, plus the total Ohmic losses in V, assuming V contains no generators. If V contains *power generators*, the additional volume integral of $\mathbf{J}_g\cdot\mathbf{E}$ over the designated *active* current sources $\mathbf{J}_g$ in the region permits writing (7-8)

$$-\oint_S \mathscr{P}\cdot d\mathbf{s} = \frac{\partial}{\partial t}\int_V \left[\frac{\mathbf{H}\cdot\mathbf{B}}{2} + \frac{\mathbf{E}\cdot\mathbf{D}}{2}\right] dv + \int_V \mathbf{J}\cdot\mathbf{E}\, dv + \int_V \mathbf{J}_g\cdot\mathbf{E}\, dv$$

If the latter is rearranged with the generated power term on the left to read

$$-\int_V \mathbf{J}_g\cdot\mathbf{E}\, dv = \frac{\partial}{\partial t}\int_V \left[\frac{\mathbf{H}\cdot\mathbf{B}}{2} + \frac{\mathbf{E}\cdot\mathbf{D}}{2}\right] dv + \int_V \mathbf{J}\cdot\mathbf{E}\, dv + \oint_S \mathscr{P}\cdot d\mathbf{s} \quad (7\text{-}11)$$

the result is interpreted physically as follows. The total instantaneous generated power in V, given by the left side of (7-11), equals the sum of the time rate of increase in electromagnetic energy in V, the Ohmic losses in V, and the outgoing power-flux passing through the surface enclosing V. This form has some interpretive advantages when applied to an antenna, for example, in which case the last term, the integral of $\mathscr{P}\cdot d\mathbf{s}$ over any surface enclosing the antenna, denotes the power-flux radiated into remote regions of space. Of special physical interest is the time average of the radiated power which is discussed in the next section.

In a *static* electromagnetic system carrying only direct currents, the operator $\partial/\partial t$ is zero, reducing Poynting's theorem (7-8) or (7-11) to

$$-\oint_S \mathscr{P}\cdot d\mathbf{s} = \int_V \mathbf{J}\cdot\mathbf{E}\, dv \text{ W} \qquad \text{Time static} \quad (7\text{-}12)$$

assuming V contains no generators. Thus, in a dc system, the net power-flux entering a closed surface S constructed about the current-carrying conductors is a measure of the Ohmic losses in those conductors. The application of (7-12) to a direct-current-carrying wire is considered in an example.

EXAMPLE 7-1. From (7-12), evaluate the total power-flux entering the closed surface S embracing a length ℓ of a long, round wire carrying a direct current I

as in (*a*) of the accompanying figure. Compare the result with the volume integral of (7-12).

The closed surface S is noted in (*b*). The Poynting vector $\mathscr{P}$ on the peripheral surface $\rho = a$ is obtained from the known **E** and **H** fields, **H** being given by

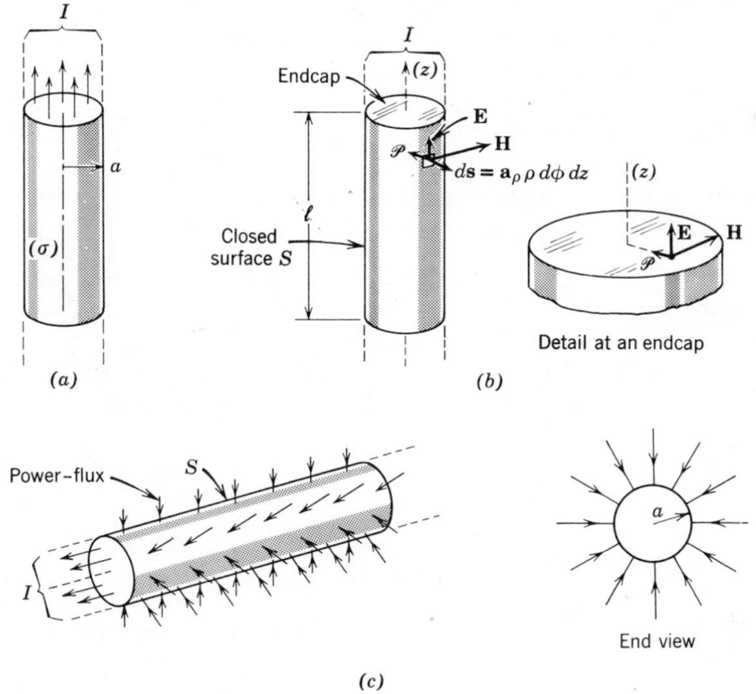

Example 7-1. (*a*) **Long, round wire carrying a static current** I. (*b*) **The E and H fields on the surface** S. (*c*) **Inward power-flux associated with direct current flow in a wire.**

(5-11) of Example 5-1, while **E** is obtained from the current density $J_z = I/A$ combined with (3-7)

$$\mathbf{E} = \mathbf{a}_z E_z = \mathbf{a}_z J_z/\sigma = \mathbf{a}_z I/\sigma A \qquad \mathbf{H} = \mathbf{a}_\phi H_\phi = \mathbf{a}_\phi I/2\pi a$$

The Poynting vector at $\rho = a$ on S is obtained from (7-1)

$$\mathscr{P} = \mathbf{E} \times \mathbf{H} = \left(\frac{\mathbf{a}_z I}{\sigma A}\right) \times \left(\frac{\mathbf{a}_\phi I}{2\pi a}\right) = -\mathbf{a}_\rho \frac{I^2}{2\pi a A \sigma}$$

As seen in (*b*), $\mathscr{P}$ on the end caps contributes nothing to the inward power-flux, making the total inward power-flux (7-9) over S

$$P \equiv -\oint_S \mathscr{P}\cdot d\mathbf{s} = -\int_{z=0}^{\ell}\int_{\phi=0}^{2\pi}\left(-\mathbf{a}_\rho \frac{I^2}{2\pi a A\sigma}\right)\cdot\mathbf{a}_\rho a\, d\phi\, dz$$

$$= \frac{I^2 2\pi a\ell}{2\pi a A\sigma} = I^2\frac{\ell}{\sigma A} = I^2 R \text{ W}$$

a result expressed in terms of the resistance (4-157) of the wire.

From (7-12), the result I^2R is also obtainable from the volume integral of $\mathbf{J}\cdot\mathbf{E}$ taken throughout the interior of S. Thus

$$\int_V \mathbf{J}\cdot\mathbf{E}\, dv = \int_V (\sigma\mathbf{E})\cdot\mathbf{E}\, dv = \int_V \sigma E_z^2\, dv = \int_{z=0}^{\ell}\int_{\phi=0}^{2\pi}\int_{\rho=0}^{a}\sigma\left(\frac{I}{\sigma A}\right)^2 \rho\, d\rho\, d\phi\, dz$$

integrating to I^2R as expected. The positive sign accounts for the actual *inward* sense of the power-flux P over S, as noted in (*c*).

Illustrations of the Poynting theorem in the time domain can be drawn from the theory of plane waves developed in Sections 2-10 and 3-6. Thus, the power-flux-density vector $\mathscr{P}$ associated with a plane wave in a region is obtained by use of (7-1) applied to the appropriate fields. In empty space, assume that a positive z traveling plane wave has electric and magnetic fields inferred from (2-121a) and (2-130a)

$$\mathbf{E} = \mathbf{a}_x E_x^+(z, t) = \mathbf{a}_x E_m^+ \cos(\omega t - \beta_0 z) \tag{7-13}$$

$$\mathbf{H} = \mathbf{a}_y H_y^+(z, t) = \mathbf{a}_y \frac{E_m^+}{\eta_0}\cos(\omega t - \beta_0 z) \tag{7-14}$$

Applying these to (7-1) obtains the time-instantaneous Poynting vector at any z position

$$\mathscr{P}(z, t) = \mathbf{E}\times\mathbf{H} = [\mathbf{a}_x E_m^+ \cos(\omega t - \beta_0 z)]\times\left[\mathbf{a}_y\frac{E_m^+}{\eta_0}\cos(\omega t - \beta_0 z)\right]$$

$$= \mathbf{a}_z\frac{(E_m^+)^2}{\eta_0}\cos^2(\omega t - \beta_0 z) \tag{7-15a}$$

The sketch of (7-15a) in Figure 7-2(*a*) shows $\mathscr{P}$ everywhere positive z directed. Denoting $\mathscr{P}(z, t)$ by $\mathbf{a}_z\mathscr{P}_z^+(z, t)$, an alternative plot of the scalar $\mathscr{P}_z^+$ is shown by the solid line in Figure 7-2(*b*). A double frequency variation of $\mathscr{P}$ with t and z produced by the squared cosine function is evident from these diagrams. Using the identity $\cos^2\theta = 1/2 + (1/2)\cos 2\theta$ permits writing

$$\mathscr{P} = \mathbf{a}_z\mathscr{P}_z^+ = \mathbf{a}_z\frac{(E_m^+)^2}{2\eta_0}[1 + \cos 2(\omega t - \beta_0 z)] \tag{7-15b}$$

a result useful when considering time-average power in the next section.

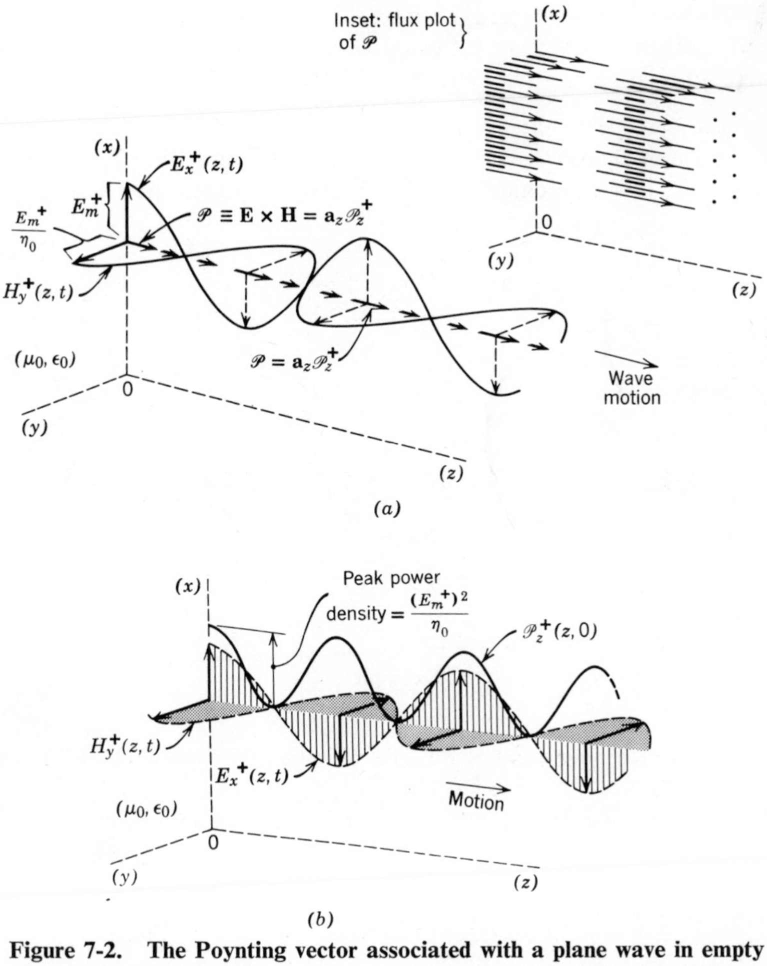

Figure 7-2. The Poynting vector associated with a plane wave in empty space. (*a*) The vector $\mathscr{P} = \mathbf{a}_z \mathscr{P}_z$ versus z at $t = 0$. (*b*) The scalar $\mathscr{P}_z(z, t)$ at $t = 0$.

The Poynting integral (7-8) applied to a region with no Ohmic losses and no generators present reduces to

$$P(t) \equiv -\oint_S \mathscr{P} \cdot d\mathbf{s} = \frac{\partial}{\partial t} \int_V \left[\frac{\mu_0 H^2}{2} + \frac{\epsilon_0 E^2}{2} \right] dv \text{ W} \tag{7-16}$$

signifying that the flux of $\mathscr{P}$ *into* a closed surface S in the lossless region is instantaneously a measure of the time rate of increase of the stored electromagnetic energy within S. In the example that follows, the validity of (7-16) is examined relative to a plane wave in free space.

EXAMPLE 7-2. Given the plane wave defined by (7-13) and (7-14), determine the net power flux $P(t)$ entering a closed, box-shaped surface S having dimensions as in the accompanying figure. Show that the time rate of increase

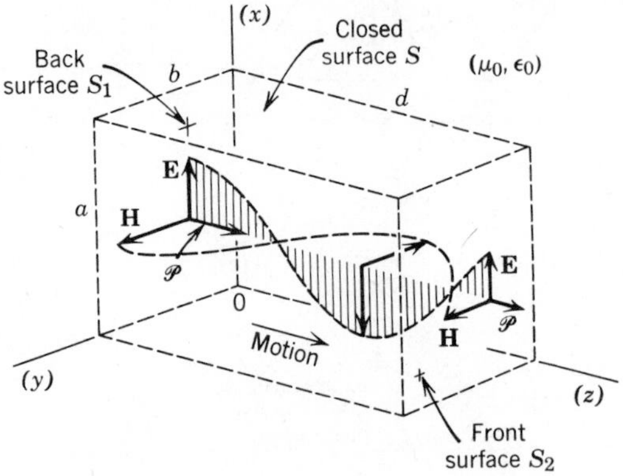

Example 7-2

of the electromagnetic energy within the volume of the box provides the same answer.

Because $\mathscr{P}$ is everywhere z directed, the only contributions to power-flux entering the box are on the ends S_1 and S_2 shown, so (7-16) yields

$$P_1(t) \equiv -\int_{S_1} \mathscr{P}\cdot d\mathbf{s} = -\int_{y=0}^{b}\int_{x=0}^{a}\left[\mathbf{a}_z \frac{(E_m^+)^2}{\eta_0}\cos^2(\omega t - \beta_0 z)\right]_{z=0}\cdot(-\mathbf{a}_z\, dx\, dy)$$

$$= \frac{(E_m^+)^2}{\eta_0} ab \cos^2 \omega t \tag{7-17a}$$

$$P_2(t) \equiv -\int_{S_2} \mathscr{P}\cdot d\mathbf{s} = -\frac{(E_m^+)^2}{\eta_0} ab \cos^2(\omega t - \beta_0 d) \tag{7-17b}$$

The net power-flux entering S is therefore

$$-\oint_S \mathscr{P}\cdot d\mathbf{s} \equiv P(t) = P_1(t) + P_2(t) = \frac{(E_m^+)^2}{\eta_0} ab[\cos^2 \omega t - \cos^2(\omega t - \beta_0 d)]$$

$$= \frac{(E_m^+)^2}{2\eta_0} ab[\cos 2\omega t - \cos 2(\omega t - \beta_0 d)] \text{ W} \tag{7-18}$$

the last being obtained by use of $\cos^2\theta = 1/2 + (1/2)\cos 2\theta$.

Equivalently, if the right side of (7-16) is integrated throughout the *volume* of the box, (7-18) should again be obtained. Substituting (7-13) and (7-14) yields

$$\frac{\partial}{\partial t}\int_V \left[\frac{\mu_0 H^2}{2} + \frac{\epsilon_0 E^2}{2}\right] dv = \frac{\partial}{\partial t}\left\{\frac{\mu_0 (E_m^+)^2}{4\eta_0^2}\int_0^d \int_0^b \int_0^a [1 + \cos 2(\omega t - \beta_0 z)]\, dx\, dy\, dz \right.$$
$$\left. + \frac{\epsilon_0 (E_m^+)^2}{4}\int_0^d \int_0^b \int_0^a [1 + \cos 2(\omega t - \beta_0 z)]\, dx\, dy\, dz\right\}$$
$$= \frac{\partial}{\partial t}\left\{\frac{\epsilon_0 (E_m^+)^2}{2}\int_0^d \int_0^b \int_0^a [1 + \cos 2(\omega t - \beta_0 z)]\, dx\, dy\, dz\right\}$$
$$= \frac{\omega \epsilon_0 (E_m^+)^2 ab}{2\beta_0}[-\cos 2(\omega t - \beta_0 z)]_0^d$$
$$= \frac{(E_m^+)^2 ab}{2\eta_0}[\cos 2\omega t - \cos 2(\omega t - \beta_0 d)] \tag{7-19}$$

agreeing with (7-18) as expected.†

For a plane wave traveling in a *conductive* (Ohmic) region, the effects of the attenuation of **E** and **H** and the phase shift between them is expected to influence the net power-flux $P(t)$ entering a closed surface S. For this case, the fields are given by real-time expressions inferred from (3-94) and (3-98c)

$$\mathbf{E} = \mathbf{a}_x E_x^+(z, t) = \mathbf{a}_x E_m^+ e^{-\alpha z} \cos(\omega t - \beta z) \tag{7-20}$$

$$\mathbf{H} = \mathbf{a}_y H_y^+(z, t) = \mathbf{a}_y \frac{E_m^+}{\eta} e^{-\alpha z} \cos(\omega t - \beta z - \theta) \tag{7-21}$$

in which θ is the angle of the wave impedance (3-99). The Poynting vector (7-1) thus becomes

$$\mathscr{P}(z, t) = \mathbf{E} \times \mathbf{H} = \mathbf{a}_z \frac{(E_m^+)^2}{\eta} e^{-2\alpha z} \cos(\omega t - \beta z)\cos(\omega t - \beta z - \theta) \tag{7-22a}$$

and the use of $\cos A \cos B = (1/2)[\cos(A + B) + \cos(A - B)]$ obtains

$$\mathscr{P} = \mathbf{a}_z \mathscr{P}_z^+(z, t) = \mathbf{a}_z \frac{(E_m^+)^2}{2\eta} e^{-2\alpha z}[\cos\theta + \cos(2\omega t - 2\beta z - \theta)] \text{ W/m}^2 \tag{7-22b}$$

A graph of $\mathscr{P}_z^+(z, t)$ versus z at $t = 0$ is shown in Figure 7-3. Not only does the attenuation of E_x^+ and H_y^+ account for a doubly attenuated power-flux density $\mathscr{P}_z^+$, but the effect of $\cos\theta$ in (7-22b), replacing the term unity in

† In the course of obtaining (7-19), note that with the substitution $(\mu_0/\eta_0^2) = \epsilon_0$, the two integrals in the first step become identical, so they combine into one. The time differentiation is taken inside the integral to eliminate the constant unity term, while in the last step, the identity $(\omega\epsilon_0/\beta_0) = \eta_0^{-1}$ is used.

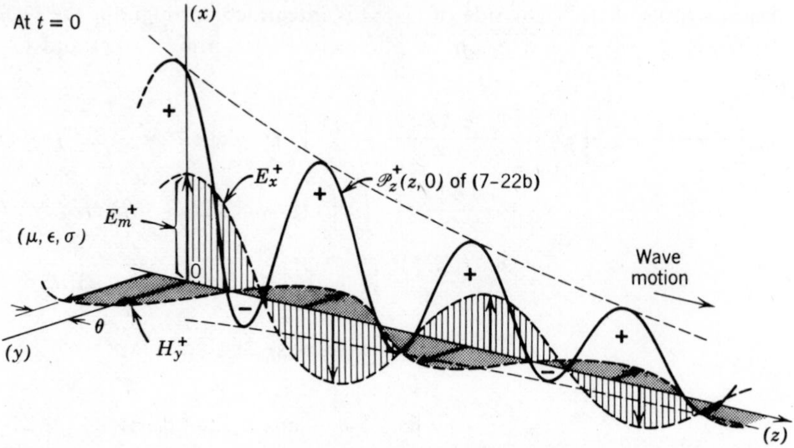

Figure 7-3. The instantaneous Poynting vector $\mathscr{P}_z^+(z, t)$ associated with a positive z traveling plane wave in a conductive region.

(7-15b) for the lossless case, is to cause $\mathscr{P}_z^+$ to go negative over a portion of each cycle, an effect associated with the phase shift θ between the electric and magnetic fields and detracting from the average power transmitted in the z direction.

EXAMPLE 7-3. If a plane wave exists in a conductive region, evaluate the net, instantaneous power-flux *entering* the box-shaped closed surface of dimensions as shown.

Integrating (7-22b) over the ends S_1 at $z = 0$ and S_2 at $z = d$ yields the instantaneous power-fluxes

$$P_1(t) \equiv -\int_{S_1} \mathscr{P} \cdot d\mathbf{s}$$

$$= -\int_{y=0}^{b} \int_{x=0}^{a} \left\{ \mathbf{a}_z \frac{(E_m^+)^2}{\eta} [\cos\theta + \cos(2\omega t - \theta)] \right\} \cdot (-\mathbf{a}_z\, dx\, dy)$$

$$= \frac{(E_m^+)^2}{2\eta} ab[\cos\theta + \cos(2\omega t - \theta)] \tag{7-23a}$$

$$P_2(t) = -\frac{(E_m^+)^2}{2\eta} e^{-2\alpha d} ab[\cos\theta + \cos(2\omega t - 2\beta d - \theta)] \tag{7-23b}$$

so that the net power-flux entering the box is their sum

$$P(t) = \frac{(E_m^+)^2}{2\eta} ab[(1 - e^{-2\alpha d})\cos\theta + \cos(2\omega t - \theta) - e^{-2\alpha d}\cos(2\omega t - 2\beta d - \theta)] \tag{7-24}$$

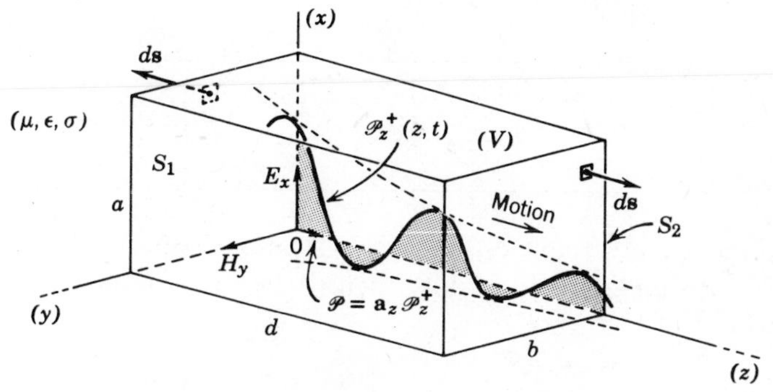

Example 7-3

From the Poynting theorem (7-8) it is evident that (7-24) is a measure of the time rate of increase of the stored electromagnetic energy within the volume plus the instantaneous Ohmic loss occurring therein. One can expect that (7-24) will reduce to (7-18) if a *lossless* region ($\sigma = 0$) is assumed.

7-2 Time-Average Poynting Vector and Power

In a consideration of the electromagnetic power delivered by sinusoidally time-varying fields to a region or system, one's interest from the point of view of practical measurements leans towards the *time average* of the power-flux rather than its instantaneous value considered in the previous section. Time-average power in electromagnetic fields is important for the same reasons as in circuit theory. The time-average power entering the terminals of a passive network, found by use of an electrodynamometer type wattmeter or from the knowledge of the amplitude and phase of the input voltage and current, is a measure of the average power dissipated as heat in all the resistive elements of the network. From the electromagnetic viewpoint, the time-average power-flux entering a closed surface containing no generators is a criterion of the same thing: the heat-producing Ohmic losses in the region.

In laboratory measurements, the time average of a time-harmonic function is customarily taken over a time interval embracing many cycles or periods. Since for steady state sinusoidal functions all periods are alike, an average over one period will yield the same result as that taken over many such periods. The time average of the Poynting vector $\mathscr{P}(u_1, u_2, u_3, t)$, denoted by $\mathscr{P}_{\text{av}}$, is defined as the area under the function $\mathscr{P}$ over a cycle, divided by the duration T (period) of the cycle, i.e.,

$$\mathscr{P}_{\mathrm{av}}(u_1, u_2, u_3) = \frac{\text{Area under } \mathscr{P} \text{ over a cycle}}{\text{Base } (T \text{ sec})} = \frac{1}{T}\int_0^T \mathscr{P}(u_1, u_2, u_3, t)\, dt \tag{7-25a}$$

if t is chosen as the variable of integration. One may alternatively choose ωt as the angular integration variable; then (7-25a) is written with 2π as the base-divisor

$$\mathscr{P}_{\mathrm{av}}(u_1, u_2, u_3) = \frac{1}{2\pi}\int_0^{2\pi} \mathscr{P}(u_1, u_2, u_3, t)\, d(\omega t) \text{ W/m}^2 \tag{7-25b}$$

It is evident that the time-average Poynting vector is a function solely of position in space, the time variable having been integrated out over definite limits (in t or ωt) in the averaging process.

Illustrations of the time-average Poynting vector can be drawn from examples in the last section. Equation (7-15a) denotes a time-instantaneous Poynting vector $\mathscr{P}(z, t) = \mathbf{a}_z \mathscr{P}_z(z, t)$ attributed to the wave of (7-13) and (7-14). Applying (7-25b) obtains its time average

$$\mathscr{P}_{\mathrm{av}}(z) = \frac{\mathbf{a}_z}{2\pi}\frac{(E_m^+)^2}{2\eta_0}\int_0^{2\pi} d(\omega t) + \frac{\mathbf{a}_z}{2\pi}\frac{(E_m^+)^2}{2\eta_0}\int_0^{2\pi} \cos 2(\omega t - \beta_0 z)\, d(\omega t) = \mathbf{a}_z \frac{(E_m^+)^2}{2\eta_0} \tag{7-26}$$

Thus the time-average result (7-26) is attributable wholly to the constant, first term of the time-instantaneous expression (7-15b). The double frequency term contributes nothing on the time average because it possesses canceling positive and negative areas over a cycle, evident from the $\mathscr{P}_z(z, t)$ diagram of Figure 7-4(a), which is just an extension of Figure 7-2(b) to successive instants in time t. The inset in Figure 7-4(a), showing the wave at the fixed $z = 0$ location, yields an average Poynting vector (area divided by the base) that is one-half the peak power density $(E_m^+)^2/\eta_0$, or (7-26).

If the region is lossy, $\mathscr{P}_{\mathrm{av}}$ becomes a function of z due to the wave attenuation produced by the losses. The time-instantaneous Poynting vector, in this case expressed by (7-22b), is depicted in Figure 7-4(b), an extension of Figure 7-3. In the insets are shown variations of $\mathscr{P}_z(z, t)$ with t at two fixed z locations ($z = 0$ and λ). Making use of (7-25b) leads to the time-average

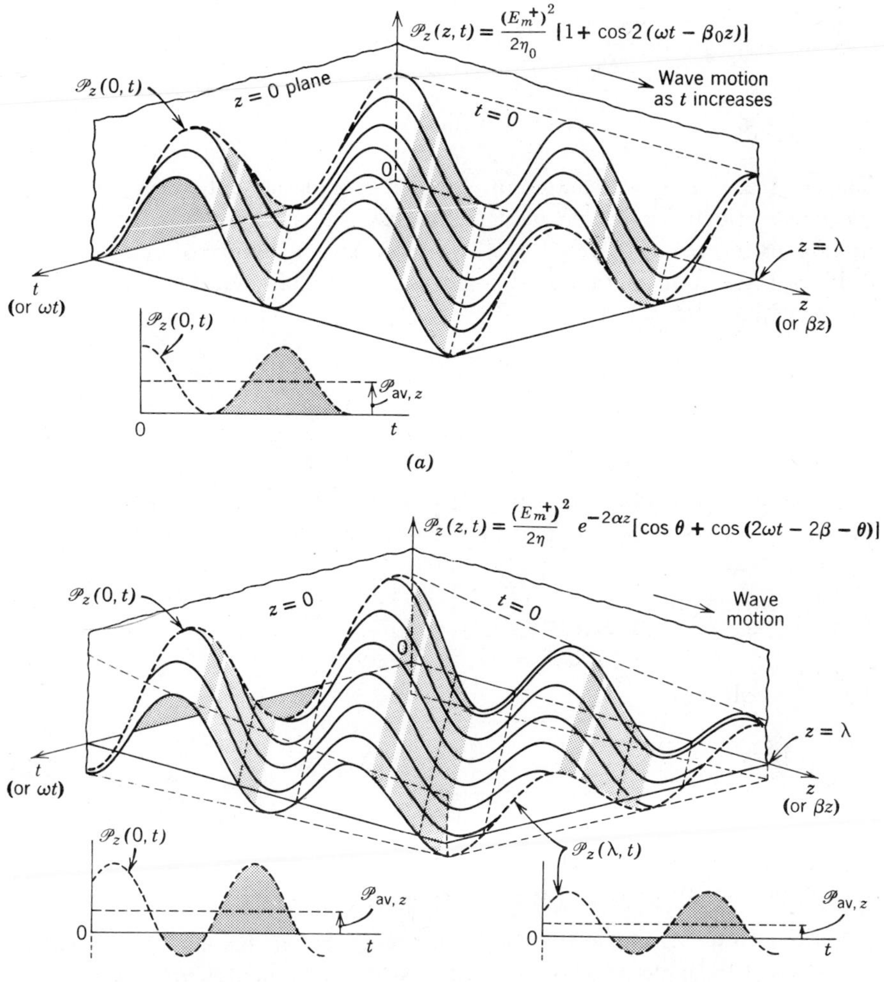

Figure 7-4. Poynting vector $\mathscr{P}_z(z, t)$ of forward z traveling plane waves in lossless and lossy regions. (*a*) Plane wave in a lossless region. Time variations at $z = 0$ are noted in the lower inset. (*b*) Plane wave in a lossy region. Below are shown time variations at $z = 0$ and $z = \lambda$.

Poynting vector

$$\mathscr{P}_{av}(z) = \frac{1}{2\pi}\int_0^{2\pi} \mathbf{a}_z \frac{(E_m^+)^2}{2\eta} e^{-2\alpha z}[\cos\theta + \cos(2\omega t - 2\beta z - \theta)]\, d(\omega t)$$

$$= \mathbf{a}_z \frac{(E_m^+)^2}{2\eta} e^{-2\alpha z} \cos\theta \ \text{W/m}^2 \tag{7-27}$$

The result is doubly attenuated in z; it also retains the factor $\cos\theta$ produced by the electric and magnetic fields being out of phase by an angle θ, a factor analogous to the *power factor* of a two-terminal impedance of circuit theory.

If the total, *time-average power-flux* emerging from a surface S is desired, one must integrate $\mathscr{P}_{av}$ to obtain

$$P_{av} = \int_S \mathscr{P}_{av}\cdot d\mathbf{s} \ \text{W} \tag{7-28a}$$

Another way to evaluate P_{av} is by averaging the total, time-instantaneous power-flux $P(t)$ through S. Thus, (7-25b) inserted in (7-28a) yields

$$P_{av} = \int_S \mathscr{P}_{av}\cdot d\mathbf{s} = \int_S \left[\frac{1}{2\pi}\int_0^{2\pi} \mathscr{P}\, d(\omega t)\right]\cdot d\mathbf{s} = \frac{1}{2\pi}\int_0^{2\pi}\left[\int_S \mathscr{P}\cdot d\mathbf{s}\right] d(\omega t)$$

whence

$$P_{av} = \frac{1}{2\pi}\int_0^{2\pi} P(t)\, d(\omega t) \ \text{W} \tag{7-28b}$$

The preference for (7-28a) or (7-28b) in evaluating P_{av} depends on the comparative convenience of the integration process.

EXAMPLE 7-4. Evaluate the net, time-average power-flux *entering* the closed surface of Example 7-2 in a free-space region containing the given wave.

The time-average power-flux entering the box is found by use of (7-28a) or (7-28b). With $d\mathbf{s}$ denoting a positive outward surface-element, (7-28a) is written with a minus sign if the net, *inward* flux is desired

$$P_{av} = -\oint \mathscr{P}_{av}\cdot d\mathbf{s} \tag{7-29}$$

With $\mathscr{P}_{av}$ given by (7-26), the average power-flux entering S_1 is

$$P_{av,1} = -\int_{S_1} \mathscr{P}_{av}\cdot d\mathbf{s} = -\int_0^b\int_0^a \mathbf{a}_z\left[\frac{(E_m^+)^2}{2\eta_0}\right]\cdot(-\mathbf{a}_z\, dx\, dy) = \frac{(E_m^+)^2}{2\eta_0} ab$$

positive because the true direction of the flux is *into* the box. A similar integration over S_2 yields the negative of that result because the flux comes *out* of the box. The net time-average power-flux entering the box is thus zero, i.e., $P_{av} = P_{av,1} + P_{av,2} = 0$, a result expected generally from closed surfaces embracing a lossless region and containing no sources.

For a sinusoidally time-varying electromagnetic field in a region possessing *losses* but no sources, the time-average power-flux P_{av} entering a closed surface is a measure of the time-average Ohmic power loss within the interior volume. This is demonstrated by beginning with the time-instantaneous integral form (7-8) of Poynting's theorem

$$P(t) \equiv -\oint_S \mathscr{P} \cdot d\mathbf{s} = \frac{\partial}{\partial t}(U_e + U_m) + \int_V \mathbf{J} \cdot \mathbf{E}\, dv \text{ W} \qquad [7\text{-}8]$$

Assuming sinusoidal fields, the time-average of the left side of (7-8), given by (7-29), equals the time average of the right side, yielding

$$-\oint_S \mathscr{P}_{av} \cdot d\mathbf{s} = \frac{1}{2\pi}\int_0^{2\pi} \frac{\partial U_e}{\partial t}\, d(\omega t) + \frac{1}{2\pi}\int_0^{2\pi} \frac{\partial U_m}{\partial t}\, d(\omega t)$$
$$+ \frac{1}{2\pi}\int_0^{2\pi}\left[\int_V \mathbf{J} \cdot \mathbf{E}\, dv\right] d(\omega t) \qquad (7\text{-}30)$$

The stored-energy quantities U_e and U_m are, from (7-10), obtained from volume integrals of E^2 and H^2 respectively, implying double frequency variations in time. Such time variations of U_e in a volume region are depicted in Figure 7-5, along with its time derivative $\partial U_e/\partial t$. Its time average, given by the first integral of the right side of (7-30), is therefore zero. Similar arguments lead to a zero time average of $\partial U_m/\partial t$, reducing (7-30) to

$$P_{av} \equiv -\oint_S \mathscr{P}_{av} \cdot d\mathbf{s} = \frac{1}{2\pi}\int_0^{2\pi}\left[\int_V \mathbf{J} \cdot \mathbf{E}\, dv\right] d(\omega t) \text{ W} \qquad (7\text{-}31)$$

One concludes that the time-average power-flux entering a closed surface S equals the average power dissipated as heat inside V bounded by S, provided there are no sources in V.

EXAMPLE 7-5. Compare the net time-average power-flux entering the box-shaped surface of Example 7-3 with the time-average Ohmic losses inside, assuming the same attenuated wave in the region.

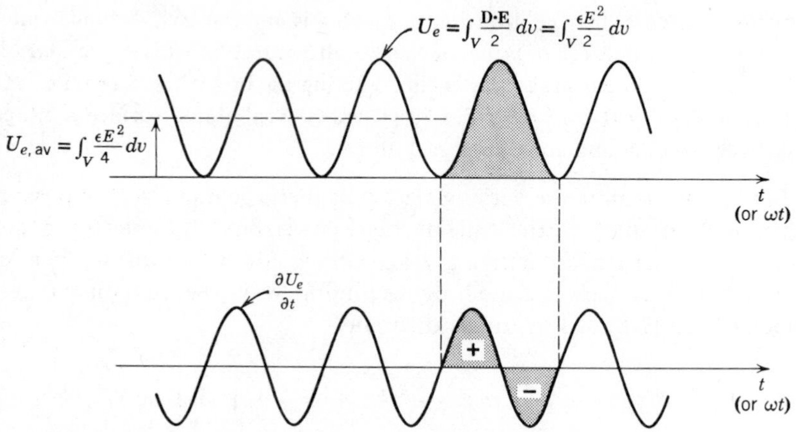

Figure 7-5. Total electric field energy and energy time rate conditions as a function of time for a volume region, assuming sinusoidal fields.

Anywhere in the region $\mathscr{P}_{\text{av}}$ is given by (7-27)

$$\mathscr{P}_{\text{av}}(z) = \mathbf{a}_z \frac{(E_m^+)^2}{2\eta} e^{-2\alpha z} \cos\theta \tag{7-32}$$

Inserting (7-32) into (7-31) yields contributions to P_{av} over only the box ends at $z = 0$ and $z = d$ as follows:

$$P_{\text{av}} \equiv -\oint_S \mathscr{P}_{\text{av}}\cdot d\mathbf{s} = -\int_{x=0}^{a}\int_{y=0}^{b}\left(\mathbf{a}_z \frac{(E_m^+)^2}{2\eta} e^{-2\alpha z}\cos\theta\right)\cdot(-\mathbf{a}_z\,dx\,dy)\Big]_{z=0}$$
$$-\int_{x=0}^{a}\int_{y=0}^{b}\left(\mathbf{a}_z \frac{(E_m^+)^2}{2\eta} e^{-2\alpha z}\cos\theta\right)\cdot(\mathbf{a}_z\,dx\,dy)\Big]_{z=d}$$
$$= \frac{(E_m^+)^2}{2\eta} ab[1 - e^{-2\alpha d}]\cos\theta \tag{7-33}$$

One can also obtain (7-33) by use of the right side of (7-31) through the time-average Ohmic losses in V. The integration simplifies if one puts

$$\frac{1}{2\pi}\int_0^{2\pi}\left[\int_V \mathbf{J}\cdot\mathbf{E}\,dv\right]d(\omega t) = \int_V\left[\frac{1}{2\pi}\int_0^{2\pi}\mathbf{J}\cdot\mathbf{E}\,d(\omega t)\right]dv \tag{7-34}$$

stating that the time average of the volume integral of $\mathbf{J}\cdot\mathbf{E}$ equals the volume integral of the time average of $\mathbf{J}\cdot\mathbf{E}$. Inserting $\mathbf{E}$ from (7-20) and making use of $\mathbf{J} = \sigma\mathbf{E}$ yields

$$\int_V\left[\frac{1}{2\pi}\int_0^{2\pi}\mathbf{J}\cdot\mathbf{E}\,d(\omega t)\right]dv = \int_V\left[\frac{1}{2\pi}\int_0^{2\pi}\sigma(E_m^+)^2 e^{-2\alpha z}\cos^2(\omega t - \beta z)\,d(\omega t)\right]dv$$
$$= \frac{\sigma(E_m^+)^2}{4\alpha} ab[1 - e^{-2\alpha d}] \tag{7-35}$$

which equals (7-33) provided that

$$\frac{\cos\theta}{\eta} = \frac{\sigma}{2\alpha} \tag{7-36}$$

It is left to the reader to prove the latter, using the appropriate definitions of α, η, and θ from Section 3-6.

7-3 Time-Average Poynting Vector and Time-Harmonic Fields

In the discussion of plane wave fields in Sections 2-10 and 3-6, it has been seen how the use of the complex forms eliminates t through the use of the factor $e^{j\omega t}$. Because in the course of problem-solving, field solutions are frequently obtained in complex form, it is useful to be able to find the time-average Poynting vector $\mathscr{P}_{\text{av}}$ directly from the complex solutions. Such results are obtained in this section, along with a version of the Poynting theorem (7-8) employing complex forms.

Revising (7-25) in terms of the complex fields requires restating the Poynting vector (7-1) in terms of the complex fields. The real-time fields **E** and **H** are related to their complex forms $\hat{\mathbf{E}}$ and $\hat{\mathbf{H}}$ by (2-74); i.e.,

$$\mathbf{E}(u_1, u_2, u_3, t) = \text{Re}\,[\hat{\mathbf{E}}(u_1, u_2, u_3)e^{j\omega t}] \tag{7-37}$$

$$\mathbf{H}(u_1, u_2, u_3, t) = \text{Re}\,[\hat{\mathbf{H}}(u_1, u_2, u_3)e^{j\omega t}] \tag{7-38}$$

$\hat{\mathbf{E}}$ and $\hat{\mathbf{H}}$ are expressed in complex polar and rectangular forms as follows:

$$\hat{\mathbf{E}}(u_1, u_2, u_3) \equiv \mathbf{a}_e\hat{E}(u_1, u_2, u_3) = \mathbf{a}_e E e^{j\theta_e} = \mathbf{a}_e[E_r + jE_i] \tag{7-39}$$

$$\hat{\mathbf{H}}(u_1, u_2, u_3) \equiv \mathbf{a}_h\hat{H}(u_1, u_2, u_3) = \mathbf{a}_h H e^{j\theta_h} = \mathbf{a}_h[H_r + jH_i] \tag{7-40}$$

and are depicted graphically in Figure 7-6(*a*), the unit vectors $\mathbf{a}_e$ and $\mathbf{a}_h$

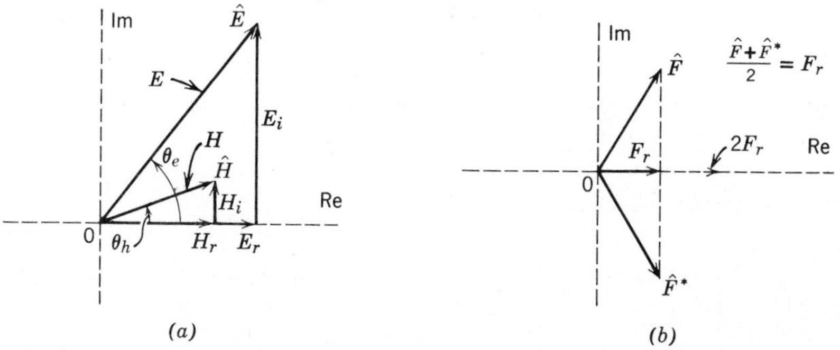

Figure 7-6. Features of fields represented in the complex plane. (*a*) Electric and magnetic field phasors, showing polar and rectangular forms. (*b*) Complex function $\hat{F}$ and its conjugate $\hat{F}^*$, showing relationship to the real part F_r.

designating the vector directions of the fields. The polar form in (7-39) substituted into (7-37) obtains the relationships between the complex and real-time forms†

$$\mathbf{E}(u_1, u_2, u_3, t) = \text{Re}\,[\hat{\mathbf{E}}(u_1, u_2, u_3)e^{j\omega t}] = \mathbf{a}_e\,\text{Re}\,[\hat{E}e^{j\omega t}] \tag{7-41}$$

$$= \mathbf{a}_e\,\text{Re}\,[Ee^{j(\omega t + \theta_e)}] = \mathbf{a}_e E \cos(\omega t + \theta_e) \tag{7-42}$$

Similarly, for the magnetic field

$$\mathbf{H}(u_1, u_2, u_3, t) = \mathbf{a}_h\,\text{Re}\,[\hat{H}e^{j\omega t}] = \mathbf{a}_h H \cos(\omega t + \theta_h) \tag{7-43}$$

The **E** field (7-41) can also be written

$$\mathbf{E}(u_1, u_2, u_3, t) = \text{Re}\,[\hat{\mathbf{E}}(u_1, u_2, u_3)e^{j\omega t}] = \mathbf{a}_e\,\text{Re}\,[\hat{E}e^{j\omega t}]$$

$$= \mathbf{a}_e \frac{\hat{E}e^{j\omega t} + (\hat{E}e^{j\omega t})^*}{2} = \mathbf{a}_e \frac{\hat{E}e^{j\omega t} + \hat{E}^*e^{-j\omega t}}{2} \tag{7-44}$$

with the latter obtained from the relationship of the real part F_r of any complex function to $\hat{F}$ and its conjugate $\hat{F}^*$ as depicted in Figure 7-6(*b*)

$$F_r = \text{Re}\,[\hat{F}] = \frac{\hat{F} + \hat{F}^*}{2} \tag{7-45}$$

Thus, the terms $\hat{E}e^{j\omega t}$ and $\hat{E}^*e^{-j\omega t}$ in (7-44) appear as two counter-rotating phasors with angular speeds ω and $-\omega$, always in conjugate positions such that half their sum produces the required real axis projection at every instant. The substitution of (7-44) and a similar expression for **H** into (7-1) obtains the real-time Poynting vector

$$\mathscr{P} \equiv \mathbf{E} \times \mathbf{H} = \mathbf{a}_e \times \mathbf{a}_h \tfrac{1}{4}(\hat{E}e^{j\omega t} + \hat{E}^*e^{-j\omega t})(\hat{H}e^{j\omega t} + \hat{H}^*e^{-j\omega t})$$

$$= \mathbf{a}_e \times \mathbf{a}_h \tfrac{1}{4}(\hat{E}\hat{H}e^{j2\omega t} + \hat{E}^*\hat{H}^*e^{-j2\omega t} + \hat{E}^*\hat{H} + \hat{E}\hat{H}^*) \tag{7-46}$$

The first two terms, by use of the complex rectangular forms (7-39) and (7-40), yield the double frequency results

$$\hat{E}\hat{H}e^{j2\omega t} + \hat{E}^*\hat{H}^*e^{-j2\omega t} = 2[(E_rH_r - E_iH_i)\cos 2\omega t - (E_iH_r + E_rH_i)\sin 2\omega t] \tag{7-47}$$

† The physical meanings of the symbols $\mathbf{a}_e$, E, and θ_e are clarified by comparing (7-41) with explicit solutions. For example, from (3-93), a positive z traveling wave in a dissipative region is

$$\mathbf{E}(z, t) = \mathbf{a}_x\,\text{Re}\,[E_m^+ e^{-\alpha z} e^{j(\omega t - \beta z)}]$$

Comparison with (7-42) shows in this case that $\mathbf{a}_e = \mathbf{a}_x$, $E = (E_m^+ e^{-\alpha z})$, and $\theta_e = -\beta z$.

while the last two terms of (7-46), independent of time, become

$$\hat{E}^*\hat{H} = (E_r - jE_i)(H_r + jH_i) = E_rH_r + E_iH_i + j(E_rH_i - E_iH_r) \quad \text{(7-48a)}$$

$$\hat{E}\hat{H}^* = (E_r + jE_i)(H_r - jH_i) = E_rH_r + E_iH_i - j(E_rH_i - E_iH_r) \quad \text{(7-48b)}$$

yielding

$$\hat{E}^*\hat{H} + \hat{E}\hat{H}^* = 2(E_rH_r + E_iH_i) \quad \text{(7-49)}$$

It is evident that of these contributions to (7-46), only the constant (7-49) contributes to the *time average* defined by (7-25b)

$$\mathscr{P}_{\text{av}} = \frac{1}{2\pi}\int_0^{2\pi} (\mathbf{E} \times \mathbf{H})\, d(\omega t) = \mathbf{a}_e \times \mathbf{a}_h \tfrac{1}{2}(E_rH_r + E_iH_i) \quad \text{(7-50)}$$

The latter is simplified upon noting that the parenthesized factor is just the real part of either (7-48a) or (7-48b), yielding

$$\mathscr{P}_{\text{av}} = \mathbf{a}_e \times \mathbf{a}_h \tfrac{1}{2}\,\text{Re}\,(\hat{E}^*\hat{H}) = \tfrac{1}{2}\,\text{Re}\,(\hat{\mathbf{E}}^* \times \hat{\mathbf{H}})\ \text{W/m}^2 \quad \text{(7-51a)}$$

or

$$\mathscr{P}_{\text{av}} = \mathbf{a}_e \times \mathbf{a}_h \tfrac{1}{2}\,\text{Re}\,(\hat{E}\hat{H}^*) = \tfrac{1}{2}\,\text{Re}\,(\hat{\mathbf{E}} \times \hat{\mathbf{H}}^*)\ \text{W/m}^2 \quad \text{(7-51b)}$$

the desired results. Equation (7-51b) states that *one-half the real part of the cross product of the complex electric field with the complex conjugate of the magnetic field yields the time-average Poynting vector at any position.* An alternative form of the same thing is (7-51a). These expressions obviate the time-averaging integration of (7-25) by making direct use of the complex forms of the electric and magnetic fields, and so they are most convenient.

If the net time-average power-flux entering a closed surface S is desired, inserting (7-51b) into (7-29) now obtains

$$P_{\text{av}} = -\oint_S \mathscr{P}_{\text{av}} \cdot d\mathbf{s} = -\oint_S \tfrac{1}{2}\,\text{Re}\,(\hat{\mathbf{E}} \times \hat{\mathbf{H}}^*) \cdot d\mathbf{s}\ \text{W} \quad \text{(7-52)}$$

EXAMPLE 7-6. Use the complex form of the attenuated plane wave fields (7-20) and (7-21) to obtain the time-average Poynting vector at any position in the region.

The complex forms of (7-20) and (7-21) are

$$\hat{\mathbf{E}}(z) = \mathbf{a}_x E_m^+ e^{-\alpha z} e^{-j\beta z} \tag{7-53}$$

$$\hat{\mathbf{H}}(z) = \mathbf{a}_y \frac{E_m^+}{\eta} e^{-\alpha z} e^{-j\theta} e^{-j\beta z} \tag{7-54}$$

and with these into (7-51b)

$$\mathscr{P}_{\text{av}} = \tfrac{1}{2}\,\text{Re}\,(\hat{\mathbf{E}} \times \hat{\mathbf{H}}^*) = \tfrac{1}{2}\,\text{Re}\left(\mathbf{a}_x \times \mathbf{a}_y \frac{(E_m^+)^2}{\eta} e^{-2\alpha z} e^{j\theta}\right)$$

$$= \mathbf{a}_z \frac{(E_m^+)^2}{2\eta} e^{-2\alpha z} \cos\theta \text{ W/m}^2 \tag{7-55}$$

which agrees with (7-27).

In the foregoing was shown how the time-average electromagnetic power-flux entering a closed surface is obtained using the complex $\hat{\mathbf{E}}$ and $\hat{\mathbf{H}}$ fields. This was seen, from the time-average Poynting theorem (7-31), to have the important interpretation of representing the time-average Ohmic power loss in the volume enclosed, assuming no sources therein. An alternative version of (7-31) is obtained directly from the complex Maxwell equations. Thus, beginning with (3-83) and (3-84), $\nabla \times \hat{\mathbf{E}} = -j\omega\mu\hat{\mathbf{H}}$ and $\nabla \times \hat{\mathbf{H}} = \hat{\mathbf{J}} + j\omega\epsilon\hat{\mathbf{E}}$, and forming the dot product of (3-83) with the conjugate of $\hat{\mathbf{H}}$, and the dot product of the conjugate of (3-84) with $\hat{\mathbf{E}}$, obtains

$$(\nabla \times \hat{\mathbf{E}})\cdot\hat{\mathbf{H}}^* = -j\omega\mu\hat{\mathbf{H}}\cdot\hat{\mathbf{H}}^* \tag{7-56}$$

$$(\nabla \times \hat{\mathbf{H}}^*)\cdot\hat{\mathbf{E}} = \hat{\mathbf{J}}^*\cdot\hat{\mathbf{E}} + (j\omega\epsilon\hat{\mathbf{E}})^*\cdot\hat{\mathbf{E}} = \hat{\mathbf{J}}^*\cdot\hat{\mathbf{E}} - j\omega\epsilon\hat{\mathbf{E}}^*\cdot\hat{\mathbf{E}} \tag{7-57}$$

Subtracting (7-57) from (7-56) yields

$$\hat{\mathbf{H}}^*\cdot(\nabla \times \hat{\mathbf{E}}) - \hat{\mathbf{E}}\cdot(\nabla \times \hat{\mathbf{H}}^*) = -j\omega\mu\hat{\mathbf{H}}\cdot\hat{\mathbf{H}}^* + j\omega\epsilon\hat{\mathbf{E}}\cdot\hat{\mathbf{E}}^* - \hat{\mathbf{J}}^*\cdot\hat{\mathbf{E}}$$

the left side of which reduces, using (16) of Table 2-2, to yield

$$\nabla\cdot(\hat{\mathbf{E}} \times \hat{\mathbf{H}}^*) = -j\omega\mu\hat{\mathbf{H}}\cdot\hat{\mathbf{H}}^* + j\omega\epsilon\hat{\mathbf{E}}\cdot\hat{\mathbf{E}}^* - \hat{\mathbf{J}}^*\cdot\hat{\mathbf{E}} \tag{7-58}$$

Integrating (7-58) throughout any volume V obtains, upon applying the divergence theorem (2-34) to the left side, the following *complex version of the Poynting theorem:*

$$-\oint_S (\hat{\mathbf{E}} \times \hat{\mathbf{H}}^*)\cdot d\mathbf{s} = j\omega\int_V [\mu\hat{\mathbf{H}}\cdot\hat{\mathbf{H}}^* - \epsilon\hat{\mathbf{E}}\cdot\hat{\mathbf{E}}^*]\,dv + \int_V \hat{\mathbf{J}}^*\cdot\hat{\mathbf{E}}\,dv \tag{7-59a}$$

If the current densities $\hat{\mathbf{J}}$ in V consist partly of *driven sources* $\hat{\mathbf{J}}_g$ (generated currents), the additional volume integral of $\hat{\mathbf{J}}_g^* \cdot \hat{\mathbf{E}}$ over those sources converts (7-59a) to a result which, when rearranged with the generated power term on the left, reads

$$-\int_V \hat{\mathbf{J}}_g^* \cdot \hat{\mathbf{E}}\, dv = j\omega \int_V [\mu \hat{\mathbf{H}} \cdot \hat{\mathbf{H}}^* - \epsilon \hat{\mathbf{E}} \cdot \hat{\mathbf{E}}^*]\, dv + \int_V \hat{\mathbf{J}}^* \cdot \hat{\mathbf{E}}\, dv + \oint_S (\hat{\mathbf{E}} \times \hat{\mathbf{H}}^*) \cdot d\mathbf{s} \tag{7-59b}$$

The choice of (7-59a) or (7-59b) thus depends on whether or not current generators $\hat{\mathbf{J}}_g$ are present in the volume under consideration. Their *real-time* counterparts are (7-8) and (7-11), developed in Section 7-1. The physical interpretations are rather different, however, as seen from the following.

Physical interpretations of the complex Poynting expressions become evident upon equating the real and the imaginary parts of (7-59a) or (7-59b). Assume a source free, dissipative volume region with μ and ϵ pure real and $\mathbf{J} = \sigma \mathbf{E}$. Equating one-half the *real* parts of (7-59a) yields the following Poynting integral expression:

$$P_{av} \equiv -\oint_S \tfrac{1}{2} \operatorname{Re} (\hat{\mathbf{E}} \times \hat{\mathbf{H}}^*) \cdot d\mathbf{s} = \int_V \frac{\sigma E^2}{2}\, dv \text{ W} \tag{7-60}$$

noting that $\operatorname{Re} (\hat{\mathbf{J}}^* \cdot \hat{\mathbf{E}}) = \operatorname{Re} (\sigma \hat{\mathbf{E}}^* \cdot \hat{\mathbf{E}}) = \sigma E^2$, in which E denotes the magnitude of $\hat{\mathbf{E}}$ according to (7-42), while from (7-52) the left integral of (7-60) is just P_{av}, the time-average power-flux entering S. Therefore, (7-60) and (7-31) are entirely *equivalent expressions*.

The equality of one-half the *imaginary* parts of (7-59a) obtains the less important result

$$-\oint_S \tfrac{1}{2} \operatorname{Im} (\hat{\mathbf{E}} \times \hat{\mathbf{H}}^*) \cdot d\mathbf{s} = 2\omega \int_V \left[\frac{\mu \hat{\mathbf{H}} \cdot \hat{\mathbf{H}}^*}{4} - \frac{\epsilon \hat{\mathbf{E}} \cdot \hat{\mathbf{E}}^*}{4} \right] dv \tag{7-61}$$

The terms $(1/4)\mu \hat{\mathbf{H}} \cdot \hat{\mathbf{H}}^*$ and $(1/4)\epsilon \hat{\mathbf{E}} \cdot \hat{\mathbf{E}}^*$, independent of time, denote the time averages of the stored energy densities of the magnetic and electric fields in V, a fact appreciated upon reexamining Figure 7-5, showing the total instantaneous field energy of a sinusoidal electric field along with its time average in a typical volume region. Thus, in a volume region containing no sources, (7-61) states that the imaginary part of the complex power-flux entering the closed surface bounding V is a measure of 2ω *times the difference* *of the time-average energies stored in the magnetic and electric fields.*† (This quantity is sometimes symbolized Q when applied to L and C energy-storage elements of circuits, details of which are given further discussion in the appendix.

The foregoing interpretations of the real and imaginary parts of the complex Poynting theorem (7-59a) can be extended to a region containing current generators of density $\hat{\mathbf{J}}_g$ by a similar consideration of (7-59b). One-half of the real parts then yields

$$-\int_V \tfrac{1}{2}\,\mathrm{Re}\,(\hat{\mathbf{J}}_g^* \cdot \hat{\mathbf{E}})\,dv = \int_V \frac{\sigma E^2}{2}\,dv + \oint_S \tfrac{1}{2}\,\mathrm{Re}\,(\hat{\mathbf{E}} \times \hat{\mathbf{H}}^*) \cdot d\mathbf{s}\ \mathrm{W} \qquad (7\text{-}62)$$

The left side denotes the time-average generated power in V, contributed by components of $\hat{\mathbf{E}}$ in phase with the current density sources $\hat{\mathbf{J}}_g$. *The time-average generated power thus equals the sum of the time-average Ohmic losses in V plus the time-average of the total power-flux leaving the closed surface S that bounds V.* This form of the Poynting theorem is useful when applied, for example, to generators of radiated power such as antennas. Thus, in free space (containing no losses), (7-62) states that the power-flux emerging (radiated) from any surface S enclosing the antenna equals the power driving the antenna terminals, or simply a statement of the conservation of energy.

REFERENCES

ELLIOTT, R. S. *Electromagnetics*. New York: McGraw-Hill, 1966.

LORRAIN, P., and D. R. CORSON. *Electromagnetic Fields and Waves*. San Francisco: W. H. Freeman, 1970.

PLONSEY, R., and R. E. COLLIN. *Principles and Applications of Electromagnetic Fields*. New York: McGraw-Hill, 1961.

STRATTON, J. A. *Electromagnetic Theory*. New York: McGraw-Hill, 1941.

PROBLEMS

7-1. Verify in detail the result I^2R obtained from the volume integral of $\mathbf{J} \cdot \mathbf{E}$ inside the direct-current-carrying wire of Example 7-1.

7-2. Prove that the instantaneous electric and magnetic field energy densities $\epsilon_0 E^2/2$ and $\mu_0 H^2/2$ of a plane wave at any point *in free space* are equal.

† The counterparts of the volume integrals of $(1/4)\mu\hat{\mathbf{H}} \cdot \hat{\mathbf{H}}^*$ and $(1/4)\epsilon\hat{\mathbf{E}} \cdot \hat{\mathbf{E}}^*$ in a series or parallel RLC *circuit* are the quantities $(1/4)L\hat{I}\hat{I}^*$ and $(1/4)C\hat{V}\hat{V}^*$, which represent the time averages of stored magnetic and electric field energies of an inductor and a capacitor. Details of the circuit implications of these energy quantities are described relative to (A-22) developed in the appendix.

7-3. A negative z traveling, x polarized uniform plane wave in free space is specified in real time by

$$\mathbf{E} = \mathbf{a}_x E_x^-(z, t) = \mathbf{a}_x E_m^- \cos(\omega t + \beta_0 z)$$

$$\mathbf{H} = \mathbf{a}_y H_y^-(z, t) = -\mathbf{a}_y \frac{E_m^-}{\eta_0} \cos(\omega t + \beta_0 z)$$

(*a*) Find the time-instantaneous Poynting vector $\mathscr{P}(z, t)$. Sketch a figure similar to Figure 7-2(*b*) applicable to this wave and its Poynting vector at $t = 0$.

(*b*) Given a closed, boxed-shaped surface like that of Example 7-2, find the instantaneous, net power-flux *entering* that surface. What is the physical meaning of this result in relation to the electromagnetic energy inside the surface?

7-4. Use the defining relation (7-25b) to determine the time-average Poynting vector $\mathscr{P}_{av}$ of the plane wave given in Problem 7-3, and show that the net, time-average power-flux entering the box-shaped closed surface of Problem 7-3(*b*) is zero. Does this result satisfy the time-average form (7-31) of the Poynting theorem?

7-5. Given the plane wave of Problem 7-3, use (7-51a) or (7-51b) to obtain its time-average Poynting vector. How much time-average power-flux passes through each end of the box-shaped surface in that problem?

7-6. Prove (7-47). Why is its time average zero?

7-7. Given the following *negative z traveling* plane wave in a lossy region as discussed in Section 3-6

$$\mathbf{E} = \mathbf{a}_x E_x^-(z, t) = \mathbf{a}_x E_m^- e^{\alpha z} \cos(\omega t + \beta z)$$

$$\mathbf{H} = \mathbf{a}_y H_y^-(z, t) = -\mathbf{a}_y \frac{E_m^-}{\eta} e^{\alpha z} \cos(\omega t + \beta z - \theta)$$

in which θ denotes the angle of the complex wave impedance (3-99a), determine the following:

(*a*) Show that the time-instantaneous Poynting vector is

$$\mathscr{P} = \mathbf{a}_z \mathscr{P}_z^-(z, t) = -\mathbf{a}_z \frac{(E_m^-)^2}{2\eta} e^{2\alpha z}[\cos\theta + \cos(2\omega t + 2\beta z - \theta)]$$

Sketch a figure similar to in Figure 7-3, showing $\mathscr{P}_z^-$ at $t = 0$.

(*b*) Given a box-shaped closed surface like that of Example 7-3, find the instantaneous power-flux *entering* that surface. Use the Poynting theorem (7-8) to give a physical interpretation to your answer.

7-8. Use (7-25b) to obtain the time-average Poynting vector of the wave of Problem 7-7. Then find the time-average power-flux entering the box-shaped surface given in that problem. Interpret the result physically by use of (7-31).

7-9. Find the time-average Poynting vector of the wave of Problem 7-7 using the complex forms of (7-51a) or (7-51b). (The answer should agree with that obtained in Problem 7-8.)

7-10. Given the plane wave of Example 2-11 in *empty space* as follows:

$$E_x^+(z, t) = 1000 \cos (\omega t - \beta_0 z) \text{ V/m}$$
$$H_y^+(z, t) = 2.65 \cos (\omega t - \beta_0 z) \text{ A/m}$$

in which $f = 20$ MHz and $\beta_0 = \omega\sqrt{\mu_0\epsilon_0} = 0.420$ rad/m, evaluate the following:

(*a*) Find the time-instantaneous Poynting vector in the region. (Express it as the sum of a constant term and a double frequency term.)

(*b*) Find the time-average Poynting vector at any position z. Obtain your answer two ways: using the definition (7-25b), and from (7-51) in terms of the complex forms of the fields.

(*c*) Determine the time-average power-flux *entering* any closed box-shaped surface like that of Example 7-2, making use of (7-31).

7-11. Given the *attenuated* plane wave of Example 3-8,

$$E_x^+(z, t) = 1000e^{-1.9z} \cos (\omega t - 4.58z) \text{ V/m}$$
$$H_y^+(z, t) = 6.29e^{-1.9z} \cos \left(\omega t - 4.58z - \frac{\pi}{8}\right) \text{ A/m}$$

in which the frequency $f = 10^8$ Hz was assumed, evaluate the following:

(*a*) Find the time-instantaneous Poynting vector in the region. Express it as the sum of a double frequency term plus another that is a function of the phase difference $\pi/8$.

(*b*) Obtain the time-average Poynting vector for this wave. Find the answer two ways: from the definition (7-25b), and using (7-51) in terms of the complex forms of the fields.

(*c*) Determine the time-average power-flux *entering* a box-shaped surface like that of Example 7-3, assuming $a = b = d = 1$ m. Interpret this answer physically, in view of the time-average form (7-31) or (7-60) of the Poynting theorem.

7-12. The time-average electromagnetic power-flux density arriving at the earth from the sun is about 1340 W/m^2. This power is contributed by many frequency components, ranging from the radio frequencies through the ultraviolet region and beyond.

(*a*) If one supposed that this electromagnetic radiation were arriving via a uniform plane wave at a single sinusoidal frequency, what electric and magnetic field amplitudes would need to be associated with this wave to supply this power density?

(*b*) By means of a suitable surface integration, calculate the total time-average power radiated from the sun, assuming the distance from the sun to the earth to be 9.2×10^7 mi.

7-13. Incident and reflected plane waves exist simultaneously in empty space such that the total electric field is

$$E_x(z, t) = E_x^+(z, t) + E_x^-(z, t)$$
$$= 100 \cos(\omega t - \beta_0 z) + 20 \cos(\omega t + \beta_0 z) \text{ V/m}$$

in which $\beta_0 = \omega\sqrt{\mu_0\epsilon_0}$.

(*a*) What standing wave ratio (SWR) exists in this region? What is the magnitude of the reflection coefficient?

(*b*) Since the accompanying magnetic field is

$$H_y(z, t) = H_y^+(z, t) + H_y^-(z, t)$$
$$= \frac{100}{120\pi} \cos(\omega t - \beta_0 z) - \frac{20}{120\pi} \cos(\omega t + \beta_0 z) \text{ A/m}$$

show that the time-instantaneous Poynting vector becomes

$$\mathscr{P}(z, t) = \mathbf{a}_z\left[\frac{(100)^2}{120\pi} \cos^2(\omega t - \beta_0 z) - \frac{(20)^2}{120\pi} \cos^2(\omega t + \beta_0 z)\right] \text{ W/m}^2$$

or just the difference of the time-instantaneous Poynting vectors $\mathscr{P}^+(z, t)$ and $\mathscr{P}^-(z, t)$ associated with the incident and reflected waves.

(*c*) Show that the time-average Poynting vector is

$$\mathscr{P}_{av} = \frac{\mathbf{a}_z}{2}\left[\frac{(100)^2}{120\pi} - \frac{(20)^2}{120\pi}\right] \text{ W/m}^2$$

or just the difference of the time-average Poynting vectors associated with the individual waves.

7-14. Incident and reflected plane waves exist simultaneously in a *lossy* region such that

$$\hat{E}_x(z) = \hat{E}_m^+ e^{-\gamma z} + \hat{E}_m^- e^{\gamma z}$$
$$\hat{H}_y(z) = \frac{\hat{E}_m^+}{\hat{\eta}} e^{-\gamma z} - \frac{\hat{E}_m^-}{\hat{\eta}} e^{\gamma z}$$

in which the complex amplitudes are given by $\hat{E}_m^+ = E_m^+ e^{j\phi+}$ and $\hat{E}_m^- = E_m^- e^{j\phi-}$, and the wave impedance is denoted by $\hat{\eta} = \eta e^{j\theta}$.

(*a*) Show that the time-average Poynting vector at any z location is

$$\mathscr{P}_{av} = \mathbf{a}_z\Bigg[\frac{(E_m^+)^2}{2\eta} e^{-2\alpha z} \cos\theta - \frac{(E_m^-)^2}{2\eta} e^{2\alpha z} \cos\theta$$
$$+ \frac{E_m^+ E_m^-}{2\eta} \{\cos[2\beta z + \theta + (\phi^- + \phi^+)]$$
$$- \cos[2\beta z - \theta + (\phi^- - \phi^+)]\}\Bigg]$$

Compare this expression with the time-average Poynting vectors associated with the incident and reflected waves taken separately. Comment on this comparison.

(*b*) Show to what result $\mathscr{P}_{av}$ converges if the region is considered lossless ($\hat{\eta}$ is the pure real η, and $\alpha \to 0$); i.e., show that

$$\mathscr{P}_{av} = \mathbf{a}_z\left[\frac{(E_m^+)^2}{2\eta} - \frac{(E_m^-)^2}{2\eta}\right] \qquad \text{Lossless region}$$

[*Note:* This is the symbolic form of the answer obtained for Problem 7-13(*c*).]

CHAPTER

8

Mode Theory of Waveguides

This chapter considers the extension of the one-dimensional wave reflection problems of Chapter 6 to the theory of waveguides, regions bounded by conducting walls parallel to the propagation direction and of uniform cross-section. Typical waveguide configurations are shown in Figure 8-1. To simplify the analysis, perfectly conducting walls are assumed, except in Section 8-6 in which the attenuative effects of wall losses are analyzed. The boundary effects of the conducting walls, producing only normal electric and tangential magnetic fields there, favors a z direction of energy flow, so the waves are said to be guided in the z direction. In this sense, the wave transmission systems are said to be *waveguides*, though this term is usually restricted to the hollow, rectangular and circular cylindrical systems of Figures 8-1(c) and (d). Two-conductor wave-guiding systems exemplified by the parallel-wire and coaxial lines of Figures 8-1(a) and (b) are commonly called *transmission lines*; in the strict sense they are also waveguides.

The mode theory of uniform waveguides is considered in this chapter, with particular emphasis on the rectangular hollow waveguides shown in Figure 8-1(c). A boundary-value-problem approach is utilized; i.e., solutions of Maxwell's equations, subject to boundary conditions, are obtained. The complex, time-harmonic forms of Maxwell's equations are used, time dependence of the fields being assumed according to the usual factor $e^{j\omega t}$, but because of the invariance of the guide cross-section with respect to the propagation direction z, an additional exponential z dependence factor $e^{\mp\gamma z}$ is assumed, with γ identified as a z direction propagation constant. With t and

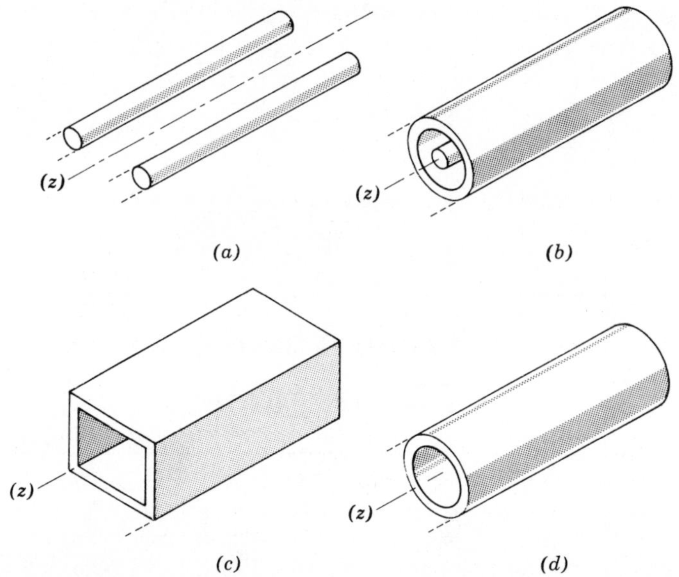

Figure 8-1. Uniform transmission line or waveguide structures of common occurrence. (*a*) Parallel-wire transmission line. (*b*) Circular cylindrical coaxial pair transmission line. (*c*) Rectangular hollow waveguide. (*d*) Circular hollow waveguide.

z thus absorbed in the factor $e^{j\omega t \mp \gamma z}$, the wave equation in terms of **E** or **H** reduces to a dependence on only the transverse variables x, y in the case of the rectangular waveguide (or in terms of ρ, ϕ for the circular cylindrical waveguide). These assumptions enable one to express the transverse field components E_x, E_y, H_x, and H_y in terms of the longitudinal components E_z and H_z. This permits a *mode* characterization of the field solutions: transverse-electric (TE) modes and transverse-magnetic (TM) modes are defined by putting $E_z = 0$ and $H_z = 0$, respectively. It is shown that an infinity of solutions belongs to each mode classification, and that a given mode propagates as a wave only if its frequency f is above a critical cutoff value f_c; below this frequency the mode becomes attenuated. In this sense, hollow waveguides act as highpass filters. The mode that will propagate at the lowest frequency is called the dominant mode of the waveguide. For hollow waveguides, the lowest frequency at which energy is propagated as waves is a function of the waveguide size, but for two-conductor transmission systems such as the coaxial line, it is seen that propagation down to zero frequency is possible by means of the dominant transverse-electromagnetic

(TEM) mode. The details of the TEM mode for two-conductor transmission lines are deferred until Chapter 9.

The chapter concludes with a discussion of the wave attenuation effects of wall losses, assuming finite conductivity of the wave-guiding walls.

8-1 Maxwell's Relations When Fields Have $e^{j\omega t \mp \gamma z}$ Dependence

It has been seen that expressing Maxwell's equations in complex, time-harmonic form through a time dependence given by the factor $e^{j\omega t}$ eliminates t from the equations. Wave-guiding systems of *uniform cross-section*, like those in Figure 8-1, permit an additional assumption of z dependence of the fields in accordance with the factor $e^{\mp\gamma z}$, inasmuch as any length ℓ of the system will influence wave propagation in exactly the same manner as any other length ℓ. The time and z dependence is therefore postulated to occur solely in accordance with the factor $e^{j\omega t \mp \gamma z}$, in which the $-$ and $+$ signs are identified with the positive z and negative z traveling wave solutions respectively. The **E** and **H** fields of Maxwell's equations are thus replaced with *complex* functions $\hat{\mathscr{E}}$ and $\hat{\mathscr{H}}$ of the transverse coordinates u_1 and u_2, multiplied by the exponential factor as follows:

$$\begin{aligned} &\mathbf{E}(u_1, u_2, z, t) \text{ is replaced by } \hat{\mathscr{E}}^{\pm}(u_1, u_2)e^{j\omega t \mp \gamma z} \\ &\mathbf{H}(u_1, u_2, z, t) \text{ is replaced by } \hat{\mathscr{H}}^{\pm}(u_1, u_2)e^{j\omega t \mp \gamma z} \end{aligned} \tag{8-1a}$$

assuming generalized *cylindrical* coordinates (u_1, u_2, z). The superscripts $\pm$ on the symbols $\hat{\mathscr{E}}$ and $\hat{\mathscr{H}}$ denote the field solutions identified with the positive z and negative z traveling waves in the waveguide. Once the complex solutions $\hat{\mathscr{E}}^{\pm}(u_1, u_2)$ and $\hat{\mathscr{H}}^{\pm}(u_1, u_2)$ are found, a restoration to their *real-time* form is obtained using

$$\begin{aligned} \mathbf{E}^{\pm}(u_1, u_2, z, t) &= \text{Re}\,[\hat{\mathscr{E}}^{\pm}(u_1, u_2)e^{j\omega t \mp \gamma z}] \\ \mathbf{H}^{\pm}(u_1, u_2, z, t) &= \text{Re}\,[\hat{\mathscr{H}}^{\pm}(u_1, u_2)e^{j\omega t \mp \gamma z}] \end{aligned} \tag{8-1b}$$

The dielectric region bounded by the waveguide conductors is assumed lossless, making $\mathbf{J} = 0$ therein, so that Maxwell's equations (3-59) and (3-77) governing the fields in the dielectric are

$$\nabla \times \mathbf{E} = -\frac{\partial \mathbf{B}}{\partial t} \tag{8-2}$$

$$\nabla \times \mathbf{H} = \frac{\partial \mathbf{D}}{\partial t} \tag{8-3}$$

With the replacement of the complex forms of (8-1a) into the latter, assuming in rectangular coördinates

$$\begin{aligned} \hat{\mathscr{E}}^{\pm}(x, y) &= \mathbf{a}_x\hat{\mathscr{E}}_x^{\pm}(x, y) + \mathbf{a}_y\hat{\mathscr{E}}_y^{\pm}(x, y) + \mathbf{a}_z\hat{\mathscr{E}}_z^{\pm}(x, y) \\ \hat{\mathscr{H}}^{\pm}(x, y) &= \mathbf{a}_x\hat{\mathscr{H}}_x^{\pm}(x, y) + \mathbf{a}_y\hat{\mathscr{H}}_y^{\pm}(x, y) + \mathbf{a}_z\hat{\mathscr{H}}_z^{\pm}(x, y) \end{aligned} \tag{8-4}$$

one obtains from (8-2)

$$\begin{vmatrix} \mathbf{a}_x & \mathbf{a}_y & \mathbf{a}_z \\ \dfrac{\partial}{\partial x} & \dfrac{\partial}{\partial y} & \dfrac{\partial}{\partial z} \\ \hat{\mathscr{E}}_x^{\pm} e^{j\omega t \mp \gamma z} & \hat{\mathscr{E}}_y^{\pm} e^{j\omega t \mp \gamma z} & \hat{\mathscr{E}}_z^{\pm} e^{j\omega t \mp \gamma z} \end{vmatrix} = -\mu \frac{\partial}{\partial t}[\mathbf{a}_x\hat{\mathscr{H}}_x^{\pm} + \mathbf{a}_y\hat{\mathscr{H}}_y^{\pm} + \mathbf{a}_z\hat{\mathscr{H}}_z^{\pm}]e^{j\omega t \mp \gamma z}$$

This expands into the three scalar relations

$$\begin{aligned} \left[\frac{\partial}{\partial y}\hat{\mathscr{E}}_z^{\pm}(x, y) - (\mp\gamma)\hat{\mathscr{E}}_y^{\pm}(x, y)\right]e^{j\omega t \mp \gamma z} &= -j\omega\mu\hat{\mathscr{H}}_x^{\pm}(x, y)e^{j\omega t \mp \gamma z} \\ \left[\mp\gamma\hat{\mathscr{E}}_x^{\pm}(x, y) - \frac{\partial}{\partial x}\hat{\mathscr{E}}_z^{\pm}(x, y)\right]e^{j\omega t \mp \gamma z} &= -j\omega\mu\hat{\mathscr{H}}_y^{\pm}(x, y)e^{j\omega t \mp \gamma z} \\ \left[\frac{\partial}{\partial x}\hat{\mathscr{E}}_y^{\pm}(x, y) - \frac{\partial}{\partial y}\hat{\mathscr{E}}_x^{\pm}(x, y)\right]e^{j\omega t \mp \gamma z} &= -j\omega\mu\hat{\mathscr{H}}_z^{\pm}(x, y)e^{j\omega t \mp \gamma z} \end{aligned}$$

in which the (x, y) dependence has been left in for clarity. The exponential factors cancel, obtaining the simplified expansion of (8-2)

$$\begin{aligned} \frac{\partial\hat{\mathscr{E}}_z^{\pm}}{\partial y} \pm \gamma\hat{\mathscr{E}}_y^{\pm} &= -j\omega\mu\hat{\mathscr{H}}_x^{\pm} \\ \mp\gamma\hat{\mathscr{E}}_x^{\pm} - \frac{\partial\hat{\mathscr{E}}_z^{\pm}}{\partial x} &= -j\omega\mu\hat{\mathscr{H}}_y^{\pm} \\ \frac{\partial\hat{\mathscr{E}}_y^{\pm}}{\partial x} - \frac{\partial\hat{\mathscr{E}}_x^{\pm}}{\partial y} &= -j\omega\mu\hat{\mathscr{H}}_z^{\pm} \end{aligned} \tag{8-5}$$

These results can be written in the compact form

$$\boxed{\nabla' \times \hat{\mathscr{E}}^{\pm} = -j\omega\mu\hat{\mathscr{H}}^{\pm}} \tag{8-6}$$

provided one defines a *modified-curl* operator, $\nabla'\times$, as follows:

$$\nabla' \times \hat{\mathscr{E}}^{\pm} \equiv \begin{vmatrix} \mathbf{a}_x & \mathbf{a}_y & \mathbf{a}_z \\ \dfrac{\partial}{\partial x} & \dfrac{\partial}{\partial y} & \mp\gamma \\ \hat{\mathscr{E}}_x^{\pm} & \hat{\mathscr{E}}_y^{\pm} & \hat{\mathscr{E}}_z^{\pm} \end{vmatrix} \tag{8-7}$$

One should regard (8-6) as the equivalent of the Maxwell equation (8-2), assuming the exponential t and z dependence of the fields noted in (8-1a). One may note that the operator $\nabla' \times$ defined by (8-7) differs from the conventional curl operator $\nabla \times$ of (2-52) to the extent of a replacement of $\partial/\partial z$ with $\mp\gamma$, a consequence of the assumption of the z dependence of the fields according to the factor $e^{\mp\gamma z}$.

In a similar manner, with the substitution of (8-1a) and (8-4), the Maxwell curl equation (8-3) yields the compact result

$$\nabla' \times \hat{\mathscr{H}}^{\pm} = j\omega\epsilon\hat{\mathscr{E}}^{\pm} \tag{8-8}$$

with the modified-curl operator $\nabla' \times$ defined once more by (8-7).

In the *generalized* cylindrical coordinate system (u_1, u_2, z) one can show that the Maxwell modified-curl relations (8-6) and (8-8), for a current-free region, become

$$\nabla' \times \hat{\mathscr{E}}^{\pm} \equiv \begin{vmatrix} \dfrac{\mathbf{a}_1}{h_2} & \dfrac{\mathbf{a}_2}{h_1} & \dfrac{\mathbf{a}_z}{h_1 h_2} \\ \dfrac{\partial}{\partial u_1} & \dfrac{\partial}{\partial u_2} & \mp\gamma \\ h_1\hat{\mathscr{E}}_1^{\pm} & h_2\hat{\mathscr{E}}_2^{\pm} & \hat{\mathscr{E}}_z^{\pm} \end{vmatrix} = -j\omega\mu\hat{\mathscr{H}}^{\pm} \tag{8-9}$$

$$\nabla' \times \hat{\mathscr{H}}^{\pm} \equiv \begin{vmatrix} \dfrac{\mathbf{a}_1}{h_2} & \dfrac{\mathbf{a}_2}{h_1} & \dfrac{\mathbf{a}_z}{h_1 h_2} \\ \dfrac{\partial}{\partial u_1} & \dfrac{\partial}{\partial u_2} & \mp\gamma \\ h_1\hat{\mathscr{H}}_1^{\pm} & h_2\hat{\mathscr{H}}_2^{\pm} & \hat{\mathscr{H}}_z^{\pm} \end{vmatrix} = j\omega\epsilon\hat{\mathscr{E}}^{\pm} \tag{8-10}$$

assuming

$$\begin{aligned} \hat{\mathscr{E}}^{\pm}(u_1, u_2) &= \mathbf{a}_1\hat{\mathscr{E}}_1^{\pm}(u_1, u_2) + \mathbf{a}_2\hat{\mathscr{E}}_2^{\pm}(u_1, u_2) + \mathbf{a}_z\hat{\mathscr{E}}_z^{\pm}(u_1, u_2) \\ \hat{\mathscr{H}}^{\pm}(u_1, u_2) &= \mathbf{a}_1\hat{\mathscr{H}}_1^{\pm}(u_1, u_2) + \mathbf{a}_2\hat{\mathscr{H}}_2^{\pm}(u_1, u_2) + \mathbf{a}_z\hat{\mathscr{H}}_z^{\pm}(u_1, u_2) \end{aligned} \tag{8-11}$$

Simplifications of the *wave equations* are also possible when field variations occur according to the factor $e^{j\omega t \mp \gamma z}$. The simultaneous manipulation of the Maxwell relations (8-2) and (8-3), applicable to a current-free region, has been seen in Section 2-9 to lead to the homogeneous vector wave equations

$$\nabla^2\mathbf{E} - \mu\epsilon\frac{\partial^2\mathbf{E}}{\partial t^2} = 0 \tag{8-12a}$$

$$\nabla^2 \mathbf{H} - \mu\epsilon \frac{\partial^2 \mathbf{H}}{\partial t^2} = 0 \tag{8-12b}$$

Using the definition (2-83) of $\nabla^2 \mathbf{E}$ applicable to the rectangular coordinate system, the vector wave equation (8-12a), for example, expands into the three scalar wave equations

$$\nabla^2 E_x - \mu\epsilon \frac{\partial^2 E_x}{\partial t^2} = 0 \tag{8-13a}$$

$$\nabla^2 E_y - \mu\epsilon \frac{\partial^2 E_y}{\partial t^2} = 0 \tag{8-13b}$$

$$\nabla^2 E_z - \mu\epsilon \frac{\partial^2 E_z}{\partial t^2} = 0 \tag{8-13c}$$

In the cartesian system, all three scalar wave equations are of identical forms, so their solutions are of the same type. From the definition (2-78) of the Laplacian of a scalar function, (8-13a) expands as follows:

$$\frac{\partial^2 E_x}{\partial x^2} + \frac{\partial^2 E_x}{\partial y^2} + \frac{\partial^2 E_x}{\partial z^2} - \mu\epsilon \frac{\partial^2 E_x}{\partial t^2} = 0 \tag{8-14}$$

If the substitution of the complex exponential form of E_x, given by (8-1a), is made into (8-14), one obtains, after canceling the exponential factors

$$\frac{\partial^2 \hat{\mathscr{E}}_x^{\pm}}{\partial x^2} + \frac{\partial^2 \hat{\mathscr{E}}_x^{\pm}}{\partial y^2} + (\gamma^2 + \omega^2 \mu\epsilon)\hat{\mathscr{E}}_x^{\pm} = 0$$

Denoting $\gamma^2 + \omega^2\mu\epsilon$ by the symbol

$$\hat{k}_c^2 \equiv \gamma^2 + \omega^2 \mu\epsilon \tag{8-15}$$

one may write the scalar wave equation

$$\frac{\partial^2 \hat{\mathscr{E}}_x^{\pm}}{\partial x^2} + \frac{\partial^2 \hat{\mathscr{E}}_x^{\pm}}{\partial y^2} + \hat{k}_c^2 \hat{\mathscr{E}}_x^{\pm} = 0 \tag{8-16a}$$

Similar substitutions of (8-1a) into (8-13b) and (8-13c) produce the simplifications

$$\frac{\partial^2 \hat{\mathscr{E}}_y^{\pm}}{\partial x^2} + \frac{\partial^2 \hat{\mathscr{E}}_y^{\pm}}{\partial y^2} + \hat{k}_c^2 \hat{\mathscr{E}}_y^{\pm} = 0 \tag{8-16b}$$

$$\frac{\partial^2 \hat{\mathscr{E}}_z^{\pm}}{\partial x^2} + \frac{\partial^2 \hat{\mathscr{E}}_z^{\pm}}{\partial y^2} + \hat{k}_c^2 \hat{\mathscr{E}}_z^{\pm} = 0 \tag{8-16c}$$

Beginning with the vector wave equation (8-12b), a procedure identical with the foregoing evidently produces three similar wave equations in terms of the components of $\mathscr{H}^{\pm}$

$$\frac{\partial^2 \hat{\mathscr{H}}_x^{\pm}}{\partial x^2} + \frac{\partial^2 \hat{\mathscr{H}}_x^{\pm}}{\partial y^2} + \hat{k}_c^2 \hat{\mathscr{H}}_x^{\pm} = 0 \tag{8-16d}$$

$$\frac{\partial^2 \hat{\mathscr{H}}_y^{\pm}}{\partial x^2} + \frac{\partial^2 \hat{\mathscr{H}}_y^{\pm}}{\partial y^2} + \hat{k}_c^2 \hat{\mathscr{H}}_y^{\pm} = 0 \tag{8-16e}$$

$$\frac{\partial^2 \hat{\mathscr{H}}_z^{\pm}}{\partial x^2} + \frac{\partial^2 \hat{\mathscr{H}}_z^{\pm}}{\partial y^2} + \hat{k}_c^2 \hat{\mathscr{H}}_z^{\pm} = 0 \tag{8-16f}$$

Any of the last six partial differential equations is useful in obtaining wave solutions for the rectangular, hollow waveguide of Figure 8-1(c), to be discussed in Section 8-3. Relationships pertaining to the mode character of the solutions are developed first.

8-2 TE, TM, and TEM Mode Relationships

A study of the expansions of the Maxwell modified-curl relationships (8-6) and (8-8) reveals that one can express the transverse components $\hat{\mathscr{E}}_1^{\pm}$, $\hat{\mathscr{E}}_2^{\pm}$, $\hat{\mathscr{H}}_1^{\pm}$, and $\hat{\mathscr{H}}_2^{\pm}$ explicitly in terms of the x and y derivatives of the longitudinal field components $\hat{\mathscr{E}}_z^{\pm}$ and $\hat{\mathscr{H}}_z^{\pm}$. These results form a basis for the *mode* description of the field solutions, relationships established in the following in rectangular coordinates.

Beginning with the expansions of (8-6) and (9-8) in rectangular coordinates

$$\frac{\partial \hat{\mathscr{E}}_z^{\pm}}{\partial y} \pm \gamma \hat{\mathscr{E}}_y^{\pm} = -j\omega\mu \hat{\mathscr{H}}_x^{\pm} \tag{8-17a}$$

$$\mp \gamma \hat{\mathscr{E}}_x^{\pm} - \frac{\partial \hat{\mathscr{E}}_z^{\pm}}{\partial x} = -j\omega\mu \hat{\mathscr{H}}_y^{\pm} \tag{8-17b}$$

$$\frac{\partial \hat{\mathscr{E}}_y^{\pm}}{\partial x} - \frac{\partial \hat{\mathscr{E}}_x^{\pm}}{\partial y} = -j\omega\mu \hat{\mathscr{H}}_z^{\pm} \tag{8-17c}$$

$$\frac{\partial \hat{\mathscr{H}}_z^{\pm}}{\partial y} \pm \gamma \hat{\mathscr{H}}_y^{\pm} = j\omega\epsilon \hat{\mathscr{E}}_x^{\pm} \tag{8-18a}$$

$$\mp \gamma \hat{\mathscr{H}}_x^{\pm} - \frac{\partial \hat{\mathscr{H}}_z^{\pm}}{\partial x} = j\omega\epsilon \hat{\mathscr{E}}_y^{\pm} \tag{8-18b}$$

$$\frac{\partial \hat{\mathscr{H}}_y^{\pm}}{\partial x} - \frac{\partial \hat{\mathscr{H}}_x^{\pm}}{\partial y} = j\omega\epsilon \hat{\mathscr{E}}_z^{\pm} \tag{8-18c}$$

one can see that the first two of each of these groups of equations contain derivative terms in only $\hat{\mathscr{E}}_z^{\pm}$ and $\hat{\mathscr{H}}_z^{\pm}$; the other terms are algebraic. This makes it possible, for example, to eliminate $\hat{\mathscr{H}}_y^{\pm}$ from (8-17b) and (8-18a) and solve for $\hat{\mathscr{E}}_x^{\pm}$, yielding

$$\hat{\mathscr{E}}_x^{\pm} = -\frac{1}{\hat{k}_c^2}\left[\pm\gamma\frac{\partial\hat{\mathscr{E}}_z^{\pm}}{\partial x} + j\omega\mu\frac{\partial\hat{\mathscr{H}}_z^{\pm}}{\partial y}\right] \tag{8-19a}$$

in which $\hat{k}_c^2$ is defined by (8-15). Similarly eliminating $\hat{\mathscr{H}}_x^{\mp}$ from (8-17a) and (8-18b) yields

$$\hat{\mathscr{E}}_y^{\pm} = \frac{1}{\hat{k}_c^2}\left[\mp\gamma\frac{\partial\hat{\mathscr{E}}_z^{\pm}}{\partial y} + j\omega\mu\frac{\partial\hat{\mathscr{H}}_z^{\pm}}{\partial x}\right] \tag{8-19b}$$

Successively eliminating $\hat{\mathscr{E}}_x^{\pm}$ and $\hat{\mathscr{E}}_y^{\pm}$ from the same pairs of relations obtains the following expressions for $\hat{\mathscr{H}}_x^{\pm}$ and $\hat{\mathscr{H}}_y^{\pm}$:

$$\hat{\mathscr{H}}_x^{\pm} = \frac{1}{\hat{k}_c^2}\left[j\omega\epsilon\frac{\partial\hat{\mathscr{E}}_z^{\pm}}{\partial y} \mp \gamma\frac{\partial\hat{\mathscr{H}}_z^{\pm}}{\partial x}\right] \tag{8-19c}$$

$$\hat{\mathscr{H}}_y^{\pm} = -\frac{1}{\hat{k}_c^2}\left[j\omega\epsilon\frac{\partial\hat{\mathscr{E}}_z^{\pm}}{\partial x} \pm \gamma\frac{\partial\hat{\mathscr{H}}_z^{\pm}}{\partial y}\right] \tag{8-19d}$$

These results permit finding the transverse field components of a rectangular waveguide whenever the longitudinal components $\hat{\mathscr{E}}_z^{\pm}$ and $\hat{\mathscr{H}}_z^{\pm}$ are known. They also serve as a basis for decomposing the field solutions into classes known as *modes*, depending on which longitudinal component, $\hat{\mathscr{E}}_z^{\pm}$ or $\hat{\mathscr{H}}_z^{\pm}$, is present.

The modes of the uniform waveguides of Figure 8-1(*b*), (*c*), (*d*) are defined as:

1. *Transverse magnetic* (TM) modes, for which $\hat{\mathscr{H}}_z^{\pm} = 0$.
2. *Transverse electric* (TE) modes, for which $\hat{\mathscr{E}}_z^{\pm} = 0$.
3. *Transverse electromagnetic* (TEM) modes, for which both $\hat{\mathscr{E}}_z^{\pm} = 0$ and $\hat{\mathscr{H}}_z^{\pm} = 0$.

Out of these definitions evolve properties of the modes as follows:

1. *TM Modes* (Transverse-Magnetic Waves). With $\hat{\mathscr{H}}_z^{\pm} = 0$, the TM mode in a waveguide has five components as noted in Figure 8-2(*a*). Putting $\hat{\mathscr{H}}_z^{\pm} = 0$ into equations (8-19) produces the following expressions for the transverse field components in rectangular coordinates:

$$\hat{\mathscr{E}}_x^{\pm} = \frac{\mp\gamma}{\hat{k}_c^2}\frac{\partial\hat{\mathscr{E}}_z^{\pm}}{\partial x} \tag{8-20a}$$

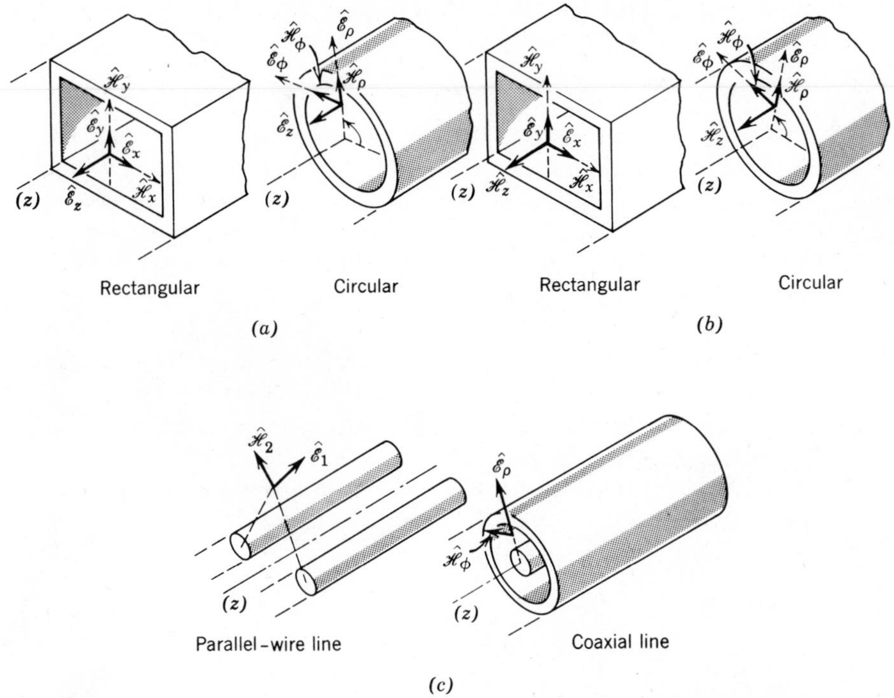

Figure 8-2. Field components of TM, TE, and TEM modes in typical waveguides or transmission lines. (*a*) Field components of TM mode, for which $\hat{\mathscr{H}}_z = 0$. (*b*) Field components of TE mode, for which $\hat{\mathscr{E}}_z = 0$. (*c*) Field components of the dominant TEM mode of two-conductor systems.

$$\hat{\mathscr{E}}_y^{\pm} = \frac{\mp\gamma}{\hat{k}_c^2}\frac{\partial\hat{\mathscr{E}}_z^{\pm}}{\partial y} \qquad \text{(8-20b)}$$

$$\hat{\mathscr{H}}_x^{\pm} = \frac{j\omega\epsilon}{\hat{k}_c^2}\frac{\partial\hat{\mathscr{E}}_z^{\pm}}{\partial y} \qquad \text{(8-20c)} \qquad \text{TM}$$

$$\hat{\mathscr{H}}_y^{\pm} = -\frac{j\omega\epsilon}{\hat{k}_c^2}\frac{\partial\hat{\mathscr{E}}_z^{\pm}}{\partial x} \qquad \text{(8-20d)}$$

in which $\hat{k}_c^2$ denotes $\gamma^2 + \omega^2\mu\epsilon$. Since the factor $\partial\hat{\mathscr{E}}_z^{\pm}/\partial x$ is common to (8-20a) and (8-20d), their ratio becomes

$$\frac{\hat{\mathscr{E}}_x^{\pm}}{\hat{\mathscr{H}}_y^{\pm}} = \pm\frac{\gamma}{j\omega\epsilon} \qquad \text{which means} \qquad \frac{\hat{\mathscr{E}}_x^{+}}{\hat{\mathscr{H}}_y^{+}} = \frac{\gamma}{j\omega\epsilon} \qquad \text{and} \qquad \frac{\hat{\mathscr{E}}_x^{-}}{\hat{\mathscr{H}}_y^{-}} = -\frac{\gamma}{j\omega\epsilon}$$

Similar results, with changes in signs, are obtained from the ratios of (8-20b) to (8-20c); calling $\gamma/j\omega\epsilon$ in each case the *intrinsic wave impedance* of TM modes, denoted by the symbol $\hat{\eta}_{\text{TM}}$, one produces the four ratios

$$\hat{\eta}_{\text{TM}} \equiv \frac{\gamma}{j\omega\epsilon} = \frac{\hat{\mathscr{E}}_x^+}{\hat{\mathscr{H}}_y^+} = -\frac{\hat{\mathscr{E}}_x^-}{\hat{\mathscr{H}}_y^-} = -\frac{\hat{\mathscr{E}}_y^+}{\hat{\mathscr{H}}_x^+} = \frac{\hat{\mathscr{E}}_y^-}{\hat{\mathscr{H}}_x^-}\ \Omega \tag{8-21}$$

The use of the latter makes it necessary to obtain only *two* of the transverse field components from $\hat{\mathscr{E}}_z^\pm$ by means of (8-20); the remaining two components are available in terms of the impedance ratios (8-21). In the detailed analysis of TM modes carried out in Section 8-3, it is seen that the propagation constant γ appearing in (8-20a, b) and (8-21) is dependent on the waveguide dimensions and the wave frequency.

Using the modified-curl relations (8-9) and (8-10) and following a procedure similar to the foregoing, modal expressions similar to (8-20) and (8-21), but applicable to waveguides in the circular cylindrical (ρ, ϕ, z) or the generalized cylindrical system (u_1, u_2, z), can be found. This is left as an exercise for the reader.

2. *TE Modes* (Transverse-Electric Waves). With $\hat{\mathscr{E}}_z^\pm = 0$, the TE mode has the five components typified in Figure 8-2(*b*), so that equations (8-19) in rectangular coordinates yield

$$\hat{\mathscr{E}}_x^\pm = -\frac{j\omega\mu}{\hat{k}_c^2}\frac{\partial\hat{\mathscr{H}}_z^\pm}{\partial y} \tag{8-22a}$$

$$\hat{\mathscr{E}}_y^\pm = \frac{j\omega\mu}{\hat{k}_c^2}\frac{\partial\hat{\mathscr{H}}_z^\pm}{\partial x} \tag{8-22b}$$

TE

$$\hat{\mathscr{H}}_x^\pm = \mp\frac{\gamma}{\hat{k}_c^2}\frac{\partial\hat{\mathscr{H}}_z^\pm}{\partial x} \tag{8-22c}$$

$$\hat{\mathscr{H}}_y^\pm = \mp\frac{\gamma}{\hat{k}_c^2}\frac{\partial\hat{\mathscr{H}}_z^\pm}{\partial y} \tag{8-22d}$$

An intrinsic wave impedance $\hat{\eta}_{\text{TE}}$ is evident from ratios of the latter as follows:

$$\hat{\eta}_{\text{TE}} \equiv \frac{j\omega\mu}{\gamma} = \frac{\hat{\mathscr{E}}_x^+}{\hat{\mathscr{H}}_y^+} = -\frac{\hat{\mathscr{E}}_x^-}{\hat{\mathscr{H}}_y^-} = -\frac{\hat{\mathscr{E}}_y^+}{\hat{\mathscr{H}}_x^+} = \frac{\hat{\mathscr{E}}_y^-}{\hat{\mathscr{H}}_x^-}\ \Omega \tag{8-23}$$

An analysis of TE modes for rectangular waveguides is pursued in Section 8-4, making use of the expressions (8-22) and (8-23).

3. *TEM Modes* (Transverse-Electromagnetic Waves). This mode, having neither $\hat{\mathscr{E}}_z$ nor $\hat{\mathscr{H}}_z$ field components, is the dominant mode of transmission lines having at least two conductors. Substituting $\hat{\mathscr{E}}_z^\pm = 0$ and $\hat{\mathscr{H}}_z^\pm = 0$ into the four relations (8-19) would appear to force all field components to vanish, thereby reducing the TEM mode to a trivial, nonexistent case. Inspection of the denominator $\hat{k}_c^2$ in these relations reveals the flaw in

this argument, for putting $\hat{k}_c^2 = 0$ simultaneously as one assumes $\hat{\mathscr{E}}_z^{\pm} = 0$ and $\hat{\mathscr{H}}_z^{\pm} = 0$ means $\gamma^2 + \omega^2\mu\epsilon = 0$, or

$$\gamma = j\omega\sqrt{\mu\epsilon} = j\beta \text{ rad/m} \tag{8-24}$$

Comparison with (3-88) shows that the transverse field components of the TEM mode comprise a wave phenomenon possessing a phase constant (8-24) identical with that of a uniform plane wave propagating in an unbounded region of parameters μ and ϵ. Substituting (8-24) into either wave impedance relation (8-21) or (8-23) further obtains the intrinsic wave impedance for the TEM mode

$$\hat{\eta}_{\text{TEM}} \equiv \frac{\gamma}{j\omega\epsilon} = \frac{j\omega\sqrt{\mu\epsilon}}{j\omega\epsilon} = \sqrt{\frac{\mu}{\epsilon}} \equiv \eta^{(0)}\ \Omega \tag{8-25}$$

Comparing (8-25) with (3-99a) reveals an intrinsic wave impedance identical with that of uniform plane waves in an unbounded region. These similarities of TEM mode characteristics with those of uniform plane waves are appreciated when one realizes that the uniform plane wave is itself TEM. The TEM mode, the *dominant mode* of energy propagation on *two-conductor lines*, is of such importance in wave transmission along open-wire or coaxial lines that it is accorded a separate detailed treatment in Chapters 9 and 10.

Generally speaking, hollow, single-conductor waveguides are capable of propagating TM and TE modes. In Section 8-4 it is shown that the so-called TE_{10} mode of the rectangular waveguide is its dominant mode, i.e., the mode propagating at the lowest frequency in that waveguide. Two-conductor systems such as coaxial lines propagate all three mode types: TEM, TM, and TE, although only the dominant TEM mode is capable of wave propagation down to zero frequency.

Signal transmission in the microwave region (at frequencies of about 1000 MHz and higher) by use of rectangular waveguides is a practical reality. Because of their intrinsically highpass characteristics, hollow waveguides become physically too large and expensive at frequencies much below this range; at lower frequencies, coaxial lines or open-wire lines may be more practical. A rectangular waveguide designed to operate with its dominant mode at about 10,000 MHz will be shown to require an interior width of about 2.5 cm; at one-hundredth this frequency (100 MHz), the guide width is required to be about 2.5 m if waves are to be propagated and not cut off. Coaxial, two-conductor lines are the obvious choice at such lower frequencies. In microwave transmission, a rectangular waveguide is usually more desirable than one of circular cross-section because the asymmetry of the rectangular cross-section provides a deliberate control of the polarization of the transmitted mode, of importance when considering the excitation of the line termination (a crystal detector, an antenna, etc.). Circular waveguide is of more limited utility, having applications to rotating joints that couple into spinning antenna dishes, to cylindrical resonant cavity frequency meters, etc.

8-3 TM Mode Solutions of Rectangular Waveguides

An analysis of the TM mode solutions of rectangular, hollow waveguides is described in the following. The cross-sectional geometry of Figure 8-3 is adopted, and the following assumptions are made:

1. The hollow, rectangular conducting pipe is assumed very long (avoiding end effects) and of uniform transverse dimensions a, b as noted in Figure 8-3.
2. The dielectric medium filling the pipe has the constant material parameters μ, ϵ and is assumed lossless, such that $\rho_v = 0$ and $\mathbf{J} = 0$ therein.
3. The waveguide walls are assumed ideal perfect conductors, simplifying the application of the boundary conditions.
4. All field quantities are assumed to vary with z and t solely in accordance with the factor $e^{j\omega t \mp \gamma z}$, in which the $-$ and $+$ signs are associated with positive z and negative z traveling wave solutions. The sinusoidal angular frequency of the fields is ω, determined by the generator frequency.

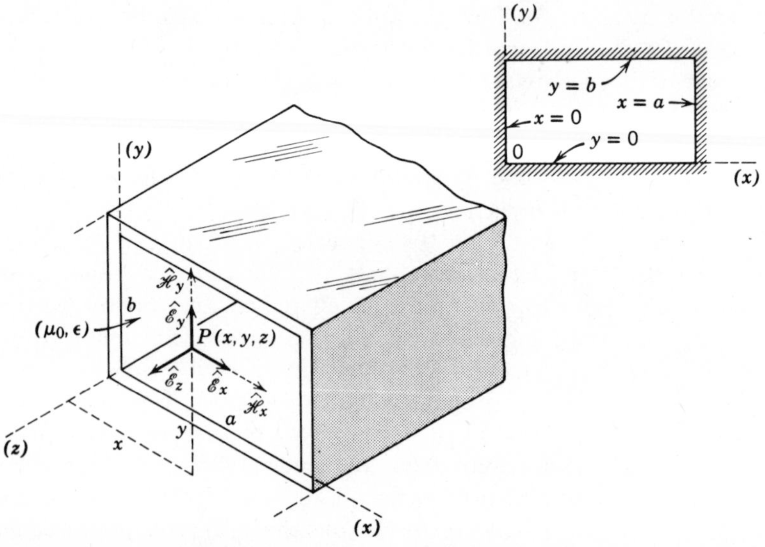

Figure 8-3. Geometry of a hollow, rectangular waveguide of uniform cross-sectional dimensions a, b.

5. For the TM modes under consideration, $\hat{\mathscr{H}}_z^{\pm} = 0$, leaving at most five field components in the pipe as noted in Figure 8-3.

Bearing in mind that the field relationships (8-20) and (8-21) are applicable to the TM case, it is convenient to begin with the wave equation (8-16c), in terms of the longitudinal field component $\hat{\mathscr{E}}_z^{\pm}$

$$\frac{\partial^2 \hat{\mathscr{E}}_z^{\pm}}{\partial x^2} + \frac{\partial^2 \hat{\mathscr{E}}_z^{\pm}}{\partial y^2} + \hat{k}_c^2 \hat{\mathscr{E}}_z^{\pm} = 0 \qquad [8\text{-}16c]$$

in which $\hat{k}_c^2 = \gamma^2 + \omega^2\mu\epsilon$. This partial differential equation is to be solved by the standard method of the separation of variables. Thus, *assume* a solution of the product form

$$\hat{\mathscr{E}}_z^{\pm}(x, y) = \hat{X}(x)\hat{Y}(y) \qquad (8\text{-}26)$$

in which $\hat{X}(x)$ and $\hat{Y}(y)$ are respectively functions of x and y only, and in general are complex. Substituting (8-26) back into (8-16c) yields

$$\hat{X}''\hat{Y} + \hat{X}\hat{Y}'' = -\hat{k}_c^2\hat{X}\hat{Y}$$

in which primes denote partial differentiations with respect to x or y. Dividing by $\hat{X}\hat{Y}$ obtains

$$\frac{\hat{X}''}{\hat{X}} + \frac{\hat{Y}''}{\hat{Y}} = -\hat{k}_c^2 \qquad (8\text{-}27)$$

If the two functions of x and y comprising the left side of (8-27) are to add up to the indicated constant for all values of x and y within the cross-section of Figure 8-3, then both of those functions must be equal to constants as well. That is, one must have

$$\frac{\hat{X}''}{\hat{X}} = -\hat{k}_x^2 \quad \text{and} \quad \frac{\hat{Y}''}{\hat{Y}} = -\hat{k}_y^2 \qquad (8\text{-}28)$$

with $-\hat{k}_x^2$ and $-\hat{k}_y$ denoting those constants. With (8-28) inserted into (8-27), it is seen that the interrelationship

$$\hat{k}_x^2 + \hat{k}_y^2 = \hat{k}_c^2 \qquad (8\text{-}29)$$

must hold among the constants. The meanings of the so-called *separation constants* $\hat{k}_x$ and $\hat{k}_y$ are ascertained later from the application of boundary conditions at the walls. Since the two differential equations (8-28) are respectively functions of x and y only, they can be written as the ordinary differential equations

$$\frac{d^2\hat{X}}{dx^2} + \hat{k}_x^2\hat{X} = 0 \quad \text{and} \quad \frac{d^2\hat{Y}}{dy^2} + \hat{k}_y^2\hat{Y} = 0 \tag{8-30}$$

These have the solutions

$$\hat{X}(x) = \hat{C}_1 \cos \hat{k}_x x + \hat{C}_2 \sin \hat{k}_x x \tag{8-31a}$$

$$\hat{Y}(y) = \hat{C}_3 \cos \hat{k}_y y + \hat{C}_4 \sin \hat{k}_y y \tag{8-31b}$$

in which $\hat{C}_1$ through $\hat{C}_4$ are constants of integration (complex, in general), to be evaluated from boundary conditions. The separated solutions (8-31), substituted back into the product expression (8-26), thus yield the desired particular solution of the wave equation (8-16c) as follows:

$$\hat{\mathscr{E}}_z^{\pm}(x, y) = (\hat{C}_1 \cos \hat{k}_x x + \hat{C}_2 \sin \hat{k}_x x)(\hat{C}_3 \cos \hat{k}_y y + \hat{C}_4 \sin \hat{k}_y y) \tag{8-32}$$

The complex constants appearing in (8-32) are evaluated from boundary conditions as follows. The component $\hat{\mathscr{E}}_z^{\pm}$ of (8-32) is tangential at the four walls $x = 0$, $x = a$ and $y = 0$, $y = b$ noted in Figure 8-3. For perfectly conducting walls the tangential $\hat{\mathscr{E}}_z$ just inside the dielectric waveguide region must vanish, so from the continuity relation (3-79) one obtains the boundary conditions

$$\begin{aligned} &1.\ \hat{\mathscr{E}}_z^{\pm}(0, y) = 0 \\ &2.\ \hat{\mathscr{E}}_z^{\pm}(a, y) = 0 \\ &3.\ \hat{\mathscr{E}}_z^{\pm}(x, 0) = 0 \\ &4.\ \hat{\mathscr{E}}_z^{\pm}(x, b) = 0 \end{aligned} \tag{8-33}$$

Boundary condition (1) applied to (8-32) yields

$$0 = (\hat{C}_1)(\hat{C}_3 \cos \hat{k}_y y + \hat{C}_4 \sin \hat{k}_y y)$$

whence $\hat{C}_1 = 0$ if the latter is to hold for all y on the wall $x = 0$. Then (8-32) becomes

$$\hat{\mathscr{E}}_z^{\pm}(x, y) = \hat{C}_2 \sin \hat{k}_x x(\hat{C}_3 \cos \hat{k}_y y + \hat{C}_4 \sin \hat{k}_y y) \tag{8-34}$$

Applying the boundary condition (2) to (8-34) obtains

$$0 = \hat{C}_2 \sin \hat{k}_x a(\hat{C}_3 \cos \hat{k}_y y + \hat{C}_4 \sin \hat{k}_y y)$$

which holds for all y on the wall $x = a$ upon setting $\sin k_x a = 0$. The latter is valid only at the zeros of the sine function, so that $\hat{k}_x a = m\pi$, in which $m = \pm 1, \pm 2, \pm 3, \ldots$, which corresponds to an infinite set of discrete values for $\hat{k}_x$ (hence to an infinite number of particular solutions, or modes) that satisfy the original wave equation. The value $m = 0$ is omitted because it produces the null, or trivial, solution. The negative values of m, moreover, add no new solutions to the set, so that one obtains for $\hat{k}_x$

$$\hat{k}_x = \frac{m\pi}{a} \qquad m = 1, 2, 3, \ldots \tag{8-35}$$

making (8-34) become

$$\hat{\mathscr{E}}_z^{\pm}(x, y) = \hat{C}_2 \sin \frac{m\pi}{a} x (\hat{C}_3 \cos \hat{k}_y y + \hat{C}_4 \sin k_y y) \quad m = 1, 2, 3, \ldots \tag{8-36}$$

The remaining boundary conditions (3) and (4) of (8-33) are next applied to (8-36); from the *similarity* of the solutions $\hat{X}(x)$ and $\hat{Y}(y)$ appearing in (8-32), together with the resemblance of the boundary conditions (3) and (4) to (1) and (2), one may infer by analogy with the preceding arguments that applying the boundary conditions (3) and (4) to (8-36) must lead to the results

$$\hat{C}_3 = 0 \qquad \text{and} \qquad \hat{k}_y = \frac{n\pi}{b} \qquad n = 1, 2, 3, \ldots \tag{8-37}$$

With these inserted into (8-36), the solution finally becomes

$$\hat{\mathscr{E}}_z^{\pm}(x, y) = \hat{C}_2 \hat{C}_4 \sin \frac{m\pi}{a} x \sin \frac{n\pi}{b} y \quad m, n = 1, 2, 3, \ldots$$

The coefficient $\hat{C}_2 \hat{C}_4$ evidently denotes the complex amplitude of any member of this solution set, which must include both positive z and negative z traveling waves. Replacing $\hat{C}_2 \hat{C}_4$ by the symbol $\hat{E}_{z,mn}^{\pm}$ yields

$$\hat{\mathscr{E}}_z^{\pm}(x, y) = \hat{E}_{z,\mathrm{mn}}^{\pm} \sin \frac{m\pi}{a} x \sin \frac{n\pi}{b} y \quad m, n = 1, 2, 3, \ldots \tag{8-38a}$$

in which $\hat{E}_{z,mn}^{+}$ or $\hat{E}_{z,mn}^{-}$ denotes the complex amplitude of any positive z or negative z traveling $\hat{\mathscr{E}}_z$ component associated with specific values of the mode numbers m and n. The solution set (8-38a) describes the z directed electric field component of the transverse-magnetic mode with mode integers m, n assigned, so the field component (8-38a) is said to belong to the TM_{mn} mode.

Solutions (8-38a) satisfying the partial differential equation (8-16c) and the boundary conditions (8-33) are also called the *eigenfunctions* (proper functions) of that boundary-value problem.

Examples of the field variations of $\hat{\mathscr{E}}_z$ predicted by (8-38a) within the waveguide cross-section are depicted in Figure 8-4, which shows how $\hat{\mathscr{E}}_z$ varies with x and y for two of the modes, TM_{11} and TM_{21}. These sketches show that the integers m and n denote the number of half-sinusoids of variations in $\hat{\mathscr{E}}_z$ occurring between the guide walls, with $\hat{\mathscr{E}}_z$ vanishing at the walls as required by the boundary conditions (8-33).

The sketches of Figure 8-4 do not show the complete field configurations of those TM_{mn} modes; the four remaining transverse field components denoted in Figure 8-3 are yet to be found. These are obtained by substituting $\hat{\mathscr{E}}_z^{\pm}$ of (8-38a) into the TM mode relationships (8-20), whence

$$\hat{\mathscr{E}}_x^{\pm}(x, y) = \left[\mp \frac{\gamma_{mn}}{\hat{k}_c^2} \frac{m\pi}{a} \hat{E}_{z,mn}^{\pm}\right] \cos \frac{m\pi}{a} x \sin \frac{n\pi}{b} y$$

$$= \hat{E}_{x,mn}^{\pm} \cos \frac{m\pi}{a} x \sin \frac{n\pi}{b} y \tag{8-38b}$$

$$\hat{\mathscr{E}}_y^{\pm}(x, y) = \left[\frac{\mp \gamma_{mn}}{\hat{k}_c^2} \frac{n\pi}{b} \hat{E}_{z,mn}^{\pm}\right] \sin \frac{m\pi}{a} x \cos \frac{n\pi}{b} y$$

$$= \hat{E}_{y,mn}^{\pm} \sin \frac{m\pi}{a} x \cos \frac{n\pi}{b} y \tag{8-38c}$$

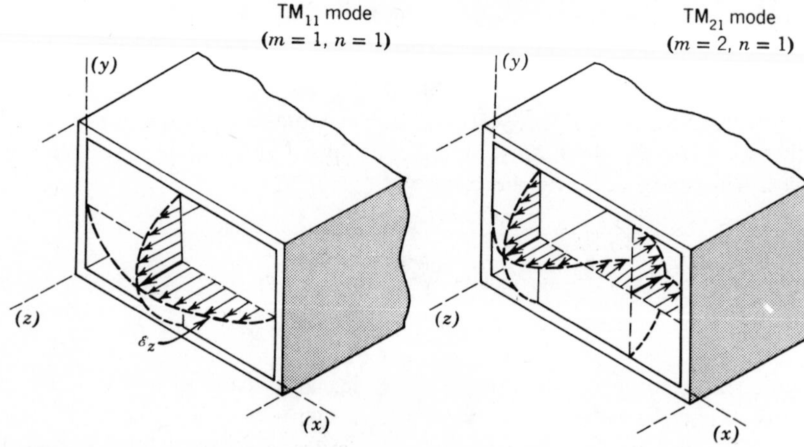

Figure 8-4. Typical cross-sectional standing-wave variations in the longitudinal electric field component $\hat{\mathscr{E}}_z$ of TM_{mn} modes, for two cases.

$$\mathscr{H}_x^{\pm}(x, y) = \mp\frac{\mathscr{E}_y^{\pm}}{\hat{\eta}_{TM_{mn}}} = \left[\frac{j\omega\epsilon}{\hat{k}_c^2}\frac{n\pi}{b}\hat{E}_{z,mn}^{\pm}\right]\sin\frac{m\pi}{a}x\cos\frac{n\pi}{b}y$$

$$= \hat{H}_{x,mn}^{\pm}\sin\frac{m\pi}{a}x\cos\frac{n\pi}{b}y \tag{8-38d}$$

$$\mathscr{H}_y^{\pm}(x, y) = \pm\frac{\mathscr{E}_x^{\pm}}{\hat{\eta}_{TM_{mn}}} = \left[\frac{-j\omega\epsilon}{\hat{k}_c^2}\frac{m\pi}{a}\hat{E}_{z,mn}^{\pm}\right]\cos\frac{m\pi}{a}x\sin\frac{n\pi}{b}y$$

$$= \hat{H}_{y,mn}^{\pm}\cos\frac{m\pi}{a}x\sin\frac{n\pi}{b}y \tag{8-38e}$$

The bracketed quantities denote the complex amplitudes $\hat{E}_{x,mn}^{\pm}$, $\hat{E}_{y,mn}^{\pm}$, $\hat{H}_{x,mn}^{\pm}$, and $\hat{H}_{y,mn}^{\pm}$ of the transverse field components, expressed in terms of the amplitude $\hat{E}_{z,mn}^{\pm}$ of the longitudinal component. Note further that the *total* electric and magnetic fields associated with any TM_{mn} mode are given by (8-4), or the appropriate vector sums of the components (8-38a) through (8-38e) (with the component $\mathscr{H}_z^{\pm}$ missing). Total field sketches in the form of flux plots are presented after a discussion of the propagation constant γ of the TM_{mn} modes, which concerns the nature of the waves along the z axis of the guide.

It is to be shown that the influence of the perfectly conducting walls of the rectangular waveguide of Figure 8-4 goes further than simply to produce sine or cosine type field variations over the guide cross-section. The walls also produce a marked frequency dependence of the z direction propagation constant γ, an effect so pronounced that at frequencies below a particular critical value (for a specified TM_{mn} mode) the mode will no longer propagate as a wave, but rather it simply attenuates with z distance. This behavior is predictable from the insertion of the proper values (8-35) and (8-37) of $\hat{k}_x$ and $\hat{k}_y$ into (8-29), yielding

$$\left(\frac{m\pi}{a}\right)^2 + \left(\frac{n\pi}{b}\right)^2 = \gamma^2 + \omega^2\mu\epsilon$$

which, upon solving for the z direction propagation constant γ, becomes†

$$\gamma_{mn} = \sqrt{\left[\left(\frac{m\pi}{a}\right)^2 + \left(\frac{n\pi}{b}\right)^2\right] - [\omega^2\mu\epsilon]} \tag{8-39}$$

in which the subscripts mn denote the dependence of γ on the choice of mode integers. Thus γ_{mn} is a function of the wall dimensions a and b, the

† Only the *positive* root of (8-39) need be chosen as the solution for γ, because the earlier assumption of t and z dependence of the fields according to the factor $\exp(j\omega t \mp \gamma z)$ accounts properly for the presence of both positive z and negative z traveling wave solutions.

frequency ω, the parameters μ and ϵ of the dielectric, and the specified mode numbers m, n. The bracketed quantities in the radicand of (8-39) are both seen to be positive real. Since the *difference* of these positive quantities is to determine γ_{mn}, it is evident that γ_{mn} becomes a *pure real* quantity (an attenuation factor α) if $(m\pi/a)^2 + (n\pi/b)^2$ is larger than $\omega^2\mu\epsilon$; with γ_{mn} becoming *pure imaginary* (a phase factor β) if the reverse is true. The transition between these two propagation conditions occurs at an angular frequency $\omega = \omega_{c,mn}$ called the *cutoff* frequency, defined where the bracketed quantities of (8-39) are equal; i.e.

$$\omega_{c,mn}^2\mu\epsilon = \left(\frac{m\pi}{a}\right)^2 + \left(\frac{n\pi}{b}\right)^2 \tag{8-40a}$$

Solving the latter for $f_{c,mn} = \omega_{c,mn}/2\pi$ yields

$$f_{c,mn} = \frac{1}{2\pi\sqrt{\mu\epsilon}}\left[\left(\frac{m\pi}{a}\right)^2 + \left(\frac{n\pi}{b}\right)^2\right]^{1/2} \tag{8-40b}$$

With (8-40a) substituted into (8-39), the propagation constant γ_{mn} is expressible in terms of $f_{c,mn}$ as follows:

$$\gamma_{mn} = \sqrt{\omega_{c,mn}^2\mu\epsilon - \omega^2\mu\epsilon} = \omega\sqrt{\mu\epsilon}\sqrt{\left(\frac{f_{c,mn}}{f}\right)^2 - 1}\ \text{m}^{-1} \tag{8-41}$$

From the latter one may infer, depending on whether a given TM$_{mn}$ mode in the rectangular guide is generated at a frequency f that is above or below the cutoff value $f_{c,mn}$ of (8-40b), that γ_{mn} becomes either pure real or pure imaginary as follows:

$$\gamma_{mn} = \alpha_{mn} \equiv \omega\sqrt{\mu\epsilon}\sqrt{\left(\frac{f_{c,mn}}{f}\right)^2 - 1}\ \text{Np/m} \qquad f < f_{c,mn} \tag{8-42a}$$

$$\gamma_{mn} = j\beta_{mn} \equiv j\omega\sqrt{\mu\epsilon}\sqrt{1 - \left(\frac{f_{c,mn}}{f}\right)^2}\ \text{rad/m} \quad f > f_{c,mn} \tag{8-42b}$$

The factor $\omega\sqrt{\mu\epsilon}$ appearing in these expressions is a phase constant $\beta^{(0)}$ identified with a uniform plane wave traveling at the frequency f in an unbounded region having the material parameters μ and ϵ, a value obtained from (3-90b) with $\sigma = 0$ or (3-110) with $\epsilon''/\epsilon' = 0$. The quantity $\beta^{(0)}$ thus serves as a convenient reference value with which the phase constant β_{mn} in the waveguide can be compared. From (8-42) it is evident that a rectangular waveguide carrying a TM$_{mn}$ mode acts as a highpass filter, allowing unattenuated wave motion characterized by the pure imaginary $\gamma_{mn} = j\beta_{mn}$

if the generator frequency f responsible for the mode exceeds the cutoff frequency $f_{c,mn}$, but attenuating the TM_{mn} mode fields with $\gamma_{mn} = \alpha_{mn}$ if $f < f_{c,mn}$.

An additional appreciation of the physical meanings of the real and imaginary results (8-42) for γ_{mn} is gained if the wave expressions for the TM_{mn} modes, including dependence on t and z, are examined. For example, multiplying the component $\hat{\mathscr{E}}_z^{\pm}$ of (8-38a) by the exponential factor $e^{j\omega t \mp \gamma_{mn} z}$ according to (8-1a) produces field solutions that depend on whether the propagation constant γ_{mn} of (8-42) is real or imaginary, as follows. If $f > f_{c,mn}$, then $\gamma_{mn} = j\beta_{mn}$ so that (8-38a), including exponential t and z dependence, becomes

$$\hat{\mathscr{E}}_z^{\pm}(x, y)e^{j\omega t \mp \gamma_{mn} z} = \hat{E}_{z,mn}^{\pm} \sin \frac{m\pi}{a} x \sin \frac{n\pi}{b} y \, e^{j(\omega t \mp \beta_{mn} z)}$$

$$m, n = 1, 2, 3, \ldots \text{ and } f > f_{c,mn} \qquad \text{(8-43a)}$$

The *traveling wave* nature of this field component is clearly specified by the factor $e^{j(\omega t - \beta_{mn} z)}$ for the positive z traveling solution, and by $e^{j(\omega t + \beta_{mn} z)}$ for the negative z traveling wave. If $f < f_c$, then $\gamma_{mn} = \alpha_{mn}$ according to (8-42a). Then the $\hat{\mathscr{E}}_z^{\pm}$ field solutions, including time and z dependence, become

$$\hat{\mathscr{E}}_z^{\pm}(x, y)e^{j\omega t \mp \gamma_{mn} z} = \hat{E}_{mn,z}^{\pm} e^{\mp \alpha_{mn} z} \sin \frac{m\pi}{a} x \sin \frac{n\pi}{b} y \, e^{j\omega t}$$

$$m, n = 1, 2, 3, \ldots \text{ and } f < f_{c,mn} \qquad \text{(8-43b)}$$

The attenuation with z provided by the factors $e^{-\alpha_{mn} z}$ or $e^{\alpha_{mn} z}$ is thus noted whenever the generator frequency f is too low. The mode will not then propagate as a wave motion; instead, the fields of the mode evanesce (vanish) with increasing distance from the generator or wave source. A mode at a frequency *below* its cutoff frequency $f_{c,mn}$ is called an *evanescent mode*. The foregoing discussion was limited to the longitudinal component $\hat{\mathscr{E}}_z^{\pm}$. The four remaining transverse components (8-38b) through (8-38e) are similarly propagated as waves along with $\hat{\mathscr{E}}_z^{\pm}$ if $f > f_{c,mn}$ or are evanescent if $f < f_{c,mn}$.

The *real-time* forms of the field components can likewise be employed to illustrate the conclusions of the foregoing discussion. Using (8-1b), the real-time form of the time-harmonic field component (8-43a) is, for $f > f_{c,mn}$ $E_z^{\pm}(x, y, z, t) = \text{Re}\,[\hat{\mathscr{E}}_z^{\pm}(x, y)e^{j(\omega t \mp \beta_{mn} z)}]$, yielding

$$E_z^{\pm}(x, y, z, t) = E_{z,mn}^{\pm} \sin \frac{m\pi}{a} x \sin \frac{n\pi}{b} y \cos(\omega t \mp \beta_{mn} z + \phi_{mn}^{\pm})$$

$$f > f_{c,mn} \qquad \text{(8-44a)}$$

the traveling wave nature of which is illustrated, in a constant-y plane, in Figure 8-5(a). Similarly, for f *below* the cutoff frequency $f_{c,mn}$:

$$E_z^{\pm}(x, y, z, t) = \text{Re}\,[\hat{\mathscr{E}}_z^{\pm}(x, y)e^{j\omega t \mp \alpha_{mn}z}]$$
$$= E_{z,mn}^{\pm}e^{\mp \alpha_{mn}z} \sin \frac{m\pi}{a} x \sin \frac{n\pi}{b} y \cos (\omega t + \phi_{mn}^{\pm})$$
$$f < f_{c,mn} \quad \text{(8-44b)}$$

Note that the complex amplitudes in these expressions may include arbitrary phase angles $\phi_{mn}^{\pm}$ according to $\hat{E}_{z,mn}^{\pm} = E_{z,mn}^{\pm}e^{j\phi_{mn}^{\pm}}$. Figure 8-5 illustrates both the wave or the evanescent behavior of the single field component

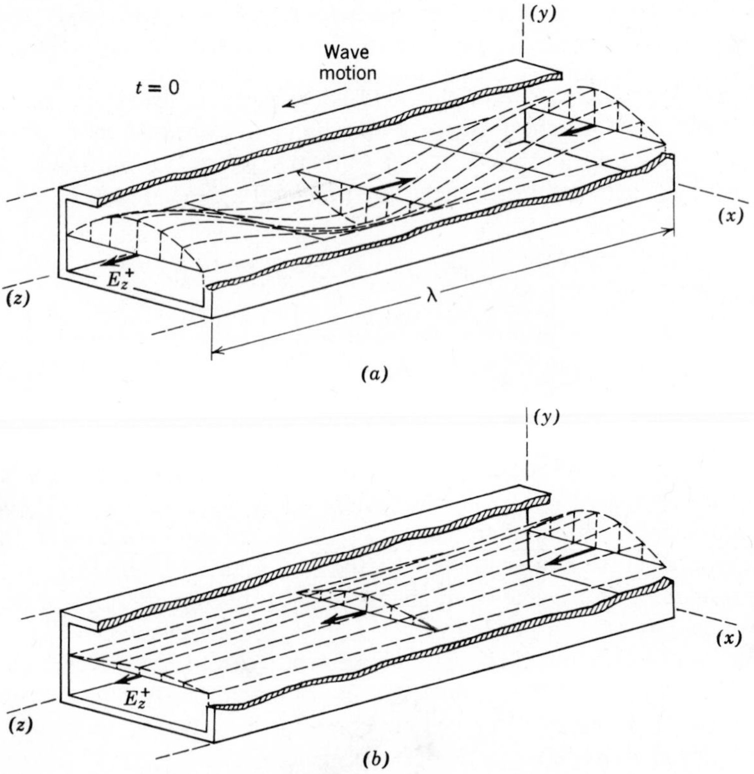

Figure 8-5. Field intensity variations of the longitudinal component $E_z^+(x, y, z, t)$ of the TM_{11} mode, shown over the plane $y = b/2$. (a) The forward z traveling wave E_z^+, if $f > f_c$. (b) The evanescence of E_z^+ with increasing z, if $f < f_c$.

$E_z^+(x, y, z, t)$, depending on whether the frequency f lies above or below $f_{c,mn}$.

Once the phase constant (8-42b) is obtained, other TM_{mn} mode properties, such as wavelength in the guide, phase velocity, and intrinsic wave impedance, can be derived. Assuming the generator frequency f of a given TM_{mn} mode to be above the cutoff value (8-40b), the wavelength λ of that mode, measured along the z axis as noted in Figure 8-5(a), for example, is found from the definition $\beta\lambda = 2\pi$. By use of (8-42b), this yields

$$\lambda_{mn} = \frac{2\pi}{\beta_{mn}} = \frac{2\pi}{\beta^{(0)}\sqrt{1 - \left(\frac{f_{c,mn}}{f}\right)^2}} = \frac{\lambda^{(0)}}{\sqrt{1 - \left(\frac{f_{c,mn}}{f}\right)^2}} \qquad f > f_{c,mn} \qquad (8\text{-}45)$$

in which $\lambda^{(0)}$ denotes the comparison wavelength $2\pi/\beta^{(0)}$ of a uniform plane wave in an unbounded region with the same dielectric parameters μ and ϵ. The z direction phase velocity is obtained using $v_p = \omega/\beta$, yielding

$$v_{p,mn} = \frac{\omega}{\beta_{mn}} = \frac{v_p^{(0)}}{\sqrt{1 - \left(\frac{f_{c,mn}}{f}\right)^2}} \qquad f > f_{c,mn} \qquad (8\text{-}46)$$

wherein $v_p^{(0)} = \omega/\beta^{(0)} = (\mu\epsilon)^{-1/2}$. The intrinsic wave impedance for TM_{mn} modes, specifying the ratios of transverse field components, is found from (8-21). If $f > f_{c,mn}$ one obtains the real result

$$\hat{\eta}_{\text{TM},mn} = \frac{j\beta_{mn}}{j\omega\epsilon} = \eta^{(0)}\sqrt{1 - \left(\frac{f_{c,mn}}{f}\right)^2} \qquad f > f_{c,mn} \qquad (8\text{-}47)$$

in which $\eta^{(0)} = \sqrt{\mu/\epsilon}$.

For a TM_{mn} mode generated at a frequency f *below* the cutoff value, the wavelength and phase velocities are not defined, in view of the purely evanescent character of the field distributions as exemplified in Figure 8-5(b). The intrinsic wave impedance for $f < f_{c,mn}$, however, from the substitution of (8-42a) into (8-21), becomes

$$\hat{\eta}_{\text{TM},mn} = \frac{\alpha}{j\omega\epsilon} = -j\eta^{(0)}\sqrt{\left(\frac{f_{c,mn}}{f}\right)^2 - 1} \qquad f < f_{c,mn} \qquad (8\text{-}48)$$

This purely reactive result implies no time-average power transfer in the z direction for an evanescent mode because of the 90° phase between the transverse electric and magnetic field components.

If the information contained in the five field expressions (8-38a) through

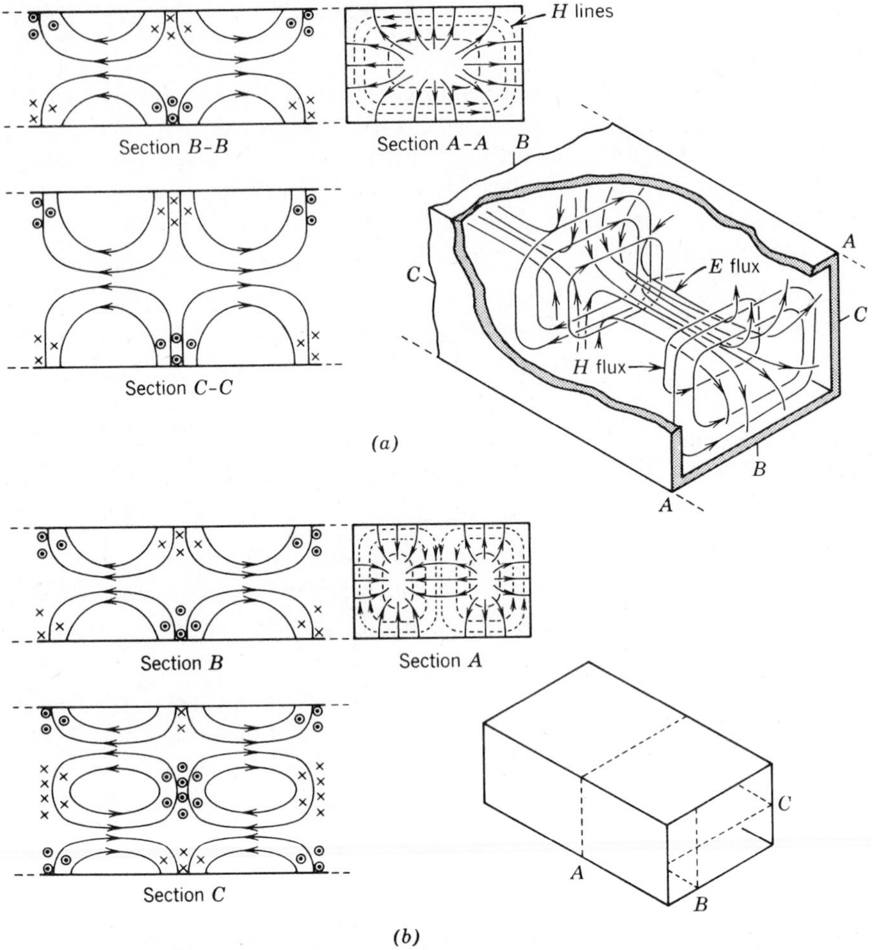

Figure 8-6. Two low-order TM modes of a rectangular waveguide. (*a*) The TM_{11} mode. (*b*) the TM_{21} mode.

(8-38e) is combined to construct the total fields **E** and **H** of the TM_{mn} modes, complete flux sketches resembling those in Figure 8-6 can be obtained. Flux sketches of two modes, TM_{11} and TM_{21}, are illustrated. A knowledge of such flux configurations is useful, for example, if the electric or magnetic fields are to be probed or linked with a short wire antenna or loop, for purposes of extracting energy from the mode.

In general, a large number of modes, propagating or evanescent, exist in the neighborhood of waveguide discontinuities such as bends and transitions.

The analysis of such nonuniformities in a waveguide is beyond the scope of this treatment. The propagation of energy in a rectangular guide is usually accomplished, at a given frequency, by selecting the dimensions a, b so that only *one mode* (the dominant mode) *propagates*, to the exclusion of all higher-order modes thus forced to become evanescent. This procedure assures a well-defined single-mode field configuration in the pipe, from which energy can be readily extracted by use of suitable transition devices (for example, a wave-guide-to-coaxial line transition). The discussion of the next section, covering TE modes, reveals that of all the modes capable of propagating in a rectangular waveguide, TM and TE, the TE_{10} mode is the dominant one.

EXAMPLE 8-1. A common air filled rectangular waveguide has the interior dimensions $a = 0.9$ in. and $b = 0.4$ in. (2.29×1.02 cm), the so-called X-band guide. (*a*) Find the cutoff frequency of the lowest-order, nontrivial TM mode. (*b*) At a source frequency that is twice the cutoff value of (*a*), determine the propagation constant for this mode. Also obtain the wavelength in the guide, the phase velocity, and the intrinsic wave impedance. (*c*) Repeat (*b*), assuming $f = f_c/2$.

(*a*) From (8-40b) it is seen that the cutoff frequency has its lowest value for TM modes if $m = 1$ and $n = 1$, the smallest integers producing non-trivial fields. Thus for the TM_{11} mode, the given dimensions yield

$$f_{c,11} = \frac{1}{2\pi\sqrt{\mu_0\epsilon_0}}\left[\left(\frac{\pi}{a}\right)^2 + \left(\frac{\pi}{b}\right)^2\right]^{1/2}$$

$$= \frac{3 \times 10^8}{2}\left[\frac{1}{(0.0229)^2} + \frac{1}{(0.0102)^2}\right]^{1/2} = 16{,}100 \text{ MHz}$$

The TM_{11} mode will thus propagate in this guide if its frequency exceeds 16,100 MHz. Below this frequency, the mode is evanescent.

(*b*) At $f = 32{,}200$ MHz, (8-42b) yields

$$\beta_{11} = \beta^{(0)}\sqrt{1 - \left(\frac{f_{c,11}}{f}\right)^2} = \frac{2\pi(32.2 \times 10^9)}{3 \times 10^8}\sqrt{1 - (\tfrac{1}{2})^2} = 585 \text{ rad/m}$$

In free space, $\lambda^{(0)} = c/f = 3 \times 10^8/32.2 \times 10^9 = 0.933$ cm, so from (8-45)

$$\lambda_{11} = \frac{\lambda^{(0)}}{\sqrt{1 - \left(\frac{f_{c,11}}{f}\right)^2}} = \frac{0.933}{\sqrt{1 - (\frac{1}{2})^2}} = \frac{0.933}{0.866} = 1.076 \text{ cm}$$

while the phase velocity and intrinsic wave impedance, from (8-46) and (8-47), are

$$v_{p,11} = \frac{3 \times 10^8}{0.866} = 3.46 \times 10^8 \text{ m/sec} \qquad \hat{\eta}_{TM,11} = 377(0.866) = 326\ \Omega$$

(*c*) At $f = 8.05$ GHz, (8-42a) obtains

$$\alpha_{11} = \beta^{(0)}\sqrt{\left(\frac{f_{c,11}}{f}\right)^2 - 1} = \frac{2\pi(8.05 \times 10^9)}{3 \times 10^8}\sqrt{2^2 - 1} = 291 \text{ Np/m}$$

Below $f_{c,11}$, wavelength and phase velocity are undefined, in view of evanescent fields, but below cutoff, from (8-48)

$$\hat{\eta}_{TM,11} = -j\eta^{(0)}\sqrt{\left(\frac{f_{c,11}}{f}\right)^2 - 1} = -j377\sqrt{2^2 - 1} = -j653\ \Omega$$

8-4 TE Mode Solutions of Rectangular Waveguides

The analysis of the TE mode solutions of rectangular waveguides proceeds essentially along the lines employed for finding the TM mode solutions in Section 8-3, so only an outline of this boundary-value problem is given. The assumptions are as follows:

1. The rectangular, hollow pipe is very long and of interior dimensions a, b as noted in Figure 8-7.
2. The lossless dielectric has the parameters μ, ϵ, with $\rho_v = 0$ and $\mathbf{J} = 0$.
3. The waveguide walls are perfectly conducting.
4. All field quantities vary as $e^{j\omega t \mp \gamma z}$.
5. $\hat{\mathscr{E}}_z^{\pm} = 0$ for TE modes.

Only the last assumption differs from those used in the derivation of TM modes in Section 8-3.

The four TE field relations (8-22) suggest that a solution for $\hat{\mathscr{H}}_z^{\pm}$ might first be obtained, whereupon (8-22) can be employed to obtain the remaining

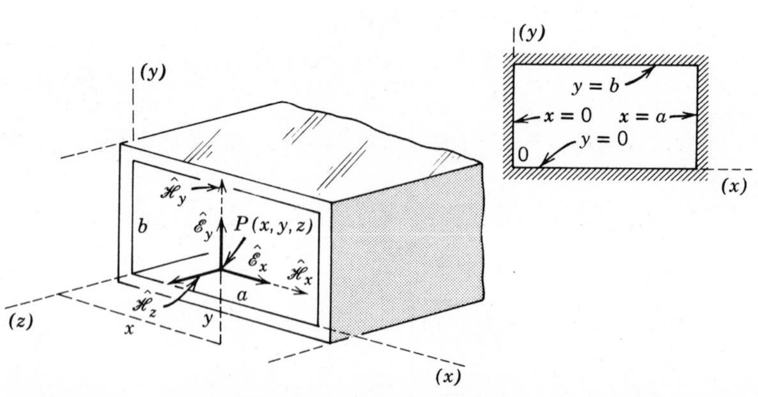

Figure 8-7. Geometry of a hollow, rectangular waveguide, showing field components corresponding to the TE modes.

transverse components. Beginning with the scalar wave equation (8-16f) in terms of $\hat{\mathscr{H}}_z^{\pm}$

$$\frac{\partial^2 \hat{\mathscr{H}}_z^{\pm}}{\partial x^2} + \frac{\partial^2 \hat{\mathscr{H}}_z^{\pm}}{\partial y^2} + \hat{k}_c^2 \hat{\mathscr{H}}_z^{\pm} = 0 \quad [8\text{-}16f]$$

by analogy with the separation-of-variables method applied to the wave equation (8-16c) in Section 8-3, a particular solution of (8-16f) is analogous with (8-32) such that

$$\hat{\mathscr{H}}_z^{\pm}(x, y) = (\hat{C}_1 \cos \hat{k}_x x + \hat{C}_2 \sin \hat{k}_x x)(\hat{C}_3 \cos \hat{k}_y y + \hat{C}_4 \sin \hat{k}_y y) \quad (8\text{-}49)$$

The boundary conditions at the perfectly conducting walls shown in Figure 8-7 demand that the tangential components of the electric field vanish there, i.e., $\hat{\mathscr{E}}_y^{\pm}(0, y) = 0$, $\hat{\mathscr{E}}_y^{\pm}(a, y) = 0$, $\hat{\mathscr{E}}_x^{\pm}(x, 0) = 0$, and $\hat{\mathscr{E}}_x^{\pm}(x, b) = 0$, but the latter are converted into equivalent boundary conditions applicable to the longitudinal component $\hat{\mathscr{H}}_z^{\pm}$ of (8-49), upon inserting them into the two TE modal relations (8-22a) and (8-22b), yielding

$$\begin{aligned} &1.\ \left.\frac{\partial \hat{\mathscr{H}}_z^{\pm}}{\partial x}\right|_{x=0} = 0 \\ &2.\ \left.\frac{\partial \hat{\mathscr{H}}_z^{\pm}}{\partial x}\right|_{x=a} = 0 \\ &3.\ \left.\frac{\partial \hat{\mathscr{H}}_z^{\pm}}{\partial y}\right|_{y=0} = 0 \\ &4.\ \left.\frac{\partial \hat{\mathscr{H}}_z^{\pm}}{\partial y}\right|_{y=b} = 0 \end{aligned} \quad (8\text{-}50)$$

Applying these boundary conditions to the appropriate x or y derivative of the $\hat{\mathscr{H}}_z^{\pm}$ solution (8-49) can be shown to obtain the following proper solutions (eigenfunctions):

$$\hat{\mathscr{H}}_z^{\pm}(x, y) = \hat{H}_{z,mn}^{\pm} \cos \frac{m\pi}{a} x \cos \frac{n\pi}{b} y \quad m, n = 0, 1, 2, \ldots \quad (8\text{-}51a)$$

in which m, n are arbitrary integers designating an infinite set of TE_{mn} modes. As in the TM mode case treated in Section 8-3, two separation constants, $\hat{k}_x = m\pi/a$ and $\hat{k}_y = n\pi/b$, are related to $\hat{k}_c^2 \equiv \gamma^2 + \omega^2\mu\epsilon$ by (8-29). The remaining field components of the TE_{mn} modes are found using (8-22), yielding

$$\hat{\mathscr{E}}_x^{\pm}(x, y) = \left[\frac{j\omega\mu}{\hat{k}_c^2}\frac{n\pi}{b}\hat{H}_{z,mn}^{\pm}\right]\cos\frac{m\pi}{a}x\sin\frac{n\pi}{b}y \equiv \hat{E}_{x,mn}^{\pm}\cos\frac{m\pi}{a}x\sin\frac{n\pi}{b}y \tag{8-51b}$$

$$\hat{\mathscr{E}}_y^{\pm}(x, y) = \left[\frac{-j\omega\mu}{\hat{k}_c^2}\frac{m\pi}{a}\hat{H}_{z,mn}^{\pm}\right]\sin\frac{m\pi}{a}x\cos\frac{n\pi}{b}y$$

$$\equiv \hat{E}_{y,mn}^{\pm}\sin\frac{m\pi}{a}x\cos\frac{n\pi}{b}y \tag{8-51c}$$

$$\hat{\mathscr{H}}_x^{\pm}(x, y) = \mp\frac{\hat{\mathscr{E}}_y^{\pm}}{\hat{\eta}_{\text{TE},mn}} = \left[\pm\frac{\gamma_{mn}}{\hat{k}_c^2}\frac{m\pi}{a}\hat{H}_{z,mn}^{\pm}\right]\sin\frac{m\pi}{a}x\cos\frac{n\pi}{b}y$$

$$\equiv \hat{H}_{x,mn}^{\pm}\sin\frac{m\pi}{a}x\cos\frac{n\pi}{b}y \tag{8-51d}$$

$$\hat{\mathscr{H}}_y^{\pm}(x, y) = \pm\frac{\hat{\mathscr{E}}_x^{\pm}}{\hat{\eta}_{\text{TE},mn}} = \left[\pm\frac{\gamma_{mn}}{\hat{k}_c^2}\frac{n\pi}{b}\hat{H}_{z,mn}^{\pm}\right]\cos\frac{m\pi}{a}x\sin\frac{n\pi}{b}y$$

$$\equiv \hat{H}_{y,mn}^{\pm}\cos\frac{m\pi}{a}x\sin\frac{n\pi}{b}y \tag{8-51e}$$

wherein

$$\hat{k}_c^2 \equiv \gamma_{mn}^2 + \omega^2\mu\epsilon = \left(\frac{m\pi}{a}\right)^2 + \left(\frac{n\pi}{b}\right)^2 \tag{8-51f}$$

implying a propagation constant γ_{mn} given by an expression identical with (8-39) for TM modes

$$\gamma_{mn} = \sqrt{\left(\frac{m\pi}{a}\right)^2 + \left(\frac{n\pi}{b}\right)^2 - \omega^2\mu\epsilon} \tag{8-52}$$

The latter implies a cutoff frequency for TE_{mn} modes in a rectangular waveguide given by an expression like (8-40b) for the comparable TM_{mn} modes

$$f_{c,mn} = \frac{1}{2\pi\sqrt{\mu\epsilon}}\sqrt{\left(\frac{m\pi}{a}\right)^2 + \left(\frac{n\pi}{b}\right)^2} \tag{8-53}$$

It therefore follows that the propagation constant γ_{mn} of (8-52) is a pure real attenuation factor α_{mn} if $f < f_{c,mn}$, or a pure imaginary phase factor $j\beta_{mn}$ if $f > f_{c,mn}$; thus

$$\gamma_{mn} = \alpha_{mn} \equiv \omega\sqrt{\mu\epsilon}\sqrt{\left(\frac{f_{c,mn}}{f}\right)^2 - 1}\ \text{Np/m} \qquad f < f_{c,mn} \tag{8-54a}$$

$$\gamma_{mn} = j\beta_{mn} \equiv j\omega\sqrt{\mu\epsilon}\sqrt{1 - \left(\frac{f_{c,mn}}{f}\right)^2} \text{ rad/m} \quad f > f_{c,mn} \quad (8\text{-}54b)$$

From (8-54b) one can infer, for a specified TE_{mn} mode, a wavelength λ_{mn} and phase velocity $v_{p,mn}$ given by expressions identical with (8-45) and (8-46) for the comparable TM_{mn} mode

$$\lambda_{mn} = \frac{\lambda^{(0)}}{\sqrt{1 + \left(\frac{f_{c,mn}}{f}\right)^2}} \text{ m} \qquad f > f_{c,mn} \quad (8\text{-}55)$$

$$v_{p,mn} = \frac{v_p^{(0)}}{\sqrt{1 - \left(\frac{f_{c,mn}}{f}\right)^2}} \text{ m/s} \qquad f > f_{c,mn} \quad (8\text{-}56)$$

in which $\lambda^{(0)}$ and $v_p^{(0)}$ are the wavelength and phase velocity associated with plane waves propagating at the frequency f in an unbounded region filled with the same dielectric with the parameters μ and ϵ. A comparison of (8-21) with (8-23) shows that the intrinsic wave impedances of TE and TM modes are not the same; from (8-23) and (8-54b) one obtains for TE_{mn} modes above cutoff

$$\eta_{TE,mn} = \frac{j\omega\mu}{j\beta_{mn}} = \frac{\eta^{(0)}}{\sqrt{1 - \left(\frac{f_{c,mn}}{f}\right)^2}} \ \Omega \qquad f > f_{c,mn} \quad (8\text{-}57)$$

which deserves comparison with expression (8-47) for $\hat{\eta}_{TM,mn}$.

If a TE_{mn} mode is generated at a frequency *below* the cutoff value specified by (8-53), the propagation constant γ_{mn} becomes the pure real α_{mn} of (8-54a), producing an evanescence of the field components (8-51) resembling that for TM_{mn} modes below cutoff as shown in Figure 8-5(*b*). While wavelength and phase velocity are undefined in the absence of wave motion for $f < f_{c,mn}$, the intrinsic wave impedance for a TE_{mn} mode below cutoff is obtained from (8-54a) and (8-23), yielding

$$\hat{\eta}_{TE,mn} = \frac{j\omega\mu}{\alpha_{mn}} = \frac{j\eta^{(0)}}{\sqrt{\left(\frac{f_{c,mn}}{f}\right)^2 - 1}} \ \Omega \qquad f < f_{c,mn} \quad (8\text{-}58)$$

From this result one may again see, as from (8-48) for TM_{mn} modes, that whenever a mode evanesces ($f < f_{c,mn}$) the wave impedance $\hat{\eta}_{TM}$ or $\hat{\eta}_{TE}$ becomes imaginary, showing that there is no power flow associated with an evanescent mode.

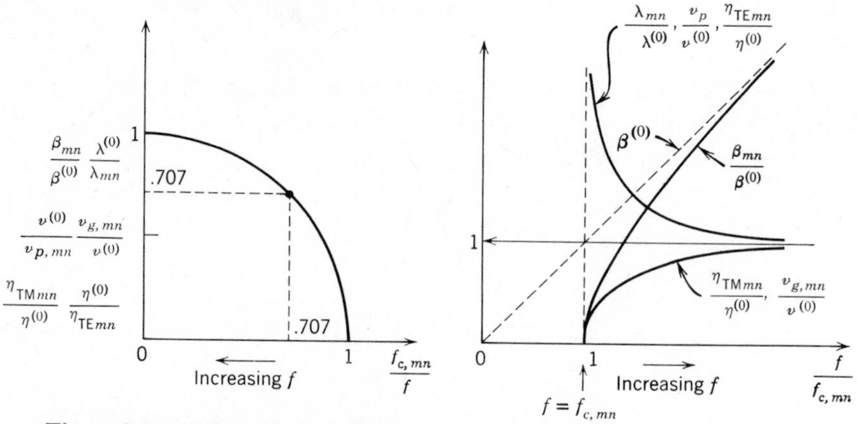

Figure 8-8. Universal circle diagram (left) and quantities plotted directly against frequency (right), for TM and TE modes.

The common factor $\sqrt{1 - (f_{c,mn}/f)^2}$ appearing in the various expressions (8-45) through (8-48) for TM_{mn} modes, together with the comparable relations (8-54) through (8-57) for TE_{mn} modes, permits graphing them as normalized quantities on the *universal circle diagram* shown in Figure 8-8. For example, the expressions (8-42b) and (8-54b) for the phase factor β_{mn} of TM or TE modes are normalized by dividing through by $\beta^{(0)} = \omega\sqrt{\mu\epsilon}$ to obtain

$$\left(\frac{\beta_{mn}}{\beta^{(0)}}\right)^2 + \left(\frac{f_{c,mn}}{f}\right)^2 = 1$$

the equation of a circle, considering $\beta_{mn}/\beta^{(0)}$ and $f_{c,mn}/f$ as the variables. A discussion of the group velocity v_g noted in the diagram is reserved for Section 8-5. To the right in the figure is also shown a graph of the same quantities plotted directly against frequency, which may have some interpretive advantages. Thus, the phase constant β_{mn} of a desired mode is seen to be zero at the cutoff frequency $f_{c,mn}$ while asymptotically approaching the unbounded space value $\beta^{(0)} = \omega\sqrt{\mu\epsilon}$ represented by the diagonal straight line as f becomes sufficiently large.

The expressions (8-51) for the five field components of the TE_{mn} modes lead to flux plots of typical modes as seen in Figure 8-9. The electric field lines are entirely transverse in any cross-section of the guide, as required for TE modes; they terminate normally at the perfectly conducting walls to satisfy the boundary conditions there. The magnetic lines, moreover, form closed loops and link electric flux (displacement currents) in the process, as required by Maxwell's equations. A comparison with Figure 8-6 points out the inherent

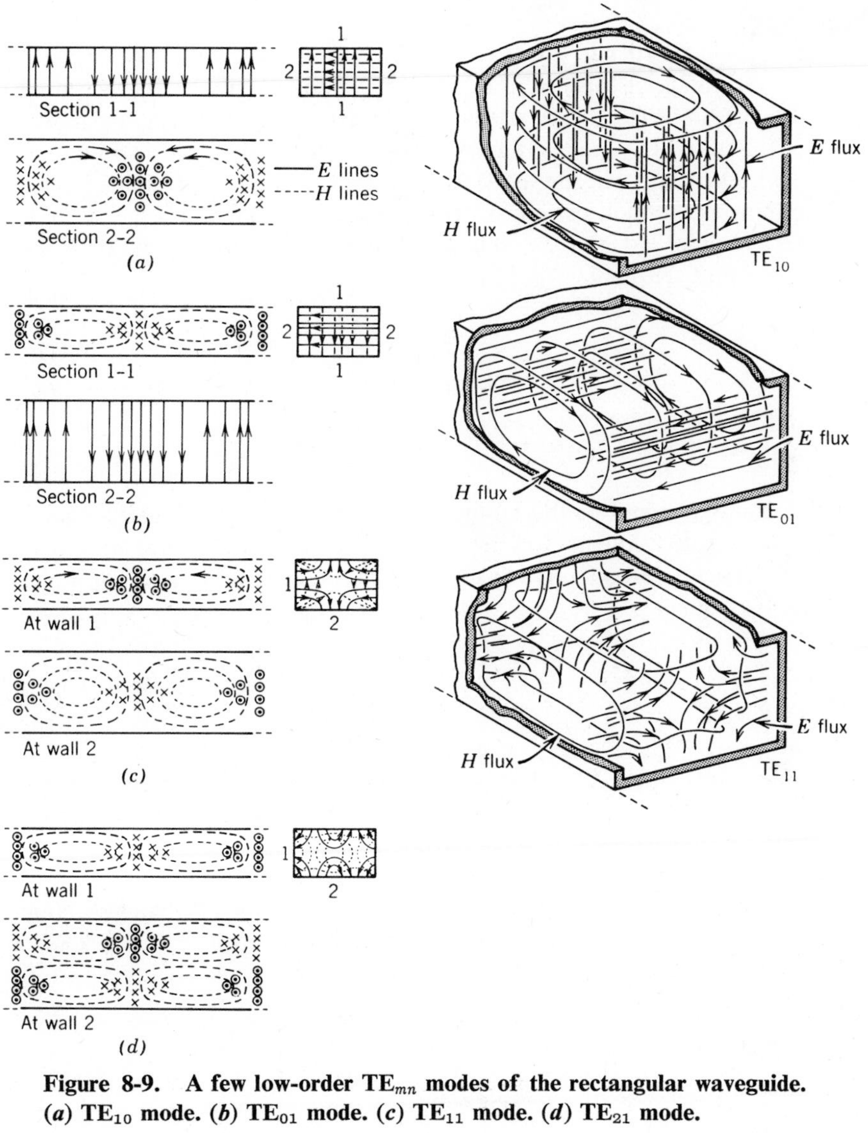

Figure 8-9. **A few low-order TE_{mn} modes of the rectangular waveguide. (*a*) TE_{10} mode. (*b*) TE_{01} mode. (*c*) TE_{11} mode. (*d*) TE_{21} mode.**

differences between TM and TE mode field configurations in a rectangular guide.

In Section 8-3, the TM mode expressions (8-38) reveal that the lowest-order nontrivial mode of this group is the TM_{11} mode. A similar inspection of the field expressions (8-51) shows that the lowest-order nontrivial TE modes are the TE_{10} and TE_{01} modes, flux plots of which are depicted in Figure 8-9(*a*) and (*b*). Of these two, the mode having the lowest cutoff frequency (8-53) is determined by which of the two transverse guide dimensions, *a* or *b*, is the larger. With $m = 1$ and $n = 0$ inserted into (8-53), the TE_{10} mode is seen to have a cutoff frequency

$$\boxed{f_{c,10} = \frac{1}{2\pi\sqrt{\mu\epsilon}}\frac{\pi}{a} = \frac{v_p^{(0)}}{2a}} \tag{8-59a}$$

a result independent of the *b* dimension because $n = 0$. Thus (8-59a) states that the cutoff frequency of the TE_{10} mode is the frequency at which the width *a* is just one-half a free-space wavelength. Similarly, the TE_{01} mode has a cutoff frequency

$$f_{c,01} = \frac{v_p^{(0)}}{2b} \tag{8-59b}$$

a value larger than $f_{c,10}$ if $a > b$, the dimensional condition depicted in Figure 8-10(*a*). From the identical cutoff frequency expressions (8-53) and (8-40b), all higher-order TE *and* TM modes exhibit cutoff frequencies higher than (8-59a), assuming $a > b$, making the TE_{10} mode the *dominant mode* of that rectangular waveguide. For example, the so-called X-band rectangular waveguide, assumed air filled and of interior dimensions $a = 0.9$ in. and $b = 0.4$ in. (0.02286×0.01016 m) has a cutoff frequency obtained from (8-59a), yielding

$$f_{c,10} = \frac{3 \times 10^8}{2(0.02286)} = 6.557 \text{ GHz} \quad X\text{-band guide} \tag{8-60}$$

while the cutoff frequency of the next higher-order mode, TE_{20}, becomes $f_{c,20} = 13.11$ GHz, from (8-53). The TE_{01} mode, from (8-59b), yields $f_{c,01} = 14.77$ GHz, while using (8-53) or (8-40b) obtains cutoff frequencies for the TE_{11} and TM_{11} modes that are even higher ($f_{c,11} = 16.10$ GHz). Their positions on a frequency scale are portrayed in Figure 8-10(*a*), showing why

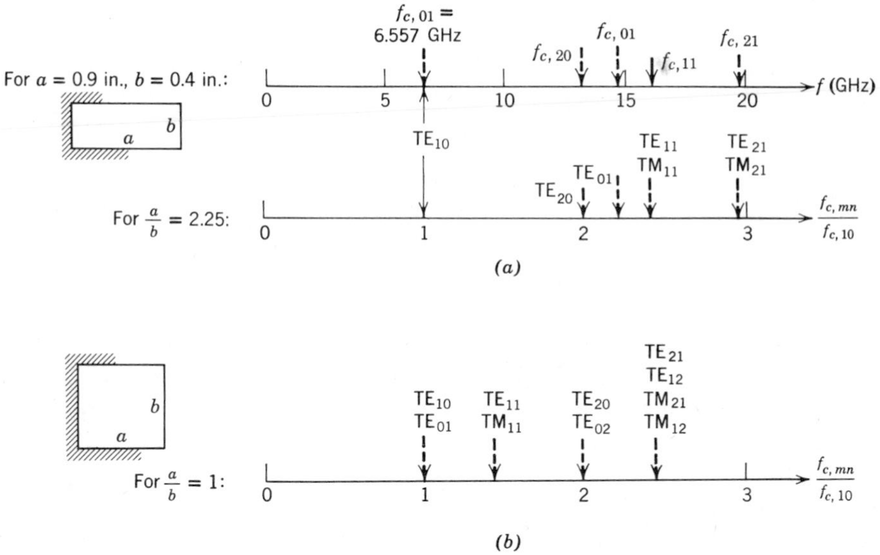

Figure 8-10. Cutoff frequencies of lower-order modes in rectangular and square waveguides. (*a*) For $a/b = 2.25$. Cutoff frequencies shown relative to $f_{c,10}$ on lower graph. (*b*) For $a/b = 1$.

the propagation of electromagnetic power via the single, dominant TE_{10} mode in a rectangular waveguide is possible by keeping the generated frequency f above the cutoff frequency of the TE_{10} mode, but below the cutoff frequencies of all other modes. This choice assures a traveling wave TE_{10} mode and the evanescence of all other modes, thereby justifying the designation dominant for the propagating TE_{10} mode. For example, the band 8.2 to 12.4 GHz is chosen as the X band; frequencies that propagate only in the dominant TE_{10} mode in a 0.4 in. × 0.9 in. rectangular waveguide.

It is evident that a *square* waveguide ($a = b$) will not possess only one dominant mode, for the TE_{10} and TE_{01} modes then have identical cutoff frequencies from (8-59). Figure 8-10(*b*) shows the positions of the cutoff frequencies of lower-order modes for a square waveguide on a relative frequency scale. A comparison of Figure 8-10(*a*) with (*b*) reveals that the use of a rectangular waveguide, with $a > b$ as in (*a*) of that figure, provides a desirable control over the E field polarization of the propagated mode. Figure 8-11 illustrates the manner in which a microwave power source (a klystron, magnetron, etc.) is connected to a waveguide by utilizing a small antenna wire protruding into the waveguide, such that the wire alignment agrees with the polarization of the dominant mode being launched. The power can similarly

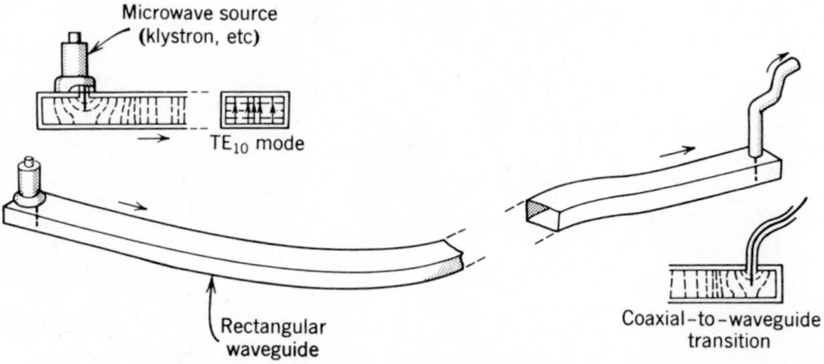

Figure 8-11. Typical waveguide transmission system, showing launching of the dominant mode, and a transition from waveguide to coaxial transmission line.

be extracted at the other end, if desired, by means of the center conductor of a coaxial line used as a receiving antenna (a waveguide-to-coax transition). The propagation of the energy down the waveguide via the dominant TE_{10} mode thus assures the known field polarization necessary to the efficient launching and retrieval of the energy.

Since signal power in a rectangular waveguide is commonly dispatched by use of the dominant TE_{10} mode, its properties are for convenience collected separately in the following. The expressions (8-51) for TE_{mn} modes, with $m = 1$, $n = 0$ inserted, reduce to three components

$$\hat{\mathscr{H}}_z^{\pm}(x) = \hat{H}_{z,10}^{\pm} \cos\frac{\pi}{a}x \tag{8-61a}$$

$$\hat{\mathscr{E}}_x^{\pm} = \hat{\mathscr{H}}_y^{\pm} = 0$$

$$\hat{\mathscr{E}}_y^{\pm}(x) = \left[\frac{-j\omega\mu a}{\pi}\hat{H}_{z,10}^{\pm}\right]\sin\frac{\pi}{a}x = \hat{E}_{y,10}^{\pm}\sin\frac{\pi}{a}x \tag{8-61b}$$

$$\hat{\mathscr{H}}_x^{\pm}(x) = \mp\frac{\hat{\mathscr{E}}_y^{\pm}}{\hat{\eta}_{\mathrm{TE},10}} = \left[\frac{\pm j\beta_{10}a}{\pi}\hat{H}_{z,10}^{\pm}\right]\sin\frac{\pi}{a}x$$

$$= \left[\frac{\pm j2a}{\lambda_{10}}\hat{H}_{z,10}^{\pm}\right]\sin\frac{\pi}{a}x = \hat{H}_{x,10}^{\pm}\sin\frac{\pi}{a}x \tag{8-61c}$$

assuming $f > f_{c,10}$. The foregoing may for some purposes be more conveniently expressed in terms of the complex amplitudes $\hat{E}_{y,10}^{\pm}$ of the y directed electric field (8-61b), yielding

$$\hat{\mathscr{E}}_y^{\pm}(x) = \hat{E}_{y,10}^{\pm} \sin \frac{\pi}{a} x \tag{8-62a}$$

$$\hat{\mathscr{H}}_x^{\pm}(x) = \mp \frac{\hat{E}_{y,10}^{\pm}}{\hat{\eta}_{\mathrm{TE},10}} \sin \frac{\pi}{a} x = \hat{H}_{x,10}^{\pm} \sin \frac{\pi}{a} x \tag{8-62b}$$

$$\hat{\mathscr{H}}_z^{\pm}(x) = j \frac{\hat{E}_{y,10}^{\pm} \lambda^{(0)}}{2\eta^{(0)} a} \cos \frac{\pi}{a} x = \hat{H}_{z,10}^{\pm} \cos \frac{\pi}{a} x \tag{8-62c}$$

The remaining properties of the TE_{10} mode are related to its cutoff frequency $f_{c,10}$ specified by (8-59a). From the latter, the ratio $f_{c,10}/f$ is

$$\frac{f_{c,10}}{f} = \frac{v_p^{(0)}}{2af} = \frac{\lambda^{(0)}}{2a} \tag{8-63}$$

to permit writing the propagation constant, wavelength in the guide, and phase velocity for the TE_{10} mode as follows:

$$\gamma_{10} = \alpha_{10} \equiv \beta^{(0)} \sqrt{\left(\frac{\lambda^{(0)}}{2a}\right)^2 - 1} \ \mathrm{Np/m} \qquad f < f_{c,10} \tag{8-64a}$$

$$\gamma_{10} = j\beta_{10} \equiv j\beta^{(0)} \sqrt{1 - \left(\frac{\lambda^{(0)}}{2a}\right)^2} \ \mathrm{rad/m} \qquad f > f_{c,10} \tag{8-64b}$$

$$\lambda_{10} = \frac{\lambda^{(0)}}{\sqrt{1 - \left(\dfrac{\lambda^{(0)}}{2a}\right)^2}} \ \mathrm{m} \qquad f > f_{c,10} \tag{8-65}$$

$$v_{p,10} = \frac{v_p^{(0)}}{\sqrt{1 - \left(\dfrac{\lambda^{(0)}}{2a}\right)^2}} \ \mathrm{m/s} \qquad f > f_{c,10} \tag{8-66}$$

in which $\beta^{(0)} = \omega\sqrt{\mu\epsilon}$, $\lambda^{(0)} = 2\pi/\beta^{(0)} = v_p^{(0)}/f$, and $v_p^{(0)} = (\mu\epsilon)^{-1/2}$ as before. The intrinsic wave impedance obtained from (8-57) or (8-58) becomes

$$\hat{\eta}_{\mathrm{TE},10} = \frac{\eta^{(0)}}{\sqrt{1 - \left(\dfrac{\lambda^{(0)}}{2a}\right)^2}} \ \Omega \qquad f > f_{c,10} \tag{8-67}$$

$$\hat{\eta}_{\mathrm{TE},10} = \frac{j\eta^{(0)}}{\sqrt{\left(\dfrac{\lambda^{(0)}}{2a}\right)^2 - 1}} \ \Omega \qquad f < f_{c,10} \tag{8-68}$$

wherein $\eta^{(0)} = \sqrt{\mu/\epsilon}$. Thus, the TE_{10} mode fields (8-61) propagating in a rectangular waveguide at a frequency above cutoff involve inphase transverse field components $\hat{\mathscr{E}}_y^{\pm}$ and $\hat{\mathscr{H}}_x^{\pm}$ related by the real impedance (8-67). Below cutoff, the imaginary result (8-68) assures no power transfer by the accompanying evanescent mode, attenuated through (8-64a).

The time and z dependence of the TE_{10} mode field expressions (8-62) are included by multiplying them by $e^{j\omega t \mp \gamma_{10} z}$. Taking the real part of the resulting products obtains the real-time traveling wave expressions as follows, assuming $f > f_{c,10}$:

$$E_y^{\pm}(x, z, t) = E_{y,10}^{\pm} \sin \frac{\pi}{a} x \cos (\omega t \mp \beta_{10} z + \phi_{10}^{\pm}) \tag{8-69a}$$

$$H_x^{\pm}(x, z, t) = \frac{\mp E_{y,10}^{\pm}}{\eta^{(0)}} \sqrt{1 - \left(\frac{\lambda^{(0)}}{2a}\right)^2} \sin \frac{\pi}{a} x \cos (\omega t \mp \beta_{10} z + \phi_{10}^{\pm}) \tag{8-69b}$$

$$H_z^{\pm}(x, z, t) = \frac{-E_{y,10}^{\pm} \lambda^{(0)}}{2a\eta^{(0)}} \cos \frac{\pi}{a} x \sin (\omega t \mp \beta_{10} z + \phi_{10}^{\pm}) \tag{8-69c}$$

in which $\phi_{10}^{\pm}$ denotes the arbitrary phase angles of the complex amplitudes $\hat{E}_{y,10}^{\pm}$, assuming $\hat{E}_{y,10}^{\pm} = E_{y,10}^{\pm} e^{j\phi_{10}^{\pm}}$. The flux plots of these fields have already been displayed in Figure 8-9(a), although the wave nature of the TE_{10} mode is perhaps more readily apparent if the field components are separately depicted as in Figure 8-12.

At the interface between the dielectric region filling a waveguide and the idealized perfectly conducting walls, the electric and magnetic fields fall abruptly from their finite values in the dielectric to zero within the conductor. This discontinuity of the fields gives rise to a surface layer of electric charges and currents at the interface, predictable from the boundary conditions (3-45) and (3-72)

$$\rho_s = \mathbf{n} \cdot \mathbf{D} \ \mathrm{C/m^2} \tag{3-45}$$

$$\mathbf{J}_s = \mathbf{n} \times \mathbf{H} \ \mathrm{A/m} \tag{3-72}$$

in which **n** denotes a normal unit vector directed into the dielectric region. The electric field of the TE_{10} mode develops a surface charge density ρ_s on only two walls of the rectangular guide, since the y directed **E** field produces a normal component $\mathbf{n} \cdot \mathbf{D}$ on only the lower ($y = 0$) and upper ($y = b$) walls. Thus, using the electric field (8-69a) in $\mathbf{D} = \epsilon \mathbf{E} = \epsilon \mathbf{a}_y E_y^{\pm}$, the boundary condition (3-45) yields the surface charge density as follows:

$$\rho_s^{\pm}]_{y=0} = \mathbf{a}_y \cdot \mathbf{a}_y \epsilon E_y^{\pm} = \epsilon E_{y,10}^{\pm} \sin \frac{\pi}{a} x \cos (\omega t \mp \beta_{10} z + \phi_{10}^{\pm}) \tag{8-70}$$

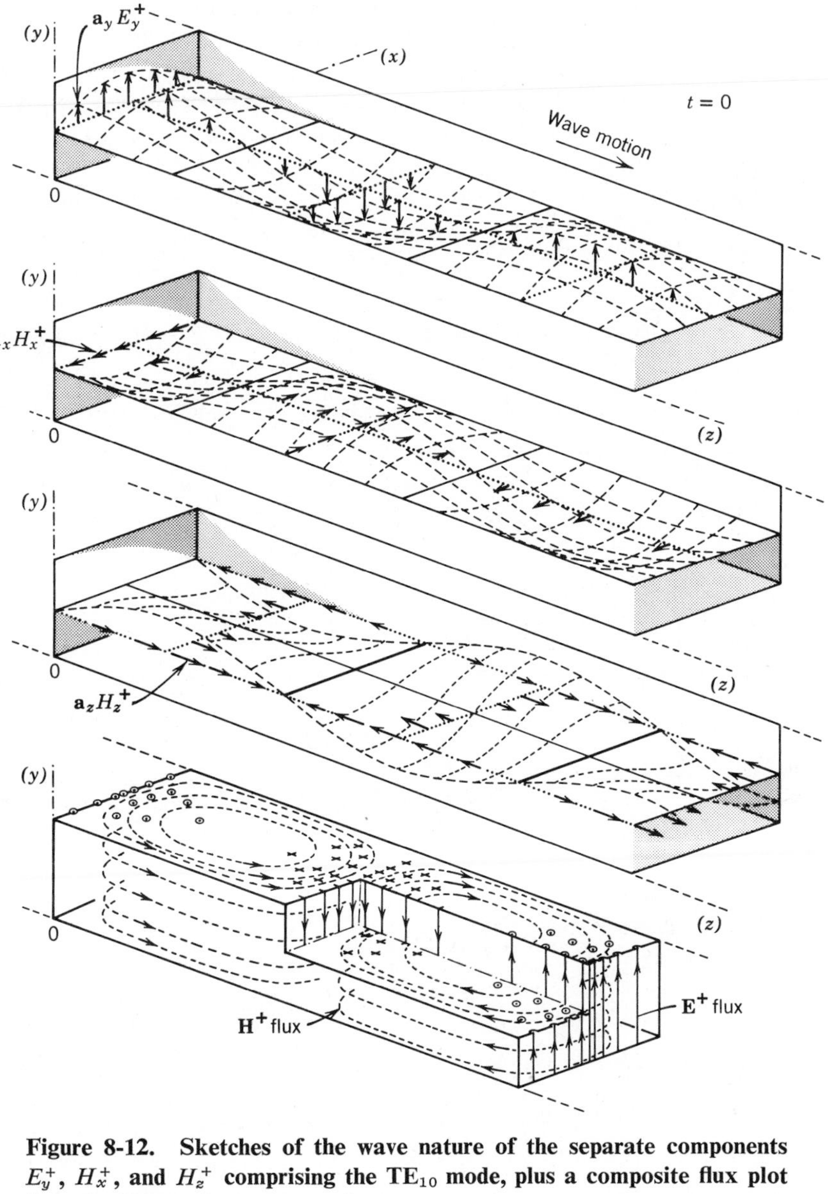

Figure 8-12. Sketches of the wave nature of the separate components E_y^+, H_x^+, and H_z^+ comprising the TE_{10} mode, plus a composite flux plot (below). All are shown at $t = 0$.

in time-instantaneous form. One may similarly show that the surface charge density on the opposite wall (at $y = b$) is the negative of (8-70).

Surface current densities given by (3-72) appear at all four walls of the guide, because tangential magnetic fields occur at every wall. For example, on the lower wall where the total magnetic field is the vector sum of (8-69b) and (8-69c), the surface current becomes

$$\mathbf{J}_s^\pm]_{y=0} = \mathbf{a}_y \times [\mathbf{a}_x H_x^\pm + \mathbf{a}_z H_z^\pm] = -\mathbf{a}_x \frac{E_{y,10}^\pm \lambda^{(0)}}{2\eta^{(0)} a} \cos\frac{\pi}{a}x \sin(\omega t \mp \beta_{10}x + \phi_{10}^\pm)$$

$$\pm\, \mathbf{a}_z \frac{E_{y,10}^\pm}{\eta^{(0)}} \sqrt{1 - \left(\frac{\lambda^{(0)}}{2a}\right)^2} \sin\frac{\pi}{a}x \cos(\omega t \mp \beta_{10}z + \phi_{10}^\pm) \qquad \text{(8-71a)}$$

On the side walls, the surface current density has but one component, being entirely y directed. Thus, on the wall at $x = 0$

$$\mathbf{J}_s^\pm]_{x=0} = \mathbf{a}_x \times (\mathbf{a}_x H_x^\pm + \mathbf{a}_z H_z^\pm) = \mathbf{a}_y \frac{\lambda^{(0)} E_{y,10}^\pm}{2\eta^{(0)} a} \sin(\omega t \mp \beta_{10}z + \phi_{10}^\pm) \qquad \text{(8-71b)}$$

The densities at $y = b$ and $x = a$ are similarly obtained. A sketch of the wall currents (8-71) is shown in Figure 8-13(a), useful if slots are to be cut in

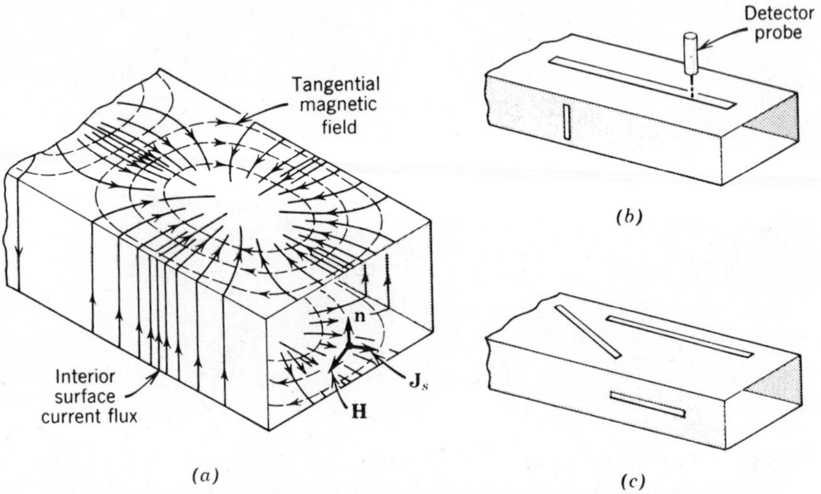

Figure 8-13. The surface currents induced by the tangential magnetic field of the TE_{10} mode on perfectly conducting inner walls of a waveguide, and wall-slot configurations. (*a*) Flux plot of surface currents on waveguide inner walls. (*b*) Slots producing negligible wall-current perturbation. (*c*) Slots producing significant wall-current perturbations.

the walls. For example, a longitudinal slot centered on the broad wall of a rectangular waveguide carrying the dominant TE_{10} mode as shown in Figure 8-13(*b*) is useful in field-probing techniques for the detection of standing waves (slotted-line measurements). A slot that does not cut across wall current flux lines produces a minimal perturbation of the waveguide fields, permitting field detection schemes that yield measurements essentially the same as those expected without the slot. In Figure 8-13(*c*), however, are shown slots that interrupt wall currents significantly, producing substantial field fringing through the slot with power radiation into the space outside the waveguide. Such configurations form the basis for *slot antennas* or arrays using waveguide fields for excitation.

EXAMPLE 8-2. An air-filled, *X*-band, rectangular waveguide carries a positive z traveling TE_{10} mode at $f = 9$ GHz. (*a*) Find the phase constant, wavelength, phase velocity, and intrinsic wave impedance associated with this mode at the given frequency. (*b*) If $\hat{\mathscr{E}}_y^+$ has the amplitude 10^4 V/m, determine the amplitudes of $\hat{\mathscr{H}}_x^+$ and $\hat{\mathscr{H}}_z^+$. What time-average power-flux is transmitted through every cross-sectional surface of the waveguide by this mode?

(*a*) At 9 GHz, the wavelength in unbounded free space is $\lambda^{(0)} = v_p^{(0)}/f = (3 \times 10^8)/(9 \times 10^9) = 3.33$ cm. With $a = 0.9$ in., the ratio $f_{c,10}/f$ given by (8-63) is $\lambda^{(0)}/2a = 0.729$, while $\beta^{(0)} = \omega\sqrt{\mu_0\epsilon_0} = 2\pi/\lambda^{(0)} = 60\pi$ rad/m. By use of (8-64b), the phase constant becomes

$$\beta_{10} = \beta^{(0)}\sqrt{1 - \left(\frac{\lambda^{(0)}}{2a}\right)^2} = 60\pi\sqrt{1 - (0.729)^2} = 60\pi(0.683) = 128.8 \text{ m}^{-1}$$

Thus, from (8-65), (8-66), and (8-67)

$$\lambda_{10} = \frac{3.33}{0.683} = 4.88 \text{ cm}$$

$$v_{p,10} = \frac{3 \times 10^8}{0.683} = 4.39 \times 10^8 \text{ m/sec}$$

$$\hat{\eta}_{\text{TE},10} = \frac{120\pi}{0.683} = 552\ \Omega$$

(*b*) With $\hat{E}_{y,10}^+ = 10^4$ V/m, the remaining amplitudes, from (8-62), are

$$\hat{H}_{x,10}^+ = -\frac{\hat{E}_{y,10}^+}{\hat{\eta}_{\text{TE},10}} = -\frac{10^4}{552} = -18.1 \text{ A/m}$$

$$\hat{H}_{z,10}^+ = j\frac{\hat{E}_{y,10}^+\lambda^{(0)}}{2\eta^{(0)}a} = \frac{j10^4(0.033)}{2(120\pi)0.0229} = j19.3 = 19.3e^{j90°} \text{ A/m}$$

The time-average power-flux transmitted through any cross-section is obtained using (7-52), in which the minus sign is omitted if it is agreed that power-flux emerging from the positive z side of the cross-section is desired. Thus, with $P_{av} = \int_S (1/2)\,\text{Re}\,[\hat{\mathbf{E}} \times \hat{\mathbf{H}}^*]\cdot d\mathbf{s}$, in which $\hat{\mathbf{E}} = \mathbf{a}_y\hat{\mathscr{E}}_y^+ e^{-j\beta_{10}z}$, $\hat{\mathbf{H}} = [\mathbf{a}_x\hat{\mathscr{H}}_x^+ + \mathbf{a}_z\hat{\mathscr{H}}_z^+]e^{-j\beta_{10}z}$, and $\hat{\mathscr{E}}_y^+$, $\hat{\mathscr{H}}_x^+$, and $\hat{\mathscr{H}}_z^+$ are supplied by (8-62), one obtains

$$P_{av} = \int_{y=0}^{b}\int_{x=0}^{a} \tfrac{1}{2}\,\text{Re}\left\{\mathbf{a}_z \frac{(\hat{E}_{y,10}^+)(\hat{E}_{y,10}^+)^*}{\hat{\eta}_{\text{TE},10}^*} \times \sin^2\frac{\pi}{a}x(e^{-j\beta_{10}z})(e^{-j\beta_{10}z})^*\right\}\cdot\mathbf{a}_z\,dx\,dy$$

$$= \frac{|\hat{E}_{y,10}^+|^2}{2\eta_{\text{TE},10}}\,b\int_0^a \sin^2\frac{\pi}{a}x\,dx = \frac{|\hat{E}_{y,10}^+|^2}{4\eta_{\text{TE},10}}\,ab$$

With $\hat{E}_{y,10}^+ = 10^4$ V/m, $\eta_{\text{TE},10} = 552\ \Omega$, $a = 0.0229$ m, and $b = 0.0102$ m, the time-average transmitted power becomes $P_{av} = 10.6$ W.

*8-5 Dispersion in Hollow Waveguides: Group Velocity

All previous discussions of wave phenomena in this text have been restricted to single-frequency, sinusoidal waves. Whether with reference to plane waves propagating in lossless or lossy unbounded regions as described in Chapter 6, or in connection with waves traveling in hollow metal tubes as considered in the present chapter, z traveling, single-frequency waves are characterized by functions of the form

$$\hat{A}e^{j(\omega t - \beta z)} \tag{8-72}$$

in which $\hat{A}$ is any complex amplitude coefficient (possibly a function of (x, y)), and in which any equiphase surface is defined by $\omega t - \beta z =$ constant. This yields the phase velocity

$$v_p = \frac{\omega}{\beta} \tag{8-73}$$

a quantity that may or may not be frequency dependent, depending on the phase factor β. Thus, in the case of plane waves traveling in unbounded free space, $\beta = \beta_0 = \omega\sqrt{\mu_0\epsilon_0}$, to yield (2-125b)

$$v_p = c \cong 3 \times 10^8\ \text{m/sec} \tag{8-74}$$

a result independent of frequency. Free space is therefore termed *dispersionless*, in view of the constant v_p regardless of the frequency. On the other hand, waves of a given TM or TE mode in a rectangular, hollow waveguide have a phase velocity given by (8-46) or (8-56)

$$v_p = \frac{v^{(0)}}{\sqrt{1 - \left(\frac{f_{c,mn}}{f}\right)^2}} \tag{8-75}$$

a decidedly frequency dependent result. While the concept of phase velocity is applicable only to steady state sinusoidal fields (constant amplitude and frequency), the Fourier superposition of any number of sinusoidal steady state field solutions having different frequencies can be used to construct *modulated* waves of variable amplitude or frequency. This important process leads to another concept known as the *group velocity*, or the velocity of the signal, or information, associated with the group of waves distributed over the spectrum of frequencies comprising the modulated signal. This is considered in the following.

No information or intelligence is transmitted by a steady state, single-frequency sinusoidal traveling wave as that illustrated in Figure 8-14(*a*). It can, however, become a carrier of information by inflicting on it the process known as *modulation*. The transmission of information via a carrier wave requires a modulation (or changing, in time), in proportion to the instantaneous value of a desired signal, of either the *amplitude* or the *frequency* of the carrier, thereby yielding an amplitude-modulated (AM) or a frequency-modulated (FM) carrier.

The present discussion is limited to the AM carrier, examples of which are illustrated in Figure 8-14(*b*) and (*c*). As suggested by the name, in this type of modulation the carrier amplitude is forced to become proportional to the signal level at every instant t. The frequency spectra of signals used to modulate a carrier typically fall within the audio range (dc to about 15 kHz) for ordinary voice or music transmission, or in the video range (dc to several megahertz) for television or coded-pulse transmission. The Fourier analysis of a high-frequency carrier, amplitude-modulated by a spectrum of lower signal frequencies, reveals what range of frequencies must be transmitted by the system containing perhaps waveguides, coaxial lines, filter circuits, antennas, and other elements. Such an analysis shows that the transmission system must be capable of passing the carrier frequency f_0 plus additional frequency components contributed by the signal spectrum of width $2\Delta f$, components appearing in two adjacent frequency bands termed side bands of width Δf just above and below f_0. For example, a 100 MHz carrier, amplitude-modulated by a video signal embracing frequency components from dc to 4 MHz, will require a transmission band from $f_0 - \Delta f$ to $f_0 + \Delta f$, namely 96 to 104 MHz, or an 8% bandwidth. On the other hand, if a 10,000 MHz carrier

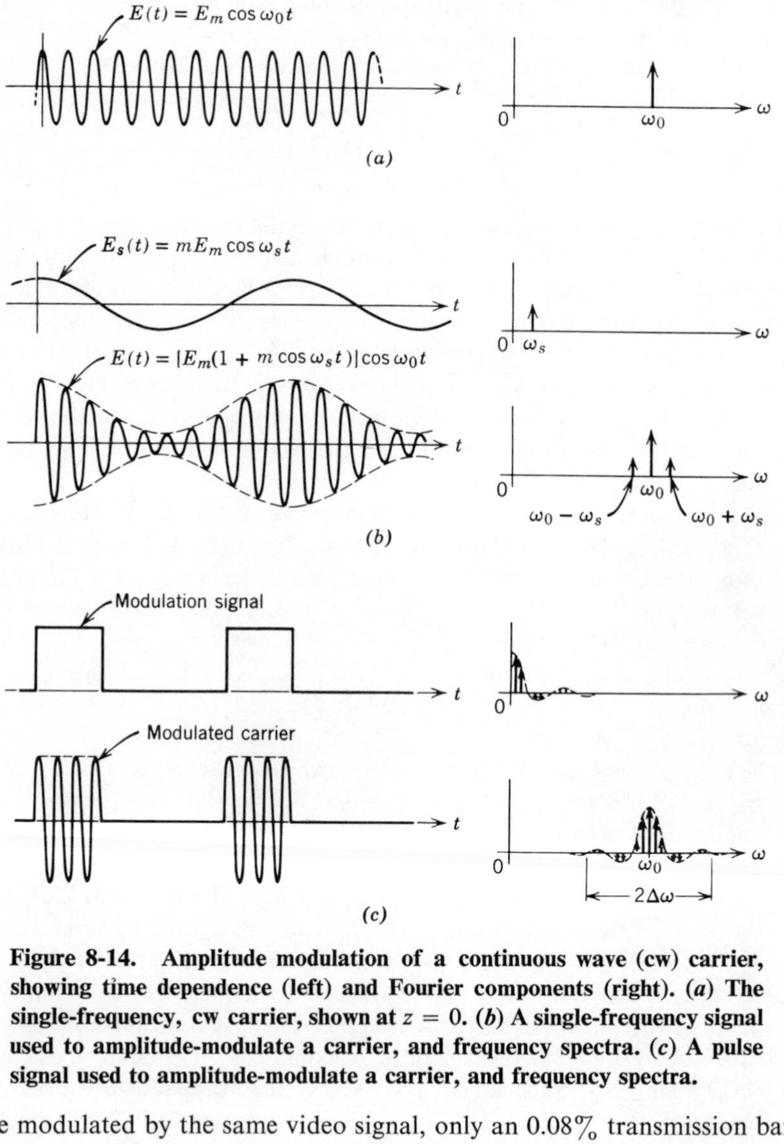

Figure 8-14. Amplitude modulation of a continuous wave (cw) carrier, showing time dependence (left) and Fourier components (right). (*a*) The single-frequency, cw carrier, shown at $z = 0$. (*b*) A single-frequency signal used to amplitude-modulate a carrier, and frequency spectra. (*c*) A pulse signal used to amplitude-modulate a carrier, and frequency spectra.

were modulated by the same video signal, only an 0.08% transmission band extending from 9,996 MHz to 10,004 MHz would be required to handle the ± 4 MHz signal spectrum. Short-pulse-communication and other high information rate systems require a correspondingly wide frequency band, therefore pulse communication systems using many channels simultaneously must operate at carrier frequencies in the uhf or microwave regions, and more recently they have even gone into the optical range of frequencies.

The generic example of amplitude modulation is illustrated in Figure 8-14(*b*), depicting the simplest case of a carrier at the sinusoidal frequency ω_0, amplitude-modulated by a time-harmonic signal at the single frequency ω_s. The carrier amplitude E_m is modulated sinusoidally in time with a signal amplitude mE_m, in which m is called the *modulation factor*, so that the real-time expression for an electric field carrier modulated in this way becomes

$$E(t) = [E_m(1 + m \cos \omega_s t)] \cos \omega_0 t \tag{8-76a}$$

The bracketed factor denotes the amplitude variations at the signal frequency ω_s. Equation (8-76a) specifies field behavior in the reference plane $z = 0$, whereas the additional z dependence needed to provide its traveling wave behavior is included momentarily. The amplitude-modulated carrier (8-76a) possesses three terms in its Fourier series expansion, or spectrum. Thus, with the substitution $\cos A \cos B = (1/2)[\cos (A + B) + \cos (A - B)]$, (8-76a) yields

$$E(t) = E_m \cos \omega_0 t + \frac{mE_m}{2} \cos (\omega_0 + \omega_s)t + \frac{mE_m}{2} \cos (\omega_0 - \omega_s)t \tag{8-76b}$$

This is a three-term (finite) Fourier series, possessing a carrier frequency term of amplitude E_m, plus just two side band terms of amplitude $mE_m/2$ at the sum and difference frequencies $(\omega_0 + \omega_s)$ and $(\omega_0 - \omega_s)$. This spectrum of three frequency components is depicted in the diagram at the right in Figure 8-14(*b*). The expressions (8-76) can be taken as the amplitude-modulated electric field of a plane wave (at $z = 0$) propagating in unbounded free-space, or denote a field component of a propagating mode inside a hollow waveguide or a coaxial line, etc. Equation (8-76b) is readily rewritten to specify the spectrum of positive z traveling waves of an amplitude-modulated carrier moving through a lossless transmission region, simply by adding in the proper phase delay terms βz as follows:

$$E^+(z, t) = E_m^+ \cos (\omega_0 t - \beta_0 z) + \frac{mE_m^+}{2} \cos [(\omega_0 + \omega_s)t - \beta_+ z]$$
$$+ \frac{mE_m^+}{2} \cos [(\omega_0 - \omega_s)t - \beta_- z] \tag{8-77}$$

in which β_0, β_+, and β_- denote the z propagation phase constants at the respective frequencies ω_0, $\omega_0 + \omega_s$, and $\omega_0 - \omega_s$. One is to examine (8-77) for its wave-envelope velocity, or so-called group velocity, for two classes of regions: a *nondispersive* region, in which all frequency components of a spectrum of waves move with the same phase velocity; and a *dispersive* region, in which the phase velocities of the spectral components are frequency dependent.

A. Group Velocity in a Nondispersive Region

Suppose the signal (8-77) denotes the amplitude-modulated field E_x of a plane wave propagating in free space. The phase velocity is then the constant $v_p = (\mu_0\epsilon_0)^{-1} = c$ given by (2-125b), making free space a nondispersive region. Therefore (8-77) written with $\beta_0 = \omega_0/c$, $\beta_+ = (\omega_0 + \omega_s)/c$, and $\beta_- = (\omega_0 - \omega_s)/c$, yielding

$$E_x^+(z, t) = E_m^+ \cos \omega_0\left(t - \frac{z}{c}\right) + \frac{mE_m^+}{2} \cos\left[(\omega_0 + \omega_s)\left(t - \frac{z}{c}\right)\right] + \frac{mE_m^+}{2} \cos\left[(\omega_0 - \omega_s)\left(t - \frac{z}{c}\right)\right] \quad (8\text{-}78)$$

Since the three Fourier terms remain in the same phase relationship no matter how far the modulated wave travels, the wave envelope must move at a velocity identical with the phase velocity in a nondispersive region. The wave envelope velocity, also called the *group velocity* (from the spectral group), is thus

$$v_g = v_p = \frac{\omega}{\beta} \qquad \text{Nondispersive} \quad (8\text{-}79)$$

for a nondispersive region. This result is correct no matter how complex the spectral structure of the wave. Hence, for the pulse-modulated signal of Figure 8-14(c), all terms of its Fourier series expansion will propagate through the medium at the same phase-velocity v_p. One thus concludes that a dispersionless region is also distortionless.

B. Group Velocity in a Dispersive Region

A hollow waveguide is an example of a wave transmission device exhibiting the phase velocity dispersion characteristic (8-75), depicted as a function of frequency in the graphs of Figure 8-8. The different phase velocities of the Fourier terms that characterize a modulated traveling wave in a waveguide result in the wave envelope appearing to slip behind the carrier appearing under the envelope. This phenomenon arises from the group velocity being *slower* than the phase velocities of the Fourier components. Thus, while the phase velocities of the Fourier terms of a modulated wave in an air filled hollow waveguide all *exceed* the speed of light, the speed of the transmission of the information (the wave envelope) at the group velocity is at a speed *less* than c.

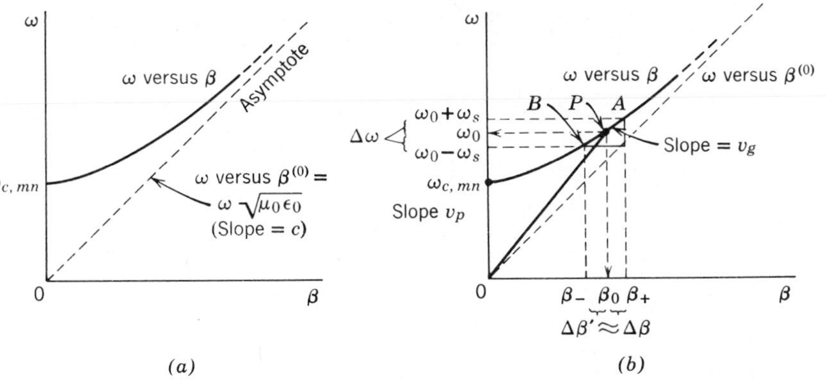

Figure 8-15. The ω-β diagram for a given mode in a hollow waveguide, and its velocity interpretations. (*a*) The ω-β diagram. (*b*) Constructions leading to the phase and group velocities.

To validate the foregoing remarks, one can appeal to the graph of β versus frequency of Figure 8-8 for the hollow waveguide. It is redrawn in Figure 8-15 as an ω-β diagram, a choice seen to yield velocities from slopes (rather than inverse slopes). Suppose the waveguide carries, for the sake of illustration, the dominate TE_{10} mode, amplitude-modulated sinusoidally at the signal frequency ω_s. The three-term expression (8-77) is readily adapted to this problem by assuming $E(z, t)$ denotes the y directed electric field component of the TE_{10} mode, given by (8-62a) in complex form or (8-69a) in real-time form. Using only the positive z traveling wave solution, and replacing the field amplitude E_m^+ in (8-77) with the x dependent amplitude factor $\mathscr{E}_y^+$ of (8-62a) as follows:

$$E_m^+ \rightarrow \mathscr{E}_y^+(x) = E_{y,10}^+ \sin \frac{\pi}{a} x$$

the expansion (8-77) for the amplitude-modulated electric field of the TE_{10} mode is written

$$E_y^+(x, z, t) = \mathscr{E}_y^+ \cos(\omega_0 t - \beta_0 z) + \frac{m\mathscr{E}_y^+}{2} \cos[(\omega_0 + \omega_s)t - \beta_+ z]$$

$$+ \frac{m\mathscr{E}_y^+}{2} \cos[(\omega_0 - \omega_s)t - \beta_- z] \qquad \text{(8-80a)}$$

Note that the phase constant β_0 of the carrier is β_{10} of (8-64b) for the TE_{10} mode.

The ω-β diagram of Figure 8-15(*b*) shows that the upper and lower sideband frequencies differ from the carrier frequency by the signal frequency

ω_s, taken to be an incremental $\Delta\omega$, whence $\omega_0 + \omega_s = \omega_0 + \Delta\omega$ and $\omega_0 - \omega_s = \omega_0 - \Delta\omega$. The phase constants β_+ and β_- associated with the upper and lower sideband terms also differ incrementally from the carrier value β_0, yielding $\beta^+ = \beta_0 + \Delta\beta$ and $\beta_- = \beta_0 - \Delta\beta'$ as noted in the diagram. If $\omega_s = \Delta\omega$ is small compared with the carrier frequency ω_0, one can put $\Delta\beta' \cong \Delta\beta$, whence (8-80a) becomes

$$E_y^+(x, z, t) = \mathscr{E}_y^+ \cos(\omega_0 t - \beta_0 z) + \frac{m\mathscr{E}_y^+}{2} \cos[(\omega_0 + \Delta\omega)t - (\beta_0 + \Delta\beta)z]$$

$$+ \frac{m\mathscr{E}_y^+}{2} \cos[(\omega_0 - \Delta\omega)t - (\beta_0 - \Delta\beta)z]$$

which can be recombined to yield

$$E_y^+(x, z, t) = \mathscr{E}_y^+[1 + m\cos(\Delta\omega\cdot t - \Delta\beta\cdot z)]\cos(\omega_0 t - \beta_0 z) \quad \text{(8-80b)}$$

A comparison with (8-76a) shows that (8-80b) describes the amplitude-modulated wave delayed in phase from the $z = 0$ reference plane by the amount $\beta_0 z$ insofar as the *carrier* at the frequency ω_0 is concerned, while the bracketed factor specifies how the *envelope* progresses down the z axis in time. Since any equiphase surface on the envelope is defined by $\Delta\omega\cdot t - \Delta\beta\cdot z =$ constant, the envelope moves down the z axis with the group velocity $v_g = \Delta\omega/\Delta\beta$. With the signal frequency $\omega_s = \Delta\omega$ small compared to the carrier frequency, $\Delta\omega/\Delta\beta$ becomes the limit

$$v_g = \frac{\partial\omega}{\partial\beta} = \left(\frac{\partial\beta}{\partial\omega}\right)^{-1} \text{ m/sec} \quad \text{(8-81)}$$

The last form, written as an inverse, is the more useful since β is usually given explicitly in terms of ω. A comparison with (8-73) shows that group and phase velocities are obtained from slope interpretations on the ω-β diagram of Figure 8-15(b). Thus v_g is given by the tangent to the ω-β curve at point P, while v_p is the slope of the line extending from the origin 0 to P. It is seen that the slope of the asymptote (ω versus $\beta^{(0)}$) is the velocity $(\mu\epsilon)^{-1}$, a value falling between v_g and v_p.

Applying the result (8-81) to the expression (8-42b) or (8-54b) for the phase constant β obtains the group velocity

$$v_{g,mn} = \left\{\frac{\partial}{\partial\omega}\left[\omega\sqrt{\mu\epsilon}\sqrt{1 - \left(\frac{\omega_{c,mn}}{\omega}\right)^2}\right]\right\}^{-1} = v^{(0)}\sqrt{1 - \left(\frac{f_{c,mn}}{f}\right)^2} \quad \text{(8-82)}$$

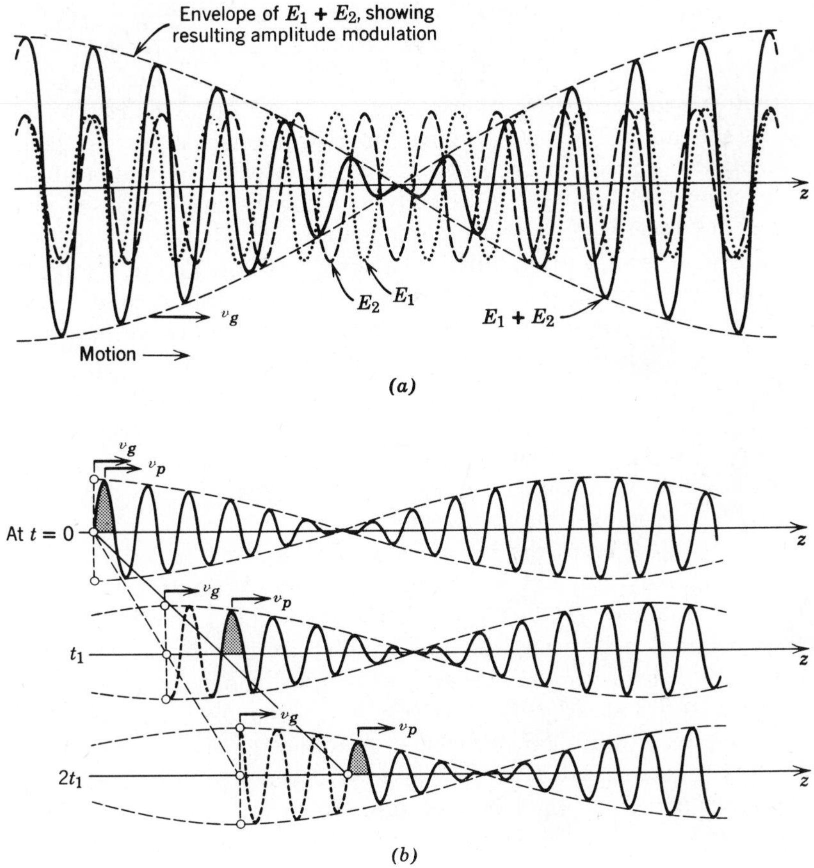

Figure 8-16. Group and phase velocities associated with an amplitude-modulated wave. (*a*) The sum of the two sideband frequencies of an amplitude-modulated wave, showing beat effects. (*b*) Depicting phase and group velocities in the wave of (*a*), as time increases. The medium is assumed normally dispersive.

for any TM or TE mode in a hollow waveguide. A comparison with the phase velocity expression (8-46) and (8-56) shows that

$$v_p v_g = [v^{(0)}]^2 \tag{8-83}$$

revealing that the unbounded-space velocity $v^{(0)}$ is the geometric mean of the phase and group velocities for hollow waveguide modes.

Figure 8-16 illustrates the phenomena of phase and group velocities relative to the upper and lower sideband frequency terms of an amplitude-modulated carrier propagating in a dispersive medium. (The carrier term is omitted to simplify the graphic addition of the waves.) Note the alternate constructive and destructive interference (i.e., amplitude modulation) produced by the sum of the waves. If the sideband components were propagating in a *nondispersive* medium, their identical phase velocities would produce the same envelope velocity (group velocity). In a *dispersive* region as shown, however, the upper sideband term has a phase velocity lower than that of the lower sideband term, as noted from the slope of $0P$ in Figure 8-15(*b*). This causes the point of constructive interference, or maximum amplitude on the diagram of Figure 8-16(*b*), to slip behind both sideband terms with the passage of time, yielding an envelope velocity (v_g) smaller than the phase velocity, i.e., smaller than $v^{(0)} = (\mu\epsilon)^{-1/2}$ by an amount such that (8-83) is satisfied.

The dispersive phenomenon in hollow waveguides, sometimes termed normal dispersion, is different from that found in the transmission of modulated signals through a conductive region such as earth or sea water, or along a lossy two-conductor cable or transmission line as considered in Chapters 9 and 10. In these cases the phase velocity is seen to be *less* than $v^{(0)} = (\mu\epsilon)^{-1/2}$, as already noted in Example 3-8 dealing with uniform plane wave propagation in a lossy region. Applying (8-81) to an example from this class of problems produces a group velocity that exceeds the phase velocities of the Fourier terms; the dispersion in such cases is called anomalous. The envelope is then seen to slip ahead of the carrier as a modulated wave moves through the lossy region. Additional comments are found in a number of sources.†

If a carrier is modulated with a pulse of very short duration, the steady state Fourier components of the spectrum, noted in Figure 8-14(*c*), may be spread over such a wide frequency band that the ratio $\Delta\omega/\Delta\beta$ of (8-81), defining a group velocity in the limit, may lose its significance. Then group velocity should be replaced with a signal velocity concept discussed in Stratton.‡ Group velocity has a precise meaning only if the carrier frequency is kept sufficiently high compared to the frequency width embracing the important terms of the Fourier spectrum.

Of historical interest concerning group velocity are the measurements made by Michelson to determine the velocity of light in carbon disulphide. Chopped (pulsed) light measurements gave a ratio of the velocity in air to that in the liquid to be 1.76, seemingly contradicting the ratio 1.64 obtained from the indices of refraction of the two media. The answer to the riddle

† See Jordan, E. C., and K. G. Balmain. *Electromagnetic Waves and Radiating Systems*, 2nd ed. Englewood Cliffs N. J.: Prentice-Hall, 1968, p. 726. The Laplace transform applied to signal velocities is discussed in J. Stratton. *Electromagnetic Theory*. New York: McGraw-Hill, 1941, pp. 330–340.

‡ Ibid.

lay in the fact that the index of refraction measurements used a continuous wave, while the velocity measured with pulsed light dealt with a group of frequencies, and so they yielded the group velocity in the dispersive liquid.

EXAMPLE 8-3. Find the group velocity associated with a suitably narrow band of Fourier components characterizing a modulated carrier traveling in the waveguide of Example 8-2, assuming a 9 GHz carrier frequency and a pure TE_{10} mode.

Using (8-82), one obtains the group velocity

$$v_{g,10} = v^{(0)}\sqrt{1 - \left(\frac{\lambda^{(0)}}{2a}\right)^2} = 3 \times 10^8\sqrt{1 - (0.729)^2} = 2.05 \times 10^8 \text{ m/sec}$$

observed to be less than the speed of light. This might be compared with the phase velocity $v_{p,10} = 4.39 \times 10^8$ m/sec obtained at the carrier frequency.

*8-6 Wall-Loss Attenuation in Hollow Waveguides

In the previous discussions of wave propagation in rectangular hollow waveguides, it was assumed that the waveguide walls were perfectly conducting. Practical waveguides are necessarily made of finitely conducting metals (e.g., brass, aluminum, silver), and waves moving down the interior will generate wall currents much like those depicted in Figure 8-12 for the dominant TE_{10} mode. In the ideal, perfectly conducting case, the wall currents are restricted to surface currents characterized by a penetration depth of zero, the tangential magnetic field being a measure of the surface current density according to the boundary condition (3-72). The fields inside the perfect conductor are zero, to make the wall power losses zero for this idealized case.

With finitely conductive walls, however, the continuity of the tangential magnetic field guarantees a time-varying magnetic field inside the conductor, producing therein an electromagnetic field rapidly diminishing with depth. The fields penetrate the conducting wall essentially at right angles to the surface. The ensuing Ohmic power loss due to the transference of a small portion of the available transmitted mode energy into the walls results in a measurable attenuation of the propagated mode. For example, the wall-loss attenuation occurring in an X-band brass waveguide carrying the TE_{10} mode at 10 GHz is of the order of 0.2 dB per meter, a significant amount for long waveguide runs. It is the purpose of this section to outline a method for the approximate analysis of the wall-loss attenuation problem for hollow guides.

In the propagation of a TM or TE mode down an ideal (lossless) waveguide, the power-flux travels unabated down the pipe, the same time-average power passing through every cross-section of the guide. As shown in

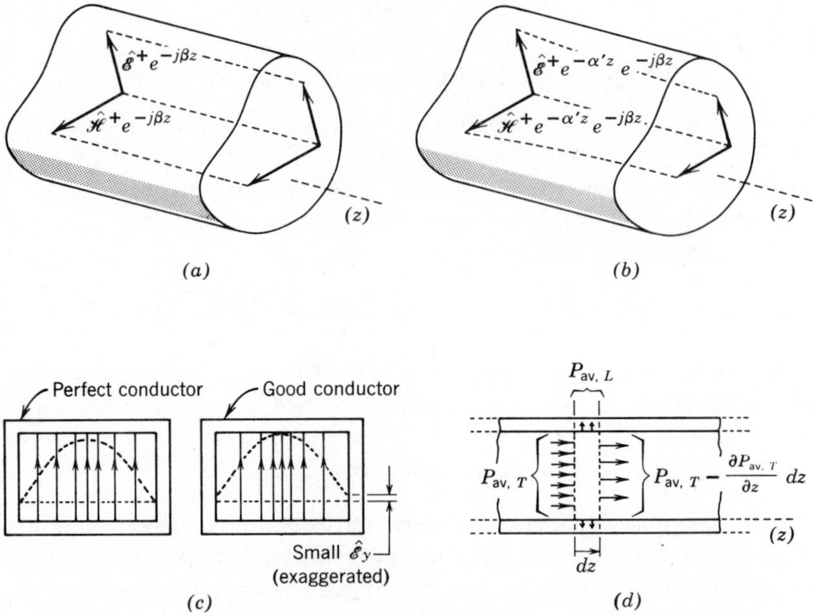

Figure 8-17. Relative to the wall-loss attenuation in a waveguide of uniform cross-section. (*a*) Unattenuated fields in a lossless, ideal waveguide. (*b*) The attenuation of the fields due to power absorption by the walls. (*c*) Showing a small tangential E_y component at the walls, compared to the lossless mode configuration. (*d*) Volume region of length dz, for comparing transmitted and wall-loss average powers.

Figure 8-17(*a*), the positive z traveling, unattenuated fields are designated in the usual complex notation

$$\hat{\mathscr{E}}^+(u_1, u_2)e^{-j\beta z} \qquad \hat{\mathscr{H}}^+(u_1, u_2)e^{-j\beta z} \tag{8-84}$$

fields defining the unperturbed mode in a loss-free waveguide. In the event of a finitely conductive wall material, a portion of the transmitted power is diverted into the walls, leading to an exponential decay of the average power through successive cross-sections of the waveguide, as suggested by Figure 8-17(*b*). The wall-loss attenuation achieved in this process is designated by α', and with the field distributions $\hat{\mathscr{E}}(u_1, u_2)$ and $\hat{\mathscr{H}}(u_1, u_2)$ assumed unchanged from (8-84), the attenuated fields are written

$$\hat{\mathscr{E}}^+(u_1, u_2)e^{-\alpha' z}e^{-j\beta z} \qquad \hat{\mathscr{H}}^+(u_1, u_2)e^{-\alpha' z}e^{-j\beta z} \tag{8-85}$$

The $\hat{\mathscr{E}}^+$ and $\hat{\mathscr{H}}^+$ factors in (8-85) will differ by a small amount from those given in (8-84), a fact appreciated upon inspecting Figure 8-17(c). Shown is the $\hat{\mathscr{E}}_y$ distribution for the TE_{10} mode of a rectangular guide, with a very small $\hat{\mathscr{E}}_y$ component existing at the $x = 0$ and $x = a$ walls due to the field penetration into the conductive material. In the first-order analysis to follow, this small perturbation of the ideal TE_{10} mode is ignored, an assumption producing acceptable wall-loss attenuation predictions if the wall-losses are small.

An expression is derived for the wall-loss attenuation factor α' in terms of the time-average transmitted power and the small fraction of this power that escapes into the walls in every length dz of the waveguide. It is shown, for a given mode, that

$$\alpha' = \frac{1}{2}\frac{\dfrac{dP_{\text{av},L}}{dz}}{P_{\text{av},T}} \text{ Np/m} \tag{8-86}$$

in which the meanings of the symbols are illustrated in Figure 8-17(d). $P_{\text{av},T}$ denotes the average power-flux transmitted by the mode through any cross-section of the waveguide, while $dP_{\text{av},L}$ is that lost into the walls through the peripheral strip of width dz.

One can derive (8-86) by noting that if the volume slice of length dz in Figure 8-17(d) contains no Ohmic losses or sources, then by (7-31) or (7-60) the *net* time-average power entering (or leaving) the surface enclosing that volume is zero. Therefore, $dP_{\text{av},L} = -P_{\text{av},T} + [P_{\text{av},T} - (\partial P_{\text{av},T}/\partial z)\, dz]$, yielding

$$dP_{\text{av},L} = -\frac{\partial P_{\text{av},T}}{\partial z}\, dz \tag{8-87}$$

The average power transmitted through the waveguide cross-section is obtained from the cross-sectional surface integral of the time-average Poynting vector (7-51b), with the fields (8-85) inserted

$$\begin{aligned} P_{\text{av},T} &= \int_{S(\text{c.s.})} \tfrac{1}{2}\,\text{Re}\,[(\hat{\mathscr{E}}^+ e^{-\alpha' z} e^{-j\beta z}) \times (\hat{\mathscr{H}}^+ e^{-\alpha' z} e^{-j\beta z})^*]\cdot d\mathbf{s} \\ &= e^{-2\alpha' z} \int_{S(\text{c.s.})} \tfrac{1}{2}\,\text{Re}\,[\hat{\mathscr{E}}^+ \times \hat{\mathscr{H}}^{+*}]\cdot d\mathbf{s} \end{aligned} \tag{8-88}$$

Differentiating (8-88) with respect to z obtains

$$\frac{\partial P_{\text{av},T}}{\partial z} = -2\alpha' e^{-2\alpha' z} \int_S \tfrac{1}{2}\,\text{Re}\,[\hat{\mathscr{E}}^+ \times \hat{\mathscr{H}}^{+*}]\cdot d\mathbf{s} = -2\alpha' P_{\text{av},T}$$

and solving for α' yields

$$\alpha' = -\frac{1}{2}\frac{\dfrac{\partial P_{\mathrm{av},T}}{\partial z}}{P_{\mathrm{av},T}}$$

but from (8-87), $\partial P_{\mathrm{av},T}/\partial z$ can be replaced with $-dP_{\mathrm{av},L}/dz$, so

$$\alpha'\,dz = \frac{1}{2}\frac{dP_{\mathrm{av},L}}{P_{\mathrm{av},T}}\ \mathrm{Np} \tag{8-89}$$

which is just (8-86), that which was to have been proved.

To illustrate the use of (8-89) in finding α' for a given waveguide mode, consider the dominant TE_{10} mode. The average transmitted power $P_{\mathrm{av},T}$ in (8-89) has already been found in Example 8-2

$$P_{\mathrm{av},T} = \frac{|\hat{E}^{+}_{y,10}|^2}{4\eta_{\mathrm{TE},10}}\,ab\ \mathrm{W} \tag{8-90}$$

The power loss $dP_{\mathrm{av},L}$ in (8-89) arises from the electromagnetic wave induced inside the conductor. Just within the walls are tangential magnetic fields, identical, by continuity, with the magnetic fields of the known components (8-62) of the unperturbed TE_{10} mode. Also appearing therein are electric fields, obtainable from the known magnetic fields by use of wave impedance expressions like (3-97), since the electromagnetic field propagates essentially at right angles into the conductors much like a localized plane wave. This fact is corroborated by the results of the skin effect analysis in Section 9.3, showing the analogy of the fields penetrating round conductors to plane wave fields at sufficiently high frequencies. In Figure 8-18(a) is shown the continuity of the known $\hat{\mathscr{H}}^{+}_{z}$ component (8-62c) of the TE_{10} mode. A small component $\hat{\mathscr{E}}_y$ is induced by $\hat{\mathscr{H}}_z$ in the metal such that $\hat{\mathscr{E}}_y = \hat{\eta}\hat{\mathscr{H}}_z$, and together they comprise essentially a plane wave traveling nearly perpendicularly into the conductor with a large attenuation. $\hat{\mathscr{H}}_z$ is maximal at the $x = 0$ and the $x = a$ walls, with a consinusoidal variation between these values existing along the $y = 0$ and the $y = b$ walls as in Figure 8-18(b). The electric fields induced just inside the $x = 0$ and $x = a$ walls thus become

$$\hat{\mathscr{E}}_y]_{x=0} = -\hat{\eta}\hat{\mathscr{H}}_z]_{x=0} \quad \hat{\mathscr{E}}_y]_{x=a} = \hat{\eta}\hat{\mathscr{H}}_z]_{x=a} \tag{8-91}$$

in which $\hat{\eta} = (\omega\mu/\sigma)^{1/2}e^{j\pi/4}$ from (3-112c), the negative sign properly accounting for the propagation of the wave *into* the metal. Similar expressions apply at the $y = 0$ and $y = b$ walls.

The time-average power loss $dP_{\mathrm{av},L}$ in (8-89) is obtained by integrating

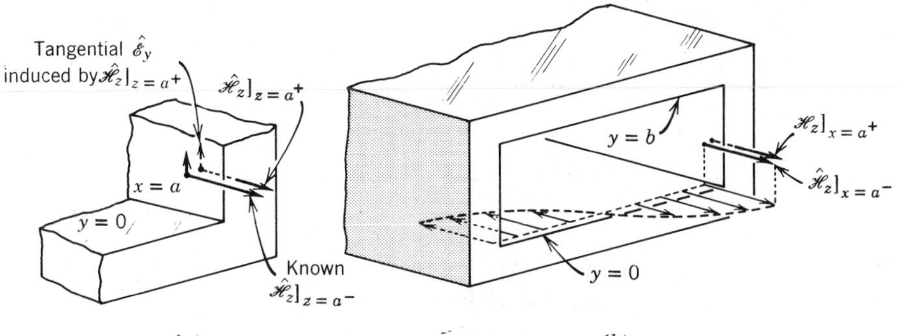

Figure 8-18. Concerning the boundary condition on the tangential magnetic field, leading to wall skin currents. (*a*) Continuity of tangential $\hat{\mathscr{H}}_z$ leads to induced $\hat{\mathscr{E}}_y$ inside conductor. (*b*) Showing cosine distribution of the tangential magnetic field on the top and bottom, uniform on side walls (TE_{10} mode).

over the four sides of the peripheral strip of length dz embracing a cross-section; thus

$$dP_{\text{av},L} = dP_{\text{av},L}]_{x=0-} + dP_{\text{av},L}]_{x=a+} + dP_{\text{av},L}]_{y=0-} + dP_{\text{av},L}]_{y=b+} \tag{8-92}$$

On the $x = 0-$ strip, for example, making use of (8-91) obtains

$$\begin{aligned} dP_{\text{av},L}]_{x=0-} &= \int_{S(x=0-)} \tfrac{1}{2}\,\text{Re}\,[(-\mathbf{a}_y\hat{\eta}\hat{\mathscr{H}}_z^+)\times \mathbf{a}_z\hat{\mathscr{H}}_z^{+*}]\cdot(-\mathbf{a}_x\,dy\,dz) \\ &= \tfrac{1}{2}\,\text{Re}\int_{y=0}^{b}\left\{\sqrt{\frac{\omega\mu}{\sigma}}\,e^{j45^\circ}\left[\frac{|\hat{E}_{y,10}|^2}{\omega^2\mu^2}\left(\frac{\pi}{a}\right)^2\cos^2\frac{\pi}{a}x\right]_{x=0-}\right\}dy\,dz \\ &= \frac{b}{2}\sqrt{\frac{\omega\mu}{2\sigma}}\,\frac{|\hat{E}_{y,10}^+|^2}{\omega^2\mu^2}\left(\frac{\pi}{a}\right)^2 dz \end{aligned} \tag{8-93}$$

Similarly, the wall-loss at $y = 0-$ becomes

$$dP_{\text{av},L}]_{y=0-} = \frac{a|\hat{E}_{y,10}^+|^2}{4\omega^2\mu^2}\sqrt{\frac{\omega\mu}{2\sigma}}\,\omega^2\mu\epsilon\,dz \tag{8-94}$$

a result accounting for both tangential components $\hat{\mathscr{H}}_x^+$ and $\hat{\mathscr{H}}_z^+$ of (8-62b) and (8-62c), and making use of the identity $(\pi/a)^2 + \beta^2 = \omega^2\mu\epsilon$ for the TE_{10}

mode. Evaluating all four wall-loss contributions of (8-92) yields

$$dP_{\text{av},L} = \frac{1}{2}\frac{|\hat{E}^+_{y,10}|^2}{\omega^2\mu^2}\sqrt{\frac{\omega\mu}{2\sigma}}\left[\frac{2\pi^2 b}{a^2} + a\omega^2\mu\epsilon\right] dz$$

$$= \frac{a}{2}\frac{|\hat{E}^+_{y,10}|^2}{\omega^2\mu^2}\sqrt{\frac{\omega\mu}{2\sigma}}\,\omega^2\mu\epsilon\left[\frac{2b}{a}\left(\frac{f_{c,10}}{f}\right)^2 + 1\right] dz \text{ W} \qquad (8\text{-}95)$$

the latter making use of $(f_{c,10}/f)^2 = \pi^2/\omega^2\mu\epsilon a^2$ from (8-53). Inserting (8-95) and (8-90) into (8-89) yields the wall-loss attenuation for the TE_{10} mode

$$\alpha'_{10} = \sqrt{\frac{\omega\mu}{2\sigma}}\,\frac{\dfrac{2b}{a}\left(\dfrac{f_{c,10}}{f}\right)^2 + 1}{\eta_0 b\sqrt{1 - \left(\dfrac{f_{c,10}}{f}\right)^2}} \text{ Np/m} \qquad (8\text{-}96)$$

The b factor in the denominator of (8-96) shows that making the height too small results in a large wall-loss attenuation. This is a consequence, at a fixed field amplitude $E^+_{y,10}$, and as seen from (8-90), of the smaller cross-sectional area through which the correspondingly smaller transmitted power $P_{\text{av},T}$ must flow, the wall-loss power remaining nearly the same as for a waveguide with a larger b height. It is also evident from (8-96) that as the cutoff frequency is approached, the wall-loss attenuation becomes indefinitely large. A graph of (8-96) versus frequency for two choices of b height is shown in Figure 8-19(a), along with the wall-loss attenuation characteristic of the TM_{11} mode in a rectangular waveguide.†

From Figure 8-19 it is evident that different modes undergo different amounts of attenuation in a given waveguide. It would appear that a way of reducing wall-loss attenuation is to minimize the exposure of the magnetic field component tangential to the wall. Nearly all modes in hollow waveguides have an increasing wall-loss attenuation with increasing frequency, with α' exhibiting a minimum value at some optimum frequency as already seen in Figure 8-19(a). It develops that a *circular waveguide* mode, the TE_{01}, deserves special attention in that it exhibits an indefinitely decreasing α' with increasing frequency, the result of a smaller and smaller component of the tangential H field at the metallic wall as the incidence of the wave becomes more nearly

† A further discussion of the wall-loss attenuation factor associated with the remaining modes of rectangular waveguides can be found in Ramo, S., J. Whinnery, and T. van Duzer. *Fields and Waves in Communication Electronics*. New York: Wiley, 1965, Chapter 8.

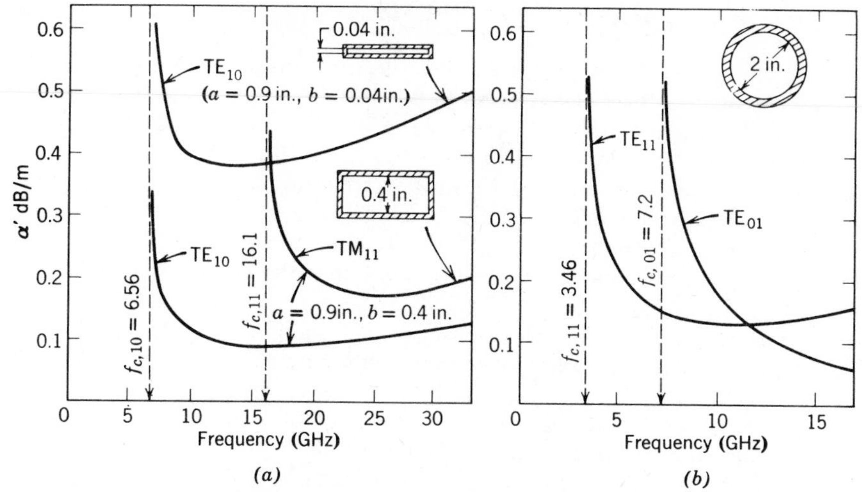

Figure 8-19. Wall-loss attenuation versus frequency for copper. (*a*) Attenuation versus frequency for modes in rectangular waveguides. (*b*) Attenuation characteristic for a circular waveguide.

grazing.† This mode, having the attenuation characteristic depicted in Figure 8-19(*b*), shows promise in long-range, low-loss transmission at superhigh frequencies in hollow, metal cylindrical pipes, though problems are posed by the fact that the TE_{11} mode, and not TE_{01}, is the dominant mode in a circular waveguide.

REFERENCES

GINZTON, E. L. *Microwave Measurements.* New York: McGraw-Hill, 1957.

JORDAN, E. C., and K. G. BALMAIN. *Electromagnetic Waves and Radiating Systems*, 2nd ed. Englewood Cliffs, N.J.: Prentice-Hall, 1968.

LANCE, A. L. *Introduction to Microwave Theory and Measurements.* New York: McGraw-Hill, 1964.

MARCUVITZ, N. *Waveguide Handbook.* M.I.T. Radiation Lab. Series, vol. 10. New York: McGraw-Hill, 1951.

RAMO, S., J. R. WHINNERY, and T. VAN DUZER. *Fields and Waves in Communication Electronics.* New York: Wiley, 1965.

† For a discussion of the theory of the circular waveguide, see Ramo, S., J. Whinnery, and T. Van Duzer, *Fields and Waves in Communication Electronics.* New York: Wiley, 1965, Chapter 8. Circular waveguides have important applications to rotating joints used for feeding movable antennas, and to tunable resonant cavities.

PROBLEMS

8-1. In the cartesian system, prove that the substitution of (8-1a) into (8-3) yields the modified-curl relation (8-8).

2. Repeat Problem 8-1, except carry out the details in the generalized cylindrical coordinate system (u_1, u_2, z) to prove (8-10).

-3. Show from the definition (2-83) of the Laplacian of a vector function that (8-12b) expands into three scalar wave equations analogous with (8-13). Then with the replacement of H_x, H_y, and H_z with their complex exponential form denoted by (8-1a), show that the scalar wave equations (8-16) are obtained.

8-4. Expressions (8-19) for the transverse field components of a hollow rectangular waveguide were obtained from the simultaneous manipulation of (8-17) and (8-18). Show in detail how (8-19a) is obtained.

8-5. Expand the modified Maxwell curl relations (8-9) and (8-10) in circular cylindrical coordinates, and from these obtain

$$\hat{\mathscr{E}}_\rho^\pm = \frac{-1}{\hat{k}_c^2}\left[\pm\gamma\frac{\partial\hat{\mathscr{E}}_z^\pm}{\partial\rho} + \frac{j\omega\mu}{\rho}\frac{\partial\hat{\mathscr{H}}_z^\pm}{\partial\phi}\right]$$

$$\hat{\mathscr{E}}_\phi^\pm = \frac{1}{\hat{k}_c^2}\left[\frac{\pm\gamma}{\rho}\frac{\partial\hat{\mathscr{E}}_z^\pm}{\partial\phi} + j\omega\mu\frac{\partial\hat{\mathscr{H}}_z^\pm}{\partial\rho}\right]$$

$$\hat{\mathscr{H}}_\rho^\pm = \frac{1}{\hat{k}_c^2}\left[\frac{j\omega\epsilon}{\rho}\frac{\partial\hat{\mathscr{E}}_z^\pm}{\partial\phi} \mp \gamma\frac{\partial\hat{\mathscr{H}}_z^\pm}{\partial\rho}\right]$$

$$\hat{\mathscr{H}}_\phi^\pm = \frac{-1}{\hat{k}_c^2}\left[j\omega\epsilon\frac{\partial\hat{\mathscr{E}}_z^\pm}{\partial\rho} \pm \frac{\gamma}{\rho}\frac{\partial\hat{\mathscr{H}}_z^\pm}{\partial\phi}\right]$$

From the results, modal expressions analogous with (8-20) through (8-25), but applicable to the TE, TM, and TEM modes of circular cylindrical waveguides and coaxial transmission lines, may be found.

8-6. Repeat Problem 8-5, except carry out the details in the generalized cylindrical coordinate system, obtaining

$$\hat{\mathscr{E}}_1^\pm = \frac{-1}{\hat{k}_c^2}\left[\frac{\pm\gamma}{h_1}\frac{\partial\hat{\mathscr{E}}_z^\pm}{\partial u_1} + \frac{j\omega\mu}{h_2}\frac{\partial\hat{\mathscr{H}}_z^\pm}{\partial u_2}\right]$$

$$\hat{\mathscr{E}}_2^\pm = \frac{1}{\hat{k}_c^2}\left[\frac{\mp\gamma}{h_2}\frac{\partial\hat{\mathscr{E}}_z^\pm}{\partial u_2} + \frac{j\omega\mu}{h_1}\frac{\partial\hat{\mathscr{H}}_z^\pm}{\partial u_1}\right]$$

$$\hat{\mathscr{H}}_1^\pm = \frac{1}{\hat{k}_c^2}\left[\frac{j\omega\epsilon}{h_2}\frac{\partial\hat{\mathscr{E}}_z^\pm}{\partial u_2} \mp \frac{\gamma}{h_1}\frac{\partial\hat{\mathscr{H}}_z^\pm}{\partial u_1}\right]$$

$$\hat{\mathscr{H}}_2^\pm = \frac{-1}{\hat{k}_c^2}\left[\frac{j\omega\epsilon}{h_1}\frac{\partial\hat{\mathscr{E}}_z^\pm}{\partial u_1} \pm \frac{\gamma}{h_2}\frac{\partial\hat{\mathscr{H}}_z^\pm}{\partial u_2}\right]$$

As a check, show that these results reduce to (8-19) in the rectangular system.

8-7. Use results given in Problem 8-5 to obtain expressions for the intrinsic wave impedances associated with the transverse field components of the TM and TE modes of uniform, circular cylindrical transmission systems, namely

$$\pm\frac{\hat{\mathscr{E}}_\rho^\pm}{\hat{\mathscr{H}}_\phi^\pm} = \mp\frac{\hat{\mathscr{E}}_\phi^\pm}{\hat{\mathscr{H}}_\rho^\pm} = \frac{\gamma}{j\omega\epsilon} \equiv \hat{\eta}_{\text{TM}}$$

$$\pm\frac{\hat{\mathscr{E}}_\rho^\pm}{\hat{\mathscr{H}}_\phi^\pm} = \mp\frac{\hat{\mathscr{E}}_\phi^\pm}{\hat{\mathscr{H}}_\rho^\pm} = \frac{j\omega\mu}{\gamma} \equiv \hat{\eta}_{\text{TE}}$$

8-8. Repeat Problem 8-7, but for generalized cylindrical coordinates.

8-9. From the substitution of (8-38a) into the TM mode relations (8-20), verify the remaining, transverse electric and magnetic fields (8-38b) through (8-38e).

8-10. Show that the lowest frequency at which a wave will just propagate (not evanesce) in the lowest-order TM_{11} mode within a rectangular, 0.9 in. by 0.4 in. air-filled waveguide is 16,100 MHz, but if the dimensions are increased by a factor of ten, the cutoff frequency falls to 1,610 MHz. If both waveguides are filled with a nonmagnetic dielectric material with $\epsilon_r = 4$, determine the effect on the cutoff frequencies.

8-11. Given two rectangular waveguides with the same inside perimeters as follows, compare their cutoff frequencies for the TM_{11} mode:
   (*a*) $a = 0.9$ in, $b = 0.4$ in.
   (*b*) $a = b = 0.65$ in.

8-12. Compare the cutoff frequencies for the TM_{12} mode in waveguides with inside dimensions:
   (*a*) $a = 0.9$ in., $b = 0.4$ in. (2.286×1.016 cm)
   (*b*) $a = 0.4$ in., $b = 0.9$ in.
   (*c*) $a = b = 0.65$ in. (1.651 cm square)

   Sketch the variations of the longitudinal field component $\hat{\mathscr{E}}_z$ in the waveguide cross-section for each case, in the manner of Figure 8-4.

8-13. Find the dimensions of the smallest, square, air-filled waveguide which will just propagate the TM_{11} mode at the frequencies:
   (*a*) 10 GHz
   (*b*) 10 MHz
   (*c*) 10 kHz

   [*Answer:* (*a*) $a = b = 2.12$ cm; (*b*) 21.2 m; (*c*) 21.2 km]

8-14. An air-filled rectangular waveguide having the interior dimensions $a = 0.9$ in. and $b = 0.4$ in. operates in the TM_{11} mode at 20 GHz. Determine for this mode:
   (*a*) The cutoff frequency
   (*b*) The phase constant β_{11}
   (*c*) The wavelength in the guide
   (*d*) The phase velocity
   (*e*) The intrinsic wave impedance
   (*f*) The propagation constant at $f = 10$ GHz (below cutoff)

Compare the answers to parts (*b*) through (*e*) with those obtained for a plane wave in unbounded free space at 20 GHz.

8-15. Rework Problem 8-14, assuming the waveguide is filled with a lossless dielectric material having $\epsilon_r = 4$.

8-16. Beginning with the wave equation (8-16f) in terms of $\mathscr{H}_z^{\pm}$, use the separation-of-variables method to obtain (8-49) for TE modes.

8-17. Applying the boundary conditions (8-50) to (8-49), show that the proper solutions (8-51a) are obtained, and that $\hat{k}_c^2 = (m\pi/a)^2 + (n\pi/b)^2$.

8-18. Using (8-51a), verify the remaining field expressions (8-51b) through (8-51e).

8-19. Utilize (8-52) for the propagation constant of TE_{mn} modes to obtain the cutoff frequency (8-53).

8-20. Given an *X*-band, air-filled, rectangular waveguide having the inside dimensions $a = 0.9$ in. and $b = 0.4$ in., compute the cutoff frequencies of the TE_{10}, TE_{01}, TE_{20}, TE_{11}, TM_{11}, TM_{21}, and TM_{12} modes. (Which of these modes will propagate as waves, and which will evanesce, at a source frequency of 10 GHz?) [*Answer:* 6.55, 14.7, 13.1, 16.1, 16.1, 19.75, and 30.1 GHz]

8-21. Show that the solutions (8-51) for the fields of TE_{mn} modes reduce to (8-61) for the TE_{10} mode.

8-22. Given six air-filled, rectangular waveguides of the following inside dimensions, compute their cutoff frequencies for the dominant TE_{10} mode:

(*a*) 6.25 in. × 3.25 in. (15.875 × 8.255 cm) (*L* band)
(*b*) 2.84 in. × 1.34 in. (7.214 × 3.404 cm) (*S* band)
(*c*) 1.872 in. × 0.872 in. (4.755 × 2.215 cm) (*C* band)
(*d*) 0.9 in. × 0.4 in. (2.286 × 1.016 cm) (*X* band)
(*e*) 0.420 in. × 0.210 in. (1.067 × 0.533 cm) (*K* band)
(*f*) 0.148 in. × 0.074 in. (0.376 × 0.188 cm) (*V* band)

[*Answer:* 945 MHz, 2.078 GHz, 3.155 GHz, 6.557 GHz, 14.067 GHz, 39.863 GHz]

8-23. If the amplitude of the field $E_y^{\pm}$ of the dominant, TE_{10} mode in an *X*-band (0.9″ × 0.4″), air-filled, rectangular waveguide is 1000 V/m, determine the amplitudes of $H_x^{\pm}$ and $H_z^{\pm}$, assuming $f = 10$ GHz. What is the time-average power transmitted?

8-24. Hollow waveguides are impractical for operating frequencies much below 1000 MHz. Compute the smallest *a* width of an air-filled, rectangular waveguide that will just propagate the TE_{10} mode at the following frequencies:

(*a*) 10 GHz
(*b*) 100 MHz
(*c*) 10 MHz
(*d*) 60 Hz

[*Answer:* 1.5 cm, 1.5 m, 15 m, 2500 km]

8-25. An *X*-band, air-filled, rectangular waveguide having the interior dimensions $a = 0.9$ in. and $b = 0.4$ in. carries the dominant TE_{10} mode at a source frequency of 10 GHz. Determine for this mode:

(*a*) The cutoff frequency

(*b*) The phase constant β_{10}

(*c*) The wavelength in the guide

(*d*) The phase velocity

(*e*) The intrinsic wave impedance

Compare the numerical answers to parts (*b*) through (*e*) with those obtained for a uniform plane wave in an unbounded air region, at the same frequency.

8-26. Assuming the waveguide of Problem 8-25 is excited with the TE_{10} mode at 5 GHz (below cutoff), determine:

(*a*) The attenuation constant

(*b*) The intrinsic wave impedance

(*c*) The z distance required to cause a diminution of the fields to e^{-1} of the reference value

8-27. Use the boundary conditions (3-45) and (3-72) to determine the surface charge and current densities on the $y = b$ and $x = a$ walls of a rectangular waveguide carrying the TE_{10} mode. Compare the results with (8-70) and (8-71), the densities on the opposite walls.

8-28. Prove (8-80b) from the preceding expression.

8-29. Show the details of the differentiation of the β expression (8-42b) or (8-54b) for rectangular waveguide modes, obtaining (8-82) for the group velocity.

8-30. For an X-band rectangular waveguide (a = 0.9 in, b = 0.4 in.) having a cutoff frequency of 6.557 GHz for the dominant mode, find its phase and group velocities at 8.2, 10, and 12.4 GHz.

8-31. Verify the TE_{10} mode wall-loss power (8-95) for a differential strip of rectangular waveguide, by carrying out the remaining details called for in (8-92). Then verify (8-96).

8-32. For sufficiently high frequencies, show that the wall-loss attenuation factor for the TE_{10} mode increases as $\sqrt{f}$.

8-33. Evaluate the wall-loss attenuation factor for the TE_{10} mode in a copper, X band rectangular waveguide ($\sigma = 5.8 \times 10^7$ ℧/m), at 8.2, 10, and 12.4 GHz.

CHAPTER
9

TEM Waves on Two-Conductor Transmission Lines

In the previous chapter were considered the TM and TE mode configurations of rectangular, hollow (single-conductor) waveguides. Omitted from detailed discussion was the TEM (transverse-electromagnetic) mode, the dominant mode of transmission lines using two (or more) conductors. The parallel-conductor line, shown in Figure 4-14(*b*) and in Example 5-16, and the circular coaxial line, depicted in Examples 4-8 and 5-13, are commonly used in the transmission of this mode. It is seen that at least two conductors are required to establish the TEM mode, transmittable over a range of frequencies extending all the way down to zero frequency (dc).

Though the TEM mode is by far the most important, TM and TE modes are also capable of propagating on two-conductor transmission lines. The latter modes, however, are evanescent below their cutoff frequencies which occur for ordinary coaxial lines in the upper microwave frequencies and beyond. The TM and TE modes on two-conductor lines thus have no useful applications to signal or power transmission, so they are omitted from detailed discussion.†

Two-conductor uniform transmission lines of the coaxial or parallel-wire type, operating in the TEM mode and illustrated in Figure 9-1(*a*) and (*b*),

† Higher-order modes on the coaxial line are discussed in Ramo, S., J. Whinnery and T. Van Duzer. *Fields and Waves in Communication Electronics*. New York: Wiley, 1965, p. 446.

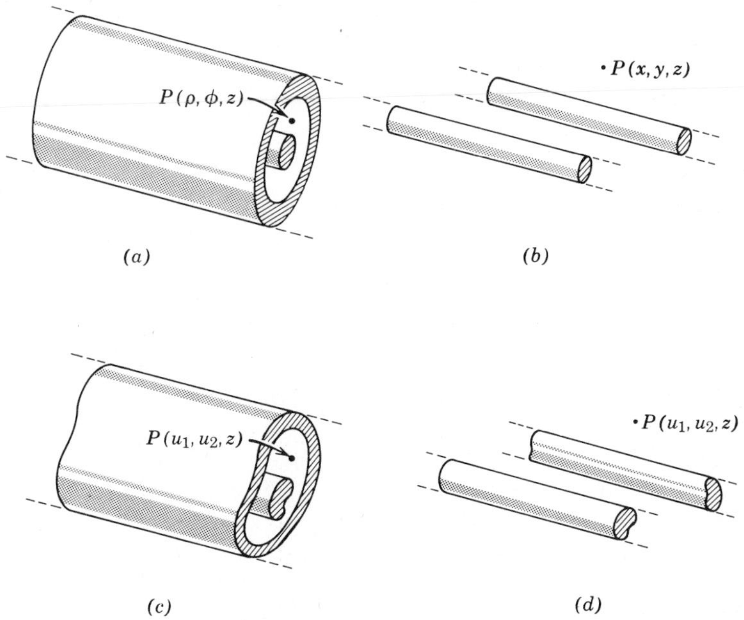

Figure 9-1. Two-conductor uniform transmission lines. (*a*) The coaxial line. Concentric conductors separated by air or a dielectric material. (*b*) Parallel-wire line. Usually separated by air. (*c*) Generalized line. One conductor inside the other. (*d*) Generalized line. Conductors externally located.

are commonly used in power distribution and signal communication systems. Power transmission lines carry power in the megawatts up to hundreds of kilometers from generating stations to urban regions. Voice and pulse-data signals are carried over telephone lines, with signal amplification applied every few tens of miles if the information is to be carried over long distances. Power lines usually operate at 50 or 60 Hz, employing parallel-wire lines suspended on poles or towers, or using buried cables. Telephone lines are seen in pairs on poles, though many buried coaxial and multiconductor cables are in use. These may carry audio signals directly, or information transmitted as a modulation of the amplitude of a carrier frequency operating up to several megahertz, permitting the transmission of several modulated carriers simultaneously over the same transmission line, or the signals may be multiplexed using pulse-code modulation at high pulse rates to increase the information-handling capacity significantly. Coaxial lines are commonly used, for example, to interconnect a radio frequency transmitter to an antenna employed for launching electromagnetic waves into the atmosphere. At the higher microwave frequencies, hollow waveguides can be employed to connect a data transmitter or perhaps a radar to an electromagnetic horn or a dish-reflector antenna. Short sections of uniform transmission line, having low losses at the higher frequencies, can be used as the high Q resonant (frequency selective) elements of filters; they may serve as reflective elements in pulse-forming networks; they may be utilized to transport pulse data from one place to another with low distortion in high-speed computers. From this partial list of applications, it becomes apparent that a detailed study of transmission line behavior can be of substantial importance to the engineer and applied scientist.

This chapter begins with a discussion of the properties of the electric and magnetic fields of the TEM mode on two-conductor lines. The related currents and voltages are developed next, to introduce the concept of characteristic impedance. The transmission line equations are deduced in terms of the distributed line parameters, first assuming ideally perfect conductors, and then for the physically realizable line employing finitely conductive elements. The time-harmonic voltage and current analysis of interconnected lines using arbitrary load impedances is deferred until Chapter 10.

*9-1 Electrostatic Potential and the TEM Mode Fields

A uniform, two-conductor transmission line is represented in generalized cylindrical coordinates in Figure 9-1(*c*) and (*d*). The pure TEM mode exists (ideally) on a line composed of perfect conductors. For conductors with finite conductivity, the z directed currents in them account for a z component of the electric field at the conductor surfaces. The small z component of the **E** field required to sustain the electric field inside even good conductors, if longitudinal currents are to flow in them, gives rise to what might be called *essentially TEM* waves. Such waves produce internal resistive and inductive effects in the conductors, considered later in Section 9.7.

A pure TEM wave, associated with perfectly conducting, uniform lines, is defined by setting

$$E_z = H_z = 0 \tag{9-1}$$

* Sections 9-1 and 9-2 cover details of the electric and magnetic *fields* of the TEM mode. Field details are important, for example, in the design of lines for which considerations of maximum field strengths, concerned with corona and voltage breakdown, may be of interest. The reader interested in a more conventional approach starting with line voltages and currents may elect to go directly to Section 9-3.

This results in completely *transverse* electric and magnetic fields between the conductors, expressed in generalized cylindrical coordinates by

$$\mathbf{E}(u_1, u_2, z, t) = \mathbf{a}_1 E_1(u_1, u_2, z, t) + \mathbf{a}_2 E_2(u_1, u_2, z, t) \qquad \text{(9-2a)}$$

$$\mathbf{H}(u_1, u_2, z, t) = \mathbf{a}_1 H_1(u_1, u_2, z, t) + \mathbf{a}_2 H_2(u_1, u_2, z, t) \qquad \text{(9-2b)}$$

At the perfectly conducting walls, the electric field (9-2a) satisfies boundary conditions obtained from (3-45) and (3-79)

$$\mathbf{n} \cdot \mathbf{D} = \rho_s \qquad \text{(9-3a)}$$

$$E_t = 0 \qquad \text{(9-3b)}$$

implying an **E** field *normal* to the conductor surfaces and terminating there in a surface charge density ρ_s. The magnetic field (9-2b) must satisfy the boundary conditions (3-50) and (3-72)

$$B_n = 0 \qquad \text{(9-4a)}$$

$$\mathbf{n} \times \mathbf{H} = \mathbf{J}_s \qquad \text{(9-4b)}$$

meaning that **H** is entirely *tangential* to the conductors, terminating there in a surface current density $\mathbf{J}_s$.

Maxwell's line-integration laws (3-66) and (3-78)

$$\oint_{\ell} \mathbf{H} \cdot d\boldsymbol{\ell} = I + \frac{d}{dt} \int_S \mathbf{D} \cdot d\mathbf{s} \qquad \text{[3-66]}$$

$$\oint_{\ell} \mathbf{E} \cdot d\boldsymbol{\ell} = -\frac{d}{dt} \int_S \mathbf{B} \cdot d\mathbf{s} \qquad \text{[3-78]}$$

can be written in the special forms

$$\oint_{\ell(\text{c.s.})} \mathbf{H} \cdot d\boldsymbol{\ell} = \int_S \mathbf{J} \cdot d\mathbf{s} \equiv I \text{ A} \qquad \text{(9-5a)}$$

$$\oint_{\ell(\text{c.s.})} \mathbf{E} \cdot d\boldsymbol{\ell} = 0 \text{ V} \qquad \text{(9-5b)}$$

if the line integrals are restricted to arbitrary closed paths ℓ within *any fixed cross-section* of the line. The simplification to (9-5) is evident from the constraint (9-1), making the flux of both **D** and **B** zero through any transverse surface S bounded by ℓ. They are interpreted as follows:

1. The integral (9-5a) is observed to be of the same form as the *time-static* Ampère's Law (5-5).

2. The integral (9-5b) means **E** is *conservative with respect to any closed path* ℓ *in a fixed cross-section* at a given instant. One can thus expect that static **E** field solutions of the Maxwell relations (4-2) and (4-6) might also serve as TEM wave solutions for transmission lines.

The variables z and t are eliminated from (9-5a) and (9-5b) by assuming field variations in accordance with the factor $\exp(j\omega t \mp \gamma z)$, as discussed in Section 8-1. The real-time fields symbolized in (9-2a) and (9-2b) are thus related to their complex time-harmonic forms by

$$\mathbf{E}(u_1, u_2, z, t) = \mathrm{Re}\,[\hat{\mathscr{E}}^+(u_1, u_2)e^{j\omega t - \gamma z} + \hat{\mathscr{E}}^-(u_1, u_2)e^{j\omega t + \gamma z}] \quad \text{(9-6a)}$$

$$\mathbf{H}(u_1, u_2, z, t) = \mathrm{Re}\,[\hat{\mathscr{H}}^+(u_1, u_2)e^{j\omega t - \gamma z} + \hat{\mathscr{H}}^-(u_1, u_2)e^{j\omega t + \gamma z}] \quad \text{(9-6b)}$$

in which $\hat{\mathscr{E}}$ and $\hat{\mathscr{H}}$ denote complex functions of the transverse coordinates u_1 and u_2 only, the + and − superscripts denoting the positive z and negative z traveling wave solutions.

Introducing the complex notations $\hat{\mathscr{E}}$, $\hat{\mathscr{H}}$, and $\hat{I}_m$ into (9-5a) and (9-5b) yields special forms as follows. With the substitution of

$$\begin{aligned} &\hat{\mathscr{H}}^{\pm}e^{j\omega t \mp \gamma z} \quad \text{for } \mathbf{H} \\ &\hat{I}_m^{\pm}e^{j\omega t \mp \gamma z} \quad \text{for } I \end{aligned} \quad \text{(9-7)}$$

(9-5a) is written

$$\oint_{\ell(\text{c.s.})} \hat{\mathscr{H}}^{\pm}\cdot d\ell = \hat{I}_m^{\pm} \text{ A} \quad \text{(9-8)}$$

after canceling the exponential factors. The *total* time-harmonic **H** field is the superposition of wave solutions

$$\hat{\mathbf{H}}(u_1, u_2, z)e^{j\omega t} = \hat{\mathscr{H}}^+(u_1, u_2)e^{j\omega t - \gamma z} + \hat{\mathscr{H}}^-(u_1, u_2)e^{j\omega t + \gamma z} \quad \text{(9-9a)}$$

written also without the factor $e^{j\omega t}$ as

$$\hat{\mathbf{H}}(u_1, u_2, z) = \hat{\mathscr{H}}^+(u_1, u_2)e^{-\gamma z} + \hat{\mathscr{H}}^-(u_1, u_2)e^{\gamma z} \quad \text{(9-9b)}$$

The total line current through any cross-section of either conductor is provided by the sum of current traveling waves

$$\hat{I}(z) = \hat{I}_m^+ e^{-\gamma z} + \hat{I}_m^- e^{\gamma z} \quad \text{(9-10a)}$$

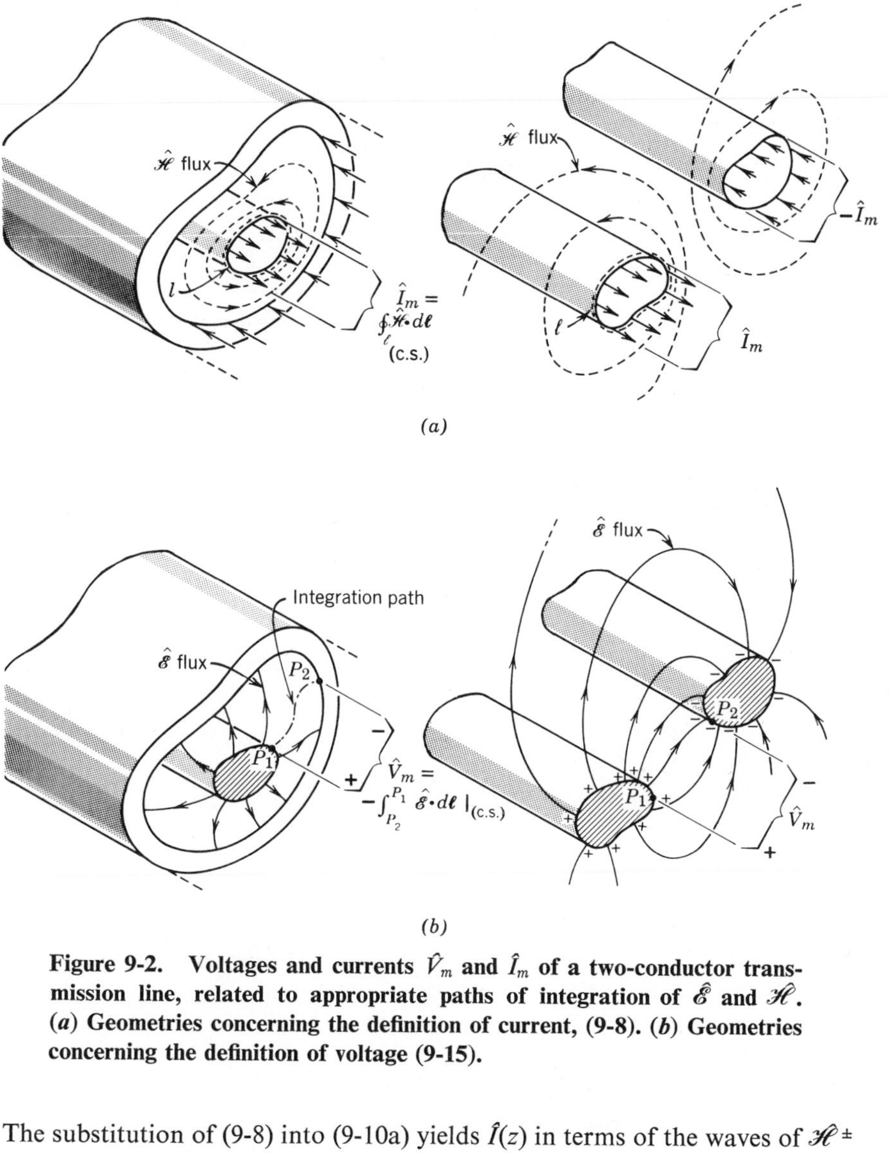

Figure 9-2. Voltages and currents $\hat{V}_m$ and $\hat{I}_m$ of a two-conductor transmission line, related to appropriate paths of integration of $\hat{\mathscr{E}}$ and $\hat{\mathscr{H}}$. (*a*) Geometries concerning the definition of current, (9-8). (*b*) Geometries concerning the definition of voltage (9-15).

The substitution of (9-8) into (9-10a) yields $\hat{I}(z)$ in terms of the waves of $\hat{\mathscr{H}}^{\pm}$

$$\hat{I}(z) = \left[\oint_{\ell(\text{c.s.})} \hat{\mathscr{H}}^{+} \cdot d\ell\right] e^{-\gamma z} + \left[\oint_{\ell(\text{c.s.})} \hat{\mathscr{H}}^{-} \cdot d\ell\right] e^{\gamma z} \qquad \text{(9-10b)}$$

In Figure 9-2(*a*) is illustrated the meaning of the integrals (9-8) appearing in (9-10b), ℓ enclosing the desired conductor within a fixed cross-section.

The real-time form of the total current (9-10a) is obtained from multiplying by $e^{j\omega t}$ and taking the real part

$$I(z, t) = \text{Re}\,[\hat{I}_m^+ e^{j\omega t - \gamma z} + \hat{I}_m^- e^{j\omega t + \gamma z}] \tag{9-10c}$$

Similarly, the substitution into (9-5b) of

$$\hat{\mathscr{E}}^{\pm} e^{j\omega t \mp \gamma z} \quad \text{for } \mathbf{E} \tag{9-11}$$

retains the *static form* of the Faraday law for TEM traveling waves

$$\boxed{\oint_{\ell(\text{c.s.})} \hat{\mathscr{E}}^{\pm} \cdot d\ell = 0} \tag{9-12}$$

after canceling the exponential factor. As with the magnetic field characterized by (9-9a), the total time-harmonic **E** field is obtained from the superposition

$$\hat{\mathbf{E}}(u_1, u_2, z)e^{j\omega t} = \hat{\mathscr{E}}^+(u_1, u_2)e^{j\omega t - \gamma z} + \hat{\mathscr{E}}^-(u_1, u_2)e^{j\omega t + \gamma z} \tag{9-13a}$$

or just

$$\hat{\mathbf{E}}(u_1, u_2, z) = \hat{\mathscr{E}}^+(u_1, u_2)e^{-\gamma z} + \hat{\mathscr{E}}^-(u_1, u_2)e^{\gamma z} \tag{9-13b}$$

The conservative property (9-12) of the transverse $\hat{\mathscr{E}}$ field of TEM waves permits writing

$$\nabla_T \times \hat{\mathscr{E}}^{\pm} = 0 \tag{9-14}$$

in which the subscript T (transverse) denotes a curl operator including only the variations $\partial/\partial u_1$ and $\partial/\partial u_2$, in view of $\hat{\mathscr{E}}^{\pm}$ being a function of only u_1 and u_2. Expanding (9-14) in generalized cylindrical coordinates yields

$$\nabla_T \times \hat{\mathscr{E}}^{\pm} \equiv \begin{vmatrix} \dfrac{\mathbf{a}_1}{h_2} & \dfrac{\mathbf{a}_2}{h_1} & \dfrac{\mathbf{a}_z}{h_1 h_2} \\ \dfrac{\partial}{\partial u_1} & \dfrac{\partial}{\partial u_2} & 0 \\ h_1\hat{\mathscr{E}}_1^{\pm} & h_2\hat{\mathscr{E}}_2^{\pm} & 0 \end{vmatrix} \equiv \mathbf{a}_z \frac{1}{h_1 h_2}\left[\frac{\partial(h_2\hat{\mathscr{E}}_2^{\pm})}{\partial u_1} - \frac{\partial(h_1\hat{\mathscr{E}}_1^{\pm})}{\partial u_2}\right] = 0 \tag{9-15}$$

Equation (9-15) specifies only the longitudinal component of the curl of $\hat{\mathscr{E}}$;

the *total* curl is given by Maxwell's modified-curl relation (8-6), which for TEM waves is

$$\nabla' \times \hat{\mathscr{E}}^{\pm} \equiv \begin{vmatrix} \dfrac{\mathbf{a}_1}{h_2} & \dfrac{\mathbf{a}_2}{h_1} & \dfrac{\mathbf{a}_z}{h_1 h_2} \\ \dfrac{\partial}{\partial u_1} & \dfrac{\partial}{\partial u_2} & \mp\gamma \\ h_1\hat{\mathscr{E}}_1^{\pm} & h_2\hat{\mathscr{E}}_2^{\pm} & 0 \end{vmatrix} = -j\omega\mu(\mathbf{a}_1\hat{\mathscr{H}}_1^{\pm} + \mathbf{a}_2\hat{\mathscr{H}}_2^{\pm}) \tag{9-16}$$

The conservative property (9-14) shows that one can relate $\hat{\mathscr{E}}^{\pm}$ to the two-dimensional *gradient* of the *potential* functions $\hat{\Phi}^{\pm}(u_1, u_2)$ by

$$\hat{\mathscr{E}}^{\pm} = -\nabla_T \hat{\Phi}^{\pm} \tag{9-17}$$

a procedure justified in Section 4-5. The total electric field given by (9-13b) may thus be written with (9-17) inserted to yield

$$\hat{\mathbf{E}}(u_1, u_2, z) = \hat{\mathscr{E}}^{+}e^{-\gamma z} + \hat{\mathscr{E}}^{-}e^{\gamma z} = (-\nabla_T\hat{\Phi}^{+})e^{-\gamma z} + (-\nabla_T\hat{\Phi}^{-})e^{\gamma z} \tag{9-18}$$

The two-dimensional (9-17) is analogous with the potential relation (4-31) of *static* electric field theory. Thus the two-dimensional scalar potential Φ for a transmission line becomes

$$\hat{\Phi}^{\pm}(u_1, u_2) = -\int_{P_0}^{P(u_1,u_2)} \hat{\mathscr{E}}^{\pm}(u_1, u_2)\cdot d\ell \Bigg]_{(\text{c.s.})} \tag{9-19a}$$

in which P_0 is an arbitrary potential reference, assuming the integration path confined to a fixed cross-section. With the conductors ideally assumed perfect, the tangential electric field vanishes on each, making them *equipotential surfaces within any cross-section.* Denoting the conductor equipotentials by $\hat{\Phi}_1$ and $\hat{\Phi}_2$, a unique potential-difference, or *voltage*, $\hat{\Phi}_1 - \hat{\Phi}_2 = \hat{V}$, is identified with the $\hat{\mathscr{E}}$-field in any cross-section such that

$$\hat{\Phi}_1^{\pm} - \hat{\Phi}_2^{\pm} \equiv \hat{V}_m^{\pm} = -\int_{P_2}^{P_1} \hat{\mathscr{E}}^{\pm}(u_1, u_2)\cdot d\ell \Bigg]_{(\text{c.s.})} \tag{9-19b}$$

in which P_2 designates the zero potential reference on one conductor as in Figure 9-2(*b*), with the line integration taken to any P_1 on the other (positive) conductor. In (9-19b) $\hat{V}_m^{\pm}$ designates *complex amplitudes* $\hat{V}_m^{+}$ *and* $\hat{V}_m^{-}$ associated with the traveling waves of voltage. Their linear superposition, after multiplying by $e^{\pm \gamma z}$, provides the total line voltage

$$\hat{V}(z) = \hat{V}_m^+ e^{-\gamma z} + \hat{V}_m^- e^{\gamma z} \tag{9-20a}$$

a form deserving comparison with the total current (9-10a). A substitution of (9-19b) into (9-20a) yields $\hat{V}(z)$ in terms of the waves of $\hat{\mathscr{E}}$

$$\hat{V}(z) = \left[-\int_{P_2}^{P_1} \hat{\mathscr{E}}^+ \cdot d\ell\right]_{(\text{c.s.})} e^{-\gamma z} + \left[-\int_{P_2}^{P_1} \hat{\mathscr{E}}^- \cdot d\ell\right]_{(\text{c.s.})} e^{\gamma z} \tag{9-20b}$$

The real-time form of (9-20a) is obtained as usual from

$$V(z, t) = \text{Re}\,[\hat{V}_m^+ e^{j\omega t - \gamma z} + \hat{V}_m^- e^{j\omega t + \gamma z}] \tag{9-20c}$$

EXAMPLE 9-1. A long, uniform, coaxial transmission line consisting of perfect conductors with the dimensions shown has a dielectric with the parameters μ, ϵ, and σ. (*a*) Use the static potential field solution Φ to obtain the time-harmonic potential $\hat{\Phi}^\pm$ in any cross-section. (Express $\hat{\Phi}^\pm$ as a function of the

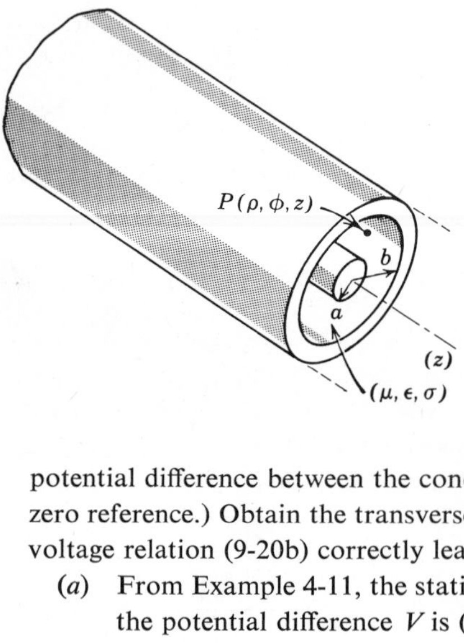

Example 9-1

potential difference between the conductors, taking the outer conductor as the zero reference.) Obtain the transverse electric field $\hat{\mathscr{E}}^\pm$. (*b*) Verify that the total voltage relation (9-20b) correctly leads to the result (9-20a).

(*a*) From Example 4-11, the static potential field of the coaxial system with the potential difference V is (4-70)

$$\Phi(\rho) = \frac{V}{\ell n \dfrac{b}{a}} \ell n \frac{b}{\rho} \tag{9-21}$$

a solution of Laplace's equation and the boundary conditions. The analogous solution applicable to TEM waves is

$$\hat{\Phi}^{\pm}(\rho) = \frac{\hat{V}_m^{\pm}}{\ln \frac{b}{a}} \ln \frac{b}{\rho} \tag{9-22}$$

in which $\hat{V}_m^{\pm}$ denotes *complex* voltage amplitudes associated with positive z and negative z traveling wave solutions. The corresponding electric field solutions are found from (9-17) to yield

$$\hat{\mathscr{E}}^{\pm} = -\nabla_T \hat{\Phi}^{\pm} = -\mathbf{a}_\rho \, \partial \hat{\Phi}^{\pm}/\partial \rho,$$

obtaining

$$\hat{\mathscr{E}}^{\pm} = \mathbf{a}_\rho \hat{\mathscr{E}}_\rho^{\pm}(\rho) = \mathbf{a}_\rho \frac{\hat{V}_m^{\pm}}{\ln \frac{b}{a}} \frac{1}{\rho} \tag{9-23}$$

The total **E** field is then the superposition (9-18)

$$\hat{\mathbf{E}}(\rho, z) = \mathbf{a}_\rho [\hat{\mathscr{E}}_\rho^{+}(\rho) e^{-\gamma z} + \hat{\mathscr{E}}_\rho^{-}(\rho) e^{\gamma z}]$$

$$= \mathbf{a}_\rho \left[\frac{\hat{V}_m^{+}}{\ln \frac{b}{a}} \frac{1}{\rho} e^{-\gamma z} + \frac{\hat{V}_m^{-}}{\ln \frac{b}{a}} \frac{1}{\rho} e^{\gamma z} \right] \tag{9-24}$$

in which γ is yet to be found, the amplitudes $\hat{V}_m^{+}$ and $\hat{V}_m^{-}$ depending on the generator and possible reflections occurring down the line.

(*b*) Inserting (9-23) into (9-20b) yields

$$\hat{V}(z) = \left[-\frac{\hat{V}_m^{+}}{\ln \frac{b}{a}} \int_b^a \frac{d\rho}{\rho} \right] e^{-\gamma z} + \left[-\frac{\hat{V}_m^{-}}{\ln \frac{b}{a}} \int_b^a \frac{d\rho}{\rho} \right] e^{\gamma z}$$

$$= \hat{V}_m^{+} e^{-\gamma z} + \hat{V}_m^{-} e^{\gamma z}$$

or just the expected result (9-20a).

9-2 Wave Impedance and Propagation Constant of the TEM Mode

Once the transverse $\hat{\mathscr{E}}^{\pm}$ have been determined for a given line, the corresponding $\hat{\mathscr{H}}^{\pm}$ can be found from an impedance result obtained from Maxwell's modified-curl equations. Thus, the expansion of (9-16) yields three scalar relations

$$\pm \gamma \hat{\mathscr{E}}_2^{\pm} = -j\omega\mu \hat{\mathscr{H}}_1^{\pm} \tag{9-25a}$$

$$\mp \gamma \hat{\mathscr{E}}_1^{\pm} = -j\omega\mu \hat{\mathscr{H}}_2^{\pm} \tag{9-25b}$$

$$\frac{1}{h_1 h_2} \left[\frac{\partial (h_2 \hat{\mathscr{E}}_2^{\pm})}{\partial u_1} - \frac{\partial (h_1 \hat{\mathscr{E}}_1^{\pm})}{\partial u_2} \right] = 0 \tag{9-25c}$$

the latter of which is just a restatement of (9-15), that $\nabla_T \times \hat{\mathscr{E}}^{\pm} = 0$. The modified-curl $\mathscr{H}^{\pm}$ relation has an analogous expansion, treated as follows. To account for a dielectric having a nonzero conductivity σ (or loss tangent ϵ''/ϵ'), (8-8) is written with the additional term $\sigma\hat{\mathscr{E}}^{\pm}$:

$$\begin{aligned}\nabla' \times \mathscr{H}^{\pm} &= \sigma\hat{\mathscr{E}}^{\pm} + j\omega\epsilon\hat{\mathscr{E}}^{\pm} \\ &= j\omega\left(\epsilon + \frac{\sigma}{j\omega}\right)\hat{\mathscr{E}}^{\pm} = j\omega\hat{\epsilon}\hat{\mathscr{E}}^{\pm}\end{aligned} \tag{9-26}$$

in which a *complex permittivity* $\hat{\epsilon}$ of (3-103) or (3-105)

$$\hat{\epsilon} \equiv \epsilon + \frac{\sigma}{j\omega} = \epsilon\left(1 - j\frac{\sigma}{\omega\epsilon}\right) \equiv \epsilon' - j\epsilon'' \tag{9-27}$$

again appears, valid also for uniform plane waves propagating in an unbounded lossy region, discussed in Section 3-6 and 3-7. The fields of TEM waves have no z components, so (9-26) becomes

$$\nabla' \times \mathscr{H}^{\pm} \equiv \begin{vmatrix} \dfrac{\mathbf{a}_1}{h_2} & \dfrac{\mathbf{a}_2}{h_1} & \dfrac{\mathbf{a}_z}{h_1 h_2} \\ \dfrac{\partial}{\partial u_1} & \dfrac{\partial}{\partial u_2} & \mp\gamma \\ h_1\mathscr{H}_1^{\pm} & h_2\mathscr{H}_2^{\pm} & 0 \end{vmatrix} = j\omega\hat{\epsilon}[\mathbf{a}_1\hat{\mathscr{E}}_1^{\pm} + \mathbf{a}_2\hat{\mathscr{E}}_2^{\pm}] \tag{9-28}$$

yielding the three scalar relations

$$\pm\gamma\mathscr{H}_2^{\pm} = j\omega\hat{\epsilon}\hat{\mathscr{E}}_1^{\pm} \tag{9-29a}$$

$$\mp\gamma\mathscr{H}_1^{\pm} = j\omega\hat{\epsilon}\hat{\mathscr{E}}_2^{\pm} \tag{9-29b}$$

$$\frac{1}{h_1 h_2}\left[\frac{\partial(h_2\mathscr{H}_2^{\pm})}{\partial u_1} - \frac{\partial(h_1\mathscr{H}_1^{\pm})}{\partial u_2}\right] = 0 \tag{9-29c}$$

The first two of this triplet, as well as those of (9-25), are algebraic and contain the wave impedance information required to find one of the fields if the other is known. Thus, (9-25b) and (9-29a) contain only $\hat{\mathscr{E}}_1^{\pm}$ and $\mathscr{H}_2^{\pm}$, making their complex ratio, an *intrinsic wave impedance* denoted by the symbol $\hat{\eta}_{\text{TEM}}$, expressible two ways

$$\frac{\hat{\mathscr{E}}_1^{\pm}}{\mathscr{H}_2^{\pm}} = \pm\frac{j\omega\mu}{\gamma} = \pm\frac{\gamma}{j\omega\hat{\epsilon}} (\equiv \pm\hat{\eta}_{\text{TEM}})\ \Omega \tag{9-30a}$$

Similarly, (9-25a) and (9-29b) yield

$$\frac{\hat{\mathscr{E}}_2^{\pm}}{\mathscr{H}_1^{\pm}} = \mp\frac{j\omega\mu}{\gamma} = \mp\frac{\gamma}{j\omega\hat{\epsilon}} (\equiv \mp\hat{\eta}_{\text{TEM}})\ \Omega \tag{9-30b}$$

Equating $j\omega\mu/\gamma$ to $\gamma/j\omega\hat{\epsilon}$ in either result yields $\gamma^2 = -\omega^2\mu\hat{\epsilon}$, obtaining the *propagating constant* for TEM waves

$$\gamma \equiv \alpha + j\beta = j\omega\sqrt{\mu\hat{\epsilon}}\ \text{m}^{-1} \tag{9-31}$$

A comparison with (3-88) reveals a propagation constant the same as for uniform plane waves in a lossy region having the same parameters μ, ϵ, σ as the transmission line dielectric. Further results from Section 3-7 yield the attenuation and phase constants α and β for TEM waves, given by (3-109) and (3-110)

$$\alpha = \frac{\omega\sqrt{\mu\epsilon}}{\sqrt{2}}\sqrt{\sqrt{1 + \left(\frac{\epsilon''}{\epsilon'}\right)^2} - 1}\ \text{Np/m} \tag{9-32}$$

$$\beta = \frac{\omega\sqrt{\mu\epsilon}}{\sqrt{2}}\sqrt{\sqrt{1 + \left(\frac{\epsilon''}{\epsilon'}\right)^2} + 1}\ \text{rad/m} \tag{9-33}$$

For a line with a lossless dielectric ($\epsilon''/\epsilon' = 0$)

$$\alpha = 0 \tag{9-34a}$$

Lossless dielectric

$$\beta = \omega\sqrt{\mu\epsilon} \tag{9-34b}$$

If (9-31) is substituted into either of the $\hat{\eta}_{\text{TEM}}$ relations (9-30), the following obtains:

$$\hat{\eta}_{\text{TEM}} = \frac{\gamma}{j\omega\hat{\epsilon}} = \frac{j\omega\sqrt{\mu\hat{\epsilon}}}{j\omega\hat{\epsilon}} = \sqrt{\frac{\mu}{\hat{\epsilon}}} \equiv \hat{\eta}\ \Omega \tag{9-35}$$

a result for TEM waves identical with (3-97) for *uniform plane waves* in the lossy region. Thus (3-99a) and (3-111) also apply to TEM waves

$$\hat{\eta} = \frac{\sqrt{\frac{\mu}{\epsilon}}}{\left[1 + \left(\frac{\epsilon''}{\epsilon'}\right)^2\right]^{1/4}} e^{j(1/2)\arctan(\epsilon''/\epsilon')} = \eta e^{j\theta}\ \Omega \tag{9-36}$$

If $\epsilon'' = 0$, (9-36) reduces to

$$\hat{\eta} \to \eta^{(0)} \equiv \sqrt{\frac{\mu}{\epsilon}} \qquad \text{Lossless dielectric} \tag{9-37}$$

having the real value $120\pi\ \Omega$ for an air dielectric line.

The wavelength is defined by $\beta\lambda = 2\pi$, making

$$\lambda = \frac{2\pi}{\beta} \text{ m} \tag{9-38}$$

to which the β expressions (9-33) or (9-34b) are applicable. The phase velocity is

$$v_p = \frac{\omega}{\beta} \text{ m/sec} \tag{9-39}$$

reducing for an ideal air filled line to $v_p = c = 3 \times 10^8$ m/sec.

An alternative expression for the impedance relations (9-30), including the vector direction of $\hat{\mathscr{E}}^\pm$ and $\hat{\mathscr{H}}^\pm$, is obtained from the expansion of (9-16)

$$\nabla' \times \hat{\mathscr{E}} \equiv \mathbf{a}_1(\pm\gamma\hat{\mathscr{e}}_2^\pm) + \mathbf{a}_2(\mp\gamma\hat{\mathscr{e}}_1^\pm) = -j\omega\mu\hat{\mathscr{H}}^\pm \tag{9-40}$$

The $\mathbf{a}_1$ and $\mathbf{a}_2$ components can be combined to yield $\mathbf{a}_1(\pm\gamma\hat{\mathscr{e}}_2^\pm) + \mathbf{a}_2(\mp\gamma\hat{\mathscr{e}}_1^\pm)$ $= \mp\gamma\mathbf{a}_z \times [\mathbf{a}_1\hat{\mathscr{e}}_1^\pm + \mathbf{a}_2\hat{\mathscr{e}}_2^\pm] = \mp\gamma\mathbf{a}_z \times \hat{\mathscr{E}}^\pm$ because $\mathbf{a}_z \times \mathbf{a}_1 = \mathbf{a}_2$ and $\mathbf{a}_z \times \mathbf{a}_2 = -\mathbf{a}_1$. So (9-40) obtains $\pm\gamma\mathbf{a}_z \times \hat{\mathscr{E}}^\pm = j\omega\mu\hat{\mathscr{H}}^\pm$, whence

$$\hat{\mathscr{H}}^\pm = \pm\frac{\gamma}{j\omega\mu} \mathbf{a}_z \times \hat{\mathscr{E}}^\pm = \pm\frac{\mathbf{a}_z \times \hat{\mathscr{E}}^\pm}{\hat{\eta}} \text{ A/m} \tag{9-41}$$

if (9-30) is utilized. Equation (9-41) is useful for obtaining the $\hat{\mathscr{H}}^\pm$ field of a TEM wave whenever $\hat{\mathscr{E}}^\pm$ is known. From it one can see that $\hat{\mathscr{E}}^\pm$ and $\hat{\mathscr{H}}^\pm$ are vectors everywhere *perpendicular* to each other. With (9-41) in (9-9b), the total field $\hat{\mathbf{H}}(u_1, u_2, z)$ is written

$$\hat{\mathbf{H}}(u_1, u_2, z) = \frac{\mathbf{a}_z \times \hat{\mathscr{E}}^+}{\hat{\eta}} e^{-\gamma z} - \frac{\mathbf{a}_z \times \hat{\mathscr{E}}^-}{\hat{\eta}} e^{\gamma z} \text{ A/m} \tag{9-42}$$

The dual of (9-42) follows from (9-28), yielding

$$\hat{\mathscr{E}}^\pm = \mp\frac{\gamma}{j\omega\hat{\epsilon}} \mathbf{a}_z \times \hat{\mathscr{H}}^\pm = \mp\hat{\eta}\mathbf{a}_z \times \hat{\mathscr{H}}^\pm \text{ V/m} \tag{9-43}$$

which is the converse of (9-41), since it enables finding $\hat{\mathscr{E}}^\pm$ whenever $\hat{\mathscr{H}}^\pm$ is known.

EXAMPLE 9-2.   Find the time-harmonic magnetic field expressions for the coaxial line of Example 9-1. Express the total electric and magnetic fields in

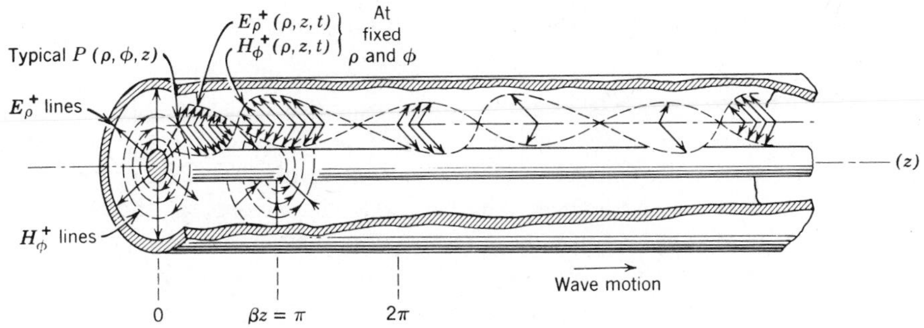

Example 9-2

complex form. Obtain their real-time forms for a line with a lossless dielectric, showing a flux sketch of only the positive z traveling fields.

The solutions (9-23) inserted into (9-41) yield

$$\mathscr{H}^{\pm} = \pm \frac{\mathbf{a}_z \times \hat{\mathscr{E}}^{\pm}}{\hat{\eta}} = \pm \mathbf{a}_\phi \frac{\hat{V}_m^{\pm}}{\hat{\eta}\, \ell n \dfrac{b}{a}} \frac{1}{\rho} \tag{9-44}$$

The *total* magnetic field is given by (9-9b), a superposition of (9-44) after multiplication by $e^{-\gamma z}$ and $e^{\gamma z}$

$$\begin{aligned}\hat{\mathbf{H}}(\rho, z) &= \mathbf{a}_\phi[\mathscr{H}_\phi^+(\rho)e^{-\gamma z} + \mathscr{H}_\phi^-(\rho)e^{\gamma z}] \\ &= \mathbf{a}_\phi\left[\frac{\hat{V}_m^+}{\hat{\eta}\, \ell n \dfrac{b}{a}} \frac{1}{\rho} e^{-\gamma z} - \frac{\hat{V}_m^-}{\hat{\eta}\, \ell n \dfrac{b}{a}} \frac{1}{\rho} e^{\gamma z}\right]\end{aligned} \tag{9-45}$$

in which $\hat{\eta}$ and γ are given by (9-36) and (9-31). From (9-34b) and (9-37), $\hat{\eta} = \eta = \sqrt{\mu/\epsilon}$ and $\gamma = j\omega\sqrt{\mu\epsilon} = j\beta$ since the dielectric is lossless. The real-time forms of (9-24) and (9-45) are found using (9-6a) and (9-6b)

$$\begin{aligned}\mathbf{E}(\rho, z, t) &= \mathbf{a}_\rho \operatorname{Re}\left[\frac{\hat{V}_m^+}{\ell n \dfrac{b}{a}} \frac{1}{\rho} e^{j(\omega t - \beta z)} + \frac{\hat{V}_m^-}{\ell n \dfrac{b}{a}} \frac{1}{\rho} e^{j(\omega t + \beta z)}\right] \\ &= \mathbf{a}_\rho\left[\frac{V_m^+}{\ell n \dfrac{b}{a}} \frac{1}{\rho} \cos(\omega t - \beta z + \phi^+) + \frac{V_m^-}{\ell n \dfrac{b}{a}} \frac{1}{\rho} \cos(\omega t + \beta z + \phi^-)\right]\end{aligned} \tag{9-46}$$

$$\mathbf{H}(\rho, z, t) = \mathbf{a}_\phi\left[\frac{V_m^+}{\eta\, \ell n \dfrac{b}{a}} \frac{1}{\rho} \cos(\omega t - \beta z + \phi^+) - \frac{V_m^-}{\eta\, \ell n \dfrac{b}{a}} \frac{1}{\rho} \cos(\omega t + \beta z + \phi^-)\right] \tag{9-47}$$

assuming complex amplitudes of the form $\hat{V}_m^{\pm} = V_m^{\pm} e^{j\phi\pm}$.

A flux sketch of the positive z traveling fields is shown.

9-3 Voltage and Current Waves; Characteristic Impedance

It is usually desirable to characterize TEM waves on a line in terms of their voltage and current waves rather than the electric and magnetic field quantities discussed in the foregoing sections. The advantages are evident from the fact that voltages and currents on a transmission line are readily measured quantities at frequencies below 1 GHz or so, whereas the electric and magnetic fields must usually be inferred from such measurements.

It has been seen from (9-10) and (9-20) that the sinusoidal voltage and current waves on a line carrying the dominant TEM mode are expressed

$$\hat{V}(z) = \hat{V}_m^+ e^{-\gamma z} + \hat{V}_m^- e^{\gamma z} \quad [9\text{-}20a]$$

$$\hat{V}(z) = \left[-\int_{P_2}^{P_1} \hat{\mathscr{E}}^+ \cdot d\ell \right]_{(\text{c.s.})} e^{-\gamma z} + \left[-\int_{P_2}^{P_1} \hat{\mathscr{E}}^- \cdot d\ell \right]_{(\text{c.s.})} e^{\gamma z} \quad [9\text{-}20b]$$

$$\hat{I}(z) = \hat{I}_m^+ e^{-\gamma z} + \hat{I}_m^- e^{\gamma z} \quad [9\text{-}10a]$$

$$\hat{I}(z) = \left[\oint_{\ell(\text{c.s.})} \hat{\mathscr{H}}^+ \cdot d\ell \right] e^{-\gamma z} + \left[\oint_{\ell(\text{c.s.})} \hat{\mathscr{H}}^- \cdot d\ell \right] e^{\gamma z} \quad [9\text{-}10b]$$

$$\hat{I}(z) = \left[\oint_{\ell(\text{c.s.})} \frac{\mathbf{a}_z \times \hat{\mathscr{E}}^+}{\hat{\eta}} \cdot d\ell \right] e^{-\gamma z} - \left[\oint_{\ell(\text{c.s.})} \frac{\mathbf{a}_z \times \hat{\mathscr{E}}^-}{\hat{\eta}} \cdot d\ell \right] e^{\gamma z} \quad (9\text{-}48)$$

The total current (9-48) follows from (9-42), expressing the forward and backward current waves in terms of the electric fields $\hat{\mathscr{E}}^{\pm}$, which are in turn proportional to the complex amplitudes $\hat{V}_m^{\pm}$ in (9-20a). (An example of the proportionality connecting $\hat{\mathscr{E}}^{\pm}$ with $\hat{V}_m^{\pm}$ has been observed in (9-23) of Example 9-1). Therefore (9-48) can be written

$$\hat{I}(z) = \frac{\hat{V}_m^+}{\hat{Z}_0} e^{-\gamma z} - \frac{\hat{V}_m^-}{\hat{Z}_0} e^{\gamma z} \quad (9\text{-}49)$$

in which $\hat{Z}_0$ is called the *characteristic impedance* of the given line. Equating the bracketed factors of (9-10b) or (9-48) to like factors of (9-49) yields the following expressions for $\hat{Z}_0$:

$$\hat{Z}_0 \equiv \pm \frac{\hat{V}_m^{\pm}}{\hat{I}_m^{\pm}} = \pm \frac{\hat{V}_m^{\pm}}{\oint_{\ell(\text{c.s.})} \hat{\mathscr{H}}^{\pm} \cdot d\ell} = \frac{\hat{V}_m^{\pm}}{\oint_{\ell(\text{c.s.})} \frac{\mathbf{a}_z \times \hat{\mathscr{E}}^{\pm}}{\hat{\eta}} \cdot d\ell} \ \Omega \quad (9\text{-}50)$$

Because $\hat{I}_m^{\pm}$ appearing in the denominators of (9-50) are proportional to the amplitudes $\hat{V}_m^{\pm}$, the latter will cancel upon carrying out the indicated integration. This makes $\hat{Z}_0$ of an ideal transmission line a function of only the conductor dimensions and the dielectric parameters μ, ϵ, and σ at a given frequency.

A summary of TEM mode relationships is given in Table 9-1.

EXAMPLE 9-3. For the coaxial line of Examples 9-1 and 9-2 (*a*) express the total current $\hat{I}(z)$ in terms of $\hat{V}_m^{\pm}$, and deduce $\hat{Z}_0$; and (*b*) sketch the electric and magnetic fields, showing the related voltage and current senses.

(*a*) To express $\hat{I}(z)$ in terms of $\hat{V}_m^{\pm}$, insert (9-44) into (9-10b), yielding

$$\hat{I}(z) = \left[\int_0^{2\pi}\left(\mathbf{a}_\phi \frac{\hat{V}_m^+}{\rho\hat{\eta}\,\ell n\frac{b}{a}}\right)\cdot\mathbf{a}_\phi\rho\,d\phi\right]e^{-\gamma z} - \left[\int_0^{2\pi}\left(\mathbf{a}_\phi \frac{\hat{V}_m^-}{\rho\hat{\eta}\,\ell n\frac{b}{a}}\right)\cdot\mathbf{a}_\phi\rho\,d\phi\right]e^{\gamma z}$$

$$= \frac{\hat{V}_m^+}{\frac{\hat{\eta}}{2\pi}\ell n\frac{b}{a}}e^{-\gamma z} - \frac{\hat{V}_m^-}{\frac{\hat{\eta}}{2\pi}\ell n\frac{b}{a}}e^{\gamma z} \tag{9-51}$$

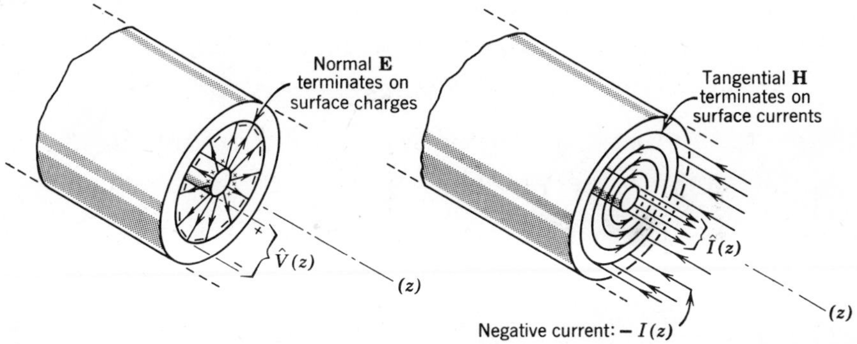

Example 9-3

A comparison with (9-49) shows that $\hat{Z}_0$ of the coaxial line with perfect conductors is

$$\hat{Z}_0 \equiv \pm\frac{\hat{V}_m^{\pm}}{\hat{I}_m^{\pm}} = \frac{\hat{\eta}}{2\pi}\ell n\frac{b}{a} \tag{9-52}$$

(*b*) The fields and related voltage and current conventions are noted in the accompanying sketch.

EXAMPLE 9-4. A lossless coaxial line has the dimensions $2a = 0.1$ in. and $2b = 0.326$ in., using a dielectric with $\epsilon_r = 2$. Assume $f = 20$ MHz. (*a*) Find its $\hat{Z}_0$, β, and v_p. (*b*) If the dielectric had the small loss tangent $(\epsilon''/\epsilon') = 0.0002$ at this frequency, determine how much $\hat{Z}_0$, β, and v_p change and how much attenuation is introduced.

(*a*) The characteristic impedance is found using (9-52), requiring (9-35) which yields $\hat{\eta} = \sqrt{\mu_0/2\epsilon_0} = 120\pi/\sqrt{2} = 266.5\ \Omega$. Thus

$$\hat{Z}_0 = \frac{\hat{\eta}}{2\pi}\,\ell n\,\frac{b}{a} = \frac{266.5}{2\pi}\,\ell n\,\frac{0.326}{0.1} = 50\ \Omega$$

From (9-34b) and (9-39)

$$\beta = \omega\sqrt{\mu_0(2\epsilon_0)} = \frac{2\pi(2 \times 10^7)\sqrt{2}}{3 \times 10^8} = 0.594\ \text{rad/m}$$

$$v_p = \frac{1}{\sqrt{\mu_0(2\epsilon_0)}} = \frac{c}{\sqrt{2}} = 2.12 \times 10^8\ \text{m/sec}$$

(*b*) The dielectric with $\epsilon''/\epsilon' = 0.0002$ has from (9-36)

$$\hat{\eta} = \frac{120\pi/\sqrt{2}}{[1 + (0.0002)^2]^{1/4}}\,e^{j(1/2)\arctan 0.0002} \cong 266.5e^{j0.0001} \cong 266.5\ \Omega$$

yielding, from (9-52), $\hat{Z}_0$ with the same negligible angle

$$\hat{Z}_0 \cong 50e^{j0.0001} \cong 50\ \Omega$$

The constants α and β are evaluated by use of the small loss approximations (3-113a) and (3-113b); thus

$$\beta \cong \omega\sqrt{\mu\epsilon}\left[1 + \frac{1}{8}\left(\frac{\epsilon''}{\epsilon'}\right)^2\right] = 0.594\left[1 + \frac{(0.0002)^2}{8}\right] \cong 0.594\ \text{rad/m}$$

$$\alpha \cong \frac{\omega\sqrt{\mu\epsilon}}{2}\left(\frac{\epsilon''}{\epsilon'}\right) = \frac{0.594}{2}(0.0002) = 5.94 \times 10^{-5}\ \text{Np/m}$$

The latter implies a wave decay to e^{-1} in a distance $d = \alpha^{-1} = (5.94 \times 10^{-5})^{-1} = 16.8$ km $= 10.4$ mi. With β essentially unchanged, v_p remains at 2.12×10^8 m/sec in the lossy dielectric.

9-4 Transmission-Line Parameters, Perfect Conductors Assumed

Maxwell's equations can be used to derive a pair of coupled differential equations expressed in terms of the voltage and current on a transmission line carrying the dominant TEM mode. The development is carried out first in the *real-time* domain. The present discussion concerns a line composed of perfect conductors separated by a dielectric with parameters (ϵ, μ, σ). It is

Table 9-1 Summary relations for TEM waves

Electric fields are found from quasi-static potentials	Magnetic fields are obtained from $\hat{\mathscr{E}}^{\pm}$ solutions by use of
$\hat{\mathscr{E}}^{\pm}(u_1, u_2) = -\nabla_T \hat{\Phi}^{\pm}$ [9-17]	$\hat{\mathscr{H}}^{\pm}(u_1, u_2) = \pm \dfrac{\mathbf{a}_z \times \hat{\mathscr{E}}^{\pm}}{\hat{\eta}}$ [9-41]
in which $\hat{\Phi}^{\pm}$ are solutions of Laplace's equation (4-68).	
The total electric field, including z dependence, is	The total magnetic field, including z dependence, is
$\hat{\mathbf{E}}(u_1, u_2, z) = \hat{\mathscr{E}}^{+}(u_1, u_2)e^{-\gamma z} + \hat{\mathscr{E}}^{-}(u_1, u_2)e^{\gamma z}$ [9-13b]	$\hat{\mathbf{H}}(u_1, u_2, z) = \hat{\mathscr{H}}^{+}(u_1, u_2)e^{-\gamma z} + \hat{\mathscr{H}}^{-}(u_1, u_2)e^{\gamma z}$ [9-9b]
$= (-\nabla_T \hat{\Phi}^{+})e^{-\gamma z} + (-\nabla_T \hat{\Phi}^{-})e^{\gamma z}$ [9-18]	$= \dfrac{\mathbf{a}_z \times \hat{\mathscr{E}}^{+}}{\hat{\eta}} e^{-\gamma z} - \dfrac{\mathbf{a}_z \times \hat{\mathscr{E}}^{-}}{\hat{\eta}} e^{\gamma z}$ [9-42]
Voltage-wave complex amplitudes are	Current wave complex amplitudes are
$\hat{V}_m^{\pm} = -\left.\int_{P_2}^{P_1} \hat{\mathscr{E}}^{\pm}(u_1, u_2) \cdot d\ell \right]_{(\text{c.s.})}$ [9-19b]	$\hat{I}_m^{\pm} = \oint_{\ell(\text{c.s.})} \hat{\mathscr{H}}^{\pm}(u_1, u_2) \cdot d\ell$ [9-8]
making the total voltage	making the total current
$\hat{V}(z) = \hat{V}_m^{+} e^{-\gamma z} + \hat{V}_m^{-} e^{\gamma z}$ [9-20a]	$\hat{I}(z) = \hat{I}_m^{+} e^{-\gamma z} + \hat{I}_m^{-} e^{\gamma z}$ [9-10a]
$= \left[-\int_{P_2}^{P_1} \hat{\mathscr{E}}^{+} \cdot d\ell\right]_{(\text{c.s.})} e^{-\gamma z} + \left[-\int_{P_2}^{P_1} \hat{\mathscr{E}}^{-} \cdot d\ell\right] e^{\gamma z}$ [9-20b]	$= \left[\oint_{\ell(\text{c.s.})}^{a_z} \dfrac{\mathbf{a}_z \times \hat{\mathscr{E}}^{+}}{\hat{\eta}} \cdot d\ell\right] e^{-\gamma z} - \left[\oint_{\ell(\text{c.s.})} \dfrac{\mathbf{a}_z \times \hat{\mathscr{E}}^{-}}{\hat{\eta}} \cdot d\ell\right] e^{\gamma z}$ [9-48]

With $\hat{\mathscr{E}}^{\pm}$ and $\hat{I}_m^{\pm}$ proportional to $\hat{V}_m^{\pm}$, (9-48) can be written

$$\hat{I}(z) = \frac{\hat{V}_m^{+}}{\hat{Z}_0} e^{-\gamma z} - \frac{\hat{V}_m^{-}}{\hat{Z}_0} e^{\gamma z} \qquad [9\text{-}49]$$

in which the characteristic impedance $\hat{Z}_0$ is

$$\hat{Z}_0 = \pm \frac{\hat{V}_m^{\pm}}{\hat{I}_m^{\pm}} = \pm \frac{\hat{V}_m^{\pm}}{\oint_{\ell(\text{c.s.})} \hat{\mathscr{H}}^{\pm} \cdot d\ell} = \frac{\hat{V}_m^{\pm}}{\displaystyle\oint_{\ell(\text{c.s.})} \frac{\mathbf{a}_z \times \hat{\mathscr{E}}^{\pm}}{\hat{\eta}} \cdot d\ell} \qquad [9\text{-}50]$$

shown that at any z cross-section on the line, the voltage and current satisfy the differential equations

$$\frac{\partial V}{\partial z} = -l_e \frac{\partial I}{\partial t} \tag{9-53}$$

$$\frac{\partial I}{\partial z} = -c \frac{\partial V}{\partial t} - gV \tag{9-54}$$

in which l_e, c, and g are distributed (per meter) inductance, capacitance, and conductance parameters to be defined. The equations are *coupled* in the sense that both dependent variables $V(z, t)$ and $I(z, t)$ appear in each.

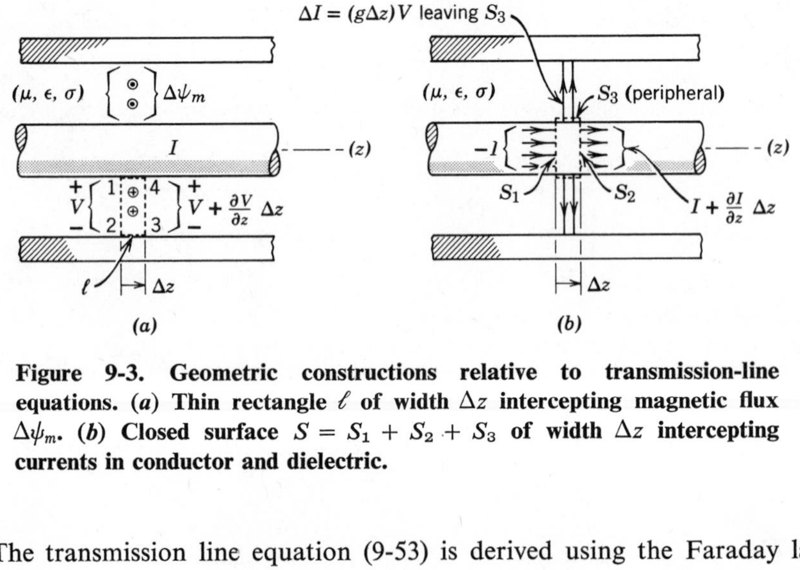

Figure 9-3. Geometric constructions relative to transmission-line equations. (*a*) Thin rectangle ℓ of width Δz intercepting magnetic flux $\Delta\psi_m$. (*b*) Closed surface $S = S_1 + S_2 + S_3$ of width Δz intercepting currents in conductor and dielectric.

The transmission line equation (9-53) is derived using the Faraday law (3-78)

$$\oint_\ell \mathbf{E}\cdot d\boldsymbol{\ell} = -\frac{\partial}{\partial t}\int_S \mathbf{B}\cdot d\mathbf{s} = -\frac{\partial \psi_m}{\partial t}\ \mathrm{V} \tag{9-55}$$

applied to a thin, closed rectangle ℓ of width $\Delta z \to dz$ in a typical cross-section as shown in Figure 9-3(a), with a magnetic flux $\Delta\psi_m$ passing through the surface bounded by ℓ. The left side of (9-55), integrated about ℓ in the 1–2–3–4–1 sense, yields voltages $-V(z, t)$ and $V + (\partial V/\partial z)\,\Delta z$ over 1–2 and 3–4, with no contributions over 2–3 and 4–1 at the perfectly conducting walls.

Thus the left side becomes

$$\oint_{\ell} \mathbf{E} \cdot d\boldsymbol{\ell} = -V + \left(V + \frac{\partial V}{\partial z} \Delta z\right) = \frac{\partial V}{\partial z} \Delta z \tag{9-56}$$

The right side of (9-55) involves the flux $\Delta\psi_m$ intercepted by ℓ. It was shown from (9-5a) that the time-varying magnetic field of the TEM mode, in a fixed cross-section of the line, satisfies the same Maxwell equations and boundary conditions as the *static magnetic field* produced by a *direct current* flowing in the line. Therefore the external inductance expression (5-88a) is applicable becoming

$$\Delta\psi_m = \Delta L_e I \text{ Wb} \tag{9-57a}$$

in which ΔL_e denotes the static external inductance associated with any Δz slice. By writing $\Delta L_e = (\boldsymbol{l}_e \, \Delta z)$, implying

$$\boldsymbol{l}_e = \frac{\Delta L_e}{\Delta z} \text{ H/m} \tag{9-57b}$$

(9-57a) becomes

$$\Delta\psi_m = (\boldsymbol{l}_e \, \Delta z) I \tag{9-57c}$$

in which $\boldsymbol{l}_e$ denotes the static *external inductance per unit length*, or external distributed inductance parameter, of the line.

With (9-56) and (9-57c) substituted into the Faraday law (9-55), one obtains

$$\frac{\partial V}{\partial z} = -\frac{\partial}{\partial t}(\boldsymbol{l}_e I) \tag{9-58}$$

The parameter $\boldsymbol{l}_e$ is a constant in a rigid line having a dielectric with a constant μ, so (9-58) becomes

$$\frac{\partial V}{\partial z} = -\boldsymbol{l}_e \frac{\partial I}{\partial t} \tag{9-59}$$

which is (9-53), that which was to have been proved.

Similarly, (9-54) is derived from the current continuity relation (4-19)

$$\oint_S \mathbf{J} \cdot d\mathbf{s} = -\frac{\partial q}{\partial t} \text{ A} \tag{4-19}$$

applied to a closed surface S of width $\Delta z \to dz$ in the same cross-section, as shown in Figure 9-3(*b*). The conductor at the assumed positive polarity is chosen for the construction, where the positive I sense is taken to be z directed. The right side of (4-19) involves a surface charge increment Δq deposited at any t on the peripheral S_3 shown, in view of the boundary condition (3-45). As seen from (9-5b) and (9-3), the time-varying electric field of the TEM mode, in a fixed cross-section of the line, satisfies the same Maxwell equations and boundary conditions as the *static* electric field between those conductors. The definition (4-47) of static capacitance can therefore be used to relate Δq to the instantaneous voltage V between the conductors as follows:

$$\Delta q = (\Delta C)V \text{ C} \qquad \text{(9-60a)}$$

with ΔC denoting the static capacitance of the Δz slice. Putting $\Delta C = (\boldsymbol{c}\,\Delta z)$, or

$$\boldsymbol{c} = \frac{\Delta C}{\Delta z} \text{ F/m} \qquad \text{(9-60b)}$$

(9-60a) becomes

$$\Delta q = (\boldsymbol{c}\,\Delta z)V \qquad \text{(9-60c)}$$

in which $\boldsymbol{c}$ denotes the static *capacitance per unit length*, or distributed capacitance parameter, of the uniform line.

The left side of (4-19) denotes the net current flux emergent from S at any t. The contributions through S_1 and S_2 in Figure 9-3(*b*) yield the net amount

$$-I + I + \frac{\partial I}{\partial z}\Delta z = \frac{\partial I}{\partial z}\Delta z \qquad \text{(9-61)}$$

An additional current increment ΔI leaves the peripheral surface S_3 and enters the region between the conductors, assuming the dielectric has a conductivity σ. From (4-138), ΔI is proportional to V, obeying

$$\Delta I = (\Delta G)V \qquad \text{(9-62a)}$$

in which ΔG denotes the conductance of the Δz slice. By putting $\Delta G = (\boldsymbol{g}\,\Delta z)$, implying

$$\boldsymbol{g} = \frac{\Delta G}{\Delta z} \text{ ℧/m} \qquad \text{(9-62b)}$$

(9-62a) becomes

$$\Delta I = (\boldsymbol{g}\,\Delta z)V \qquad \text{(9-62c)}$$

in which g defined by (9-62b) denotes the *conductance per unit length*, or distributed conductance parameter of the line. It is evident from (4-140), from which $\Delta G = (\sigma/\epsilon)\,\Delta C$, that g is not an independent quantity; it is related to the distributed capacitance parameter† c upon making use of (3-108)

$$g = \frac{\sigma}{\epsilon} c \left(= \frac{\sigma}{\omega\epsilon} \omega c = \frac{\epsilon''}{\epsilon'} \omega c \right) \ \mho/\text{m} \tag{9-63}$$

Inserting (9-60c), (9-61), and (9-62c) into (4-19) yields

$$\frac{\partial I}{\partial z} \Delta z + (g\,\Delta z)V = -\frac{\partial}{\partial t}(c\,\Delta z)V$$

reducing to the differential equation

$$\frac{\partial I}{\partial z} = -\frac{\partial}{\partial t}(cV) - gV$$

If the parameter c is not a function of time, the latter becomes

$$\boxed{\frac{\partial I}{\partial z} = -c\frac{\partial V}{\partial t} - gV} \tag{9-64}$$

or just (9-54).

Many *dielectric materials* used in transmission lines have parameters μ, ϵ, and σ that may be functions of the sinusoidal frequency ω of the fields, as seen from Table 3-3. From this point of view, the time-harmonic forms of the transmission-line equations may be of greater interest than the real-time forms (9-53) and (9-54). Thus, if into the latter

$$\begin{aligned} &V(z, t) \text{ is replaced with } \hat{V}(z)e^{j\omega t} \\ &I(z, t) \text{ is replaced with } \hat{I}(z)e^{j\omega t} \end{aligned} \tag{9-65}$$

one obtains the time-harmonic transmission-line equations

$$\frac{d\hat{V}}{dz} = -j\omega l_e \hat{I} \tag{9-66a}$$

$$\frac{d\hat{I}}{dz} = -(g + j\omega c)\hat{V} \tag{9-66b}$$

† The last forms for g in (9-63) involve the frequency ω and so apply to the time-harmonic case only.

These are also written

$$\frac{d\hat{V}}{dz} = -\hat{z}\hat{I} \tag{9-67a}$$

$$\frac{d\hat{I}}{dz} = -\hat{y}\hat{V} \tag{9-67b}$$

upon taking $\hat{z}$ and $\hat{y}$ to mean

$$\hat{z} = j\omega \boldsymbol{l}_e \ \Omega/\text{m} \qquad \text{Series distributed impedance} \tag{9-68}$$

$$\hat{y} = \boldsymbol{g} + j\omega \boldsymbol{c} \ \mho/\text{m} \qquad \text{Shunt distributed admittance} \tag{9-69}$$

With the substitution of (9-63), $\hat{y}$ is written in terms of the dielectric loss tangent as follows:

$$\hat{y} = \boldsymbol{g} + j\omega\boldsymbol{c} = \left(\frac{\epsilon''}{\epsilon'} + j\right)\omega\boldsymbol{c} \ \mho/\text{m} \tag{9-70}$$

For example, using a dielectric with a loss tangent of 0.001 yields $\hat{y} = (0.001 + j)\omega\boldsymbol{c}$, or very nearly $j\omega\boldsymbol{c}$.

The wave solutions for $\hat{V}(z)$ and $\hat{I}(z)$ of the transmission-line equations (9-67) have been supplied by (9-20a) and (9-49). They yield expressions for γ and $\hat{Z}_0$ in terms of the parameters $\hat{z}$ and $\hat{y}$ as follows. If, into (9-53) and (9-54)

$$\begin{aligned} &V(z, t) \text{ is replaced with } \hat{V}_m^{\pm} e^{j\omega t \mp \gamma z} \\ &I(z, t) \text{ is replaced with } \hat{I}_m^{\pm} e^{j\omega t \mp \gamma z} \end{aligned} \tag{9-71}$$

one obtains purely algebraic results

$$\mp\gamma\hat{V}_m^{\pm} = -j\omega \boldsymbol{l}_e\hat{I}_m^{\pm} = -\hat{z}\hat{I}_m^{\pm} \tag{9-72}$$

$$\mp\gamma\hat{I}_m^{\pm} = -(\boldsymbol{g} + j\omega\boldsymbol{c})\hat{V}_m^{\pm} = -\hat{y}\hat{V}_m^{\pm} \tag{9-73}$$

But from (9-50), the ratio $\pm\hat{V}_m^{\pm}/\hat{I}_m^{\pm}$ is $\hat{Z}_0$, obtaining from (9-72) and (9-73)

$$\hat{Z}_0 \equiv \pm\frac{\hat{V}_m^{\pm}}{\hat{I}_m^{\pm}} = \frac{\hat{z}}{\gamma} = \frac{\gamma}{\hat{y}} \tag{9-74}$$

The last equality enables expressing γ in terms of the distributed parameters (9-68) and (9-69)

$$\gamma \equiv \alpha + j\beta = \sqrt{\hat{z}\hat{y}} \tag{9-75a}$$

$$= \sqrt{(j\omega \boldsymbol{l}_e)(\boldsymbol{g} + j\omega\boldsymbol{c})} \ \text{m}^{-1} \tag{9-75b}$$

also written

$$\gamma = \sqrt{-\omega^2 l_e c + j\omega l_e g} = j\omega\sqrt{l_e c\left(1 + \frac{g}{j\omega c}\right)} \tag{9-75c}$$

γ can also be expressed in terms of the *dielectric* parameters as

$$\gamma = j\omega\sqrt{\mu\epsilon\left(1 + \frac{\sigma}{j\omega\epsilon}\right)} \tag{9-75d}$$

obtained by substituting $\hat{\epsilon} = \epsilon(1 + \sigma/j\omega\epsilon)$ from (9-27) into (9-31). If $g = (\sigma/\epsilon)c$ of (9-63) is combined with (9-75c) and (9-75d), one obtains the special result

$$l_e c = \mu\epsilon \tag{9-76}$$

The use of (9-63) and (9-76) permits finding g and l_e of a line once c, for example, has been obtained, presuming the constants μ, ϵ, and σ (or the loss tangent) of the dielectric are known.

Inserting (9-75a) back into (9-74) obtains also $\hat{Z}_0$ in terms of the distributed parameters

$$\hat{Z}_0 \equiv \pm\frac{\hat{V}_m^{\pm}}{\hat{I}_m^{\pm}} = \sqrt{\frac{\hat{z}}{\hat{y}}} \tag{9-77a}$$

$$= \sqrt{\frac{j\omega l_e}{g + j\omega c}}\ \Omega \tag{9-77b}$$

Putting (9-63) and (9-76) into (9-77b), one finds that $\hat{Z}_0$ of a line with perfect conductors and dielectric losses can be expressed

$$\hat{Z}_0 = \hat{\eta}\frac{l_e}{\mu} = \hat{\eta}\frac{\epsilon}{c} \tag{9-77c}$$

If the line is completely idealized by assuming no dielectric losses, (9-75c) reduces to $\gamma = j\omega\sqrt{l_e c} = j\omega\sqrt{\mu\epsilon}$, implying

$$\beta = \omega\sqrt{l_e c} = \omega\sqrt{\mu\epsilon} \qquad \text{Lossless} \tag{9-78}$$

Then (9-77c) yields the *pure real* characteristic impedance

$$Z_0 = \eta^{(0)}\frac{l_e}{\mu} = \eta^{(0)}\frac{\epsilon}{c} = \sqrt{\frac{l_e}{c}} \qquad \text{Lossless} \tag{9-79}$$

in which $\eta^{(0)} = \sqrt{\mu/\epsilon}$ is the intrinsic wave impedance associated with the lossless dielectric. Equations (9-78) and (9-79) are useful for short transmission lines used at high frequencies, for which neglecting the small losses may not entail serious errors.

Note that (9-77c), applied to the special cases of coaxial and parallel-wire lines with perfect conductors, produces the following results upon making use of the static capacitances (4-51) and (4-107):

$$\hat{Z}_0 = \frac{\hat{\eta}}{2\pi} \ell n \frac{b}{a} \quad \text{Coaxial line, perfect conductors} \qquad \text{(9-80a)}$$

and

$$\hat{Z}_0 = \frac{\hat{\eta}}{\pi} \ell n \left[\frac{h}{R} + \sqrt{\left(\frac{h}{R}\right)^2 - 1} \right] \quad \text{Parallel-wire line, perfect conductors} \qquad \text{(9-80b)}$$

If dielectric losses are neglected, the wave impedance becomes $\hat{\eta} \to \eta^{(0)} = 120\pi/\sqrt{\epsilon_r}$, with the reasonable assumption of a nonmagnetic dielectric. Then (9-80a) and (9-80b) yield the real results

$$Z_0 = \frac{60}{\sqrt{\epsilon_r}} \ell n \frac{b}{a} \quad \text{Lossless coaxial line} \qquad \text{(9-80c)}$$

$$Z_0 = \frac{120}{\sqrt{\epsilon_r}} \ell n \left[\frac{h}{R} + \sqrt{\left(\frac{h}{R}\right)^2 - 1} \right] \quad \text{Lossless parallel-wire line} \qquad \text{(9-80d)}$$

and are graphed in Figure 9-4. They are useful approximations for transmission lines at high frequencies, for which impedive effects due to the field penetration into the conductors, described in Section 9-7 under the topic of skin effect, are neglected.

EXAMPLE 9-5. A coaxial line has perfect conductors but not necessarily a lossless dielectric. (*a*) Adapt the static capacitance of the line to find its parameter $\boldsymbol{c}$. (*b*) Find $\boldsymbol{g}$. (*c*) Use (9-57c) to derive $\boldsymbol{l}_e$, the distributed (external) inductance parameter.

(*a*) The static capacitance is given by (4-51): $C/\ell = 2\pi\epsilon/\ell n\,(b/a)$. This ratio is also $\Delta C/\Delta z$, whence from (9-60b)

$$\boldsymbol{c} = \frac{2\pi\epsilon}{\ell n \dfrac{b}{a}} \qquad \text{(9-81a)}$$

(*b*) The distributed conductance parameter is given by (9-63)

$$\boldsymbol{g} = \frac{\sigma}{\epsilon}\boldsymbol{c} = \frac{2\pi\sigma}{\ell n \dfrac{b}{a}} \qquad \text{(9-81b)}$$

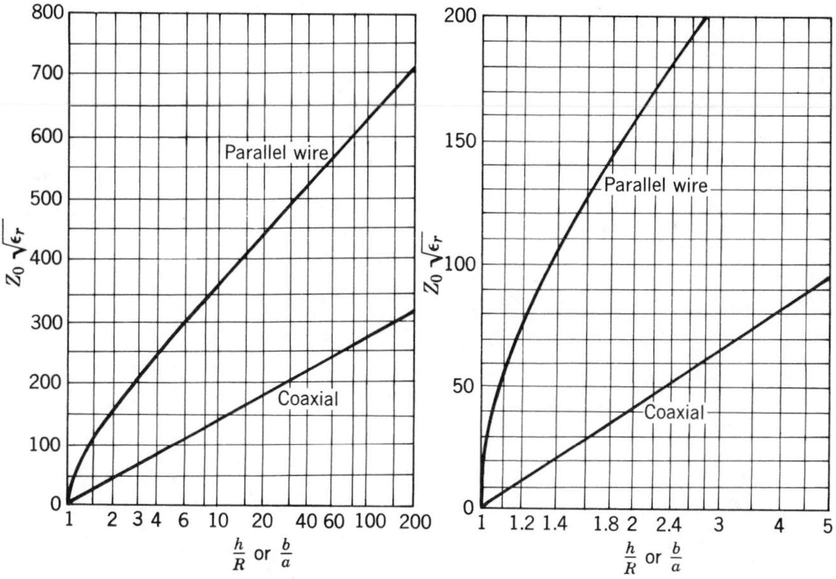

Figure 9-4. Characteristic impedance of lossless coaxial and parallel-wire lines.

The *shunt distributed admittance* (9-70) is therefore

$$\hat{y} = g + j\omega c = \left(\frac{\epsilon''}{\epsilon'} + j\right)\omega c = \left(\frac{\epsilon''}{\epsilon'} + j\right)\omega \frac{2\pi\epsilon}{\ln \dfrac{b}{a}} \tag{9-82}$$

(*c*) The defining relation for the distributed external inductance parameter l_e is (9-57c)

$$l_e \,\Delta z = \frac{\Delta\psi_m}{I} \tag{9-83}$$

in which $\Delta\psi_m$ is the flux intercepted by a thin rectangle as in Figure 9-3(*a*). Thus $\Delta\psi_m$ intercepted by ℓ is $\int_S \mathbf{B}\cdot d\mathbf{s}$ in which $\mathbf{B} = \mu\mathbf{H}$, with (9-44) providing the solutions $\mathscr{H}^{\pm}$ for the coaxial line. Thus (9-83) in complex form becomes, as $\Delta z \to dz$

$$l_e \,dz = \frac{\int_S \mu\mathscr{H}^{\pm}\cdot d\mathbf{s}}{\hat{I}_m^{\pm}} = \frac{\mu \displaystyle\int_S \left(\pm\mathbf{a}_\phi \frac{\hat{V}_m^{\pm}}{\hat{\eta}\rho \ln \dfrac{b}{a}}\right)\cdot \mathbf{a}_\phi \, d\rho \, dz}{\hat{I}_m^{\pm}} = \pm\frac{\mu \, dz \hat{V}_m^{\pm}}{\hat{\eta}\hat{I}_m^{\pm}}$$

but $\pm \hat{V}_m^{\pm}/I_m^{\pm} = \hat{Z}_0$ from (9-50), so canceling dz obtains

$$\boldsymbol{l}_e = \mu \frac{\hat{Z}_0}{\hat{\eta}} = \frac{\mu}{2\pi} \ln \frac{b}{a} \tag{9-84}$$

a result also directly obtainable by use of (9-76). Then from (9-68)

$$\boldsymbol{z} = j\omega \boldsymbol{l}_e = j\omega \frac{\mu}{2\pi} \ln \frac{b}{a} \tag{9-85}$$

*9-5 Circuit Model of a Line with Perfect Conductors

The transmission-line equations (9-53) and (9-54) can be used to establish a circuit model of a two-conductor line. Such a model employing lumped circuit elements L, C, and R exhibits the same voltage and current characteristics as the line being modelled. It assumes the configuration as in Figure 9-5, depicting the model of some length increment Δz of the line to consist of a series inductive element $L = \boldsymbol{l}_e \, \Delta z$ and shunt capacitance and conductance elements $C = \boldsymbol{c} \, \Delta z$ and $G = \boldsymbol{g} \, \Delta z$, in which $\boldsymbol{l}_e$, $\boldsymbol{c}$, and $\boldsymbol{g}$ are the line distributed parameters. As many of such sections as are needed to model a line length d are cascaded as shown. This procedure permits scaling a transmission line of any length into a circuit model suitable for tests within the confines of a laboratory. Transmission-line network analyzers, useful in the performance

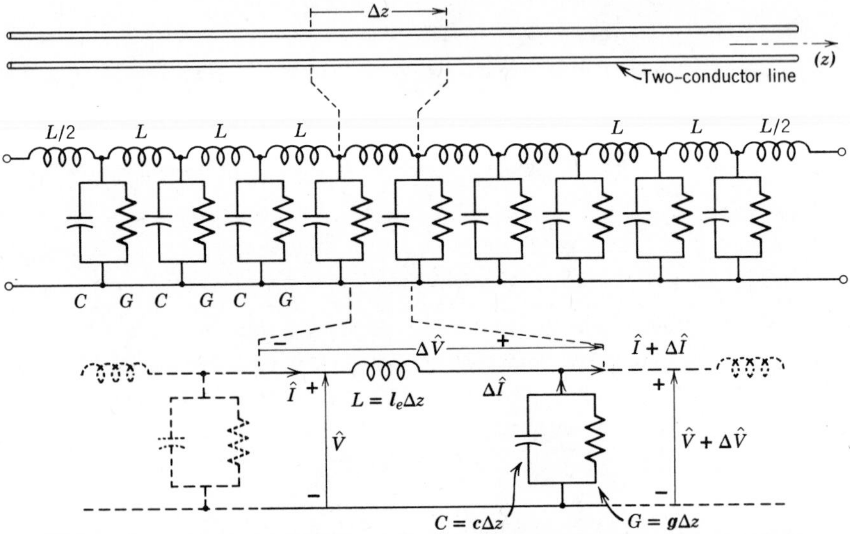

Figure 9-5. Circuit model of a line with perfect condúctors and dielectric losses.

prediction of power line and telephone systems, can be built according to such a modelling technique. Another application is to pulse-forming circuits, utilizing the wave delay and reflection properties of a transmission line.

To show that the model of Figure 9-5 essentially obeys the transmission-line equations (9-53) and (9-54), note that I flowing through the series inductive element $L = \mathbf{l}_e \Delta z$ produces an incremental voltage drop given by $\Delta V = -\partial/\partial t[\mathbf{l}_e \Delta z I]$, in which the minus sign accounts for the polarity relative to the assumed positive sense of I. This is written

$$\frac{\Delta V}{\Delta z} = -\mathbf{l}_e \frac{\partial I}{\partial t} \tag{9-86}$$

if $\mathbf{l}_e$ is not time-dependent. The finite difference form of (9-53) is (9-86), one of the desired results.

Similarly, the current increment diverted through the shunt elements $C = \mathbf{c}\,\Delta z$ and $G = \mathbf{g}\,\Delta z$ in Figure 9-5 is $\Delta I = -\partial/\partial t[(\mathbf{c}\,\Delta z)(V + \Delta V)] - (\mathbf{g}\,\Delta z)(V + \Delta V)$. Assuming $\mathbf{c}$ a constant and neglecting the ΔV terms yields

$$\frac{\Delta I}{\Delta z} = -\mathbf{c}\frac{\partial V}{\partial t} - \mathbf{g}V \tag{9-87}$$

the finite-difference counterpart of (9-54).

The accuracy with which the voltage and current characteristics of a transmission line can be modelled depends in part on the length increment Δz chosen; for some purposes it is desirable to pick Δz such that twenty or more circuit sections per wavelength are employed to simulate the actual line at the highest frequency of operation. Problems affecting the modelling accuracy are the possibility of mutual couplings among the series inductors, and the resistance and distributed capacitance effects in the physical inductors.

EXAMPLE 9-6. A 50-Ω coaxial line with a polyethylene dielectric ($\epsilon_r = 2.26$) is used at a maximum frequency of 100 kHz. Neglecting dielectric losses and assuming perfect conductors, design a circuit model to simulate 1000 m of line, assuming 20 sections per wavelength is adequate at 100 kHz.

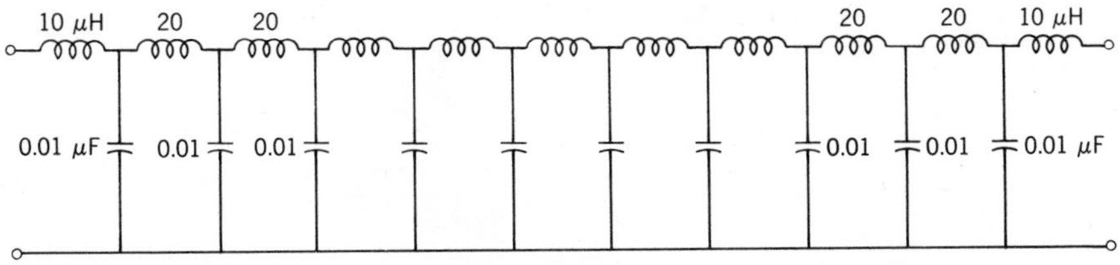

Example 9-6

At 100 kHz, the wavelength in the cable is given by (9-38), yielding $\lambda = 2\pi/\beta = c/f\sqrt{\epsilon_r} = 2000$ m. The line is 1000 m long, so 10 sections, each equivalent to $\Delta z = 100$ m of line are called for (or nine sections puts a half-section on each end of the line). The parameter $\boldsymbol{l}_e$ is found using (9-77c)

$$\boldsymbol{l}_e = \mu \frac{\hat{Z}_0}{\hat{\eta}} = 4\pi \times 10^{-7} \frac{50}{80\pi} = 0.2\ \mu\text{H/m}$$

and if each section is to model $\Delta z = 100$ m of line, one obtains $L = (0.2)100 = 20\ \mu$H. The parameter $\boldsymbol{c}$ is found by inserting $\boldsymbol{l}_e$ into (9-76) or (9-77c)

$$\boldsymbol{c} = \frac{\mu\epsilon}{\boldsymbol{l}_e} = \epsilon \frac{\hat{\eta}}{\hat{Z}_0} = \frac{(8.85 \times 10^{-12})(2.26)80\pi}{50} = 100\ \text{pF/m}$$

obtaining $C = (100)100 = 0.01\ \mu$F. The model of this line is shown in the accompanying figure.

*9-6 Wave Equations for a Line with Perfect Conductors

The voltage and current on a transmission line obey wave equations analogous to those derived in Chapter 8 for the fields **E** and **H** of TM and TE modes. In the present case of TEM modes, manipulating the transmission-line equations (9-53) and (9-54) leads to wave equations in terms of V or I. Thus, if (9-53) is differentiated with respect to z and (9-54) substituted into the result, a wave equation in terms of V is obtained, while reversing the procedure yields a wave equation in terms of I. These become

$$\frac{\partial^2 V}{\partial z^2} - \boldsymbol{l}_e \boldsymbol{c} \frac{\partial^2 V}{\partial t^2} - \boldsymbol{l}_e \boldsymbol{g} \frac{\partial V}{\partial t} = 0 \quad \text{(9-88a)}$$

$$\frac{\partial^2 I}{\partial z^2} - \boldsymbol{l}_e \boldsymbol{c} \frac{\partial^2 I}{\partial t^2} - \boldsymbol{l}_e \boldsymbol{g} \frac{\partial I}{\partial t} = 0 \quad \text{(9-88b)}$$

The third term of each equation is attributable to dielectric losses, for setting $\boldsymbol{g} = 0$ reduces them to

$$\frac{\partial^2 V}{\partial z^2} - \boldsymbol{l}_e \boldsymbol{c} \frac{\partial^2 V}{\partial t^2} = 0 \quad \text{(9-89a)}$$

Wave equations for lossless line

$$\frac{\partial^2 I}{\partial z^2} - \boldsymbol{l}_e \boldsymbol{c} \frac{\partial^2 I}{\partial t^2} = 0 \quad \text{(9-89b)}$$

The complex, time-harmonic forms of the wave equations (9-88) are obtained by the usual substitution of the exponential functions (9-65), yielding

$$\frac{d^2\hat{V}}{dz^2} + (\omega^2 l_e c - j\omega l_e g)\hat{V} = 0 \tag{9-90a}$$

$$\frac{d^2\hat{I}}{dz^2} + (\omega^2 l_e c - j\omega l_e g)\hat{I} = 0 \tag{9-90b}$$

also written

$$\frac{d^2\hat{V}}{dz^2} - \hat{z}\hat{y}\hat{V} = 0 \tag{9-90c}$$

$$\frac{d^2\hat{I}}{dz^2} - \hat{z}\hat{y}\hat{I} = 0 \tag{9-90d}$$

if $\hat{z} = j\omega l_e$ and $\hat{y} = g + j\omega c$ according to (9-68) and (9-69). Solving (9-90c) and (9-90d) produces the wave solutions

$$\hat{V}(z) = \hat{V}_m^+ e^{-\gamma z} + \hat{V}_m^- e^{\gamma z} \tag{9-91a}$$

$$\hat{I}(z) = \hat{I}_m^+ e^{-\gamma z} + \hat{I}_m^- e^{\gamma z} \tag{9-91b}$$

with the propagation constant γ given by

$$\gamma \equiv \alpha + j\beta = \sqrt{\hat{z}\hat{y}} = \sqrt{j\omega l_e(g + j\omega c)} = \sqrt{-\omega^2 l_e c + j\omega l_e g} \tag{9-92}$$

These results are consistent, as expected, with the solutions (9-10a) and (9-20a) for $\hat{V}(z)$ and $\hat{I}(z)$, and with (9-75) for γ, all obtained previously via a rather different route.

*9-7 Transmission-Line Parameters, Conductor-Impedance Included

In previous sections, the transmission line carrying the TEM mode was discussed, assuming possible losses within the dielectric but with the idealization of perfect conductors ($\sigma_c \to \infty$). In the real world, conductors with a high

* The reader wishing to omit the development of internal resistance and inductance parameters in this section, to conserve time, may just refer to the results (9-103) and (9-105) for γ and $\hat{Z}_0$ and proceed to the next chapter.

(though finite) conductivity σ_c are used in the fabrication of lines, introducing two new problems into the study of this mode:

1. The distortion of the electric field in the dielectric from a true *transverse* (TEM) condition, brought about by the presence of an E_z component at the conductor walls.
2. The current penetration into the conductor interiors, given the name skin effect.

The longitudinal current $I(z, t)$ in the conductors gives rise to a z directed electric field component along the conductor wall whenever its conductivity σ_c is finite, in view of the relation (3-7), $\mathbf{J} = \sigma_c \mathbf{E}$. The presence of such a longitudinal component at the dielectric conductor interface would appear to contradict the assumption (9-1) defining the TEM mode for the line. With a sufficiently high conductivity of the conductors, however, an almost negligible E_z component is developed at the walls, compared with the transverse component present there. This condition is shown in Figure 9-6(*a*), using a

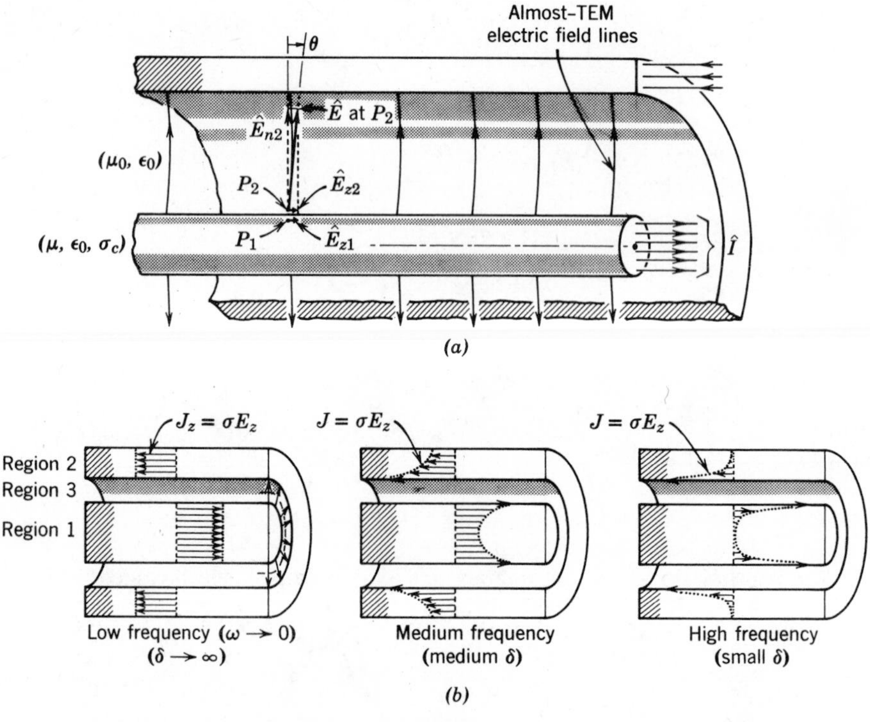

Figure 9-6. Effects of finite conductivity in a coaxial line. (*a*) Influence of the E_z field on the distortion of the TEM mode (shown exaggerated). (*b*) Illustrating current density variations at various frequencies.

coaxial line for illustrative purposes. A slight curvature of the electric flux between the conductors is produced by E_z.

The *current penetration* into the conductors of a line complicates the derivation of suitable transmission-line equations resembling (9-67a) and (9-67b). Infinitely conductive lines possess only surface currents on the conductor walls, but with a finite conductivity, current conditions like those of Figure 9-6(*b*) are obtained, with a relatively small penetration occurring at high frequencies, while a uniform current density prevails over the cross-section under dc conditions. An exact solution entails satisfying the boundary conditions for the fields, i.e., matching the normal and tangential components of the fields at the interfaces separating the dielectric region from the two conductors. Some of these solutions, derived for particular line geometries, are discussed in the following.

In deriving the transmission-line equation (9-59) or its time-harmonic form (9-67a), $d\hat{V}/dz = -\hat{z}I$, it was seen that the Faraday law (3-78) yielded only an external inductance contribution, $\hat{z} = j\omega l_e$. No resistive term was obtained because of the assumption of perfect conductors, allowing no tangential $\mathbf{E}$ component along the short sides of the rectangle. For physical conductors with a finite conductivity, the time-varying tangential $\hat{\mathbf{H}}$-field at the conductor wall, required by (3-71) to be continuous into the interior, generates an associated $\hat{\mathbf{E}}$ component propagating along with $\hat{\mathbf{H}}$ *into* the conductors as suggested by Figure 9-6(*b*). A current density field $\mathbf{J}$ accompanies this electromagnetic wave, in view of (3-7). An illustration of this process was found in the propagation of wall-loss currents into the metal boundaries of a hollow waveguide, as in Figure 8-18. In the following development it is shown that the form of transmission-line equations (9-67) are also applicable to a line whose conductors have finite conductivity, the series distributed parameter $\hat{z}$ acquiring two *internal impedance* contributions $\hat{z}_{i1}$ and $\hat{z}_{i2}$, attributable to the tangential components $\hat{E}_{z1}$ and $\hat{E}_{z2}$ developed along the conductor walls. To show this, Faraday's law (3-78), $\oint_\ell \mathbf{E}\cdot d\boldsymbol{\ell} \triangleq -d\psi_m/dt$, is applied again to a thin rectangle ℓ constructed as in Figure 9-3(*a*) in a cross-section of the line. The finite conductivity produces additional voltage terms $\Delta\hat{V}_1$ and $\Delta\hat{V}_2$ at the z segments as depicted in Figures 9-7(*a*) and (*b*). Along the edge 4–1, utilizing time-harmonic quantities,

$$\Delta\hat{V}_1 = -\hat{E}_z\,\Delta z \tag{9-93}$$

is obtained from the integral of $-\hat{\mathbf{E}}\cdot d\boldsymbol{\ell}$ along Δz. Continuous into the conductor surface is $\hat{E}_z$, where it is proportional to the current density $\hat{J}_z$, through (3-7), while $\hat{J}_z$ is in turn proportional to the current $\hat{I}$ carried in the conductor, related to the $\mathbf{H}$ field by (9-5a)

$$\hat{I} = \oint \hat{\mathbf{H}}\cdot d\boldsymbol{\ell} \tag{9-94}$$

One can therefore write (9-93) as

$$\Delta\hat{V}_1 = -\hat{E}_z\,\Delta z = -\hat{z}_{i1}\hat{I}\,\Delta z \tag{9-95}$$

in which the proportionality constant $\hat{z}_{i1}$ is called the *internal distributed impedance* (a surface impedance) parameter, in view of its units (ohm per meter). A similar argument concerns the voltage along the edge 2–3, so define internal impedance parameters for each conductor by

$$\hat{z}_{i1} = \frac{-\Delta\hat{V}_1/\Delta z]_{4-1}}{\hat{I}} = \frac{\hat{E}_z]_{4-1}}{\hat{I}} \qquad \hat{z}_{i2} = \frac{-\Delta\hat{V}_2/\Delta z]_{2-3}}{\hat{I}} = \frac{\hat{E}_z]_{2-3}}{\hat{I}}\ \Omega/\text{m} \tag{9-96}$$

The integral (3-78) taken about the four sides of ℓ in Figure 9-7(*a*) or (*b*) thus obtains

$$-\hat{V} + \left(\hat{V} + \frac{\partial\hat{V}}{\partial z}\Delta z\right) + (\hat{z}_{i1}\,\Delta z)\hat{I} + (\hat{z}_{i2}\,\Delta z)\hat{I}$$

The right side of (3-78), moreover, in time-harmonic form, becomes $-j\omega(\boldsymbol{l}_e\,\Delta z)\hat{I}$, based on the external magnetic flux $\Delta\psi_m = (\boldsymbol{l}_e\,\Delta z)\hat{I}$ linking ℓ as already considered for Figure 9-3(*a*).

Thus (3-78) becomes

$$-\hat{V} + \left(\hat{V} + \frac{\partial\hat{V}}{\partial z}\Delta z\right) + (\hat{z}_{i1}\,\Delta z)\hat{I} + (\hat{z}_{i2}\,\Delta z)\hat{I} = -j\omega(\boldsymbol{l}_e\,\Delta z)\hat{I}$$

which, after canceling Δz, yields the differential equation

$$\frac{d\hat{V}}{dz} = -(\hat{z}_{i1} + \hat{z}_{i2} + j\omega\boldsymbol{l}_e)\hat{I} = -\hat{z}\hat{I} \tag{9-97}$$

The *total* series distributed impedance $\hat{z}$ is therefore

$$\hat{z} = \hat{z}_{i1} + \hat{z}_{i2} + j\omega\boldsymbol{l}_e \tag{9-98a}$$

the sum of the internal parameters (9-96) plus the reactance of the external inductance parameter $\boldsymbol{l}_e$ defined by (9-57b). In the discussion to follow it develops that $\hat{z}_{i1}$ and $\hat{z}_{i2}$ consist of resistive and inductive parts, to permit writing (9-98a)

$$\begin{aligned}\hat{z} &= \hat{z}_{i1} + \hat{z}_{i2} + j\omega\boldsymbol{l}_e \\ &= (\boldsymbol{r}_{i1} + j\omega\boldsymbol{l}_{i1}) + (\boldsymbol{r}_{i2} + j\omega\boldsymbol{l}_{i2}) + j\omega\boldsymbol{l}_e \\ &= \boldsymbol{r} + j\omega\boldsymbol{l}\ \Omega/\text{m}\end{aligned} \tag{9-98b}$$

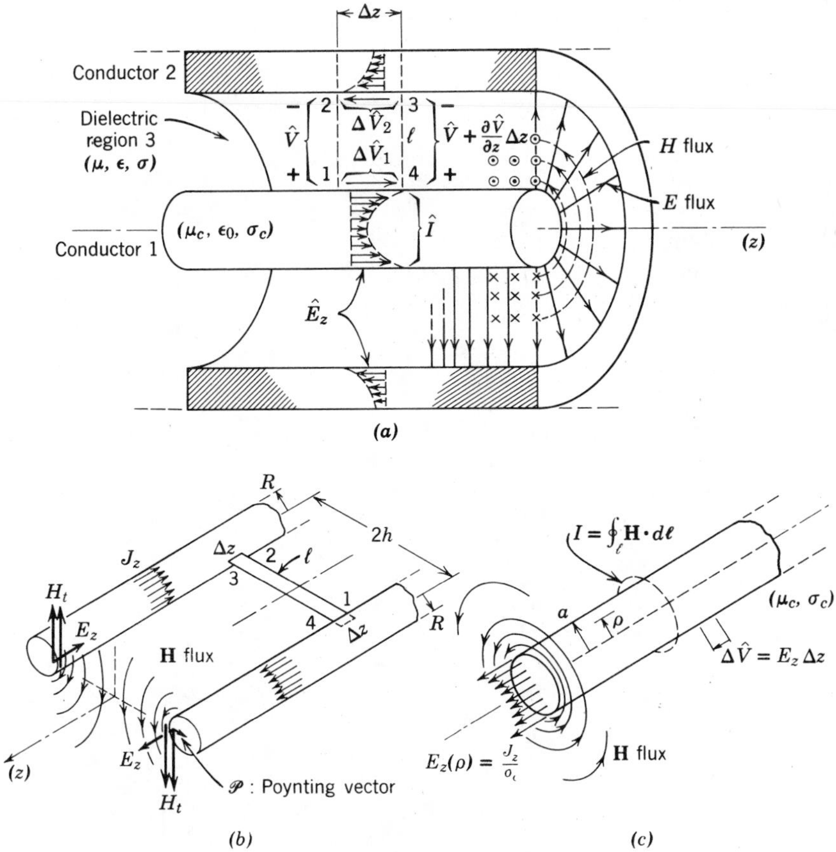

Figure 9-7. Relative to internal conductor-impedance effects in lines. (*a*) The coaxial line, showing closed line ℓ (1234) and voltages along its edges. (*b*) Parallel-wire line, with rectangle ℓ. (*c*) An isolated conductor, with axial symmetry.

in which $\boldsymbol{r}$ and $\boldsymbol{l}$ denote the *total* series distributed resistance and inductance parameters of the line.

The other transmission-line equation, obtainable from the continuity relation (4-19), maintains the same forms (9-64) or (9-67b) as for the line with perfect conductors

$$\frac{d\hat{I}}{dz} = -\hat{y}\hat{V} \tag{9-99}$$

in which the shunt distributed parameter $\hat{y}$ denotes

$$\hat{y} = g + j\omega c = \left(\frac{\epsilon''}{\epsilon'} + j\right)\omega c \text{ ℧/m} \tag{9-100}$$

with g and c defined as usual by (9-60b) and (9-63), or by (9-70).

The simultaneous manipulation of (9-97) and (9-99) further leads to wave equations comparable to (9-90) obtained for the perfect conductor case, becoming

$$\frac{d^2\hat{V}}{dz^2} - \hat{z}\hat{y}\hat{V} = 0 \tag{9-101a}$$

$$\frac{d^2\hat{I}}{dz^2} - \hat{z}\hat{y}\hat{I} = 0 \tag{9-101b}$$

Their solutions are evidently

$$\hat{V}(z) = \hat{V}_m^+ e^{-\gamma z} + \hat{V}_m^- e^{\gamma z} \tag{9-102a}$$

$$\hat{I}(z) = \hat{I}_m^+ e^{-\gamma z} + \hat{I}_m^- e^{\gamma z} \tag{9-102b}$$

if γ denotes the complex propagation constant

$$\gamma \equiv \alpha + j\beta = \sqrt{\hat{z}\hat{y}} = \sqrt{(r + j\omega l)(g + j\omega c)} \text{ m}^{-1} \tag{9-103a}$$

and $\hat{V}_m^{\pm}$ and $\hat{I}_m^{\pm}$ are the usual complex amplitudes of the forward and backward waves.

If the line parameters r and g are *small* compared with $j\omega l$ and $j\omega c$ respectively, a useful simplification of (9-103a) can be shown to yield $\gamma = (1/2)(r\sqrt{c/l} + g\sqrt{l/c}) + j\omega\sqrt{lc}$, obtaining the attenuation and phase constants

$$\alpha \cong \frac{r}{2}\sqrt{\frac{c}{l}} + \frac{g}{2}\sqrt{\frac{l}{c}} \text{ Np/m} \quad \text{Low-loss line} \tag{9-103b}$$

$$\beta \cong \omega\sqrt{lc} \text{ rad/m} \qquad r \ll \omega l \quad g \ll \omega c \tag{9-103c}$$

The form of (9-103b) is seen to separate the attenuative effects into contributions due to the series and shunt loss-parameters r and g.

That (9-102b) can be written in terms of $\hat{V}_m^+$ and $\hat{V}_m^-$, according to

$$\hat{I}(z) = \frac{\hat{V}_m^+}{\hat{Z}_0} e^{-\gamma z} - \frac{\hat{V}_m^-}{\hat{Z}_0} e^{\gamma z} \tag{9-104}$$

is demonstrated by inserting the solution (9-102a) back into (9-97), yielding

$$\hat{I}(z) = -\frac{1}{\hat{z}} \frac{\partial \hat{V}}{\partial z} = -\frac{1}{\hat{z}} [-\gamma \hat{V}_m^+ e^{-\gamma z} + \gamma \hat{V}_m^- e^{\gamma z}]$$

This becomes (9-104) upon inserting (9-103a), provided the characteristic impedance $\hat{Z}_0$ is written

$$\hat{Z}_0 \equiv \pm \frac{\hat{V}_m^\pm}{\hat{I}_m^\pm} = \sqrt{\frac{\hat{z}}{\hat{y}}} \tag{9-105a}$$

$$= \sqrt{\frac{r + j\omega l}{g + j\omega c}} \ \Omega \tag{9-105b}$$

The circuit model illustrated in Figure 9-8 can be shown to be valid for the transmission line having both dielectric and conductor losses. Applying the

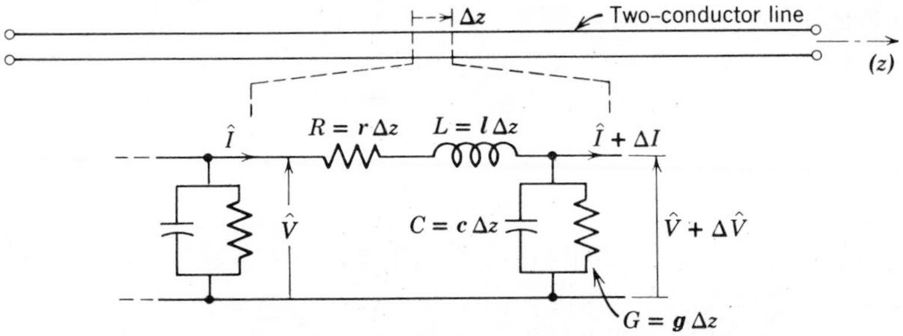

Figure 9-8. Circuit model of a line with dielectric (shunt) and conductor (series) losses.

method of Section 9-5, finite difference versions of the transmission-line equations (9-97) and (9-99) hold for the model. A comparison with Figure 9-4 reveals that the effect of conductor losses is to insert the series resistance element $r\,\Delta z$ into each section of the model, while the series inductance element $l\,\Delta z$ must include internal inductance contributions.

In the following, two examples of transmission lines are analyzed for their line parameters, (*a*) the parallel-wire line, assuming a separation sufficient to neglect proximity effects, and (*b*) the coaxial line, examined only for its high-frequency and dc behavior. The analysis of the internal distributed impedance of an *isolated wire* is useful for simplifying both problems, so this is taken up first.

A. Current Penetration in Round Wire (Skin Effect); Internal Distributed Parameters

The isolated round conductor shown in Figure 9-7(*c*) is to be regarded as an element of a two-conductor transmission line, so to the associated time-harmonic electric and magnetic fields are attributed the usual factors $e^{j\omega t \mp \gamma z}$. Only positive z traveling waves need be considered, thus requiring only $e^{j\omega t - \gamma z}$. The internal impedance contribution $\hat{z}_i$ defined by (9-96) is desired for this conductor. Considered as one of a pair of conductors carrying the TEM mode, its magnetic field is assumed the axially symmetric component $\hat{\mathscr{H}}_\phi e^{j\omega t - \gamma z}$. The continuity of $\hat{\mathscr{H}}_\phi$ into the conductor interior generates an electric field component therein, related to its axial current density $\hat{J}_z$ by (3-7)

$$\hat{J}_z = \sigma_c \hat{\mathscr{E}}_z \tag{9-106}$$

The internal impedance $\hat{z}_i$, defined by (9-96), denotes the voltage drop per unit length $\Delta\hat{V}/\Delta z = \hat{\mathscr{E}}_z$ divided by the total current $\hat{I}$ in a given cross-section as in Figure 9-7(*c*); i.e.,

$$\hat{z}_i = \frac{-\dfrac{\Delta \hat{V}}{\Delta z}}{\hat{I}} = \frac{\hat{\mathscr{E}}_z(z)}{\oint_{(\text{c.s.})} \hat{\mathscr{H}} \cdot d\ell} = \frac{\hat{\mathscr{E}}_z(a)}{\oint_{(\text{c.s.})} \hat{\mathscr{H}}_\phi a \, d\phi} \tag{9-107}$$

The evaluation of $\hat{z}_i$ is facilitated if $\hat{\mathscr{H}}_\phi$ is expressed in terms of $\hat{\mathscr{E}}_z$. The desired relationship is obtained from the modified-curl expression (8-6), reducing to the following from axial symmetry and only $\hat{\mathscr{E}}_z$ and $\hat{\mathscr{H}}_\phi$ present

$$\hat{\mathscr{H}} = \mathbf{a}_\phi \hat{\mathscr{H}}_\phi = -\frac{1}{j\omega\mu_c} \begin{vmatrix} \dfrac{\mathbf{a}_\rho}{\rho} & \mathbf{a}_\phi & \dfrac{\mathbf{a}_z}{\rho} \\ \dfrac{\partial}{\partial \rho} & 0 & -\gamma \\ 0 & 0 & \hat{\mathscr{E}}_z \end{vmatrix} = \mathbf{a}_\phi \frac{1}{j\omega\mu_c} \frac{\partial \hat{\mathscr{E}}_z}{\partial \rho} \tag{9-108}$$

In the latter, $\hat{\mathscr{E}}_z$ satisfies a wave equation of the form of (2-96) which, for the conductor with constants $(\mu_c, \epsilon_c, \sigma_c)$ and in time-harmonic form, becomes

$\nabla^2\hat{\mathbf{E}} + \omega^2\mu_c\epsilon_c\hat{\mathbf{E}} = j\omega\mu_c\sigma_c\hat{\mathbf{E}}$. Combining terms in $\hat{\mathbf{E}}$ yields

$$\nabla^2\hat{\mathbf{E}} + \omega^2\mu_c\epsilon_c(1 - j\sigma_c/\omega\epsilon_c)\hat{\mathbf{E}} = 0$$

but in a good conductor, $\sigma_c/\omega\epsilon_c \gg 1$, reducing it to $\nabla^2\hat{\mathbf{E}} - j\omega\mu_c\sigma_c\hat{\mathbf{E}} = 0$. With only $\hat{E}_z$ present, one obtains the scalar wave equation $\nabla^2\hat{E}_z - j\omega\mu_c\sigma_c\hat{E}_z = 0$. One expresses $\nabla^2\hat{E}_z$ in circular cylindrical coordinates by (2-80), symmetry requiring $\partial/\partial\phi = 0$, and with dependence on z given by $e^{-\gamma z}$, yields

$$\frac{1}{\rho}\frac{\partial}{\partial\rho}\left(\rho\frac{\partial\hat{\mathscr{E}}_z}{\partial\rho}\right) + (\gamma^2 - j\omega\mu_c\sigma_c)\hat{\mathscr{E}}_z = 0 \tag{9-109}$$

With the z propagation constant in the *dielectric* given by (9-31) (essentially $j\omega\sqrt{\mu\hat{\epsilon}}$), one has $|\gamma^2| \ll \omega\mu_c\sigma_c$, so discarding the γ^2 term in (9-109) yields

$$\frac{d^2\hat{\mathscr{E}}_z}{d\rho^2} + \frac{1}{\rho}\frac{d\hat{\mathscr{E}}_z}{d\rho} - j\omega\mu_c\sigma_c\hat{\mathscr{E}}_z = 0 \tag{9-110}$$

With $\hat{J}_z/\sigma_c$ for $\hat{\mathscr{E}}_z$, multiplying by ρ^2 obtains

$$\rho^2\frac{d^2\hat{J}_z}{d\rho^2} + \rho\frac{d\hat{J}_z}{d\rho} - j\omega\mu_c\sigma_c\rho^2\hat{J}_z = 0 \tag{9-111}$$

a result known as the Bessel differential equation of zero order. Its solution is obtainable by assuming a power series solution,† leading to the current density

$$\hat{J}_z(\rho) = \hat{C}_1 J_0(\hat{k}\rho) + \hat{C}_2 N_0(\hat{k}\rho) \tag{9-112}$$

in which the factor $-j\omega\mu_c\sigma_c$ in (9-111), denoted by $\hat{k}^2$, implies

$$\hat{k} = \sqrt{-j\omega\mu_c\sigma_c} = j^{-1/2}\sqrt{\omega\mu_c\sigma_c} = j^{-1/2}\frac{\sqrt{2}}{\delta} \tag{9-113}$$

The symbol δ used in the latter evidently means

$$\delta = \sqrt{\frac{2}{\omega\mu_c\sigma_c}}\ \text{m} \quad \text{Uniform plane wave} \tag{9-114}$$

† For details of the power series method and Bessel functions, see Wylie, C. R. *Advanced Engineering Mathematics*, 2nd ed. New York: McGraw-Hill, 1960, Chapter 10; or Ramo, S., J. Whinnery, and T. van Duzer. *Fields and Waves in Communication Electronics*. New York: Wiley, 1965, pp. 203–215.

which from (3-95) and (3-112a) specifies the skin depth of penetration of a *plane wave* into a conductor with the same parameters μ_c, σ_c and is used for comparison purposes in what follows. The symbols J_0 and N_0 appearing in (9-112) denote zero order Bessel functions of *first kind* and *second kind*, respectively. One represents $J_0(u)$ by the following power series:

$$J_0(u) = \sum_{m=0}^{\infty} \frac{(-1)^m (u/2)^{2m}}{(m)!^2} = 1 - \frac{u^2}{2^2} + \frac{u^4}{(2!)^2 2^4} - \frac{u^6}{(3!)^2 2^6} + \cdots \tag{9-115}$$

If $u = \hat{k}\rho = j^{\;1/2}\sqrt{2}\rho/\delta$, and with the latter into (9-115) and grouping the real and imaginary terms, one obtains

$$J_0\left(j^{-1/2}\frac{\sqrt{2}}{\delta}\rho\right) = 1 - \frac{\left(\sqrt{2}\frac{\rho}{\delta}\right)^4}{(2!)^2 2^4} + \frac{\left(\sqrt{2}\frac{\rho}{\delta}\right)^8}{(4!)^2 2^8} - \cdots$$
$$+ j\left[\frac{\left(\sqrt{2}\frac{\rho}{\delta}\right)^2}{(1!)^2 2^2} - \frac{\left(\sqrt{2}\frac{\rho}{\delta}\right)^6}{(3!)^2 2^6} + \frac{\left(\sqrt{2}\frac{\rho}{\delta}\right)^{10}}{(5!)^2 2^{10}} - \cdots\right] \tag{9-116a}$$

$$= \text{ber}\left(\sqrt{2}\frac{\rho}{\delta}\right) + j\,\text{bei}\left(\sqrt{2}\frac{\rho}{\delta}\right) \tag{9-116b}$$

The symbols ber and bei (Bessel-real and Bessel-imaginary) thus denote the real and imaginary parts of J_0, and are tabulated in numerous references.†

The Bessel function $N_0(u)$ in (9-112), the second solution of (9-111), is discarded here because of a singularity‡ at $\rho = 0$ ($N_0(0) \to -\infty$); i.e., the wire center $\rho = 0$ is within the region of discussion of the current density $\hat{J}_z$, so $N_0(u)$ is of no physical use. Therefore (9-112) can be written with $\hat{C}_2 = 0$ and (9-116b) inserted, obtaining

$$\hat{J}_z(\rho) = \hat{C}_1 J_1\left(j^{-1/2}\sqrt{2}\frac{\rho}{\delta}\right) = \hat{C}_1\left[\text{ber}\left(\sqrt{2}\frac{\rho}{\delta}\right) + j\,\text{bei}\left(\sqrt{2}\frac{\rho}{\delta}\right)\right] \tag{9-117}$$

Putting $\rho = a$ into (9-117) expresses $\hat{C}_1$ in terms of the density $\hat{J}_z(a)$ at the wire surface, obtaining $\hat{C}_1 = \hat{J}_z(a)/[\text{ber}(\sqrt{2}\,a/\delta) + j\,\text{bei}(\sqrt{2}\,a/\delta)]$. The latter into (9-117) yields the current density distribution in the wire

$$\hat{J}_z(\rho) = \hat{J}_z(a)\frac{\text{ber}\left(\sqrt{2}\frac{\rho}{\delta}\right) + j\,\text{bei}\left(\sqrt{2}\frac{\rho}{\delta}\right)}{\text{ber}\left(\sqrt{2}\frac{a}{\delta}\right) + j\,\text{bei}\left(\sqrt{2}\frac{a}{\delta}\right)} \text{ A/m}^2 \tag{9-118}$$

† For example, see Dwight, H. B. *Tables of Integrals and Other Mathematical Data*, revised ed. New York: Macmillan, 1961.

‡ See Ramo, S., et al., op. cit., p. 208, for a graph.

Using tabulations of ber and bei, or resorting to the series definitions (9-116), one can plot (9-118) as the skin effect curves shown as solid lines in Figure 9-9(*a*). The curves are universalized by expressing the argument $\sqrt{2}\rho/\delta$ in (9-118)

$$\sqrt{2}\frac{\rho}{\delta} = \sqrt{2}\frac{a}{\delta}\cdot\frac{\rho}{a} \tag{9-119}$$

This permits using the *normalized* radius ρ/a with a range (0, 1) from the wire center to the surface, while a/δ expresses the wire radius a in terms of the plane wave comparison skin depth δ given by (9-114). Both graphs show that the axial current density is depressed noticeably below the surface value if the wire radius exceeds a plane wave skin depth or so ($a/\delta > 1$). For example, if $a/\delta = 2$, $|\hat{J}_z(0)|$ is seen to be about 62% of the value at the surface.

Figure 9-9(*a*) shows (dashed) the amplitude attenuation of plane waves in a conductor, for comparison with the round wire current density (solid curves) given by (9-118). Plane wave attenuation was described in Chapter

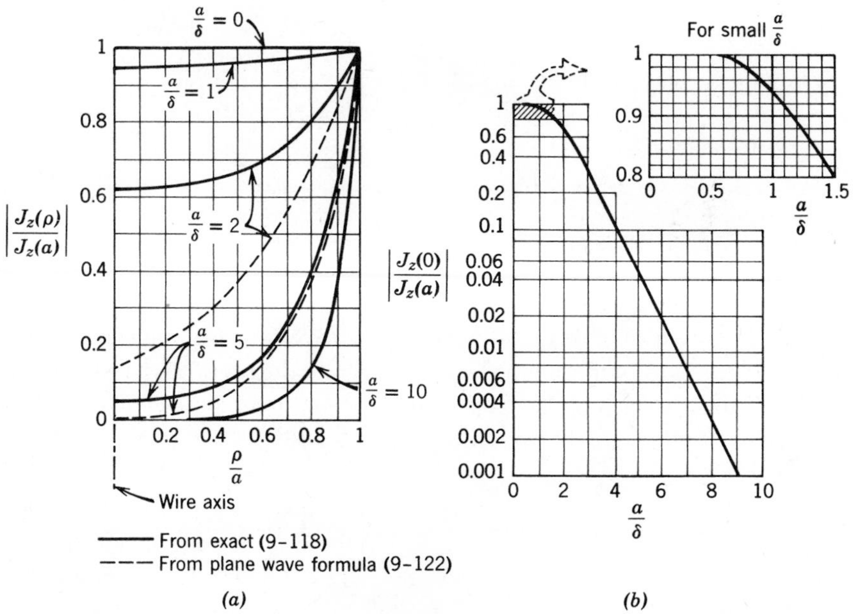

Figure 9-9. **Current density in round wires. (*a*) Current density magnitude versus normalized radial distance ρ/a. Dashed lines show plane wave approximations. (*b*) Ratio of current density magnitude at the center to that at the surface, as a function of wire radius in plane wave skin depths.**

3 with reference to Figure 3-17, implying that a z directed current density plane wave, traveling in the x direction in a good conductor, is given by

$$\hat{J}_z(x) = J_m e^{-\alpha x} e^{-j\beta x} = J_m e^{-x/\delta} e^{-jx/\delta} \tag{9-120}$$

in which J_m denotes a reference amplitude at $x = 0$, and attenuation and phase factors $\alpha = \beta = \delta^{-1}$ are given by (3-112a) and (3-112b). That $\hat{J}_z(\rho)$ of (9-118) in a round wire reduces essentially to (9-120) is shown using the asymptotic limits of the Bessel functions. Assuming the argument sufficient large ($u > 10$ or so), the following asymptotic form holds:†

$$J_0(u) \to \sqrt{\frac{2}{\pi u}} \cos\left(u - \frac{\pi}{4}\right) \tag{9-121a}$$

becoming exact as $u \to \infty$. With the complex argument $u = j^{-1/2}\sqrt{2}\rho/\delta$, (9-121a) becomes

$$J_0\left(j^{-1/2}\sqrt{2}\frac{\rho}{\delta}\right) \to \frac{e^{\rho/\delta}}{\sqrt{\dfrac{2\pi\sqrt{2}\rho}{\delta}}} e^{j(\rho/\delta - \pi/8)} \tag{9-121b}$$

for large ρ/δ. Using (9-121b) in the numerator of (9-118) and the same with $\rho = a$ in the denominator yields the following approximate current density in a round conductor:

$$\hat{J}_z(\rho) \to \hat{J}_z(a)\sqrt{\frac{a}{\rho}}\, e^{-(a-\rho)/\delta} e^{-j(a-\rho)/\delta} \qquad \frac{a}{\delta} \text{ large} \tag{9-122}$$

essentially correct if the conductor radius is several plane wave skin depths ($a/\delta > 10$ or so). Comparing (9-122) with (9-120) thus verifies that if $a/\delta > 10$ or so, δ can be used to denote the *depth of penetration in a round conductor*. The solid and dashed curves in Figure 9-9(a) indicate how well the plane wave expression (9-120) approximates the exact round wire relation (9-118).

EXAMPLE 9-7. (a) At 60 Hz, what maximum radius should a copper wire have if J_z on the wire axis is not to be less than 90% of the surface value? (b) If f is raised to 6 GHz (in the microwave range), what maximum radius meets the same criterion?

† For a tabulation of asymptotic forms, see for example, Ramo, S., J. Whinnery and T. van Duzer. *Fields and Waves in Communication Electronics*. New York: Wiley, 1965, p. 212.

(*a*) From Figure 9-9(*b*), if $|J_z(0)/J_z(a)| = 0.9$ then $a/\delta = 1.16$, implying $a = 1.16$ plane wave skin depths, but at 60 Hz, δ from (9-114) becomes

$$\delta = \sqrt{\frac{2}{\omega\mu_c\sigma_c}} = \sqrt{\frac{2}{2\pi(60)4\pi \times 10^{-7}(5.8 \times 10^7)}} = 0.00853 \text{ m} = 8.53 \text{ mm}$$

so the wire radius must not exceed $a = 1.16\delta = 9.89$ mm $= 0.389$ in.

(*b*) If f were increased by 10^8 to 6×10^9 Hz, δ of (9-114) becomes 0.00853×10^{-4}, yielding $a = 1.16\delta = 0.989 \times 10^{-3}$ mm $= 0.989\ \mu$m $= 0.389$ mil, a very small wire. Barretter wires, used as resistance elements in microwave power bridges, are made this thin to yield dc and rf resistances that are *essentially the same*, permitting substitution methods (rf for dc power) to be utilized.

EXAMPLE 9-8. A copper conductor is of $a = 1$ mm radius. (*a*) At what frequency is the radius 10 plane wave skin depths? (*b*) What is δ if f is increased by 10^4?

(*a*) Putting $a = 10\delta$ (or $a/\delta = 10$) means from (9-114) that $a = 10\sqrt{2/\omega\mu_c\sigma_c}$, so solving for f yields

$$f = \frac{100(2)}{2\pi a^2\mu_c\sigma_c} = \frac{200}{2\pi(10^{-6})4\pi \times 10^{-7}(5.8 \times 10^7)} = 0.432 \times 10^6 = 432 \text{ kHz}$$

At or above this frequency, the skin depth in the wire is essentially that for a *plane wave* in this conducting material.

(*b*) Increasing f to 4.32 GHz decreases δ, from (9-114), by $\sqrt{10^{-4}} = 10^{-2}$, yielding $\delta = 10^{-3}$ mm $= 1\ \mu$m.

The impedance parameter $\hat{z}_i$ of an isolated round wire is now found by use of (9-107). Substituting (9-118) into (9-106) yields

$$\hat{\mathscr{E}}_z(\rho) = \frac{\hat{J}_z(\rho)}{\sigma_c} = \frac{\hat{J}_z(a)}{\sigma_c}\,\frac{\operatorname{ber}\left(\sqrt{2}\,\frac{\rho}{\delta}\right) + j\operatorname{bei}\left(\sqrt{2}\,\frac{\rho}{\delta}\right)}{\operatorname{ber}\left(\sqrt{2}\,\frac{a}{\delta}\right) + j\operatorname{bei}\left(\sqrt{2}\,\frac{a}{\rho}\right)} \tag{9-123}$$

Also needed is $\hat{\mathscr{H}}_\phi$ at the wire surface, obtained from (9-123) by use of (9-108); hence

$$\hat{\mathscr{H}}_\phi(a) = \frac{1}{j\omega\mu_c}\left.\frac{\partial\hat{\mathscr{E}}_z(\rho)}{\partial\rho}\right]_{\rho=a} = \frac{1}{j\omega\mu_c\sigma_c}\left.\frac{\partial\hat{J}_z(\rho)}{\partial\rho}\right]_{\rho=a} = \frac{1}{j\omega\mu_c\sigma_c}\left(\frac{\sqrt{2}}{\delta}\right)\left.\frac{\partial\hat{J}_z(\rho)}{\partial u}\right]_{\rho=a}$$

$$= \frac{\hat{J}_z(a)}{j\sqrt{\omega\mu_c\sigma_c}}\,\frac{\operatorname{ber}'\left(\sqrt{2}\,\frac{a}{\delta}\right) + j\operatorname{bei}'\left(\sqrt{2}\,\frac{a}{\delta}\right)}{\operatorname{ber}\left(\sqrt{2}\,\frac{a}{\delta}\right) + j\operatorname{bei}\left(\sqrt{2}\,\frac{a}{\delta}\right)} \tag{9-124}$$

In (9-124) a change to the variable $u = \sqrt{2}\rho/\delta$ has been made, the primes signifying differentiations with respect to u. With (9-123) and (9-124) into (9-107), the symmetry of $\mathscr{H}_\phi$ permits a simple integration around the wire to yield $\hat{I}$, obtaining

$$\hat{z}_i = \frac{\hat{\mathscr{E}}_z(a)}{\oint_{\ell(a)} \mathscr{H}_\phi(a) \cdot a\, d\phi} = \frac{j}{\sqrt{2}\pi a}\sqrt{\frac{\omega\mu_c}{2\sigma_c}}\,\frac{\text{ber}\left(\sqrt{2}\frac{a}{\delta}\right) + j\,\text{bei}\left(\sqrt{2}\frac{a}{\delta}\right)}{\text{ber}'\left(\sqrt{2}\frac{a}{\delta}\right) + j\,\text{bei}'\left(\sqrt{2}\frac{a}{\delta}\right)}\ \Omega/\text{m} \tag{9-125}$$

One can decompose $\hat{z}_i$ into $r_i + j\omega l_i$ by rationalizing the denominator, obtaining internal resistance and inductive contributions as follows:

$$r_i = \frac{1}{2\pi a}\sqrt{\frac{\omega\mu_c}{\sigma_c}}\,\frac{\text{ber}\left(\sqrt{2}\frac{a}{\delta}\right)\text{bei}'\left(\sqrt{2}\frac{a}{\delta}\right) - \text{bei}\left(\sqrt{2}\frac{a}{\delta}\right)\text{ber}'\left(\sqrt{2}\frac{a}{\delta}\right)}{\left[\text{ber}'\left(\sqrt{2}\frac{a}{\delta}\right)\right]^2 + \left[\text{bei}'\left(\sqrt{2}\frac{a}{\delta}\right)\right]^2}\ \Omega/\text{m} \tag{9-126a}$$

$$\omega l_i = \frac{1}{2\pi a}\sqrt{\frac{\omega\mu_c}{\sigma_c}}\,\frac{\text{ber}\left(\sqrt{2}\frac{a}{\delta}\right)\text{ber}'\left(\sqrt{2}\frac{a}{\delta}\right) + \text{bei}\left(\sqrt{2}\frac{a}{\delta}\right)\text{bei}'\left(\sqrt{2}\frac{a}{\delta}\right)}{\left[\text{ber}'\left(\sqrt{2}\frac{a}{\delta}\right)\right]^2 + \left[\text{bei}'\left(\sqrt{2}\frac{a}{\delta}\right)\right]^2}\ \Omega/\text{m} \tag{9-126b}$$

That the latter are correct is appreciated from the zero frequency limits. Noting that $\sqrt{2}a/\delta \equiv w$ approaches zero as $\omega \to 0$, all but the first term of (9-116) can be discarded, yielding ber $w \to 1$ and bei $w \to w^2/4$ as $\omega \to 0$. Similarly, derivatives of the power series obtain ber$'$ $w \to w^3/16$ and bei$'$ $w \to w/2$ as $\omega \to 0$. These into (9-126a) yield the dc resistance

$$r_{i,\text{dc}} = \frac{1}{2\pi a}\sqrt{\frac{\omega\mu_c}{\sigma_c}}\frac{4}{w^2} = \frac{1}{\sigma_c \pi a^2}\ \Omega/\text{m} \tag{9-127a}$$

a result seen to agree with the static result (4-157)

$$\frac{R_{\text{dc}}}{\ell} = \frac{1}{\sigma A} = \frac{1}{\sigma\pi a^2}$$

The zero frequency inductance obtained from (9-126b) is

$$l_{i,\text{dc}} = \frac{1}{2\pi a\omega}\sqrt{\frac{\omega\mu_c}{\sigma_c}}\,\frac{-\frac{w^3}{16} + \frac{w^3}{8}}{\frac{w^2}{4}} = \frac{\mu_c}{8\pi}\ \text{H/m} \tag{9-127b}$$

agreeing with the static result L_i/ℓ of (5-82).

A graph of r_i and l_i for the isolated conductor, expressed as ratios to the dc values (9-127), is shown as solid curves in Figure 9-10, along with a short table for improved accuracy. Also shown dashed are high-frequency approximations to the internal parameters, approaching the exact curves for a/δ sufficiently large as discussed in the following.

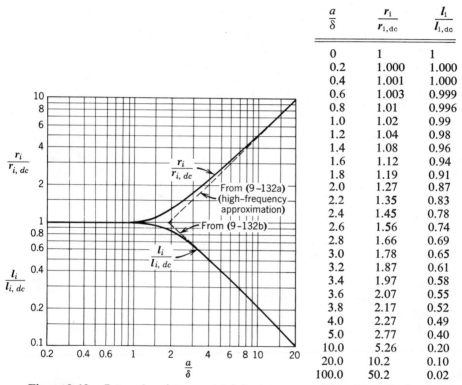

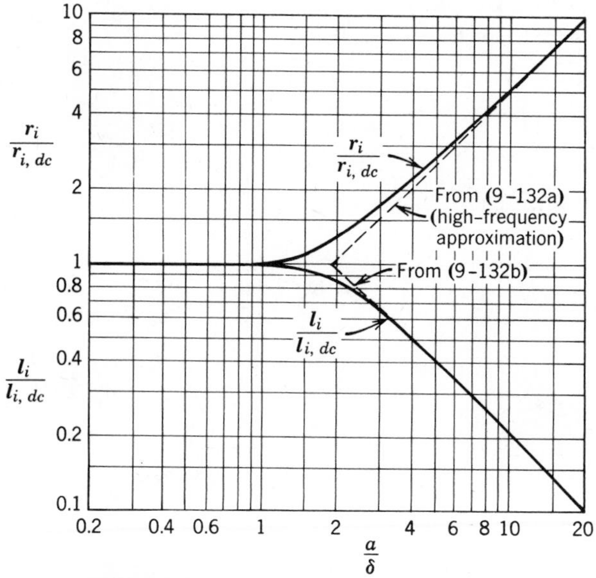

$\frac{a}{\delta}$	$\frac{r_i}{r_{i,dc}}$	$\frac{l_i}{l_{i,dc}}$
0	1	1
0.2	1.000	1.000
0.4	1.001	1.000
0.6	1.003	0.999
0.8	1.01	0.996
1.0	1.02	0.99
1.2	1.04	0.98
1.4	1.08	0.96
1.6	1.12	0.94
1.8	1.19	0.91
2.0	1.27	0.87
2.2	1.35	0.83
2.4	1.45	0.78
2.6	1.56	0.74
2.8	1.66	0.69
3.0	1.78	0.65
3.2	1.87	0.61
3.4	1.97	0.58
3.6	2.07	0.55
3.8	2.17	0.52
4.0	2.27	0.49
5.0	2.77	0.40
10.0	5.26	0.20
20.0	10.2	0.10
100.0	50.2	0.02

Figure 9-10. Internal resistance and inductance parameters for an isolated round wire.

Asymptotic approximations can be found for $\hat{z}_i$ from (9-107), but it is convenient to reexpress $\hat{\mathscr{H}}_\phi$ in terms of $J_0(\hat{k}\rho)$, instead of using the ber and bei functions. From (9-108) and (9-117)

$$\hat{\mathscr{H}}_\phi(\rho) = \frac{1}{j\omega\mu_c}\frac{\partial \hat{\mathscr{E}}_z(\rho)}{\partial \rho} = \frac{1}{j\omega\mu_c\sigma_c}\frac{\partial \hat{J}_z(\rho)}{\partial \rho} = \frac{1}{j\omega\mu_c\sigma_c}\frac{\hat{J}_z(a)}{J_0(\hat{k}a)}\frac{\partial}{\partial \rho}J_0(\hat{k}\rho) \qquad \text{(9-128a)}$$

wherein $\hat{k} = \sqrt{-j\omega\mu\sigma} = j^{-1/2}\sqrt{2}/\delta$ and δ is given by (9-114). Also

$$\frac{\partial}{\partial \rho}J_0(\hat{k}\rho) = \frac{\partial}{\partial(\hat{k}\rho)}J_0(\hat{k}\rho)\frac{\partial(\hat{k}\rho)}{\partial \rho} = \hat{k}_0 J'(\hat{k}\rho)$$

in which the prime denotes the derivative with respect to the argument $\hat{k}\rho = j^{-1/2}\sqrt{2}\rho/\delta$ Then (9-128a) yields, at $\rho = a$

$$\hat{\mathscr{H}}_\phi(a) = \frac{\hat{k}\hat{J}_z(a)\, J_0'(\hat{k}a)}{j\omega\mu_c\sigma_c\, J_0(\hat{k}a)} = \frac{j^{1/2}\hat{J}_z(a)\, J_0'(\hat{k}a)}{\sqrt{\omega\mu_c\sigma_c}\, J_0(\hat{k}a)} \tag{9-128b}$$

Using (9-128b) in (9-107), the internal distributed impedance becomes

$$\hat{z}_i = \frac{\hat{\mathscr{E}}_z(a)}{\hat{I}} = \frac{\hat{J}_z(a)}{2\pi a\sigma_c\hat{\mathscr{H}}_\phi(a)} = \frac{-j^{-1/2}}{\sqrt{2}\pi a}\sqrt{\frac{\omega\mu_c}{2\sigma_c}}\frac{J_0(\hat{k}a)}{J_1(\hat{k}a)} \tag{9-129}$$

in which the Bessel function of order unity is obtained from the identity† $J_0'(v) = -J_1(v)$. The asymptotic form for $J_1(\hat{k}a)$ is

$$J_1\left(j^{-1/2}\sqrt{2}\frac{a}{\delta}\right) \rightarrow \frac{e^{a/\delta}}{\sqrt{2\pi\sqrt{2}\dfrac{a}{\delta}}}\, e^{j(a/\delta - 5\pi/8)} \tag{9-130}$$

and the latter with (9-121b) into (9-129) yields

$$\hat{z}_i \rightarrow \frac{j^{-1/2}}{\sqrt{2}\pi a}\sqrt{\frac{\omega\mu_c}{2\sigma_c}}\frac{e^{j(a/\delta - \pi/8)}}{e^{j(a/\delta - 5\pi/8)}} = \frac{1}{2\pi a}(1 + j)\sqrt{\frac{\omega\mu_c}{2\sigma_c}}\ \Omega/\text{m} \tag{9-131}$$

valid for sufficiently large a/δ. The real and imaginary parts of (9-131) yield

$$r_i \rightarrow \frac{1}{2\pi a}\sqrt{\frac{\omega\mu_c}{2\sigma_c}} = \frac{1}{2\pi a\sigma_c\delta} = \frac{a}{2\delta}\, r_{i(\text{dc})}\ \Omega/\text{m} \qquad \frac{a}{\delta}\ \text{large} \tag{9-132a}$$

$$l_i \rightarrow \frac{1}{2\pi a\omega}\sqrt{\frac{\omega\mu_c}{2\sigma_c}} = \frac{1}{2\pi a\sigma_c\delta\omega} = \frac{2\delta}{a}\, l_{i(\text{dc})}\ \text{H/m} \qquad \frac{a}{\delta}\ \text{large} \tag{9-132b}$$

For instance, if $a/\delta = 5$, the asymptotic expression (9-132a) can be used in lieu of (9-126a) with an error of about 10%, decreasing to zero error for a/δ sufficiently larger.

The asymptotic result (9-131) is seen to contain the quantity $\hat{\eta}$ of (3-112c), i.e., the intrinsic wave impedance for a plane wave in a conductive region. Thus (9-131) yields the ratio for a round wire

$$\frac{\hat{\mathscr{E}}_z(a)}{\hat{\mathscr{H}}_\phi(a)} \rightarrow (1 + j)\sqrt{\frac{\omega\mu_c}{2\sigma_c}} \equiv \hat{\eta}\ \Omega \qquad \frac{a}{\delta}\ \text{large} \tag{9-133}$$

† See Ramo, S., J. Whinnery and T. van Duzer. *Fields and Waves in Communications Electronics.* New York: Wiley, 1965, p. 213.

One concludes that for current penetration small, the impedance ratio $\hat{\mathscr{E}}_z/\hat{\mathscr{H}}_\phi$ at the surface of a round wire becomes $\hat{\eta}$, the same as the ratio of electric to magnetic *plane wave* fields in a conductor.

B. Distributed Parameters of a Parallel-Wire Line, Conductor Impedances Included

The isolated wire, internal impedance results of the previous discussion can be applied directly to the parallel-wire line of Figure 9-7(*b*), with the object of finding (9-98b), the distributed parameter $\hat{\boldsymbol{z}}$. Proximity effects† are neglected, which assumes fields in each wire undisturbed from the axially symmetric configuration attained when isolated, a reasonable assumption if the axial separation is greater than about 10 conductor diameters.

With effects of proximity neglected, the *internal* parameters of the parallel-wire line of Figure 9-7(*b*) are double the results (9-126) obtained for the isolated conductor, in view of the impedance encountered twice along the edges Δz of the rectangle ℓ. The series parameter (9-98b) therefore becomes

$$\begin{aligned} \hat{\boldsymbol{z}} &= 2\boldsymbol{r}_i + j\omega(2\boldsymbol{l}_i + \boldsymbol{l}_e) \\ \hat{\boldsymbol{z}} &= \boldsymbol{r} + j\omega \boldsymbol{l}\ \Omega/\text{m} \end{aligned} \qquad (9\text{-}134)$$

in which $\boldsymbol{r}_i$ and $\boldsymbol{l}_i$ are given by (9-126) or (9-132). In (9-134) $\boldsymbol{l}_e$ is related to the magnetic field exterior to the wires, permitting the use of that obtained in Problem 9-5(*a*) for the perfect conductor case

$$\boldsymbol{l}_e = \frac{\mu}{\pi} \ln \frac{h + d}{a}\ \text{H/m} \qquad (9\text{-}135)$$

wherein a is the wire radius, $2h$ the separation, and $d = \sqrt{h^2 - a^2}$.

The expression (9-100) for the shunt parameter $\hat{\boldsymbol{y}} = \boldsymbol{g} + j\omega\boldsymbol{c}$ is the same whether or not the conductors are perfectly conducting; thus, from Problem 9-4

$$\hat{\boldsymbol{y}} = \boldsymbol{g} + j\omega\boldsymbol{c} = \left(\frac{\epsilon''}{\epsilon'} + j\right)\omega\boldsymbol{c} = \left(\frac{\epsilon''}{\epsilon'} + j\right)\frac{\omega\pi\epsilon}{\ln \dfrac{h + d}{a}}\ \mho/\text{m} \qquad (9\text{-}136)$$

† An analysis of the effects of the proximity of the wires on the increase in internal resistance is found in Arnold, A. H. M. "The alternating-current resistance of parallel conductors of circular cross-section," *Jour. I.E.E.*, **77**, 1935, p. 49.

For an air dielectric, the assumption $g = 0$ is appropriate. In telephone lines using poles for support, the insulator leakage is often reduced to an equivalent distributed loss effect along the line, yielding a parameter g determined by the number of poles used per mile or kilometer.

EXAMPLE 9-9. A telephone line consists of 0.104 in. (0.264 cm) diameter hard-drawn copper wires ($\sigma_0 = 5.14 \times 10^7$ ℧/m) separated 12 in. (30.5 cm) in air. Neglect leakage due to the supporting insulators. (*a*) Compute the distributed constants r, l, g, and c at 1 kHz. (*b*) Find $\hat{Z}_0$, α, β, λ, and v_p at 1 kHz.

(*a*) With $2a = 0.104$ in. and $2h = 12$ in., $h/a = 115.5$, so (9-136) yields

$$c = \frac{\pi\epsilon}{\ln\left[\frac{h}{a} + \sqrt{\left(\frac{h}{a}\right)^2 - 1}\right]} = \frac{10^{-9}/36}{\ln[231]} = 5.10 \text{ pF/m} = 0.00822\ \mu\text{F/mi}$$

in which the conversion 1609 m/mi is used. If leakage is neglected $g = 0$.

With $\hat{z}$ given by (9-134), evaluating r_i and l_i requires first expressing the wire radius as a function of δ given by (9-114)

$$\delta = \sqrt{\frac{2}{\omega\mu_c\sigma_c}} = \sqrt{\frac{2}{2\pi(10^3)4\pi \times 10^{-7}(5.63 \times 10^7)}} = 0.00213 \text{ m}$$

making $a/\delta = 0.622$. From Figure 9-10, $r_i/r_{i,\text{dc}} = 1.003$, $l_i/l_{i,\text{dc}} = 0.999$. Thus, with $r_i = 1.003 r_{i,\text{dc}}$, (9-127a) in (9-134) yields

$$r = 2r_i = 2\left(1.003\frac{1}{\sigma_c\pi a^2}\right) = \frac{2 \times 1.003}{5.14 \times 10^7\pi(0.00132)^2} = 0.00712\ \Omega/\text{m}$$

or 11.47 Ω/mi. Similarly, using $l_{i,\text{dc}}$ of (9-127b) in (9-134) obtains $2l_i = 2(0.999)\mu_0/8\pi = 0.999\ \mu\text{H/m} = 0.161$ mH/mi., and from (9-135)

$$l_e = \frac{\mu_0}{\pi}\ln\frac{h+d}{a} = \frac{4\pi \times 10^{-7}}{\pi}\ln(231) = 2.18\ \mu\text{H/m} = 3.51 \text{ mH/mi}$$

so the total distributed inductance in (9-134) becomes

$$l = 2l_i + l_e = 2.28\ \mu\text{H/m} = 3.67 \text{ mH/mi}$$

(*b*) One obtains $\hat{Z}_0$ using (9-105)

$$\hat{Z}_0 = \sqrt{\frac{r + j\omega l}{g + j\omega c}} = \sqrt{\frac{11.47 + j2\pi(10^3)3.67 \times 10^{-3}}{0 + j2\pi(10^3)0.00822}} = 703e^{-j13.2^\circ}\ \Omega$$

By use of (9-103a)

$$\gamma = \sqrt{(r + j\omega l)(g + j\omega c)} = \sqrt{(25.7e^{j63.5^\circ})(j51.6 \times 10^{-6})}$$
$$= 0.0083 + j0.035 \text{ mi}^{-1}$$

yielding $\alpha = 0.0083$ Np/mi, $\beta = 0.035$ rad/mi. From (9-38) and (9-39)

$$\lambda = \frac{2\pi}{\beta} = \frac{2\pi}{0.035} = 179 \text{ mi} = 288 \text{ km}$$

$$v_p = \frac{\omega}{\beta} = f\lambda = 179{,}000 \text{ mi/sec} = 2.88 \times 10^8 \text{ m/sec}$$

From the value of α, a wave on the line attenuates to $e^{-1} = 36.8\%$ of its input value in $d = \alpha^{-1} = (0.0083)^{-1} = 120$ mi at $f = 1$ kHz.

C. Distributed Parameters of a Coaxial Line, Conductor Impedances Included

In Figure 9-6 is shown the current distributions obtained in the outer conductor of a coaxial line at low, medium, and high frequencies. At the higher frequencies the currents concentrate towards the TEM fields responsible for those currents, namely, toward the outer wall of the inner conductor and towards the *inner* wall of the *outer* conductor. This latter property of the coaxial line, like hollow waveguides, makes it an excellent *shielding* device at high frequencies, the essentially zero currents on the outer wall eliminating the possibility of a small tangential electric field being coupled outside the outer conductor.

The isolated wire, internal impedance results (9-126) are, in view of the axial symmetry, directly applicable with no approximation to the *center* conductor of the coaxial line. An additional internal impedance is associated with the continuity of $\mathscr{H}_\phi$ on the inner wall of the outer conductor (at $\rho = b$), as suggested by Figure 9-11(a). Defined by (9-107), it is written

$$\mathfrak{z}_{i2} = \frac{\hat{\mathscr{E}}_z(b)}{\hat{I}} = \frac{\hat{\mathscr{E}}_z(b)}{\oint_\ell \hat{\mathscr{H}}_\phi b\,d\phi} = \frac{\hat{\mathscr{E}}_z(b)}{2\pi b \hat{\mathscr{H}}_\phi(b)} \tag{9-137}$$

which requires expressing $\hat{\mathscr{H}}_\phi$ in terms of the induced $\hat{\mathscr{E}}_z$ just inside the outer conductor, already accomplished through (9-108). The solution for $\hat{\mathscr{E}}_z(\rho)$ in the outer conductor, moreover, must satisfy the differential equation (9-111), yielding once again solutions (9-112). *Both* Bessel functions $J_0(\hat{k}\rho)$ and $N_0(\hat{k}\rho)$ must be retained to satisfy boundary conditions on the outer conductor. (Thus, from the range $b < \rho < c$ within the outer conductor, the singularity at $\rho = 0$ of N_0 does not become troublesome.) The exact field solution in the outer conductor is, however, not pursued further here; only a high-frequency approximation to the internal impedance for the outer conductor is employed in the following.

In determining the distributed impedance parameters of the coaxial line,

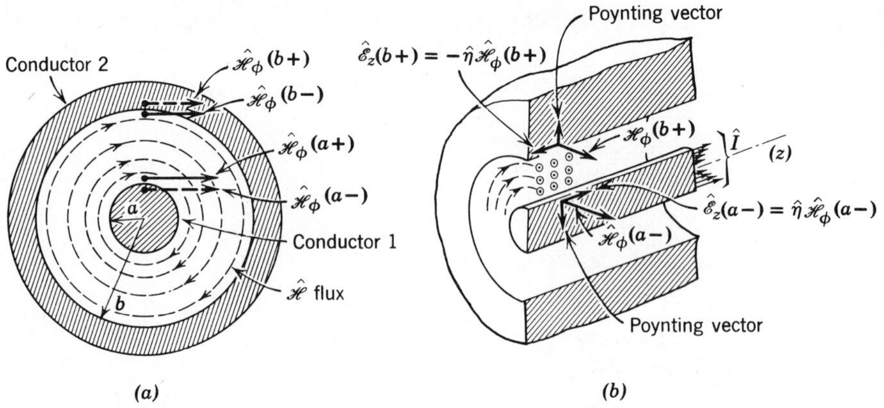

Figure 9-11. Relative to the distributed internal impedance of a coaxial line. (*a*) Continuity of tangential magnetic fields into conductors. (*b*) Longitudinal electric field induced by $\hat{\mathscr{H}}_\phi$, producing skin effect.

from axial symmetry the high-frequency approximation (9-131) for an isolated conductor is the same for the *inner* conductor of the coaxial line; hence

$$\hat{z}_{i1} \to \frac{1}{2\pi a}(1+j)\sqrt{\frac{\omega\mu_c}{2\sigma_c}} \tag{9-138}$$

Thus, the internal resistance and inductance contributions are:

$$r_{i1} \to \frac{1}{2\pi a}\sqrt{\frac{\omega\mu_c}{2\sigma_c}} \tag{9-139a}$$

High-frequency approximations

$$l_{i1} \to \frac{1}{2\pi a\omega}\sqrt{\frac{\omega\mu_c}{2\sigma_c}} \tag{9-139b}$$

The outer conductor internal impedance $\hat{z}_{i2}$ defined by (9-137) can be heuristically expressed in the high-frequency limit by use of the wave impedance (3-112c)

$$\frac{\hat{\mathscr{E}}_z(b)}{\hat{\mathscr{H}}_\phi(b)} \to \hat{\eta} = (1+j)\sqrt{\frac{\omega\mu_c}{2\sigma_c}} \tag{9-140}$$

a ratio approaching that for *plane wave* fields in a good conductor, at frequencies sufficiently high to make δ small compared to the radius b. Thus (9-140) applied to (9-137) yields

$$\hat{z}_{i2} \to \frac{\hat{\mathscr{E}}_z(b)}{2\pi b_\phi \hat{\mathscr{H}}(b)} = \frac{1+j}{2\pi b}\sqrt{\frac{\omega\mu_c}{2\sigma_c}} \tag{9-141}$$

with real and imaginary parts analogous with (9-139)

$$\mathbf{r}_{i2} \to \frac{1}{2\pi b}\sqrt{\frac{\omega\mu_c}{2\sigma_c}} \quad \text{(9-142a)}$$

High-frequency approximations

$$\mathbf{l}_{i2} \to \frac{1}{2\pi b\omega}\sqrt{\frac{\omega\mu_c}{2\sigma_c}} \quad \text{(9-142b)}$$

Adding (9-139), (9-142), and (9-84) thus obtains, from (9-98)

$$\hat{z} = \hat{z}_{i1} + \hat{z}_{i2} + j\omega \mathbf{l}_e = \mathbf{r}_{i1} + \mathbf{r}_{i2} + j\omega(\mathbf{l}_{i1} + \mathbf{l}_{i2} + \mathbf{l}_e)$$

$$\to \frac{1}{2\pi}\left(\frac{1}{a} + \frac{1}{b}\right)\sqrt{\frac{\omega\mu_c}{2\sigma_c}} + j\omega\left[\frac{1}{\omega 2\pi}\left(\frac{1}{a} + \frac{1}{b}\right)\sqrt{\frac{\omega\mu_c}{2\sigma_c}} + \frac{\mu}{2\pi}\ln\frac{b}{a}\right]$$

$$= \mathbf{r} + j\omega \mathbf{l}\ \Omega/\text{m} \qquad \text{High-frequency approximations} \quad \text{(9-143)}$$

The series distributed parameters of a coaxial line, from (9-143), have the following properties in the high-frequency approximation:

1. The resistive part $\mathbf{r}$ increases as the square root of the frequency, and it decreases inversely with $\sqrt{\sigma_c}$.
2. The inductive part has two contributions. The first is the inductive part of the *internal* impedance, behavior like $\mathbf{r}$. The second is the *external* inductance (9-84), providing the major contribution to $\mathbf{l}$ in practical coaxial lines.

Finally, the shunt parameter (9-100), $\hat{y} = \mathbf{g} + j\omega \mathbf{c}$, remains unaltered from (9-82), applicable to the line with perfect conductors. This conclusion follows as usual from the dependence of $\mathbf{g}$ and $\mathbf{c}$ on only the electric field in the dielectric region. Thus,

$$\hat{y} = \mathbf{g} + j\omega \mathbf{c} = \left(\frac{\epsilon''}{\epsilon'} + j\right)\omega \mathbf{c} = \left(\frac{\epsilon''}{\epsilon'} + j\right)\omega \frac{2\pi\epsilon}{\ln\frac{b}{a}}\ \mho/\text{m} \quad \text{(9-144)}$$

valid at all frequencies, and not just a high-frequency approximation.

The general derivation of the series parameters of the coaxial line at an arbitrary frequency ω is omitted here, but at low frequencies they are readily obtained in the absence of the skin effect. Then the dc inductance results obtained in Example 5-13 are applicable, while the resistance parameters are obtained from an adaptation of (4-157), assuming uniform current densities over the conductor cross-sections. It is left for the reader to carry out the details.

EXAMPLE 9-10. Assume the cable in Example 9-4 has the dielectric loss tangent of 0.0002, but this time copper conductors are used. Recalculate $\hat{Z}_0$, α, and β from the distributed parameters $\hat{y}$ and $\hat{z}$, at $f = 20$ MHz.

Note first the plane wave δ obtained from (3-114) is $\delta = \sqrt{2/\omega\mu_c\sigma_c} = 1.48 \times 10^{-5}$ m $= 0.0148$ mm, sufficiently smaller than $a = 0.05$ in. $= 1.27$ mm such that the high-frequency approximation assumed for (9-143) applies. The series $\boldsymbol{r}$ parameter then becomes

$$\boldsymbol{r} = \frac{1}{2\pi}\left(\frac{1}{a} + \frac{1}{b}\right)\sqrt{\frac{\omega\mu_c}{2\sigma_c}} = \frac{1}{2\pi}\left(\frac{10^3}{1.27} + \frac{10^3}{4.14}\right)\sqrt{\frac{2\pi(2 \times 10^7)4\pi \times 10^{-7}}{2 \times 5.8 \times 10^7}}$$
$$= 0.191\ \Omega/\text{m} \tag{1}$$

The inductance parameter has internal and external contributions

$$j\omega\boldsymbol{l} = j\left[\boldsymbol{r} + \omega\frac{\mu_0}{2\pi}\ln\frac{b}{a}\right] = j\left[0.191 + 2\pi(2 \times 10^7)\frac{4\pi \times 10^{-7}}{2\pi}\ln 3.26\right]$$
$$= j29.8\ \Omega/\text{m} \tag{2}$$

in which $\boldsymbol{l}_e$ contributes almost wholly to $j\omega\boldsymbol{l}$. Thus, $\hat{z} \equiv \boldsymbol{r} + j\omega\boldsymbol{l} = 0.191 + j29.8\ \Omega/\text{m}$.

The shunt parameter $\hat{y}$ is found from (9-144), whence $\boldsymbol{c} = 2\pi\epsilon/\ln(b/a) = 2\pi(2 \times 10^{-9}/36\pi)/1.18 = 94.2$ pF/m, yielding $\hat{y} = \boldsymbol{g} + j\omega\boldsymbol{c} = (\epsilon''/\epsilon') + j)\omega\boldsymbol{c} = (0.0002 + j)\omega\boldsymbol{c} = j1.184 \times 10^{-2}$ ℧/m. Dielectric losses are negligible in this example.

The characteristic impedance is found by use of (9-105)

$$\hat{Z}_0 = \sqrt{\frac{\boldsymbol{r} + j\omega\boldsymbol{l}}{\boldsymbol{g} + j\omega\boldsymbol{c}}} = \sqrt{\frac{0.191 + j29.8}{j1.184 \times 10^{-2}}} = \sqrt{\frac{29.8e^{j89.7°}}{j1.184 \times 10^{-2}}} \cong 50\ \Omega \tag{3}$$

or essentially that of Example 9-4 for the lossless case, expected since $\boldsymbol{r} \ll \omega\boldsymbol{l} \cong \omega\boldsymbol{l}_e$ and $\boldsymbol{g} \ll \omega\boldsymbol{c}$.

The propagation constant obtained from (9-103a) is $\gamma = \sqrt{\hat{y}\hat{z}} = \sqrt{(0.191 + j29.8)j1.184 \times 10^{-2}} = 0.595e^{j89.85°}$ m^{-1}, yielding from the imaginary part

$$\beta = 0.595 \sin 89.85° \cong 0.595\ \text{rad/m} \tag{4}$$

The latter is also obtainable using (9-103c), since $\boldsymbol{r} \ll \omega\boldsymbol{l}$ and $\boldsymbol{g} \ll \omega\boldsymbol{c}$ are satisfied. To find α, (9-103b) provides good accuracy

$$\alpha \cong \frac{\boldsymbol{r}}{2}\sqrt{\frac{\boldsymbol{c}}{\boldsymbol{l}}} + \frac{\boldsymbol{g}}{2}\sqrt{\frac{\boldsymbol{l}}{\boldsymbol{c}}} = \frac{\boldsymbol{r}}{2Z_0} + \frac{\boldsymbol{g}Z_0}{2} = \frac{0.191}{2 \times 50} + \frac{2.37 \times 10^{-6}}{2}50 \tag{5}$$
$$\cong 1.97 \times 10^{-3}\ \text{Np/m} = 1.71 \times 10^{-2}\ \text{dB/m} \tag{5}$$

and while still a low value, it represents an increase of over 30 times that obtained in Example 9-4 with *perfect* conductors assumed. Waves thus attenuate to 36.8% of the input value of a length $d = \alpha^{-1} = (1.97 \times 10^{-3})^{-1} = 508$ m of this line at $f = 20$ MHz.

At 20 MHz, (9-38) and (9-39) yield $\lambda = 2\pi/\beta = 10.55$ m and $v_p = \omega/\beta = 2.12 \times 10^8$ m/sec, results essentially those of Example 9-4, in view of the small losses.

REFERENCES

JORDAN, E. C., and K. G. BALMAIN. *Electromagnetic Waves and Radiating Systems*, 2nd ed. Englewood Cliffs, N.J.: Prentice-Hall, 1968.

MAGID, L. M. *Electromagnetic Fields, Energy and Waves.* New York: Wiley, 1972.

RAMO, S., J. R. WHINNERY, and T. VAN DUZER. *Fields and Waves in Communication Electronics.* New York: Wiley, 1965.

PROBLEMS

9-1. A long parallel-wire line in air consists of perfect conductors of radius R and separation $2h$ as shown. It carries the TEM mode.

(*a*) Borrowing from the static potential solution (4-96), show that the time-harmonic potentials associated with TEM waves are (making only the necessary changes in notation)

$$\hat{\Phi}^{\pm}(x, y) = \frac{\hat{Q}_m^{\pm}}{2\pi\epsilon\ell} \ln \frac{\sqrt{(x+d)^2 + y^2}}{\sqrt{(x-d)^2 + y^2}}$$

$$= \frac{\hat{V}_m^{\pm}/2}{\ln \dfrac{h+d}{R}} \ln \frac{\sqrt{(x+d)^2 + y^2}}{\sqrt{(x-d)^2 + y^2}}$$

in which $d = \sqrt{h^2 - R^2}$. [*Note:* The latter form utilizes (4-105) to express Φ in terms of the voltage difference V between assumed

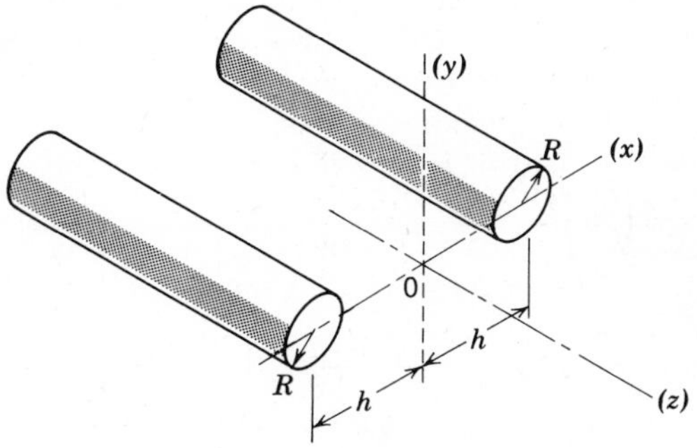

Problem 9-1

positive conductor and the reference at the origin, or only *half* the voltage between the conductors, so replacing V with $\hat{V}_m^{\pm}/2$ becomes necessary.]

(*b*) Show from (*a*) that the electric fields become

$$\hat{\mathscr{E}}^{\pm}(x, y) = \frac{\hat{V}_m^{\pm}}{2\,\ell n \dfrac{h+d}{R}}\left\{\mathbf{a}_x\left[\frac{x-d}{(x-d)^2+y^2} - \frac{x+d}{(x+d)^2+y^2}\right] + \mathbf{a}_y\left[\frac{y}{(x-d)^2+y^2} - \frac{y}{(x+d)^2+y^2}\right]\right\}$$

$$= \frac{\hat{V}_m^{\pm}}{2\,\ell n \dfrac{d+h-R}{d-h+R}}\left\{\mathbf{a}_x\left[\frac{x-d}{(x-d)^2+y^2} - \frac{x+d}{(x+d)^2+y^2}\right] + \mathbf{a}_y\,\frac{y}{(x-d)^2+y^2} - \frac{y}{(x+d)^2+y^2}\right]\right\}$$

The latter is verified by use of

$$(h + d)/R = (d + h - R)/(d - h + R),$$

with $d = \sqrt{h^2 - R^2}$.

(*c*) Write the *total* $\hat{\mathbf{E}}(x, y, z)$ as a superposition of $\hat{\mathscr{E}}^+$ and $\hat{\mathscr{E}}^-$. Express it in real-time form, $\mathbf{E}(x, y, z, t)$, assuming $\hat{V}_m^{\pm} = V_m^{\pm} e^{j\theta \pm}$.

9-2. Show for the line of Problem 9-1 that the magnetic fields are given by

$$\hat{\mathscr{H}}^{\pm}(x, y) = \frac{\pm\hat{V}_m^{\pm}}{2\eta_0\,\ell n \dfrac{d+h-R}{d-h+R}}\left\{\mathbf{a}_x\left[\frac{y}{(x+d)^2+y^2} - \frac{y}{(x-d)^2+y^2}\right] + \mathbf{a}_y\left[\frac{x-d}{(x-d)^2+y^2} - \frac{x+d}{(x+d)^2+y^2}\right]\right\}$$

in which $d = \sqrt{h^2 - R^2}$. Superpose these to obtain the total field $\hat{\mathbf{H}}(x, y, z)$, as well as the real-time form, assuming $\hat{V}_m^{\pm} = V_m^{\pm} e^{j\theta \pm}$.

9-3. (*a*) Show that the integrals of (9-20b), applied to the $\hat{\mathscr{E}}_m^{\pm}$ fields obtained in Problem 9-1, yield the expected relation (9-20a) for $\hat{V}(z)$ in terms of the voltage amplitudes $\hat{V}_m^{\pm}$. A path of integration ℓ connecting the conductors as shown in (*a*) of the figure is suggested.

(*b*) Integrate (9-10b) to express $\hat{I}(z)$ in terms of the complex voltage amplitudes $\hat{V}_m^{\pm}$, making use of $\hat{\mathscr{H}}^{\pm}$ obtained in Problem 9-2. [*Note:* The path of integration should encompass one conductor as in (*b*) of the figure, showing two possibilities ℓ_1 and ℓ_2. The last is suggested for ease of integration; it is valid to assume no contribution to the integral about the semicircle at infinity.] [*Answer:*]

$$\hat{I}(z) = \frac{\hat{V}_m^+}{\dfrac{\eta_0}{\pi}\,\ell n \dfrac{h+d}{R}} e^{-\gamma z} - \frac{\hat{V}_m^-}{\dfrac{\eta_0}{\pi}\,\ell n \dfrac{h+d}{R}} e^{\gamma z} \text{ A}$$

in which $d = \sqrt{h^2 - R^2}$.

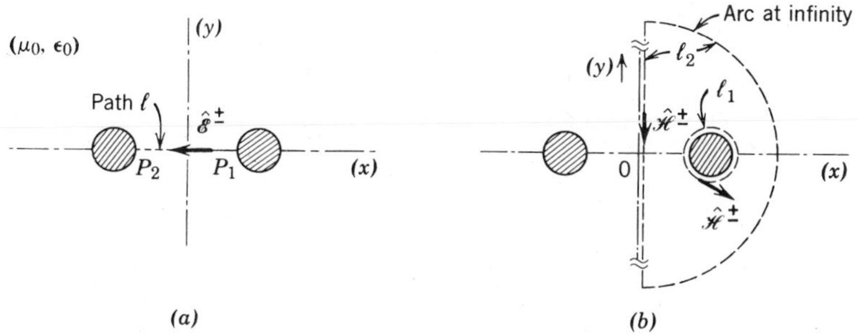

Problem 9-3. ***(a)* Suggested integration path ℓ, for obtaining voltage from the electric field. *(b)* Integration paths ℓ_1 and ℓ_2 surrounding a conductor, for obtaining current from the magnetic field.**

(*c*) From (*b*), show that the characteristic impedance of the parallel-wire line is

$$\hat{Z}_0 = \frac{\eta_0}{\pi} \ln \frac{h + d}{R} = \frac{\eta_0}{\pi} \cosh^{-1} \frac{h}{R} \cong 120 \ln \frac{h + d}{R} \, \Omega$$

with $d = \sqrt{h^2 - R^2}$. Show that if three air lines have h/R ratios of 2, 10, and 100, Z_0 becomes 158, 360, and 635 Ω, respectively.

9-4. (*a*) Adapt the static capacitance of a parallel-wire line in Section 4-11 to obtain the distributed capacitance parameter for the line carrying the TEM mode

$$c = \frac{\pi \epsilon_0}{\ln \dfrac{h + d}{R}}$$

if $d = \sqrt{h^2 - R^2}$. Show that three air lines with the h/R ratios 2, 10, and 100 have c parameters of 21, 9.3, and 5.25 pF/m, respectively.

(*b*) Show that the distributed conductance parameter of the same line would be

$$g = \frac{\pi \sigma}{\ln \dfrac{h + d}{R}}$$

if the uninsulated wires were immersed in a region of conductivity σ.

9-5. (*a*) Using the defining expression (9-57c), $\Delta\psi_m = (l_e \Delta z)I$, derive the external inductance parameter of an air dielectric, parallel-wire line

$$l_e = \frac{\mu_0}{\pi} \ln \frac{h + d}{R}$$

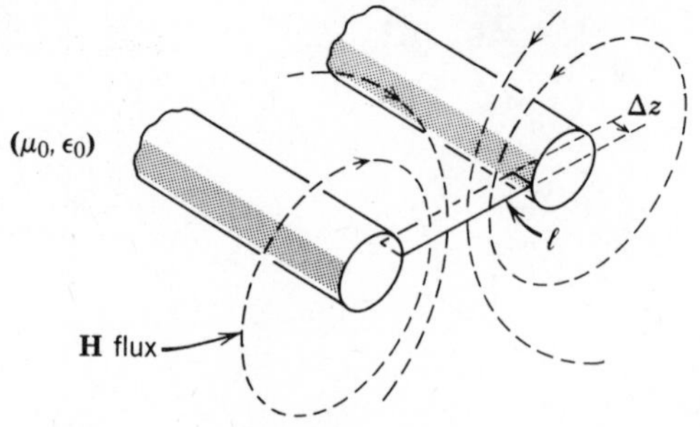

Problem 9-5

in which $d = \sqrt{h^2 - R^2}$. [*Note:* The integration for $\Delta\psi_m$ is simplified by using a rectangle in the plane of the wires as in the figure.]

(*b*) Check the answer to (*a*) using (9-77c) and the c parameter obtained in Problem 9-4. Find Z_0 for the three lines of Problem 9-4(a).

(*c*) Show that three different lines with h/R ratios of 2, 10, and 100 have the l_e parameters 0.528, 1.84, and 2.12 μH/m, respectively.

9-6. Verify that Z_0 obtained in Problem 9-3(*c*) satisfies (9-77c) upon using l_e and c obtained in Problems 9-4 and 9-5.

9-7. Find the parameters c, g, and l_e of a coaxial line, assuming perfect conductors, $b/a = 3$, and a polyethylene dielectric with $\epsilon_r = 2.26$ and a loss tangent of 0.0002 at $f = 1$ MHz. Find also γ and $\hat{Z}_0$.

9-8. If three coaxial lines with an air dielectric are to have characteristic impedances of 50, 70, and 100 Ω, show that their b/a ratios must be about 2.3, 3.2, and 5.3, respectively. How are these ratios affected if polyethylene dielectric is used ($\epsilon_r = 2.26$)? Ignore losses.

9-9. Based solely on the knowledge that a particular coaxial line has an air dielectric and a characteristic impedance of 50 Ω, determine its distributed parameters c, g, and l_e assuming perfect conductors. [*Answer:* $c = 66.8$ pF/m, $l_e = 0.167$ μH/m]

9-10. At $f = 1$ MHz, compare the wavelength in any air dielectric cable carrying the TEM mode with that in the cable using a polyethylene dielectric ($\epsilon_r = 2.26$). Ignore losses.

9-11. Show that for a uniform line carrying TEM waves and assuming perfect conductors separated by a lossy dielectric, α and β given by (9-32) and (9-33) can be expressed

$$\alpha = \frac{\omega\sqrt{l_e c}}{\sqrt{2}}\sqrt{\sqrt{1 + \left(\frac{g}{\omega c}\right)^2} - 1} \qquad \beta = \frac{\omega\sqrt{l_e c}}{\sqrt{2}}\sqrt{\sqrt{1 + \left(\frac{g}{\omega c}\right)^2} + 1}$$

To what do these reduce if $g = 0$?

9-12. Illustrated is a configuration known as the *microstrip* transmission line, a two-conductor line with more confined electromagnetic field energies

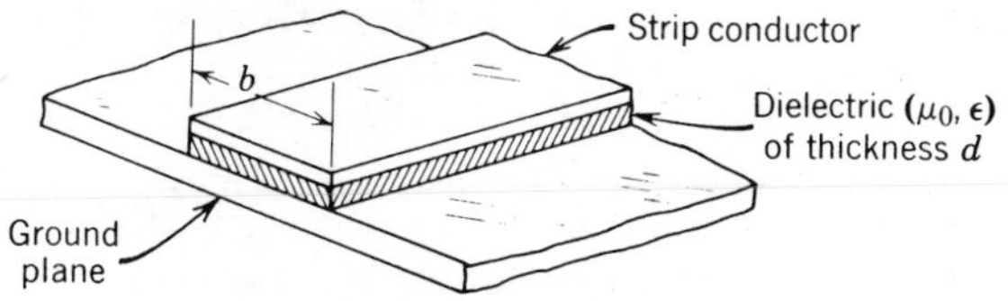

Problem 9-12

than is provided by the conventional parallel-wire line. Consisting of a thin metal strip separated from *chassis* ground by a thin dielectric layer as shown, the microstrip line is amenable to printed-circuit methods for inexpensive and compact layouts applicable to microwave circuitry, for example.

(*a*) Neglecting losses and electric field-fringing at the strip sides, adapt the static capacitance (4-52) of a parallel-plate system to show that the distributed capacitance of the stripline is approximately $\boldsymbol{c} = \epsilon(b/d)$. Use (9-76) to show that its distributed inductance is about $\boldsymbol{l}_e = \mu_0(d/b)$, assuming a nonmagnetic dielectric strip. Then find an expression for its characteristic impedance.

(*b*) What b/d ratio is required to make a stripline a 50 Ω line, assuming a polyethylene dielectric ($\epsilon_r = 2.26$)? Find the strip width b, if the polyethylene strip is 0.1 mm thick.

(*c*) Suggest how the approximate expression for $\boldsymbol{c}$ in part (*a*) might be improved by field-plotting methods of Section 4-13, accounting for the different permittivities of the dielectric strip and the surrounding air.

9-13. Beginning with the transmission-line equations (9-53) and (9-54), derive the wave equations (9-88a) and (9-88b) applicable to a line with dielectric losses and perfect conductors.

9-14. Show that (9-103a) for the propagation constant reduces to (9-103b) and (9-103c), assuming $\boldsymbol{r} \ll \omega \boldsymbol{l}$ and $\boldsymbol{g} \ll \omega \boldsymbol{c}$. [*Hint:* Make use of the binomial theorem, retaining only two terms to yield $(a^2 + b^2)^{1/2} \simeq a[1 + b^2/2a^2]$, if $b \ll a$.] Show also that the characteristic impedance becomes the pure real $\hat{Z}_0 = (\boldsymbol{l}/\boldsymbol{c})^{1/2}$ Ω.

9-15. Verify the limiting expressions (9-127) for the internal distributed resistance and inductance of a wire, beginning with (9-126).

9-16. Verify the high-frequency ($a/\delta \ll 1$) expressions (9-132) for the internal distributed resistance and inductance of a wire, beginning with (9-129).

9-17. (*a*) For the telephone line of Example 9-9, using hard-drawn copper wire 0.104 in. in diameter separated 12 in. in air, show that its distributed parameters at 10 kHz become, neglecting leakage ($\boldsymbol{g} = 0$): $\boldsymbol{c} = 0.00822$ μF/mi, $\boldsymbol{r} = 12$ Ω/mi, $\boldsymbol{l} = 3.64$ mH/mi.

(*b*) At 100 kHz, show that $\boldsymbol{c} = 0.00\,822$ μF/mi, $\boldsymbol{r} = 35.1$ Ω/mi, $\boldsymbol{l} = 3.55$ mH/mi.

9-18. For a parallel-wire line spaced $2h$ in air (neglecting proximity effects) explain why a reduction in wire size results in an increase in the parameter l_e and a reduction in c. How is the magnitude of $\hat{Z}_0$ affected?

9-19. (*a*) Evaluate r, l, g, and c for the cable described in Example 9-10, for a frequency ten times as large (f = 200 MHz). Compare with results obtained at 20 MHz, explaining changes observed.
(*b*) Use the parameter computed in (*a*) to determine $\hat{Z}_0$, α, β, v_p, and λ on this line at 200 MHz. What line length yields a wave attenuaation to 36.8% of input values at this frequency?

9-20. A high-power coaxial line uses a polyethylene dielectric with $\epsilon_r = 2.26$ and a loss tangent of 0.0002, and it has the conductor dimensions a = 0.0975 in. = 0.248 cm, b = 0.340 in. = 0.864 cm. Conductors are copper.
(*a*) At 10 MHz, find $\hat{z} = r + j\omega l$ using the high-frequency approximation (9-143), verifying that the skin depth is sufficiently small to allow the approximation. Determine also $\hat{y}$.
(*b*) From parameters obtained in (*a*), show that $\hat{Z}_0$ is essentially 50 Ω, and find β and α at 10 MHz. (Express α in nepers per meter, decibels per meter, decibels per 100 ft.) What length of this line yields an attenuation to 36.8% of input values at this frequency?

9-21. The maximum allowable average power input into the high-power 50-Ω cable of Problem 9-19, yielding a safe inner conductor temperature not in excess of 80°C, can be shown to be 7 kW at 10 MHz, and about 2 kW at 100 MHz, for an ambient temperature of 40°C. What sinusoidal voltage and current amplitudes are associated with the traveling waves at these power levels?

9-22. For the cable of Example 9-10, what percentage of the attenuation factor α is attributable to conductor wall-losses and what part is due to losses in the dielectric at 20 MHz?

9-23. Use (9-103b) to express α as a function of the dimensions a, b of a coaxial cable operated at high frequencies, assuming g negligible. Holding b constant, minimize this expression with respect to a, showing that a minimum wall-loss attenuation is obtained if b/a is about 3.6. Show that the characteristic impedance of this line is about $77/\sqrt{\epsilon_r}$ Ω.

CHAPTER

10

Analysis of Reflective Transmission Lines

In the introductory paragraphs of Chapter 9 were cited applications of two-conductor transmission lines, embracing the transmission of *power* at the lower end of the frequency spectrum to *signal* transmission at frequencies of many megahertz. In that chapter were covered the determination of line parameters and propagation characteristics from the line geometry and materials, in addition to relating the electric and magnetic fields of the line to its voltage and current waves.

This chapter continues with the analysis of such transmission lines when terminated in arbitrary load impedances. In engineering practice, a communication line used for signal transmission is usually terminated in its characteristic impedance, unless the load value is fixed by the physical nature of the load (e.g., an antenna). Then it may be necessary to employ a *load-matching* scheme to adjust the input impedance of the combination to the value of the line characteristic impedance. Power transmission lines, on the other hand, invariably operate under load-mismatch conditions, in view of the variable loading depending on power demand. At their low operating frequency (usually between 50 and 400 Hz), however, power lines are usually electrically short ($\ell \ll \lambda$), so the analysis can often be simplified through lumped-element, equivalent circuit methods. These techniques are omitted from discussion here.

This chapter begins with analytical methods for determining voltage, current, and line impedance conditions on a two-conductor transmission line fed from a sinusoidal source and terminated in an arbitrary load impedance.

Use is made of the reflection coefficient and line impedance technique, developed in Chapter 6 for uniform plane waves at normal incidence to plane interfaces. The logical application of the Smith chart follows, with emphasis on both the impedance and admittance versions of the chart. There follows an analysis of standing waves of current and voltage on mismatched lines, making further use of the Smith chart. Analytical expressions for line input impedance under arbitrary termination conditions are developed next. Impedance matching of a mismatched line by use of reactive stubs is considered. The chapter concludes with a discussion in the time domain of nonsinusoidal waves on lossless lines, a topic of interest to designers of pulse-shaping circuits, to experimentalists making use of the time domain reflectometer in the analysis of the effects of mismatches on lossless lines, etc.

10-1 Voltage and Current Calculation on Lines with Reflection

In the present section, a connection is established between the forward and backward wave amplitudes $\hat{V}_m^+$ and $\hat{V}_m^-$. It is found that the reflected voltage amplitude $\hat{V}_m^-$ relative to an incident amplitude $\hat{V}_m^+$ depends on the disparity between the load value $\hat{Z}_L$ terminating the line and its characteristic impedance $\hat{Z}_0$, no reflection occurring on a line properly terminated with $\hat{Z}_L = \hat{Z}_0$. In a manner analogous to the methods of Chapter 6 concerned with plane wave reflections in multilayer systems, the cascaded line system of Figure 10-1(a) is analyzed by using reflection coefficient and impedance concepts. A simple extension of these ideas permits analyzing transmission-line systems such as the branched arrangement of Figure 10-1(b).

From the developments of Sections 9-3 and 9-7, the total voltage and current on a line in time-harmonic form are (9-102a) and (9-104)

$$\hat{V}(z) = \hat{V}_m^+ e^{-\gamma z} + \hat{V}_m^- e^{\gamma z} \qquad [9\text{-}102a]$$

$$\hat{I}(z) = \frac{\hat{V}_m^+}{\hat{Z}_0} e^{-\gamma z} - \frac{\hat{V}_m^-}{\hat{Z}_0} e^{\gamma z} \qquad [9\text{-}104]$$

The propagation constant and characteristic impedance are related to the line parameters by (9-103a) and (9-105)

$$\gamma(\equiv \alpha + j\beta) = \sqrt{\hat{z}\hat{y}} = \sqrt{(\boldsymbol{r} + j\omega \boldsymbol{l})(\boldsymbol{g} + j\omega \boldsymbol{c})} \qquad [9\text{-}103a]$$

$$\hat{Z}_0\left(\equiv \pm\frac{\hat{V}_m^\pm}{\hat{I}_m^\pm}\right) = \sqrt{\frac{\hat{z}}{\hat{y}}} = \sqrt{\frac{\boldsymbol{r} + j\omega \boldsymbol{l}}{\boldsymbol{g} + j\omega \boldsymbol{c}}} \qquad [9\text{-}105]$$

A comparison of (9-102a) and (9-104) with the electric and magnetic fields (6-29) and (6-31) reveals the analogy of the waves of voltage and current on a transmission line with plane waves normally incident on multilayer systems

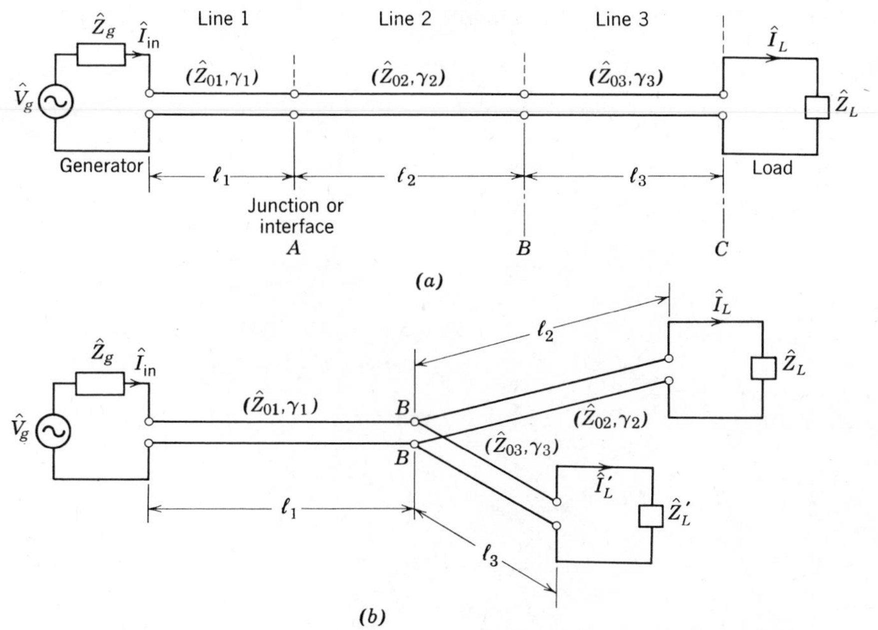

Figure 10-1. Generators connected to loads by use of lines with different $\hat{Z}_0$ and γ values. (*a*) Cascaded system of transmission lines of different γ and $\hat{Z}_0$, connected between a generator and load $\hat{Z}_L$. (*b*) Generator feeding two loads through a branched system.

as described in Section 6-6. Thus the cascaded line systems shown in Figure 10-1(*a*) can be analyzed by techniques already described in Chapter 6. Equations (9-102a) and (9-104), applicable to any line section of Figure 10-1, can be written in terms of a reflection coefficient $\hat{\Gamma}(z)$ as follows:

$$\hat{V}(z) = \hat{V}_m^+ e^{-\gamma z} + \hat{V}_m^- e^{\gamma z} = \hat{V}_m^+ e^{-\gamma z}[1 + \hat{\Gamma}(z)] \tag{10-1}$$

$$\hat{I}(z) = \frac{\hat{V}_m^+}{\hat{Z}_0} e^{-\gamma z} - \frac{\hat{V}_m^-}{\hat{Z}_0} e^{\gamma z} = \frac{\hat{V}_m^+}{\hat{Z}_0} e^{-\gamma z}[1 - \hat{\Gamma}(z)] \tag{10-2}$$

with $\hat{\Gamma}(z)$ defined in a manner analogous with (6-30)

$$\hat{\Gamma}(z) = \frac{\hat{V}_m^-}{\hat{V}_m^+} e^{2\gamma z} \tag{10-3}$$

A transmission-line impedance, defined as the ratio of the line voltage (10-1) to the line current (10-2), is analogous with (6-32)

$$\hat{Z}(z) \equiv \frac{\hat{V}(z)}{\hat{I}(z)} = \hat{Z}_0 \frac{1 + \hat{\Gamma}(z)}{1 - \hat{\Gamma}(z)} \tag{10-4}$$

Solving for $\hat{\Gamma}(z)$, one has conversely

$$\hat{\Gamma}(z) = \frac{\hat{Z}(z) - \hat{Z}_0}{\hat{Z}(z) + \hat{Z}_0} \tag{10-5}$$

the analog of (6-33). The reflection coefficient at any other location z', in terms of $\hat{\Gamma}(z)$, is obtained from (10-3) in the way used to obtain (6-34), yielding

$$\hat{\Gamma}(z') = \hat{\Gamma}(z)e^{2\gamma(z' - z)} \tag{10-6}$$

The item completing the analogy concerns the continuity of the impedance $\hat{Z}(z)$ of (10-4) at the junction of different lines (e.g., in Figure 10-1(a), the junctions A, B, and C). From the continuity of the voltage and the current to either side of the common plane between the lines, as in Figure 10-2, it is required that

$$\hat{Z}(z-) = \hat{Z}(z+) \tag{10-7}$$

the analog of (6-35). While from (10-7) $\hat{Z}(z)$ is *continuous* at a junction of two lines, it is evident from (10-5) that $\hat{\Gamma}(z)$ is *not*, since $\hat{Z}_0$ is different on the two sides of the interface.

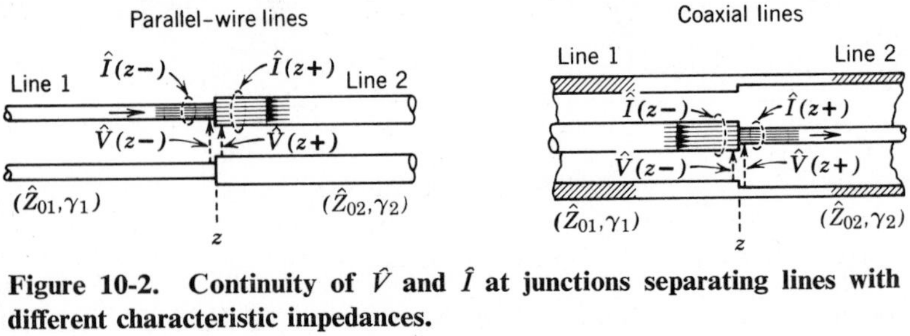

Figure 10-2. Continuity of $\hat{V}$ and $\hat{I}$ at junctions separating lines with different characteristic impedances.

In Table 10-1 is given a summary of the foregoing relations, along with analogous relations of Chapter 6 for plane waves. The application of (10-1) through (10-7) to line problems is illustrated in examples to follow.

In transmission-line systems it is noteworthy that, besides the dominant TEM mode, TM and TE modes can also exist on two-conductor lines. The latter modes, however, as for hollow waveguides treated in Chapter 8, are ordinarily highly attenuated below their cutoff frequencies, occurring for typical coaxial lines at the upper microwave frequencies and beyond. When a desired signal is dispatched in the dominant mode down a two-conductor line or a waveguide, a partial conversion of the signal power into higher-order modes will occur at discontinuities (sudden dimensional changes, sharp bends, etc.). Because of their high attenuation (evanescence), the higher modes vanish to negligible levels a short distance away. Accompanying the presence of the higher modes, however, is the development of an unwanted reflection in the dominant mode. For example, in joining two coaxial lines of different radial dimensions, the discontinuity at the junction may be shown to generate a reflection equivalent to that produced by a small capacitance C shunted across the lines at their junction,† even if their characteristic impedances are the same. At all frequencies except those approaching the microwave region, however, this effect is ordinarily very small (C is of the order of a few picofarads). It is ignored in the present treatment.

EXAMPLE 10-1. A transmitter, operated at 20 MHz and developing $\hat{V}_g = 100e^{j0^\circ}$ V with 50 Ω internal impedance, is connected to an antenna load through 6.33 m of the line described in Example 9-10. The antenna impedance at 20 MHz measures $\hat{Z}_L = 36 + j20\ \Omega$. (*a*) What are $\hat{Z}_0$, α, and β of this line, and how long is it in wavelengths? (*b*) Determine the input impedance of the line when terminated with $\hat{Z}_L$. (*c*) How much power is delivered to the line? (*d*) Compute the load current and time-average power absorbed by $\hat{Z}_L$. (*e*) If $\hat{Z}_L = 50\ \Omega$, what is the input impedance and how much average power is delivered to $\hat{Z}_L$?

(*a*) From Example 9-10, $\hat{Z}_0 = 50\ \Omega$, $\alpha = 1.97 \times 10^{-3}$ Np/m, $\beta = 0.595$ rad/m. With $\lambda = 10.55$ m, ℓ in wavelengths becomes $\ell/\lambda = 6.33/10.55 = 0.6$.

(*b*) $\hat{Z}_{\text{in}}$ is obtained by first finding $\hat{\Gamma}$ at the load using (10-5)

$$\hat{\Gamma}(0) = \frac{\hat{Z}(0) - \hat{Z}_0}{\hat{Z}(0) + \hat{Z}_0} = \frac{\hat{Z}_L - \hat{Z}_0}{\hat{Z}_L + \hat{Z}_0} = \frac{(36 + j20) - 50}{36 + j20 + 50} = 0.2765e^{j111.9^\circ} \quad (1)$$

† For the application of higher modes to coaxial-line discontinuities, see Whinnery, J. R., and H. W. Jamieson. "Equivalent circuits for discontinuities in transmission lines," *Proc. I.R.E.*, **32**, February 1944, p. 98; or Ghose, R. N. *Microwave Circuit Theory and Analysis*. New York: McGraw-Hill, 1963, Chapter 11.

Table 10-1 Transmission-line analog of plane wave propagation in multilayered regions

---

Multilayer regions with plane waves

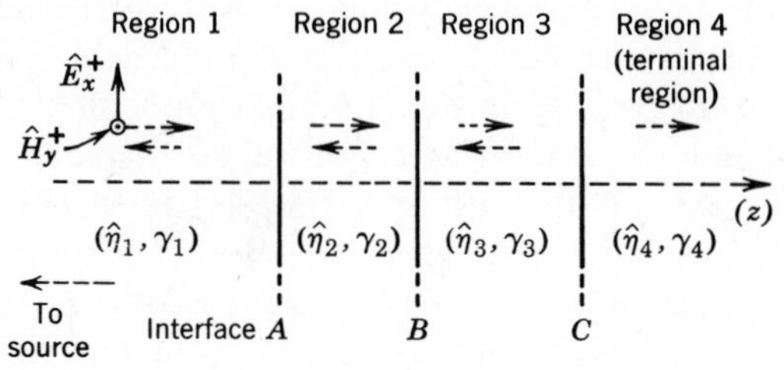

Total fields:

$$\hat{E}_x(z) = \hat{E}_m^+ e^{-\gamma z}[1 + \hat{\Gamma}(z)] \qquad [6\text{-}29]$$

$$\hat{H}_y(z) = \frac{\hat{E}_m^+}{\hat{\eta}} e^{-\gamma z}[1 - \hat{\Gamma}(z)] \qquad [6\text{-}31]$$

with

$$\hat{\Gamma}(z) \equiv \frac{\hat{E}_m^-}{\hat{E}_m^+} e^{2\gamma z} \qquad [6\text{-}30]$$

Total transverse field impedance:

$$\hat{Z}(z) \equiv \frac{\hat{E}_x(z)}{\hat{H}_y(z)} = \hat{\eta}\,\frac{1 + \hat{\Gamma}(z)}{1 - \hat{\Gamma}(z)} \qquad [6\text{-}32]$$

making

$$\hat{\Gamma}(z) = \frac{\hat{Z}(z) - \hat{\eta}}{\hat{Z}(z) + \hat{\eta}} \qquad [6\text{-}33]$$

At another location z':

$$\hat{\Gamma}(z') = \hat{\Gamma}(z)e^{2\gamma(z' - z)} \qquad [6\text{-}34]$$

Continuity of $\hat{Z}(z)$:

$$\hat{Z}(z-) = \hat{Z}(z+) \qquad [6\text{-}35]$$

Smith-chart use, normalizing (6-32):

$$\hat{z}(z) \equiv \frac{\hat{Z}(z)}{\hat{\eta}} = \frac{1 + \hat{\Gamma}(z)}{1 - \hat{\Gamma}(z)} \qquad [6\text{-}36]$$

Cascaded transmission lines

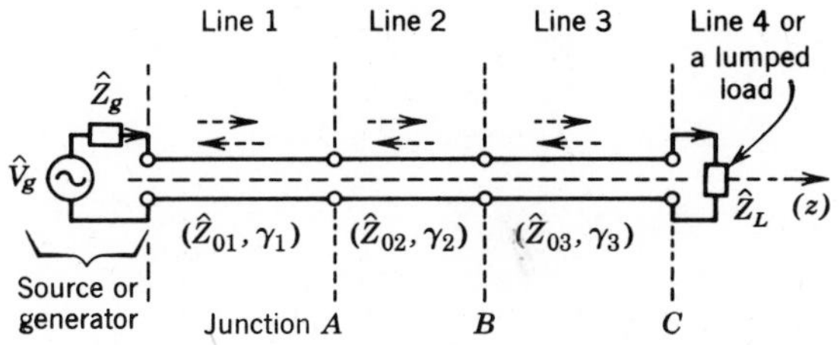

Total voltage and current:

$$\hat{V}(z) = \hat{V}_m^+ e^{-\gamma z}[1 + \hat{\Gamma}(z)] \qquad [10\text{-}1]$$

$$\hat{I}(z) = \frac{\hat{V}_m^+}{\hat{Z}_0} e^{-\gamma z}[1 - \hat{\Gamma}(z)] \qquad [10\text{-}2]$$

with

$$\hat{\Gamma}(z) \equiv \frac{\hat{V}_m^-}{\hat{V}_m^+} e^{2\gamma z} \qquad [10\text{-}3]$$

Line impedance:

$$\hat{Z}(z) \equiv \frac{\hat{V}(z)}{\hat{I}(z)} = \hat{Z}_0 \frac{1 + \hat{\Gamma}(z)}{1 - \hat{\Gamma}(z)} \qquad [10\text{-}4]$$

making

$$\hat{\Gamma}(z) = \frac{\hat{Z}(z) - \hat{Z}_0}{\hat{Z}(z) + \hat{Z}_0} \qquad [10\text{-}5]$$

At another location z':

$$\hat{\Gamma}(z') = \hat{\Gamma}(z) e^{2\gamma(z' - z)} \qquad [10\text{-}6]$$

Continuity of $\hat{Z}(z)$:

$$\hat{Z}(z-) = \hat{Z}(z+) \qquad [10\text{-}7]$$

Smith-chart use, normalizing (10-4):

$$\hat{z}(z) \equiv \frac{\hat{Z}(z)}{\hat{Z}_0} = \frac{1 + \hat{\Gamma}(z)}{1 - \hat{\Gamma}(z)} \qquad [10\text{-}8]$$

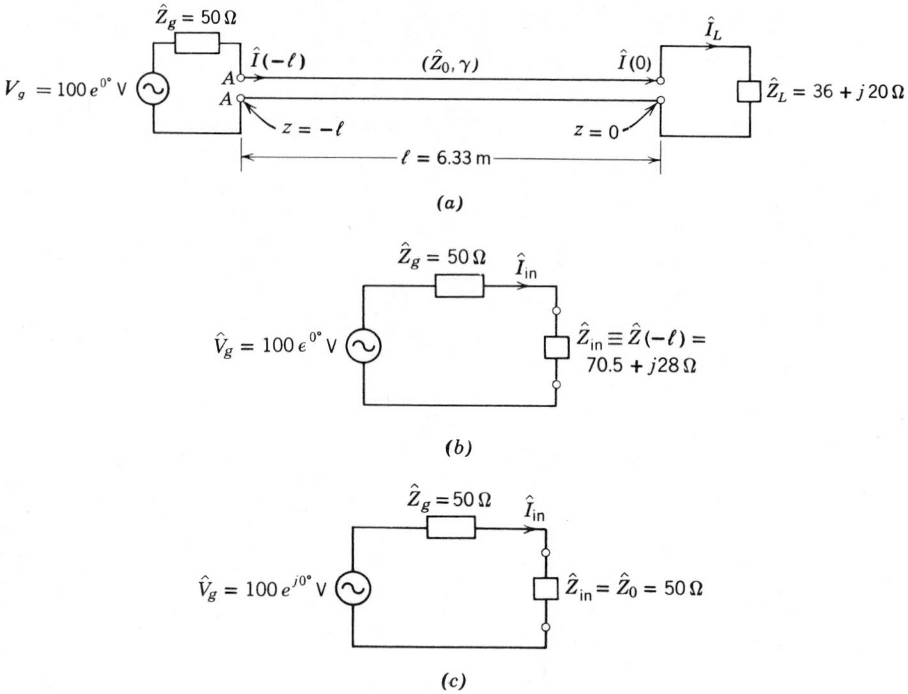

Example 10-1

Then $\hat{\Gamma}$ at $A - A$, by use of (10-6), becomes

$$\hat{\Gamma}(-\ell) = \hat{\Gamma}(0)e^{-2\alpha\ell}e^{-j2\beta\ell} = (0.2765e^{j111.9°})e^{-2(2\times10^{-3})6.33}e^{-j2(2\pi/\lambda)0.6\lambda}$$

$$= 0.2765e^{j111.9°}e^{-0.0253}e^{-j72°} = 0.212 + j0.177 \qquad (2)$$

in which the factor $e^{-0.00253}$ due to the low attenuation of this short line is essentially unity. Equation (2) into (10-4) yields the desired answer

$$\hat{Z}_{in} = \hat{Z}(-\ell) = \hat{Z}_0 \frac{1 + \hat{\Gamma}(-\ell)}{1 - \hat{\Gamma}(-\ell)} = 50\frac{1.212 + j0.177}{0.788 - j0.177}$$

$$= 75.8e^{j21°} = 70.5 + j28\ \Omega \qquad (3)$$

(c) $\hat{I}_{in}$ is obtained from the equivalent input circuit in figure (b)

$$\hat{I}_{in} = \hat{I}(-\ell) = \frac{V_g}{\hat{Z}_g + \hat{Z}(-\ell)} = \frac{100}{50 + 70.5 + j28} = 0.808e^{-j13.1°}\ \text{A} \qquad (4)$$

Thus the average input power, from (A-17) in the appendix, is

$$P_{av,in} = \tfrac{1}{2}\,\text{Re}\,(\hat{V}_{in}\hat{I}_{in}^*) = \tfrac{1}{2}\,\text{Re}\,(\hat{Z}_{in}\hat{I}_{in}\hat{I}_{in}^*) = \tfrac{1}{2}\,\text{Re}\,[75.8e^{j21°}(0.808)^2]$$

$$= \tfrac{1}{2}75.8(0.808)^2\cos 21° = 23\ \text{W} \qquad (5)$$

(*d*) Using (10-2), $\hat{I}_{in}$ is expressed in terms of $\hat{I}_m^+$; hence, $\hat{I}_{in} = \hat{I}(-\ell) = \hat{I}_m^+ e^{\alpha\ell} e^{j\beta\ell}[1 - \hat{\Gamma}(-\ell)]$, in which all quantities are known except $\hat{I}_m^+$. Solving for it yields

$$\hat{I}_m^+ = \frac{\hat{I}(-\ell)e^{-\alpha\ell}e^{-j\beta\ell}}{1 - \hat{\Gamma}(-\ell)} = \frac{0.808e^{-j13.1°}(1)e^{-(j2\pi/\lambda_0)(0.6\lambda_0)}}{1 - (0.212 + j0.177)} = 1e^{-j215.4°} \text{ A} \quad (6)$$

$\hat{I}_L$, written in terms of $\hat{I}_m^+$ by use of (10-2), thus becomes

$$\hat{I}_L = \hat{I}(0) = \hat{I}_m^+[1 - \hat{\Gamma}(0)] = 1e^{-j215.4°}[1 - (-0.103 + j0.2565)]$$
$$= 1.133e^{-j228.5°} \text{ A} \quad (7)$$

The average load power is found from (A-17) in the appendix

$$P_{av,L} = \tfrac{1}{2} \text{Re} [\hat{Z}_L \hat{I}_L \hat{I}_L^*] = \tfrac{1}{2} [41.2(1.133)^2 \cos 29°] = 23 \text{ W} \quad (8)$$

agreeing with (5) because of negligible losses.

(*e*) With $\hat{Z}_L = \hat{Z}_0 = 50\ \Omega$, $\hat{Z}_{in} = \hat{Z}_0 = 50\ \Omega$ also. Then from the equivalent input circuit, $\hat{I}_{in} = \hat{V}_g/(\hat{Z}_g + \hat{Z}_0) = 1$ A, yielding from figure (c)

$$P_{av,in} = \tfrac{1}{2} \text{Re} [50(1)^2] = 25 \text{ W} \quad (9)$$

From a well-known theorem of circuit theory, (9) represents the maximum power available from $\hat{V}_g$.

EXAMPLE 10-2. Sixty miles of the line of Example 9-9 are connected between a generator (developing, at 1 kHz, $\hat{V}_g = 20e^{j0°}$ V with $\hat{Z}_g = 700\ \Omega$) and the 1000-Ω load shown. (*a*) What are $\hat{Z}_0$, α, and β and what is ℓ in wavelengths? (*b*) Determine $\hat{Z}_{in}$ at $A - A$. (*c*) What is P_{av} into $A - A$? (*d*) Determine $\hat{I}_L$ and $P_{av,L}$. (*e*) With the line terminated in $\hat{Z}_0$ and the generator adjusted for a conjugate match ($\hat{Z}_g = \hat{Z}_0^*$), how much power is delivered to $A - A$ and the load?

(*a*) From Example 9-9, $\hat{Z}_0 = 703e^{-j13.2°}\ \Omega$, $\alpha = 0.0083$ Np/mi, and $\beta = 0.035$ rad/mi. The latter yields $\lambda = 179$ mi, obtaining $\ell/\lambda = 60/179 = 0.335$.

(*b*) $\hat{Z}_{in}$ is obtained by finding first $\hat{\Gamma}$ at the load from (10-5), yielding $\hat{\Gamma}(0) = (\hat{Z}_L - \hat{Z}_0)/(\hat{Z}_L + \hat{Z}_0) = 0.209e^{j32.5°}$. Applying (10-6), $\hat{\Gamma}$ at $A - A$ becomes

$$\hat{\Gamma}(-\ell) = \hat{\Gamma}(0)e^{2\gamma(-\ell-0)} = \hat{\Gamma}(0)e^{-2\alpha\ell}e^{-j2\beta\ell}$$
$$= (0.209e^{j32.5°})e^{-2(0.0083 \times 60)}e^{-j2(0.035 \times 60)}$$
$$= -0.0675 + j0.0371 \quad (1)$$

yielding the input impedance by use of (10-4)

$$\hat{Z}(-\ell) = \hat{Z}_0 \frac{1 + \hat{\Gamma}(-\ell)}{1 - \hat{\Gamma}(-\ell)} = 703e^{-j13.2°} \frac{1 + (-0.0675 + j0.0371)}{1 - (-0.0675 + j0.0371)}$$
$$= 615e^{-j8.8°}\ \Omega \quad (2)$$

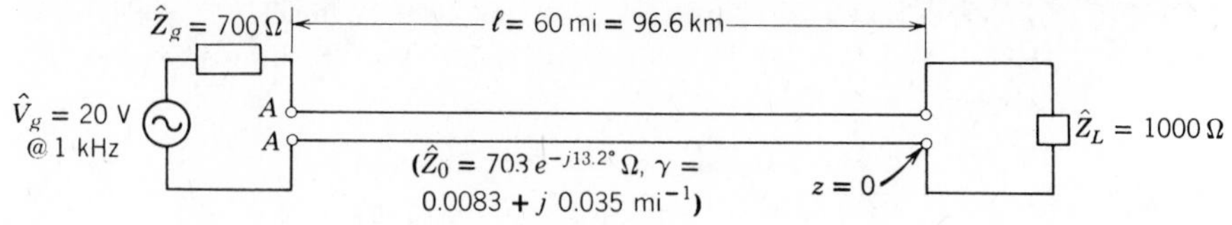

(a)

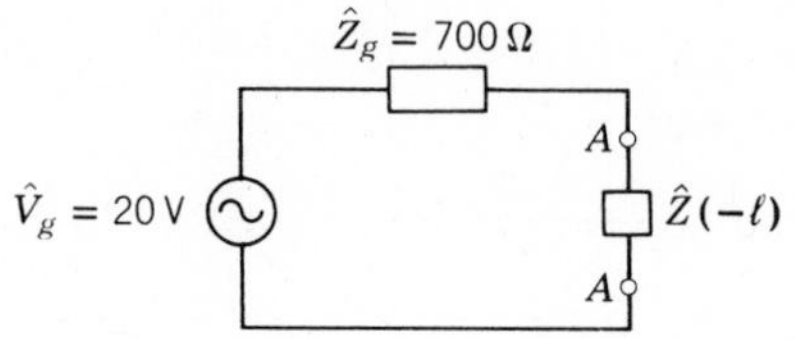

(b)

Example 10-2

(*c*) The power input at $A - A$ is found using the equivalent input circuit, whence $\hat{I}_{\text{in}} = 20/(700 + 607 - j93) = 0.0153e^{j4.1°}$ A. Then

$$P_{\text{av,in}} = \tfrac{1}{2}[61\,5(0.0153)^2 \cos 8.8°] = 71.8 \text{ mW} \tag{3}$$

(*d*) To find $\hat{I}_L$ from the input current, apply (10-2), obtaining $\hat{I}_{\text{in}} = \hat{I}_m^+ e^{\gamma\ell}[1 - \hat{\Gamma}(-\ell)]$; solving it yields $\hat{I}_m^+ = 0.0087e^{-j114.6°}$ A. Then by use of (10-2) at $z = 0$

$$\hat{I}(0) = \hat{I}_m^+ e^{-0}[1 - \Gamma(0)] = 0.0087e^{-j114.6°}[1 - 0.177 - j0.111]$$
$$= 0.0072e^{-j122.3°} \text{ A} \tag{4}$$

obtaining the average load power

$$P_{\text{av},L} = \tfrac{1}{2}\,\text{Re}\,[\hat{Z}_L\hat{I}_L\hat{I}_L^*] = \tfrac{1}{2}[1000(7.2 \times 10^{-3})^2 \cos 0°] = 25.9 \text{ mW} \tag{5}$$

(*e*) With $\hat{Z}_L = \hat{Z}_0$, the input impedance becomes $\hat{Z}_0$; so $\hat{Z}_g = \hat{Z}_0^* = 685 + j161$ obtains $\hat{I}_{\text{in}} = 20/(685 + j161 + 685 - j161) = 0.0146$ A. Then

$$P_{\text{av,in}} = \tfrac{1}{2}[703(0.0146)^2 \cos 13.2°] = 72.9 \text{ mW} \tag{6}$$

To find $\hat{I}_L$ from $\hat{I}_{\text{in}}$ when $\hat{\Gamma}(-\ell) = 0$, (10-2) yields $\hat{I}_m^+ = \hat{I}(-\ell)e^{-\alpha\ell - j\beta\ell} = 0.0146e^{-0.498}e^{-j120.7°} = 0.00887e^{-j120.7°}$ A, which is also $\hat{I}_L$, since (10-2) with $\hat{\Gamma}(0) = 0$ yields $\hat{I}(0) = \hat{I}_m^+$. Then

$$P_{\text{av},L} = \tfrac{1}{2}\,\text{Re}\,[\hat{Z}_0\hat{I}_L\hat{I}_L^*] = \tfrac{1}{2}[703(0.887 \times 10^{-2})^2 \cos 13.2°] = 27 \text{ mW} \tag{7}$$

10-2 Smith Chart Analysis of Transmission Lines

From Section 6-7 one recalls the normalized total transverse field impedance (6-36), used for analyzing plane wave behavior in multilayer regions by use of the Smith chart. Upon normalizing the line impedance (10-4) using the characteristic impedance $\hat{Z}_0$, one obtains the normalized impedance analogous with (6-36)

$$\frac{\hat{Z}(z)}{\hat{Z}_0} \equiv \hat{z}(z) = \frac{1 + \hat{\Gamma}(z)}{1 - \hat{\Gamma}(z)} \tag{10-8}$$

The latter, together with (10-6)

$$\hat{\Gamma}(z') = \hat{\Gamma}(z)e^{2\gamma(z'-z)} \qquad [10\text{-}6]$$

provide the ingredients for using the Smith chart in transmission-line applications. Examples follow showing details of the method. Though some accuracy is admittedly lost, time can be saved in problem-solving using the Smith chart. More significantly perhaps, the Smith chart displays graphically many transmission-line solutions all at once, providing a valuable tool for indicating the range of possible answers with variations in the available parameters.

EXAMPLE 10-3. Rework (*b*) of Example 10-1, using the Smith chart to obtain the input impedance.

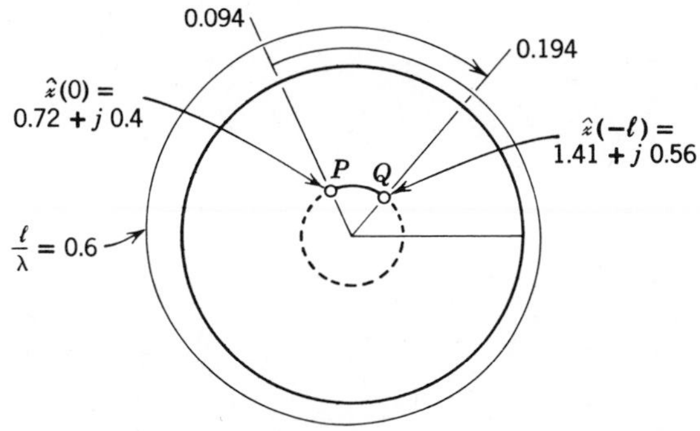

Example 10-3

The load impedance $\hat{Z}_L = \hat{Z}(0)$ is normalized using (10-8)

$$\hat{z}(0) \equiv \frac{\hat{Z}(0)}{\hat{Z}_0} = \frac{36 + j20}{50} = 0.72 + j0.40 \tag{1}$$

a result entered into the chart at P in the figure. The rotation by 0.6 λ, the line length, towards the generator obtains the normalized input impedance $\hat{z}(-\ell) =$

$1.41 + j0.56$, shown at Q. Denormalizing obtains

$$\hat{Z}(-\ell) = \hat{z}(-\ell)\hat{Z}_0 = (1.41 + j0.56)50 = 70.5 + j28 = 75.8e^{j21.7°}\ \Omega$$

the same as (4) in Example 10-1.

If desired, values of $\hat{\Gamma}$ at the output and input of the line can also be read at P and Q, obtaining

$$\hat{\Gamma}(0) = 0.28e^{j112°} \qquad \hat{\Gamma}(-\ell) = 0.28e^{j39.5°}$$

agreeing with (1) and (2) obtained analytically in Example 10-1.

EXAMPLE 10-4. Rework (*b*) of Example 10-2, using the Smith chart to obtain $\hat{Z}_{in}$ of the *lossy* line.

Normalizing $\hat{Z}_L \equiv \hat{Z}(0)$ by use of (10-8) yields $\hat{z}(0) = \hat{Z}(0)/\hat{Z}_0 = 1000/703e^{-j13.2°} = 1.384 + j0.324$, entered onto the chart at P in the figure. One obtains $\hat{z}(-\ell)$ from a phase rotation from P to Q *plus* a decrease in reflection coefficient amplitude in accordance with (10-6)

$$\hat{\Gamma}(-\ell) = \hat{\Gamma}(0)e^{2(\alpha + j\beta)(-\ell - 0)} = \hat{\Gamma}(0)e^{-2\alpha\ell}e^{-j2\beta\ell} = \hat{\Gamma}(0)e^{-2(0.0083 \times 60)}e^{-j2\beta\ell}$$
$$= 0.369\hat{\Gamma}(0)e^{-j2\beta\ell}$$

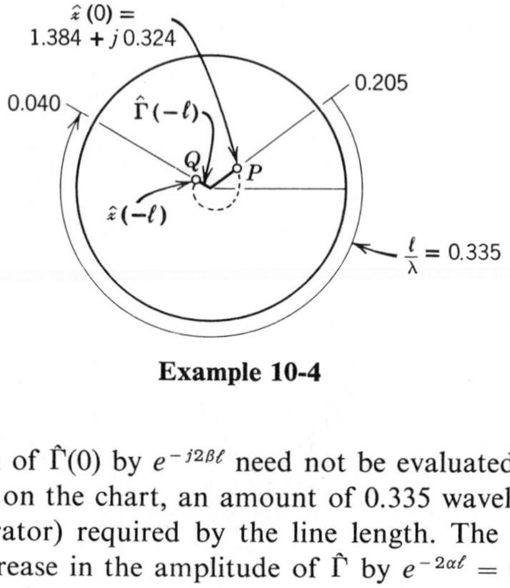

Example 10-4

The phase rotation of $\hat{\Gamma}(0)$ by $e^{-j2\beta\ell}$ need not be evaluated, being obtained from the *rim scale* on the chart, an amount of 0.335 wavelengths clockwise (towards the generator) required by the line length. The real factor $e^{-2\alpha\ell}$ accounts for a decrease in the amplitude of $\hat{\Gamma}$ by $e^{-2\alpha\ell} = 0.369$ in rotating from P to Q. Then $\hat{z}(-\ell)$ is read off the chart at Q, it is $\hat{z}(-\ell) = 0.87 + j0.07$. Denormalizing yields

$$\hat{Z}(-\ell) \equiv \hat{Z}_{in} = \hat{z}(-\ell)\hat{Z}_0 = 0.87e^{j4.5°}(703e^{-j13.2°}) = 613e^{-j8.7°}\ \Omega$$

in agreement with (2) of Example 10-2(*b*).

EXAMPLE 10-5. An antenna with a measured impedance $72 + j40\ \Omega$ at 20 MHz is to be driven by a transmitter 27 ft away. All that is available for this purpose are two coaxial lines with the characteristics:

$$\text{Line 1: } \hat{Z}_{01} = 70\ \Omega,\ \ell_1 = 15 \text{ ft} = 4.57 \text{ m, air dielectric}$$

$$\text{Line 2: } \hat{Z}_{02} = 50\ \Omega,\ \ell_2 = 12 \text{ ft} = 3.66 \text{ m, dielectric } \epsilon_r = 2$$

Ignore losses for these relatively short lines, connected as depicted in (*a*). (*a*) Express line lengths in terms of wavelength on each line. (*b*) With $\hat{V}_g = 100e^{j0°}$ and $\hat{Z}_g = 100\ \Omega$ use the Smith chart to find the impedance at $A - A$, and the average power delivered to $\hat{Z}_L$.

(*a*) Line 1 is lossless, so by use of (9-34b) and (9-38), $\lambda_1 = 2\pi/\beta_0 = c/f =$

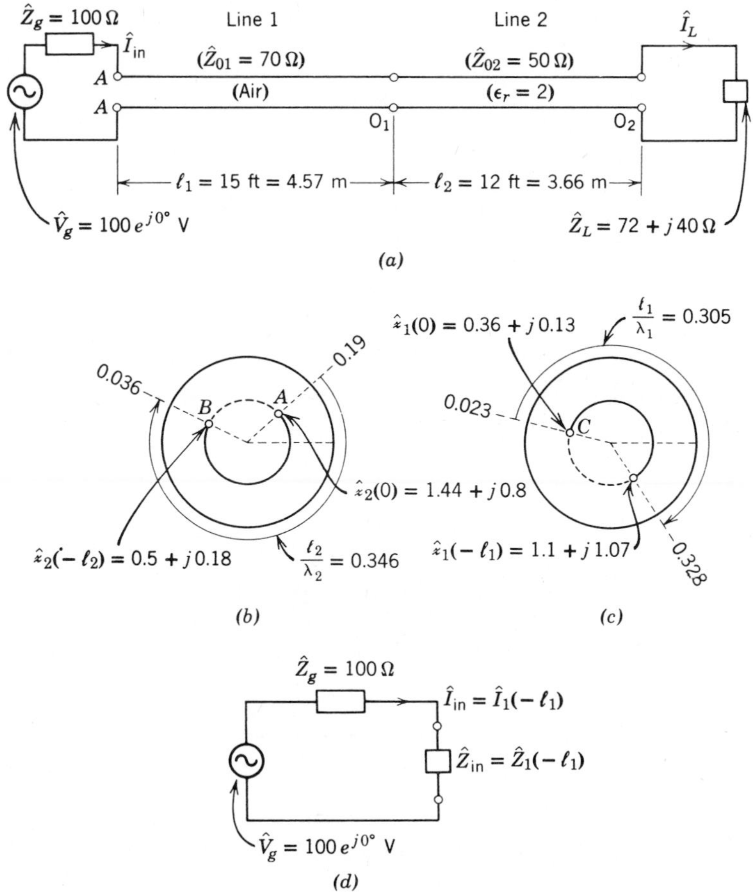

Example 10-5

15 m, whence $\ell_1 = 4.57$ m $= (4.57/15)\,\lambda_1 = 0.305\,\lambda_1$. Similarly, for line 2, $\lambda_2 = 10.6$ m, yielding $\ell_2 = 0.346\,\lambda_2$.

(*b*) The origins 0_1 and 0_2 are located as shown in (*a*). The normalized load impedance is $\hat{z}_2(0) = \hat{Z}_2(0)/\hat{Z}_{02} = (72 + j40)/50 = 1.44 + j0.8$, which entered at A in figure (*b*) and rotated $0.346\,\lambda_2$ towards the generator yields $\hat{z}_2(-\ell_2) = 0.5 + j0.18$, shown at B. From (10-7) it is the actual line impedance (not normalized) that is continuous at the junction, so denormalizing $\hat{z}_2(-\ell_2)$ obtains $\hat{Z}_2(-\ell_2) = \hat{z}_2(-\ell_2)\hat{Z}_{02} = (0.5 + j0.18)50 = 25 + j9\ \Omega = \hat{Z}_1(0)$. Normalizing $\hat{Z}_1(0)$ yields

$$\hat{z}_1(0) \equiv \frac{\hat{Z}_1(0)}{\hat{Z}_{01}} = \frac{25 + j9}{70} = 0.358 + j0.128 \tag{1}$$

entered at point C in figure (*c*). Rotating $0.305\,\lambda_1$ yields the normalized input impedance $\hat{z}_1(-\ell_1) = 1.1 - j1.07$, whence

$$\begin{aligned}\hat{Z}_{\text{in}} &= \hat{Z}_1(-\ell_1) = \hat{z}_1(-\ell_1)\hat{Z}_{01} = (1.1 - j1.07)70 \\ &= 78 - j73.5 = 107.2e^{-j43.3^\circ}\ \Omega\end{aligned} \tag{2}$$

From the equivalent input circuit one obtains $\hat{I}_{\text{in}} = 100/(178 - j73.5) = 0.518e^{j22.4^\circ}$ A, yielding from figure (d)

$$P_{\text{av,in}} = \tfrac{1}{2}[107.2(0.518)^2 \cos 43.3^\circ] = 10.5\ \text{W} \tag{3}$$

With both lines lossless, this is also $P_{\text{av},L}$ delivered to the antenna.

In the cascaded system of Figure 10-1(*a*), the line impedance appearing on the load side of any opened-up junction is the impedance seen by the line at the other side of the junction, evident from the continuity relation (10-7). If the junction at 0_1 in Example 10-5 were opened, for example, the impedance $\hat{Z}_2(-\ell_2) = 25 + j9\ \Omega$ obtained looking into line 2 is that seen by line 1 upon closing the junction. Similarly, $\hat{Z}_1(-\ell_1) = 78 - j73.5\ \Omega$ into line 1 is seen by the generator when connected to those terminals. In a system with branched lines as in Figure 10-1(*b*), the impedance seen by the line 1 at $B - B$, where line 2 and line 3 are parallel-connected, is just the parallel combination of their input impedances $\hat{Z}_2(-\ell_2)$ and $\hat{Z}_3(-\ell_3)$.

To find the impedance of two elements $\hat{Z}_1$ and $\hat{Z}_2$ connected in *parallel*, the expression

$$\hat{Z}_{\parallel} = \frac{\hat{Z}_1\hat{Z}_2}{\hat{Z}_1 + \hat{Z}_2}\ \Omega \tag{10-9a}$$

applies. By use of their admittances $\hat{Y}_1 = \hat{Z}_1^{-1}$ and $\hat{Y}_2 = \hat{Z}_1^{-1}$, the simpler relation

$$\hat{Y}_{\parallel} \equiv \hat{Z}_{\parallel}^{-1} = \hat{Y}_1 + \hat{Y}_2\ \mho \tag{10-9b}$$

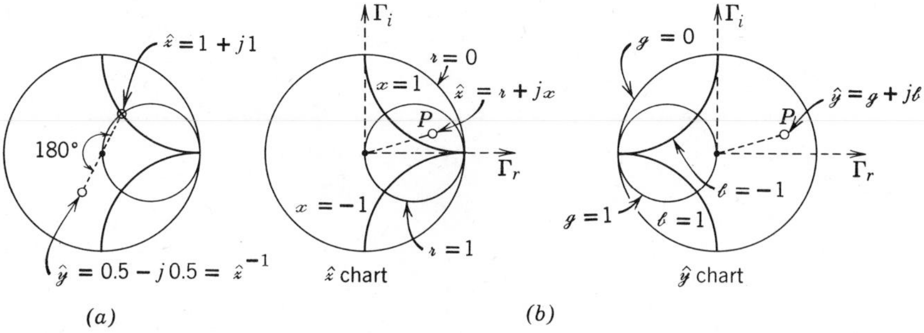

Figure 10-3. Reciprocal aspects of the Smith chart. (*a*) Reciprocal of $\hat{z}$ through 180° rotation. (*b*) Normalized impedance and admittance charts.

can be employed. If an attainable 1% graphic accuracy is adequate, the Smith chart is useful for finding $\hat{Y} \equiv \hat{Z}^{-1}$. This is possible through a property of the chart yielding the reciprocal of any complex number from its 180° rotation about the chart.† Thus, entering $\hat{z} = 1 + j1$ onto the chart and rotating it 180° as in Figure 10-3(*a*) yields $\hat{y} = 0.5 - j0.5$, the reciprocal of $\hat{z}$. In the case of a complex number with a large or small magnitude compared with unity, however, an arbitrary normalization forcing its magnitude near unity improves the accuracy of this process. For example $\hat{Z} = 150 + j100$ is very close to the point ∞ on the Smith chart; thus its reciprocal falls near the diametrically opposite zero point. Normalizing $\hat{Z}$ through a division by 100 to obtain $\hat{z} = 1.5 + j1$, however, yields from the chart its reciprocal $\hat{y} = \hat{z}^{-1} = 0.46 - j0.31$, and denormalizing yields $\hat{Y} = 0.0046 - j0.0031$, the desired reciprocal of $\hat{Z}$.

The foregoing reciprocal property leads to another version of the Smith chart, the normalized admittance chart, shown in Figure 10-3(*b*) alongside the more usual normalized impedance form. By simply *relabeling* the r and x circles of the $\hat{z}$ chart with g and b respectively, and rotating the chart 180°, one obtains the normalized admittance chart on which $\hat{y} = g + jb = \hat{z}^{-1}$. Thus, a given reflection coefficient $\hat{\Gamma}(z)$ depicted at P yields $\hat{z} = r + jx$ there on the $\hat{z}$ chart, while simultaneously displaying the corresponding $\hat{y} = g + jb$ at the same point on the admittance chart. The $\hat{y}$ chart is especially useful in the analysis of systems involving parallel-connected components requiring the use of (10-9b); e.g., in the application of stub sections of line to impedance matching taken up in Section 10-5.

EXAMPLE 10-6. Suppose the line and load described in Example 10-1 are connected to the lines of Example 10-5, forming the branched system in (*a*). With the aid of a Smith chart determine the following. (*a*) Find the impedance

† This property can be proved by use of (10-31) in Section 10.4.

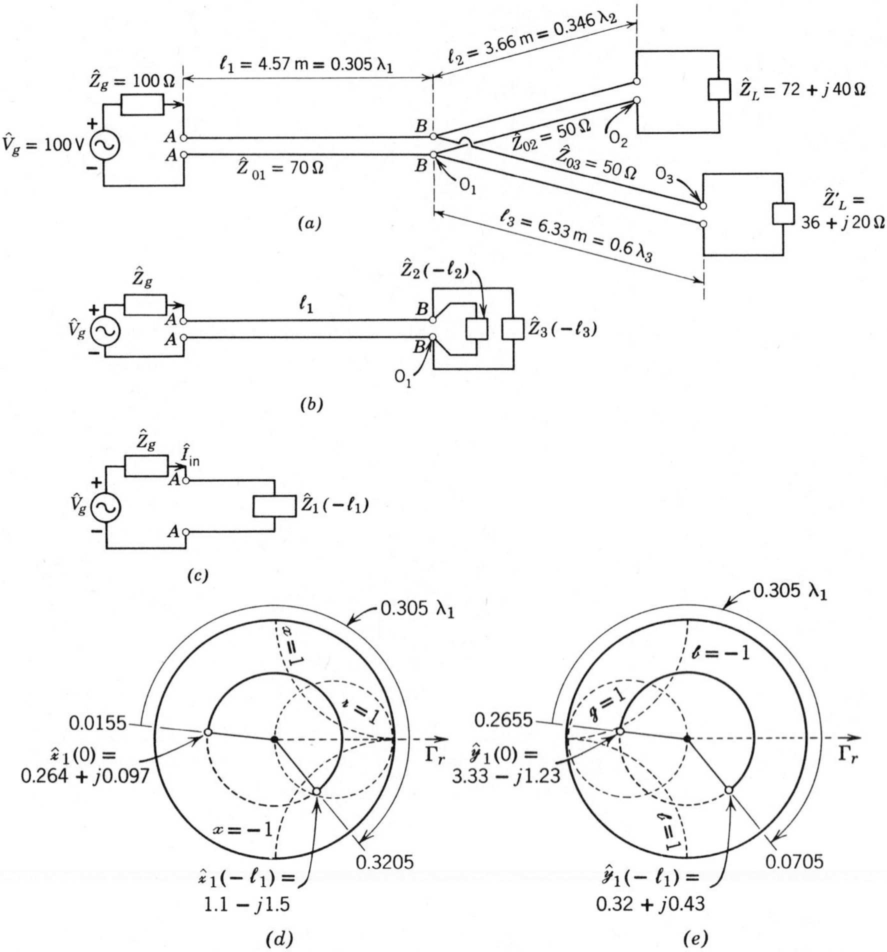

Example 10-6. (*a*) Branched system. (*b*) Equivalence at *BB*. (*c*) Equivalence at *AA*. (*d*) Using the $\hat{z}$ chart. (*e*) Using the $\hat{y}$ chart.

produced at $B - B$ by the parallel-connected lines and seen by line 1. Determine the impedance seen by the generator at $A - A$. (*b*) What power is delivered into $A - A$?

(*a*) From Example 10-5 and 10-3, $\hat{Z}_2(-\ell_2) = 25 + j9\ \Omega$ and $\hat{Z}_3(-\ell_3) = 70.5 + j28\ \Omega$. Line 1 is terminated with the parallel combination as in (*b*)

$$\hat{Z}_1(0) = \frac{\hat{Z}_2(-\ell_2)\hat{Z}_3(-\ell_3)}{\hat{Z}_2(-\ell_2) + \hat{Z}_3(-\ell_3)} = \frac{(25 + j9)(70.5 + j28)}{95.5 + j37}$$

$$= \frac{(26.6e^{j19.8^\circ})(75.7e^{j21.7^\circ})}{102.3e^{j21.2^\circ}}$$

$$= 19.7e^{j20.3^\circ} = 18.5 + j6.83\ \Omega \qquad (1)$$

Normalizing by use of $\hat{Z}_{01} = 70\ \Omega$ yields $\hat{z}_1(0) = 0.264 + j0.097$ which, entered onto the chart and rotated $0.305\ \lambda_1$ as in (*d*), yields $\hat{z}_1(-\ell_1) = 1.1 - j1.5$, whence

$$\hat{Z}_{\text{in}} = \hat{Z}_1(-\ell_1) = \hat{z}_1(-\ell_1)\hat{Z}_{01} = (1.1 - j1.5)70 = 77 - j105$$
$$= 130e^{-j53.7^\circ}\ \Omega \qquad (2)$$

A result equivalent to (2), expressed as the input admittance $\hat{Y}_1(-\ell_1)$, might have been found by beginning with $\hat{Y}_1(0) = [\hat{Z}_1(0)]^{-1} = 0.0476 - j0.0176$ loading line 1. Normalizing the latter by use of $\hat{Y}_{01} = \hat{Z}_{01}^{-1} = 0.0143\ \mho$ obtains $\hat{y}_1(0) = 3.33 - j1.23$. Entering it onto the $\hat{y}$ chart and rotating $0.305\ \lambda_1$ as in (*e*) obtains $\hat{y}_1(-\ell_1) = 0.32 + j0.43$, yielding

$$\hat{Y}_1(-\ell_1) = 0.0143(0.32 + j0.43) = 0.0046 + j0.0062\ \mho \qquad (3)$$

the reciprocal of (2) as expected.

(*b*) The generator delivers into line 1 $\hat{I}_{\text{in}} = 100/(177 - j105) = 0.485e^{j30.7^\circ}$ A, obtaining

$$P_{\text{av,in}} = \tfrac{1}{2}\,\text{Re}\,[\hat{Z}_{\text{in}}\hat{I}_{\text{in}}\hat{I}_{\text{in}}^*] = \tfrac{1}{2}[130(0.485)^2 \cos 53.7^\circ] = 9.05\ \text{W} \qquad (4)$$

The lines being lossless, (4) is divided (unequally) between the loads $\hat{Z}_L$ and $\hat{Z}_L'$. The power division can be obtained from the fact that a common voltage supplies the lines at their junction.

10-3 Standing Waves on Transmission Lines

The reflected waves of voltage and current occurring at the mismatched termination ($\hat{Z}_L \neq \hat{Z}_0$) of a line produce *standing waves* in a manner analogous to that process described for plane waves in Section 6-8. Voltage and current waves on a line with reflections are given by (10-1) and (10-2)

$$\hat{V}(z) = \hat{V}_m^+ e^{-\gamma z} + \hat{V}_m^- e^{\gamma z} = \hat{V}_m^+ e^{-\gamma z}[1 + \hat{\Gamma}(z)] \qquad (10\text{-}10)$$

$$\hat{I}(z) = \frac{\hat{V}_m^+}{\hat{Z}_0} e^{-\gamma z} - \frac{\hat{V}_m^-}{\hat{Z}_0} e^{\gamma z} = \frac{\hat{V}_m^+}{\hat{Z}_0} e^{-\gamma z}[1 - \hat{\Gamma}(z)] \qquad (10\text{-}11)$$

The presence of forward and backward waves gives rise to standing waves of voltage and current magnitudes as depicted in Figure 10-4. In (*a*) is seen the effect of the factors $e^{-\alpha z}$ and $e^{\alpha z}$ on the forward and backward voltage or current waves with losses present. This produces the standing-wave behavior

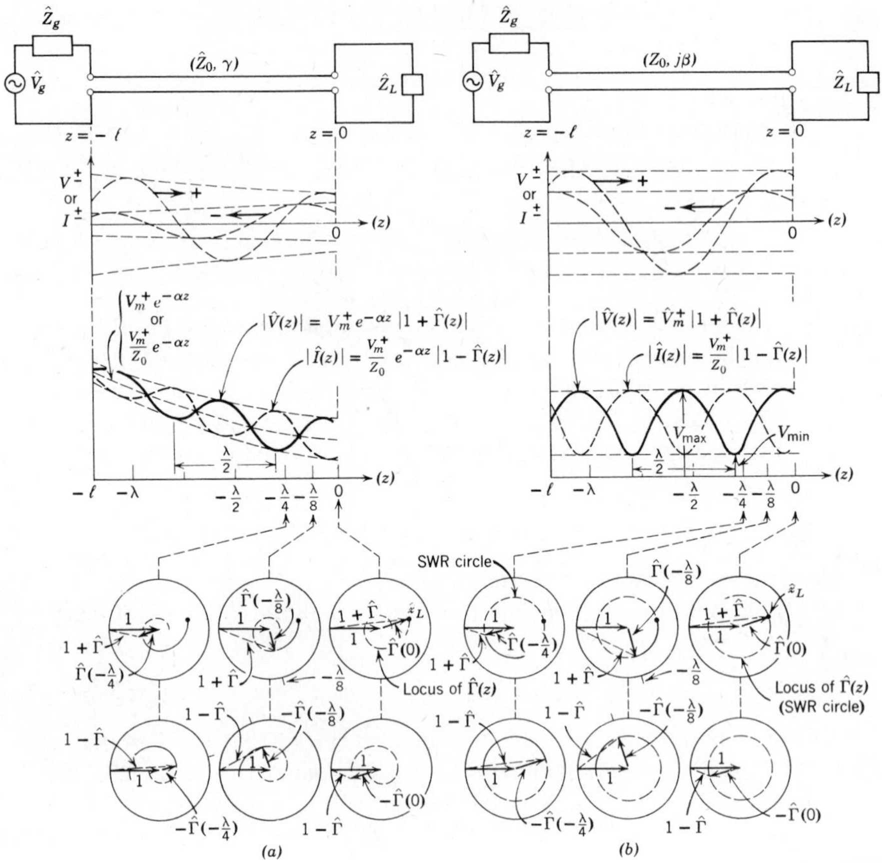

Figure 10-4. Depicting incident and reflected voltage and current waves on a lossy and lossless line, their magnitudes, and $1 + \hat{\Gamma}$ and $1 - \hat{\Gamma}$ from the Smith chart. (*a*) A line with losses. (*b*) A lossless line.

shown as curves of the magnitudes $|\hat{V}(z)|$ and $|\hat{I}(z)|$ plotted against distance with the undulations becoming smaller as the reflected wave diminishes with increasing distance from the load. The current standing wave dips where the voltage standing wave is maximal, a direct result of the minus sign in the reflected term of (10-11). Upon examining the magnitudes of (10-10) and (10-11)

$$|\hat{V}(z)| = V_m^+ e^{-\alpha z}|1 + \hat{\Gamma}(z)| \tag{10-12}$$

$$|\hat{I}(z)| = \frac{V_m^+}{Z_0} e^{-\alpha z}|1 - \hat{\Gamma}(z)| \tag{10-13}$$

it is seen that their graphs are readily obtained with the aid of a Smith chart. Thus, with the evaluation of $\hat{\Gamma}$ at the load position by use of (10-5) the quantities $|1 + \hat{\Gamma}|$ and $|1 - \hat{\Gamma}|$ appearing in (10-12) and (10-13) are obtained directly from the Smith chart as shown in Figure 10-4(*a*). Thus $\hat{\Gamma}$ is retarded in phase by $e^{j2\beta\ell}$ and diminished in length by $e^{2\alpha\ell}$ according to (10-6) in going to $z = -\ell$ towards the generator. Then with $|1 + \hat{\Gamma}|$ and $|1 - \hat{\Gamma}|$ multiplied by $e^{-\alpha z}$ and scaled by the factors V_m^+ and V_m^+/Z_0 according to (10-12) and (10-13), $|\hat{V}(z)|$ and $|\hat{I}(z)|$ follow curves typified in Figure 10-4(*a*). For a lossy line of sufficient length, the spiraling of $\hat{\Gamma}(z)$ towards the chart center reduces the undulations nearest the generator to practically zero, in view of the reduction of the reflected wave amplitude. The input impedance of a lossy line of sufficient length thus tends towards $\hat{Z}_0$, regardless of the load termination used.

If the line has sufficiently low over-all wave attenuation that it is essentially *lossless*, the standing-wave behavior is simplified. With $\gamma = j\beta$ and $\hat{Z}_0$ a pure resistance, the amplitudes of (10-10) and (10-11) become

$$|\hat{V}(z)| = V_m^+ |1 + \hat{\Gamma}(z)| \tag{10-14}$$

$$|\hat{I}(z)| = \frac{V_m^+}{Z_0} |1 - \hat{\Gamma}(z)| \tag{10-15}$$

The latter are analogous with (6-51) relative to the reflection of plane waves discussed in Chapter 6. In the absence of attenuation, $\hat{\Gamma}(z)$ varies only in *phase* along the line in accordance with the lossless version of (10-6): $\hat{\Gamma}(z) = \hat{\Gamma}(0)e^{j2\beta z}$. This provides the familiar *circular* locus of $\hat{\Gamma}(z)$ on the Smith chart shown typically in Figure 10-4(*b*) and termed the SWR circle. From it one can find the voltage and current along the line, obtainable by use of (10-14) and (10-15), with $|1 + \hat{\Gamma}(z)|$ and $|1 - \hat{\Gamma}(z)|$ found graphically from the charts of Figure 10-4(*b*). The analogy of this process with that of Figure 6-13 for a region with plane waves is evident. In such lossless systems, the standing-wave maxima and minima occur 90° (or $\lambda/4$) apart, with V_{max} (or equivalently E_{max}) at the location of I_{min} (or H_{min}), and *vice versa*.

The standing-wave ratio (SWR) associated with the voltage and current magnitude diagrams of Figure 10-4(*b*) is defined for a *lossless* line by

$$\text{SWR} \equiv \frac{|\hat{V}(z)|_{max}}{|\hat{V}(z)|_{min}} \equiv \frac{V_{max}}{V_{min}} = \frac{I_{max}}{I_{min}} \tag{10-16a}$$

analogous with (6-50). From the Smith chart representations in Figure 10-4(*b*), V_{max} and V_{min} on the line are seen to occur respectively at the locations of $1 + |\hat{\Gamma}|$ and $1 - |\hat{\Gamma}|$ on the SWR circle. Thus (10-16a) can be written in terms of the reflection coefficient magnitude as follows:

$$\text{SWR} = \frac{1 + |\hat{\Gamma}(z)|}{1 - |\hat{\Gamma}(z)|} \tag{10-16b}$$

having the converse

$$|\hat{\Gamma}(z)| = \frac{\text{SWR} - 1}{\text{SWR} + 1} \tag{10-17}$$

By arguments analogous with those used for plane waves, the SWR circle of a lossless line with reflections is centered on the Smith chart such that it passes through the SWR $= r$ point on the real axis of the chart, as depicted in Figure 6-12(c) for plane waves. Additional details concerning graphic interpretations of the forward and backward voltage and current waves can be developed from figures analogous with the electric and magnetic field diagrams shown in Figure 6-13.

Example 10-7. Find the SWR on the lossless line 2 in Example 10-5. Where are V_{max} and V_{min} located?

The magnitude of the reflection coefficient obtained from the Smith chart in Example 10-5 is $|\hat{\Gamma}| = 0.36$. The SWR using (10-16b) is therefore

$$\text{SWR} = \frac{1 + |\hat{\Gamma}|}{1 - |\hat{\Gamma}|} = \frac{1 + 0.36}{1 - 0.36} = \frac{1.36}{0.64} = 2.12$$

an answer also obtained from the r value intercepted by the SWR circle along the positive real axis of the chart as shown in (a). V_{max} occurs at N on the SWR circle, located $d = 0.06\ \lambda_2$ towards the generator, or $d = 0.06(10.6) = 0.636$ m from the load as shown in (b). V_{min} is at M, an additional quarter wave towards the generator as shown.

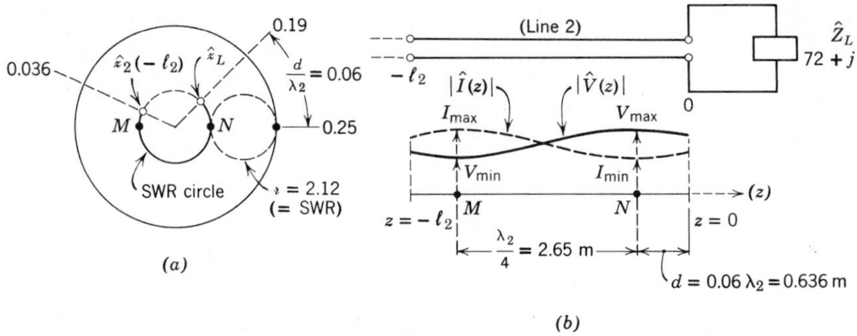

Example 10-7

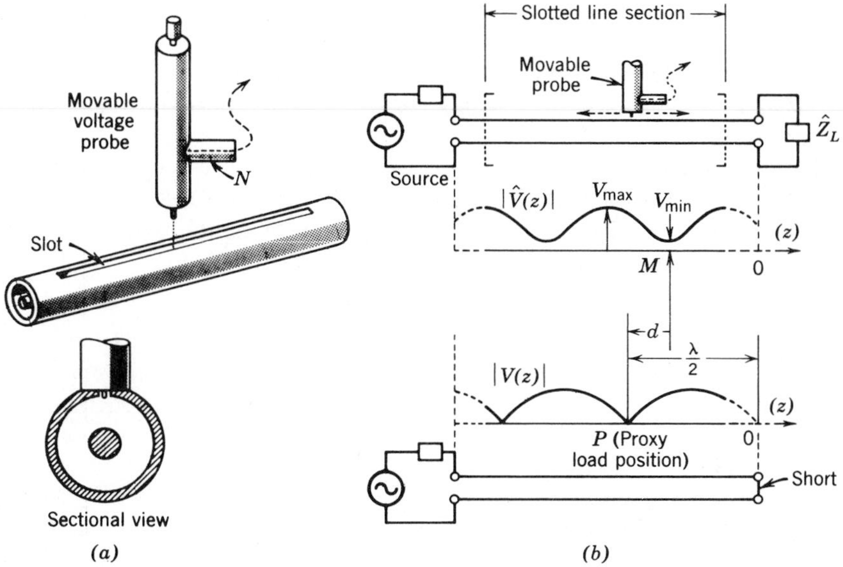

Figure 10-5. Slotted line and impedance measurements. (*a*) A coaxial slotted-line section and voltage (electric field) probe. (*b*) Determination of an unknown impedance from standing-wave measurements.

At the higher frequencies, above 100 MHz or so, impedances can be inferred from standing-wave data obtained from an instrument known as the *slotted line*, illustrated in Figure 10-5(*a*). A slotted line may be a rigid section of air dielectric line having a precision slot milled lengthwise through the outer conductor to accept a movable voltage-sensing probe. The latter travels along an externally mounted carriage to permit measuring, usually by use of a detector and amplifier system, the relative voltage anywhere along the slot. Position measurements are facilitated by use of an attached scale. The probe is permitted to penetrate only a short distance into the slot to minimize the distortion of the electric field being measured. When the slotted line is connected between a generator and a load as in Figure 10-5(*b*), the voltage standing wave developed within the slotted section is measured by the detector output. The impedance of the load can be inferred from measurements on the standing wave as follows.

With an unknown load $\hat{Z}_L$ attached to the slotted line, the SWR and the V_{min} position are recorded. The corresponding SWR circle is drawn on a Smith chart, with V_{min} (denoted M) occurring at the intersection of the SWR circle and the negative real axis. If $\hat{Z}_L$ is replaced with a short circuit in the load plane, the standing wave produced has nulls spaced by half wavelengths as shown in Figure 10-5(*b*). Each null location can be regarded as a *proxy load position*, a position where the load impedance is replicated when the line is once again terminated in $\hat{Z}_L$. (This property of a lossless line reproducing an impedance every half wavelength is evident from the Smith chart, since moving a half wavelength corresponds to a full rotation about the SWR circle.) Thus, if the proxy load position P were located a distance d from the V_{min} location M as in Figure 10-5(*b*), the impedance at P would be obtained from the Smith chart by a rotation d/λ on the SWR circle from M to P. Denormalizing $\hat{z}$ there by use of the line Z_0 obtains the unknown impedance at P, and hence $\hat{Z}_L$.

EXAMPLE 10-8. An unknown impedance $\hat{Z}_L$ is to be measured at 500 MHz by use of a 50 Ω slotted air line. Because of the location of $\hat{Z}_L$, it is connected to the slotted line using an additional length of lossless 50 Ω cable as in (*a*). With $\hat{Z}_L$ in place, the measured SWR is 3.2, V_{min} occurring at the scale position 19.4 cm along the slotted line. Replacing $\hat{Z}_L$ with a short, a null is observed at the position 11.2 cm. Determine $\hat{Z}_L$.

Drawing the SWR = 3.2 circle on the Smith chart as in (*b*), V_{min} is at M. The shift from V_{min} to the proxy load position P is $d = 19.4 - 11.2 = 8.2$ cm

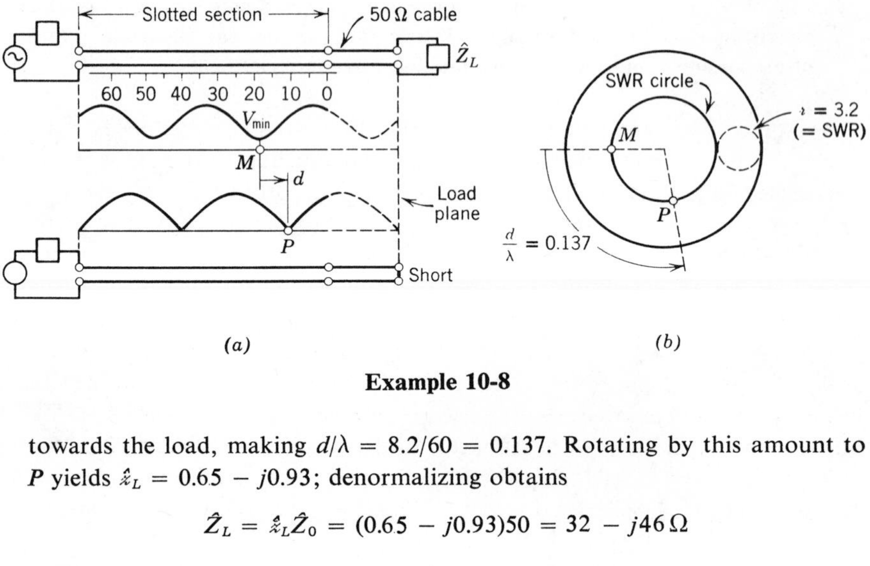

Example 10-8

towards the load, making $d/\lambda = 8.2/60 = 0.137$. Rotating by this amount to P yields $\hat{z}_L = 0.65 - j0.93$; denormalizing obtains

$$\hat{Z}_L = \hat{z}_L \hat{Z}_0 = (0.65 - j0.93)50 = 32 - j46\,\Omega$$

10-4 Analytical Expressions for Line Impedance

From previous sections it was seen that the input impedance of a section of line is a function of the complex termination $\hat{Z}_L$, the line length, and the line parameters that determine γ and $\hat{Z}_0$. One can consolidate these effects into a single expression for input impedance, if desired, noting first that the input impedance of the terminated line illustrated in Figure 10-4 is expressed by (10-4)

$$\hat{Z}_{\text{in}} \equiv \hat{Z}(-\ell) = \hat{Z}_0 \frac{1 + \hat{\Gamma}(-\ell)}{1 - \hat{\Gamma}(-\ell)} \tag{10-18}$$

The input reflection-coefficient $\hat{\Gamma}(-\ell)$ is written in terms of the load-value $\hat{\Gamma}(0)$ using (10-6), such that

$$\hat{\Gamma}(-\ell) = \hat{\Gamma}(0)e^{2\gamma(-\ell-0)} = \hat{\Gamma}(0)e^{-2\gamma\ell} \tag{10-19}$$

in which $\hat{\Gamma}(0)$ is specified in terms of the load $\hat{Z}(0) \equiv \hat{Z}_L$ by (10-4)

$$\hat{\Gamma}(0) = \frac{\hat{Z}_L - \hat{Z}_0}{\hat{Z}_L + \hat{Z}_0} \tag{10-20}$$

The substitution of the latter and (10-19) into (10-18) obtains

$$\hat{Z}_{\text{in}} \equiv \hat{Z}_0 \frac{1 + \dfrac{\hat{Z}_L - \hat{Z}_0}{\hat{Z}_L + \hat{Z}_0} e^{-2\gamma\ell}}{1 - \dfrac{\hat{Z}_L - \hat{Z}_0}{\hat{Z}_L + \hat{Z}_0} e^{-2\gamma\ell}} = \hat{Z}_0 \frac{(\hat{Z}_L + \hat{Z}_0) + (\hat{Z}_L - \hat{Z}_0)e^{-2\gamma\ell}}{(\hat{Z}_L + \hat{Z}_0) - (\hat{Z}_L - \hat{Z}_0)e^{-2\gamma\ell}} \tag{10-21a}$$

$$\hat{Z}_{\text{in}} = \hat{Z}_0 \frac{(\hat{Z}_L + \hat{Z}_0)e^{\gamma\ell} + (\hat{Z}_L - \hat{Z}_0)e^{-\gamma\ell}}{(\hat{Z}_L + \hat{Z}_0)e^{\gamma\ell} - (\hat{Z}_L - \hat{Z}_0)e^{-\gamma\ell}} \ \Omega \tag{10-21b}$$

Upon collecting like terms, (10-21b) can also be written in terms of the hyperbolic cosine and sine functions, obtaining

$$\hat{Z}_{\text{in}} = \hat{Z}_0 \frac{\hat{Z}_L \cosh \gamma\ell + \hat{Z}_0 \sinh \gamma\ell}{\hat{Z}_0 \cosh \gamma\ell + \hat{Z}_L \sinh \gamma\ell} \ \Omega \tag{10-21c}$$

if the definitions

$$\cosh \gamma\ell \equiv \frac{e^{\gamma\ell} + e^{-\gamma\ell}}{2} \qquad \sinh \gamma\ell \equiv \frac{e^{\gamma\ell} - e^{-\gamma\ell}}{2} \tag{10-22}$$

are employed. Note that if each of the expressions (10-21) is examined for the input impedance obtained if the load impedance equals $\hat{Z}_0$, the expected result $\hat{Z}_{\text{in}} = \hat{Z}_0$ obtains.

EXAMPLE 10-9. Use one of the expressions (10-21) to find the input impedance of the 60 mi of line terminated in 1000 Ω described in Example 10-2.

Substituting into (10-21a) the values of $\hat{Z}_0$, $\hat{Z}_L$, α, β, and ℓ obtained from Example 10-2 yields

$$\hat{Z}_{\text{in}} = \hat{Z}_0 \frac{(1000 + 685 - j161) + (1000 - 685 + j161)e^{-2(0.0082)60}e^{-j2(0.035)60}}{(1000 + 685 - j161) - (1000 - 685 + j161)e^{-2(0.0082)60}e^{-j2(0.035)60}}$$

$$= 703e^{-j13.2^\circ}\frac{1580e^{-j3^\circ}}{1807e^{-j7.6^\circ}} = 615e^{-j8.6^\circ}\ \Omega$$

which agrees with (2) of Example 10-2.

One can simplify (10-21) for the special case of a *lossless* line. With $\gamma = j\beta$ and the pure real characteristic impedance Z_0, (10-21c) reduces to

$$\hat{Z}_{\text{in}} \equiv \hat{Z}(-\ell) = Z_0 \frac{\hat{Z}_L \cos\beta\ell + jZ_0 \sin\beta\ell}{Z_0 \cos\beta\ell + j\hat{Z}_L \sin\beta\ell}\ \Omega \quad \text{Lossless line} \quad (10\text{-}23)$$

noting from (10-22) that

$$\cosh(j\beta\ell) = \cos\beta\ell \qquad \sinh(j\beta\ell) = j\sin\beta\ell \qquad (10\text{-}24)$$

In impedance calculations for lossless coaxial or parallel-wire lines, particularly at high frequencies, one is reminded that the Z_0 expressions (9-80c) and (9-80d), graphed in Figure 9-4, are useful in lossless line expressions such as (10-23).

Additional special cases of (10-21) can be generated for particular line lengths and loads. Of interest are the short-circuit and open-circuit load cases. If $\hat{Z}_L = 0$, the input impedance (10-21c) reduces to

$$\hat{Z}_{\text{in,sc}} = \hat{Z}_0 \tanh\gamma\ell\ \Omega \qquad \text{Short-circuit load} \quad (10\text{-}25)$$

with the hyperbolic tangent function defined by

$$\tanh\gamma\ell \equiv \frac{\sinh\gamma\ell}{\cosh\gamma\ell} = \frac{e^{\gamma\ell} - e^{-\gamma\ell}}{e^{\gamma\ell} + e^{-\gamma\ell}} \qquad (10\text{-}26)$$

With the line *lossless* ($\gamma = j\beta$, $\hat{Z}_0$ pure real), (10-25) becomes

$$\hat{Z}_{\text{in,sc}} = jZ_0 \tan\beta\ell \quad \text{Lossless line, short-circuit load} \quad (10\text{-}27)$$

since from (10-24) and (10-26), $\tanh(j\beta\ell) = j\tan\beta\ell$. Equation (10-27)

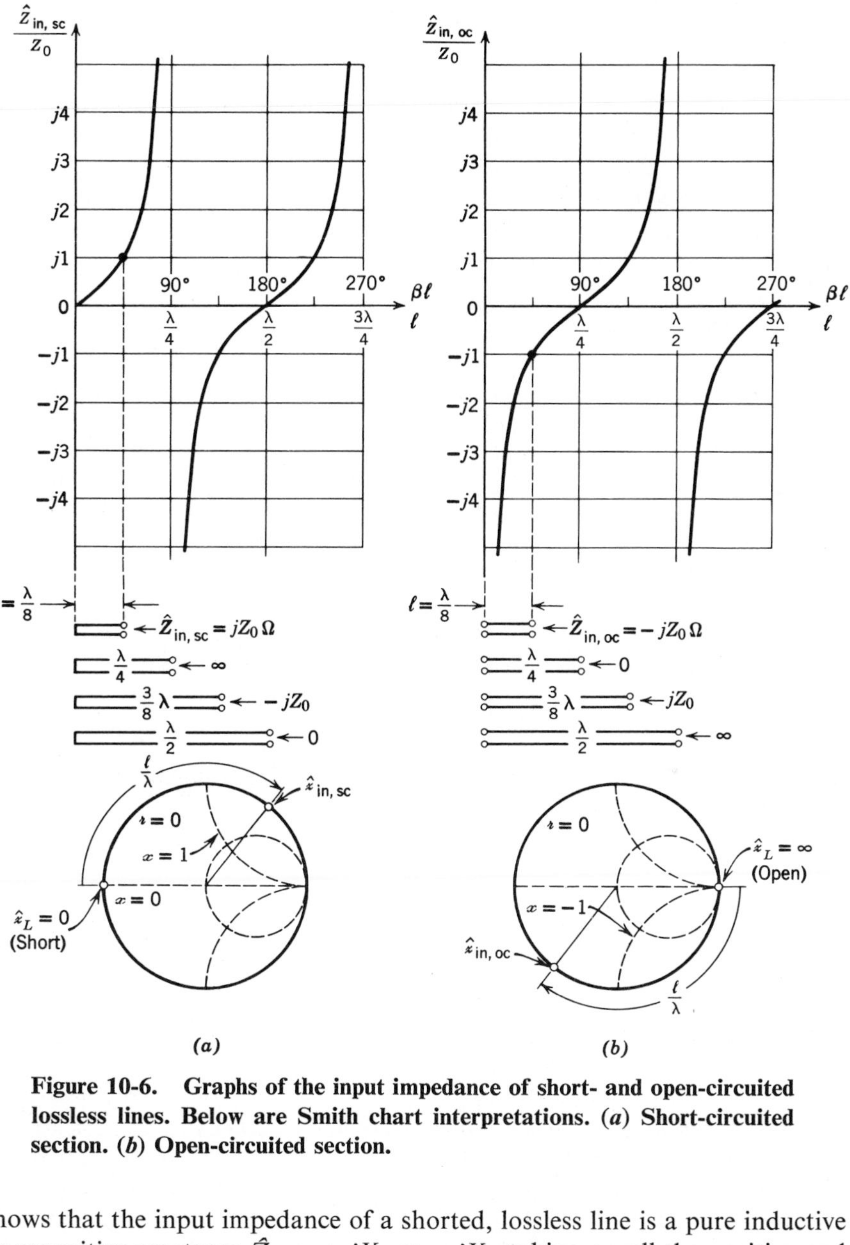

Figure 10-6. Graphs of the input impedance of short- and open-circuited lossless lines. Below are Smith chart interpretations. (*a*) Short-circuited section. (*b*) Open-circuited section.

shows that the input impedance of a shorted, lossless line is a pure inductive or capacitive reactance $\hat{Z}_{\mathrm{in,sc}} = jX_L$ or $-jX_c$, taking on all the positive and negative values of the tangent function with varying line length. In Figure 10-6(*a*) is shown a graph of (10-27), together with its Smith chart interpretation. Entering the Smith chart at the shorted load value $\hat{z}_L = 0$ (corresponding to the reflection coefficient $\hat{\Gamma} = -1$ there), a rotation on the *rim* by ℓ/λ provides the desired input reactance predicted by (10-27). Thus, a quarter wave, lossless, shorted line has an infinite input impedance. A shorted length of low-loss line is called a stub; if its length is variable through the use of telescoping conductors, it is an *adjustable stub*. Stubs are often used at high frequencies as the reactive elements in narrow band impedance-matching schemes, as described in the next section.

A section of transmission line with *open* load terminals ($Z_L \to \infty$) has an input impedance obtained from (10-21c)

$$\hat{Z}_{\text{in,oc}} = \hat{Z}_0 \coth \gamma\ell \ \Omega \quad \text{Open-circuit load} \quad (10\text{-}28)$$

wherein the hyperbolic cotangent function, $\coth \gamma\ell$, means $(\tanh \gamma\ell)^{-1}$. Equation (10-28) reduces, for a lossless line, to

$$\hat{Z}_{\text{in,oc}} = -jZ_0 \cot \beta\ell \quad \text{Lossless line, open-circuit load} \quad (10\text{-}29)$$

The unlimited range of capacitive and inductive reactance values provided by an open-circuited stub is depicted similarly in Figure 10-6(*b*).

Of additional interest is the input impedance of a lossless, one-half or one-quarter wave line with an arbitrary load $\hat{Z}_L$. For the line one-half wave long ($\beta\ell = \pi = 180°$), (10-23) reduces to

$$\hat{Z}_{\text{in}} = \hat{Z}_L \qquad \text{Lossless line, } \frac{\lambda}{2} \text{ long} \quad (10\text{-}30)$$

This result is not unexpected, since from the Smith chart, impedances on the SWR circle of a lossless line repeat themselves every half wavelength along the line, corresponding to a full rotation around the chart.

For a line one-quarter wavelength long ($\beta\ell = \pi/2 = 90°$), (10-23) becomes

$$\hat{Z}_{\text{in}} = \frac{Z_0^2}{\hat{Z}_L} \qquad \text{Lossless line, } \frac{\lambda}{4} \text{ long} \quad (10\text{-}31)$$

In view of (10-30), adding any integral number of half-wave sections of lossless line to the input of a quarter wave line still yields (10-31), making the latter valid for any line length totaling an odd number of quarter wavelengths. A lossless line obeying (10-31) is called a *quarter wave transformer*, a name arising from its use in matching a high or a low impedance load to a transmission line, from the insertion of a quarter wave section of lossless line having a properly chosen characteristic impedance. Thus, if a given load $\hat{Z}_L$ is to be fed from a line with a $\hat{Z}_0$ different from $\hat{Z}_L$, a quarter wave transformer connected to $\hat{Z}_L$ will have an input impedance $\hat{Z}_{\text{in}}$ that matches the feed-line characteristic impedance if the transformer section has the characteristic impedance, from (10-31), given by

$$Z_0 = \sqrt{\hat{Z}_{\text{in}}\hat{Z}_L} \tag{10-32}$$

In practice, a quarter wave transformer is used at high frequencies to connect a resistive load to a lossless line (with a pure real $\hat{Z}_0$). Because the method depends on the transformer section being a quarter wave long, the degree of impedance match is necessarily frequency-dependent. The frequency *bandwidth* of a matching scheme is conveniently specified in terms of the frequency deviation, to either side of the design frequency, at which the SWR established on the feed line departs from unity by more than some specified amount; a limit of 1.5 or so is often an acceptable criterion.

It can be shown that an increased bandwidth of the quarter wave matching scheme is realizable if the impedance transformation is made in two or more stages, with transformations made to intermediate resistive values. The limit of stepped systems such as this is the *tapered transmission line*, made several wavelengths long, providing a slowly varying characteristic impedance starting at the Z_0 of the input line and tapering to the load resistance value. The result is an extremely broadband matching device. Details of the bandwidth analysis of relatively narrow band matching devices such as the quarter wave transformer and of stub-matching systems described in the next section are found in a number of sources.†

EXAMPLE 10-10. A dipole antenna having a measured terminal impedance of 72 Ω at 150 MHz is driven from a parallel-wire line having a 300 Ω characteristic impedance. The feed-line conductors are spaced $2h = 0.75$ in. Design a quarter wave section of parallel-wire air line that will match the 72 Ω load to the 300 Ω line at this frequency.

The characteristic impedance of the quarter wave transformer is obtained from its load impedance and required input impedance, $\hat{Z}_{\text{in}} = 300\ \Omega$, by use

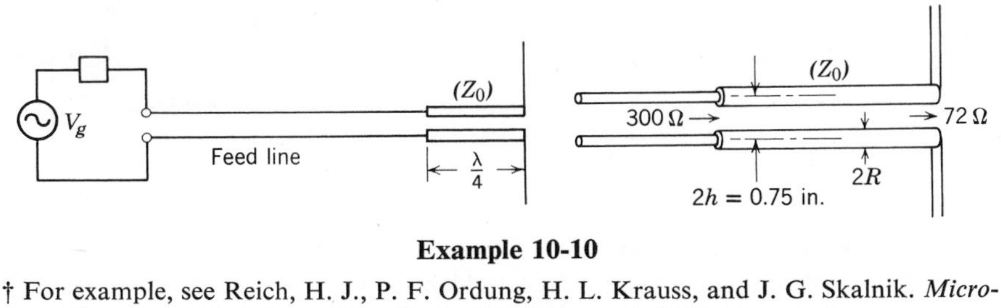

Example 10-10

† For example, see Reich, H. J., P. F. Ordung, H. L. Krauss, and J. G. Skalnik. *Microwave Theory and Techniques*. New York: Van Nostrand, 1953, Chapter 4.

of (10-32)

$$Z_0 = \sqrt{(72)(300)} = 147\ \Omega \tag{1}$$

Using the graph of Figure 9-4, $(h/R) = 1.85$ yields $Z_0 = 147\ \Omega$ for a lossless, parallel-wire line in air. Choosing the spacing $2h = 0.75$ in. for the quarter wave transformer, the conductor diameter becomes $2R = 0.405$ in. At 150 MHz, λ on the air dielectric transformer section, assumed lossless, is obtained using (9-34b) and (9-38); i.e., $\lambda = c/f = 2$ m, yielding the required length $\lambda/4 = 0.5$ m for the quarter wave section.

10-5 Impedance-Matching: Stub-Matching of Lossless Lines

In communication systems, the matching of line terminations to line characteristic impedances results in no power reflections, important in maximizing the power transfer to the load. Just as vital to system performance is the consideration that, if a load impedance were not matched to a line, the generator at the feed end would see different impedances at the various frequencies within the information-carrying band, a result of the frequency sensitivity exhibited generally by the input impedance (10-21) of an improperly terminated line. Thus, if pulse data were being transmitted, an improper termination would yield different reflections at the various frequencies within the Fourier spectrum of the pulse. The result is pulse-shape distortion, correctible by properly matching the load over the desired frequency band.

Practical impedance-matching arrangements are shown in Figure 10-7. At

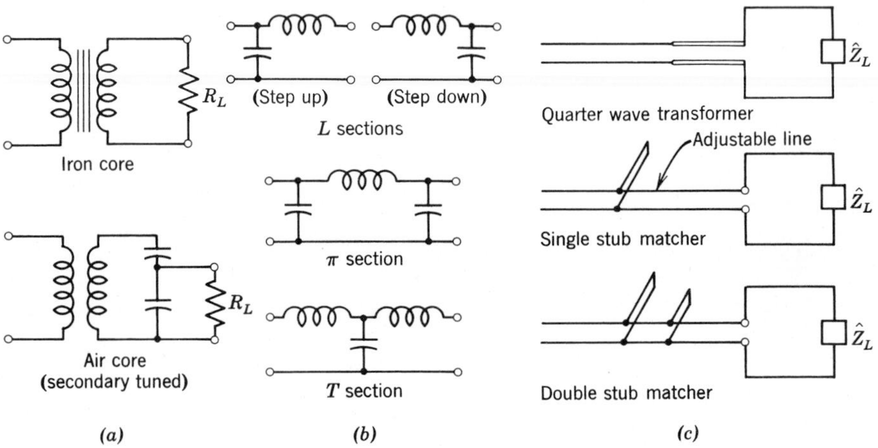

Figure 10-7. Impedance-matching schemes. (*a*) Transformers as impedance matchers. (*b*) Narrow band matching sections using lumped reactors. (*c*) High-frequency matchers using transformers or stubs.

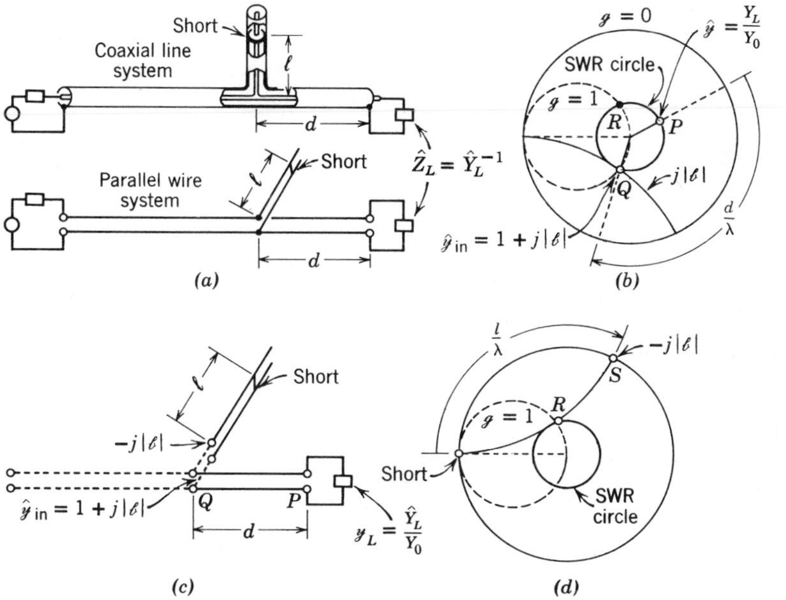

Figure 10-8. Details of single-stub-matching on a lossless line. (*a*) Use of a shorted stub in impedance-matching. (*b*) Use of the $\hat{y}$ chart in impedance-matching. (*c*) Adjustment of *d* and ℓ for a match. (*d*) Determining the stub length ℓ.

lower frequencies, transformers as shown in (*a*) of that figure can be employed for impedance-matching, with untuned, iron core transformers useful at power or audio frequencies. At radio frequencies not too high to eliminate the use of lumped or printed circuit elements, the *L*, *T*, and π configurations of Figure 10-7(*b*) are useable. Thus, antenna impedances can be matched with such schemes well into the VHF band (up to frequencies of the order of 100 MHz or so). At still higher frequencies (at wavelengths up to a few meters), lumped elements are physically too small or inefficient (low *Q*), so replacing them with low-loss transformer or stub sections as depicted in Figure 10-7(*c*) might be desirable. The object of the present section is to examine an impedance-matching technique making use of reactive stub elements introduced along the transmission line.

The single-stub-matching arrangement of Figure 10-8(*a*) is analyzed. Similar schemes employ double or triple-stub combinations. With the proper adjustment of the length ℓ of the stub and its position *d* from the arbitrary load $\hat{Z}_L$, it is shown how a low loss line can be matched to the impedance produced by the parallel combination of the stub and the remaining length *d* of line terminated in the mismatched load $\hat{Z}_L$. The Smith chart is an important time-saver in the analysis.

Because of the parallel connection of the stub and the transmission line, it is advantageous to employ the *admittance* form of the Smith chart described relative to Figure 10-3(*b*). The line and stub are both considered lossless, each having the same pure real characteristic admittance $\hat{Y}_0 = \hat{Z}_0^{-1}$. With the known load $\hat{Y}_L = \hat{Z}_L^{-1}$, normalized it becomes $\hat{y}_L = \hat{Y}_L/Y_0$, yielding an SWR circle passing through $\hat{y}_L$ at some point such as P in Figure 10-8(*b*). Moving towards the generator a distance d such that the intersection with the $g = 1$ circle at Q is obtained, the input admittance into that d length becomes $\hat{y} = 1 + j|b|$ as depicted in Figure 10-8(*c*). Another intersection with $g = 1$ is obtained farther towards the generator at R on the SWR circle; there the line admittance is $1 - j|b|$. If at either Q or R the line is shunted with the susceptance $\mp j|b|$ (provided by an adjustable shorted stub), there results a cancellation of the susceptive part $\pm j|b|$ of the line admittance, yielding the parallel admittance $\hat{y} = 1$ at Q or R. A matched impedance is thus obtained at Q or R upon reattaching the line to the left.

The remaining task is to find the stub length ℓ needed to provide $-j|b|$ at Q, or $+j|b|$ at R. If the stub is attached to Q as in Figure 10-8(*c*), the positive (capacitive) susceptance of the input admittance $\hat{y} = 1 + j|b|$ must be canceled by the negative (inductive) susceptance $-j|b|$ of the shorted stub of length ℓ. Its length is obtained as the distance ℓ/λ shown in Figure 10-8(*d*), measured as a rotation towards the generator from the susceptance $b \to \infty$ at the short to the required susceptance $-|b|$ at S on the chart rim.

Example 10-11. A transmitter operated at 150 MHz ($\lambda_0 = 2$ m) feeds a 72 Ω antenna load through 12 m of a lossless, 300 Ω, parallel-wire, air dielectric line. Determine the position d from the load at which a shorted stub should be connected and the required stub length, to match the load to the line as shown in (*a*). Assume the stub to be made of the same 300 Ω line.

The load admittance being $(72)^{-1} = 0.014$ and with the line $Y_0 = (300)^{-1} = 0.00333$, the normalized admittance becomes $y_L = (0.014)/(0.00333) = 4.17$,

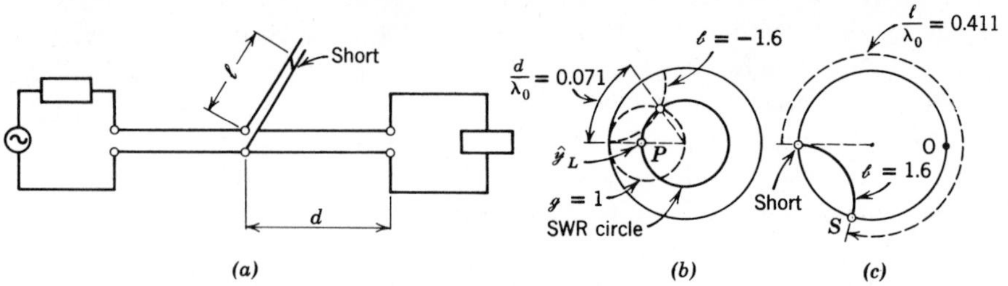

Example 10-11

shown at P on the Smith chart in (*b*). Rotating the latter by 0.071 λ on the SWR circle provides an intersection with $g = 1$ at $\hat{y} = 1 - j1.6$, so the stub must be located $d = 0.071\ \lambda_0 = 14.2$ cm from the load. To cancel the inductive susceptance $-j|b| = -j1.6$ there, the required stub length ℓ is given by the clockwise rotation ℓ/λ_0 from its short to the normalized susceptance $b = 1.6$ as shown in (*c*), yielding $\ell/\lambda_0 = 0.411$, so $\ell = (0.411)2 = 0.822$ m. If an open-circuited stub had been used, its length producing the same input susceptance is $\lambda/4$ shorter than that of the shorted stub, corresponding to the distance 0 to S on the rim.

*10-6 Nonsinusoidal Waves on Lossless Lines

The preceding sections dealt almost exclusively with single-frequency, time-harmonic voltage and current waves on transmission lines. The present section is concerned with waves produced by nonsinusoidal or pulse sources applied to such lines. As discussed in Section 8-5 dealing with group velocity, nonsinusoidal signals involving amplitude-modulated or frequency-modulated carriers are encountered extensively in systems making use of transmission lines to transport the signals over suitable paths. Their occurrence in computers, radar, television, and telephone and telegraph multiplexing systems involving data, voice, and video signals are but a few representative examples. The waves on a line fed by a nonsinusoidal generator are analyzed readily from the superposition of the harmonic frequency terms making up the generated waveshape, applying to each the time-harmonic, single-frequency methods already developed in foregoing sections. An alternative method is to attempt direct solutions of the time domain wave equations such as (9-88) or (9-89), shown to be readily obtained if the line is *lossless*.

The phase velocity of a sinusoidal transmission-line wave with the angular frequency ω is always (9-39)

$$v_p = \frac{\omega}{\beta} \text{ m/sec} \qquad [9\text{-}39]$$

as depicted in Figure 10-9(*a*). On a lossless line, β is specified by (9-34b) or (9-103c), yielding

$$v_p = \frac{1}{\sqrt{\mu\epsilon}} = \frac{1}{\sqrt{\boldsymbol{l}_e \boldsymbol{c}}} \text{ m/sec} \qquad (10\text{-}33)$$

a value independent of the frequency whenever μ and ϵ (or $\boldsymbol{l}_e$ and $\boldsymbol{c}$) can be regarded as constants over the spectrum of Fourier frequency components of interest within the transmission band. A Fourier superposition of sinusoidal harmonic waves is useful for representing the nonsinusoidal recurrent

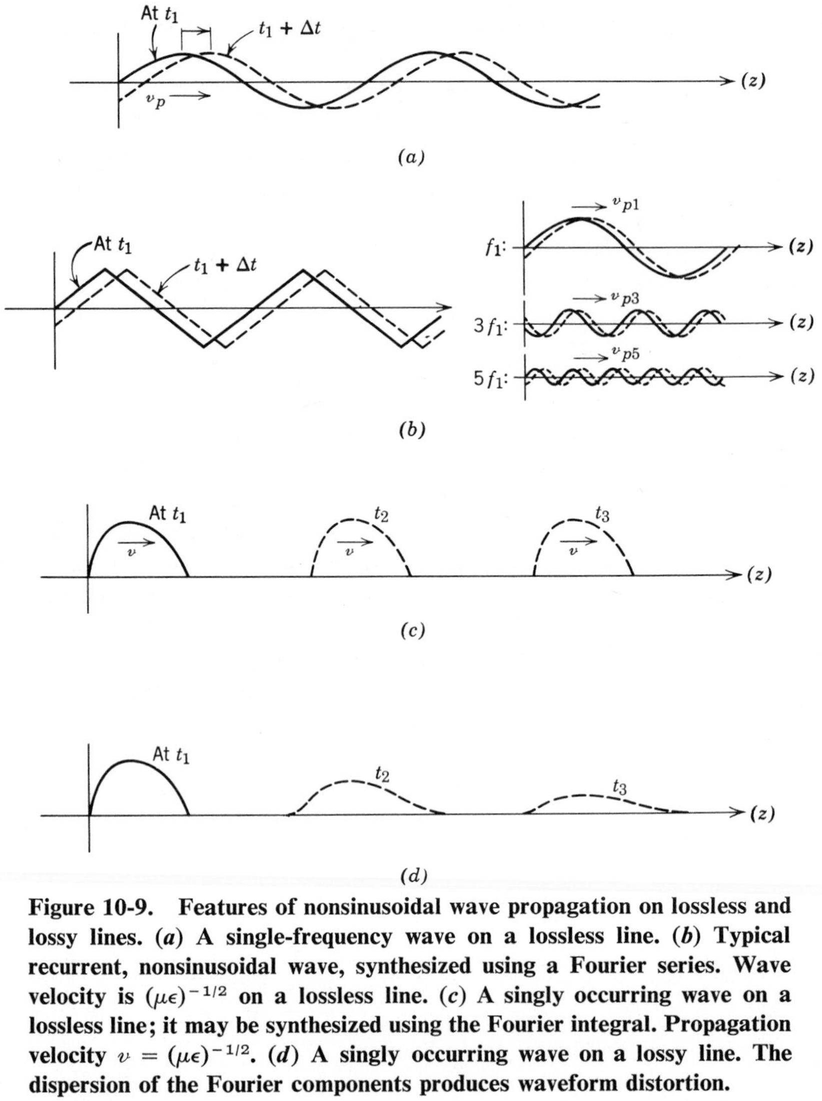

Figure 10-9. Features of nonsinusoidal wave propagation on lossless and lossy lines. (*a*) A single-frequency wave on a lossless line. (*b*) Typical recurrent, nonsinusoidal wave, synthesized using a Fourier series. Wave velocity is $(\mu\epsilon)^{-1/2}$ on a lossless line. (*c*) A singly occurring wave on a lossless line; it may be synthesized using the Fourier integral. Propagation velocity $v = (\mu\epsilon)^{-1/2}$. (*d*) A singly occurring wave on a lossy line. The dispersion of the Fourier components produces waveform distortion.

wave typified in Figure 10-9(*b*), or even singly occurring pulse waves as shown in Figure 10-9(*c*). Whenever each component proceeds down a line with exactly the same phase velocity and unattenuated amplitude, no wave-shape distortion can occur, in view of the precisely maintained phase relationship among all the constituent waves. A lossless line, for which v_p is constant at all frequencies, is termed nondispersive, yielding the distortionless transmission of nonsinusoidal or pulse signals. Free space is a nondispersive region, in view of the constant phase velocity $(\mu_0\epsilon_0)^{-1/2} = c$ given by (2-125b) for plane waves. An example of the dispersionless propagation of a nonsinusoidal wave in free space was illustrated by the amplitude-modulated carrier (8-78), depicted in Figure 8-14(*b*). Its Fourier series representation (8-78) reveals three time-harmonic plane waves propagating in a fixed phase relationship regardless of how far the modulated wave travels.

On a line with losses, the phase velocities of the harmonic Fourier terms vary with frequency, in view of the phase constant β given by the imaginary part of (9-103a). Waveform distortion is expected on a TEM line with losses, becoming more severe with longer line lengths because of the increase in the phase disparity among the Fourier terms. This phenomenon is depicted in Figure 10-9(*d*).

From the foregoing it has been seen, in principle at least, that the time-harmonic solution of a wave transmission problem contains all the necessary information about its performance under nonsinusoidal conditions. Even so, a direct attack on the transient wave problem in the time domain can often be more instructive, especially from the physical pictures it provides. The direct solution is readily obtained for the lossless transmission line, and it is to this dispersionless case that the remainder of this section is devoted.

The wave equations for lossless lines have been found to be (9-89a) and (9-89b)

$$\frac{\partial^2 V}{\partial z^2} - l_e c \frac{\partial^2 V}{\partial t^2} = 0 \qquad [9\text{-}89a]$$

$$\frac{\partial^2 I}{\partial z^2} - l_e c \frac{\partial^2 I}{\partial t^3} = 0 \qquad [9\text{-}89b]$$

The solutions of (9-89a) are in general a linear sum of forward- and backward-traveling waves

$$V(z, t) = V^+\left(t - \frac{z}{v}\right) + V^-\left(t + \frac{z}{v}\right) \text{ V} \qquad (10\text{-}34)$$

in which V^+ and V^- denote arbitrary functions of the traveling wave variables $t - z/v$ and $t + z/v$ respectively, v being the velocity of V^+ and V^- in their respective directions, given by (10-33)

$$v = \frac{1}{\sqrt{\mu\epsilon}} = \frac{1}{\sqrt{l_e c}} \text{ m/sec} \qquad (10\text{-}35)$$

That (10-35) is the correct velocity of a wave with an arbitrary waveshape is

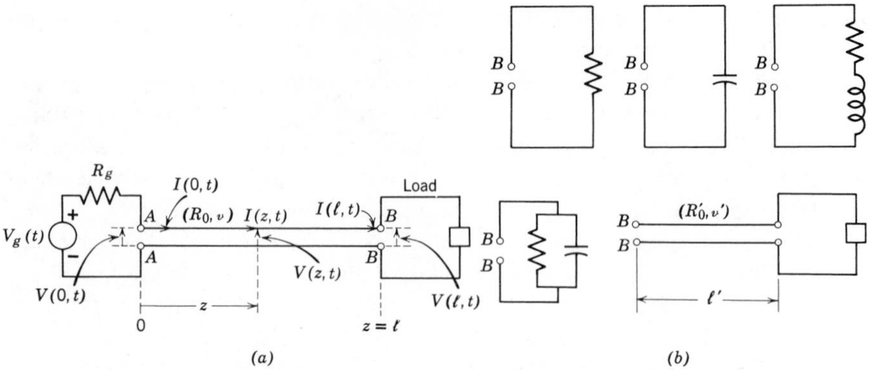

Figure 10-10. A lossless line fed by a nonsinusoidal generator. (*a*) Lossless line terminated arbitrarily. (*b*) Typical load terminations.

appreciated from the fact that the Fourier sinusoidal time-harmonic wave terms required to synthesize it must all have the same phase velocity v_p on a dispersionless line. The arbitrary waveshapes of the traveling wave terms in (10-34) are shown for specific problems to be governed by the waveform of the generated voltage at the sending end, the effect of the generator internal resistance, and the load conditions at the receiving end.

The nonsinusoidal wave denoted by $V^+(t - z/v)$ in (10-34) is launched onto the line upon applying the generated voltage to the input terminals, giving rise to that wave traveling down the line with the velocity v. If the line were properly terminated in its characteristic impedance, the pure resistance value given by (9-79)

$$\hat{Z}_0 = R_0 = \sqrt{\frac{l_e}{c}}\ \Omega \tag{10-36}$$

then the incident wave $V^+(t - z/v)$ would be totally absorbed at the load without reflection, since all the Fourier harmonics making up V^+ are absorbed by the proper termination. If any reflected wave $V^-(t - z/v)$ is to exist in (10-34), the line termination must differ from R_0.

The current waves accompanying (10-34) can be deduced from the fact that all Fourier voltage terms needed to construct (10-34) are related to their corresponding currents by R_0 of (10-36); therefore

$$I(z, t) = I^+\left(t - \frac{z}{v}\right) + I^-\left(t + \frac{z}{v}\right) \tag{10-37a}$$

$$I(z, t) = \frac{V^+\left(t - \frac{z}{v}\right)}{R_0} - \frac{V^-\left(t + \frac{z}{v}\right)}{R_0} \tag{10-37b}$$

Consider a class of problems typified in Figure 10-10(*a*). The source $V_g(t)$ of arbitrary waveshape (i.e., a pulse, a ramp function, etc.) has an internal resistance R_g. The details are given in the following in terms of (*A*) the input conditions resulting in positive z traveling waves of voltage and current and (*B*) the reflected waves obtained from the load conditions.

A. The Line Input Conditions and the Forward-Propagated Waves

Upon applying $V_g(t)$ to the system of Figure 10-10(*a*), the voltage and current waves consist initially of only the forward-traveling terms V^+ and I^+ of (10-34) and (10-37), from the physical fact that, with active sources only at the generator end, backward waves V^- and I^- cannot appear until the incident waves V^+ and I^+ reach the load (and then only if a load mismatch exists there). So before reflections appear, (10-34) and (10-37) are written

$$V(z, t) = V^+\left(t - \frac{z}{v}\right) \tag{10-38a}$$

$$t < \frac{\ell}{v}$$

$$I(z, t) = I^+\left(t - \frac{z}{v}\right) = \frac{V^+\left(t - \frac{z}{v}\right)}{R_0} \tag{10-38b}$$

The analytical form of $V^+(t - z/v)$ depends on the input voltage $V(0, t)$ developed at A-A, found by writing (10-38) in a form valid at the input $z = 0$

$$V(0, t) = V^+\left(t - \frac{0}{v}\right) \tag{10-39a}$$

$$z = 0 \text{ and } 0 < t < \frac{\ell}{v}$$

$$I(0, t) = \frac{V^+\left(t - \frac{0}{v}\right)}{R_0} \tag{10-39b}$$

but Kirchhoff's voltage law around the generator circuit and across A-A yields

$$V_g(t) = R_g I(0, t) + V(0, t) \tag{10-40}$$

so combining (10-40) with (10-39) yields the input voltage and current at $z = 0$

$$V(0, t) = V^+\left(t - \frac{0}{v}\right) = V_g(t)\frac{R_0}{R_g + R_0} \tag{10-41a}$$

$$I(0, t) = I^+\left(t - \frac{0}{v}\right) = \frac{V^+\left(t - \frac{0}{v}\right)}{R_0} = \frac{V_g(t)}{R_g + R_0} \tag{10-41b}$$

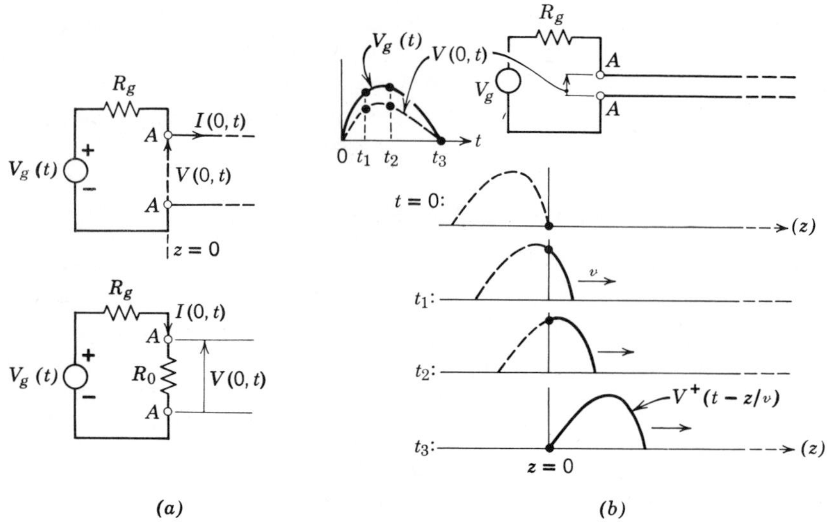

Figure 10-11. Nonsinusoidal input phenomena for a lossless line. (*a*) Equivalent input circuit for determining forward waves. (*b*) Launching of forward voltage wave produced by $V(0, t)$.

(10-41) suggests the equivalent input circuit in Figure 10-11(*a*). Thus $V^+(t - 0/v)$ developed at *A-A* sees the characteristic resistance R_0 of the line, to permit finding the input current $I(0, t) \equiv I^+(t - 0/v)$ by use of (10-41b).

The forward waves $V^+(t - z/v)$ and $I^+(t - z/v)$ launched on the line are a direct consequence of (10-41) appearing at *A-A*; i.e., they are simply (10-41) delayed by the retardation-time z/v. For example, if the applied voltage $V(0, t) = V^+(t - 0/v)$ obtained from (10-41a) were a steady voltage obtained by switching a battery onto the line at $t = 0$, the traveling wave $V^+(t - z/v)$ would be a wavefront moving down the line with the velocity $v = (\mu\epsilon)^{-1/2}$, spreading the constant voltage distribution on the entire line after a time lapse of ℓ/v seconds. For an arbitrary applied voltage $V(0, t)$ like that of Figure 10-11(*b*), the resulting wave $V^+(t - z/v)$ is a consequence of the voltage $V(0, t)$ appearing at *A-A* and moving down the line with the velocity v, illustrated at successive instants 0, t_1, t_2, t_3 in the figure.

EXAMPLE 10-12. The generated voltage $V_g(t)$ of 1 μsec duration and having the trapezoidal shape shown is applied at $t = 0$ to the lossless 50 Ω air line. The generator has 20 Ω resistance. Find the voltage and current developed at *A-A* and the forward waves established after $t = 0$. Sketch the results.

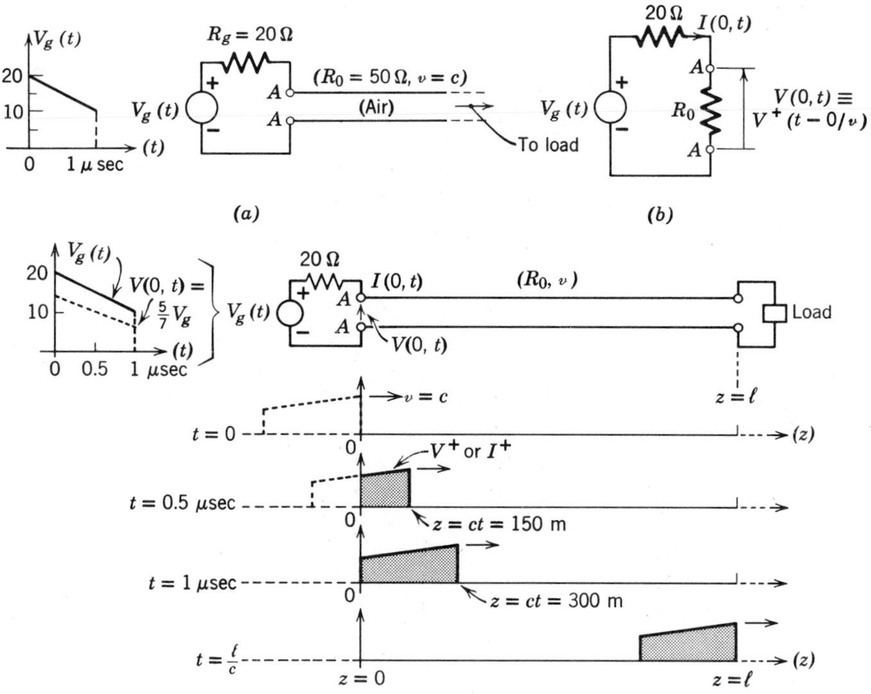

(c)

Example 10-12

At A-A, by use of (10-41) and the input circuit of (b)

$$V(0, t) = V^+\left(t - \frac{0}{v}\right) = V_g(t)\,\frac{50}{20 + 50} = \tfrac{5}{7}V_g(t)\text{ V} \tag{1}$$

$$I(0, t) = I^+\left(t - \frac{0}{v}\right) = \frac{V^+\left(t - \frac{0}{v}\right)}{50}\text{ A} \tag{2}$$

The resulting waves are (1) and (2) delayed by z/v sec (with $v = c$ in the air line), yielding

$$V(z, t) = V^+\left(t - \frac{z}{c}\right) = \tfrac{5}{7}V_g\left(t - \frac{z}{c}\right)\text{ V} \tag{3}$$

$$I(z, t) = \frac{V^+\left(t - \frac{z}{c}\right)}{50}\text{ A} \tag{4}$$

The latter are sketched at typical instants in (c).

B. Reflected Waves from Load Boundary Conditions

Next are found the reflections produced when the incident waves arrive at the load. The general solutions (10-34) and (10-37) incorporating both incident and reflected waves are required. At $z = \ell$ these are

$$V(\ell, t) = V^+\left(t - \frac{\ell}{v}\right) + V^-\left(t + \frac{\ell}{v}\right) \tag{10-42a}$$

$$I(\ell, t) = \frac{V^+\left(t - \frac{\ell}{v}\right)}{R_0} - \frac{V^-\left(t + \frac{\ell}{v}\right)}{R_0} \tag{10-42b}$$

The reflected waves in (10-42) depend on the load, examples of which are suggested in Figure 10-10(*b*). The two relations (10-42) contain three unknowns, $V(\ell, t)$, $I(\ell, t)$, and $V^-(t + \ell/v)$, since the forward wave $V^+(t - \ell/v)$ is presumed known from part A. A third equation is thus required at $z = \ell$. It can be established once the load configuration is assigned. For the representative loads shown in Figure 10-10(*b*), their Kirchhoff voltage relationships are

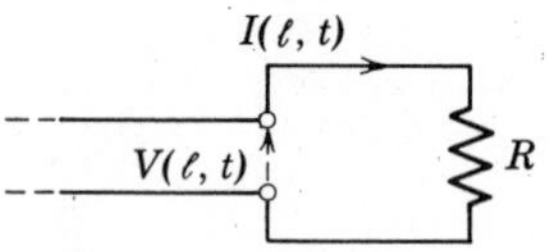

$$V(\ell, t) = RI(\ell, t) \tag{10-43a}$$

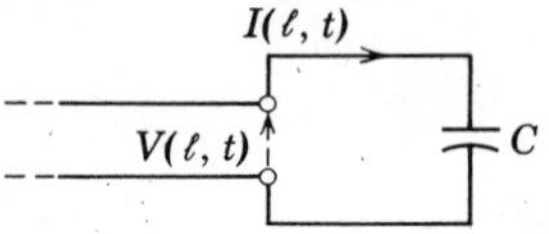

$$I(\ell, t) = C\frac{dV(\ell, t)}{dt} \tag{10-43b}$$

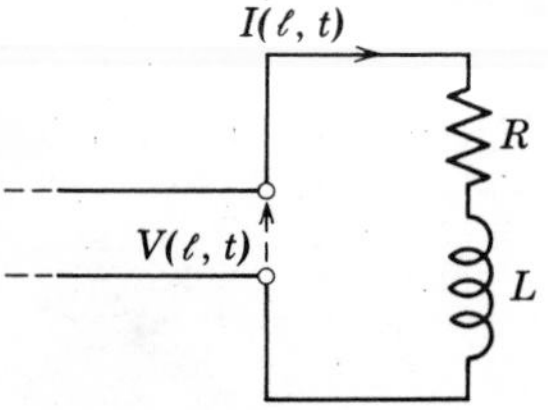

$$V(\ell, t) = RI(\ell, t) + L\frac{dI(\ell, t)}{dt} \tag{10-43c}$$

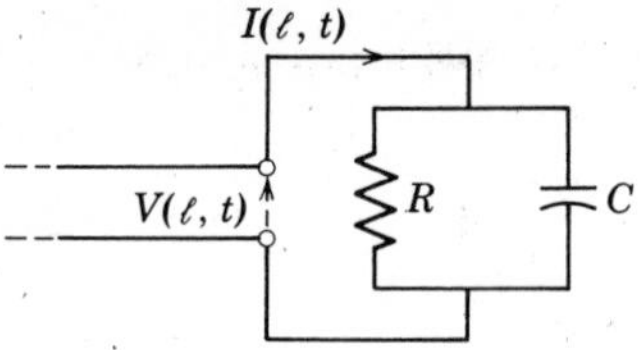

$$I(\ell, t) = \frac{V(\ell, t)}{R} + C\frac{dV(\ell, t)}{dt} \tag{10-43d}$$

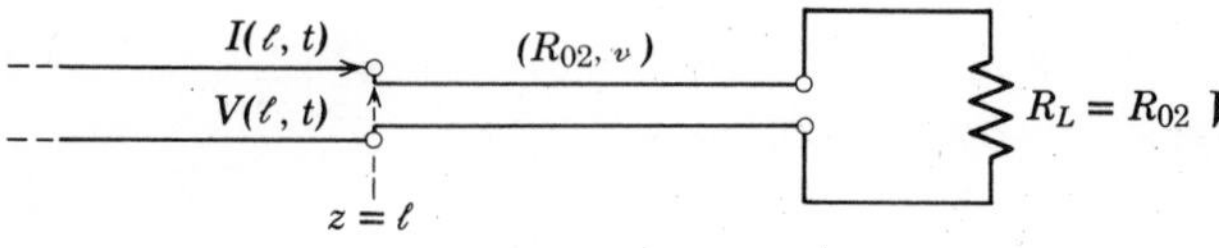

$$V(\ell, t) = R_{02}I(\ell, t) \tag{10-43e}$$

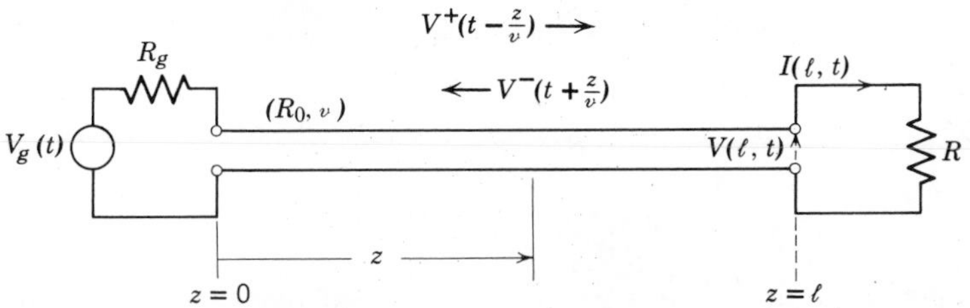

Figure 10-12. Resistive terminated line.

Considered first is the simple though important case of a resistive load, depicted in Figure 10-12. At the load position $z = \ell$, (10-42a), (10-42b) and (10-43a) must be satisfied, so with the incident waves $V^+(t - z/v)$ and $I^+(t - z/v)$ arriving when $t = \ell/v$, one may write for any t after that instant

$$V(\ell, t) = V^+\left(t - \frac{\ell}{v}\right) + V^-\left(t + \frac{\ell}{v}\right) \tag{10-44a}$$

$$I(\ell, t) = \frac{V^+\left(t - \dfrac{\ell}{v}\right)}{R_0} - \frac{V^-\left(t + \dfrac{\ell}{v}\right)}{R_0} \qquad z = \ell, t \geqslant \frac{\ell}{v} \tag{10-44b}$$

$$V(\ell, t) = RI(\ell, t) \tag{10-44c}$$

Defining a *real time domain reflection coefficient* Γ at the load as the ratio of instantaneous reflected to incident voltages

$$\Gamma(\ell, t) = \frac{V^-\left(t + \dfrac{\ell}{v}\right)}{V^+\left(t - \dfrac{\ell}{v}\right)} \tag{10-45}$$

permits writing (10-44a) and (10-44b)

$$V(\ell, t) = V^+\left(t - \frac{\ell}{v}\right)[1 + \Gamma(\ell, t)] \tag{10-46a}$$

$$I(\ell, t) = \frac{V^+\left(t - \dfrac{\ell}{v}\right)}{R_0}[1 - \Gamma(\ell, t)] \tag{10-46b}$$

From (10-44c), the ratio of the load voltage $V(\ell, t)$ to the current $I(\ell, t)$ is R, yielding from the ratio of (10-46a) to (10-46b)

$$R = R_0 \frac{1 + \Gamma(\ell)}{1 - \Gamma(\ell)} \tag{10-47}$$

Solving for $\Gamma(\ell)$ obtains

$$\Gamma(\ell) = \frac{R - R_0}{R + R_0} \equiv \frac{V^-\left(t + \dfrac{\ell}{v}\right)}{V^+\left(t - \dfrac{\ell}{v}\right)} \tag{10-48}$$

in which the notation $\Gamma(\ell, t)$ is altered to read just $\Gamma(\ell)$. Equation (10-48) is useful for finding $V^-(t + \ell/v)$ at the resistive load whenever the incident wave $V^+(t - \ell/v)$ is known. It is emphasized that the pure real $\Gamma(\ell)$ in (10-48) is a consequence of the output expression (10-43a) being purely *algebraic*. A time domain reflection coefficient is undefined for reactive loads corresponding to (10-43b, c, and d), except in the asymptotic limits for which the derivative terms in the load differential equations become negligible.

EXAMPLE 10-13. If the line of Example 10-12 is terminated in a short circuit ($R = 0$), what reflected wave is produced by the incident trapezoidal voltage wave? Sketch the results.

From (10-48), $\Gamma(\ell) = -1$. Then from (10-45)

$$V^-\left(t + \frac{\ell}{c}\right) = -V^+\left(t - \frac{\ell}{c}\right) \tag{1}$$

holding for all $t > \ell/c$ after $V^+(t - z/c)$ first appears at the load. To obtain the desired $V^-(t + z/c)$ reflected to the left, the reflected wave (1) must be delayed in time by $(\ell - z)/c$ (a time delay incurred by wave motion from the

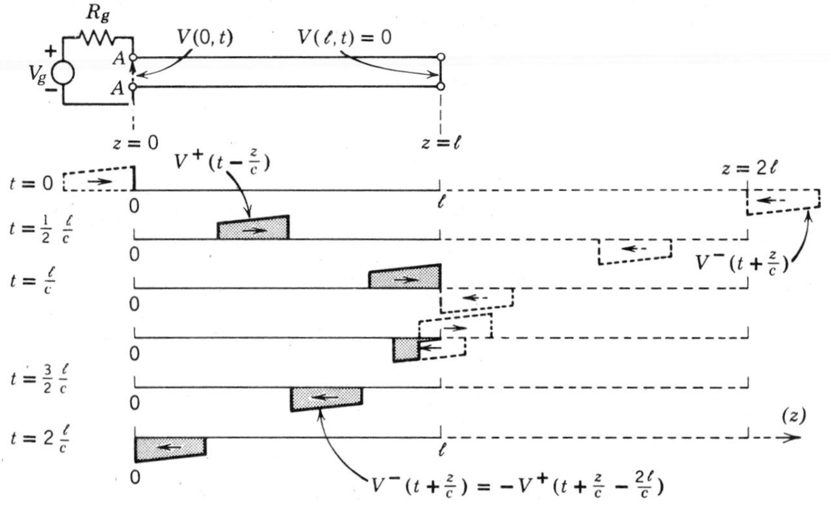

Example 10-13

load *back* to any z location towards the generator), yielding

$$V^-\left(t + \frac{z}{c}\right) = -V^+\left(t - \frac{\ell}{c} - \frac{\ell - z}{c}\right) = -V^+\left(t + \frac{z}{c} - \frac{2\ell}{c}\right) \quad t \geq \frac{\ell}{c} \tag{2}$$

As a check, observe that (2) becomes $-V^+(t - 2\ell/c)$ at the input $z = 0$, the reflection arriving there after a delay of $2\ell/c$. These results, shown in the sketch, reveal the echo $V^-(t + z/c)$ as a mirrored replica of the incident V^+, inverted by the effect of the minus sign in (2) (maintaining zero volts across the shorted load). The echo apparently originates from an image location $z = 2\ell$, due to the time-delay $2\ell/c$ needed for the reflection to reach to the input. The corresponding reflected current wave $I^-(t + z/c)$ is obtained by substituting (2) into the second term of (10-37b), yielding

$$I^-\left(t + \frac{z}{c}\right) = -\frac{V^-\left(t + \frac{z}{c}\right)}{R_0} = 0.02V^+\left(t + \frac{z}{c} - \frac{2\ell}{c}\right) \text{ A} \tag{3}$$

In the foregoing example, a second echo (re-reflection) occurs when $V^-(t + z/v)$ reaches the input at $z = 0$. It can be found by use of the arguments of that example applied to the generator circuit. With a generator internal resistance $R_g = R_0$, no re-reflection occurs. In general, an internal resistance R_g provides the time domain reflection coefficient at the line input ($z = 0$)

$$\Gamma(0) = \frac{V_2^+\left(t - \frac{0}{v}\right)}{V^-\left(t + \frac{0}{v}\right)} = \frac{R_g - R_0}{R_g + R_0} \tag{10-49}$$

with V_2^+ denoting the forward wave re-reflected from the generator to the load. These processes repeat when the wave V_2^+ in its turn arrives at the load, causing a third echo $V_2^-(t + \ell/v)$, and so on. The total wave solution is the superposition of all waves obtained in this way.

EXAMPLE 10-14. A lossless, 50-Ω coaxial line 200 m long using dielectric with $\epsilon_r = 2.25$ is terminated in 100 Ω and fed from a 150-V dc source having $R_g = 25$ Ω as shown. With the source switched on at $t = 0$, $V_g(t) = 150u(t)$ V, a step function. Find the voltage and current waves on the line after $t = 0$.

The equivalent input circuit of (*b*) yields at $A - A$

$$V(0, t) = V^+\left(t - \frac{0}{v}\right) = 150u(t)\frac{50}{25 + 50} = 100u(t) \text{ V} \tag{1}$$

$$I(0, t) = I^+\left(t - \frac{0}{v}\right) = \frac{V^+\left(t - \frac{0}{v}\right)}{50} = 2u(t) \text{ A} \tag{2}$$

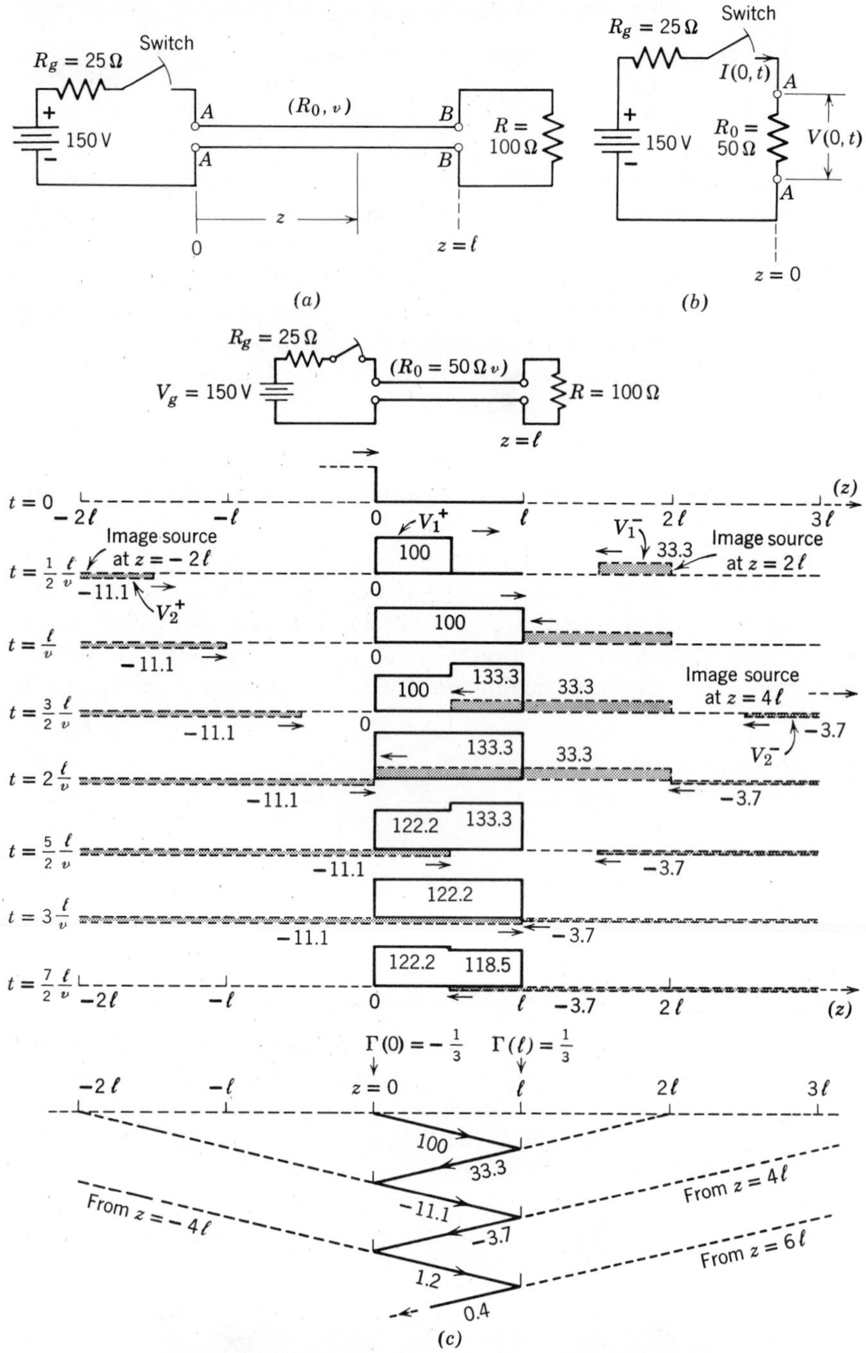

Example 10-14. ***(a)*** **Transmission-line system.** ***(b)*** **Equivalent input circuit.** ***(c)*** **Echo diagram.**

The incident waves are (1) and (2) delayed z/v sec

$$V(z, t) = V_1^+\left(t - \frac{z}{v}\right) = 100u\left(t - \frac{z}{v}\right) \text{ V} \qquad t < \frac{\ell}{v} \tag{3}$$

$$I(z, t) = I_1^+\left(t - \frac{z}{v}\right) = 2u\left(t - \frac{z}{v}\right) \text{ A} \tag{4}$$

wherein $v = (\mu_0\epsilon_0\epsilon_r)^{-1/2} = c/\sqrt{\epsilon_r} = 2 \times 10^8$ m/sec.

After a delay of $\ell/v = 200/2 \times 10^8 = 1\ \mu$sec, (3) and (4) arrive at the load. From (10-48), $\Gamma(\ell) = 1/3$, yielding from (10-45) the reflected voltage at $z = \ell$

$$V_1^-\left(t + \frac{\ell}{v}\right) = \Gamma(\ell)V_1^+\left(t - \frac{\ell}{v}\right) = 33.3u\left(t - \frac{\ell}{v}\right) \tag{5}$$

Equation (5) is delayed by $(\ell - z)/v$ to produce the first reflection towards the generator:

$$V_1^-\left(t + \frac{z}{v}\right) = 33.3u\left(t - \frac{\ell}{v} - \frac{\ell - z}{v}\right) = 33.3u\left(t + \frac{z}{v} - \frac{2\ell}{v}\right) \tag{6}$$

incorporating the delay $2\ell/v = 2\ \mu$sec. The accompanying wave diagram shows that (6) appears to originate from the image position $z = 2\ell$, arriving at the load after 1 μsec and reaching the generator after 2 μsec. When (6) reaches the generator, (10-49) yields $\Gamma(0) = (25 - 50)/(25 + 50) = -1/3$, yielding from (6) at $z = 0$ the second reflection:

$$V_2^+\left(t - \frac{0}{v}\right) = \Gamma(0)V_1^-\left(t + \frac{0}{v}\right) = (-\tfrac{1}{3})33.3u\left(t + \frac{0}{v} - \frac{2\ell}{v}\right)$$

$$= -11.1u\left(t - \frac{2\ell}{v}\right) \tag{7}$$

The latter, specified at $z = 0$, must be delayed z/v seconds as it moves towards the load

$$V_2^+\left(t - \frac{z}{v}\right) = -11.1u\left(t - \frac{z}{v} - \frac{2\ell}{v}\right) \tag{8}$$

The accompanying diagram shows how echo (8) originates from the image position $z = -2\ell$.

Extending the preceding details, (8) produces at the load a third echo, appearing to originate from $z = 4\ell$. This process continues indefinitely, approaching the steady state 120 V (obtainable from the generator-load circuit in the absence of the line). The *echo diagram* shown catalogs these results. The total voltage (10-34), given by the sum of all the positive z and negative z

traveling waves, thus becomes

$$\begin{aligned}
V(z,t) &= V^+\left(t - \frac{z}{v}\right) + V^-\left(t + \frac{z}{v}\right) \\
&= \left[V_1^+\left(t - \frac{z}{v}\right) + V_2^+\left(t - \frac{z}{v}\right) + \cdots\right] \\
&\quad + \left[V_1^-\left(t + \frac{z}{v}\right) + V_2^-\left(t + \frac{z}{v}\right) + \cdots\right] \\
&= \left[100u\left(t - \frac{z}{v}\right) - 11.1u\left(t - \frac{z}{v} - \frac{2\ell}{v}\right) + 1.2u\left(t - \frac{z}{v} - \frac{4\ell}{v}\right) - \cdots\right] \\
&\quad + \left[33.3u\left(t + \frac{z}{v} - \frac{2\ell}{v}\right) - 3.7u\left(t + \frac{z}{v} - \frac{4\ell}{v}\right)\right. \\
&\qquad \left. + 0.4u\left(t + \frac{z}{v} - \frac{6\ell}{v}\right) - \cdots\right] \qquad (9)
\end{aligned}$$

In the preceding examples, resistive loads were assumed, permitting the use of the time domain reflection coefficient (10-48). More generally, loads with capacitive or inductive elements as illustrated in Figure 10-10(b) may prevail. Analyzing their effects requires satisfying (10-42a) and (10-42b) in addition to the appropriate Kirchhoff relationship (10-43) (in general a differential or integro-differential equation). Eliminating the reflected voltage term $V^-(t + \ell/v)$ from the load relations (10-42) obtains

$$V(\ell, t) = 2V^+\left(t - \frac{\ell}{v}\right) - R_0 I(\ell, t) \qquad (10\text{-}50)$$

seen to correspond to the load terminal equivalent circuit, illustrated in Figure 10-13. Upon combining (10-50) with a load relation selected from (10-43), the resulting differential or integro-differential equation obtained is solved for the unknown load voltage or current, yielding the reflected wave. An example illustrates this procedure.

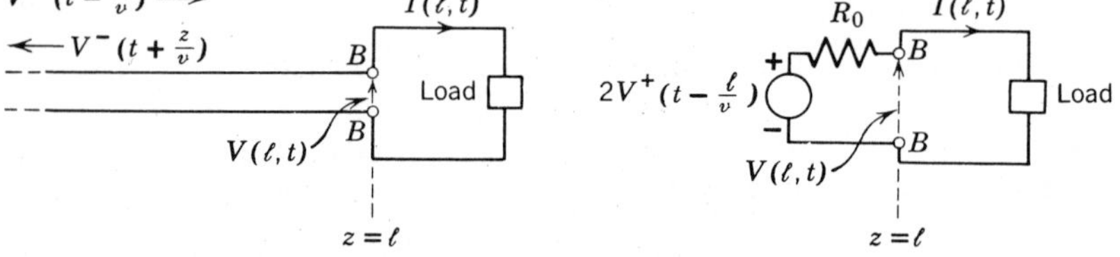

Figure 10-13. Equivalent load circuit corresponding to an arriving voltage wave $V^+(t - \ell/v)$.

EXAMPLE 10-15. The 50-Ω lossless line in (*a*) is terminated in a capacitor and fed from a 150-V dc source, assumed switched on at $t = 0$. To eliminate reflections at the generator, it is in series with a resistance such that $R_g = R_0 = 50\ \Omega$. Find the waves on the system after $t = 0$.

The input equivalent circuit of Figure 10-11(*a*) obtains, at $A - A$, $V(0, t) = 75u(t - 0/v)$. Delayed z/v, it yields the forward wave

$$V(z, t) = V^+\left(t - \frac{z}{v}\right) = 75u\left(t - \frac{z}{v}\right)\ \mathrm{V} \qquad 0 < t < \frac{\ell}{v} \quad (1)$$

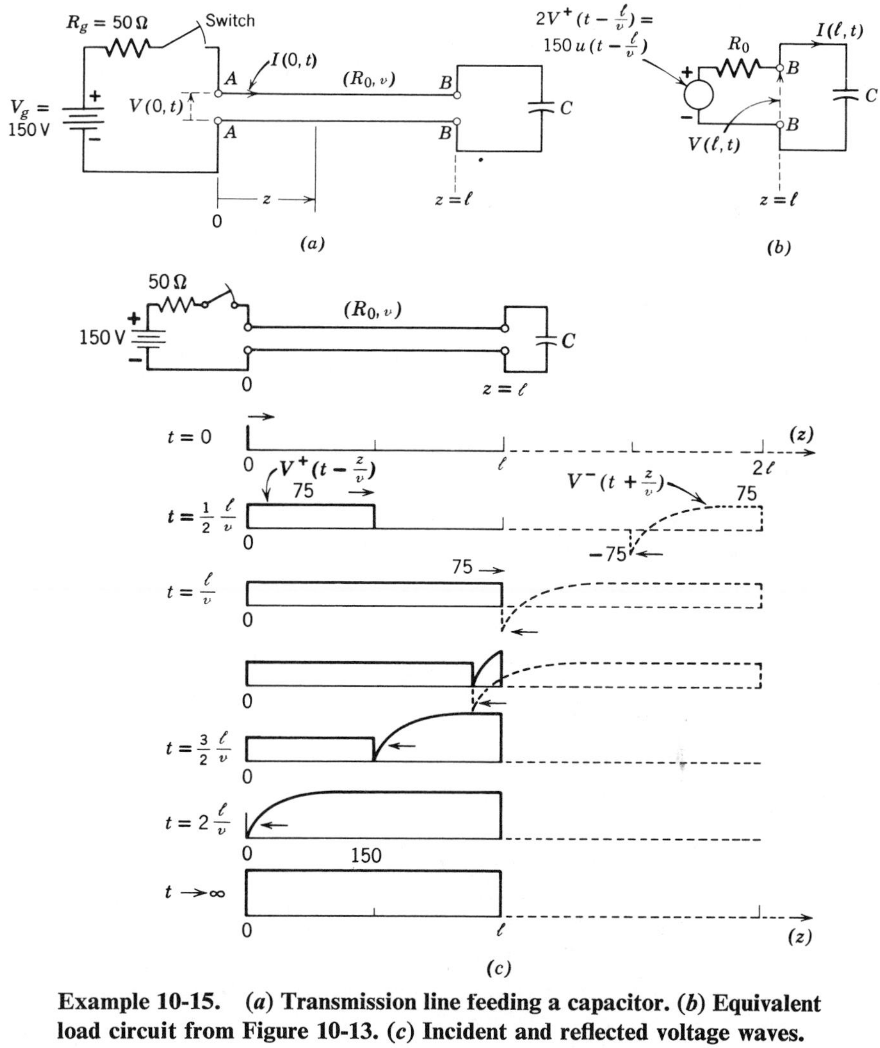

Example 10-15. (*a*) Transmission line feeding a capacitor. (*b*) Equivalent load circuit from Figure 10-13. (*c*) Incident and reflected voltage waves.

At the load position $z = \ell$, $V(\ell, t)$ of (10-42a) consists of (1) plus an unknown reflected voltage $V^-(t + \ell/v)$. To find the latter, combine (10-50),

$$V(\ell, t) = 2V^+\left(t - \frac{\ell}{v}\right) - R_0 I(\ell, t) \tag{2}$$

applicable to the equivalent load circuit of (*b*), with the current voltage relationship (10-43b) for the capacitive load

$$C\frac{dV(\ell, t)}{dt} = I(\ell, t) \tag{3}$$

(1) and (3) into (2), obtains the differential equation

$$150u\left(t - \frac{\ell}{v}\right) = R_0 C\frac{dV(\ell, t)}{dt} + V(\ell, t) \tag{4}$$

the solution of which is

$$V(\ell, t) = 150u\left(t - \frac{\ell}{v}\right)[1 - e^{-(t-\ell/v)/R_0C}] \tag{5}$$

the voltage across C. Equation (5) into (10-42a) then yields the reflected voltage at $z = \ell$

$$\begin{aligned} V^-\left(t + \frac{\ell}{v}\right) &= V(\ell, t) - V^+\left(t - \frac{\ell}{v}\right) \\ &= 150u\left(t - \frac{\ell}{v}\right)[1 - e^{-(t-\ell/v)/R_0C}] - 75u\left(t - \frac{\ell}{v}\right) \\ &= 75u\left(t - \frac{\ell}{v}\right) - 150u\left(t - \frac{\ell}{v}\right)e^{-(t-\ell/v)/R_0C} \end{aligned} \tag{6}$$

The reflection at any z is (6) delayed by $(\ell - z)/v$:

$$V^-\left(t + \frac{z}{v}\right) = 75u\left(t + \frac{z}{v} - \frac{2\ell}{v}\right) - 150u\left(t + \frac{z}{v} - \frac{2\ell}{v}\right)e^{-[t+(z/v)-(2\ell/v)]/R_0C} \tag{7}$$

The total voltage on the line is thus the superposition of (1) and (7), shown in (*c*).

REFERENCES

JOHNSON, W. C. *Transmission Lines and Networks.* New York: McGraw-Hill, 1950.

JORDAN, E. C., and K. G. BALMAIN. *Electromagnetic Waves and Radiating Systems*, 2nd ed. Englewood Cliffs, N.J.: Prentice-Hall, 1968.

Reich, H., P. F. Ordung, H. L. Krauss, and J. G. Skalnik. *Microwave Theory and Techniques*. Princeton, N.J.: D. Van Nostrand, 1953.

Stevenson, W. D., Jr. *Elements of Power Systems Analysis*. New York: McGraw-Hill, 1962.

PROBLEMS

10-1. The generator and load of Example 10-1 are connected using the same line except that its length is doubled, making $(\ell/\lambda) = 1.2$ at 20 MHz.

(*a*) Find the input impedance of the line, making use of (10-5) and (10-4). (Justify ignoring the attenuation in the calculations.)

(*b*) What voltage and current are developed by the generator at the input? Compute the average power delivered by the generator to the line. How much is delivered to the load, assuming a lossless line?

(*c*) If the line is terminated in a matched load, what is the line input impedance, and how much power is delivered to the load? How do the voltages and currents compare at the input and the output terminals under matched conditions?

10-2. At $f = 1$ kHz, the generator and load of Example 10-2 are connected to the same line, but its length is reduced to 40 mi., making $(\ell/\lambda) = 0.223$ at 1 kHz.

(*a*) By use of (10-4) and (10-5), determine the line input impedance.

(*b*) What voltage and current are provided by the generator at the line input? How much average power is delivered to the line?

(*c*) Find the load current, load voltage, and power absorbed by the load.

(*d*) Assuming the line terminated in $\hat{Z}_L = \hat{Z}_0$, and adjusting the generator for a conjugate match $\hat{Z}_g^* = \hat{Z}_0$, determine the power delivered to the line and the load.

10-3. Rework (*a*) of Problem 10-1 this time using the Smith chart to find the input impedance.

10-4. Rework (*a*) of Problem 10-2 using the Smith chart to find the input impedance.

10-5. The antenna load of Example 10-5 is fed at 20 MHz by the same generator and sections of line given, except the positions of the lines 1 and 2 are interchanged. With the aid of a Smith chart, find the line input impedance seen by the generator. Determine the average power delivered by the generator to the input. Why is this also the power delivered to the load?

10-6. The input impedance into the line of Example 10-1 was found to be $\hat{Z}_{\text{in}} = 70.5 + j28\ \Omega$. Use the Smith chart to convert this quantity to $\hat{Y}_{\text{in}}$.

10-7. Use the Smith chart to convert the load impedance terminating the line of Example 10-1 to its normalized load admittance $\hat{y}_L$. Then apply the

Smith chart (as a $\hat{y}$ chart) to obtain the input admittance of the line. Show that the latter agrees with the input impedance obtained in Example 10-1.

10-8. From the input power obtained in (*b*) of Example 10-6 compute the power absorbed by each of the loads $\hat{Z}_L$ and $\hat{Z}_L'$ terminating the branched lines.

10-9. Suppose the branched, lossless system of Example 10-6 is rearranged by a simple interchange of the lines 1 and 2, the generator and load termination being left as is.

(*a*) Use the admittance chart to find the input admittances of lines 1 and 3. Then find the input admittance seen by the generator looking into the feeder line 2.

(*b*) Compute the time-average power delivered by the generator to the system.

10-10. Based on results obtained in Example 10-3 for the essentially lossless line of Example 10-1 find the following:

(*a*) The SWR on the line. Use the Smith chart to locate the positions of V_{max} and V_{min} on the line. Where are I_{max} and I_{min} relative to the latter?

(*b*) Use the magnitudes of $1 + \hat{\Gamma}$ and $1 - \hat{\Gamma}$ obtained from the Smith chart as a basis for a graph of the voltage and current magnitudes on the line. Label the positions of V_{max}, V_{min}, I_{max}, I_{min}. [*Note: The graphs should agree with the input current and voltage magnitudes* obtained in part (*c*) of Example 10-1.]

10-11. An unknown impedance $\hat{Z}_L$ connected to a 50-Ω lossless slotted air line operated at 250 MHz produces an SWR = 2.7, with V_{min} occurring at the scale position 2.4 cm measured from a zero reference placed somewhere near the load. When the load is replaced by a short circuit, a null occurs at the scale position 47.1 cm. Determine $\hat{Z}_L$.

10-12. A 50 Ω lossless air line terminated in a resistive load has an SWR of 3. What two possible resistance values might the load have? What distance exists between the load and the first V_{min} on the standing wave in each case?

10-13. A 50 Ω lossless line is consecutively terminated in the following load impedances:

(*a*) 50 Ω
(*b*) 36 Ω
(*c*) 70 Ω
(*d*) $j36$ Ω
(*e*) $-j70$ Ω
(*f*) $30 + j40$ Ω
(*g*) $30 - j40$ Ω
(*h*) 0 Ω (a short)
(*i*) ∞ Ω (open)

With the aid of a Smith chart, determine the magnitude of the reflection coefficient and the SWR on the line for each termination. What is invariably the SWR on the line terminated in a pure reactance?

10-14. If a mismatched load is connected to a lossless line, it yields a standing wave with V_{min} located according to one of the four cases:

(*a*) V_{min} is at the load-plane

(*b*) V_{min} is a quarter wave away from the load

(*c*) V_{min} is between the load plane and a quarter wave away from the load plane

(*d*) V_{min} is between a quarter wave and a half wave away from the load plane.

Use the Smith chart to deduce for each case the type of load that must exist; i.e., whether the load is an inductive or a capacitive impedance, or whether it is a pure resistance (specifying in that event whether it is larger or smaller than Z_0).

10-15. A low-loss line with $\epsilon_r = 2.26$ and $Z_0 = 50\ \Omega$ is terminated in the abnormally low resistance of 1 Ω. Suppose the line is cut to exactly one-quarter wavelength at its operating frequency of 100 MHz.

(*a*) What is the line length?

(*b*) Determine the reflection coefficient at the load position, obtained both from (10-5) and by use of the Smith chart. What is the SWR on this line, and where is V_{min} located? Comment on the accuracy of the chart whenever a high SWR prevails on the line, as for this system.

(*c*) Determine the line input impedance assuming no line losses. Find the answer using the Smith chart and by analytical means, by use of the impedance expression (10-23) or equivalently the reflection coefficient and impedance expressions (10-4), (10-5), and (10-6).

10-16. Use the definitions (10-22) of the hyperbolic functions to obtain (10-21c) from (10-21b). Then show how (10-23) for a lossless line follows from (10-21c).

10-17. Use the exponential definitions (10-22) of the hyperbolic functions to evaluate cosh z, sinh z, and tanh z for the real arguments $z = 0$, $z = 0.5$, and $z = 1$.

10-18. Use the definitions (10-22) of the hyperbolic functions to prove that

$$\sinh (a + b) = \sinh a \cosh b + \cosh a \sinh b$$

and

$$\cosh (a + b) = \cosh a \cosh b + \sinh a \sinh b.$$

Show to what results the latter reduce if $a + b$ is replaced with the complex number $z = a + jb$; i.e., show that $\sinh z = \sinh a \cos b + j \cosh a \sin b$, and $\cosh z = \cosh a \cos b + j \sinh a \sin b$.

10-19. Deduce the input impedance expression (10-25) from (10-21c), and further show how the lossless version (10-27) follows from (10-25).

10-20. Prove the input impedance expressions (10-30) and (10-31) for lossless lines a half wave and a quarter wave long.

10-21. A piece of 50 Ω coaxial line of negligible losses is three-eighths wavelength long and terminated in a short circuit. Find the input impedance by use of (10-27), verifying the answer with the Smith chart. Neglecting losses, what is its input impedance if the line length is a quarter wave? A half wave?

10-22. An open-circuited, 75 Ω coaxial line with negligible losses is 0.3 wavelength long. Using (10-29), find its input impedance, verifying the result by use of the Smith chart. What is the input impedance if the line is a quarter wave long? A half wave?

10-23. A 70 Ω lossless line is a quarter wave long. With what impedance must this line be terminated to yield a 50 Ω input impedance?

10-24. A particular monopole antenna (a coaxial-fed wire extending above a conductive ground plane) has a measured impedance of 36 Ω at $f = 100$ MHz. This antenna is to be fed from a 70 Ω coaxial line through a quarter wave, lossless section of line with its Z_0 chosen to match the feed-line impedance. Find the required Z_0 of the quarter wave transformer and its physical length, assuming a polyethylene dielectric ($\epsilon_r = 2.26$).

10-25. An unknown impedance is connected to a 75 Ω lossless air line, operated at 600 MHz, producing a measured SWR = 3.2 on the line. To match the load to the line, find by use of a Smith chart where a short-circuited stub should be placed relative to the V_{min} on this line, and determine the required stub length. The line and the stub have the same characteristic impedance.

10-26. Prove by direct substitution that (10-34) and (10-37a), linear superpositions of arbitrary functions of $t \mp z/v$, are solutions of the wave equations (9-89a) and (9-89b).

10-27. A 50 Ω lossless line is 100 m long and has a polyethylene dielectric ($\epsilon_r = 2.26$), A single, square-topped pulse of 50 V amplitude and 0.1 μsec duration is switched into it at $t = 0$ from a generator with 50 Ω internal resistance. Find the amplitudes of the forward voltage and current waves on the line. What is the velocity of propagation of the pulse? Sketch the waveforms on the line at successive instants t when the leading edge of the wave:

(*a*) Just leaves the generator terminals
(*b*) Is halfway toward the load
(*c*) Just reaches the load

10-28. Assume the line of Problem 10-27 is terminated in a short circuit. What is the time domain reflection coefficient at the load? Sketch the forward and backward voltage waves on the line, as well as their superpositions, shown at typical instants after the forward pulse reaches the short.

10-29. Repeat Problem 10-28 assuming the line terminated in an open circuit.

10-30. An instrument known as a time domain reflectometer utilizes a step voltage generator developing V V and having 50 Ω internal impedance, the terminal voltage of which is monitored by use of a fast rise time calibrated oscilloscope. Suppose a section of 50 Ω lossless cable ($\epsilon_r = 2.26$) were connected to the generator terminals. With the cable load terminals shorted, what display would be observed on the oscilloscope, assuming the step voltage to be switched on at $t = 0$? What is the oscilloscope display with the cable load terminals open-circuited?

10-31. A direct voltage $V = 100$ V is connected in series with a 600 Ω resistor to a lossless air line 100 m long with a characteristic resistance of 200 Ω, open-circuited at its load end, What is the time domain reflection coefficient at the load and at the generator terminals? With the voltage switched on at $t = 0$, find the load voltage as a function of time during the interval required to establish three successive reflections from the load. What asymptotic value is approached? Use an echo diagram to depict the results.

10-32. The transmission line system of Example 10-15 is terminated in an inductor L. Neglecting the resistance of L, determine the voltage waves developed on the system after switching the dc source into the input at $t = 0$.

CHAPTER
11
Radiation from Antennas in Free Space

The problem of the radiation of electromagnetic energy from a transmitting antenna to a receiving system is of considerable interest to the communications engineer. Transmitting antennas are devices used in terminating a transmission line or waveguide with the intent of efficiently launching electromagnetic waves into space, and they may be regarded as sources of such waves in space. This chapter is concerned with the analysis of the radiation fields obtained from typical antenna sources, important examples of which are the linear wire antenna and the electromagnetic horn illustrated in Figure 11-1. First the physical **E** and **B** fields are described in terms of the scalar and vector auxiliary potentials Φ and **A**, which are in turn shown to satisfy wave equations. A solution of the wave equation in **A** is next obtained in the form of an integral over the antenna currents resembling (5-28), except that the time retardation effect due to the finite velocity of electromagnetic wave propagation is included. The **E** and **H** field solutions of an elementary oscillating current element are derived as a special case, with an integration leading to the radiated fields of a thin, center-fed antenna of arbitrary length. The extension of Maxwell's equations to a symmetrical set using postulated magnetic charges and currents, together with their boundary conditions, forms the basis for predicting the radiation fields of electromagnetic horns and related aperture-type antennas.

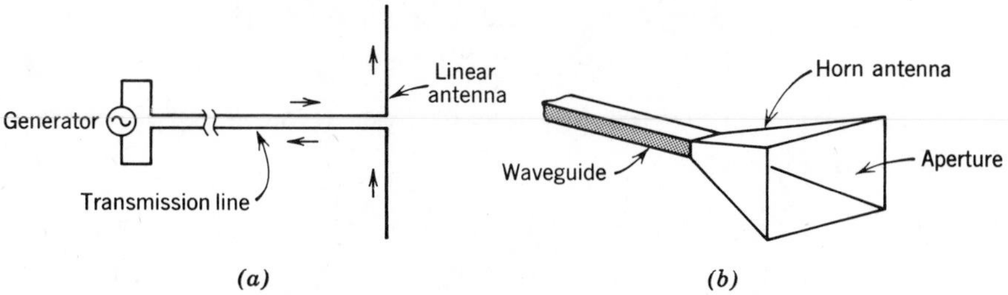

Figure 11-1. Examples of antennas. (*a*) Linear wire antenna, center-fed by a transmission line. (*b*) Electromagnetic horn, excited by use of a waveguide.

11-1 Wave Equations in Terms of Electromagnetic Potentials

As an aid in computing the radiation fields of antennas, frequently auxiliary functions (potential fields) are helpful in systematizing the mathematics. In particular, one may recall that the electric and magnetic fields of charge and current sources have already been related to potential functions by (5-22) and (5-48)

$$\mathbf{B} = \nabla \times \mathbf{A} \tag{11-1}$$

$$\mathbf{E} = -\nabla \Phi - \frac{\partial \mathbf{A}}{\partial t} \tag{11-2}$$

in which Φ is the scalar electric potential and $\mathbf{A}$ the vector magnetic potential field at any point in a region. It is recalled from Chapter 5 that the potential fields at any field point in free space may be expressed by integrations over the charge and current sources as given by (4-35) and (5-28) as follows:

$$\Phi(x, y, z) = \int_V \frac{\rho_v(x', y', z')\, dv'}{4\pi\epsilon_0 R} \qquad [4\text{-}35]$$

$$\mathbf{A}(x, y, z) = \int_V \frac{\mu_0 \mathbf{J}(x', y', z')\, dv'}{4\pi R} \qquad [5\text{-}28a]$$

assuming quasi-static conditions, i.e., that the source densities ρ_v and $\mathbf{J}$ vary sufficiently slowly in time that the finite velocity of propagation of the field effects (time retardation) can be neglected.

For radiation problems in which the fields many wavelengths from the sources are usually desired, the time retardation effects are of such significance that the potential integrals (4-35) and (5-28) become useless. Revised forms can be obtained by showing that Φ and $\mathbf{A}$ in general satisfy inhomogeneous wave equations, and from these are found integral solutions for free space.

One must first find wave equations satisfied by the potentials Φ and $\mathbf{A}$, by use of (11-1) and (11-2) substituted into the Maxwell relations (3-24), (3-48), (3-59), and (3-77)

$$\nabla \cdot \mathbf{D} = \rho_v \tag{11-3}$$

$$\nabla \cdot \mathbf{B} = 0 \tag{11-4}$$

$$\nabla \times \mathbf{E} = -\frac{\partial \mathbf{B}}{\partial t} \tag{11-5}$$

$$\nabla \times \mathbf{H} = \mathbf{J} + \frac{\partial \mathbf{D}}{\partial t} \tag{11-6}$$

Note that there is no further need to use (11-4) and (11-5) here, for the potential relationships (11-1) and (11-2) were obtained from these two Maxwell equations originally. (Putting (11-1) and (11-2) into (11-4) and (11-5) merely leads to identities.) With ϵ assumed a constant in what follows, write (11-3) as $\nabla \cdot \mathbf{E} = \rho_v/\epsilon$, and substituting (11-2) into the latter yields $\nabla \cdot (-\nabla\Phi - \partial\mathbf{A}/\partial t) = \rho_v/\epsilon$. This becomes

$$\nabla^2\Phi + \frac{\partial(\nabla \cdot \mathbf{A})}{\partial t} = -\frac{\rho_v}{\epsilon} \tag{11-7}$$

if $\nabla \cdot (\nabla\Phi) = \nabla^2\Phi$ is utilized from (18) in Table 2-2. From Section 2-8 it is recalled that the specification of both the divergence and the curl of a vector function assures its uniqueness (within an arbitrary constant), but only the curl of $\mathbf{A}$, by (11-1), has thus far been established. The divergence of $\mathbf{A}$ is therefore still arbitrary, so put

$$\nabla \cdot \mathbf{A} = -\mu\epsilon \frac{\partial \Phi}{\partial t} \tag{11-8}$$

whereupon substituting the latter into (11-7) obtains

$$\nabla^2\Phi - \mu\epsilon \frac{\partial^2 \Phi}{\partial t^2} = -\frac{\rho_v}{\epsilon} \tag{11-9}$$

Comparing this result with the form of (2-100) reveals that (11-9) is a scalar *wave equation* in terms of the potential Φ. Thus it is seen that the direct consequence of assuming (11-8), called the Lorentz condition, is to force the potential Φ to obey the wave equation (11-9).

A wave equation in terms of **A** is similarly derived if both (11-1) and (11-2) are substituted into the remaining Maxwell equation (11-6). First multiply (11-6) by μ, assumed constant, yielding $\nabla \times \mathbf{B} = \mu \mathbf{J} + \mu\epsilon\, \partial \mathbf{E}/\partial t$. Inserting (11-1) and (11-2) then obtains

$$\nabla \times (\nabla \times \mathbf{A}) = \mu \mathbf{J} + \mu\epsilon\left[-\nabla\left(\frac{\partial \Phi}{\partial t}\right) - \frac{\partial^2 \mathbf{A}}{\partial t^2}\right] \qquad (11\text{-}10)$$

The identity (21) of Table 2-2 permits writing $\nabla \times (\nabla \times \mathbf{A}) = \nabla(\nabla \cdot \mathbf{A}) - \nabla^2 \mathbf{A}$, and further substituting the Lorentz condition (11-8) produces a cancellation of terms containing Φ in (11-10) to yield

$$\nabla^2 \mathbf{A} - \mu\epsilon \frac{\partial^2 \mathbf{A}}{\partial t^2} = -\mu \mathbf{J} \qquad (11\text{-}11)$$

the desired vector wave equation expressed in terms of **A**.

A comparison of the wave equations (11-9) and (11-11) with the Poisson-type differential equations (4-67) and (5-26)

$$\nabla^2 \Phi = -\frac{\rho_v}{\epsilon} \qquad [4\text{-}67]$$

$$\nabla^2 \mathbf{A} = -\mu \mathbf{J} \qquad [5\text{-}26]$$

shows that the latter are just special cases of the wave equations, subject to the time-static assumption $\partial/\partial t = 0$. The integrals (4-35) and (5-28) given earlier are the time-static solutions of (4-67) and (5-26) in free space. In the next section, comparable integral solutions of the wave equations (11-9) and (11-11) are derived. Complex, time-harmonic fields are utilized to accomplish this. Thus, with $\mathbf{A}(u_1, u_2, u_3, t)$ in the time domain replaced with $\hat{\mathbf{A}}(u_1, u_2, u_3)e^{j\omega t}$ in the manner of (2-67), and similarly for the remaining fields, the Lorentz condition (11-8) becomes $\nabla \cdot \hat{\mathbf{A}} = -j\omega\mu\epsilon\hat{\Phi}$, yielding

$$\hat{\Phi} = -\frac{\nabla \cdot \hat{\mathbf{A}}}{j\omega\mu\epsilon} \qquad \text{Lorentz condition} \quad (11\text{-}12)$$

The collected results (11-1), (11-2), (11-9), and (11-11) in time-harmonic form become

---

$$\hat{\mathbf{B}} = \nabla \times \hat{\mathbf{A}} \qquad (11\text{-}13)$$

$$\hat{\mathbf{E}} = -\nabla\hat{\Phi} - j\omega\hat{\mathbf{A}} = \frac{\nabla(\nabla \cdot \hat{\mathbf{A}})}{j\omega\mu\epsilon} - j\omega\hat{\mathbf{A}} \qquad (11\text{-}14)$$

---

in which $\hat{\Phi}$ and $\hat{\mathbf{A}}$ are solutions of the wave equations

$$\nabla^2\hat{\Phi} + \omega^2\mu\epsilon\hat{\Phi} = -\frac{\hat{\rho}_v}{\epsilon} \tag{11-15}$$

$$\nabla^2\hat{\mathbf{A}} + \omega^2\mu\epsilon\hat{\mathbf{A}} = -\mu\hat{\mathbf{J}} \tag{11-16}$$

11-2 Integration of the Inhomogeneous Wave Equation in Free Space

It is shown that a solution of the vector wave equation (11-16) in free space can be represented as the following integral over the time-harmonic current sources $\hat{\mathbf{J}}(u_1', u_2', u_3')$

$$\hat{\mathbf{A}}(u_1, u_2, u_3) = \int_V \frac{\mu_0\hat{\mathbf{J}}(u_1', u_2', u_3')e^{-j\beta_0 R}}{4\pi R}\, dv' \tag{11-17}$$

a result closely resembling the integral (5-28a) for direct current sources in free space, except for the additional time retardation factor $\exp(-j\beta_0 R)$. The geometry of a generalized system of current densities in free space is shown in Figure 11-2. The integral (11-17) over a system of such current sources leads to the vector magnetic potential $\hat{\mathbf{A}}$ at the field point P, whence $\hat{\mathbf{B}}$ and $\hat{\mathbf{E}}$ fields are then obtained using (11-13) and (11-14).

The formal proof that (11-17) is a solution of (11-16) is given here. It is convenient to expand (11-16) in rectangular coordinates into the scalar wave equations

$$\begin{aligned}\nabla^2\hat{A}_x + \omega^2\mu_0\epsilon_0\hat{A}_x &= -\mu_0\hat{J}_x \\ \nabla^2\hat{A}_y + \omega^2\mu_0\epsilon_0\hat{A}_y &= -\mu_0\hat{J}_y \\ \nabla^2\hat{A}_z + \omega^2\mu_0\epsilon_0\hat{A}_z &= -\mu_0\hat{J}_z\end{aligned} \tag{11-18}$$

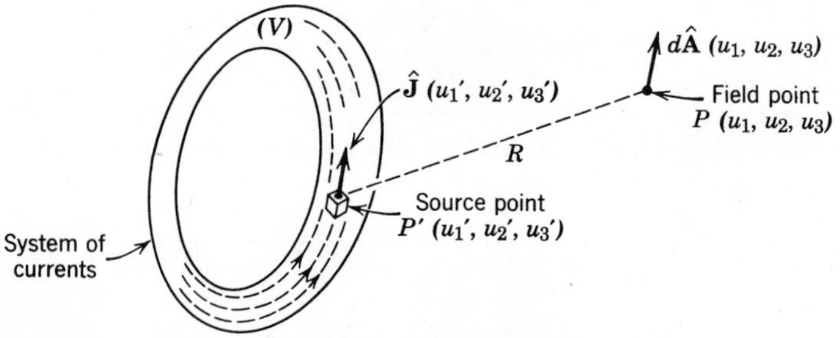

Figure 11-2. Generalized current source distribution in free space, and field point P at which the potential $\hat{\mathbf{A}}$ is found using (11-17).

making use of (2-83). Denote any of (11-18) in the unsubscripted form

$$\nabla^2 \hat{A} + \omega^2 \mu_0 \epsilon_0 \hat{A} = -\mu_0 \hat{J} \tag{11-19}$$

Recall the Green's symmetrical theorem (2-92)

$$\int_V [f\nabla^2 g - g\nabla^2 f]\, dv = \oint_S \left(f\frac{\partial g}{\partial n} - g\frac{\partial f}{\partial n} \right) ds \tag{11-20}$$

correct for *any* pair of well-behaved functions f and g in and on a volume V bounded by the closed surface S. It is shown that (11-20) leads directly to the solution (11-17) if suitable choices are made for the functions f and g. Choose f as the field $\hat{A}$ in (11-19), assuming $\hat{A}$ means any component of $\hat{\mathbf{A}}$ located at a source point P'. Thus, $\hat{A} \equiv \hat{A}(u_1', u_2', u_3') = \hat{A}(\mathbf{r}')$, in which $\mathbf{r}'$ denotes the position vector of P'. The wave equation (11-19) is certainly satisfied by $\hat{A}(\mathbf{r}')$ at all such points in space, and the Green's theorem (11-20) is written

$$\int_V [\hat{A}(\mathbf{r}')\nabla^2 g - g\nabla^2 \hat{A}(\mathbf{r}')]\, dv' = \oint_S \left[\hat{A}(\mathbf{r}')\frac{\partial g}{\partial n} - g\frac{\partial \hat{A}}{\partial n} \right] ds' \tag{11-21}$$

in which dv and ds are primed, since they are identified with locations P' in V and on S. For reasons about to be clarified, the function g in (11-21) is chosen as

$$g = \frac{e^{-j\beta_0 R}}{R} \tag{11-22}$$

In (11-22), for the free-space region to which the constants μ_0, ϵ_0 apply

$$\beta_0 = \omega\sqrt{\mu_0 \epsilon_0} \tag{11-23}$$

and $R = |\mathbf{r} - \mathbf{r}'|$ denotes the distance between the source point $P'(\mathbf{r}')$ and any fixed field point $P(\mathbf{r})$, at which $\hat{A}$ is desired to be expressed in terms of the sources. The geometry is shown in Figure 11-3(*a*). The integration (11-21) in V and on S being taken over all points P' and with P an arbitrary fixed point in the region, it is seen that $R = 0$ at P (i.e., P is the origin of R). Moreover, a property of the function g is that it satisfies the scalar wave equation,

$$\nabla^2 g + \beta_0^2 g = 0 \tag{11-24}$$

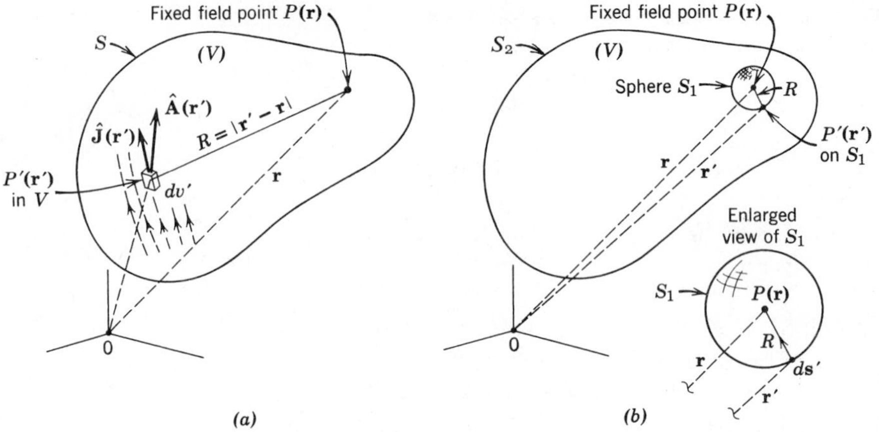

Figure 11-3. Geometries relative to the derivation of (11-17). (*a*) Current distribution at source points P' and a fixed field point P enclosed by surface S. (*b*) Showing small sphere S_1 excluding $P(\mathbf{r})$ from the region of integration.

With (11-22) inserted into (11-21), the latter becomes

$$\int_V \left[\hat{A}(\mathbf{r}')\nabla^2\left(\frac{e^{-j\beta_0 R}}{R}\right) - \frac{e^{-j\beta_0 R}}{R}\nabla^2\hat{A}(\mathbf{r}')\right] dv' = \oint_S \left[\hat{A}(\mathbf{r}')\frac{\partial}{\partial n}\left(\frac{e^{-j\beta_0 R}}{R}\right) - \frac{e^{-j\beta_0 R}}{R}\frac{\partial \hat{A}(\mathbf{r}')}{\partial n}\right] ds' \quad (11\text{-}25)$$

From (11-24), $-\beta_0^2 g$ can replace $\nabla^2 g$ in the first term of (11-25), obtaining for the volume integral

$$\int_V \left[-\hat{A}(\mathbf{r}')\beta_0^2\frac{e^{-j\beta_0 R}}{R} - \frac{e^{-j\beta_0 R}}{R}\nabla^2\hat{A}(\mathbf{r}')\right] dv' = \int_V -\frac{e^{-j\beta_0 R}}{R}[\nabla^2\hat{A}(\mathbf{r}') + \beta_0^2\hat{A}(\mathbf{r}')]\, dv' = \int_V \frac{\mu_0\hat{J}(\mathbf{r}')e^{-j\beta_0 R}}{R}\, dv' \quad (11\text{-}26)$$

the latter following from the use of (11-19).

Next, an inspection of the right side of (11-25) reveals that if the field point $P(\mathbf{r})$ is to be inside the volume region V of integration as in Figure 11-3(*a*), then the requirement of Green's theorem that the function $g = e^{-j\beta_0 R}/R$ be *well-behaved* in V is violated at P (where $R = 0$), in view of the singularity in g there. The point P is excluded from the volume region of integration by constructing a small sphere S_1 of radius $R = R_1$ about P as shown in Figure

11-3(*b*). Then V is bounded by the closed surface $S = S_1 + S_2$ as noted in the figure, whence (11-25) is written symbolically

$$\int_V [\ ]\, dv' = \int_{S_1} [\ ] ds' + \int_{S_2} [\ ]\, ds' \tag{11-27}$$

in which the brackets denote the volume and surface integrands of (11-25).

The contribution of the integral on S_1 in (11-27) is now shown to approach the value $4\pi\hat{A}(\mathbf{r})$, or just 4π times the potential $\hat{A}$ at the field point P, as the sphere S_1 vanishes. From Figure 11-3(*b*) it is evident that $\partial/\partial n = -\partial/\partial R$ on S_1, since the normal is directed *outward* with respect to the volume region V (meaning in the negative R sense). Thus

$$\int_{S_1} [\ ]\, ds' = \int_{S_1} \left[-\hat{A}(\mathbf{r}') \frac{\partial}{\partial R}\left(\frac{e^{-j\beta_0 R}}{R}\right) + \frac{e^{-j\beta_0 R}}{R} \frac{\partial \hat{A}(\mathbf{r}')}{\partial R} \right]_{R=R_1} ds'$$

$$= \int_{S_1} \left\{ \hat{A}(\mathbf{r}')\left(\frac{1}{R_1} + j\beta_0\right) \frac{e^{-j\beta_0 R_1}}{R_1} + \frac{e^{-j\beta_0 R_1}}{R_1} \frac{\partial \hat{A}(\mathbf{r}')}{\partial R}\right]_{R=R_1} \right\} ds' \tag{11-28}$$

By definition $A(\mathbf{r}')$ is well-behaved in the vicinity of the fixed point $P(\mathbf{r})$. Allowing the radius R_1 of the sphere S_1 to become arbitrarily small, the value of $\hat{A}(\mathbf{r}')$ on S_1 can be replaced by its mean value $\langle \hat{A}(\mathbf{r}')\rangle$, becoming $\hat{A}(\mathbf{r})$ as $R_1 \to 0$. Extracting the mean values of $\hat{A}(\mathbf{r}')$ and $\partial\hat{A}/\partial R$ from the integral (11-28) yields

$$\int_{S_1} [\ ]\, ds' = \left\{ \langle \hat{A} \rangle \left(\frac{1}{R_1^2} + \frac{j\beta_0}{R_1}\right) e^{-j\beta_0 R_1} + \frac{e^{-j\beta_0 R_1}}{R_1} \left\langle \frac{\partial \hat{A}}{\partial R} \right\rangle_{R=R_1} \right\} \int_{S_1} ds'$$

The integral $\int_{S_1} ds'$ is just $4\pi R_1^2$, so as $R_1 \to 0$, (11-28) becomes

$$\lim_{R_1 \to 0} \int_{S_1} [\ ]\, ds' \to 4\pi \hat{A}(\mathbf{r}) \tag{11-29}$$

or just 4π times the potential $\hat{A}$ at the field point $P(\mathbf{r})$, that which was to be proved. Upon putting (11-26) and (11-29) back into (11-27) and solving for $\hat{A}(\mathbf{r})$, the desired *integral solution* for the scalar wave equation (11-19) is obtained

$$\hat{A}(\mathbf{r}) = \int_V \frac{\mu_0 \hat{J}(\mathbf{r}') e^{-j\beta_0 R}}{4\pi R}\, dv' + \frac{1}{4\pi} \oint_S \left[\frac{\partial \hat{A}(\mathbf{r}')}{\partial n} \frac{e^{-j\beta_0 R}}{R} - \hat{A}(\mathbf{r}') \frac{\partial}{\partial n}\left(\frac{e^{-j\beta_0 R}}{R}\right)\right] ds' \tag{11-30}$$

in which the subscript on S_2 is discarded since S_1 is no longer present. Since

(11-30) is a solution of each scalar wave equation in (11-18), three such solutions, for $\hat{A}_x(\mathbf{r})$, $\hat{A}_y(\mathbf{r})$, and $\hat{A}_z(\mathbf{r})$, can be added vectorially to provide the desired solution for the vector wave equation (11-16)

$$\hat{\mathbf{A}}(\mathbf{r}) = \int_V \frac{\mu_0 \hat{\mathbf{J}}(\mathbf{r}')e^{-j\beta_0 R}}{4\pi R} dv' + \frac{1}{4\pi} \oint_S \left[\frac{\partial \hat{\mathbf{A}}(\mathbf{r}')}{\partial n} \frac{e^{-j\beta_0 R}}{R} - \hat{\mathbf{A}} \frac{\partial}{\partial n} \left(\frac{e^{-j\beta_0 R}}{R} \right) \right] ds' \tag{11-31}$$

a result due originally to Helmholtz.

The meaning of the integral solution (11-31) is clarified upon referring to Figure 11-4, and three cases of physical interest can be identified.

CASE A. Suppose no current densities exist inside the region V, bounded by a finite closed surface S as in Figure 11-4(a). With $\mathbf{J} = 0$, (11-31) yields only the surface integral

$$\hat{\mathbf{A}}(\mathbf{r}) = \frac{1}{4\pi} \oint_S \left[\frac{\partial \hat{\mathbf{A}}(\mathbf{r}')}{\partial n} \frac{e^{-j\beta_0 R}}{R} - \hat{\mathbf{A}}(\mathbf{r}') \frac{\partial}{\partial n} \left(\frac{e^{-j\beta_0 R}}{R} \right) \right] ds' \tag{11-32a}$$

With no sources inside V, the potential $\hat{\mathbf{A}}(\mathbf{r}')$ on the closed surface S must be

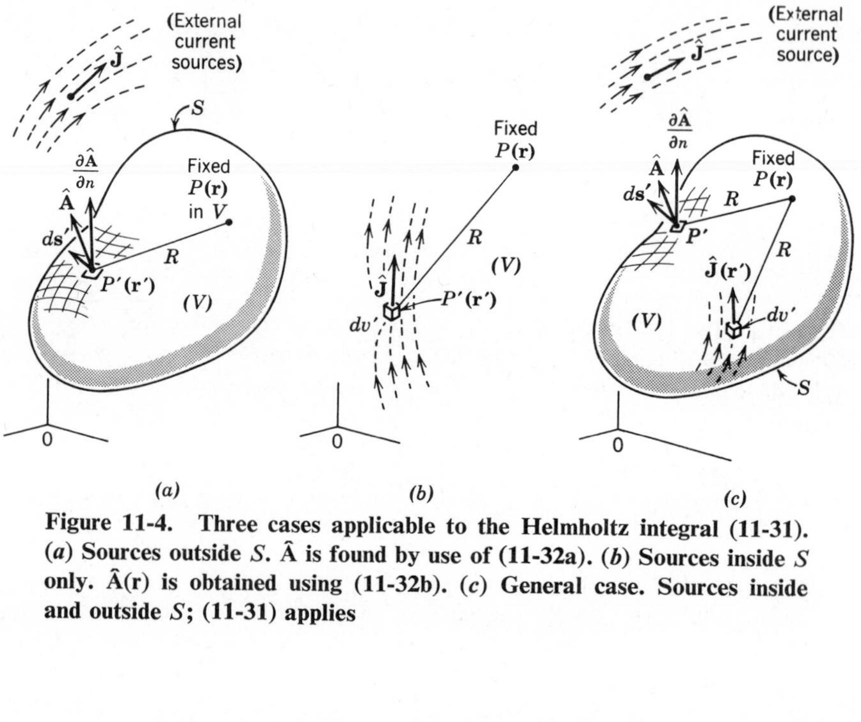

Figure 11-4. Three cases applicable to the Helmholtz integral (11-31). (*a*) Sources outside S. $\hat{\mathbf{A}}$ is found by use of (11-32a). (*b*) Sources inside S only. $\hat{\mathbf{A}}(\mathbf{r})$ is obtained using (11-32b). (*c*) General case. Sources inside and outside S; (11-31) applies

attributed to current sources lying entirely *outside* S. This type of integral was extensively investigated by Kirchhoff in optical diffraction problems.

CASE B. Suppose current densities exist within a finite distance from the origin. Then the closed surface S in (11-30) can be expanded indefinitely towards infinity, to reduce the contributions to the potential at the field point P to the volume integral (11-17)

$$\hat{\mathbf{A}}(\mathbf{r}) = \int_V \frac{\mu_0 \hat{\mathbf{J}}(\mathbf{r}')e^{-j\beta_0 R}}{4\pi R}\, dv' \tag{11-32b}$$

as depicted by the geometry in Figure 11-4(*b*). This is the reduced form of (11-31) commonly encountered in free-space antenna radiation problems. Examples are considered in the next section.

CASE C. If current density sources exist both inside the closed surface S and outside it, the general form (11-31) is applicable, amounting to a superposition of (11-32a) and (11-32b). This case is largely academic, for the better expedient is usually to expand S until all current sources are enclosed, in which event Case B applies.

In a manner analogous to the arguments leading to (11-32b), one can show

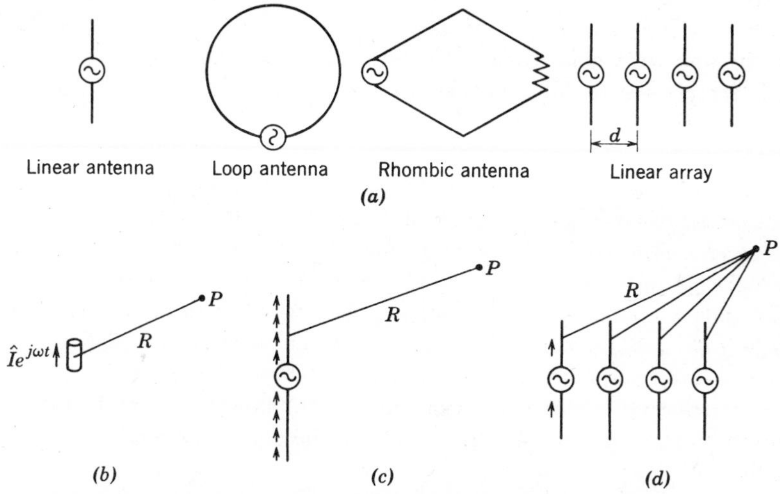

Figure 11-5. Antenna configurations amenable to field analysis by use of (11-32b). (*a*) A few antenna configurations of physical interest. (*b*) An infinitesimal oscillating current-element. (*c*) Linear antenna as a superposition of current-elements. (*d*) Array of linear wire antennas.

that the scalar electric potential $\hat{\Phi}$, developed at the field point $P(\mathbf{r})$ by a time-harmonic charge density distribution $\hat{\rho}_v(\mathbf{r}')$ in free space, becomes

$$\hat{\Phi}(r) = \int_V \frac{\rho_v(\mathbf{r}')e^{-j\beta_0 R}}{4\pi\epsilon_0 R}\, dv' \qquad (11\text{-}33)$$

assuming the closed surface S to have been expanded indefinitely towards infinity. In antenna radiation problems to follow, no use is made of (11-33), since the Lorentz condition (11-8) permits expressing $\hat{\mathbf{E}}$ and $\hat{\mathbf{B}}$, via (11-13) and (11-14), solely in terms of the potential $\hat{\mathbf{A}}$.

The integral (11-32b) has extensive applications to the determination of the radiation fields of current distributions on conductors in free space. Examples are shown in Figure 11-5(*a*). Insight into the radiation fields of such devices can be acquired initially from a study of the infinitesimal dipole element illustrated in Figure 11-5(*b*), since an end-to-end superposition of such elements can be used as a basis for constructing any of the antennas in Figure 11-5(*a*), and hence their fields as well. The field integral (11-32b) is simply an expression of such a superposition.

11-3 Radiation from the Infinitesimal Current-Element

The usefulness of (11-32b) lies in the fact that it yields, at any field point in free space, the total potential $\hat{\mathbf{A}}$ due to a system of time-harmonic current sources. The physical $\hat{\mathbf{E}}$ and $\hat{\mathbf{H}}$ fields of such sources can then be derived from the known $\hat{\mathbf{A}}$ field by use of (11-13) and (11-14).

It is instructive to find the fields of the most basic current source: the *infinitesimal current-element* (or elementary dipole) illustrated in Figures 11-5(*b*) and 11-6. The current-element is defined by $\hat{\mathbf{J}}\, dv'$ appearing in the potential integral (11-32b), with $\hat{\mathbf{J}}$ denoting the complex, time-harmonic, vector current density at some volume-element dv'. For present purposes, the transverse area ds' of the volume-element is assumed to vanish as suggested in Figure 11-6(*a*), with the current source carrying a finite (rather than infinitesimal) current $\hat{I}$. This permits expressing the volume current-element $\hat{\mathbf{J}}\, dv'$ as $\hat{I}\, d\ell'$, a *linear* current-element. It is seen from Figure 11-6(*a*) that the current $\hat{I}$ is accompanied by charge accumulations $\pm\hat{q}$ at the ends. The relation connecting a time-instantaneous current flow i with real-time charge accumulations $\pm q$ is, by (4-19), $i = dq/dt$. The corresponding time-harmonic form is

$$\hat{I} = j\omega\hat{q} \qquad (11\text{-}34)$$

Because the elementary current source involves charge displacements $\pm\hat{q}$ to the ends of the element, it is often called an oscillating *electric dipole*.

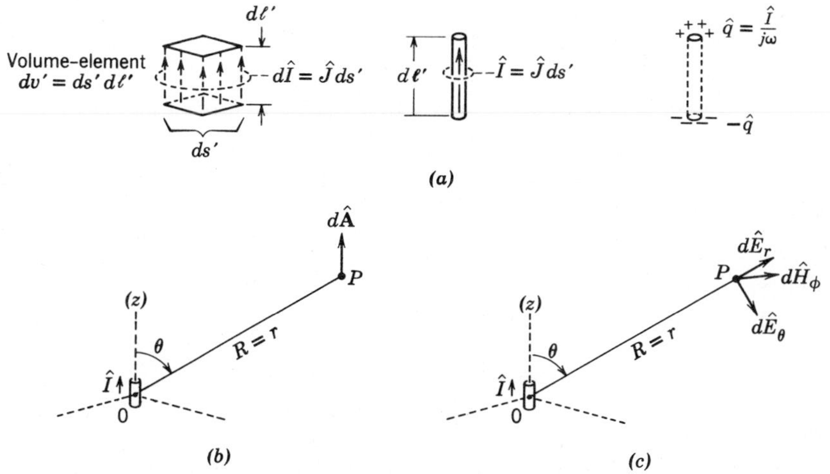

Figure 11-6. **The oscillating current-element. (*a*) Limit of the volume current-element $\hat{\mathbf{J}}\, dv'$, becoming the linear current-element $\hat{I}\, d\ell'$ as $ds' \to 0$. (*b*) Geometry relative to a current-element at the origin. (*c*) Components of electric and magnetic fields at *P*.**

The vector magnetic potential of an oscillating current-element located at the origin of a spherical coordinate system is obtained with reference to Figure 11-6(*b*). Since only an infinitesimal current-element is assumed present, no integration of (11-32b) is required, to yield the differential potential at *P*

$$d\hat{\mathbf{A}} = \mathbf{a}_z\, d\hat{A}_z = \mathbf{a}_z \frac{\mu_0 \hat{I} e^{-j\beta_0 r}}{4\pi r}\, dz' \tag{11-35}$$

The electric and magnetic fields corresponding to (11-35) are found by use of (11-1) and (11-2). Equation (11-35) is given in mixed coordinate systems, so it becomes desirable to express the z directed potential in spherical coordinates. From the geometry in Figure 11-6(*b*)

$$d\hat{A}_r = d\hat{A}_z \cos\theta \qquad d\hat{A}_\theta = -dA_z \sin\theta \tag{11-36}$$

Then the $\hat{\mathbf{H}}$ field of the elementary dipole becomes, from (11-1)

$$d\hat{\mathbf{H}} = \frac{d\hat{\mathbf{B}}}{\mu_0} = \frac{\nabla \times (d\hat{\mathbf{A}})}{\mu_0} = \frac{1}{\mu_0} \begin{vmatrix} \dfrac{\mathbf{a}_r}{r^2 \sin\theta} & \dfrac{\mathbf{a}_\theta}{r \sin\theta} & \dfrac{\mathbf{a}_\phi}{r} \\ \dfrac{\partial}{\partial r} & \dfrac{\partial}{\partial \theta} & 0 \\ d\hat{A}_r & r\, d\hat{A}_\theta & 0 \end{vmatrix}$$

$$= \mathbf{a}_\phi \frac{\hat{I}\,dz'}{4\pi} e^{-j\beta_0 r}\left[\frac{j\beta_0}{r} + \frac{1}{r^2}\right] \sin\theta \text{ A/m} \tag{11-37}$$

The electric field **E** is obtainable two ways. One is by use of (11-14)

$$d\hat{\mathbf{E}} = \frac{\nabla(\nabla\cdot d\hat{\mathbf{A}})}{j\omega\mu_0\epsilon_0} - j\omega\, d\hat{\mathbf{A}} \tag{11-38}$$

into which the substitution of (11-35) and (11-36) yields the desired $d\hat{\mathbf{E}}$ at P. An alternate method involves the use of the Maxwell equation (3-85), which in the free-space region becomes

$$d\hat{\mathbf{E}} = \frac{1}{j\omega\epsilon_0} \nabla\times(d\hat{\mathbf{H}}) \tag{11-39}$$

and into which the substitution of (11-37) obtains the desired electric field. Either method yields

$$d\hat{\mathbf{E}} = \mathbf{a}_r\, d\hat{E}_r + \mathbf{a}_\theta\, d\hat{E}_\theta \text{ V/m} \tag{11-40a}$$

in which

$$d\hat{E}_r = \frac{\hat{I}\,dz'}{4\pi} e^{-j\beta_0 r}\left[\frac{2\eta_0}{r^2} + \frac{2}{j\omega\epsilon_0 r^3}\right] \cos\theta \tag{11-40b}$$

$$d\hat{E}_\theta = \frac{\hat{I}\,dz'}{4\pi} e^{-j\beta_0 r}\left[\frac{j\omega\mu_0}{r} + \frac{\eta_0}{r^2} + \frac{1}{j\omega\epsilon_0 r^3}\right] \sin\theta \tag{11-40c}$$

and η_0 denotes the intrinsic impedance $\sqrt{\mu_0/\epsilon_0}$ for free space encountered in Chapter 2 in connection with uniform plane waves. In Figure 11-6(*c*) are shown the vector field components of the oscillating current element at the typical field point $P(r, \theta, \phi)$ and given by (11-37) and (11-40).

The real-time forms of the fields of an oscillating current-element are found by use of (2-74); thus, from (11-37) and (11-40)

$$dE_r = \text{Re}\,[d\hat{E}_r e^{j\omega t}] = \text{Re}\left[\frac{\hat{I}\,dz'}{4\pi} e^{j(\omega t - \beta_0 r)}\left(\frac{2\eta_0}{r^2} + \frac{2e^{-j90^\circ}}{\omega\epsilon_0 r^3}\right) \cos\theta\right]$$

$$= \frac{2I\,dz'}{4\pi}\left[\frac{\eta_0}{r^2}\cos(\omega t - \beta_0 r) + \frac{1}{\omega\epsilon_0 r^3}\sin(\omega t - \beta_0 r)\right]\cos\theta \tag{11-41a}$$

$$dE_\theta = \frac{I\,dz'}{4\pi}\left[\frac{-\omega\mu_0}{r}\sin(\omega t - \beta_0 r) + \frac{\eta_0}{r^2}\cos(\omega t - \beta_0 r) + \frac{1}{\omega\epsilon_0 r^3}\sin(\omega t - \beta_0 r)\right]\sin\theta \tag{11-41b}$$

$$dH_\phi = \frac{I\,dz'}{4\pi}\left[\frac{1}{r^2}\cos(\omega t - \beta_0 r) - \frac{\beta_0}{r}\sin(\omega t - \beta_0 r)\right]\sin\theta \quad \text{(11-41c)}$$

assuming the current amplitude $\hat{I}$ to be the pure real I. These real-time results are useful in sketching the flux fields of the oscillating dipole, depicted in Figure 11-7. Only the electric field lines are shown, since their components dE_θ and dE_r lie in the plane of the paper; the flux of dH_ϕ, from (11-41c), consists of an azimuthally oriented system of circles about the z axis of the figure. The fields close to the dipole, termed the *nearzone fields*, resemble the electric flux of a static charge dipole discussed in Example 4-7, in contrast with the *farzone*, or radiation, *fields* that become important at distances of a few wavelengths or more from the source.

The complexity of the foregoing electric dipole fields warrants a look at the simplifications in the so-called farzone and nearzone regions, distinctions made possible by a comparison of the terms dependent on r^{-1}, r^{-2}, and r^{-3} in the field expressions and facilitated by forming ratios of the terms. For example, from the magnetic field expression (11-37), the following ratio of the inverse r to inverse r^2 terms is obtained

$$\frac{\text{Magnitude of } 1/r \text{ term}}{\text{Magnitude of } 1/r^2 \text{ term}} = \frac{\beta_0/r}{1/r^2} = \beta_0 r = 2\pi\frac{r}{\lambda_0} \quad \text{(11-42)}$$

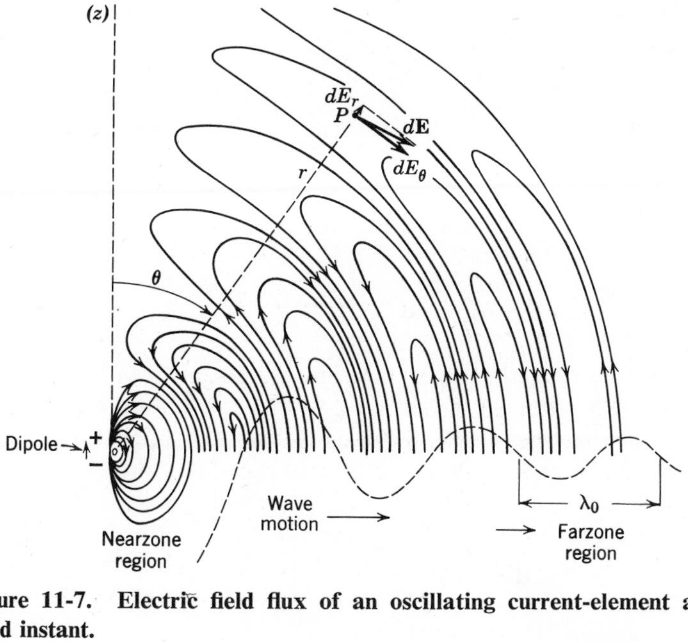

Figure 11-7. Electric field flux of an oscillating current-element at a fixed instant.

From this it is inferred that many wavelengths from the dipole source ($r \gg \lambda_0$) the $1/r$ term predominates. Conversely, the $1/r^2$ term of (11-37) is far more important *near* the source ($r \ll \lambda_0$). Similar comparisons applied to the electric field expressions (11-40b) and (11-40c) yield ratios of the magnitudes of the $1/r$ to the $1/r^2$ terms as well as the $1/r^2$ to the $1/r^3$ terms, comparable to the result (11-42). One concludes that in the nearzone region ($r \ll \lambda_0$), the fields of the elementary dipole are well approximated by only the $1/r^2$ term of $d\hat{H}_\phi$ and the $1/r^3$ terms of $d\hat{E}_r$ and $d\hat{E}_\theta$ as follows:

$$d\hat{\mathbf{E}} \cong \frac{\hat{I}\,dz'}{j\omega 4\pi\epsilon_0 r^3} e^{-j\beta_0 r}[\mathbf{a}_r 2\cos\theta + \mathbf{a}_\theta \sin\theta] \tag{11-43a}$$

Nearzone: $r \ll \lambda_0$

$$d\hat{\mathbf{H}} \cong \mathbf{a}_\phi \frac{\hat{I}\,dz'}{4\pi r^2} e^{-j\beta_0 r} \sin\theta \tag{11-43b}$$

Comparing (11-43a) with the field (4-44) of a static charge dipole shows that (11-43a) converges precisely to (4-44) in the static limit ($\omega \to 0$), upon substituting $\hat{q}$ for $\hat{I}/j\omega$ into (11-43a) from (11-34). Similarly, the magnetic field (11-43b) reduces, as $\omega \to 0$, to the static result obtained from the Biot–Savart law (5-35b) applied to a differential current-element.

With the assumption $r \gg \lambda_0$, defining the *farzone* region, the important field terms are only those varying as $1/r$, reducing (11-37), (11-40b), and (11-40c) to

$$d\hat{\mathbf{E}} \cong \mathbf{a}_\theta \frac{j\omega\mu_0 \hat{I}dz'}{4\pi r} e^{-j\beta_0 r} \sin\theta \tag{11-44a}$$

Farzone: $r \gg \lambda_0$

$$d\hat{\mathbf{H}} \cong \mathbf{a}_\phi \frac{j\beta_0 \hat{I}dz'}{4\pi r} e^{-j\beta_0 r} \sin\theta \tag{11-44b}$$

These fields are in-phase, so they become important in the radiation of energy into remote regions. Their ratio is the real intrinsic wave impedance

$$\frac{d\hat{E}_\theta}{d\hat{H}_\phi} = \sqrt{\frac{\mu_0}{\epsilon_0}} \equiv \eta_0 \approx 377\ \Omega \tag{11-45}$$

identical with (2-130), the wave impedance associated with uniform plane waves in free space treated in Section 2-10. This is not an unexpected result if one realizes that the spherical waves (11-44) are TEM waves. They are essentially uniform plane waves over a small portion of the surface of a large sphere of radius r centered at the radiating elementary dipole. The factor $\sin\theta$ appearing in the farzone field expressions (11-44) is called the *field pattern* factor of the elementary dipole. It accounts for maximum field intensities in a direction *broadside* to the elementary dipole as shown in Figure 11-8(a), tapering to zero along the dipole axis.

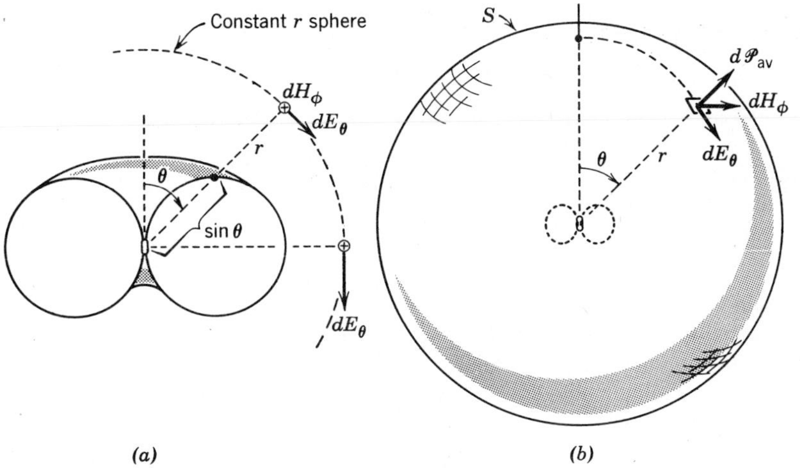

Figure 11-8. Relative to the elementary dipole. (*a*) Field pattern of the elementary dipole. Plot of sin θ versus θ. Note the axial symmetry. (*b*) Large spherical surface $S(r \gg \lambda_0)$ enclosing dipole, for finding the radiated power by use of farzone fields.

The time-average power radiated from any surface S enclosing the elementary dipole, depicted in Figure 11-8(*b*), is obtained by use of the Poynting theorem (7-62). No Ohmic dissipation occurs in the free-space region enclosed by S, making (7-62)

$$P_{av} = -\int_V \tfrac{1}{2} \operatorname{Re} [\hat{\mathbf{J}}^* \cdot \hat{\mathbf{E}}]\, dv = \oint_S \tfrac{1}{2} \operatorname{Re} [\hat{\mathbf{E}} \times \hat{\mathbf{H}}^*] \cdot d\mathbf{s} \text{ W} \qquad (11\text{-}46)$$

The volume integral denotes the time-average power generated by active sources driving the elementary dipole, seen to equal the time-average power-flux leaving (radiated from) the enclosing surface S. It should be realized that *any* surface S whatsoever may be used to enclose the dipole source, but by use of a sphere of large radius r, requiring only the farzone fields (11-44), one eliminates the need for incorporating all the terms of (11-37) and (11-40). The additional contributions to the time-average power are found to be zero anyway because of the phase condition of the nearzone terms. Inserting (11-44) into (11-46), one may show that the time-average power radiated from the elementary dipole becomes

$$P_{av} = \oint_S \tfrac{1}{2} \operatorname{Re} [d\hat{\mathbf{E}} \times d\hat{\mathbf{H}}^*] \cdot d\mathbf{s} = \frac{\eta_0 \pi I^2}{3} \left(\frac{dz'}{\lambda_0} \right)^2 \text{ W} \qquad (11\text{-}47)$$

Upon increasing the differential length dz' of the current element to a value h

(though yet small compared to the wavelength λ_0) it is seen from (11-47) that the radiated power is proportional to the square of the length. Even so, an electrically short current-element is incapable of radiating much power. For example, a wire antenna 3 cm long operated at 100 MHz has a length $h = 0.01\ \lambda_0$, making $(h/\lambda_0) = 0.01$. If one could excite the wire with 1 A of current, its radiated power, from (11-47), would be only 50 mW. Aside from the difficulty of exciting an electrically short antenna with very much current, this is still substantially less than the power obtainable from a half wave linear antenna carrying the same current, as seen from the next section. A detailed consideration of the nearzone fields and antenna impedance, a subject not considered in this text, reveals that considerable capacitive impedance to current flow is offered by electrically short antennas, with little resulting radiation.

11-4 Radiation Fields of a Linear, Center-Fed, Thin-Wire Antenna

The thin-wire antenna, fed from a voltage source applied at a gap located at some point along the wire, has been in common use for years. The prediction of the fields radiated at remote distances from such an antenna, making use of the known fields of an oscillating current-element from the previous section, is the subject of this discussion. It involves the integration, or summing, of either the differential radiation potential (11-32b), or the electric and magnetic fields (11-44a) or (11-44b), over the finite length of the antenna, accounting for the dependence of the differential field contributions at a fixed field point P, upon the variable distance R to source points on the antenna. This is suggested by the geometry of Figure 11-9(a). The field integrals are seen to contain the current $\hat{I}$, a quantity that must be known or specified at each source position z' along the antenna if the integration is to be carried out. The antenna current distribution $\hat{I}(z')$ may be found experimentally or by analytical means. Experiments in which antenna currents are probed show that their amplitudes along the antenna wire are very nearly sinusoidal standing waves. Qualitatively, at least, an open-ended linear antenna may be regarded as an opened-out section of an open-circuited parallel-wire transmission line, possessing sinusoidal standing waves of current tapering to zero at the wire ends as depicted in Figure 11-9(b). Similarly, the circular loop antenna shown in Figure 11-9(c) has a current distribution on the wire that deviates only slightly from the current standing wave on the conductors of a shorted parallel-wire transmission line, as noted in the same figure. While the effects of power radiation from an antenna tend to produce deviations from these simplified standing-wave pictures, a comparison of measured antenna field patterns with calculated results based on assumed sinusoidal current standing waves reveals that the assumption is quite suitable for farzone field calculations. A much better current distribution approximation is required, on the other hand, for predicting the *terminal impedance* of a linear wire antenna; this latter task is omitted from the present discussion.

An analytical proof of the fact that a sinusoidal standing wave of current is a reasonable assumption for linear wire antennas was originally provided

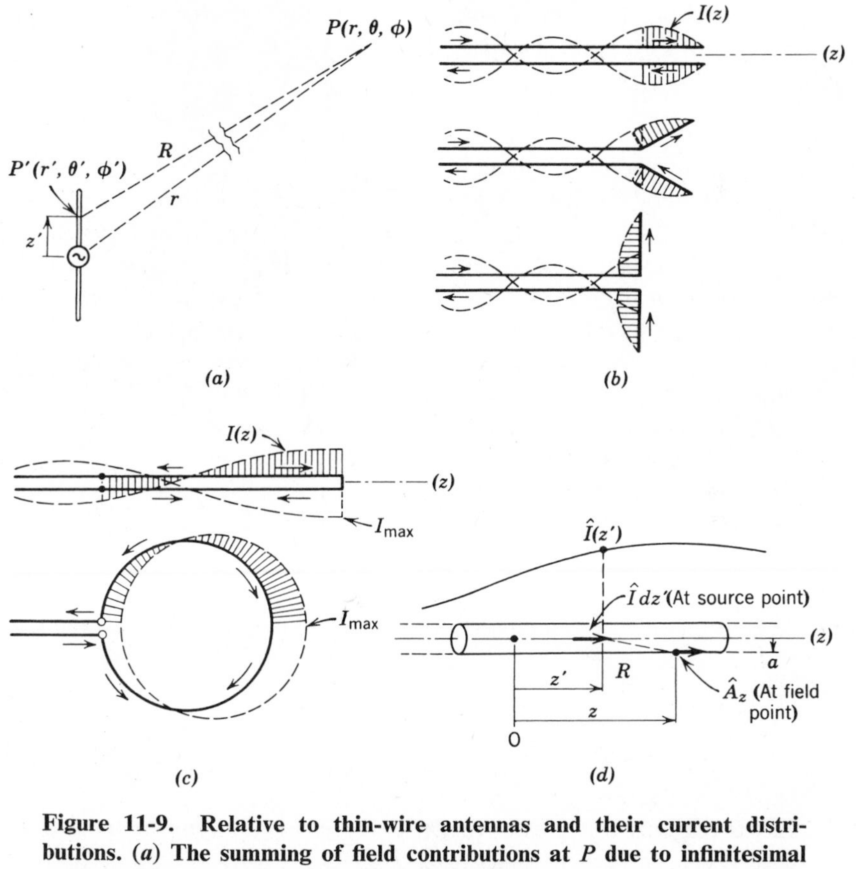

Figure 11-9. Relative to thin-wire antennas and their current distributions. (*a*) The summing of field contributions at *P* due to infinitesimal current-elements along an antenna. (*b*) Linear antenna current standing wave, obtained from a deformation of an open-circuited transmission line. (*c*) Loop antenna current standing wave, obtained from a deformation of a shorted transmission line. (*d*) Pertaining to the distribution of a current standing wave along a thin wire, as a function of *z*.

by Pocklington.† His conclusion is demonstrated from an expression for the vector magnetic potential **A** developed at the surface of a thin wire, using the integral (11-32b). Assuming the antenna current $\hat{I}(z')$ to be concentrated wholly on the wire axis as in Figure 11-9(*d*) does not significantly alter the potential $\hat{\mathbf{A}}(a, z)$ at the typical, fixed field point P *on the wire surface.* $\hat{\mathbf{A}}(a, z)$ is obtained from (11-32b) with $\hat{I}\, d\ell = \mathbf{a}_z \hat{I}(z')\, dz'$

$$\hat{\mathbf{A}}(a, z) = \mathbf{a}_z \int_{z'=z-\ell}^{z+\ell} \frac{\mu_0 \hat{I} e^{-j\beta_0 R}}{4\pi R}\, dz' \tag{11-48}$$

in which the distance R from a source point $P'(0, z')$ on the wire axis to the fixed field point $P(a, z)$ on the wire surface is

$$R = \sqrt{a^2 + (z' - z)^2} \tag{11-49}$$

It is sufficient to assign the integration limits in (11-48) over only a rather small neighborhood $(z - \ell, z + \ell)$ of the field point P, in view of the close proximity of the field point to the axial current sources. In other words, current sources located at more remote points $(z' - z) \gg a$ are too far away to alter the integral appreciably. Then the phase factor $e^{-j\beta_0 R}$ in (11-48) takes on values near unity, since $\beta_0 R$ will not depart greatly from zero. Also, the current $\hat{I}(z')$ being investigated acquires an average value $\hat{I}(z)$ over the neighborhood $(z - \ell, z + \ell)$ about the *fixed* field point P, or essentially a uniform value over that z' range. Then $\hat{I}(z)$ can be removed from the integral of (11-48) to obtain

$$\hat{A}_z(a, z) \simeq \hat{I}(z) \int_{z'=z-\ell}^{z+\ell} \frac{\mu_0\, dz'}{4\pi\sqrt{a^2 + (z' - z)^2}} = \hat{I}(z) \frac{\mu_0}{4\pi} \ln \frac{\sqrt{a^2 + \ell^2} + \ell}{\sqrt{a^2 + \ell^2} - \ell}$$

Thus $\hat{A}(z)$ at any field point close to the wire axis is *proportional* to the local current $\hat{I}(z)$; i.e.,

$$\hat{A}_z(a, z) = K\hat{I}(z) \tag{11-50}$$

In an essentially axially symmetric system such as this, the field $\hat{A}_z$ at a fixed radial position $\rho = a$ on the wire is dependent on z only. The wave equation (11-18), written only in terms of the $\hat{A}_z$ component, therefore reduces to

$$\frac{\partial^2 \hat{A}_z}{\partial z^2} + \omega^2 \mu_0 \epsilon_0 \hat{A}_z = 0 \tag{11-51}$$

† Pocklington, H. E. "Electrical oscillations in wires," *Cambr. Phil. Soc. Proc.*, **9**, Oct. 25, 1897, pp. 324–332.

for any field point $P(a, z)$ on the wire surface, but (11-50) substituted into (11-51) yields a wave equation in terms of the current $\hat{I}(z)$ on the wire axis

$$\frac{\partial^2 \hat{I}}{\partial z^2} + \omega^2 \mu_0 \epsilon_0 \hat{I} = 0 \tag{11-52}$$

This homogeneous second-order differential equation evidently has the solution

$$\hat{I}(z) = \hat{I}_m^+ e^{-j\beta_0 z} + \hat{I}_m^- e^{j\beta_0 z} \tag{11-53}$$

a sum of forward and backward-traveling waves of current on the wire. With the current at the open ends of the linear antenna of Figure 11-9(b) required to vanish, (11-53) agrees with the experimental evidence that the time-harmonic current distribution on a thin-wire antenna is essentially a *sinusoidal standing wave.* The phase factor β_0 applicable to the current standing wave on the wire implies that the wavelength associated with the current distribution is the same as that for the fields in the free space surrounding the antenna.

In Figure 11-10(a) are shown examples of linear antennas driven by a sinusoidal source and developing sinusoidal standing waves of current in accordance with the foregoing remarks. The following rules are applicable:

1. The current through the driving source must be continuous. This follows from the requirement that just as much current must leave one generator terminal as enters the other terminal.
2. If the antenna ends are open (the wire does not form a continuous loop), the current at the ends must vanish. This follows from the conservation of charge.
3. The current distributions to either side of the generator must be sinusoidal standing waves with a free-space phase constant $\beta_0 = \omega\sqrt{\mu_0\epsilon_0}$. The distribution is specified by (11-53), satisfying the boundary conditions of rules 1 and 2 above.

The foregoing rules, applied to the *center-fed,* straight-wire antenna of Figure 11-10(a) lead to the following current distributions on the upper and lower halves of the antenna:

$$\begin{aligned} \hat{I}(z) &= \hat{I}_m \sin \beta_0(\ell - z) \qquad & 0 < z < \ell \\ \hat{I}(z) &= \hat{I}_m \sin \beta_0(\ell + z) \qquad & -\ell < z < 0 \end{aligned} \tag{11-54}$$

in which $\hat{I}_m$ denotes the current amplitude occurring at the maximum along the standing wave. In general $\hat{I}_m$ may be assumed *complex*; i.e., $\hat{I}_m = I_m e^{j\phi}$

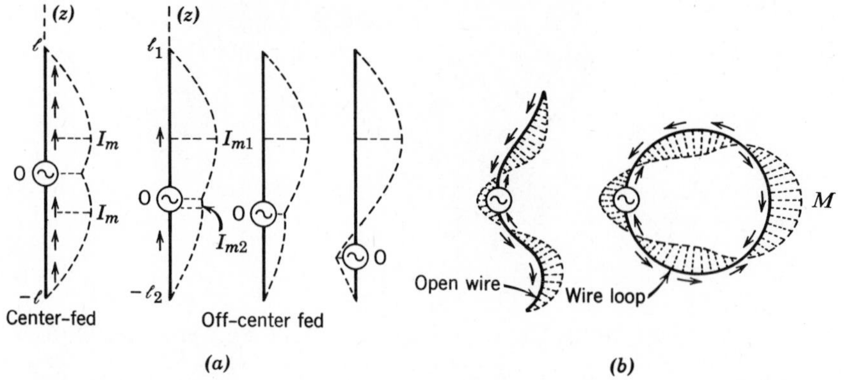

Figure 11-10. Standing-wave current distributions found on thin, linear antennas driven by a single sinusoidal generator. (*a*) Straight linear antennas: center-fed (left) and off-center-fed. (*b*) Examples of curved linear antennas: open-wire (left) and loop antenna.

if an arbitrary phase angle ϕ is desired to be included. The standing waves (11-54) are seen to be continuous at the generator location $z = 0$ in accordance with rule 1; while they further vanish at the antenna ends $z = \pm\ell$ as required by rule 2. If the generator is placed off center as in Figure 11-10(*a*), nonsymmetrical current standing waves are obtained as shown. The rules applied to this case yield current standing waves

$$\begin{aligned} \hat{I}(z) &= \hat{I}_{m1} \sin \beta_0(\ell_1 - z) \qquad 0 < z < \ell_1 \\ \hat{I}(z) &= \hat{I}_{m2} \sin \beta_0(\ell_2 + z) \qquad -\ell_2 < z < 0 \end{aligned} \tag{11-55}$$

The reader may verify what additional relationship between the standing-wave amplitudes $\hat{I}_{m1}$ and $\hat{I}_{m2}$ needs to be satisfied if current continuity through the generator is to prevail.

The current standing waves are illustrated for center-fed antennas of typical lengths in Figure 11-11(*a*). Shown in (*b*) is the variation of the antenna currents with time; the standing character of the waves is evident from the real-time behavior.

Having seen from the preceding development that only local or neighborhood current-elements appreciably affect the potential $\hat{\mathbf{A}}$ near the surface of a linear antenna, one should also expect sinusoidal current standing waves to exist on a linear antenna even when it is *curved* as depicted in Figure 11-10(*b*). For the open-wire antenna of length 2ℓ in that figure, the distribution (11-54) may still be taken as a good approximation to the current standing wave, if z is replaced by a variable ζ denoting distance measured along the curved

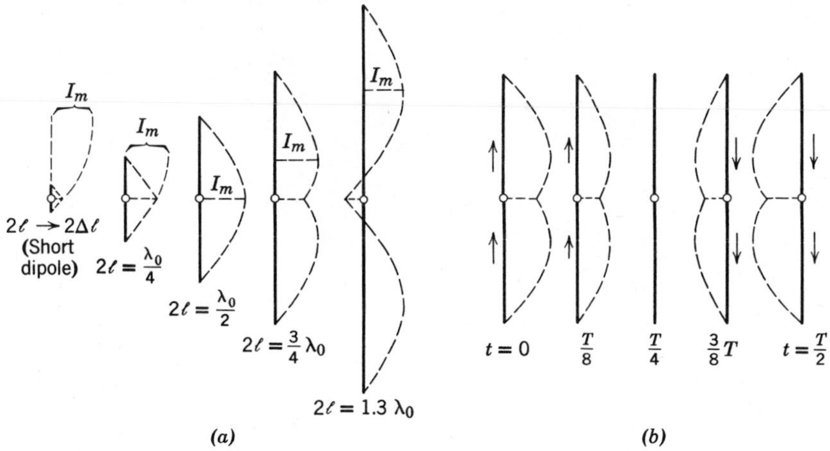

Figure 11-11. Currents on center-fed, linear antennas. (*a*) Current standing waves on antennas of different lengths, operated at the same frequency. (*b*) Variations of the current standing wave with time *t*, for a three-quarters wavelength antenna.

wire axis. A departure from the sinusoidal standing wave would be expected at sharp corners or regions of wire curvature small compared with a free-space wavelength. Finally, one may note the effect of bringing the ends of the antenna together, forming the "loop antenna" of Figure 11-10(*b*). The current standing wave in this case is predictable from a change in rule 2: instead of the current vanishing at the previously open ends, it must be continuous and well-behaved at the midpoint M, and symmetric about M.

Armed with the knowledge of a reasonable approximation (11-54) of the current distribution on the center-fed, linear antenna of Figure 11-11(*a*), one can proceed with the evaluation of its farzone electromagnetic field. Two methods are available at this stage of the development: (*a*) the integration for the vector magnetic potential $\hat{\mathbf{A}}$ at any farzone field point by use of (11-35), whence $\hat{\mathbf{E}}$ and $\hat{\mathbf{H}}$ may be found using (11-31) and (11-14); and (*b*) the direct integration for the farzone electric field by means of (11-44a).

Approach (*b*) is employed, appealing to the geometry of Figure 11-12. The contribution to the total electric field at the field point P, offered by a typical current-element $\hat{I}\,dz'$ located at P' on the antenna wire, is given by (11-44a)

$$d\hat{E}_\theta = \frac{j\eta_0\beta_0\hat{I}\,dz'}{4\pi R}\,e^{-j\beta_0 R}\sin\theta \tag{11-56}$$

With $P(r, \theta, \phi)$ in the farzone and the antenna centered at the origin, the

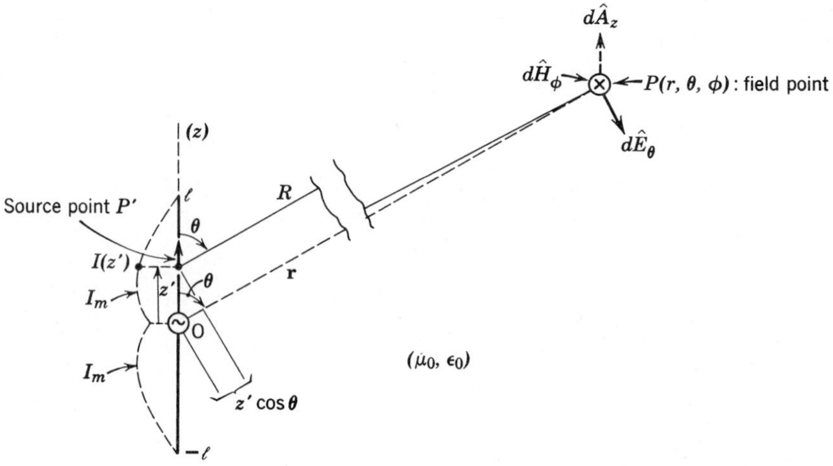

Figure 11-12. Geometry of center-fed linear antenna in relation to the determination of its farzone electromagnetic fields.

distance R from the source point P' to P is essentially parallel to r, so that

$$R \cong r - z' \cos \theta \tag{11-57}$$

assuming R very large compared with the antenna length 2ℓ. Using (11-57), the electric field at P is found by integrating (11-56)

$$\begin{aligned} \hat{E}_\theta = \int d\hat{E}_\theta &= \frac{j\beta_0\eta_0 \sin\theta}{4\pi} \int_{z'=-\ell}^{\ell} \frac{\hat{I}(z')e^{-j\beta_0(r - z'\cos\theta)}}{r - z'\cos\theta}\,dz' \\ &= \frac{j\beta_0\eta_0 \sin\theta}{4\pi} e^{-j\beta_0 r} \int_{z'=-\ell}^{\ell} \frac{\hat{I}(z')e^{j\beta_0 z'\cos\theta}}{r - z'\cos\theta}\,dz' \end{aligned} \tag{11-58}$$

The z' dependence of the current is expressed by (11-54)

$$\begin{aligned} \hat{I}(z') &= \hat{I}_m \sin\beta_0(\ell - z') \qquad 0 < z' < \ell \\ \hat{I}(z) &= \hat{I}_m \sin\beta_0(\ell + z') \qquad -\ell < z' < 0 \end{aligned} \tag{11-59}$$

The phase factor $\exp(j\beta_0 z' \cos\theta)$ appearing in (11-58) accounts for the out-of-phase condition of the contributions $d\hat{E}_\theta$ arriving at P, making the integral quite sensitive to the variable phase delay due to the variable distance R. In contrast, the factor $r - z'\cos\theta$ in the denominator in (11-58) affects only the amplitudes of the $d\hat{E}_\theta$ contribution at P, permitting there the substitution $r - z' \cos\theta \cong r$. Therefore (11-58) simplifies to

$$\hat{E}_\theta = \frac{j\beta_0\eta_0\hat{I}_m \sin\theta e^{-j\beta_0 r}}{4\pi r} \times \left\{\int_{-\ell}^{0} e^{j\beta_0 z'\cos\theta} \sin\beta_0(\ell + z')\,dz' + \int_0^{\ell} e^{j\beta_0 z'\cos\theta} \sin\beta_0(\ell - z')\,dz'\right\}$$

in which the integration is by parts or by substituting the exponential expression $\sin\alpha = (2j)^{-1}[e^{j\alpha} - e^{-j\alpha}]$ for the sine functions. Thus

$$\hat{E}_\theta = \frac{j\eta_0\hat{I}_m}{2\pi r} e^{-j\beta_0 r}\left[\frac{\cos(\beta_0\ell\cos\theta) - \cos\beta_0\ell}{\sin\theta}\right] = \eta_0\hat{H}_\phi \text{ V/m} \qquad (11\text{-}60)$$

From the latter are deduced the following properties of the radiation fields of a center-fed, linear dipole:

1. The fields $\hat{E}_\theta$ and $\hat{H}_\phi$ in the farzone region are outgoing TEM spherical waves, related by the real intrinsic wave impedance η_0 like the fields of uniform plane waves in free space. The fields have spherical equiphase surfaces (observed from putting r = constant in the spherical wave phase factor $e^{-j\beta_0 r}$), while the amplitudes vary as r^{-1}.
2. The fields are directly proportional to the excitation current amplitude $\hat{I}_m$.
3. $\hat{E}_\theta$ and $\hat{H}_\phi$ are independent of the azimuthal angle ϕ (from the axial symmetry). The θ dependence is contained in the bracketed factor of (11-60), designated by $F(\theta)$ as follows:

$$F(\theta) = \frac{\cos(\beta_0\ell\cos\theta) - \cos\beta_0\ell}{\sin\theta} \qquad (11\text{-}61)$$

   $F(\theta)$ is called the field pattern of the center-fed linear antenna. Some features of this factor are discussed in the following.

The pattern-factor $F(\theta)$ specifies how the farzone fields vary with θ over a sphere. In Figure 11-13(a), (b), and (c) are shown three ways to depict the field pattern of a center-fed antenna a half wavelength long (the so-called half wave dipole). The pattern is axially symmetric, so the sectional-cut pattern of Figure 11-13(a) adequately represents $F(\theta)$. From (d) and (e), as the antenna length 2ℓ is increased, the pattern acquires additional lobes, between which nulls or dead spots in the transmitted fields occur. This is a consequence of phase interference effects, more pronounced for longer

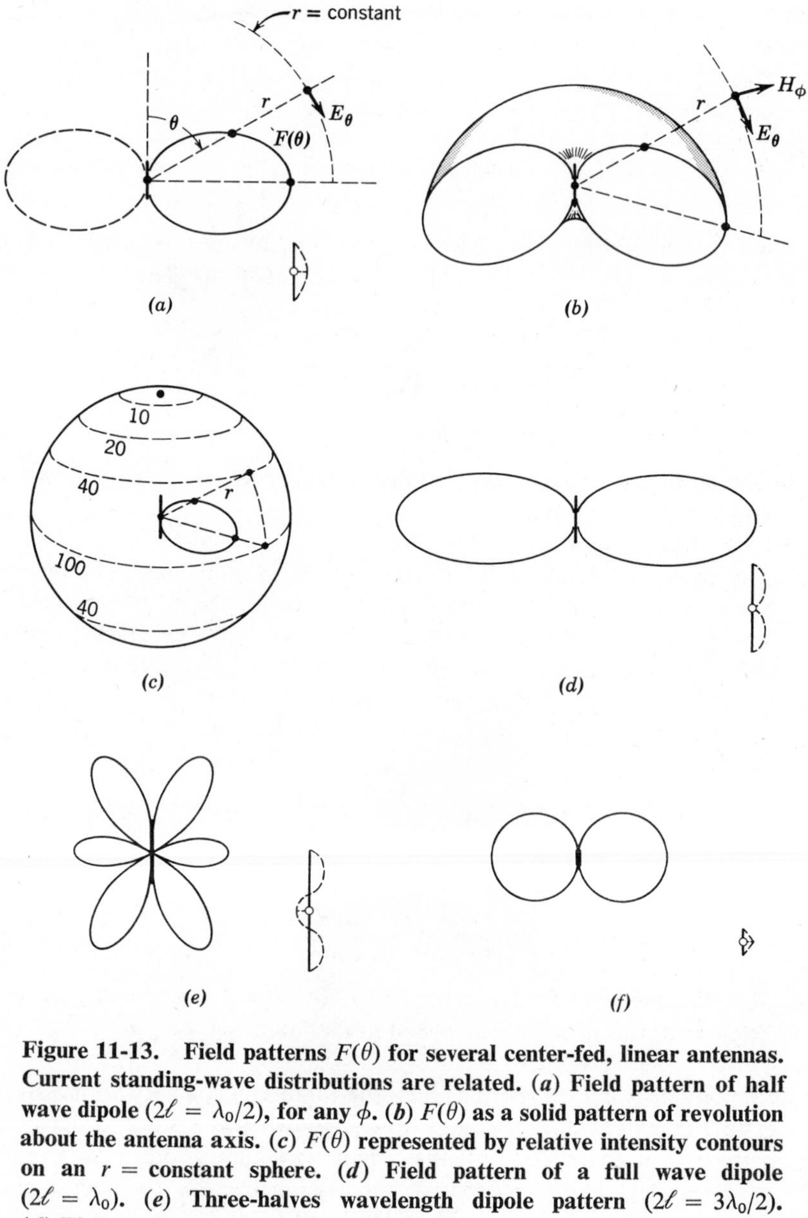

Figure 11-13. **Field patterns $F(\theta)$ for several center-fed, linear antennas. Current standing-wave distributions are related. (*a*) Field pattern of half wave dipole ($2\ell = \lambda_0/2$), for any ϕ. (*b*) $F(\theta)$ as a solid pattern of revolution about the antenna axis. (*c*) $F(\theta)$ represented by relative intensity contours on an r = constant sphere. (*d*) Field pattern of a full wave dipole ($2\ell = \lambda_0$). (*e*) Three-halves wavelength dipole pattern ($2\ell = 3\lambda_0/2$). (*f*) Field pattern of a very short dipole ($2\ell \to dz'$).**

antennas. Thus the pattern of the three-halves wavelength antenna in Figure 11-13(*e*) shows how the phase effects of the field contributions in the broadside ($\theta = 90°$) direction yields a relatively small lobe maximum there, a consequence of partial phase cancellation. For the limiting case of a very short dipole having essentially a triangular standing-wave distribution (the *tips* of sine waves), $F(\theta)$ reduces to the simple $\sin\theta$ function of Figure 11-13(*f*). In all cases, a null occurs in the pattern along the antenna axis.

The total time-average power radiated from a linear antenna may be found in the same way as for the infinitesimal dipole, by making use of (11-46). The farzone fields (11-60) thus yield the radially outward-directed, time-average Poynting vector

$$\mathscr{P}_{\text{av}} = \tfrac{1}{2}\,\text{Re}\left\{\frac{j\eta_0\hat{I}_m}{2\pi r}\,e^{-j\beta_0 r}\left(\frac{\cos(\beta_0\ell\cos\theta)-\cos\beta_0\ell}{\sin\theta}\right)\mathbf{a}_\theta \times \mathbf{a}_\phi\left[\frac{-j\hat{I}_m^*}{2\pi r}\,e^{j\beta_0 r}\left(\frac{\cos(\beta_0\ell\cos\theta)-\cos\beta_0\ell}{\sin\theta}\right)\right]\right\}$$

$$= \mathbf{a}_r\,\frac{\eta_0 I_m^2}{8\pi^2 r^2}\left(\frac{\cos(\beta_0\ell\cos\theta)-\cos\beta_0\ell}{\sin\theta}\right)^2\ \text{W/m}^2 \qquad (11\text{-}62)$$

The latter substituted into (11-46) and integrated over a sphere of radius r, on which $d\mathbf{s} = \mathbf{a}_r\,ds = \mathbf{a}_r r^2 \sin^2\theta\,d\theta\,d\phi$, obtains the time-average radiated power

$$P_{\text{av}} = \oint_S \mathscr{P}_{\text{av}}\cdot d\mathbf{s} = \int_{\theta=0}^{\pi}\int_{\phi=0}^{2\pi} \mathscr{P}_{\text{av}}\cdot\mathbf{a}_r r^2 \sin\theta\,d\theta\,d\phi$$

$$= \frac{\eta_0 I_m^2}{4\pi}\int_0^{\pi}\frac{[\cos(\beta_0\ell\cos\theta)-\cos\beta_0\ell]^2}{\sin\theta}\,d\theta\ \text{W} \qquad (11\text{-}63)$$

This expression is not integrable in closed form, though it can be evaluated by power series substitutions or by use of the tabulated sine and cosine integral functions† $Si(x)$ and $Ci(x)$; also it is readily computerized or solved by graphical methods.

Closely related to the time-average radiated power (11-63) of a linear, center-fed antenna is its so-called *radiation resistance*, R_{rad}. It is defined such that, upon multiplying it by the square of the rms value of the current amplitude I_m associated with the antenna current distribution (10-54), the radiated power (11-63) is obtained; i.e., $R_{\text{rad}} I_m^2/2 = P_{\text{av}}$, making

$$R_{\text{rad}} = \frac{2P_{\text{av}}}{I_m^2}\ \Omega \qquad (11\text{-}64)$$

† For example, see Ramo, S., J. Whinnery, and T. van Duzer. *Fields and Waves in Communication Electronics*. New York: Wiley, 1965, p. 648.

Thus R_{rad} denotes a fictitious resistance that, upon carrying the rms current $I_m/\sqrt{2}$, dissipates the same amount of power as that radiated by a center-fed linear antenna possessing a standing wave of current with the amplitude $\hat{I}_m$. It is emphasized that the radiation resistance of a linear antenna is in general *not* the same as the resistance part of the antenna impedance $\hat{Z}_a$ seen by a generator attached to the antenna terminals. This fact is related in part to the current amplitude $\hat{I}_m$ appearing elsewhere along the z axis, generally not at the antenna terminals, as seen in Figure 11-10(a).

Of considerable practical interest is the time-average power radiated from the center-fed, *half wave dipole*, the field pattern of which may be recalled from Figures 11-13(a), (b), and (c). Besides having a favorable two-lobed pattern without an excessive number of interference nulls, the half wave dipole also has a desirable terminal impedance, $\hat{Z}_a$. It develops that this impedance is essentially a pure resistance, virtually the same as its radiation resistance R_{rad}, in view of the maximum value $\hat{I}_m$ of its standing wave of current appearing at the dipole terminals as noted in Figure 11-13(a). The time-average power radiated from a half wave dipole can be calculated by use of (11-63), becoming, with $2\ell = \lambda_0/2$ $(\beta_0\ell = \pi/2)$

$$P_{\text{av}} = \frac{\eta_0 I_m^2}{4\pi}\int_0^\pi \frac{\cos^2\left(\dfrac{\pi}{2}\cos\theta\right)}{\sin\theta}\,d\theta \cong 36.5 I_m^2 \text{ W} \quad \text{Half wave dipole} \qquad (11\text{-}65)$$

The radiation resistance from (11-64) is, therefore

$$R_{\text{rad}} = \frac{2P_{\text{av}}}{I_m^2} \cong \frac{2(36.5 I_m^2)}{I_m^2} = 73\ \Omega \quad \text{Half wave dipole} \qquad (11\text{-}66)$$

*11-5 Symmetric Maxwell's Equations and Their Vector Potentials; the Field Equivalence Theorem

In Section 11-4 was demonstrated how the current distribution $\hat{I}(z')$ over a linear antenna leads to the electromagnetic fields in the surrounding region of free space, regarding electric current elements as the field sources. The external fields of *aperture antennas* such as the electromagnetic horn of Figure 11-1(b), can similarly be found from the integration over the electric currents, these occurring mainly on the inner conductive surfaces of the horn. The radiation fields of parabolic reflectors might be obtained in the same way. However, because the irregularly shaped surfaces occupied by the currents are generally not specifiable in one of the common coordinate systems, the integration could be a tedious process. It is simpler and more natural, for aperture antennas such as horns and reflecting paraboloids, to regard the *electromagnetic fields in the aperture plane* as the *sources* of the exterior fields.

In this section is described a field equivalence theorem due to Schelkunoff† putting this idea to use. A brief discussion of the history of this subject, which had its beginnings in the theory of optical diffraction, is related first to help provide some additional insight into this problem.

The radiation fields of aperture antennas may be explained by adopting a point of view not unlike that proposed by the Dutch physicist Christian Huygens in 1678. He regarded each differential surface-element of a wavefront in an electromagnetic (optical) aperture as the source of a spherical wavelet disturbance, the total effect of which, at any field point further along, was simply regarded as the phased superposition, or sum, of all the spherical wave contributions there.⁑ This approach was refined later by Augustin Fresnel (1788–1827). The Huygens–Fresnel principle was placed on a firm mathematical basis for acoustical wave phenomena by H. von Helmholtz in 1859 and for optical waves by G. Kirchhoff in 1882. Kirchhoff utilized essentially the Green's identity (2-92) to derive a field integral solution of the appropriate scalar wave equation.†† A generalization of the Kirchhoff method, extended to the vector electromagnetic wave equation, was provided by Love‡ and subsequently extended by Stratton and Chu.§ In the latter method, the use of potential fields is avoided by employing a vector version of Green's identity to yield direct integrals for the **E** and **H** fields, in terms of electric and magnetic current and charge densities in an arbitrary volume region in free space.

A vector field integral approach deduced earlier by Schelkunoff¶ is entirely equivalent to the Stratton-Chu method, but it possesses the possible advantage of additional physical insight. Schelkunoff's method is to be described here and applied to problems involving aperture-type sources of electromagnetic radiation fields. It makes use of field potentials (both magnetic *and* electric vector potentials), and it incorporates a concept termed *field equivalence*, involving the replacement of electric and magnetic fields on an arbitrary

† Schelkunoff, S. A. "Some equivalence theorems of electromagnetics and their application to radiation problems," *Bell. Syst. Tech. Jour.*, **15**, January 1936, pp. 92–112.

⁑ In this regard, one may see that the integral (11-36b) for the vector potential field of a system of current sources represents nothing more than a superposition of elementary spherical wave functions, $e^{-j\beta_0 R}/R$, proportional in strength to the current sources $\mathbf{J}\, dv'$, and summed up (integrated) in proper phase at the field point P of Figure 11-4(*b*). The Huygens picture differs, however, in that it deals only with scalar wave phenomena.

†† An account of Kirchhoff's method is given in Born, M., and E. Wolf. *Principles of Optics*. Elmsford, N.Y.: Pergamon, 1964, pp. 375–382.

‡ Love, A. E. H. "The integration of the equations of propagation of electric waves," *Phil. Trans.*, (A), **197**, 1901, p. 45.

§ Stratton, J. A., and L. J. Chu. "Diffraction theory of electromagnetic waves," *Phys. Rev.*, **56**, 1939, p. 99.

¶ Ibid.

closed surface (principally on the aperture) with equivalent electric and magnetic currents and charges. The concept of field equivalence is seen to make it desirable to postulate fictitious magnetic charges and currents in the region in question.

A. Symmetric Maxwell's Equations and Their Vector Potentials

Calling the magnetic charge density ρ_m and the magnetic current density $\mathbf{J}_m$ provides the following *symmetrical* form of Maxwell's differential equations for free space, written here in complex time-harmonic form

$$\nabla \cdot \hat{\mathbf{D}} = \hat{\rho}_v \ \mathrm{C/m^3} \tag{11-67a}$$

$$\nabla \cdot \hat{\mathbf{B}} = \hat{\rho}_m \ \mathrm{Wb/m^3} \tag{11-67b}$$

$$\nabla \times \hat{\mathbf{E}} = -\hat{\mathbf{J}}_m - j\omega\mu_0\hat{\mathbf{H}} \ \mathrm{V/m^2} \tag{11-67c}$$

$$\nabla \times \hat{\mathbf{H}} = \hat{\mathbf{J}} + j\omega\epsilon_0\hat{\mathbf{E}} \ \mathrm{A/m^2} \tag{11-67d}$$

Suppose that the electric and magnetic fields $\hat{\mathbf{D}} = \epsilon_0\hat{\mathbf{E}}$ and $\hat{\mathbf{B}} = \mu_0\hat{\mathbf{H}}$ of the latter are resolved into the contributions $\hat{\mathbf{D}}_e = \epsilon_0\hat{\mathbf{E}}_e$ and $\hat{\mathbf{B}}_e = \mu_0\hat{\mathbf{H}}_e$ attributable solely to the *electric currents and charges* $\hat{\rho}_v$ and $\hat{\mathbf{J}}$, plus the additional contributions $\hat{\mathbf{D}}_m = \epsilon_0\hat{\mathbf{E}}_m$ and $\hat{\mathbf{B}}_m = \mu_0\hat{\mathbf{H}}_m$ related to only the *fictitious magnetic currents and charges* $\hat{\rho}_m$ and $\hat{\mathbf{J}}_m$ in the region. The total fields in (11-67) can then be written as superpositions of the two sets of fields

$$\hat{\mathbf{D}} = \hat{\mathbf{D}}_e + \hat{\mathbf{D}}_m = \epsilon_0(\hat{\mathbf{E}}_e + \hat{\mathbf{E}}_m) = \epsilon_0\hat{\mathbf{E}} \tag{11-68}$$

$$\hat{\mathbf{B}} = \hat{\mathbf{B}}_e + \hat{\mathbf{B}}_m = \mu_0(\hat{\mathbf{H}}_e + \hat{\mathbf{H}}_m) = \mu_0\hat{\mathbf{H}} \tag{11-69}$$

This bifurcation permits separating the Maxwell equations (11-67) into two groups, attributable to the two kinds of sources as follows:

$$\left.\begin{array}{lll|lll}
\nabla \cdot \hat{\mathbf{D}}_e = \hat{\rho}_v & & (a) & \nabla \cdot \hat{\mathbf{D}}_m = 0 & & (e) \\
\nabla \cdot \hat{\mathbf{B}}_e = 0 & & (b) & \nabla \cdot \hat{\mathbf{B}}_m = \hat{\rho}_m & & (f) \\
\nabla \times \hat{\mathbf{E}}_e = -j\omega\mu_0\hat{\mathbf{H}}_e & & (c) & \nabla \times \hat{\mathbf{E}}_m = -\hat{\mathbf{J}}_m - j\omega\mu_0\hat{\mathbf{H}}_m & & (g) \\
\nabla \times \hat{\mathbf{H}}_e = \hat{\mathbf{J}} + j\omega\epsilon_0\hat{\mathbf{E}}_e & & (d) & \nabla \times \hat{\mathbf{H}}_m = j\omega\epsilon_0\hat{\mathbf{E}}_m & & (h)
\end{array}\right. \tag{11-70}$$

It is seen that the sums of these paired equations yield simply (11-67) again. Equations (*a*) through (*d*) are the conventional Maxwell equations, identical with (3-24, 3-48, 3-59, and 3-77). They specify the electric and magnetic fields associated with distributions of electric currents and charges of densities $\hat{\mathbf{J}}$ and $\hat{\rho}_v$. Equations (e) through (h) specify the additional electric and magnetic fields developed in a region presuming that fictitious magnetic current and charge densities $\hat{\mathbf{J}}_m$ and $\hat{\rho}_m$ are also present. In the presence of *both* types of currents and charges, electric and magnetic, the superposed fields (11-68) and (11-69) are needed, and these satisfy the Maxwell relations (11-67).

An integral solution of the system of equations (11-70) (*a*) through (*d*) has already been discussed in Section 11-2 by use of the vector magnetic potential $\hat{\mathbf{A}}$. Recall the applicable time-harmonic relations (11-13), (11-14), and (11-12) in this regard

$$\hat{\mathbf{B}}_e = \nabla \times \hat{\mathbf{A}} \tag{11-71}$$

$$\hat{\mathbf{E}}_e = -\nabla\hat{\Phi} - j\omega\hat{\mathbf{A}} \tag{11-72}$$

$$\hat{\Phi} = -\frac{\nabla\cdot\hat{\mathbf{A}}}{j\omega\mu_0\epsilon_0} \tag{11-73}$$

in which $\hat{\Phi}$ and $\hat{\mathbf{A}}$ satisfy the wave equations

$$\nabla^2\hat{\Phi} + \omega^2\mu_0\epsilon_0\hat{\Phi} = -\frac{\hat{\rho}_v}{\epsilon_0} \tag{11-74}$$

$$\nabla^2\hat{\mathbf{A}} + \omega^2\mu_0\epsilon_0\hat{\mathbf{A}} = -\mu_0\hat{\mathbf{J}} \tag{11-75}$$

It was shown moreover that (11-32b) is an integral of the wave equation (11-75)

$$\hat{\mathbf{A}}(\boldsymbol{r}) = \int_V \frac{\mu_0\hat{\mathbf{J}}(\boldsymbol{r}')e^{-j\beta_0 R}}{4\pi R}\,dv' \tag{11-76}$$

The additional effects of fictitious magnetic current and charge distributions in the free-space region can be deduced by analogy for the system of equations (11-70), (*e*) through (*h*), upon noting the *dualities* that exist between pairs of quantities in the first and second columns of Maxwell equations (11-70). Analogous pairs of equations are (*a*) and (*f*), (*b*) and (*e*), (*c*) and (*h*), and (*d*) and (*g*); so there exist the following dualities:

$$\begin{aligned}
&\hat{\mathbf{E}}_e \text{ is the dual of } \hat{\mathbf{H}}_m\\
&\hat{\mathbf{H}}_e \text{ is the dual of } -\hat{\mathbf{E}}_m\\
&\epsilon_0 \text{ is the dual of } \mu_0\\
&\mu_0 \text{ is the dual of } \epsilon_0\\
&\hat{\rho}_v \text{ is the dual of } \hat{\rho}_m\\
&\hat{\mathbf{J}} \text{ is the dual of } \hat{\mathbf{J}}_m
\end{aligned} \tag{11-77}$$

Since the fields $\hat{\mathbf{E}}_e$ and $\hat{H}_e$ are related to the vector magnetic potential $\hat{\mathbf{A}}$

and the scalar electric potential $\hat{\Phi}$ through (11-71) and (11-72), the corresponding fields $\hat{\mathbf{H}}_m$ and $\hat{\mathbf{E}}_m$ are analogously connected to the so-called *vector electric potential*, $\hat{\mathbf{F}}$, and *scalar magnetic potential*, $\hat{\Psi}$, such that

$$\begin{aligned} &\hat{\mathbf{A}} \text{ is the dual of } \hat{\mathbf{F}} \\ &\hat{\Phi} \text{ is the dual of } \hat{\Psi} \end{aligned} \tag{11-78}$$

A substitution of the quantities in the second columns of (11-77) and (11-78) for their duals in (11-71) through (11-75) leads to the potential relations

$$\hat{\mathbf{D}}_m = -\nabla \times \hat{\mathbf{F}} \tag{11-79}$$

$$\hat{\mathbf{H}}_m = -\nabla\hat{\Psi} - j\omega\hat{\mathbf{F}} \tag{11-80}$$

$$\hat{\Psi} = -\frac{\nabla \cdot \hat{\mathbf{F}}}{j\omega\mu_0\epsilon_0} \tag{11-81}$$

in which $\hat{\Psi}$ and $\hat{\mathbf{F}}$ satisfy wave equations analogous with (11-74) and (11-75)

$$\nabla^2\hat{\Psi} + \omega^2\mu_0\epsilon_0\hat{\Psi} = -\frac{\hat{\rho}_m}{\mu_0} \tag{11-82}$$

$$\nabla^2\hat{\mathbf{F}} + \omega^2\mu_0\epsilon_0\hat{\mathbf{F}} = -\epsilon_0\hat{\mathbf{J}}_m \tag{11-83}$$

Then the vector electric potential $\hat{\mathbf{F}}$ produced by the electric current density source distribution $\hat{\mathbf{J}}_m(r')$ at $P(\mathbf{r})$ is

$$\hat{\mathbf{F}}(\mathbf{r}) = \int_V \frac{\epsilon_0\hat{\mathbf{J}}_m(\mathbf{r}')e^{-j\beta_0 R}}{4\pi R}\,dv' \tag{11-84}$$

analogous with (11-76).

From the foregoing it is seen that if *both* electric and magnetic currents of densities $\hat{\mathbf{J}}$ and $\hat{\mathbf{J}}_m$ exist simultaneously in a free-space region, the total field $\hat{\mathbf{E}}$ produced at any field point $P(\mathbf{r})$ becomes the sum (11-68) of $\hat{\mathbf{E}}_e$ and $\hat{\mathbf{E}}_m$; from (11-72) added to (11-79) this becomes

$$\begin{aligned} \hat{\mathbf{E}}(\mathbf{r}) &= \hat{\mathbf{E}}_e(\mathbf{r}) + \hat{\mathbf{E}}_m(\mathbf{r}) = -\nabla\hat{\Phi} - j\omega\hat{\mathbf{A}} - \frac{1}{\epsilon_0}\nabla \times \hat{\mathbf{F}} \\ &= \frac{\nabla(\nabla\cdot\hat{\mathbf{A}})}{j\omega\mu_0\epsilon_0} - j\omega\hat{\mathbf{A}} - \frac{1}{\epsilon_0}\nabla \times \hat{\mathbf{F}} \end{aligned} \tag{11-85}$$

The total field $\hat{\mathbf{H}}(\mathbf{r})$ is similarly obtained from (11-71) and (11-80) substituted

into (11-69), yielding

$$\hat{\mathbf{H}}(\mathbf{r}) = \frac{\nabla(\nabla\cdot\hat{\mathbf{F}})}{j\omega\mu_0\epsilon_0} - j\omega\hat{\mathbf{F}} + \frac{1}{\mu_0}\nabla\times\hat{\mathbf{A}} \tag{11-86}$$

In practice, there is no need to use the latter, for once $\hat{\mathbf{E}}(\mathbf{r})$ has been found using (11-85), substituting that result into the free-space version of Maxwell curl relation (11-67c) (with $\hat{\mathbf{J}}_m = 0$) yields $\hat{\mathbf{H}}(\mathbf{r})$. For aperture sources to which the foregoing potential expressions (11-85) and (11-86) are applicable, moreover, the electric and magnetic current sources $\hat{\mathbf{J}}$ and $\hat{\mathbf{J}}_m$ are shown to take the form of *surface* densities $\hat{\mathbf{J}}_s$ and $\hat{\mathbf{J}}_{sm}$, in which case (11-76) and (11-84) become

$$\hat{\mathbf{A}}(\mathbf{r}) = \int_S \frac{\mu_0\hat{\mathbf{J}}_s(\mathbf{r}')e^{-j\beta_0 R}}{4\pi R}\,ds' \tag{11-87}$$

$$\hat{\mathbf{F}}(\mathbf{r}) = \int_S \frac{\epsilon_0\hat{\mathbf{J}}_{sm}(\mathbf{r}')e^{-j\beta_0 R}}{4\pi R}\,ds' \tag{11-88}$$

The latter prove useful in the field equivalence theorem described in the following.

B. Field Equivalence Theorem, and Fields of Aperture Antennas

The development of the field equivalence theorem to be described requires the use of the generalized boundary conditions associated with the symmetric Maxwell equations (11-67). The integral versions of the latter, in real-time form, are

$$\oint_S \mathbf{D}\cdot d\mathbf{s} = q \ \mathrm{C} \tag{11-89a}$$

$$\oint_S \mathbf{B}\cdot d\mathbf{s} = q_m \ \mathrm{Wb} \tag{11-89b}$$

$$\oint_\ell \mathbf{E}\cdot d\boldsymbol{\ell} = -\int_S \mathbf{J}_m\cdot d\mathbf{s} - \frac{\partial}{\partial t}\int_S \mathbf{B}\cdot d\mathbf{s} \ \mathrm{V} \tag{11-89c}$$

$$\oint_\ell \mathbf{H}\cdot d\boldsymbol{\ell} = \int_S \mathbf{J}\cdot d\mathbf{s} + \frac{\partial}{\partial t}\int_S \mathbf{D}\cdot d\mathbf{s} \ \mathrm{A} \tag{11-89d}$$

The corresponding boundary conditions are derived by the methods of Chapter 3, using the devices of the Gaussian pillbox or the thin, closed rectangle constructions typified in Figures 3-4 and 3-10. The results are

$$\mathbf{n}\cdot(\mathbf{D}_1 - \mathbf{D}_2) = \rho_s \text{ C/m}^2 \tag{11-90a}$$

$$\mathbf{n}\cdot(\mathbf{B}_1 - \mathbf{B}_2) = \rho_{sm} \text{ Wb/m}^2 \tag{11-90b}$$

$$\mathbf{n}\times(\mathbf{E}_1 - \mathbf{E}_2) = -\mathbf{J}_{sm} \text{ V/m} \tag{11-90c}$$

$$\mathbf{n}\times(\mathbf{H}_1 - \mathbf{H}_2) = \mathbf{J}_s \text{ A/m} \tag{11-90d}$$

in which **n** denotes as usual the normal unit vector directed from region 2 into region 1. A comparison of (11-90b) and (11-90c) with (3-50) and (3-79) reveals the additional effects of fictitious magnetic surface charge and surface current densities ρ_{sm} and $\mathbf{J}_{sm}$ at an interface. The field equivalence theorem is concerned with a boundary interface, to one side of which the fields are assumed *nullified.* With the assumption of null fields in region 2, the boundary conditions (11-90) specialize to

$$\mathbf{n}\cdot\mathbf{D}_1 = \rho_s \text{ C/m}^2 \tag{11-91a}$$

$$\mathbf{n}\cdot\mathbf{B}_1 = \rho_{sm} \text{ Wb/m}^2 \tag{11-91b}$$

$$-\mathbf{n}\times\mathbf{E}_1 = \mathbf{J}_{sm} \text{ V/m} \tag{11-91c}$$

$$\mathbf{n}\times\mathbf{H}_1 = \mathbf{J}_s \text{ A/m} \tag{11-91d}$$

(11-91a) and (11-91b) state that the *normal* components of electric and magnetic fields may undergo an abrupt jump to zero from region 1 to 2 only if a surface electric charge ρ_s and a surface magnetic charge ρ_{sm} are present in strength equal to the normal D_{n1} and B_{n1} components respectively. According to (11-91c) and (11-91d), moreover, abrupt transitions from finite values to zero, of the tangential components E_{t1} and H_{t1}, are allowable only if the respective magnetic current and electric current surface densities $\mathbf{J}_{sm}$ and $\mathbf{J}_s$ prevail at the interface.

While the free magnetic charge and magnetic current densities of (11-91c) and (11-91d) have not been proved to exist physically, they are an important mathematical concept in the field equivalence theorem. Suppose first that known distributions of electric current and charge densities exist in some portion of a free-space region, as in Figure 11-14(*a*). The fields at $P(\mathbf{r})$ could, as usual, be found by use of the potential integral (11-32b), whence $\hat{\mathbf{E}}$ and $\hat{\mathbf{H}}$ follow from (11-13) and (11-14). The fields at $P(\mathbf{r})$ can also be obtained, however, from equivalent currents established over an arbitrary surface S_1 enclosing all the sources, as depicted in Figure 11-14(*b*), the enclosed sources producing the fields $\hat{\mathbf{E}}_1$ and $\hat{\mathbf{H}}_1$ on S_1 as shown. Suppose the sources inside

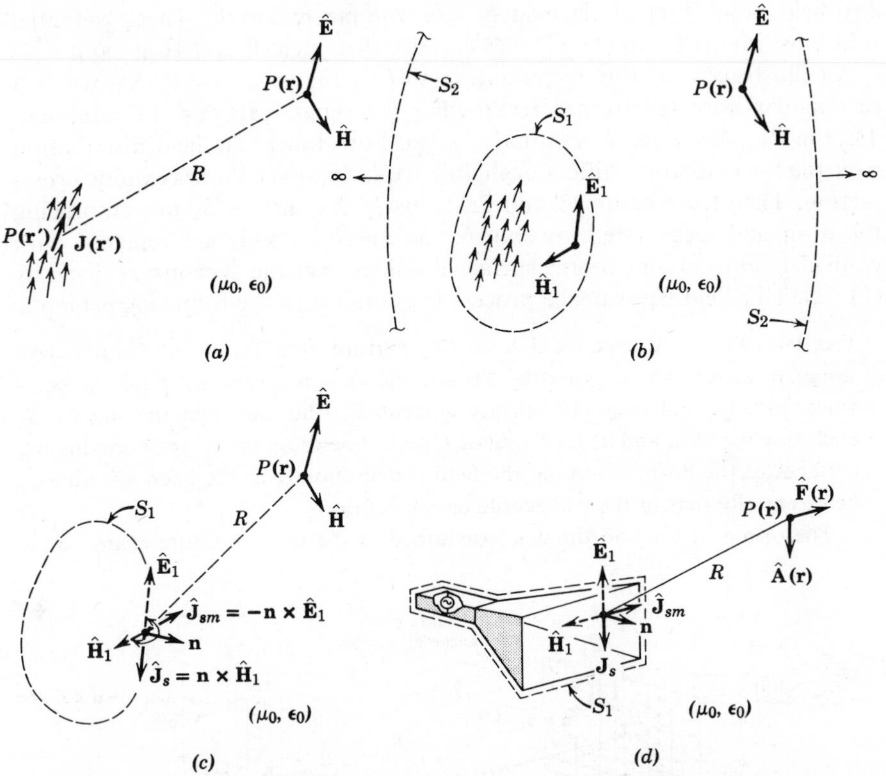

Figure 11-14. **Development of the field equivalence principle. (*a*) Physical sources establish $\hat{\mathbf{E}}_1$, $\hat{\mathbf{H}}_1$, on S_1. (*b*) Arbitrary closed surface $S = S_1 + S_2$ excludes physical sources from free-space region V. The sources establish $\hat{\mathbf{E}}_1$, $\hat{\mathbf{H}}_1$, on S_1. (*c*) Sources $\hat{\mathbf{J}}$ assumed nullified in S_1 with fields $\hat{\mathbf{E}}_1$, $\hat{\mathbf{H}}_1$ on S_1 intact. Equivalent currents $\hat{\mathbf{J}}_s$, $\hat{\mathbf{J}}_{sm}$ required on S_1 account for $\hat{\mathbf{E}}$, $\hat{\mathbf{H}}$ at $P(\mathbf{r})$. (*d*) Application of field equivalence to electromagnetic horn. Known fields on aperture S_1 replaced by equivalent surface currents $\hat{\mathbf{J}}_s$, $\hat{\mathbf{J}}_{sm}$ yield potentials $\hat{\mathbf{A}}$, $\hat{\mathbf{F}}$ at $P(\mathbf{r})$.**

S_1 are considered *nullified*, but the fields $\hat{\mathbf{E}}_1$ and $\hat{\mathbf{H}}_1$ just outside S_1 are maintained at their previous values. This condition is mathematically allowable only if the four boundary conditions (11-91) are upheld, implying the simultaneous existence of electric and magnetic charge and current densities ρ_s, ρ_{sm}, $\mathbf{J}_{sm}$, and $\mathbf{J}_s$ on the surface S_1, as suggested by Figure 11-14(*c*); with the establishment of the *equivalent sources* $\hat{\mathbf{J}}_{sm} = -\mathbf{n} \times \hat{\mathbf{E}}_1$ and $\hat{\mathbf{J}}_s = \mathbf{n} \times \hat{\mathbf{H}}_1$ on S_1, the integrals (11-87) and (11-88) can be employed to find $\hat{\mathbf{A}}$ and $\hat{\mathbf{F}}$ at any field point $P(\mathbf{r})$ in the source free volume region V. These potential solutions, inserted into (11-85) and (11-86), then yield $\hat{\mathbf{E}}$ and $\hat{\mathbf{H}}$ at $P(\mathbf{r})$.

An illustration of this technique relates to Figure 11-14(*d*). Shown is a rectangular horn fed from a rectangular waveguide carrying the dominant TE_{10} mode. Assuming a reasonably small horn taper, the field distribution over the horn aperture differs negligibly from that over the waveguide cross-section. Then from assumed null fields inside the surface S_1 just embracing the horn and feed system as shown, the aperture fields are replaced with equivalent current and charge surface densities over the aperture as given by (11-91). This field equivalence process is examined in the following example.

EXAMPLE 11-1. A pyramidal horn of aperture area ab is fed from a rectangular waveguide carrying the TE_{10} mode as in diagram (*a*). Find the equivalent electric and magnetic surface current distributions over the surface S_1 enclosing the horn and its feed system. (Neglect fields on the exterior conducting surfaces of the horn, assuming the field distribution over the horn aperture to be essentially that in the waveguide cross-section.)

The origin of the coordinates is assumed at the horn aperture *center*, as in

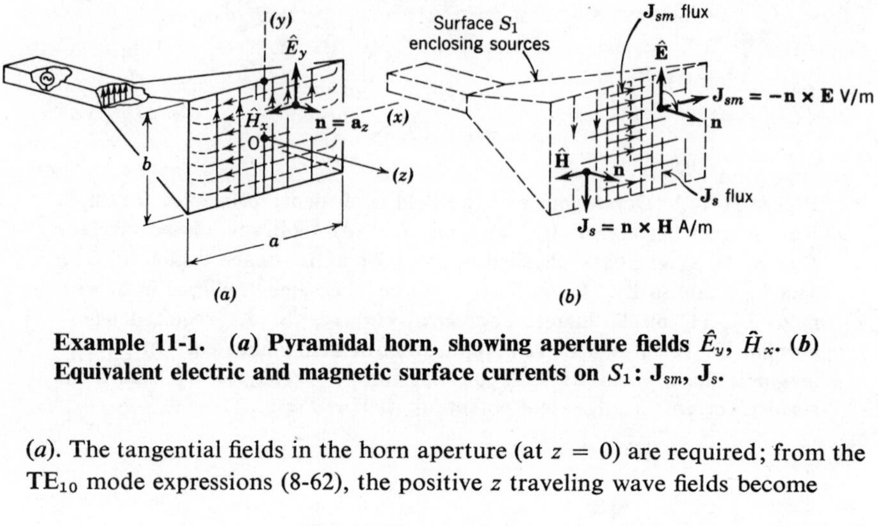

Example 11-1. (*a*) Pyramidal horn, showing aperture fields $\hat{E}_y$, $\hat{H}_x$. (*b*) Equivalent electric and magnetic surface currents on S_1: $\mathbf{J}_{sm}$, $\mathbf{J}_s$.

(*a*). The tangential fields in the horn aperture (at $z = 0$) are required; from the TE_{10} mode expressions (8-62), the positive z traveling wave fields become

$$\hat{E}_y^+(x, 0) = \hat{E}_{y,10}^+ \cos \frac{\pi}{a} x \tag{1}$$

$$\hat{H}_x^+(x, 0) = -\frac{\hat{E}_y^+(x, 0)}{\eta_{TE,10}} = -\frac{\hat{E}_{y,10}^+}{\eta_0} \sqrt{1 - \left(\frac{f_{c,10}}{f}\right)^2} \cos \frac{\pi}{a} x \tag{2}$$

in which a denotes the horn width in the H plane as shown. (The cosine distributions are the result of placing 0 at the aperture center.) Thus, the equivalent magnetic and electric surface current densities over the aperture become, by use of (11-91c) and (11-91d)

$$\hat{\mathbf{J}}_{sm} = -\mathbf{n} \times \hat{\mathbf{E}}_1 = -\mathbf{a}_z \times \mathbf{a}_y \hat{E}_y^+(x, 0) = \mathbf{a}_x \hat{E}_{y,10}^+ \cos\frac{\pi}{a}x \tag{3}$$

$$\hat{\mathbf{J}}_s = \mathbf{n} \times \hat{\mathbf{H}}_1 = \mathbf{a}_z \times \mathbf{a}_x \hat{H}_x^+(x, 0) = -\mathbf{a}_y \frac{\hat{E}_{y\,10}^+}{\eta_0}\sqrt{1 - \left(\frac{f_{c,10}}{f}\right)^2} \cos\frac{\pi}{a}x \tag{4}$$

flux sketches of which are depicted in (*b*). The reader may verify what equivalent magnetic charge density $\hat{\rho}_{sm}$ exists over the aperture, using (11-91b).

The electromagnetic fields exterior to the boundary surface S_1 enclosing the electromagnetic horn in the previous example can now be obtained by use of the potential integrals (11-87) and (11-88) taken over the aperture equivalent source currents (3) and (4), from which the fields $\hat{\mathbf{E}}$ and $\hat{\mathbf{H}}$ are found by use of (11-85) and (11-86). This process is illustrated in the somewhat different example following, which considers the radiation (diffraction) fields of a rectangular aperture in an absorbing screen illuminated with a uniform plane wave.

EXAMPLE 11-2. Suppose a rectangular aperture of dimensions a and b is cut into a thin, flat, perfectly absorbing (black) screen excited with a uniform, plane wave as in the accompanying figure. Assume that the fields are perfectly absorbed (without reflection) everywhere on the screen except in the aperture, and that the tangential fields are zero on the screen just to the right of it (at $z = 0+$). (*a*) Find the equivalent surface current distributions over the closed surface S_1 (consisting of the plane $z = 0+$ located just to the right of the screen

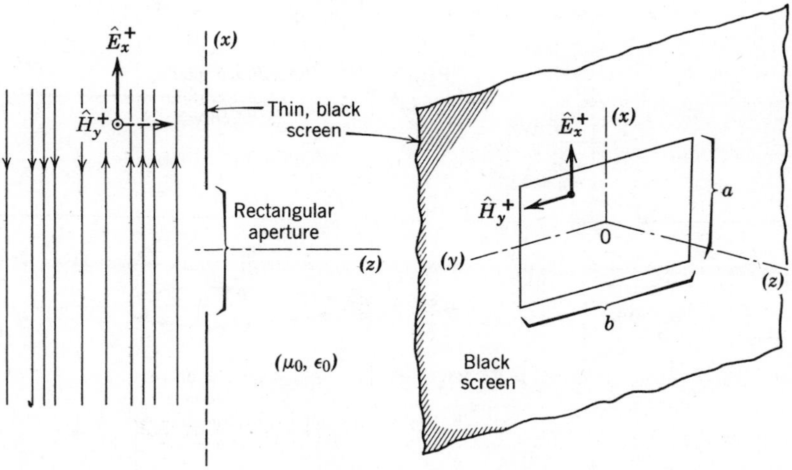

(*a*) (*b*)

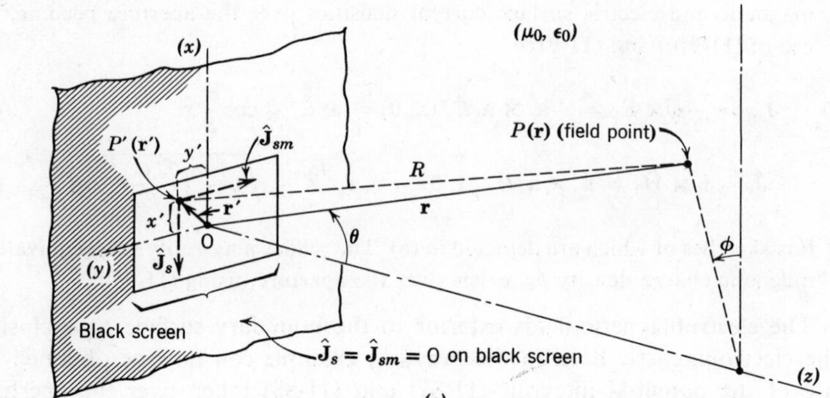
(x)
(μ0, ε0)
P′(r′)
y′
Ĵsm
x′
r′
0
(y)
Ĵs
θ
R
r
P(r) (field point)
φ
(z)
Black screen
Ĵs = Ĵsm = 0 on black screen

(c)

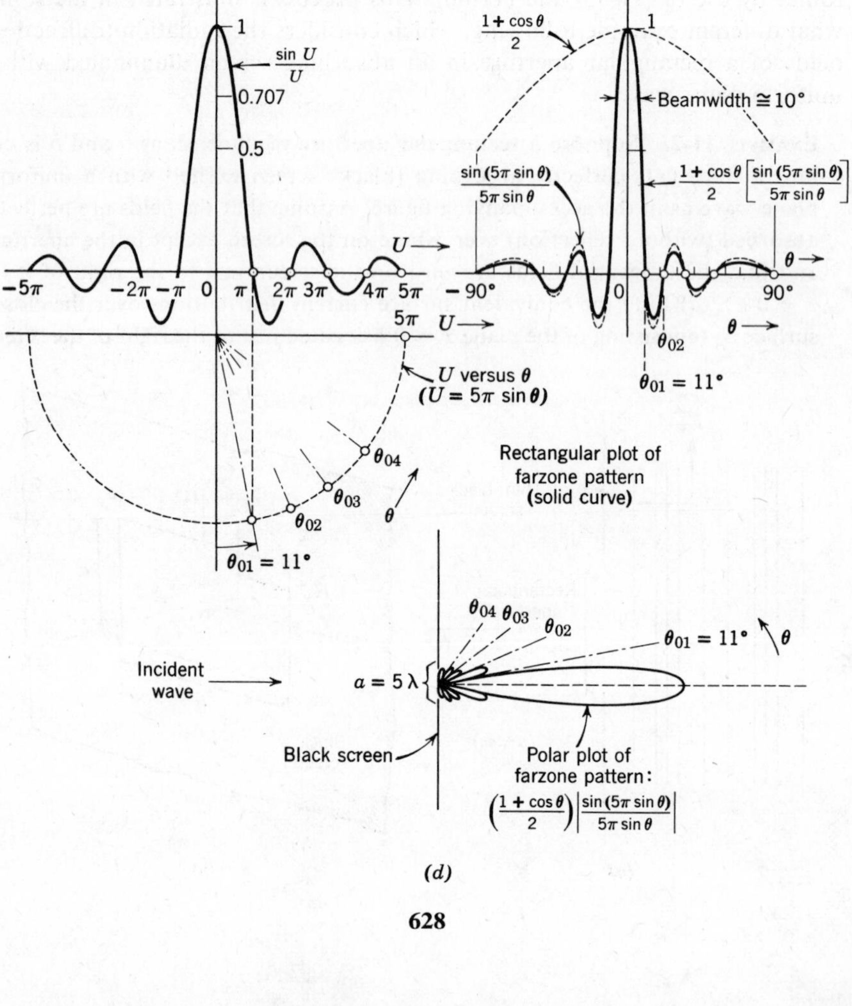
1
sin U / U
0.707
0.5
U →
−5π
−2π
−π
0
π
2π
3π
4π
5π
5π
U →
U versus θ
(U = 5π sin θ)
θ04
θ03
θ02
θ
θ01 = 11°
(1 + cos θ)/2
1
Beamwidth ≅ 10°
sin(5π sin θ) / 5π sin θ
(1 + cos θ)/2 [sin(5π sin θ) / 5π sin θ]
θ →
− 90°
0
90°
θ →
θ02
θ01 = 11°
Rectangular plot of
farzone pattern
(solid curve)
θ04
θ03
θ02
θ01 = 11°
θ
Incident
wave
a = 5 λ
Black screen
Polar plot of
farzone pattern:
((1 + cos θ)/2) |sin(5π sin θ) / 5π sin θ|

(d)

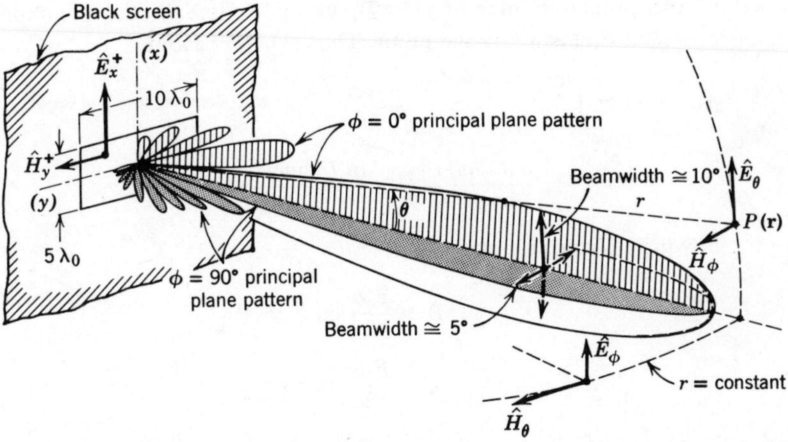

(e)

Example 11-2. **(*a*) Edge view of aperture in black screen. (*b*) Showing uniform plane wave field in the rectangular aperture. (*c*) Equivalent source currents in the aperture, and field point geometry in spherical coordinates. (*d*) Graphic constructions leading to the normalized field pattern of a rectangular aperture, in the $\phi = 0^\circ$ principal plane. (*e*) The field pattern of a rectangular, $5\lambda_0$ by $10\lambda_0$ aperture. Side lobes are shown only in the principal planes.**

and a hemisphere of indefinitely large radius encompassing the entire right-half space). (*b*) Evaluate $\hat{\mathbf{A}}(\mathbf{r})$ and $\hat{\mathbf{F}}(\mathbf{r})$ at any farzone field point $P(\mathbf{r})$ in the region $z > 0$. Use spherical coordinates. (*c*) Derive $\hat{\mathbf{E}}$ and $\hat{\mathbf{H}}$ from the potentials at $P(\mathbf{r})$ in the farzone. (*d*) Sketch the farzone diffraction field patterns in the principal planes $\phi = 0$ and $\phi = \pi/2$, if $a = 5\ \lambda_0$ and $b = 10\ \lambda_0$.

(*a*) Using the boundary conditions (11-91c) and (11-91d), the equivalent currents in the aperture plane ($z = 0$) become

$$\hat{\mathbf{J}}_{sm} = -\mathbf{n} \times \hat{\mathbf{E}}_1 = -\mathbf{a}_z \times \mathbf{a}_x \hat{E}_m^+ \tag{1}$$

$$\hat{\mathbf{J}}_s = \mathbf{n} \times \hat{\mathbf{H}}_1 = \mathbf{a}_z \times \mathbf{a}_y \frac{\hat{E}_m^+}{\eta_0} \tag{2}$$

(*b*) $\hat{\mathbf{A}}(\mathbf{r})$ is found by use of (11-87). The geometry in figure (*c*) shows $P(\mathbf{r})$ in a spherical coordinate frame of reference, with the origin at the center of the aperture. The distance between $P(\mathbf{r})$ and the source point $P(\mathbf{r}')$ is approximated by

$$R \cong r - \sin\theta(x'\cos\phi + y'\sin\phi) \tag{3}$$

The difference between R and r in (3) is significant in the phase exponent of the potential integral (11-87), but negligible in the denominator provided $P(\mathbf{r})$ is a farzone point. Thus (11-87) becomes

$$\hat{\mathbf{A}}(r, \theta, \phi) \cong \int_{y'=-b/2}^{b/2} \int_{x'=-a/2}^{a/2} \frac{-\mu_0 \mathbf{a}_x \hat{E}_m^+}{4\pi\eta_0 r} e^{-j\beta_0[r - \sin\theta(x'\cos\phi + y'\sin\phi)]} \, dx' \, dy'$$

$$= -\mathbf{a}_x \frac{\mu_0 \hat{E}_m^+ ab}{4\pi\eta_0 r} e^{-j\beta_0 r} \frac{\sin U}{U} \frac{\sin V}{V} \tag{4}$$

in which U and V signify

$$U \equiv \frac{\beta_0 a}{2} \sin\theta \cos\phi \tag{5}$$

$$V \equiv \frac{\beta_0 b}{2} \sin\theta \sin\phi \tag{6}$$

By (11-88), $\hat{\mathbf{F}}$ similarly becomes

$$\hat{\mathbf{F}}(r, \theta, \phi) \cong -\mathbf{a}_y \frac{\epsilon_0 \hat{E}_m^+ ab}{4\pi r} e^{-j\beta_0 r} \frac{\sin U}{U} \frac{\sin V}{V} \tag{7}$$

(*c*) The physical field $\hat{\mathbf{E}}$ is obtained by use of (11-85) into which (4) and (7) are substituted. If only the terms decreasing with r no faster than $1/r$ are retained, the result becomes

$$\hat{\mathbf{E}}(r, \theta, \phi) = \frac{j\beta_0 \hat{E}_m^+}{4\pi r} ab(1 + \cos\theta) e^{-j\beta_0 r} \frac{\sin U}{U} \frac{\sin V}{V} [\mathbf{a}_\theta \cos\phi - \mathbf{a}_\phi \sin\phi] \tag{11-92}$$

showing that $\mathbf{E}(\mathbf{r})$ has only the *transverse* components $\hat{E}_\theta$ and $\hat{E}_\phi$ in the farzone.

One can find $\hat{\mathbf{H}}(\mathbf{r})$ by use of the free-space version of (11-67c). If only the terms decreasing as $1/r$ are retained, the results are

$$\hat{H}_\phi = \frac{\hat{E}_\theta}{\eta_0} \qquad \hat{H}_\theta = -\frac{\hat{E}_\phi}{\eta_0} \tag{11-93}$$

The components of (11-93) are thus related by the intrinsic wave impedance η_0 of the source free environment.

(*d*) Sketches of the farzone field patterns of the given rectangular aperture can be obtained for the principal planes $\phi = 0$ and $\phi = \pi/2$ by use of (11-92). For $\phi = 0$ (the vertical principal plane), (11-92) becomes

$$\hat{\mathbf{E}}(r, \theta, \phi) \cong \mathbf{a}_\theta \frac{j\beta_0 \hat{E}_m^+}{2\pi r} abe^{-j\beta_0 r} \left[\left(\frac{1 + \cos\theta}{2} \right) \frac{\sin(5\pi \sin\theta)}{5\pi \sin\theta} \right] \tag{11-94}$$

The bracketed factor is the *normalized field pattern* of the rectangular aperture in the $\phi = 0$ principal plane. In figure (*d*) is shown the graphical construction of the factor $(\sin U)/U$ specialized to $\phi = 0$, with the field pattern plotted versus the physical angle θ. The physical, or visible, range of θ for an aperture in a black screen is the range $-90° < \theta < 90°$, seen to correspond to a visible range of U between the limits $(-5\pi, 5\pi)$. There is a slight tapering effect of the factor $(1 + \cos\theta)/2$, called the *Huygens factor*, which has the effect of reducing the side lobe amplitudes somewhat below the values predictable from the basic $(\sin U)/U$ function.

The field pattern in the other principal plane ($\phi = \pi/2$) is analyzed in a similar manner. The aperture width $b = 10\,\lambda$ applies in this case, yielding 18 side lobes (instead of the eight obtained in the $\phi = 0$ principal plane).

The beamwidth of the principal beam of such an aperture is usually defined as the angular width measured between the 70.7% amplitude points. Since $(\sin U)/U$ has the value 0.707 at $U_0 = 1.39$, one can write an expression for beamwidth $(2\theta_0)$ in a principal plane of the cophasal, uniformly illuminated rectangular aperture in the form

$$2\theta_0 = 2\sin^{-1}\left(\frac{U_0}{\frac{\beta_0 a}{2}}\right) = 2\sin^{-1} 0.443\left(\frac{\lambda_0}{a}\right)\ \text{rad} \tag{11-95}$$

If the aperture width is sufficiently large, (11-95) is approximated by

$$2\theta_0 \cong 0.88\frac{\lambda_0}{a}\ \text{rad} \cong 50\frac{\lambda_0}{a}\ \text{deg} \tag{11-96}$$

from which it is seen that an aperture width $a = 10\,\lambda_0$ produces a beamwidth $2\theta_0 = 5°$, while $a = 100\,\lambda_0$ provides an associated beamwidth $2\theta_0 = 0.5°$, etc. This result shows that the beamwidth in the principal plane of a uniformly, cophasally excited aperture is inversely proportional to the aperture width measured in that principal plane.

Another important characteristic of the diffraction pattern of an aperture source is the relative strength of its side lobes in relation to the level of the principal beam. The $(\sin U)/U$ diffraction pattern of the uniformly illuminated case treated in the preceding example and shown in figure (*d*) has a first side lobe level that is 21.7% of the mainbeam maximum, or about 13 dB down. The side lobe level achieved in a given aperture antenna design is dependent on the functional nature of the aperture field distribution. In particular, if the aperture field excitation remains cophasal but has a cosinusoidally tapered amplitude over the aperture, the first side lobe will be 23 dB below the mainbeam maximum, or 10 dB lower than that obtained with uniform aperture excitation. Side lobe suppression by means of aperture illumination tapering is usually obtained, however, only at the expense of an increase in the beamwidth. For example, a cophasally and uniformly excited aperture of 10 λ width has a beamwidth of about 5° in the principal plane which includes that width. With cosinusoidal tapering, the beamwidth increases to about 6.8°.

A summary of the effects that aperture amplitude tapering has on beamwidth and side lobe level is given in Table 11-1 for the cases of rectangular apertures and axially symmetrically excited circular apertures. The proof of these results can be established in the same manner as described in Example 11-2 for the case of a cophasally and uniformly illuminated rectangular aperture. The equivalent sources over a horn aperture with a cosinusoidal tangential field distribution in one dimension has been considered in Example 11-1. The characteristics of its farzone field pattern in the principal x-z plane are found in (2) of Table 11-1. Additional details of the circular-aperture diffraction problem are found on p. 192 of the book by Silver listed in the references.

Table 11–1 Field-pattern characteristics of large[a] apertures

Type of aperture illumination	*Sketch of amplitude*	*Pattern beamwidth (deg)*	*First sidelobe level (dB)*
	Rectangular aperture (a = width)		
(1) Uniform	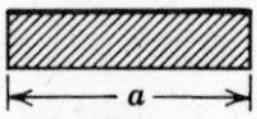	$\frac{50°}{a/\lambda_0}$	13.2
(2) Cosinusoidal		$\frac{68°}{a/\lambda_0}$	23
(3) Cosine squared		$\frac{82°}{a/\lambda_0}$	32
	Circular aperture (D = diameter)		
(4) Uniform		$\frac{58°}{D/\lambda_0}$	17.6
(5) Parabolic		$\frac{72°}{D/\lambda_0}$	24.6
(6) Parabolic squared	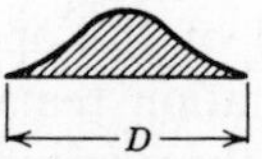	$\frac{84°}{D/\lambda_0}$	30.6

[a] By large aperture is meant one whose principal dimensions are large compared to λ_0.

REFERENCES

JORDAN, E. C., and K. G. BALMAIN. *Electromagnetic Waves and Radiating Systems*, 2nd ed. Englewood Cliffs, N.J.: Prentice-Hall, 1968.

RAMO, S., J. R. WHINNERY, and T. VAN DUZER. *Fields and Waves in Communication Electronics*. New York: Wiley, 1965.

SILVER, S. *Microwave Antenna Theory and Design*. New York: McGraw-Hill, 1949.

PROBLEMS

11-1. Suppose that, to the vector magnetic potential **A**, one adds another vector $\nabla\psi$, the gradient of an arbitrary scalar function ψ. Also, to the scalar electric potential Φ is added the function $-\partial\psi/\partial t$. Assuming that **A** and Φ satisfy (11-1) and (11-2), show that the new potentials $\mathbf{A}' = \mathbf{A} + \nabla\psi$ and $\Phi' = \Phi - \partial\psi/\partial t$ also satisfy them. Comment on the uniqueness of **A** and ψ.

11-2. Using the expressions (11-35) and (11-36) for the vector magnetic potential, derive in detail the result (11-37) for the magnetic field of the elementary electric dipole in free space.

11-3. Derive in detail the electric field (11-40) for the elementary electric dipole, using either (11-38) in terms of the vector magnetic potential or the Maxwell relation (11-39).

11-4. Obtain in detail the real-time field expressions (11-41) of the elementary electric dipole from their complex forms (11-37) and (11-40). Explain why the functions $\cos(\omega t - \beta_0 r)$ and $\sin(\omega t - \beta_0 r)$ denote outward-traveling waves.

11-5. Prove in detail (11-47) for the time-average power radiated from an elementary electric dipole, using an enclosing sphere of sufficient size that only the farzone radiation terms (11-44) need be used.

11-6. Prove the result (11-47), this time using an enclosing sphere of arbitrary size (not necessarily large), thus utilizing all the components (11-37) and (11-40) of the dipole field.

11-7. Show sketches of, and in the manner of (11-54) express analytically, the current distributions $\hat{I}(z')$ on the following thin-wire antennas:

(*a*) A center-fed antenna, one-quarter wavelength long ($2\ell = \lambda_0/4$ or $\beta_0\ell = 45°$).

(*b*) A center-fed antenna of length $2\ell = \lambda_0/2$ (a half wave dipole).

(*c*) A center-fed antenna, three-quarters wavelength long.

(*d*) An off-center-fed antenna with $\ell_1 = 3\,\lambda_0/4$, $\ell_2 = \lambda_0/4$. Specify one current maximum in terms of the other.

(*e*) An off-center-fed antenna with $\ell_1 = \lambda_0/8$, $\ell_2 = \lambda_0/4$. Specify $\hat{I}_{m2}$ in terms of $\hat{I}_{m1}$.

11-8. A circular loop antenna like that of Figure 11-10(*b*) has a circumferential length of one-quarter free-space wavelength. Show a sketch of, and find an expression for (as a function of the angle ϕ generated about the loop center), the current distribution on the wire loop. What is the effect on the current distribution of making the loop diameter small compared to λ_0? Large?

11-9. Verify the farzone field expression (11-60) for the linear antenna, by integrating (11-58) in detail.

11-10. Calculate and plot the polar field pattern factors $F(\theta)$ for the following center-fed linear antennas:

(*a*) A half wave dipole
(*b*) A three-halves wavelength dipole

[*Hint:* Use angular intervals no greater than 15°, plotting only from 0° to 90° in view of the symmetry.]

11-11. For a half wave, center-fed dipole in free space, what is the magnitude of $\hat{E}_\theta$ at a broadside range of 1 mi. (1609 m), if the current amplitude is 1 A?

11-12. Integrate the average radiated power expression (11-65) either graphically or by use of a computer, to find the radiation resistance (11-66) of a half wave, center-fed dipole.

11-13. Justify the expression (11-86) for the $\hat{\mathbf{H}}$ field by making use of (11-85) and duality.

11-14. Enclosing the thin, center-fed dipole of Figure 11-12 inside a close-fitting, cylindrical closed surface S_1, what equivalent surface currents and charges must exist over S_1 if the fields at $P(r, \theta, \phi)$ are to remain unchanged while the sources inside S_1 are nullified? [*Hint:* Assume a current distribution on the wire given by (11-59).]

11-15. Prove the result (4) in Example 11-2 by carrying out the integration in detail.

11-16. Show that the result (11-92) in Example 11-2 is correct, by substituting the potential functions (4) and (7) into (11-85).

11-17. Sketch the normalized field pattern in the $\phi = 90°$ principal plane of the rectangular aperture illuminated with a uniform plane wave as described in Example 11-2, assuming the aperture dimensions $a = 10\ \lambda_0$ and $b = 20\ \lambda_0$. Find the locations of the nulls θ_{01}, θ_{02}, etc. of that pattern. Determine its beamwidth analytically, and express its first side lobe level in decibels below the maximum value of the principal beam. Compare your results with those given in Table 11-1.

11-18. Using the equivalent aperture fields determined in Example 11-1 (assuming only the forward-traveling, dominant TE_{10} mode in the aperture), find the following for the rectangular horn illustrated, assuming negligible currents induced on the outer metallic walls.

(*a*) Show that the vector electric and magnetic potentials at any farzone point become

$$\hat{\mathbf{A}}(r, \theta, \phi) = -\mathbf{a}_y \frac{\mu_0 \hat{E}_m^+ ab}{8r\eta_{\text{TE},10}} e^{-j\beta_0 r} \frac{\sin B}{B} \frac{\cos A}{A^2 - \left(\frac{\pi}{2}\right)^2}$$

$$\hat{\mathbf{F}}(r, \theta, \phi) = \mathbf{a}_x \frac{\epsilon_0 \hat{E}_m^+ ab}{8r} e^{-j\beta_0 r} \frac{\sin B}{B} \frac{\cos A}{A^2 - \left(\frac{\pi}{2}\right)^2}$$

in which

$$A = \frac{\beta_0 a}{2} \sin\theta \cos\phi$$

$$B = \frac{\beta_0 b}{2} \sin\theta \sin\phi$$

(*b*) Use the latter potential solutions to derive the electric and magnetic fields in the farzone, obtaining

$$\hat{\mathbf{E}}(r, \theta, \phi) = \left[\left(\mathbf{a}_\theta \sin\phi \left(\frac{\beta_{10}}{\beta_0}\cos\theta + 1\right) + \mathbf{a}_\phi \cos\phi\left(\frac{\beta_{10}}{\beta_0} + \cos\theta\right)\right]\right.$$
$$\times \frac{j\beta_0 E_{y,10}}{8} \frac{e^{-j\beta_0 r}}{r} \frac{\sin B}{B} \frac{\cos A}{A^2 - \left(\frac{\pi}{2}\right)^2}$$

(*c*) Compare the farzone field results for the rectangular horn with those found in Example 11-2 for the uniform plane wave excited rectangular aperture having the same dimensions *a*, *b*. Comment especially on the similarities or differences between the field patterns in the principal planes.

11-19. From (11-95) was derived in effect a "5 and 10 rule" for a rectangular aperture illuminated uniformly in amplitude and phase. This rule implies that a 5 λ_0-wide aperture has a 10° beamwidth in that principal plane while making it 10 λ_0 wide decreases the beamwidth to 5°, and so on. Using the results compiled in Table 11-1 for other types of illumination (cophasal, but with the amplitude tapered in the manner shown) over rectangular and circular apertures, determine what comparable rule applies to the remaining cases in the table.

11-20 Use Table 11-1 to compare the beamwidths of the farzone field patterns of each of the following circular apertures. Assume uniform aperture illuminations.

(*a*) A parabolic dish having an aperture diameter of 10 wavelengths.
(*b*) The same as (*a*), except 100 wavelengths diameter.
(*c*) An optical laser having an aperture of 5 mm diameter, transmitting red light at a wavelength of about 7000 Å $= 7 \times 10^{-7}$ m $= 0.7\ \mu$m. (What is the aperture diameter in wavelengths?)
(*d*) By what factor will each beamwidth in the foregoing increase, if the aperture illumination is assumed to taper parabolically in each case?

APPENDIX

Energy and Power Concepts of Circuit Theory

Under time-harmonic conditions, the voltages and currents of a circuit have settled down to their sinusoidal steady state. Then the time rate of energy transfer, or instantaneous power, into a specified terminal pair is obtained from the product of the instantaneous, sinusoidal voltage and current there, leading to the time-average power over an integral number of periods of the voltage and current waveforms. It is the time-average power that is associated with the net transfer of energy from the sources to the circuit elements. The reactive elements store energy over one portion and release it over another portion of a cycle, to contribute nothing to the time-average power absorption or dissipation (even though a time-average *energy* storage by these elements does exist over any cycle). The resistive elements, on the other hand, account for the total time-average power absorption. Some aspects of these processes, from the points of view of both the real-time and the complex forms of the voltage, current, and power expressions, are examined in the following.

A single terminal pair linear network driven by a sinusoidal time-harmonic source is shown in Figure A-1, and it is completely characterized by a complex impedance or admittance. The sinusoidal terminal voltage and current v and i are related to their complex forms $\hat{V}$ and $\hat{I}$ as follows:

$$v(t) = \text{Re}\,[\hat{V}e^{j\omega t}] = V\cos(\omega t + \theta_v)\ \text{V} \qquad \text{(A-1)}$$

$$i(t) = \text{Re}\,[\hat{I}e^{j\omega t}] = I\cos(\omega t + \theta_i)\ \text{A} \qquad \text{(A-2)}$$

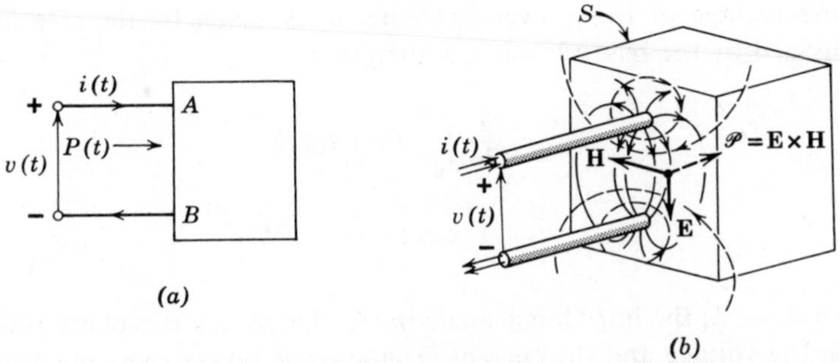

Figure A-1. A two-terminal network and its input fields. (*a*) Terminal pair network connected to a sinusoidal source. (*b*) D, H, and 𝒫 fields associated with input voltage and current.

in which V and I denote the positive, real sinusoidal amplitudes, with θ_v and θ_i representing the phases relative to an arbitrary time origin. The complex, time-harmonic forms are evidently

$$\hat{V} = Ve^{j\theta_v} \tag{A-3}$$

$$\hat{I} = Ie^{j\theta_i} \tag{A-4}$$

The complex impedance $\hat{Z}$ looking into the terminals AB is the ratio of (A-3) to (A-4)

$$\hat{Z} = \frac{V}{I} e^{j(\theta_v - \theta_i)}\ \Omega \tag{A-5}$$

Supposing the positive current and voltage conventions are as shown in Figure A-1(*a*), the instantaneous power, $P = vi$, flows into the circuit as denoted by the arrow. This fact is evident from field theory by noting that the fields **E** and **H**, associated quasi-statically with the voltage v and the current i, respectively, as depicted in Figure A-1(*b*), produce a Poynting vector $\mathcal{P} = \mathbf{E} \times \mathbf{H}$ that accounts for a net power-flux $P = \oint_S \mathcal{P} \cdot d\mathbf{s}$ flowing into the surface S encompassing the circuit as shown. It is evident that the reversal of *either v or i* during a portion of a cycle accounts for an *outflow* of power-flux from S at that instant.

With the voltage and current conventions of Figure A-1, the instantaneous power flow to the right is

$$\begin{aligned} P(t) = v(t)i(t) &= VI\cos(\omega t + \theta_v)\cos(\omega t + \theta_i) \\ &= \frac{VI}{2}[\cos(\theta_v - \theta_i) + \cos(2\omega t + \theta_v + \theta_i)] \end{aligned} \tag{A-6}$$

The time-average of this power-flux entering S, given by the area under a cycle divided by the base (2π rad), is therefore

$$P_{av} = \frac{1}{2\pi}\int_0^{2\pi} P(t)\, d(\omega t) \tag{A-7}$$

$$= \frac{VI}{2}\cos(\theta_v - \theta_i)\ \text{W} \tag{A-8}$$

in which $\theta_v - \theta_i$, the impedance angle in (A-5), denotes the phase difference between the voltage and the current. The average power over any cycle thus is determined by the constant term of (A-6). The double frequency term does not contribute to average power because its positive and negative fluctuations are equal over any cycle. In Figure A-2 are depicted the quantities $v(t)$, $i(t)$, and $p(t)$ associated with a typical network. It is significant that if the network of Figure A-1 were *passive*, its impedance angle $(\theta_v - \theta_i)$ of (A-5) would lie in the right half of the complex $\hat{Z}$ plane; i.e., $\text{Re}(\hat{Z}) > 0$. This condition means that the time-average power flow, a positive quantity from (A-8), is *into* the network. A negative real part of the impedance (A-5) would on the other hand imply an *active* circuit (containing generators), meaning that the network is the source of time-average *outward* power-flux.

It is useful at this point to discuss the energies stored by the reactive elements L and C of a circuit. The instantaneous magnetic field energy stored in the field of an ideal (resistanceless) inductor, carrying the time-harmonic current $i(t) = I\cos(\omega t + \theta_L)$, is

$$U_m(t) = \tfrac{1}{2}Li^2 = \tfrac{1}{2}LI^2\cos^2(\omega t + \theta_L)$$

$$= \tfrac{1}{4}LI^2[1 + \cos 2(\omega t + \theta_L)]\ \text{J} \tag{A-9}$$

Borrowing from the definition (A-7) of time average, the stored energy

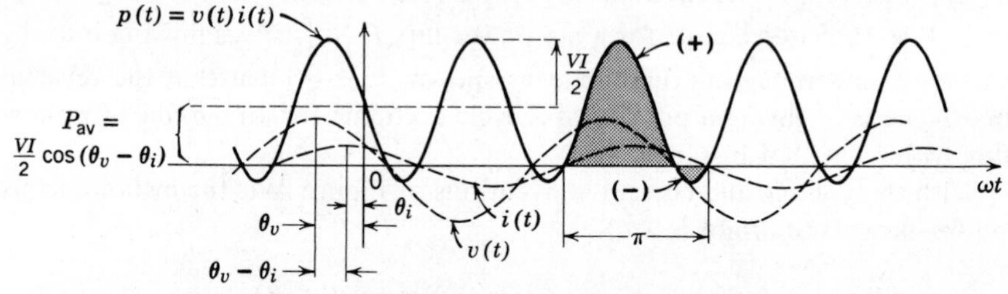

Figure A-2. Time-instantaneous voltage, current, and power, showing negative excursions of $p(t)$ over the intervals $\theta_v - \theta_i$, and the time average P_{av}.

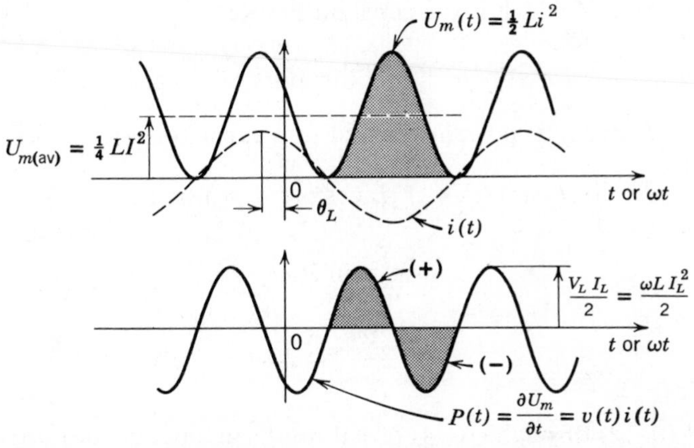

Figure A-3. Time-instantaneous energy and power input conditions for a pure inductance.

(A-9) averaged over a cycle is the constant amount

$$U_{m,\mathrm{av}} = \tfrac{1}{4}LI^2 = \tfrac{1}{4}L\hat{I}\hat{I}^* \text{ J} \tag{A-10}$$

in which the equivalence $\hat{I}\hat{I}^* \equiv I^2$ is obtained from the complex inductor current, $\hat{I} = Ie^{j\theta}$. Graphs of the instantaneous and time-average magnetic energies of a pure inductor are shown in the upper diagram of Figure A-3. The instantaneous power delivered to the inductor is found from the time derivative of the instaneous energy (A-9); equivalently, it is the product of the instantaneous inductor voltage and current

$$\begin{aligned} P(t) = v(t)i(t) &= V_L I_L \cos(\omega t + \theta_L) \sin(\omega t + \theta_L) \\ &= \frac{V_L I_L}{2} \sin 2(\omega t + \theta_L) \\ &= \frac{\omega L I_L^2}{2} \sin 2(\omega t + \theta_L) \text{ W} \end{aligned} \tag{A-11}$$

implying a power surge to and from the inductive element every quarter cycle, noted in the lower diagram of Figure A-3. The time-average power is thus zero.

In a similar fashion, the instantaneous energy stored in the electric field of a capacitor, across which the voltage $v_c(t) = V\cos(\omega t + \theta_c)$ appears, is

$$\begin{aligned} U_e(t) &= \tfrac{1}{2}Cv^2 = \tfrac{1}{2}CV^2 \cos^2(\omega t + \theta_c) \\ &= \tfrac{1}{4}CV^2[1 + \cos 2(\omega t + \theta_c)] \text{ J} \end{aligned} \tag{A-12}$$

the time average of which is the constant first term

$$U_{e,\mathrm{av}} = \tfrac{1}{4}CV^2 = \tfrac{1}{4}C\hat{V}\hat{V}^* \text{ J} \tag{A-13}$$

The time-instantaneous power delivered to a capacitor is

$$\begin{aligned} P_c(t) = v(t)i(t) &= V_c I_c \cos(\omega t + \theta_c) \sin(\omega t + \theta_c) \\ &= \frac{V_c I_c}{2} \sin 2(\omega t + \theta_c) \\ &= \frac{\omega C V_c^2}{2} \sin 2(\omega t + \theta_c) \text{ W} \end{aligned} \tag{A-14}$$

Graphs of (A-12) through (A-14) reveal results analogous to those of Figure A-3 for the inductor. The foregoing energy and power expressions are meaningful in the concept of complex power applied to networks, considered next.

The product $\hat{V}\hat{I}$, obtained by multiplying together the complex voltage (A-3) by the current (A-4) has no physical meaning, but if the *conjugate* of either the voltage or the current is employed in this product, the result has important physical interpretations. Thus, defining a *complex power* as follows:

$$\hat{P} = \tfrac{1}{2}\hat{V}\hat{I}^* \tag{A-15}$$

and applying (A-3) and (A-4), one obtains

$$\begin{aligned} \hat{P} = \tfrac{1}{2}\hat{V}\hat{I}^* &= \frac{VI}{2} e^{j(\theta_v - \theta_i)} \\ &= \frac{VI}{2} \cos(\theta_v - \theta_i) + j\frac{VI}{2} \sin(\theta_v - \theta_i) \end{aligned} \tag{A-16}$$

The real part of $\hat{P}$ evidently describes the time-average power, P_{av}, deduced previously in (A-6) using the time-instantaneous forms of the voltage and current; hence

$$P_{\mathrm{av}} = \operatorname{Re}(\hat{P}) = \operatorname{Re}(\tfrac{1}{2}\hat{V}\hat{I}^*) \text{ W} \tag{A-17}$$

Denoting the imaginary part of $\hat{P}$ by Q, (A-16) is written†

$$\hat{P} = P_{\mathrm{av}} + jQ \tag{A-18}$$

† The quantity Q, used here to define the imaginary part of complex power, should not be confused with the quality factor Q defined in (4-144).

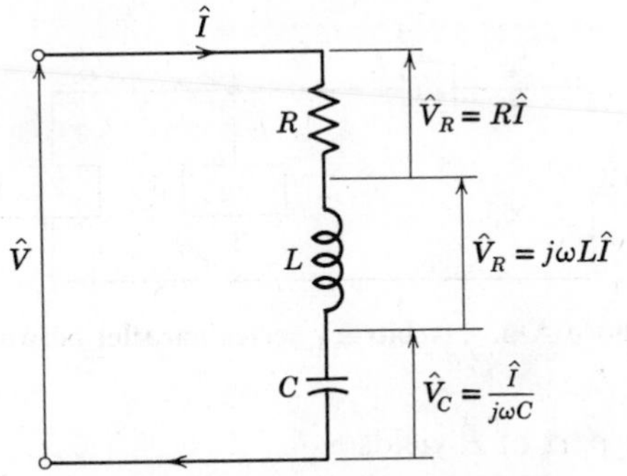

Figure A-4. Series *RLC* branch.

wherein

$$Q = \text{Im}(\hat{P}) = \text{Im}\left(\frac{\hat{V}\hat{I}^*}{2}\right) = \frac{VI}{2}\sin(\theta_v - \theta_i) \tag{A-19}$$

Thus Q may be positive or negative, depending on the sign of the impedance angle $(\theta_v - \theta_i)$, and Q is positive for an inductive impedance looking into the terminals AB of Figure A-1, negative if capacitive. It is customary to express Q in the units of volt-amperes (reactive), or vars, rather than watts, in view of its energy storage, rather than dissipative, implications.

The imaginary part Q of complex power is given a physical interpretation by considering its meaning for the series LRC branch of Figure A-4. The complex power delivered to the series branch can be expanded

$$\begin{aligned}\hat{P} &= \tfrac{1}{2}\hat{V}\hat{I}^* = \tfrac{1}{2}(\hat{V}_R + \hat{V}_L + \hat{V}_c)\hat{I}^* \\ &= \tfrac{1}{2}\left(R\hat{I} + j\omega L\hat{I} - j\frac{\hat{I}}{\omega C}\right)\hat{I}^* \\ &= \frac{R\hat{I}\hat{I}^*}{2} + j\left[\frac{\omega L\hat{I}\hat{I}^*}{2} - \frac{\omega C\hat{V}_c\hat{V}_c^*}{2}\right]\end{aligned} \tag{A-20}$$

which is the sum of the complex powers $\hat{P}_R + \hat{P}_L + \hat{P}_c$ delivered to each element.† Because $\hat{P} = P_{\text{av}} + jQ$ from (A-18), it follows that P_{av} dissipated by the RLC branch is attributable to the resistance through

$$P_{\text{av}} = \text{Re}(\hat{P}) = \frac{R\hat{I}\hat{I}^*}{2} = \frac{RI^2}{2}\text{ W} \tag{A-21}$$

† The last term of (A-20), denoting the complex power delivered to the capacitor, is obtained from $(1/2)\hat{V}_c\hat{I}^*$ by relating the capacitor current and voltage through $\hat{I} = j\omega C\hat{V}_c$ to make $I^* = \omega C(j)^* V_c^* = -j\omega C V_c^*$, so the result follows. In forming the conjugate of the product of two (or more) complex numbers, note that each is conjugated according to the identity $(\hat{z}_1\hat{z}_2)^* = \hat{z}_1^*\hat{z}_2^*$.

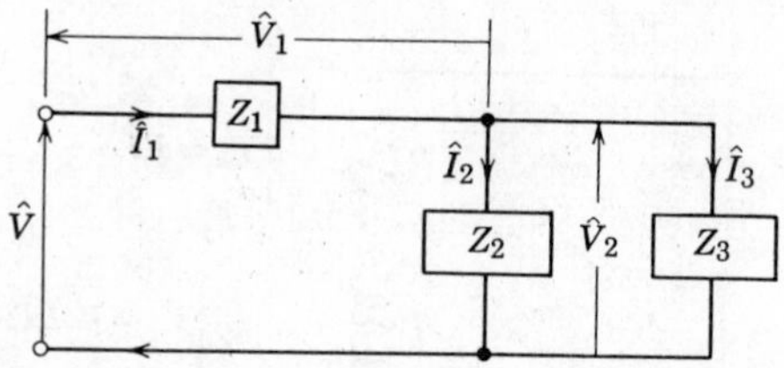

Figure A-5. Arbitrary series-parallel network.

while the imaginary part of $\hat{P}$ yields

$$Q = \text{Im}\,(\hat{P}) = \frac{\omega L\hat{I}\hat{I}^*}{2} - \frac{\omega C\hat{V}_c\hat{V}_c^*}{2} = \frac{\omega LI^2}{2} - \frac{\omega CV_c^2}{2} \tag{A-22a}$$

$$Q = \text{Im}\,(\hat{P}) = 2\omega\left[\frac{LI^2}{4} - \frac{CV_c^2}{4}\right] \text{var} \tag{A-22b}$$

A comparison of the terms of (A-22a) with the instantaneous power expressions (A-11) and (A-14) reveals that $Q = \text{Im}\,(\hat{P})$ represents the *difference of the peak values of the power delivered to the L and C elements*. Equivalently, the form (A-22b) shows that Q is 2ω *times the difference of the time averages of the energies stored in L and C*, as seen from (A-10) and (A-13). The same conclusions apply to a parallel-connected *RLC* network.

Superposition is an important property of the complex powers associated with the branches of a linear network. It is readily shown that the complex power $\hat{P}$ delivered to the input terminals of an arbitrary network is also the *sum* of the complex powers delivered to the separate branches. To prove this, the series parallel network of Figure A-5 is of sufficient generality. Suppose each impedance element $\hat{Z}_1, \hat{Z}_2$, and $\hat{Z}_3$ consists of arbitrary, series connected *RLC* elements as in Figure A-4. It is seen that the input complex power may be expanded as follows:

$$\begin{aligned}\hat{P} = \hat{V}\hat{I}_1^* = (\hat{V}_1 + \hat{V}_2)\hat{I}^* &= \hat{V}_1\hat{I}_1^* + \hat{V}_2(\hat{I}_2^* + \hat{I}_3^*)\\ &= \hat{V}_1\hat{I}_1^* + \hat{V}_2\hat{I}_2^* + \hat{V}_3\hat{I}_3^*\\ &= \hat{P}_1 + \hat{P}_2 + \hat{P}_3\end{aligned}$$

which is the sum of the complex powers delivered to each impedance element of the network. Generally, for an n branch network, the input complex power is

$$\hat{P} = \sum_{i=1}^{n} \hat{P}_i = \sum_{i=1}^{n} \frac{\hat{V}_i\hat{I}_i^*}{2} \tag{A-23}$$

If each branch contains an arbitrary series combination of R, L, and C, with the complex power delivered to each branch given by (A-20), using (A-23) obtains the total complex power into the system

$$\hat{P} = \sum_{i=1}^{n} \frac{R_i \hat{I}_i \hat{I}_i^*}{2} + j \sum_{i=1}^{n} \left[\frac{\omega L_i \hat{I}_i \hat{I}_i^*}{2} - \frac{\omega C_i \hat{V}_{ci} \hat{V}_{ci}^*}{2} \right]$$

$$= \sum_{i=1}^{n} P_{\text{av},i} + j \sum_{i=1}^{n} Q_i \tag{A-24}$$

The real and reactive powers of the branches of a network thus combine linearly to affect the complex power at the input terminals accordingly. Applications of these concepts to specific network problems are found in appropriate sources.†

† See Brenner, E., and M. Javid. *Analysis of Electric Circuits*. New York: McGraw-Hill, 1959, pp. 354–365. Applications to three-phase networks are found in Pearson, S. and G. Maler. *Introductory Circuit Analysis*. New York: Wiley, 1965, p. 368.

Index